Exotic Nuclear Spectroscopy

Exotic Nuclear Spectroscopy

Edited by

William C. McHarris
Michigan State University
East Lansing, Michigan

Springer Science+Business Media, LLC

Library of Congress Cataloging-in-Publication Data

Symposium on Exotic Nuclear Spectroscopy (1989 : Miami Beach, Fla.)
Exotic nuclear spectroscopy / edited by William C. McHarris.
p. cm.
"Proceedings of an American Chemical Society Division of Nuclear Chemistry and Technology Symposium on Exotic Nuclear Spectroscopy, held September 11-15, 1989, in Miami Beach, Florida"--T.p. verso.
Includes bibliographical references and index.
ISBN 978-0-306-43882-0 ISBN 978-1-4615-3684-0 (eBook)
DOI 10.1007/978-1-4615-3684-0
1. Nuclear spectroscopy--Congresses. I. McHarris, William C. II. American Chemical Society. Division of Nuclear Chemistry and Technology. III. Title.
QC454.N8S96 1989
543'.0858--dc20 91-8182
CIP

Proceedings of an American Chemical Society Division of Nuclear Chemistry and Technology Symposium on Exotic Nuclear Spectroscopy, held September 11-15, 1989, in Miami Beach, Florida

Preface

Several hundred nuclear scientists from more than a dozen countries met in Miami Beach, Florida, in September 1989 for a *Symposium on Exotic Nuclear Spectroscopy*, sponsored by the Division of Nuclear Chemistry and Technology of the American Chemical Society. Braving five days of beckoning beaches, they presented, listened to, and discussed a series of invited papers covering the renaissance of nuclear spectroscopy—both experimental and theoretical—that has occurred during the past decade and which promises to continue well into the decade to come. This book contains the Proceedings of that *Symposium on Exotic Nuclear Spectroscopy*. But it is much more: During the ensuing six months, most of the papers were rewritten, polished, and/or expanded; so the resulting book is a much more finished, comprehensive, and up-to-date product than a general proceedings.

Chapter sequences follow the original symposium program, which, with a few exceptions, juxtaposes like topics. Each chapter can stand more or less on its own—although, as in any complex, interrelated scientific field, there are many cross-references among chapters.

The manuscripts were edited and then uniformly typed in the Department of Chemistry at Michigan State University. I tried to keep the editing to a minimum so as to preserve the flavor of individual styles.

A great debt of gratitude goes to those who helped make this volume possible, especially to Vada O'Donnell, who bore the largest burden of typing (and clarifying) the manuscripts. Special thanks also goes to the Division of Nuclear Chemistry and Technology of ACS for its financial and moral support—J. Robb Grover, who as Chairman of the Division initiated the project, worked especially hard to help bring it about. And, finally, I thank the forty-one authors (or sets of authors) who took time out of their busy schedules to come to Miami Beach for their presentations, then took even more time out of increasingly busy schedules to write the finished papers. I hope and think this volume should be a useful reference for the next few years, indicating, as it does, the directions nuclear spectroscopy —and almost all nuclear spectroscopy nowadays can qualify as "exotic"—seems to be taking.

Wm. C. McHarris

East Lansing, Michigan
July 1990

Contents

1. Atomic Mass Measurements with Radioactive Beams and/or Targets: Where to Start — 1
P. E. Haustein

2. A Recoil Mass Spectrometer for the HHIRF Facility — 11
J. D. Cole, T. M. Cormier, and J. H. Hamilton

3. The Study of Exotic $N \approx 82$ Nuclei Using the Daresbury Recoil Mass Separator — 23
J. H. McNeill, A. A. Chishti, W. Gelletly, M. A. C. Hotchkis, B. J. Varley, J. Blomqvist, P. J. Daly, M. Piiparinen, and P. J. Woods

4. Spectroscopic Techniques in the Study of Radioactive Nuclei Far from Stability: Results for the $Z = 82$ Region — 39
E. F. Zganjar, J. L. Wood, and J. Schwarzenberg

5. Nuclear Decay Studies Far from Stability — 57
R. B. Firestone, J. M. Nitschke, P. A. Wilmarth, K. S. Vierinen, R. M. Chasteler, J. Gilat, and A. A. Shihab-Eldin

6. Accelerated Radioactive Nuclear Beams (Low Energy) — 83
John M. D'Auria

7. A Technique for Proton Drip-Line Studies via Fusion-Evaporation Reactions — 101
M. F. Mohar, W. Benenson, D. J. Morrissey, R. M. Ronningen, B. Sherrill, J. Stevenson, J. S. Winfield, J. Yurkon, J. Görres, and K. Subotic

8. Nuclear Structure in the Neutron-Rich Doubly-Magic ^{78}Ni Region — 111
John C. Hill, F. K. Wohn, J. A. Winger, E. K. Warburton, R. L. Gill, and R. B. Schuhmann

9. Nuclear Spectroscopy in the Rare-Earth Region Near the Proton Drip Line — 127
K. S. Toth, J. M. Nitschke, P. A. Wilmarth, and K. S. Vierinen

10. Shapes and Moments of Odd-A Pt Nuclei ($177 \leq A \leq 195$) 139
J. Rikovska

11. Towards Superheavy Nuclei—Status and Prospects for the Production and Investigation of Heavy Elements 181
G. Münzenberg and P. Armbruster

12. Reflection-Asymmetric Shapes in Nuclei 205
I. Ahmad, M. P. Carpenter, H. Emling, R. Holzmann, R. V. F. Janssens, T. L. Khoo, E. F. Moore, L. R. Morse, J. L. Durell, J. B. Fitzgerald, A. S. Mowbary, M. A. C. Hotchkiss, W. R. Phillips, M. W. Drigert, D. Ye, and Ph. Benet

13. Reflection Asymmetry and Fission Yields in the Mass-145 Region and a Possible Route to Superheavy-Element Synthesis 217
J. D. Robertson and W. B. Walters

14. Quadrupole and Octupole Shapes in Nuclei 229
D. Cline

15. GAMMASPHERE 245
I.-Y. Lee

16. Evidence for Octupole-Octupole and Quadrupole-Octupole Excitations in Spherical Nuclei 259
S. W. Yates, R. A. Gatenby, E. M. Baum, E. L. Johnson, J. R. Vanhoy, T. Belgya, B. Fazekas, and G. Molnar

17. Spin-Stabilized Deformation in Transitional Nuclei 271
P. Chowdhury, B. Crowell, P. J. Ennis, C. J. Lister, Ch. Winter, H. R. Andrews, D. Horn, D. C. Radford, D. Ward, J. K. Johansson, J. C. Waddington, and S. Pilotte

18. Shape Evolution Studies in the $A = 70$ Region 283
A. V. Ramayya and J. H. Hamilton

19. Shape and Structural Changes in Nuclei at High Spin and Temperature 293
T. L. Khoo, I. Ahmad, R. Holzmann, R. V. F. Janssens, E. F. Moore, F. L. H. Wolfs, M. W. Drigert, K. B. Beard, D.-Z. Ye, U. Garg, Ph. Benet, P. J. Daly, and Z. Grabowski

20. Shell-Model Calculations for Exotic Nuclei 295
B. A. Brown, E. K. Warburton, and B. H. Wildenthal

21. Recent Results on Superdeformation 321
M.-A. D. Deleplanque

22. Nuclear Collectivity and Complex Alignment Mechanisms in Light W and Os Nuclei 339
N. R. Johnson

23. Discrete-Line Spectroscopy in Superdeformed Nuclei 355
V. P. Janzen, J. K. Johansson, J. A. Kuehner, D. Radford, J. C. Waddington, D. Ward, and the 8π Collaboration

24. High-Spin Properties of Doubly-Odd Nuclei of Mass ≈130 369
E. S. Paul, C. W. Beausang, D. B. Fossan, Y. Liang, R. Ma, W. F. Piel, Jr., S. Shi, and N. Xu

25. Rotational-Band Structure in Odd-Odd Re Nuclei 379
W. A. Olivier, W.-T. Chou, A. Rios, W. C. McHarris, and R. Aryaeinejad

26. Double Decoupling in Deformed Doubly-Odd Nuclei 393
A. J. Kreiner

27. Deformed Odd-Odd Nuclei: Matrix Elements for the Residual *p-n* Interaction and Patterns of Alternating Perturbations in Level Spacings 413
R. W. Hoff, A. K. Jain, J. Kvasil, P. C. Sood, and R. K. Sheline

28. Proton-Neutron Interactions in the A = 100 Nuclides 427
R. A. Meyer, I. Morrison, and W. B. Walters

29. Dynamical Symmetries for Fermions 437
M. Guidry

30. α Particles as Probes of Nuclear Shape in the Rare Earths and Structure Effects on Proton Emission in the Mass-80 Region 457
D. G. Sarantites, N. G. Nicolis, V. Abenante, Z. Majka, T. M. Semkow, C. Baktash, J. R. Beene, G. Garcia-Bermudez, M. L. Halbert, D. C. Hensley, N. R. Johnson, I.-Y. Lee, F. K. McGowan, M. A. Riley, A. Virtanen, and H. C. Griffin

31. Fast β Transitions and Intrinsic Structures in Deformed Nuclei 473
P. C. Sood and R. K. Sheline

32. Intruder States, Coexistence, and Approaches to Deformation: The Study of ^{120}Xe and the N = 66 Isotones 495
P. F. Mantica, Jr., B. E. Zimmerman, C. E. Ford, W. B. Walters, D. Rupnik, E. F. Zganjar, H. K. Carter, J. Rokovska, and N. J. Stone

33. Global Trends and Structural Consequences of the Proton-Neutron Interaction 511
D. S. Brenner, R. F. Casten, C. Wesselborg, D. D. Warner, and J.-Y. Zhang

34. IBFFA Calculations of Odd-Odd Nuclei 523
W.-T. Chou, W. C. McHarris, and O. Scholten

35. Delayed-Neutron Emission Probabilities of Li-F Nuclides 535
P. L. Reeder, R. A. Warner, W. K. Hensley, D. J. Vieira, and J. M. Wouters

36. Exotic Decays at the Proton Drip Line 549
D. M. Moltz, T. F. Lang, J. Cerny, J. D. Robertson, and J. E. Reiff

37. Delayed Fission of Light Am Isotopes 561
H. L. Hall and D. C. Hoffman

38. High-Spin Spectroscopy for Odd-Z Nuclei with $A \approx 160$ 587
C.-H. Yu, J. Gascon, G. B. Hagemann, and J. D. Garrett

39. Fast Chemical Separations for the Study of Short-Lived Nuclides 609
K. Rengan

40. Rotational Population Patterns and Searches for the Nuclear SQUID 625
L. F. Canto, R. J. Donangelo, A. R. Farhan, M. W. Guidry, J. O. Rasmussen, M. A. Stoyer, and P. Ring

41. Time-Resolved and Time-Integral Studies of Nuclear Relaxation: An Extension of the On-Line Nuclear Orientation Technique to Shorter Half-Lives 637
N. J. Stone

42. Exotic Nuclear Spectroscopy—Remembrance of Past Futures 657
Wm. C. McHarris

Author Index 665

Nuclide Index 667

Subject Index 673

1. Atomic Mass Measurements with Radioactive Beams and/or Targets: Where to Start

Peter E. Haustein
Department of Chemistry
Brookhaven National Laboratory
Upton, NY 11973 USA

ABSTRACT

Radioactive beams or radioactive targets (or both) can significantly increase the yields of exotic isotopes, allowing studies to be performed in regions which are currently inaccessible. An important goal to pursue with these exotic species is a broad program of nuclidic mass measurements. This is motivated by the observation that mass-model predictions generally diverge from one another in regions far from β-decay stability where well-measured masses are sparse or nonexistent. Stringent tests of mass models are therefore possible, and these can highlight important features in the mass models that affect the quality of their short-range and long-range extrapolation properties. Selection of systems to study can be guided, in part, by a desire to probe those regions where distinctions among mass models are most apparent and where exotic isotope yields will be optimal. Several examples will be presented to highlight future opportunities in this area.

1.1 INTRODUCTION

Nuclei far from stability have traditionally been a fertile ground for tests of models of nuclear structure, nuclear masses, and for the discovery of new nuclear decay modes and structural dynamics. Several illustrative examples are: (super)-deformation regions and shape evolution; fissility variation and fission modes; β-delayed particle emission; direct proton radioactivity; synthesis of the heaviest elements; and the interplay between nuclear decay properties and *r*- and *rp*-process nucleosynthesis.

One can confidently expect this to continue as new methods that employ radioactive beams and/or radioactive targets allow even greater excursions to be made from the valley of β stability. Indeed, the increase of available decay energy alone opens more exotic decay channels, many of which may prove to be the doorways to the study of new features of nuclear structure.

It is instructive to look in detail at one selected area—masses of nuclei far from stability. This is motivated by the fact that mass measurements of new nuclei were not part of the database which had been used to refine the parameters used in the construction of the mass model. Therefore, well-measured masses of these new nuclides serve as impartial tests of the predictive quality among diverse sets of mass models. Such tests frequently serve to identify "good" mass models, i.e., those models whose predictions on either a global basis or in restricted mass regions prove to be of highest quality. Having identified a model

Exotic Nuclear Spectroscopy, Edited by W. C. McHarris
Plenum Press, New York, 1990

(or models) in this way, the selection of additional experiments can be guided with greater assurance.

1.2 SHORT REVIEW OF MASS MODELS

A short review of the methods used in the construction of mass models is useful in the context of the analysis which will follow. A more complete description of methods can be found in the most recent compilation of mass-model predictions and related material.[1] Construction of a mass model generally involves several steps. The model's basic content is first selected, i.e., *ab initio,* phenomenological, semi-empirical, liquid drop(let), shell model, or mass-relations based. The next step is the selection and evaluation of each mathematical component of the model that is thought to characterize the relevant physical features of the mass surface. From this step a prototype mass equation is generated. The third step involves attempts to fit the prototype mass equations to the known mass surface. This generally involves the use of iterative least-squares methods wherein mass equation terms and their coefficients are adjusted to minimize residuals with respect to the body of known masses. Once "best fit" parameters are obtained, the final step consists of the calculation of predicted masses for the measured (input) masses and the calculation of predicted (output) masses for isotopes beyond the known mass surface. This is done generally out to the neutron and proton drip-lines and into the region of superheavy elements, unless features of the model preclude such calculations or there are reasons to believe that the predictions become instrinsically unreliable in these regions.

1.3 CHARACTERISTICS OF MASS MODELS IN REGIONS WHERE EXPERIMENTAL DATA EXIST

The "goodness of fit" of a particular model can be quantified by the average and root-mean-square (rms) deviations of the predicted masses relative to the known masses. Average deviation indicates whether, on a global basis, the calculation has predicted slight over-binding or slight under-binding relative to the isotopic database. RMS deviation signals the degree of conformation of the predicted mass surface to the measured one. In general, smaller rms deviations are achieved through the use of larger numbers of adjustable parameters in the models. An interesting correlation between the numbers of adjustable parameters in mass models and the rms deviations of these models to the known mass surface was noted by Tondeur.[2] He observed that models which have most successfully and proficiently incorporated the relevant physical features of mass surface achieve the smallest rms deviations with a minimum number of adjustable parameters.

In the 1986–1987 Atomic Mass Predictions,[1] residuals for each mass model were plotted in the format of the Chart of the Nuclides. This allows one to gauge the reliability of the predictions in regions that border the experimentally known surface. The smallest collections of residuals are generally associated with those models employing the largest number of adjustable parameters. This *does not* necessarily mean that such models will have the best extrapolation properties. Analyses[3-5] of earlier (1975–81) sets of predictions[6-9] showed that for those models there is only a rough correlation between the quality of fits to the known mass surface and the quality of mass predictions for new isotopes that were subsequently measured.

1.4 RADIOACTIVE BEAMS AND TARGETS IN THE PRODUCTION OF NEW EXOTIC ISOTOPES

It is anticipated that one or more large national research centers will be established to provide users with access to radioactive beam facilities. The use of these radioactive beams (possibly in combination with radioactive targets) will permit the production of many new exotic nuclear species with yields that will be substantially higher than are obtainable with more conventional production methods. For example, fusion-evaporation or deep-inelastic-scattering reactions could start from projectile-target combinations where the reaction partners possess N/Z ratios that are significantly different from those of stable nuclides. This feature and careful control of reaction excitation energies can dramatically increase the yields of exotic isotopes far from stability. Projectile fragmentation of medium-energy to high-energy heavy-ion beams is known to yield neutron-rich species. Capture, deceleration, and cooling of such fragments will yield neutron-rich secondary beams. Nuclear structure information can be obtained directly from these beams, or reactions of such beams with stable relatively neutron-rich targets will provide large improvements in the yields of new neutron-rich species. Even larger yields will be possible when radioactive targets that are even more neutron rich are used. Eventual synthesis of the postulated superheavy elements may be achieved this way.

1.5 CHARACTERISTICS OF MASS MODELS IN REGIONS WHERE DATA DO NOT PRESENTLY EXIST

It is instructive to examine the predictions of atomic models in regions beyond the body of measured masses, since these regions will be the ones where exploratory studies with radioactive beams will begin. Previous experience has shown that mass-model predictions (as well as predictions of other nuclear properties) diverge from one another in regions far from stability and that this divergence frequently becomes large and occurs rapidly on excursions from the valley of β stability. Where these divergencies occur and how fast they occur are useful pieces of information. They can point to specific mass regions where selected projectile/target combinations can be employed to produce optimal yields of new isotopes for broad-ranging studies of nuclear structure and masses at those places where distinctions among models are most apparent.

For the specific case of the divergences in mass-model predictions in regions far from stability, a detailed analysis of the 1986–1987 Atomic Mass Predictions has been made. Two sets of predictions (Pape and Antony; Dussel, Caurier, and Zuker) do not contain predictions that cover large regions far from stability. Of the remaining eight sets of predictions (Möller-Nix, Möller et al., Comay et at., Satpathy-Nayak, Tachibana et al., Spanier-Johannson, Jänecke-Masson, and Masson-Jänecke), each includes predictions on both the proton-rich and neutron-rich sides of stability from the lightest elements up to the heaviest actinides (Spanier-Johannson predict only for $Z > 51$).

Root-mean-square divergences in the mass predictions were computed using these eight sets of calculations in the following way: 1) Predictions for any isotope (measured or predicted by systematics) in the 1986 Wapstra-Audi-Hoekstra mass table were excluded; 2) any isotope which had been predicted by any model to be on or beyond the proton or neutron drip lines was excluded; and 3) at least six predictions had to exist for the remaining cases (approximately 1,100 nuclides beyond the known mass surface)—in most cases seven or all eight predictions were used. Once the rms deviations of the predictions were calculated they were binned according to size and plotted (Figs. 1–5) in the format of the chart of the nuclides.

Two additional levels of analyses were then employed. In the first the plots were inspected to identify trends, i.e., the size and rate of the spread of the mass predictions. The trends were divided into two classes. The first was where the divergences in the mass-model predictions tended to be small and to grow slowly. The second consisted of mass regions with large and/or rapidly growing divergences in the mass predictions. Results of these analyses were then used in a final analysis step to identify appropriate (radioactive) beam/(radioactive) target combinations to reach these mass regions.

1.5.A Regions Where Divergences in Mass Model Predictions Are Small

Several regions where the divergences in mass-model predictions are small have been identified. While it may be comforting that a consensus exists among different sets of predictions in these regions, it is important nevertheless to perform experimental tests of theory in each of these regions. These tests can generally be made with modest extensions of existing methodologies; they will not require the more "heroic" efforts outlined below in Sect. 1.5B. Also, in some cases the origins of the small divergences among the mass models can be rationalized. Each case will be examined in order of increasing mass.

Z = 27 to 29, neutron rich: The first case is shown in Fig. 1 with the cluster of squares, signaling smallest rms deviations between 0 and 0.5 MeV, located for neutron-rich nuclei with $Z = 27 - 29$, just beyond the known mass surface. This case seems to be clearly associated

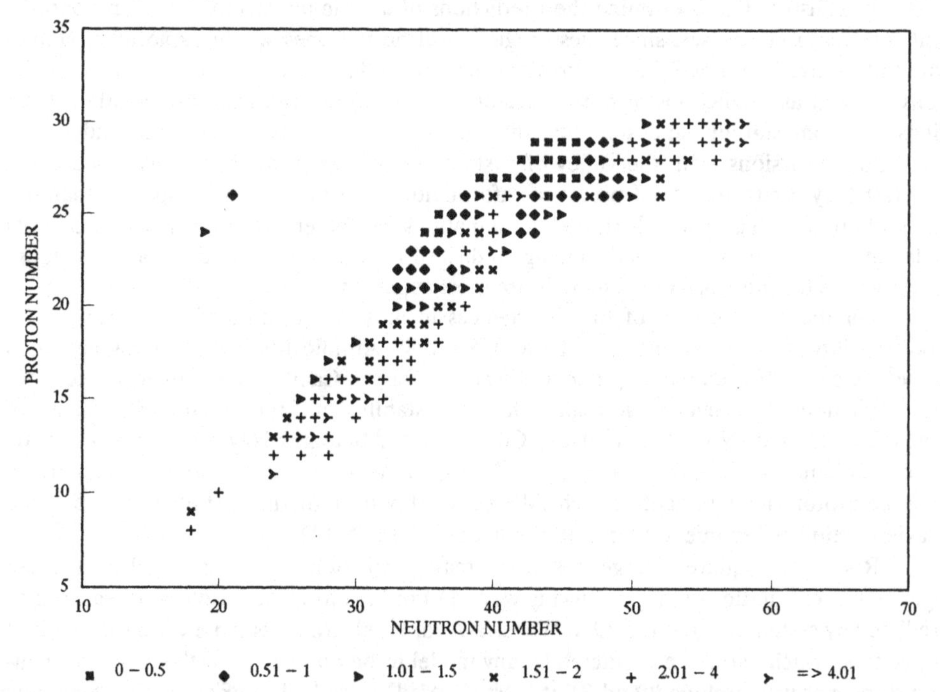

Figure 1. Root-mean-square (rms) divergences of atomic mass predictions, plotted in the format of the Chart of the Nuclides for $Z = 8$ to 30. The rms deviations (in MeV) are binned by energy. Symbols at the bottom of the figure indicate the ranges of the energy bins.

with the proton shell closure at $Z = 28$ and implies that the mass models which contain quite diverse "physics" nonetheless yield nearly identical predictions. It is also quite probable that the database of known nuclei to the left of this cluster contains both important mass and shell-correction information which were essential input for predictions in this mass region.

$Z = 44$ to 48, neutron rich: A broad region from $Z = 35$ to 48 on the neutron-rich side has mass model predictions with relatively small divergences (frequently less than 1 MeV). Between the $Z = 44$ to 48 (squares in Fig. 2) quite small divergences occur. This seems to be associated with a grouping of nuclei that are thought to have small and slowing changing deformation. Adjacent known (fission product) nuclei have been well studied, and many well determined masses have been made in this region. These features are presumed to aid accurate predictions for these nuclides.

$Z = 37$ to 43, along $N = Z$: There are several nuclides in this region, on or very close to the $N = Z$ line that have small spreads in their mass predictions. As one moves toward greater proton richness, the divergences grow rapidly (i.e., right-pointing arrows, plus signs). Adjacent nuclei that do not appear in Fig. 2 have been studied via reactions like $^{40}Ca + ^{40}Ca$ and $^{40}Ca + ^{58}Ni$, and masses have been measured. Variants of these reactions with radioactive beams or targets should allow these presently unmeasured isotopes to be produced in useful yield for mass measurements (see below).

$Z = 73$ to 76, neutron rich: Within the $Z = 62$ to 84 region shown in Fig. 4 there is a small cluster of neutron-rich nuclei with $Z = 73$ to 76 that exhibit small divergences in their pre-

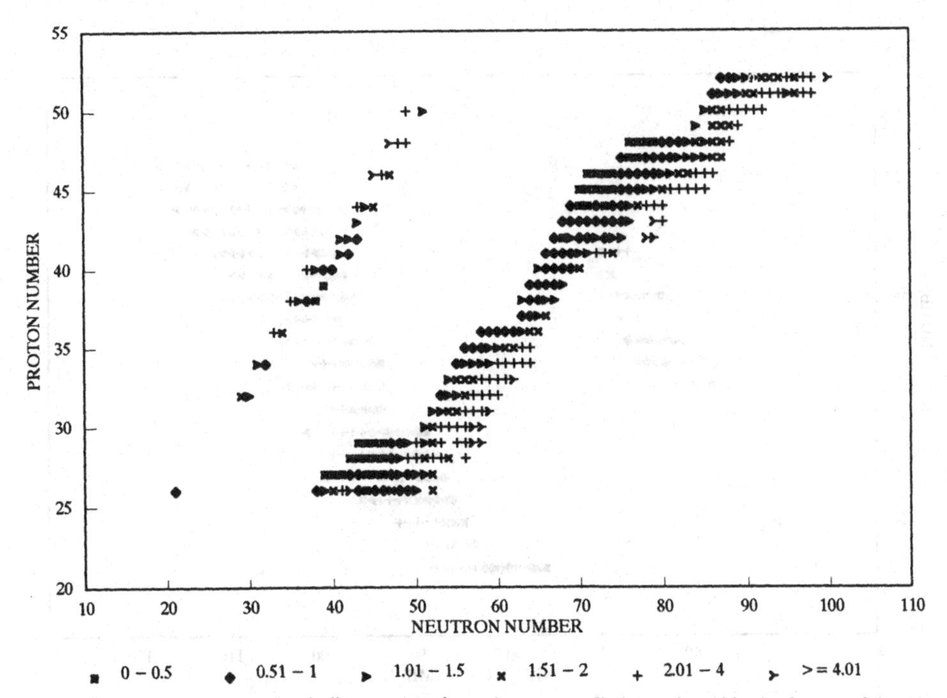

Figure 2. Root-mean-square (rms) divergences of atomic mass predictions, plotted in the format of the Chart of the Nuclides for $Z = 26$ to 52. See Fig. 1 for symbol definitions.

dicted masses. It is interesting to note that this particular region is no longer accessible by fission and that just a few sparse new mass measurements among these nuclei or adjacent ones would be quite useful here as tests of nuclear structure and mass models and to guide additional measurements in the future.

Z = 81, starting near/beyond N = 126: Figure 3 also shows an interesting feature for Z = 81 (Tl). The line of diamond symbols (rms deviations between 0.5 and 1 MeV) that start just beyond N = 126 extends to the right much farther than for adjacent elements. Eventually the trend breaks for Z = 81, and divergences in the mass predictions for very neutron-rich Tl isotopes mirror those of Hg and Pb. The "well-behaved" nature of the double shell closure at ^{208}Pb and a large well-determined mass database for neutron-rich Hg, Tl, and Pb isotopes (principally determined from the terminating mass of the natural decay chains) are supporting aspects in the mass models that appear to yield consistent sets of predictions. What is unclear is why this is more prominent in the Tl isotopes.

Z = 86, neutron rich and Z = 100 to 104, along N = 156 & 157: As a general feature it should be noted that the mass prediction divergences are rather gentle on both sides of stability in the heavy element region (Fig. 5). Two areas within this region exhibit small divergences in the mass-model predictions. The first of these is in neutron-rich Rn (Z = 86) isotopes, and it appears to be associated with the extensive nuclear structure and nuclidic-mass databases of slightly less neutron-rich Rn isotopes that have been accumulated from studies of the natural radioactivity chains and studies of Rn isotopes done with on-line separators. The second region occurs for N = 156 and 157 in elements 100 to 104. This feature appears to be connected to the fact that these nuclides represent

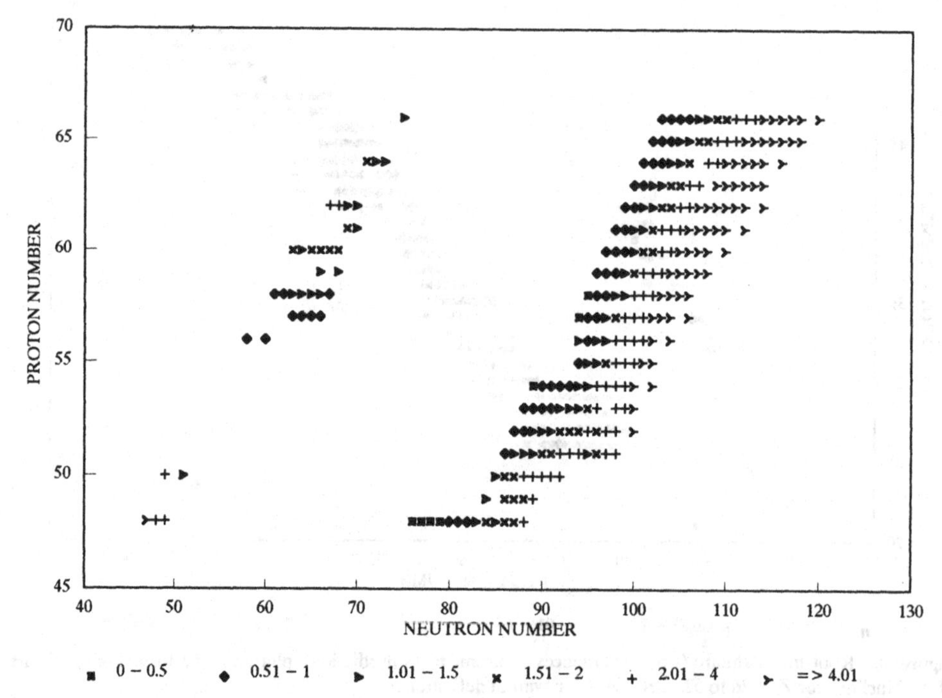

Figure 3. Root-mean-square (rms) divergences of atomic mass predictions, plotted in the format of the Chart of the Nuclides for Z = 48 to 66. See Fig. 1 for symbol definitions.

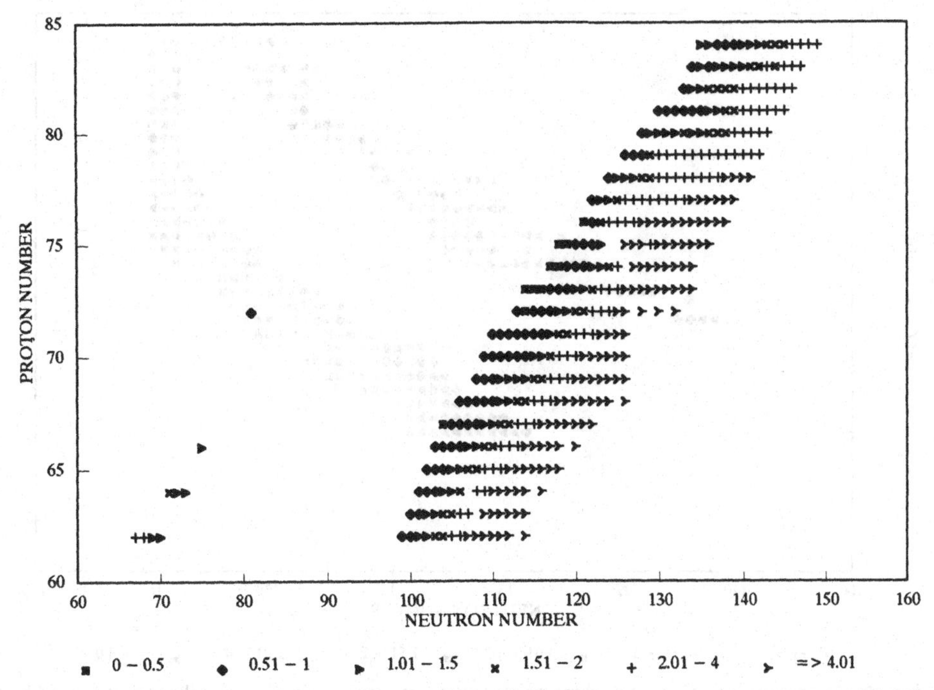

Figure 4. Root-mean-square (rms) divergences of atomic mass predictions, plotted in the format of the Chart of the Nuclides for $Z = 62$ to 84. See Fig. 1 for symbol definitions.

the heaviest long-lived actinide isotopes and that extensive nuclear spectroscopy and mass measurements have been performed in this region that have fleshed out the mass surface and quantified the details of the microscopic nuclear structure.

1.5.B Regions Where Divergences in Mass Predictions Are Large: Mass Regions to Study With Radioactive Beams and/or Targets

We now turn to those regions where divergences in the mass predictions are large and/or grow rapidly on excursion from the stability line. In general, each of these regions lies more remote from stability than those discussed in the preceeding section. In almost all cases, radioactive beams or targets will have to be used to reach these regions.

Z = 10 to 20, neutron rich: Inspection of Fig. 1 reveals that the predictions of the mass models are widely divergent in the light neutron-rich nuclei, ranging in some cases to rms deviations of more than 4 MeV. Any well measured masses of isotopes in this region will clearly be of great interest. The isotopes just beyond the known mass surface are already under study at the TOFI facility at Los Alamos and the LISE spectrometer facility at GANIL.[11] The advent of radioactive beam facilities that could capture, store, and cool light neutron-rich projectile fragments such as, for example ^{40}S or ^{36}Si from the projectile fragmentation of ^{48}Ca, will materially aid additional studies in this region. Inverse reactions of such beams on lighter targets are indicated.

Z = 28 to 36, neutron-rich, especially beyond N = 50: For $N > 50$ in the elements with Z between about 28 and 34 one sees clusters of symbols (pulses, crosses, etc.) in Fig. 2 that

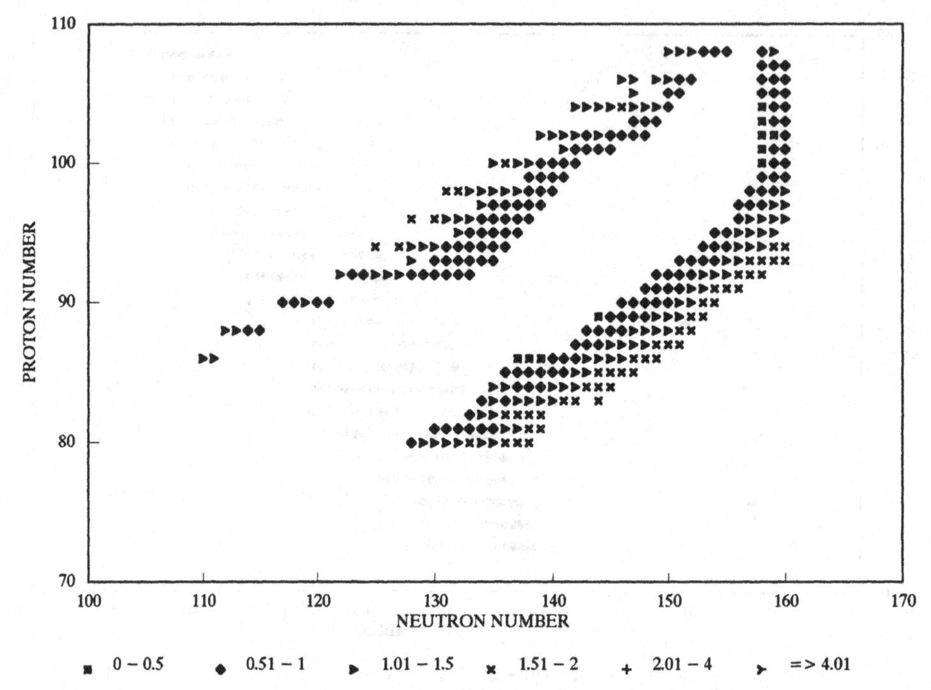

Figure 5. Root-mean-square (rms) divergences of atomic mass predictions, plotted in the format of the Chart of the Nuclides for Z = 80 to 108. See Fig. 1 for symbol definitions.

indicate large divergences in the mass model predictions. For N < 50 in the lighter elements of this region the divergences are much smaller. This signals the influence of the N = 50 neutron shell closure. It also suggests that these nuclei, which are intimately involved in r-process nucleosynthesis beginning in lighter seed nuclei, are rather poorly predicted. Careful mass measurements in this region could identify the model (or models) that best reproduce the mass surface. This could improve significantly the astrophysical aspects which are intimately connected to atomic masses. The best prospects for reaching these nuclei is by utilizing beams of projectile fragments from isotopes somewhat above this mass region, e.g., ^{96}Zr, ^{100}Mo, ^{104}Ru; fission-product 369-d ^{106}Ru also looks quite attractive.

Z = 44 to 50, along N = Z: Relatively few cases are represented in this region, shown in Fig. 2. However the few that are indicate that the small divergences in mass-model predictions that were characteristic of nuclei on or near the N = Z line with Z between 37 and 43 change to much larger ones for elements up to Sn (Z = 50). The heaviest compound nucleus with N = Z that can be made with stable beams and targets is ^{80}Zr (via ^{40}Ca + ^{40}Ca). Thus, further studies of heavier N = Z nuclei via fusion-evaopration channels must employ radioactive beams and/or targets. Several candidate species that possess N = Z and long lifetimes are ^{44}Ti, ^{48}Cr,and ^{56}Ni. Each can be made in high yield by spallation of elements a few atomic numbers higher; conventional methods would be employed in either their acceleration or in their fabrication as targets.

Z = 52 to 95, very neutron rich: Inspection of Fig. 3, 4, and 5 for these nuclides shows a consistent pattern wherein the divergences of the mass model predictions are small at the edge of the known surface but grow steadily. As the neutron drip-line is approached (right side cut off of plotted points at each Z in Figs. 3 and 4), the divergences have grown to more than 4 MeV. An equivalent statement is that the neutron drip line is poorly predicted throughout these elements. Most of the nuclei plotted in these figures will remain inaccessible even with the availability of radioactive beams or targets. The ones that will be reachable, however, are well worth studying. If enough of them are characterized, a pattern will develop that will identify the more-successful mass models in this region. Some additional progress may be made using fast fission-product separators or by extending fission-product measurements into the low-yield wings of the fission-product distribution. Alternative methods to achieve adequate isotopic yields in these regions would be either projectile fragmentation or possibly the combination of very neutron-rich radioactive beams and very neutron-rich radioactive targets.

1.6 SUMMARY

The analysis presented above has highlighted features of atomic mass predictions in regions beyond the experimentally known mass surface. Regions where predictions are generally similar have been noted. Other regions where atomic mass predictions diverge strongly from one another have been identified also. In the former cases modest extension of present day techniques commonly employed in studies of nuclear reactions and nuclear spectroscopy will permit the production and characterization of new isotopes. In the latter cases radioactive beams or radioactive targets (or both) will have to be employed to reach exotic new nuclides and study their properties. In both cases, well-determined nuclear properties such as masses will serve to test theory and to provide guidance for additional investigations.

ACKNOWLEDGMENT

This research was carried out at Brookhaven National Laboratory under Contract DE-AC02-76CH00016 with the U.S. Department of Energy and supported by its Division of High Energy and Nuclear Physics.

REFERENCES

1. The 1986 - 1987 Atomic Mass Predictions, P. E. Haustein, Special Editor, *At. Data Nucl. Data Tables* **39**, 185 (1988).
2. F. Tondeur "Self-consistent study of nuclei far from stability with the energy density method," in *Proc. 4th Int. Conf. on Nuclei Far from Stability,* P. G. Hansen and O. B. Nielsen, eds., CERN Report 81-09, Helsingor, Denmark, 1981, p. 81.
3. P. E. Haustein, "A Comprehensive and Critical Review of the Predictive Properties of the Various Mass Models" in *Proc. 7th Int. Conf. on Atomic Masses and Fundamental Constants (AMCO 7),* O. Klepper, ed., Technische Hochschule Darmstadt, Darmstadt-Seeheim, West Germany, p. 413 (1984).
4. P. E. Haustein, (*Am. Chem. Soc. Symp..* **324**) *Nuclei Off the Line of Stability*, R. A. Meyer and D. S. Brenner, eds., pp. 126-131 (1986).
5. P. E. Haustein, "Predictive Properties of Atomic Mass Models: Their Relationship to Nuclear Decay Modes by Spontaneous Charged-Particle Emission" in *Particle Emission from Nuclei*, **1**, D. N. Poenaru and M. S. Ivascu, eds., CRC Press, Boca Raton, FL, (1989).

6. 1975 Mass Predictions, S. Maripuu, Special Editor, *At. Data Nucl. Data Tables* **17**, 411 (1976).
7. P. Möller and J. R. Nix, *At. Data Nucl. Data Tables* **26**, 165 (1981).
8. J. E. Monahan and F. J. D. Serduke, *Phys. Rev.* C **17**, 1196 (1978).
9. M. Uno and M. Yamada, Report INS-NUMA-40, Waseda University, Tokyo, Japan (1982).
10. J. M. Wouters et al., Los Alamos National Laboratory Report LA-UR-87-4293 (1987).
11. A. C. Mueller et al. *Z. Phys.* **A330**, 63 (1988).

2. A Recoil Mass Spectrometer for the HHIRF Facility

J. D. Cole
Idaho National Engineering Laboratory
E. G. & G.
Idaho Falls, ID 83415

T. M. Cormier
Department of Physics and Cyclotron Institute
Texas A & M University
College Station, TX 77843

J. H. Hamilton
Department of Physics and Astronomy
Vanderbilt University
Nashville, TN 37230

ABSTRACT

A Recoil Mass Spectrometer (RMS) is to be built that will carry out a broad research program in heavy-ion science. The RMS will make possible the study of otherwise inaccessible exotic nuclei. Careful attention has been given to match the RMS to all the beams available from the HHIRF accelerators, including those beams with the highest energy, as well as massive particles for use in inverse reactions. The RMS is to be a momentum achromat followed by a split electric-dipole mass spectrometer of the type operating at NSRL at the University of Rochester. The RMS is essential for many of the proposed experiments on short-lived and/or low cross-section products. The spectrometer design is discussed, with examples and comparisons with other spectrometers given. Detector arrays to be used with the RMS are also discussed.

2.1 INTRODUCTION

For many years scientists have studied nuclei far off stability, both via their radioactive decays and in-beam γ-ray spectroscopy. (Several reviews[1-3] have examples of work relevant to this discussion.) As the regions of known nuclei are pushed out to more neutron deficient nuclei, the cross-sections for their production in heavy-ion reactions are so small ($\lesssim$ a few mb) that they are difficult or impossible to study by traditional in-beam γ-ray techniques. However, there are important physics questions to be answered by studies of these lighter nuclei. New ways have been sought to identify them.

Several groups have responded by recognizing the power of a Recoil Mass Spectrometer (RMS). The Daresbury RMS[4] is operational (yielding exciting results); the RMS at Legnaro[5] (LNL), Italy, and the RPMS[6] at MSU are becoming operational; and others, such as Argonne National Laboratory's Fragment Mass Analyzer[7] (FMA) and Texas A&M University's MARS project,[8] are being built.

The importance of these spectrometers is clear. At the University of Rochester,

Exotic Nuclear Spectroscopy, Edited by W. C. McHarris
Plenum Press, New York, 1990

where the first successful RMS of the current design was developed,[9] studies of nuclei far from stability that have production cross-sections ≈1 mb have been underway for several years. This spectrometer was combined with a segmented neutron detector[10] and Ge detectors to produce a powerful tool for in-beam spectroscopy. These in-beam γ-ray recoil-mass coincidence studies were the first ones done. As noted by the Daresbury group when these results were first reported,[11] the results clearly justified their large effort. Included in the initial studies were the identification for the first time of levels[12] in ^{73}Br and the extension to higher spins of the known bands in a number of nuclei[11,12] in this region. The success of these studies strongly encouraged the expansion of the research with the Rochester RMS.

Nevertheless, the Rochester RMS has limitations, including the energy of the accelerator, which limits the range of the nuclei far off stability that could be reached. This has been improved with the upgraded energy of the Rochester tandem, but there are still regions of heavy nuclei far from stability that will not be accessible because of beam energy. More importantly, the rigidity of the Rochester RMS was not designed to cover inverse reactions, where one uses a heavy projectile on a light target. The inverse reactions are important when very-low cross-section products are to be studied, because kinematic focusing can increase their intensity through the RMS by large factors and because the high-velocity recoils allow Z identification in a ΔE detector, both of which can make the difference between success and failure in an experiment. Thus, the idea for an RMS, which would be connected to the higher-energy accelerators at the Holifield Heavy Ion Research Facility and which would be capable of separating products in inverse reactions, was born.

The key design features which were established for the RMS to be operated at HHIRF were the following: 1) To match as well as possible the RMS rigidity to the beams and energies available from the HHIRF accelerators, especially including those for inverse reactions. 2) To make the solid angle as large as possible to study very-weak reaction channels. 3) To make the spectrometer flexible to cover broad ranges of different research areas, both for now and for the future. The RMS achieves these goals, and it will be an important facility in the world for research with such devices.

2.2 SPECTROMETER DEVELOPMENT

The Recoil Mass Spectrometer described here is designed to analyze heavy nuclear products from a heavy-ion-induced reaction. Several papers[13-15] give reviews of spectrometers used in nuclear physics, including recoil mass spectrometers. The intent here is not to reproduce those discussions but to point to the important factors that make an RMS, and in particular the RMS being constructed for HHIRF, unique in its abilities. Comparisons with specific spectrometers are made to emphasize particular points.

Various spectrometers (principally magnetic in character) have been used since the discovery that radioactive decay involved the emission of particles that could be manipulated by fields. Although the need for high-quality spectrometers dates from the early development of accelerators, it was not until the late 1940's that modern spectrometers began to be designed. There has always been particular interest in determining the Z or element number of the reaction products. The Z cannot be determined by electric or magnetic deflection alone; however, in an inverse reaction, detectors capable of determining Z within one unit can be used at the focal plane of the spectrometer. Omitting Z from further discussion, the properties that can be determined are the mass number (A), the kinetic energy (E), and in some cases the mass (m). To focus recoils from a reaction, the charge of the recoil ($q = Qe$), its velocity (v), and its mass (m) are the parameters that determine its path through a spectrometer. Magnetic devices disperse in momentum (p/q), and electric ones disperse in energy (E/q). A combination of these two elements leads to a focusing of lines of constant m/q on the focal plane of a spectrometer. This is, in essence, the approach taken with the development of the current family of recoil mass spectrometers.

The direction of the development has been influenced by several problems that must be solved to provide good identification of the nuclear products. The direct identification of the products has been limited by two problems. First, the reaction products and the elastically-scattered beam particles usually both lie in the forward direction. For a simple particle detector, this gives a high gross count rate that masks the low count rate of the reaction products. The second problem for detectors alone is that, for the more-massive recoils ($A \approx 100$), the energy resolution of the detectors is insufficient to determine mass using time-of-flight and direct-energy measurements. Highly-developed magnetic spectrometers, such as those discussed in the review[13] by Enge, allow analysis of the heavier fragments by dispersing across the focal plane, depending on various properties, but they still have inadequate beam rejection at 0°. Devices such as beam filters or velocity selectors have been developed to separate the products of interest from the beam, but they leave the balance of the analysis to downstream detectors, which still lack energy resolution for heavy fragments. Only in recent years has there been an effort to combine the beam-rejection function and the analysis function into a single "recoil spectrometer."

Some discussion of the most current efforts in the field are needed for comparison to the spectrometer discussed here. A velocity filter (SHIP)[16] is included in the discussion. SHIP is included for its success in the limited class of experiments for which it was designed. Three full-recoil spectrometers [University of Rochester,[9] Daresbury RMS,[4] and the Laboratori Nazionali di Legnaro (LNL) RMS at Padova, Italy[5]] and a reaction separator [RPMS at NSCL, MSU[7]] will be discussed, as they are working systems or are under construction. The FMA[7] at ANL will be discussed in conjunction with the Legnaro device, as they are virtually identical. Several spectrometers will not be discussed, but are important to the current work in nuclear physics. LARA[17] is being built at Munich, MARS[8] at Texas A&M; also, the LISE[18] and SPEG[19] spectrometers are in use at GANIL.

2.2.A Designs with Velocity Filters

The present SHIP is a velocity filter (*QQQEDDDDEQQQ*) at the Gesellschaft für Schwerionenforschung, Darmstadt, FRG. It is the most outstanding example of a beam-separation device for heavy-ion physics. The beam rejection ranges from 10^{12}–10^{8} with an acceptance of 2.7 msr and a dispersion of 2.2 mm/%. As can immediately be seen, the final identification of the reaction products is almost totally dependent upon the detector at the focal plane.[20] SHIP has rigidity ($E\rho$ = 20 MV = 10 MeV/q, $B\rho$ = 12 kG-m = 69.5 MeV-nucleon/q^2) in the range being discussed for the proposed spectrometer. The outstanding success of the work at GSI with SHIP is an excellent illustration of the matching of the spectrometer to the facility and the program for which it is planned. SHIP was built primarily to search for new elements, including super-heavy ones, and to explore particle-decay modes of nuclei very far from stability. These could be identified by implanting the recoil products in a charged-particle detector and studying their particle-decay modes. Inverse reactions cannot be used at SHIP.

The RPMS at MSU[6] is by far the most rigid of the systems to be discussed. The RPMS uses as its first separation element for beam rejection a wein filter which has the advantage that it can be "tuned" to pass a certain velocity. The condition on this velocity is $v = E/B$; thus, the electric field is not designed to have one value per product to produce an orbit through the machine. High-rigidity ions will pass with a low dispersion, and low-rigidity ions will pass with a high dispersion. This initially seems to be an excellent device, but some high prices are paid for these properties. Following the wein filter are nine quadrupoles and a dipole (*QQQWQQQQQQDQQQ*). The chromatic aberration in this system is large, and, if completely cancelled, the higher-order terms become a problem. This spreads the line widths to the point that the mass resolution is 100 for 8% and 200 for ±4% in ΔE. This is for a recoil energy of a maximum of 30 MeV/nucleon. The solid-angle acceptance is 1 msr. Although a good system for fragmentation or reaction studies, it is limited for other studies.

The RMS Daresbury[4] (*QQQWWQQQSSDQQQ*) is very similar to the MSU design. The velocity selector is separated into two parts, allowing the primary beam to be dumped without striking the electric plates. The solid-angle acceptance is about 1 msr, and sextupoles have been added to correct some of the aberrations. The velocity acceptance is ±2% with a mass dispersion of 10 mm/%. The use of an elegant and complex detector at the focal plane enhances the recoil identification, so the low mass resolution is somewhat misleading. For spectroscopy work, the target is surrounded by an array of compton-suppressed Ge detectors. Future plans call for the use of a Ge ball arrangement of twenty detectors called Poly-TESSA. However, the 35-cm distance from the target to the first element may present problems.

2.2.B Zero-Energy-Dispersion Designs

The discussion will now turn to the three other spectrometers and what can be gained from their design and operation. The design of the RMS at Rochester,[9] the RMS at Legnaro,[5] and the FMA at ANL[6] are all based upon the same premise. This premise is that by having a system with no spatial or angular energy dispersion, the energy aberrations will not be present or be very large for the system. Thus, by having an energy focus, both (x/δ_E) and (θ/δ_E) vanish (δ_E is the fractional energy dispersion $\Delta E/E$). The selection of an electric dipole before and after a magnetic dipole (*EDE*) is the configuration used to produce an energy focus.

The differences between these spectrometers involve how the higher-order corrections are applied and how the intermediate optic constraints result in different focusing and beam rejection. All designs use a split-cylindrical electrostatic deflector. The magnetic dipole separates the two parts of the electric deflector. The high separation of the reaction products and the elastically-scattered beam particles is achieved because of this configuration. The momenta of the beam particles and the fusion reaction products (or p/q) are very close, while their (E/q) are quite different. The electric deflector disperses based upon energy (E/q), thus allowing beam and reaction-product separation to be a maximum. The immediate criticism of this design is that the primary beam strikes the first positive-potential deflection plate in the system. However, the Rochester RMS has been working in this manner for several years with no problems, so this is not considered to be an issue.

The Legnaro spectrometer has been designed to minimize and correct for higher-order aberrations as much as possible. (This is not true of the Rochester spectrometer.) The Legnaro design (*QQESDSE*) uses only the two quadrupoles to focus through the entire spectrometer. No intermediate focus is formed. There is an energy focus, but this does not occur with any spatial focus. With this approach, there is a limit to what one can do with the beam extent and also keep the higher-order aberrations as small as possible. In the dispersive plane the focus is quite good, but in the vertical plan the beam size is approximately 6 cm. The second-order corrections that are made are used in part to make the mass-focal plane tilt go to zero. The Rochester design uses a quadrupole triplet for both entrance and exit to the spectrometer (*QQQEDEQQQ*). The final triplet both magnifies the mass focus and the dispersion preceding it and also brings about a spatial focus at the focal plane position.

The FMA at Argonne is virtually identical to the one at Legnaro in that it is a *QQEDEQQ* configuration. The sextupoles present in the Legnaro design are replaced by a curved poleface boundary on the central dipole. The final quadrupole doublet is used to reduce the vertical beam size to approximately 1.5 cm.

Several general comments on all three of these spectrometers can be grouped together. First, the high-beam rejection is based on multiple scattering within the spectrometer if the beam strikes the first positive plate of the split electric dipole. For so-called normal reactions (projectile particle lighter than target nuclei), this will always work. If the beam does not strike the first electric dipole plates, this rejection is lost. If the energy of the beam particles (E_b) and the energy of the reaction recoils (E_r) both lie within the energy acceptance of the spectrometer, both types of particles will be focused on the focal plane if the rigidity of the spectrometer is sufficiently high to handle the recoils. Obviously, the energy of the recoils can never exceed the energy of the beam particles, so this case need not be considered. Between the two extremes of $E_b >> E_r$ and $E_b \approx E_r$, a very large variety of reactions, focusing conditions, and beam rejection levels can be devised. The net result is that not all or necessarily even a large number of reactions, particularly inverse reactions, can be used in a spectrometer of this type.

Second, the rigidity of these spectrometers is relatively low at 6 MeV/Q for the Rochester RMS and 9 MeV/Q for Legnaro and Argonne machines. This puts an upper limit on the reactions that can be studied with these spectrometers. Any reaction producing a recoil of $\approx$5 MeV/nucleon cannot be studied in these systems (even assuming $Z = N$ and complete stripping). Thus, rigidity is important if the spectrometer is to cover a broad range of reactions.

The discussion has been centered on comparing the design of velocity filters or other recoil mass spectrometers to the design presented here. Some comments comparing a RMS with more traditional magnetic spectrometers (BRS, *Q3D*) or isotope separators need to be made. The rigidity and dispersion desired can be obtained in these spectrometers, but the beam rejection at 0° and large solid angle are problems that cannot be overcome together. Although mass separators can be made to have large acceptance, the problems of different ionization efficiency for different reaction products, the loss of correlation between information at the target location and the separator focal plane, and the long hold-up time of the species in the ion-source are not present (and therefore not problems) in the RMS. Considering these spectrometers, separators, and the other RMS designs discussed, the design effort here is for a spectrometer based upon the Rochester approach, but with significant improvements. Effort was made to remove the second-order aberrations but without the loss in spatial focusing. Rigidity will be high to match the HHIRF accelerators as well as possible.

2.3 DESIGN INTRODUCTION

Careful attention was given to march the RMS to all the beams available from HHIRF accelerators, including those with the highest energy, and also massive particles for use in inverse reactions. For the tandem, the highest energy for fully stripped nuclei would be about 25 MeV/q for $N = Z$. For the electrostatic deflectors used in this design, the maximum energy is about 15 MeV/q or 7.5 MeV/nucleon for $N = Z$ and fully-stripped recoils. This shows the need for the high rigidity. This last consideration is in contrast to some present and proposed RMS facilities.

In designing any spectrometer, close attention must be given to the broad view of what experiments are to be performed with the system. Initially, the primary expectation for the RMS was to study exotic nuclei from normal reactions with a beam that was lighter than the target nuclei. Some attention was given to inverse reactions, but more in the nature of those that would have the beam focused onto the focal plane. Since that original design effort, inverse reactions have been reconsidered. The current view is that "all" inverse reactions must be handled by the RMS, and the case of the primary beam reaching the focal plane is not acceptable. This is to say that the high beam rejection must work equally well for inverse reactions. The kinematic advantage in using inverse reactions is simply too large to lose for any reaction.

A completely new consideration for the spectrometer was the GAMMASPHERE project.[21] At the time the idea for GAMMASPHERE was formulated, none of the recoil spectrometers was considered to be important to its mission. After some consideration and the appearance of early results from experiments at Rochester and Daresbury, the importance of the RMS to GAMMASPHERE was seen. The problem was that none of the existing or proposed spectrometers could operate at the large image or object distances required by GAMMASPHERE without significant loss of performance.

With these new conditions in mind, a modification to the RMS was sought that would allow use of all inverse reactions, have large image and object distances, and meet the high rigidity needs of HHIRF. The result that is presented here is the spectrometer that is currently being built for use at HHIRF.

The general description of the modification is that a momentum achromat has been added to the front of the original RMS design. This is shown in the element layout in Fig. 1.

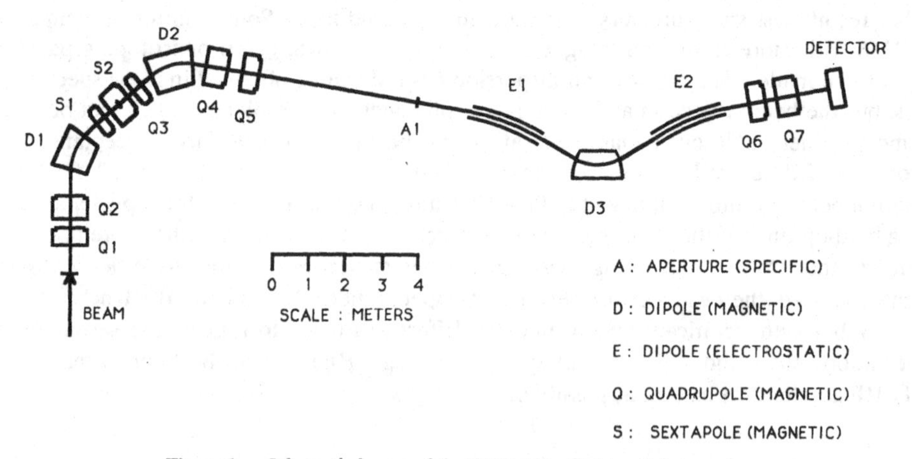

Figure 1. Schematic layout of the RMS. The flight path is to scale.

The optics of the achromat are parallel-to-parallel in the dispersive plane. At the focal plane of the achromat, the center of quadrupole #3 (Fig. 1), the dispersion is ≈10 mm/% and the resolving power is ≈350 at the full solid angle, Ω = 25 msr. The achromat is composed of three quadrupoles and two dipoles. The dipoles are 50° in bend angle with both dispersive in the same direction. The two entrance quadrupoles have 20-cm apertures, and the third quadrupole has an aperture of 30 cm. This achromat precedes a Rochester-style RMS (i.e., a split electric dipole) that is basically the original design.

The achromat is designed such that there is a focus formed between the quadrupoles and the first electrostatic deflector. At this point the beam is completely removed and an image of the target spot is reformed. Since the distance between the quadrupole and the electrostatic deflector is 6 m, with this focus occurring 4.5 m after the quadrupole, the achromat can be used independently of the rest of the RMS. This provides a momentum achromat of rigidity of 25 MeV/nucleon and a solid angle acceptance of 25 msr.

To understand the advantage and power of the addition of the achromat, consider the following example of an inverse reaction at 5 MeV/nucleon: ^{156}Gd on ^{12}C. In this case, ^{164}Yb is the reaction product of interest. For a RMS of the Rochester style, both the primary beam and the reaction products pass through the first electrostatic deflector unscattered. In the case of the ^{156}Gd on ^{12}C, the most probable charge states of the beam actually result in M/Q ratios that are focused at the focal plane. This is shown in Fig. 2 as calculated for the original design without the achromat. Charge states 47 and 48 reach the focal plane of the RMS if they are not stopped in the achromat.

Figure 3 shows the beam particles and the reaction products for the ^{156}Gd on ^{12}C case at the focal plane of the achromat. The most probable charge states are shown but without being weighted. That is to say, an equal number of rays are used in each charge state calculation, and the figure does not represent the most probable population of each charge state. The reaction products clearly fill the space available at the focal plane, while the beam particles are focused into lines. The change in height in the y plane for each mass re-

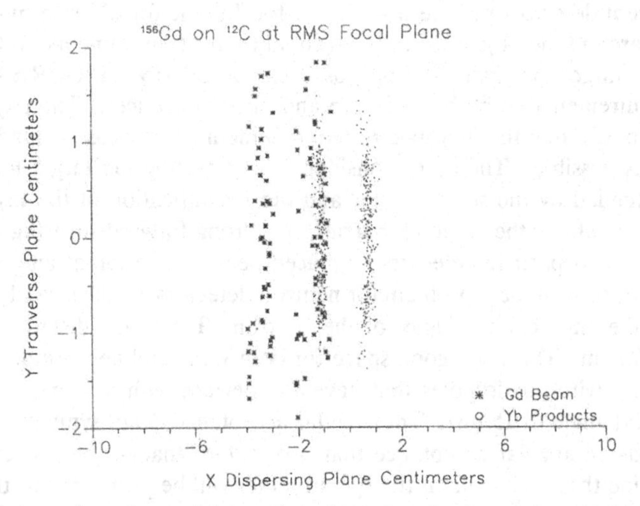

Figure 2. Both beam and reaction particles reach the focal plane for the reaction, ^{154}Gd(^{12}C,4n)^{164}Yb, without the achromat preceding the first electrostatic deflector.

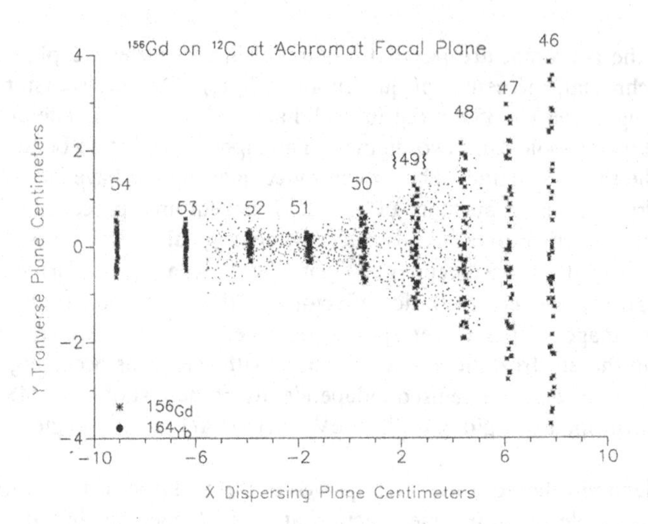

Figure 3. The reaction products and beam particles for the reaction, $^{154}Gd(^{12}C,4n)^{164}Yb$, at the focal plane the achromat. The most probable charge state for the beam is {49}. A percent of the ^{164}Yb would be stopped with the beam, but this is a negligible amount.

sults from the focal plane being at a 70° angle to the beam axis, whereas the calculation is for a plane perpendicular to the beam axis. The beam particles can easily be stopped, while the reaction products are passed with only a small loss.

2.3.A Target Location and First Quadrupoles

The general view of an experiment with an RMS is that the RMS is but one element in the experiment. Detectors (Ge, Si, plastic, NaI, liquid scintillators, gas proportional counters) in singles or various array configurations will be used with the spectrometer. A variety of different detectors may be used at the focal plane for different purposes. However, detectors around the target itself are essential to most experiments, so the initial consideration is the target location. The proposed use of GAMMASPHERE with the RMS puts severe requirements on both the image and object distances. The physical distance from the target position to the first spectrometer element (in this case a quadrupole) needs to be as large as possible. The other considerations affecting the target location are the solid angle subtended by the spectrometer and the magnification of the instrument. For energetic reactions above the coulomb barrier, the strong forward focusing of the emitted particles requires that particle detectors be placed between the target and the spectrometer. In particular, this can be a problem for neutron detectors, as the size of the detector is determined by the interaction distance of the neutron. The target distance chosen in the present case is 75 cm. This gives good space for both γ-ray and neutron detectors. Moreover, this distance with quadrupoles that have an aperture with a diameter of 20 cm still gives a large solid angle of 15 msr. The angular acceptance is not symmetrical, as the dispersive plane has an angular acceptance that is one-third that of the vertical plane. Any slits used to define the solid angle of the spectrometer will be positioned at the entrance to the first quadrupole. The first lens is diverging in the dispersive plane, and the second lens is converging.

2.3.B Electric-Magnetic-Electric Dipoles

The choice made for the present design is an electric-magnetic-electric dipole configuration following the achromat. The physical dimensions of these elements are 20° electrostatic deflectors with a radius of 600 cm and a separation of 10 cm. The deflectors are separated by a magnetic dipole with a 50.0° bend angle and a radius of 140 cm. The gap of the magnet is 10 cm, with the pole faces having no curvature. The magnet is weakly focusing with shim angles, $a = b = 15°$. The electric deflectors are planned to have initially a maximum field of 40 kV/cm, which would yield an electric rigidity, $E\rho = 24$ MV $\equiv 12$ MeV/q. The final goal is to condition the plates to hold 50 kV/cm. This would give $E\rho =$ 30 MV $\equiv$ 15 MeV/q.

2.3.C Final Quadrupole Doublet

At this point, the system is a mass spectrometer. At the exit of the second electric deflector, a triple focus occurs in the dispersive plane. By adding the quadrupole doublet, a vertical focus can be obtained, the mass dispersion increased by changing the magnification. These improvements have their price in other aberrations becoming larger, but some of the worst of these can be corrected in the focal-plane detector.

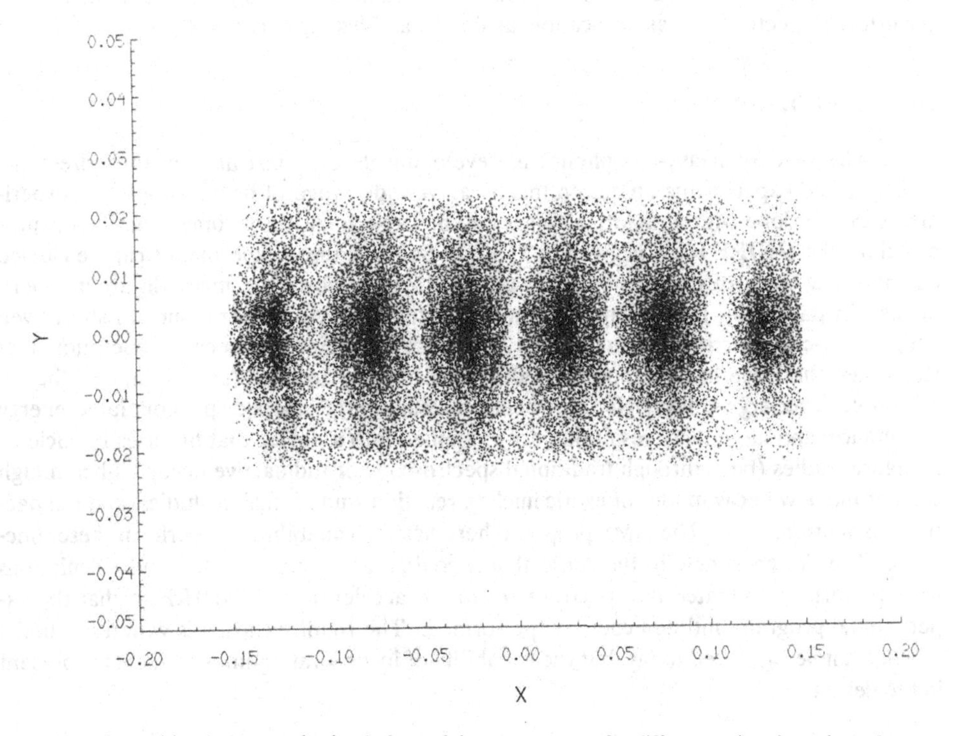

Figure 4. The position in X, Y at the focal plane of the spectrometer for different m/q values about the central mass. The central mass is $A = 100$ and the charge state is 15. The solid angle for this plot is 6.8 msr, and the mass resolution ($m/\Delta m$) is $\approx$360.

In summary, the RMS will have the following characteristics:

a) a large-acceptance solid angle of up to 15 msr
b) an energy range (ΔE) of 3.5%
c) a mass-to-charge ratio range (m/q) of $\pm 5\%$
d) a good mass resolution ($m/\Delta m \approx 770$) (FWHM) at 10.0 msr and ≈ 540 (FWHM) at 15.0 msr (Fig. 4 shows the focal plane resolution for 10.0 msr.)
e) a mass dispersion (x/δ_m) of ≈ 43 mm/% ($\delta_m = \Delta m/m$ or the fractional mass deviation in percent)
f) a magnification at the focal plane for the reaction products
g) an excellent primary beam rejection ($\approx 10^{13}$ in most cases) at 0°
h) a large target-to-first-quadrupole distance of 75 cm

The spectrometer is to be comprised of seven quadrupoles, two electrostatic deflectors, and three magnetic dipoles in a *QQDSQSDQQEDEQQ* configuration. The electrical rigidity ($E\rho$) is initially to be 12 MeV/q, with the design goal being 15 MeV/q. The magnetic rigidity ($B\rho$) for the dipoles is 20 kG-m (2 Tm) maximum with corresponding rigidity planned for the quadrupoles to match closely the spectrometer to the accelerators at ORNL.

The total transmission of the spectrometer is also very good, with approximately 90% of the reaction products reaching the focal plane at the 10.0-msr solid angle. Continuing efforts to improve the spectrometer will occur during the engineering-design phase and with the establishment of the specifications. Final adjustments and refinements will occur to the spectrometer as fabrication, assembly, and testing occur.

2.4 CONCLUSION

The field of heavy-ion physics is developing its own instruments to address the problems and experiments unique to this area. A wide range of heavy-ion-physics experiments can be done with an RMS, as suggested in Table I. A list of some experiments proposed for the HHIRF RMS is given in Table I. Some of these experiments can be carried out on other types of devices, but only an RMS can perform the many different experiments. In particular, it is essential for many of the proposed experiments on short-lived and/or low-cross-section products. To document this, since it became operational at Rochester, the RMS has been used in 50-70% of all experimental running time of the accelerator. The large solid angle, high beam rejection, large mass dispersion, large energy acceptance, and large mass resolution are combined into a device that has uses in nuclear-structure studies (be it through traditional spectroscopy of radioactive decay, in-beam high spin, or the new decay modes of exotic nuclei), reaction studies, fusion studies, and radioactive-beam production. The RMS proposed here has the capabilities to perform these functions. The design is new in the sense that a configuration was selected and calculations were performed to match the spectrometer to the accelerators at HHIRF so that the experimental program outlined could be performed. The rigidity chosen is within the limits of what can be produced today, but the capability of incorporating improvements is present in the design.

Table I. Research with an RMS

Radioactive decay of proton- and neutron-rich exotic nuclei.
- Generally inaccessible nuclei
  - Weak reaction channels ($\approx$1 mb)
  - Short half-lives ($\approx$100 ms)
  - Difficultly-ionized species (e.g., Zr, La, W, Ta)
- Exotic decay modes—at (or past) proton/neutron drip lines
- β-delayed particle emission
- Super-heavy nuclei
- Low-lying, excited-, and ground-state properties

High-Spin studies of neutron- and proton-rich nuclei
- Nuclei with weak reaction channels
- Continuum γ-ray studies
- Band structure of weakly-populated states
- Alignment at high spins
- Average internal conversion

Fusion studies
- Measure fusion cross-sections
- Fusion resonance

Reaction studies
- Quasi-elastic
- Fragmentation
- Massive transfer reactions
- Resonances

Radioactive beams
- Coulomb re-excitation of reaction products
- NMR studies
- Implantation at lattice sites
- Characterize transport techniques
- Atomic spectroscopy of highly-ionized atoms

ACKNOWLEDGMENTS

The funding for our RMS is being provided by the U.S. Department of Energy, a major contribution from the State of Tennessee, significant contributions from Idaho National Engineering Laboratory, Vanderbilt University, the University of Tennessee-Knoxville, Oak Ridge Associated Universities, and the Director's Office of Oak Ridge National Laboratory; also, contributions from UNISOR and its members, Georgia Institute of Technology, Mississippi State University, and the University of Maryland. This work was supported in part by the U. S. Department of Energy under contracts No. DE-AC07-76ID01570, DE-FG05-86ER40256, and DE-FG05-88ER40407.

REFERENCES

1. J. H. Hamilton et al., *Rep. Prog. Phys.* **48**, 631 (1985).
2. J. H. Hamilton et al., *High Angular Momentum Properties of Nuclei*, Ed. by N. R. Johnson (Harwood, New York, 1982), p. 227.
3. J. H. Hamilton, *Treatise on Heavy Ion Science*, **8**, Ed. by A. Bromley (Plenum, New York, 1984), p. 2.
4. H. G. Price, Daresbury Lab Report, DL/NUCL/R19 (1979).
5. P. Spolaore et al., *Nucl. Instr. Meth.* **A238**, 381 (1985); P. Spolaore et al., *Nuovo Cim.* **81A**, 351 (1984).
6. L. H. Harwood and J.A. Nolen, Jr., *Nucl. Instr. Meth.* **186**, 435 (1981).
7. C. N. Davids and J.D. Larson, *Nucl. Instr. Meth.* **B40-41**, 1224 (1989).
8. T. M. Cormier, private communication.
9. T. M. Cormier and P.M. Stwertka, *Nucl. Instr. Meth.* **184**, 423 (1981); T. M. Cormier et al., *Nucl. Instr. Meth.* **212**, 185 (1983).
10. R. B. Piercy et al., "Conference on Instrumentation for Heavy-Ion Nuclear Research," ORNL Report CONF-841005, 27 (1984).
11. J. H. Hamilton, *Fundamental Problems in Heavy-ion Collisions*, Ed. by N. Cindro, W. Greiner, and R. Caplar (World Scientific Publishing, Singapore, 1984), p. 111.
12. S. Wen et al., *J. Phys. G* **11** L173 (1985); S. J. Robinson et al., *Bull. Am. Phys. Soc.* **30**, 726 (1985).
13. H. A. Enge, *Nucl. Instr. Meth.* **186**, 413 (1981).
14. H. A. Enge, *Nucl. Instr. Meth.* **162**, 161 (1979).
15. K. Sistemich, *Nucl. Instr. Meth.* **139**, 203 (1976).
16. G. Münzenberg et al., *Nucl. Instr. Meth.* **161**, 65 (1979).
17. W. Wilhelm, private communication.
18. D. Guillemaud-Mueller et al., *Proceedings of the 5th International Conference on Nuclei Far From Stability, Rosseau Lake, September 1987*, Ed. by I.S. Towner, (AIP Conference Proceedings **164**, New York, 1988), p. 757.
19. Zahn Wemlong et al., *Nouvelle de GANIL* **25**, 22 (1988).
20. G. Münzenberg et al., *Nucl. Instr. Meth.* **186**, 423 (1981).
21. I.-Y. Lee, Chapter 15 in This Book.

3. The Study of Exotic $N \approx 82$ Nuclei Using the Daresbury Recoil Mass Separator

J. H. McNeill*†
Oak Ridge National Laboratory
and Joint Institute of Heavy-Ion Research
Oak Ridge, TN 37831, USA
and
Department of Physics
University of Manchester
Manchester M13 9PL, England

A. A. Chishti, W. Gelletly,‡ M. A. C. Hotchkis, and B. J. Varley
Department of Physics, Schuster Laboratory
University of Manchester
Manchester M13 9PL, England

J. Blomqvist
Manne Siegbahn Institute of Physics
S-10405 Stockholm, Sweden

P. J. Daly
Department of Chemistry
Purdue University
West Lafayette, IN 47907, USA

M. Piiparinen
Department of Physics
Jyväsklä University,
SF-40100 Jyväskylä, Finland

P. J. Woods
Department of Physics
University of Edinburgh
EH9 3JZ, Edinburgh, Scotland

*Present address: S. E. Grand Gulf Nuclear Station, Port Gibson, MS 39150, U. S. A.

† Oak Ridge National Laboratory is operated by Martin Marietta Energy Systems, Inc., under Contract No. DE-AC05-84OR21400 with the U. S. Department of Energy.

‡ Present address: SERC-Daresbury Laboratory, Daresbury, Warrington, Cheshire, WA4 4AD, England, U. K.

ABSTRACT

Experiments using the Daresbury Recoil Mass Separator have identified microsecond isomers in the exotic $N = 82, 83$ nuclei, ^{153}Yb, ^{153}Lu, ^{154}Hf, and have established their decay schemes. The results for ^{153}Lu and ^{154}Hf, together with those for lighter $N = 82$ isotones, provide an outstanding illustration of the dependence of $E2$ transition rates between J^π states on the subshell occupation, and demonstrate that half-filling of the $\pi h_{11/2}$ subshell in the $N = 82$ series occurs just below $Z = 71$ ^{153}Lu. The result for the isomer in the even-odd $N = 83$ nucleus ^{153}Yb shows that the isomerism results from a low-energy $E2$ transition, rather than an $E3$ transition as in lighter $N = 83$ even-odd isotones, which is another consequence of the $\pi h_{11/2}$ subshell being about half-filled. Furthermore, the long-lived isomer observed in the odd-odd $N = 83$ nucleus ^{154}Lu also reflects that the $\pi h_{11/2}$ subshell is close to being half-filled.

3.1 FILLING OF THE $\pi h_{11/2}$ SUBSHELL: $N = 82$ ISOTONES ^{153}Lu and ^{154}Hf

Some ten years ago it was found[1] that the $Z = 64$, $N = 82$ species ^{146}Gd displays properties resembling those of a doubly-magic nucleus. This discovery stimulated spectroscopic studies of many neighbouring nuclei that have a few valence nucleons outside the ^{146}Gd core and may be well described in terms of shell-model configurations. Particularly interesting are the proton-rich $N = 82$ isotones above ^{146}Gd, where the three nearly degenerate proton subshells $h_{11/2}$, $s_{1/2}$, and $d_{3/2}$ are being filled. The high-j unique-parity $\pi h_{11/2}$ orbitals should be especially important in the formation of high-spin states, and one could expect yrast excitations of the type $\pi h_{11/2}{}^n$ to figure prominently in the spectra of $N = 82$ isotones with Z–64 valence protons. Experiments[2-7] have identified and characterized the decays of related $\pi h_{11/2}{}^n$ $E2$ isomers below 3 MeV in the two- to six-valence-proton $N = 82$ nuclei ^{148}Dy, ^{149}Ho, ^{150}Er, ^{151}Tm, and ^{152}Yb (Fig. 1). In the even-A nuclei, the isomeric transitions are between 10^+ and 8^+ $\pi h_{11/2}{}^n$ states of seniority $v = 2$; in the

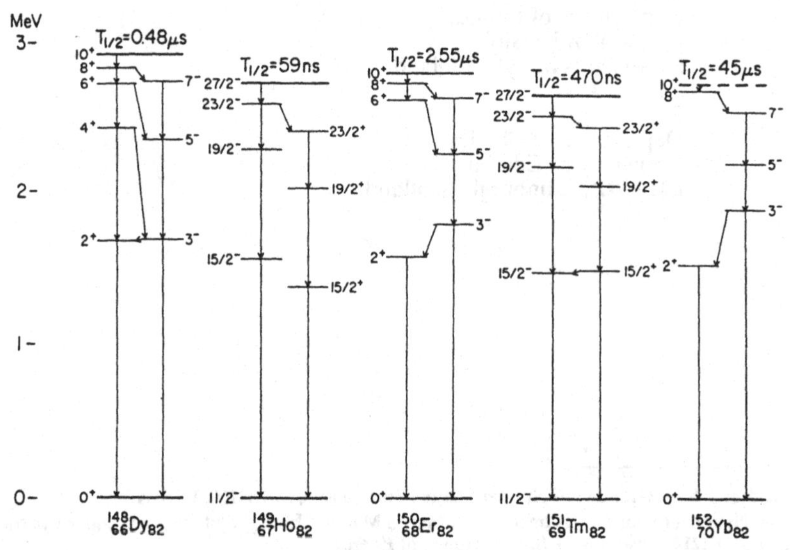

Figure 1. Systematics of the yrast levels up to 3 MeV in the $N = 82$ isotones, ^{148}Dy, ^{149}Ho, ^{150}Er, ^{151}Tm, and ^{152}Yb.

odd-A nuclei, they are $27/2^- \rightarrow 23/2^-$ transitions between $\pi h_{11/2}{}^n$ $\nu = 3$ states. These isomeric decays also populate 3^- octupole, $[\pi h_{11/2}{}^{n-1} s_{1/2}]5^-$, and $[\pi h_{11/2}{}^{n-1} d_{3/2}]7^-$ $\nu = 2$ excitations in the even-A nuclei, and corresponding $15/2^+$, $19/2^+$, and $23/2^+$ $\nu = 3$ states, formed by coupling with an additional $h_{11/2}$ proton, in the odd-A nuclei.

The results for ^{148}Dy include a complete spectrum[2] of $\pi h_{11/2}{}^2$ levels (0^+, 2^+, 4^+, 6^+, 8^+, 10^+) whose energies provide empirical two-body matrix elements for calculating the $\pi h_{11/2}{}^n$ spectra in the heavier isotones. Moreover, Lawson[8] has used the ^{148}Dy energies to show that the condition for no seniority mixing is very nearly fulfilled. When seniority is a good quantum number, the reduced $E2$ transition rates between corresponding $\pi h_{11/2}{}^n$ states of the same seniority should be proportional to $(6-n)^2$, where n is the $\pi h_{11/2}$ occupation number, and should become zero when $n = 6$ at the half-filled subshell.[9,10] The relevant expressions for the $\nu = 2$ and $\nu = 3$ transitions are[8]:

$$B(E2; 10^+ \rightarrow 8^+) = \left[\frac{6-n}{4}\right]^2 \frac{2025}{35321\pi} \cdot (e_{\text{eff}} < r^2 >)^2 \tag{1}$$

and

$$B(E2; 27/2^- \rightarrow 23/2^-) = \left[\frac{6-n}{3}\right]^2 \cdot \frac{32400}{265837\pi} (e_{\text{eff}} < r^2 >)^2 \tag{2}$$

Using the fixed value for the effective charge $e_{eff} = 1.5\,e$, and $\langle r^2 \rangle = 32$ fm^2 for the $\pi h_{11/2}$ orbital, the measured B(E2) values for the five $N = 82$ nuclei are very well reproduced[11,12] (Fig. 2) *by assuming that in every case n equals Z–64, the number of valence protons*. For ^{152}Yb with six valence protons, the experimental $B(E2)$ is not exactly zero as in the ideal picture, but its very small value (0.02 Weisskopf units) demonstrates that the $\pi h_{11/2}$ subshell is close to being half-filled in the 8^+ and 10^+ states.

It is difficult to see why this simple model is so successful, since one would expect that in the relevant states of the $Z > 67$ nuclei, the $s_{1/2}$ and $d_{3/2}$ subshells should also be partly occupied by proton pairs, reducing the occupation of $\pi h_{11/2}$. On the other hand, the single-particle energy gap at $Z = 64$ is significantly smaller than the gaps at traditional magic numbers, and scattering of proton pairs from $g_{7/2}$ or $d_{5/2}$ across the gap into $h_{11/2}$ may compensate in some measure for the depletion in $s_{1/2}$ and $d_{3/2}$. Calculations[11,13] which take account of these two effects suggest that half-filling of the $\pi h_{11/2}$ subshell might be postponed to ^{153}Lu$_{82}$ or ^{154}Hf$_{82}$, but there have been no data for these nuclei up to now.

The $N = 82$ isotones ^{153}Lu and ^{154}Hf lie very close to the proton drip line. Previous attempts to study these nuclei have been unfruitful, mainly because they can be produced only in reactions with very low cross-sections. In the present experiments, the sensitivity for detection of weak isomeric decays was considerably enhanced by the use of the Daresbury Recoil Mass Separator.[14] The ionization chamber, normally located behind the separator focal plane, was replaced by an aluminium catcher foil, which was surrounded by four large Ge detectors and a LEPS. Fusion-evaporation product recoils from the reaction, ^{102}Pd + ^{54}Fe $\rightarrow$ ^{156}Hf*, were mass analyzed, and, after passing through the position-sensitive focal plane detector,[14] were deposited on the catcher foil for γ-ray measurements. Time relationships between signals from the position-sensitive detector and from the γ-ray detectors were used for mass, half-life, and γ-γ coincidence determinations. Since the separator transit time is ≈ 1 μs, only long-lived isomeric species were observed, but they could be studied under low-background conditions. The measurements were performed over a three-day period using 10-15 particle-nA beams of 240- to 245-MeV ^{54}Fe ions on a 1-mg/cm^2 ^{102}Pd target.

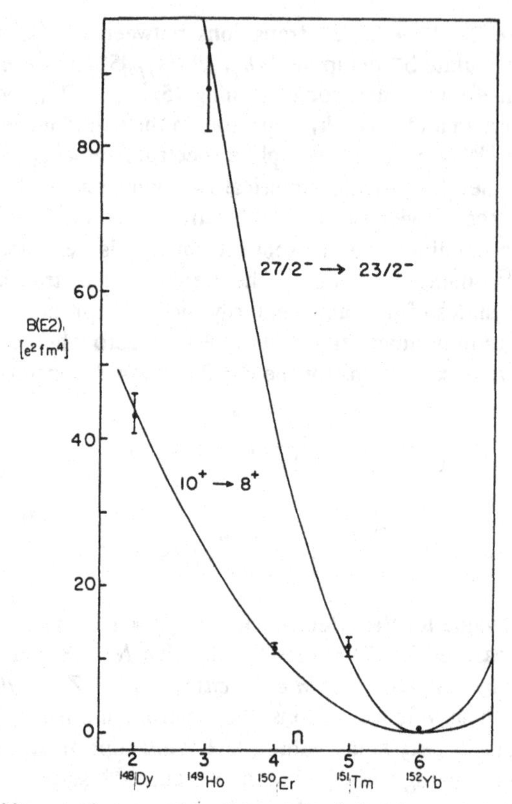

Figure 2. The $E2$ transition rates between 10^+ and 8^+ states of the $\pi h_{11/2}{}^n$ $v = 2$ spectra in the even-A nuclei (^{148}Dy, ^{150}Er, ^{152}Yb) and between $27/2^-$ and $23/2^-$ states of the $\pi h_{11/2}{}^n$ $v = 3$ spectra in the odd-A nuclei (^{149}Ho, ^{151}Tm). The measured $B(E2)$'s are shown with error bars, and the error bar for ^{152}Yb is within the plotted point. Two parabolae show the calculated $B(E2)$ transition rates of pure $\pi h_{11/2}{}^n$ configurations using $\langle e_{eff} r^2 \rangle \ \pi h_{11/2} = 49\ e\ \mathrm{fm}^2$.

Production cross-sections predicted by the code CASCADE[15] for the reaction products, ^{153}Yb, ^{153}Lu, ^{154}Lu, and ^{154}Hf, all of which were expected to have long-lived isomers, are summarized in Table I, which also lists relative yields at the catcher foil for the isomeric species identified in these experiments. In line with the predictions, by far the strongest family of γ rays was observed in the $A = 153$ mass window (Fig. 3) and to be coincident with Yb x rays. The twenty-three γ rays thus assigned to ^{153}Yb have been placed[16] in the decay scheme (Fig. 9) of a new 15-μs isomer in that nucleus. A second, much weaker (<2%) γ-ray family, including 130-, 174-, and 217-keV lines, was also observed in the A = 153 mass window. These γ rays are assigned to the $p2n$ evaporation product, ^{153}Lu, because they appeared in coincidence with Lu K x rays (Fig. 5a), whereas the ^{153}Yb lines were suppressed by more than a factor of 50 in the same spectrum. Other transitions in ^{153}Lu were identified by γ-γ coincidences (for example, Fig. 5b). The ^{153}Lu γ rays were observed to decay with $t_{1/2} = 15 \pm 3\ \mu$s, and they are all placed in the isomeric decay scheme shown in Fig. 4 on the basis of the γ-γ coincidence results. Table II lists the energies and intensities for all γ rays placed in the ^{153}Lu decay scheme. Spins and parities are assigned mainly by analogy with the similar decay schemes of the shorter-lived $27/2^-$

Table I. Production Cross-Sections for the Reaction ^{102}Pd + ^{54}Fe → ^{156}Hf* predicted by the Code CASCADE at ^{54}Fe Beam Energies of 235 and 245 MeV. The right column gives relative experimental yields (in arbitrary units) for the isomeric species identified.

Exit Channel	Product	σ (235 MeV) (mb)	σ (245 MeV) (mb)	Isomeric Yields[a]
2pn	^{153}Yb	73	60	70
p2n	^{153}Lu	2	3.5	1.0
pn	^{154}Lu	3	1.2	2.5
2n	^{154}Hf	0.2	0.1	0.06

[a]Uncorrected for differences in the separator transmission for different products.

isomers[3,4,6] in ^{149}Ho and ^{151}Tm. The total conversion coefficient of the 130 keV transition, α_{tot} (130 keV) = 1.1 ± 0.2 derived from intensity balance, is consistent with *E*2 character, and the measured half-life gives $B(E2$; 130-keV, ^{153}Lu$)$ = 0.45 ± 0.09 e^2fm^4, the smallest value determined in the *N* = 82 series. The γ-γ time distributions for ^{153}Lu indicated a lower-lying isomer with $t_{1/2}$ > 0.1 μs, which we associate with the 23/2$^-$ state.

As Fig. 4 shows, the spectrum of $\pi h_{11/2}{}^7$ levels calculated using the empirical $\pi h_{11/2}{}^2$ interactions from ^{148}Dy is in very good agreement with experiment, especially for

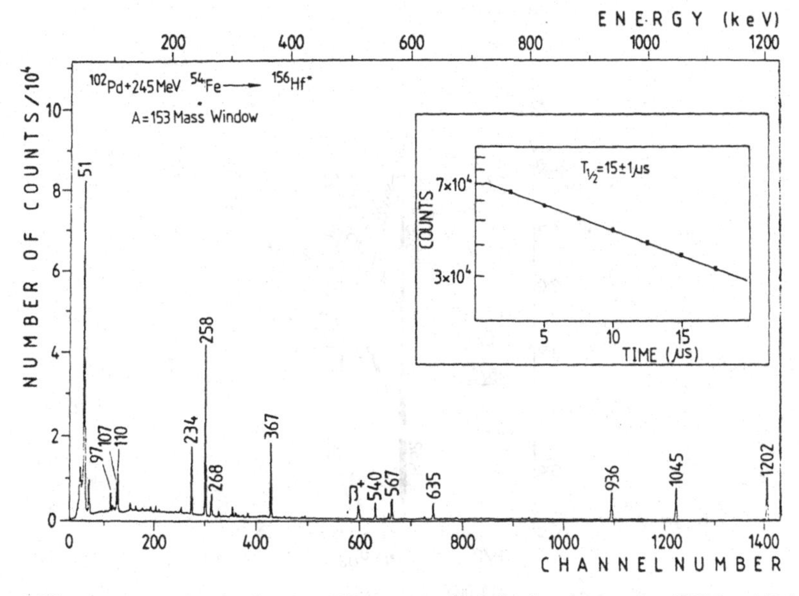

Figure 3. Off-beam γ-ray spectrum for *A* = 153 mass recoils following the reaction ^{102}Pd + 240-245-MeV ^{54}Fe. Lifetime data are shown in the inset.

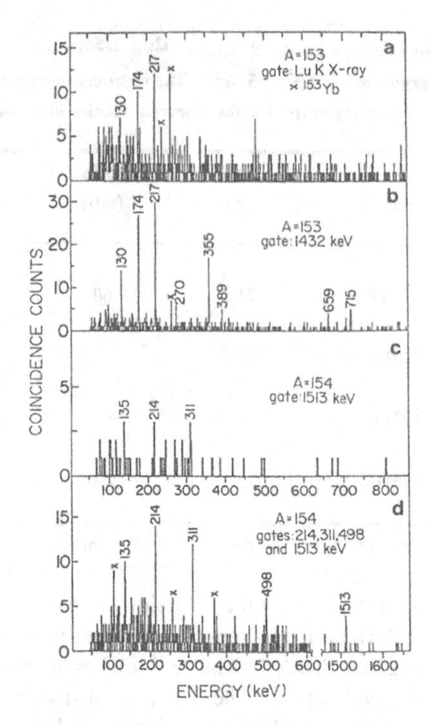

Figure 4. Isomeric decay scheme for ^{153}Lu. Transition arrow widths are proportional to the measured intensities, with internal conversion contributions unshaded. The calculated $\pi h_{11/2}{}^7$ spectrum, with the 27/2⁻ level energy matched to experiment, is shown to the left.

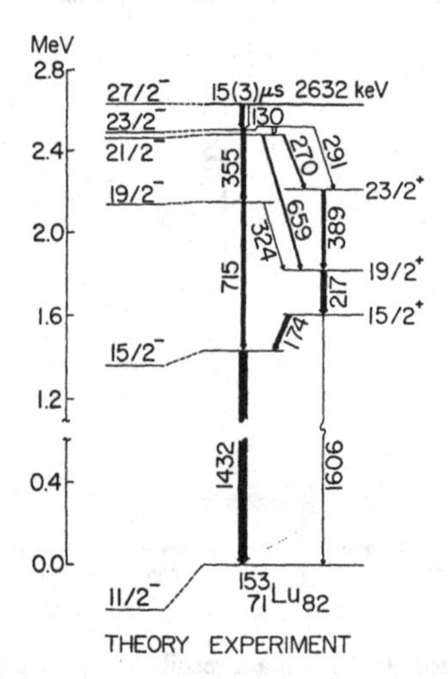

Figure 5. Key γ-ray coincidence spectra for the specified mass windows and gating transitions.

Table II. Energies, Intensities, and Placements of γ Rays Following the 15-μs Isomer in ^{153}Lu

Energy (keV)	Off-Beam Intensity	Placement (keV)
130.4(2)	54(4)	2633.0
174.0(2)	46(4)	1606.1
216.5(2)	65(6)	1822.6
270.0(3)	24(2)	2147.2
291.0(3)	8(2)[a]	2502.6
323.9(3)	10(2)[a]	2147.2
355.4(3)	57(4)	2502.6
389.1(3)	35(3)	2211.7
658.8(3)	22(3)	2481.4
715.1(3)	35(4)	2147.2
1432.1(3)	100(9)	1432.1
1606.1(3)	≈2	1601.1

[a]Obtained from coincidence gates.

the four topmost levels. The 21/2$^-$ level in ^{153}Lu, populated by an unobserved 21-keV transition from the 23/2$^-$ level, has no observed counterparts in ^{149}Ho and ^{151}Tm. Theory[8] predicts such a level ≈26 keV below 23/2$^-$ in each of the three nuclei, but the 23/2$^-$ → 21/2$^-$ $M1$ transition is strongly forbidden. It appears that the 23/2$^-$ → 19/2$^-$ $E2$ decay must be extremely retarded, as it is in ^{153}Lu, before the 23/2$^-$ → 21/2$^-$ transition becomes competitive. An interesting note is that the 27/2$^-$ 15-μs isomeric state in ^{153}Lu has a positive Q value of 3.3 ± 0.7 MeV (Ref. 17) for proton decay to the ^{152}Yb ground state, but this decay mode must be suppressed by the centrifugal barrier associated with $l = 13$ nucleon transfer and by the complicated nucleon rearrangements involved.

The possible $A = 154$ products of the reaction were ^{154}Yb, ^{154}Lu, and ^{154}Hf, and of these the $N = 84$ nucleus ^{154}Yb has no microsecond isomers. The most intense $A = 154$ γ rays were found to be coincident with Lu x rays, and they are assigned[18] to a new 35-μs isomer in ^{154}Lu (see below, Fig. 10, 11). Careful scrutiny of the $A = 154$ γ-ray spectrum for a possible ^{154}Hf $2^+ \rightarrow 0^+$ transition in the 1490-to-1530-keV range suggested by systematics revealed a weak 1513-keV γ ray that decayed with $t_{1/2} = 9 \pm 4$ μs. A gate on the 1513-keV transition (Fig. 5*c*) indicated coincidence peaks (each with 4 to 6 counts) at 135, 214, and 311 keV, and additional gates on these low-energy lines identified another γ ray of 498 keV in the same nucleus (Fig. 4*d*). Although the data are statistically poor and provide no Z identification, the firm $A = 154$ mass determination, the $N = 82$ systematics, and the observed reaction yield (Table I) all favour the assignment of these five γ rays to the decay of a 9-μs isomer in ^{154}Hf. When the transitions are ordered as shown in Fig. 6, the smooth energy-level systematics[2-7] of the even-A $N = 82$ isotones are extended to $Z = 72$ in a convincing manner. The value of the ^{152}Yb 10^+ half-life shown in Fig. 6 was determined during the present studies. By assuming that the energies of the unobserved $10^+ \rightarrow 8^+$ $E2$ transitions in ^{152}Yb and ^{154}Hf lie within the range 14 to 70 keV, as systematics indicate, the following results are obtained: $B(E2;\ 10^+ \rightarrow 8^+,\ ^{152}\mathrm{Yb}) = 0.9 \pm 0.1\ e^2\mathrm{fm}^4$ and $B(E2;\ 10^+ \rightarrow 8^+,\ ^{154}\mathrm{Hf}) = 2.9 \pm 1.4\ e^2\mathrm{fm}^4$.

The faster transition rate for ^{154}Hf is a clear signal that in this nucleus the $\pi h_{11/2}$ subshell is more than half-filled.

For both ^{153}Lu and ^{154}Hf, the $B(E2)$ results are much smaller than the values pre-

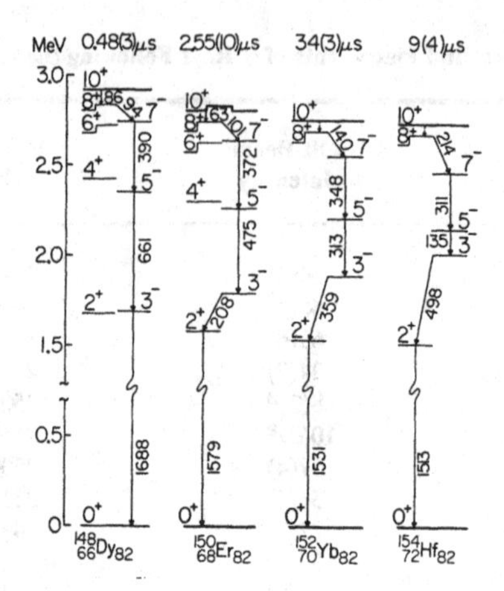

Figure 6. Energy systematics of even-A $N = 82$ isotones, showing the main decay pathways of the 10^+ isomers.

dicted using the equations and model assumptions specified earlier, and they imply that the $\pi h_{11/2}$ occupation numbers for this nuclei are, in fact, somewhat less than Z–64. The overall results are illustrated in Fig. 7, where the transition amplitudes $\sqrt{B(E2)}$ for the seven $N = 82$ isotones are plotted versus Z–64. The square root leaves an ambiguity about the sign of the $E2$ matrix element, which should change at the point of half-filling, but the data strongly indicate that the sign change occurs just before Lu. The fact that the smooth curves through the data points both intersect the zero axis at the same value of Z–64 provides reassurance that correct signs have been chosen for the amplitudes. The variation of $E2$ transition amplitude with number of valence protons is not exactly linear, but nearly so, and it is clear that the $\pi h_{11/2}$ subshell is closest to being half-filled in the $Z = 71$ isotone ^{153}Lu. It is less easy to specify the magnitude of the $\pi h_{11/2}$ $E2$ effective charge in this region, but it may be significantly larger[11] than 1.52 e, the value suggested previously.[2]

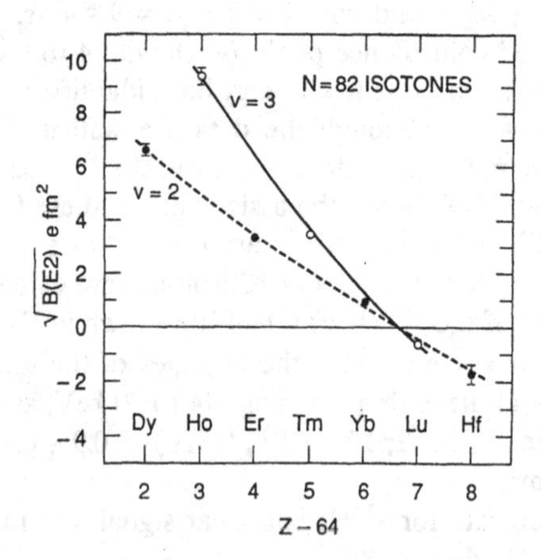

Figure 7. Measured $E2$ transition amplitudes for $\nu = 2$ and $\nu = 3$ isomeric transitions in $N = 82$ isotones as a function of $Z - 64$. Where error bars are not shown, they lie within the plotted point.

3.2 THE 15-μs ISOMER IN THE Z = 70, N = 83 NUCLEUS ^{153}Yb

An intriguing problem in the structure of proton-rich N = 83 isotones has recently come to light.[19] Three-particle $27/2^-$ isomers, having the aligned configuration, $(\pi h_{11/2}{}^n)_{10}\ \nu f_{7/2}$, are well established[2,20,21] in the Z = 66,68 isotones, ^{149}Dy and ^{151}Er. The main decay in each case is by a retarded low-energy $E3$ transition to a $21/2^+$ state of dominant $\pi h_{11/2}{}^{n-1} d_{3/2} \nu f_{7/2}$ character (Fig. 8). Consequently, the isomeric half-lives are unusually long (~0.5s), so much so that Barden et al.[19] were able to study the competitive β-decay branching from the $27/2^-$ isomers to levels in N = 84 daughter nuclei. However, when these workers went on to search for an analogous long-lived isomer in the Z = 70 isotone ^{153}Yb, they found none and concluded that no $27/2^-$ $E3$ isomer exists in that nucleus. The observation of the strong ^{153}Yb γ rays, which were observed to decay with the half-life, $t_{1/2}$ = 15 ± 1 μs, elucidates the systematics of proton-rich N = 83 nuclei discussed in the previous paragraph. Comprehensive γ-γ coincidence results allowed us to place all the observed ^{153}Yb γ rays in the level scheme in Fig. 9, and Table III lists the energies and intensities for all the γ rays known to follow the decay of the ^{153}Yb 15-μs isomer. Total conversion coefficients for low-energy transitions, determined from intensity-balance requirements, yielded key information about multipolarities. The strong 51-keV transition is determined to have the total conversion coefficient value, α_{tot}(51 keV) = 0.49 ± 0.12, which is indicative of $E1$ character. From an intensity-balance condition with the 234- and 328-keV γ rays, the conversion coefficient for the 97-keV transition is calculated to be α_{tot}(97 keV) = 3.86 ± 0.08, indicating $M1$ or $E2$ character. The 107- and 110-keV transitions are found to have the total conversion coefficient values: α_{tot}(107 keV) = 2.81 ± 0.08 and α_{tot}(110 keV) = 2.5 ± 0.4, which indicate $M1$, and $E2$ or $M1$ character, respectively. Since

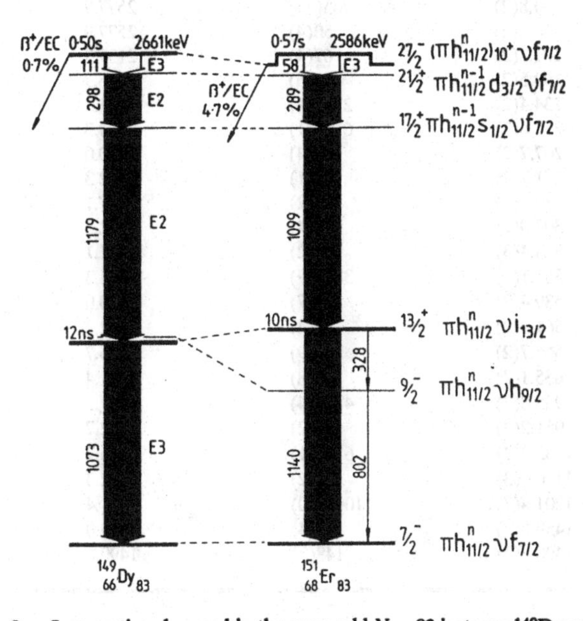

Figure 8. Systematics observed in the even-odd N = 83 isotones ^{149}Dy and ^{151}Er.

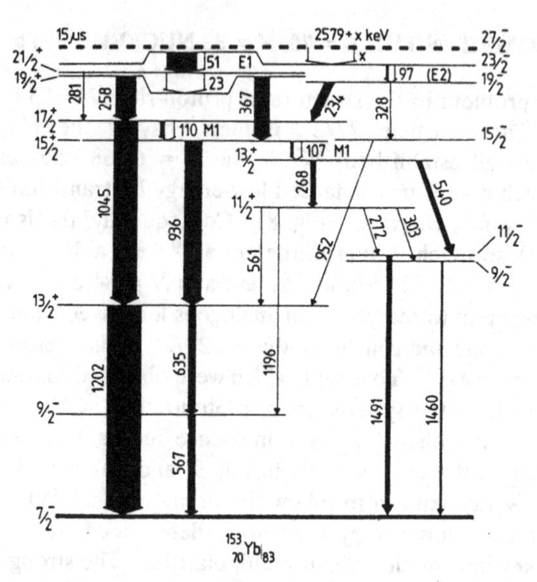

Figure 9. The ^{153}Yb level scheme. Arrow widths are proportional to the transition intensities, with internal conversion contributions unshaded.

Table III. Energies, Intensities, and Placements of γ Rays Following the 15-μs Isomer in ^{153}Yb

Energy (keV)	Off-Beam Intensity	Placement (keV)
50.8(2)	765(76)	2577.9
96.8(2)	50(4)	2577.9
107.2(2)	67(4)	2137.2
109.5(2)	98(5)	2246.7
234.4(2)	226(11)	2481.1
257.6(2)	616(31)	2504.3
267.7(2)	90(4)	2030.0
271.7(4)	6(1)	1762.3
280.5(3)	12(2)	2527.1
303.0(2)	44(4)	1762.3
328.4(3)	17(2)	2481.1
367.1(2)	384(19)	2504.3
539.4(2)	142(7)	2030.0
561.0(2)	41(4)	1762.3
566.7(2)	189(9)	566.7
635.1(2)	160(8)	1201.4
935.8(2)	473(24)	2137.2
951.2(3)	13(2)	2152.7
1045.3(2)	612(31)	2246.7
1196.0(3)	22(2)	1762.3
1201.4(2)	1000(50)	1201.4
1459.0(2)	47(4)	1459.0
1490.6(2)	148(7)	1490.6

discounted. On the basis of this information, together with the measured branching intensities and the $N = 83$ systematics, J values are assigned with confidence (Fig. 9).

The half-life of the 1202-keV level, interpreted as a $13/2^+$ state of $\nu i_{13/2} \times 3^-$ character, has been determined to be ≈ 6 ns. The energy reduced branching to the 567-keV and ground states of ^{153}Yb is almost identical to the known $M2/E3$ branching from the $13/2^+$ isomer in ^{151}Er to the $\nu h_{9/2}$ and $\nu f_{7/2}$ single particle states (Fig. 8). We note, in passing, that measurements of radioactivity from the catcher showed 567-keV γ rays coincident with 511-keV annihilation radiation, indicating that the radioisotope ^{153}Lu decays mainly via β^+ by the expected fast Gamow-Teller transition, $\pi h_{11/2} \rightarrow \nu h_{9/2}$.

From systematics, the strongly-populated $15/2^+$, $17/2^+$ and $19/2^+$, $21/2^+$ levels in ^{153}Yb are interpreted as members of $\pi h_{11/2}{}^{n-1} s_{1/2} \nu f_{7/2}$ and $\pi h_{11/2}{}^{n-1} d_{3/2} \nu f_{7/2}$ multiplets. The intense $21/2^+ \rightarrow 19/2^+$ 23-keV $M1$ was not detected, but it is clearly indicated by the coincidence results. The $21/2^+$ level is fed by the strong 51-keV $E1$ from a $23/2^-$ level that is assigned to the coupling of a $[\pi h_{11/2}{}^n]8^+$ with $\nu f_{7/2}$. This level also deexcites through a $23/2^- \rightarrow 19/2^- \rightarrow 15/2^-$ cascade, where the 97- and 329-keV transition energies closely resemble known $8^+ \rightarrow 6^+ \rightarrow 4^+$ spacings between $\pi h_{11/2}{}^n$ states in $N = 82$ nuclei such as ^{148}Dy and ^{150}Er. Furthermore, the 1460- as well as the 1491-keV transition energies resemble the interpreted $\pi h_{11/2}{}^n$ $2^+ \rightarrow 0^+$ spacing in ^{152}Yb. The 15-μs isomeric state almost certainly has the aligned configuration, $[\pi h_{11/2}{}^n \nu f_{7/2}]27/2^-$. The $27/2^- \rightarrow 23/2^-$ isomeric transition was not observed; our measurements place an upper limit of 50 keV on its energy, and the deduced $B(E2)$ is ≈ 2 e^2fm^4. This transition, like the $\pi h_{11/2}{}^n$ $10^+ \rightarrow 8^+$ $E2$ transition in ^{152}Yb, is highly retarded because in these nuclei the $\pi h_{11/2}$ subshell is close to being half-filled.[7,12] The present study thus provides a solution to the $N = 83$ systematics problem outlined above. As Barden et al.[19] have discussed, the long-lived $E3$ isomers occur in ^{149}Dy and ^{151}Er because the strongly-attractive neutron-proton interaction in the $\pi h_{11/2} \nu f_{7/2}$ 9^+ coupling lowers the $\pi h_{11/2}{}^n \nu f_{7/2}$ $27/2^-$ aligned state below the $23/2^-$ multiplet member. With increasing Z, the n-p attraction diminishes as the half-filled $\pi h_{11/2}$ subshell is approached, and one consequence is that in ^{153}Yb the $23/2^-$ is now found below the $27/2^-$ state, giving rise to $E2$ rather than $E3$ isomerism. The structure of other multiplets observed in ^{153}Yb are also influenced by the reduced n-p interaction.

3.3 THE 35-μs ISOMER IN THE Z = 71, N = 83 NUCLEUS ^{154}Lu

As was discussed previously in Sections 3.1 and 3.2, $27/2^-$ isomers of $\pi h_{11/2}{}^n$ $\nu = 3$ character occur at ≈ 2.7 MeV in the proton-rich $N = 82$ isotones, ^{149}Ho, ^{151}Tm, and ^{153}Lu. As both Z and the $\pi h_{11/2}$ occupation number increase, the observed $B(E2; 27/2^- \rightarrow 23/2^-)$ decreases sharply, reaching a value of only 0.45 e^2fm^4 (or 0.01 Wu) in ^{153}Lu, where the $\pi h_{11/2}$ subshell is approximately half-filled. Corresponding $[\pi h_{11/2}{}^n \nu f_{7/2}]17^+$ isomers have been located[22,23] in the $N = 83$ neighbouring nuclei, ^{150}Ho and ^{152}Tm, and by extrapolation a 17^+ half-life much longer than 1 μs has been predicted[23] for the ^{154}Lu isotone, which has not been studied up to now.

For the recent experiment discussed in Section 3.1, the strongest family of γ rays observed in the $A = 154$ mass window was found to be coincident with Lu x rays and therefore assigned to the pn evaporation product ^{154}Lu. The isomeric half-life was determined to be 35 ± 3 μs. Although the yield of the ^{154}Lu isomer at the catcher foil was only 4% of the ^{153}Yb isomeric yield, the ^{154}Lu γ-γ coincidence data recorded were of excellent quality, as is illustrated in Fig. 10.

The 130-keV transition was found to be in prompt coincidence with all the other ^{154}Lu γ rays (Fig. 10*a*), and its total conversion coefficient was determined from intensity balance to be 1.14 ± 0.21, in agreement with the theoretical *E*2 value. Accordingly, it is interpreted as the isomeric $17^+ \to 15^+$ transition. A key result, illustrated in Fig. 10*b*, was the identification of the 130-, 350-, 678-, and 1402-keV γ-ray cascade, which closely resembles established $17^+ \to 15^+ \to 13^+ \to 11^+ \to 9^+$ *E*2 cascades in the ^{150}Ho and ^{152}Tm isotones.[23] All the strong ^{154}Lu γ rays, except the 678- and 1402-keV lines, appeared in coincidence with the 215- keV transition (Fig. 10*c*).

The ^{154}Lu isomeric decay scheme shown in Fig. 11 is based on the comprehensive γ-γ coincidence results and on N = 83 level systematics, and Table IV lists the energies and intensities for all the γ rays that follow the decay of the ^{154}Lu 35-μs isomer. Particularly, the spin-parity and suggested configuration assignments rest largely on systematics arguments. The $\pi h_{11/2}{}^n \nu f_{7/2}$ assignment for the two lowest ^{154}Lu levels extends a regular trend observed in the ^{148}Tb, ^{150}Ho, and ^{152}Tm isotones, where the 8^+ multiplet members are found 315, 217, and 114 keV, respectively, above the $[\pi h_{11/2}{}^n \nu f_{7/2}]\ 9^+$ ground states. The decay patterns of the 9^+ and 10^+ levels in ^{154}Lu support their interpretation as $\pi h_{11/2}{}^n \nu h_{9/2}$ states, corresponding to similar states identified in ^{152}Tm[23]; the lower ener-

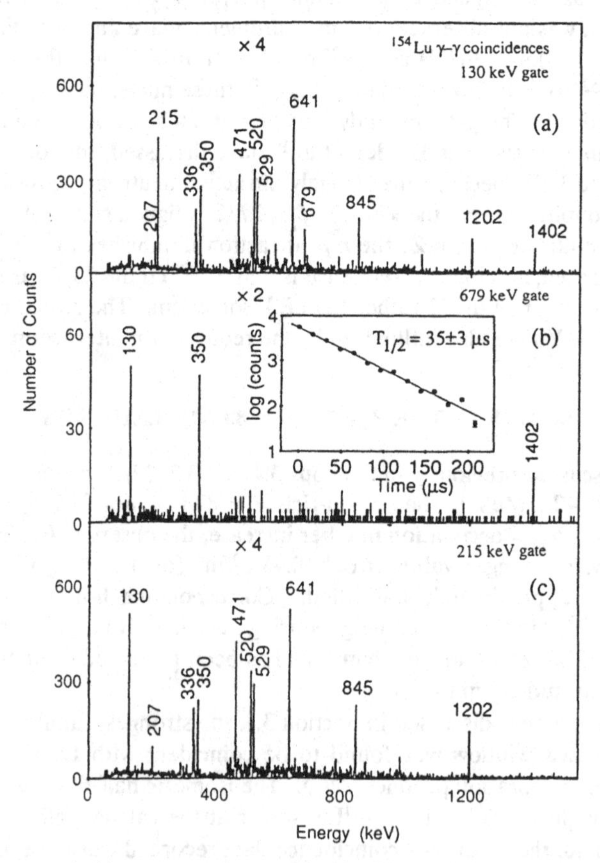

Figure 10. Key γ-ray coincidence spectra for transitions in ^{154}Lu. Lifetime data are shown in the inset.

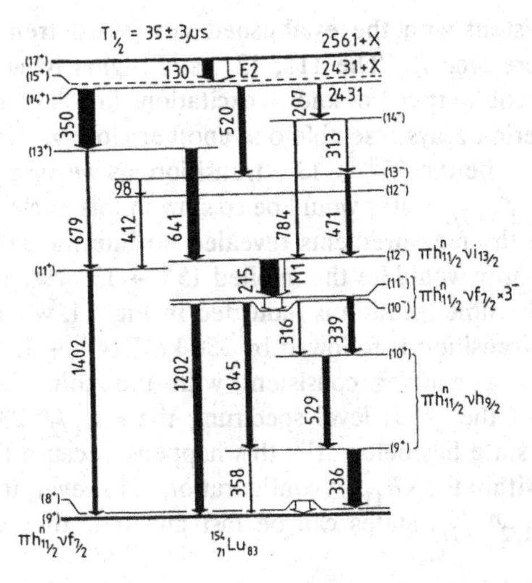

Figure 11. The ^{154}Lu level scheme. Arrow widths are proportional to the transition intensities with internal conversion contributions unshaded.

Table IV. Energies, Intensities, and Placements of γ Rays Following the 35-μs Isomer in ^{154}Lu.

Energy (keV)	Off-Beam Intensity	Placement (keV)
98.0(6)	7(4)	1911.6
130.4(3)	72(7)	2561.5+x
206.8(4)	18(2)	2431.1
215.4(3)	100(10)	1440.5
313.4(5)	10(2)	2224.3
316.2(5)	9(2)	1202.5
335.9(3)	75(8)	357.3
338.7(4)	39(4)	1225.1
350.0(3)	99(10)	2431.1
358.1(6)	5(2)	358.1
411.6(5)	10(2)	1813.9
470.7(4)	31(3)	1911.6
519.5(4)	37(4)	2431.1
529.1(4)	34(5)	886.4
640.6(4)	70(7)	2081.1
678.8(4)	18(3)	2081.1
784.3(6)	8(2)	2224.3
845.2(4)	43(4)	1202.5
1202.5(5)	50(9)	1202.5
1402.3(4)	31(3)	1402.3

gies in ^{154}Lu are consistent with the established downward trend in the $\nu h_{9/2}$ single-particle energy with increasing Z. The 11^-, 12^-, and higher-lying, negative-parity ^{154}Lu levels are also probable counterparts of known excitations in ^{150}Ho and ^{152}Tm, and in most respects the three isomeric decays resemble one another closely. However, the 350-keV γ ray in ^{154}Lu is unlikely to be the $15^+ \rightarrow 13^+$ transition, as we first assumed. Such an $E2$ transition between $\pi h_{11/2}{}^n \nu f_{7/2}$ states would be so slow in this nucleus that the parent level should be isomeric, but the measurements revealed no intermediate isomers in the level scheme. A further difficulty would be the implied $15^+ \rightarrow 13^-$ assignment for the 520-keV transition. The most plausible solution is indicated in Fig. 11, where an unobserved low-energy $15^+ \rightarrow 14^+$ $M1$ transition is followed by 350-keV $14^+ \rightarrow 13^+$ and 520-keV $14^+ \rightarrow 13^-$ decay branches in a manner consistent with the coincidence results. In the corresponding portion of the ^{153}Lu level spectrum, the $\pi h_{11/2}{}^n$ $23/2^-$ state is isomeric, even though the $21/2^-$ state lies below it[18]; this happens because the $23/2^- \rightarrow 21/2^-$ $M1$ transition is forbidden within the $\pi h_{11/2}{}^n$ configuration. However, in ^{154}Lu the $15^+ \rightarrow 14^+$ transition between $\pi h_{11/2}{}^n \nu f_{7/2}$ states can be fast and may thus carry the entire decay intensity.

While the Fig. 11 level scheme may not be correct in all details, there seems little doubt about the interpretation of the 130-keV $E2$ as the expected $17^+ \rightarrow 15^+$ isomeric transition. The measured half-life gives 0.19 ± 0.02 e^2fm^4, much smaller than the values, 182 ± 18 and 44 ± 2 e^2fm^4, determined from the analogous $17^+ \rightarrow 15^+$ decays in ^{150}Ho and ^{152}Tm. Following Ref. 23, we plot in Fig. 12 the isomeric $E2$ transition amplitudes, $\sqrt{B(E2)}$ with Z is seen to be appropriately linear for each set of the isotones. The shift in Fig. 12 between the $N = 82$ and $N = 83$ lines may be attributed to the $(\pi h_{11/2}{}^n)_{27/2} \nu f_{7/2}$ components of the 15^+ states, which give rise to similar contributions to the $17^+ \rightarrow 15^+$ transition amplitudes in each of the $N = 83$ isotones. As mentioned above, this result has been anticipated in Ref. 23, where a $B(E2;\ 17^+ \rightarrow 15^+)$ very close to zero was predicted for ^{154}Lu.

ACKNOWLEDGMENTS

We thank R. Broda and C. J. Lister for useful discussions. We also thank G. Reed from the University of Manchester and the Daresbury Laboratory staff for technical assis-

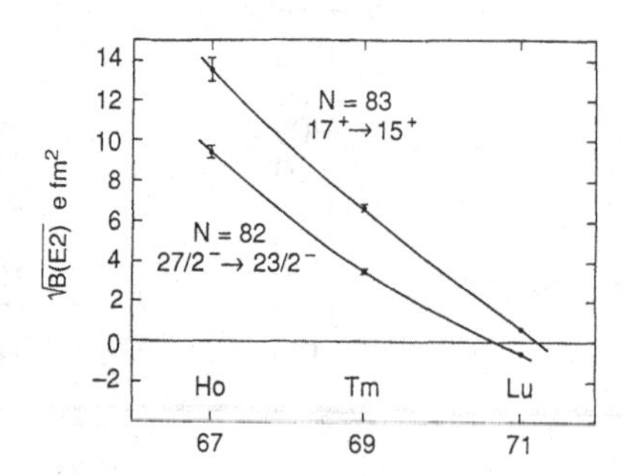

Figure 12. Measured $E2$ transition amplitudes for $\nu = 3$ and $\nu = 4$ isomeric transition in $N = 82$ and $N = 83$ isotones as a function of Z. Where error bars are not shown, they lie within the plotted point.

tance, and also special thanks go to D. Blunt for the making of the targets and to J. C. Lisle for assistance in the data analysis. This work has been funded by the U. K. Science and Engineering Research Council, during the period of which the first author, J. H. McNeill, was a postdoctoral research associate at the University of Manchester. The co-authors P. J. Daly and M. Piiparinen acknowledge support from the U.S. Department of Energy, and M. Piiparinen also acknowledges support from the Academy of Finland.

REFERENCES

1. P. Kleinheinz, M. Ogawa, R. Broda, P. J. Daly, D. Haenni, H. Beuscher, and A. Kleinrahm, *Z. Phys.* **A286**, 27 (1978).
2. P. J. Daly, P. Kleinheinz, R. Broda, S. Lunardi, H. Backe, and J. Blomqvist, *Z. Phys.* **A298**, 173 (1980).
3. J. Wilson, S. R. Faber, P. J. Daly, I. Ahmad, J. Borggreen, P. Chowdhury, T. L. Khoo, R. D. Lawson, R. K. Smither, and J. Blomqvist, *Z. Phys.* **A296**, 185 (1980).
4. H. Helppi, Y. H. Chung, P. J. Daly, S. R. Faber, A. Pakkanen, I. Ahmad, P. Chowdhury, Z. W. Grabowski, T. L. Khoo, R. D. Lawson, and J. Blomqvist, *Phys. Let*, **115B**, 11 (1982); K. S. Toth, D. C. Sousa, J. M. Nitschke, and P. A. Wilmarth, *Phys. Rev. C* **35**, 620 (1987).
5. Y. H. Chung, P. J. Daly, H. Helppi, R. Broda, Z. W. Grabowski, M. Kortelahti, J. McNeill, A. Pakkanen, P. Chowdhury, R. V. F. Janssens, T. L. Khoo, and J. Blomqvist, *Phys. Rev. C* **29**, 2153 (1984).
6. J. H. McNeill, Ph.D. Thesis, Purdue University, 1986.
7. E. Nolte, G. Korschinek, and Ch. Setzensack, *Z. Phys.* **A309**, 33 (1982); E. Nolte, S. Z. Gui, G. Colombo, G. Korschinek, and K. Eskola, *Z. Phys.* **A306**, 223 (1982).
8. R. D. Lawson, *Z. Phys.* **A303**, 51 (1981).
9. A. de Shalit and I. Talmi, *Nuclear Shell Theory* (Academic, New York, 1963).
10. R. D. Lawson, *Theory of the Nuclear Shell Model* (Clarendon, Oxford, 1980).
11. J. Blomqvist, *Int. Rev. of Nucl. Phys.* **2** (World Scientific, Signapore, 1984) pp. 1-32.
12. P. J. Daly, Y. H. Chung, Z. W. Grabowski, H. Helppi, M. Kortelahti, P. Chowdhury, R. V. F. Janssens, T. L. Khoo, R. D. Lawson, and J. Blomqvist, in *Proceedings of the International Conference on Nuclear Physics, Florence* **1**, 73 (1983).
13. R. R. Chasman, *Phys. Rev.* C **28**, 1374 (1983).
14. A. N. James, T. P. Morrison, K. L. Ying, K. A. Connell, H. G. Price, and J. Simpson, *Nucl. Inst. Meth.* **A267**, 144 (1988).
15. F. Puhlhofer, *Nucl. Phys.* **A280**, 267 (1977).
16. J. H. McNeill, A. A. Chishti, P. J. Daly, M. A. C. Hotchkis, M. Piiparinen, and B. J. Varley, *Z. Phys.* **A332**, 105 (1989).
17. Evaluated using ground state masses from A. H. Wapstra and G. Audi, *Nucl. Phys.* **A432**, 1 (1985).
18. J. H. McNeill, J. Blomqvist, A. A. Chishti, P. J. Daly, W. Gelletly, M. A. C. Hotchkis, M. Piiparinen, B. J. Varley, and P. J. Woods, *Z. Phys.* **A335**, 247 (1990).
19. R. Barden, A. Plochocki, D. Schardt, B. Rubio, M. Ogawa, P. Kleinheinz, R. Kirchner, O. Klepper, and J. Blomqvist, *Z. Phys.* **A329**, 11 (1988).
20. A. M. Stefanini, P. J. Daly, P. Kleinheinz, M. R. Maier, and R. Wagner, *Nucl. Phys.* **A258**, 34 (1976); A. M. Stefanini, P. Kleinheinz, and M. R. Maier, *Phys. Lett.* **62B**, 405 (1976).
21. J. Jastrzebski, R. Kossakowski, J. Lukasiak, M. Moszynski, Z. Preibisz, S. Andre, J. Genevey, A. Gizon, and J. Gizon, *Phys. Lett.* **97B**, 50 (1980).
22. J. Wilson, J. H. Chung, S. R. Faber, A. Pakkanen, P. J. Daly, I. Ahmad, P. Chowdhury, T. L. Khoo, R. D. Lawson, and R. K. Smither, *Phys. Lett.* **103B**, 413 (1981).
23. J. McNeill, R. Broda, Y. H. Chung, P. J. Daly, Z. W. Grabowski, H. Helppi, M. Kortelahti, R. V. F. Janssens, T. L. Khoo, R. D. Lawson, D. C. Radford, and J. Blomqvist, *Z. Phys.* **A325**, 27 (1986).

4. Spectroscopic Techniques in the Study of Radioactive Nuclei Far from Stability: Results for the $Z = 82$ Region

E. F. Zganjar
Department of Physics and Astronomy
Louisiana State University
Baton Rouge, LA 70803

J. L. Wood
School of Physics
Georgia Institute of Technology
Atlanta, GA 30332

and

J. Schwarzenberg
School of Chemistry
Georgia Institute of Technology
Atlanta, GA 30332

ABSTRACT

Techniques that have been used at UNISOR for the study of the structure of very neutron-deficient Pt, Au, and Hg isotopes are described. In particular, the aspects of this experimental program which lead to the establishment of a new class of nuclear-structure phenomenon (low-lying, low-energy, electric-monopole transitions) are described. At the core of this program lies the almost unused technique of observing coincidences between γ rays and conversion electrons. Additional techniques developed at UNISOR for radioactive-decay spectroscopy far from stability are also described. Finally, the broader range of options for studying nuclear structure far from stability, with emphasis on the future, are outlined.

4.1 INTRODUCTION

Our understanding of the structure of heavy nuclei is purely phenomenological. Thus, extending our knowledge of nuclear phenomena beyond the borders of known nuclei provides critical tests of the models developed to describe nuclear structure. It has also provided surprises. The very neutron-deficient Pt, Au, Hg, Tl, Pb, and Bi isotopes have contributed to this process extensively following the unexpected discovery, by Bonn et al. in 1972, that there is a sudden change in the nuclear charge distribution of the very light Hg isotopes.[1] The picture that has emerged is an extensive region of the mass surface within which very different shapes coexist in nuclei (see, for example, Refs. 2-5). However, although gross features are crudely understood,[4,5] extensive work is needed to provide information upon which a unified and detailed theory can be built. The aim of the present

Exotic Nuclear Spectroscopy, Edited by W. C. McHarris
Plenum Press, New York, 1990

report is to describe the contribution of the UNISOR program of radioactive decay-scheme studies to this process. It is intended to provide examples of the difficulties encountered in making detailed studies of nuclear excited states far from stability and to show how many of these problems are overcome. The unsolved problems then provide directions for future development in experimental techniques.

4.2 BASIC TECHNIQUES OF CHARACTERIZATION

The first step in a nuclear-spectroscopy study far from stability is to assign the radiations to the correct mass number and element. On-line isotope separators can, in favorable cases of elemental species, isolate isobars with half-lives as short as 100 ms (for example, 75-ms ^{32}Ar; Ref. 6). Recoil-mass spectrometers can separate even shorter half-lives (for example, 3.2-μs ^{76}Rb^{m}; Ref. 7). The time limitation of standard on-line isotope separator techniques is *hold-up* in the ion source. Higher ion-source temperatures generally help, but there is a fundamental limitation to this approach. A novel method that circumvents this *hold-up time* is the ion-guide isotope-separator on-line (IGISOL).[8] Elemental identification is less directly made, but some selectivity of elements within the ion source has been achieved.[9] Direct separation of isobars by element requires an isotope separator mass resolution of $m/\Delta m \gg 10{,}000$. Devices exceeding this capability operate on-line at CERN and Chalk River. Velocity filters, such as SHIP at GSI, Darmstadt, can also achieve mass and charge separation simultaneously.[10] However, detailed spectroscopy requires a finite counting time with the inevitable growth and decay of daughter (and grand-daughter) isotopes. Thus, the separation of isobars into elements at the isotope separator is not a complete solution to correct elemental identification. The most reliable method involves x-ray-gated coincidence spectra. An example of the degree of separation achievable at UNISOR using this technique is shown in Fig. 1. Certainly, in many instances, the half-lives of different isobars are sufficiently different that correct identification can be made by spectrum multi-scaling. However, when unresolved doublets occur, with the components belonging to different isotopes, weak lines can easily be missed in multi-scaling analyses.

Coincidence spectroscopy often leads in a straightforward way to the elucidation of the excited states populated in the decay. However, far-from-stability radioactive decay in odd-mass isotopes often populates 50 to 100 excited states, and 200 to 800 transitions may be involved, yielding many unresolved doublets and even unresolved triplets and higher multiplets. Note the situation for a small part of the ^{187}Tl $\rightarrow$ ^{187}Hg^{g} $\rightarrow$ ^{187}Au decay shown in Fig. 2. Part a) shows the portion of the spectrum from 150 to 500 keV, which contains 122 of the 747 γ-ray lines observed between 36 and 2555 keV. Although the very high resolution of bent-crystal spectrometers can overcome much of the difficulty involving multiplets, their efficiency is very low and thus they require long running times. To our knowledge, on-line γ-ray spectroscopy following radioactive decay using a bent-crystal spectrometer has never been used far from stability. Coincidence analysis can, however, reveal and quantitatively separate most multiplets. A successful example, from work done at UNISOR, is shown in Fig. 3. This case was resolved using *running* coincidence gates whereby a suspected multiplet was scanned in the gate-pulling analysis, a few channels at a time, and the corresponding coincidences carefully replotted. This is a tedious task, but, in this case, since neither of the coincident pairs exhibited other separate and distinct coincidences, it was the only way to resolve the transitions.

A nuclear decay scheme is of little value, in terms of nuclear structure information, unless the spin and parity of most of the low-lying levels can be determined. This has been

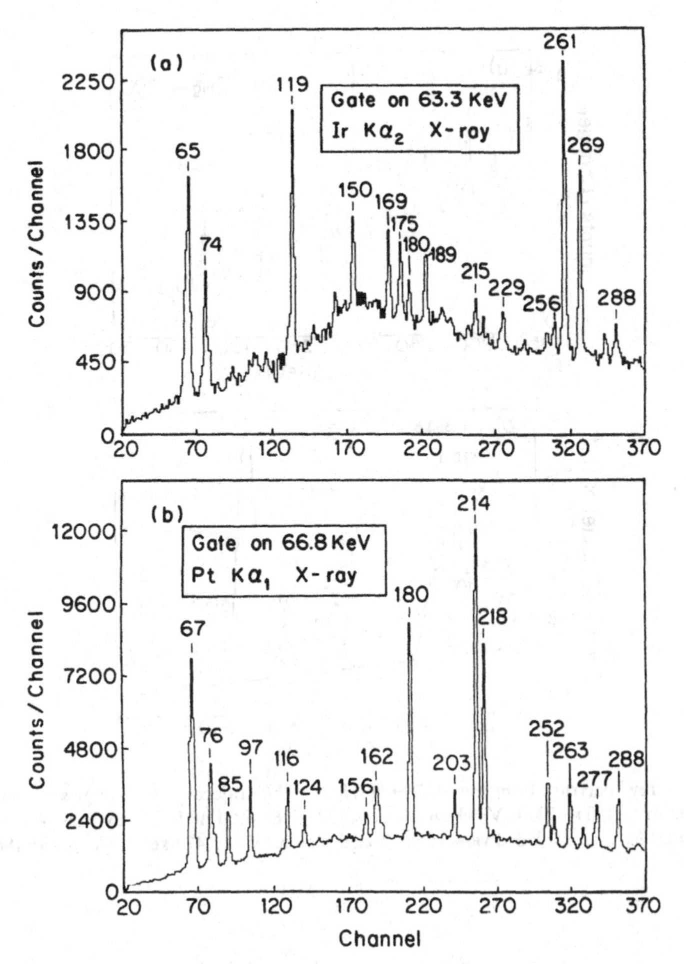

Figure 1. γ-ray spectra for $A = 183$ gated by Ir $K_{\alpha 2}$ x rays *a*) and Pt $K_{\alpha 1}$ x rays *b*). Note the absence of the 199- and 261-keV lines in *b*), and the presence of lines at 256 and 288 keV in both *a*) and *b*). The data are from Ref. 11.

done for many ground states (and isomers) when they have half-lives long enough to permit isolation and measurement. A measurement of the spin of a level may involve the relatively direct technique of atomic-beam magnetic resonance[13] or the less-direct technique of optical hyperfine spectroscopy.[14] However, the vast majority of states have sub-μs half-lives, and spin/parity information must be obtained from the emitted radiations. Three basic approaches are in use: Internal-conversion-coefficient measurements, γ-γ angular-correlation measurements, and measurements of the angular distribution of γ rays emitted from oriented nuclei. The measurement of γ-γ angular correlations far from stability has been used only to a limited degree. Because it is a coincidence technique and requires sizeable source-to-detector distances to define correlation angles adequately, intense sources are needed, and this becomes increasingly more difficult as one proceeds further from the stability region. The measurement of angular distributions of γ rays emitted from oriented nuclei is a newly-applied technique[15] in the study of nuclei far from stability. The

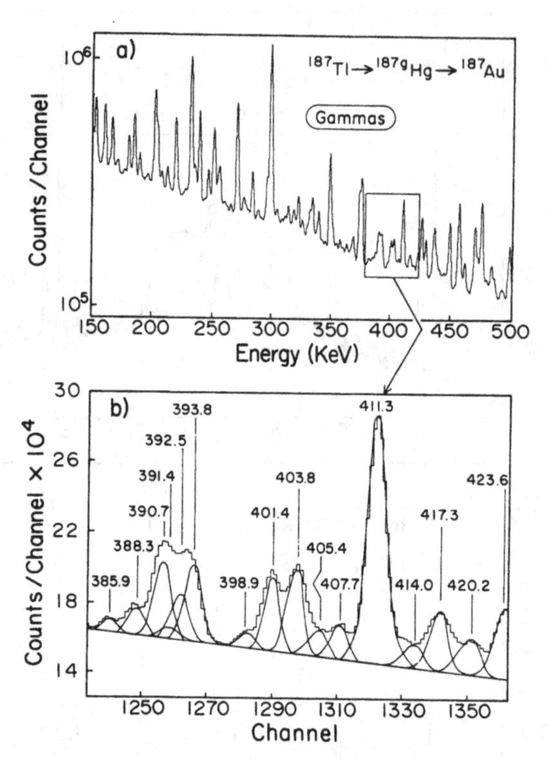

Figure 2. The γ-ray spectrum from the ^{187}Tl → ^{187}Hgg → ^{187}Au decay. All energies are in keV. Part a) shows the portion from 150 to 500 keV, which contains 122 of the total of 747 γ-ray lines observed between 36 and 2555 keV. Part b) is an exploded view of the boxed region (383 to 424 keV), along with the results of the fitting procedure.

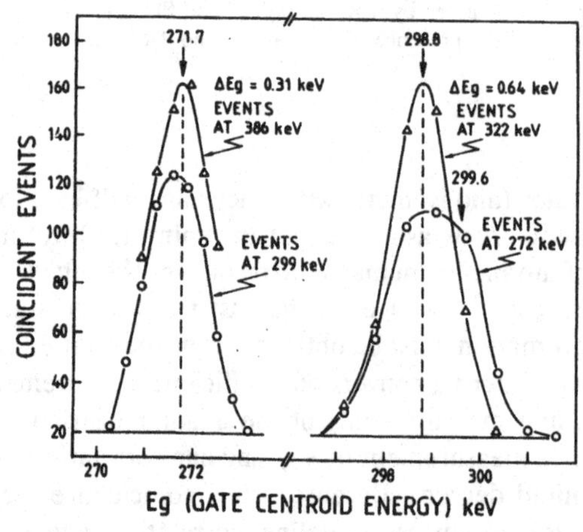

Figure 3. Coincidence scans at ≈272 and 299 keV for the γ-γ-t data from ^{187}Hg decay. Each point on the left side represents an advance of 0.31 keV of the gate set on γ_1 at the energy denoted on the horizontal axis; the number of coincident events recorded for γ_2 in the regions of 386 and 299 keV is denoted on the vertical axis. The right side represents a coarser (0.64 keV/step) example of the reverse of the same procedure. The data reveal coincident pairs of γ rays at 298.8 - 271.7 and 299.6 - 271.2 keV. The data are from Ref. 12.

difficulty that had to be overcome in such a measurement is the necessity of continuous direct implantation of the short-lived radioactive species into the $T \approx 10$ mK environment needed for orienting the nuclei. The attractiveness of the technique is that it is a singles measurement and provides unique multipolarities (even when mixed) in most situations. This technique is discussed further in the final section.

Internal-conversion-electron spectroscopy has been the major source of information on spins and parities of excited states in nuclei far from stability. The most common device in use for on-line studies of internal conversion spectra is the Si(Li) detector, which has a resolution for conversion electrons comparable to Ge detector resolution for γ rays. However, conversion-electron spectroscopy differs from γ-ray spectroscopy in that for each transition there are conversion-electron lines corresponding to at least the *K*, *L*, and *M* atomic shells. This gives rise to a much higher line density and a much higher chance of unresolved multiplets. An example from the ^{187}Tl → ^{187}Hg^{g} → ^{187}Au decay which corresponds to the 150- to 500-keV range of Fig. 3, if one subtracts 80 keV (the Au *K*-shell binding energy), is shown in Fig. 4. Additionally, internal-conversion spectroscopy is often the only means by which very low-energy ($\Delta E \leq 50$ keV) transitions can be directly observed, because such transitions are highly converted. In general, Si(Li) detectors are unable to detect electrons below energies of about 20 keV because of dead-layer effects in the detector crystal. To our knowledge, the only work that has been done far from stability on

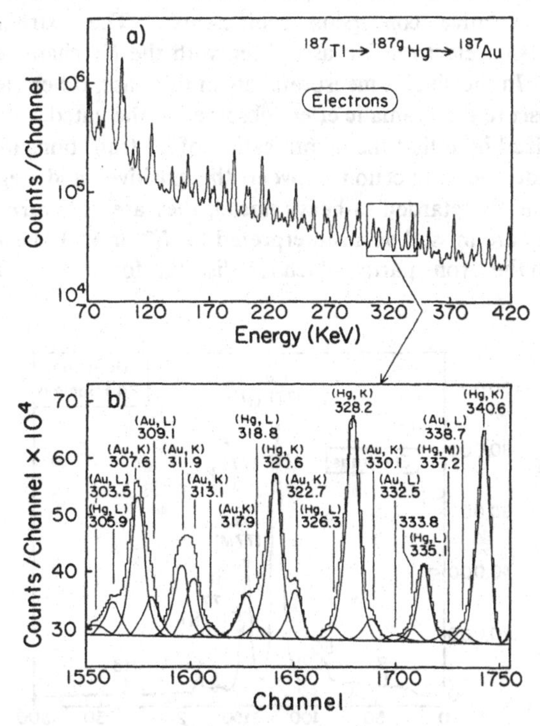

Figure 4. The conversion-electron spectrum from the ^{187}Tl → ^{187}Hg^{g} → ^{187}Au decay. All energies are in keV. Part *a*) shows the portion from 70 to 420 keV, which contains 152 lines. The energy range spanned by part *a*) exactly matches the 150-to-500-keV range of Fig. 1*a* if one subtracts 80 keV. Part *b*) is an exploded view of the boxed region (303 to 344 keV) along with the results of the fitting procedure. The *K*-shell binding energy is 80.7 keV in Au and 83.1 keV in Hg. Thus, the regions in parts *b*) of Figs. 2 and 4 enable one to make a 1:1 comparison for conversion in the *K* shell. The *L*- and *M*-subshell lines from lower-energy transitions are also indicated.

the measurement of conversion electrons with energies well below 20 keV is that of the ISOCELE group,[16] who employed considerable ingenuity in the construction of a special system, which permitted the slowing down of the incoming beam of radioactive ions as well as the acceleration of the emitted electrons into a flat magnetic spectrograph. As important as such a device is in locating the very low energy transitions, it cannot be used in the coincidence mode. The use of an ultra-low-noise, ion-implanted Si detector for coincidence work with conversion electrons, albeit with much poorer resolution than the magnetic spectrograph, is discussed in the final section.

It is evident that if a decay scheme can only be elucidated by careful γ-γ coincidence techniques because of unresolved transitions, then comparable coincidence techniques must be applied also to conversion-electron spectroscopy. In fact, because of the multiplicity of conversion-electron subshell lines, coincidence techniques are highly desirable even if the decay scheme in question is relatively simple. The use of coincidences between γ rays and conversion electrons has played a major role in the UNISOR program. A number of examples taken from work at UNISOR illustrate the importance of such coincidences to the general study of decay schemes far from stability: Figure 5 shows the ability of γ-ray-gated conversion-electron spectra to provide detailed subshell-ratio information. This usually leads to unambiguous multipolarity assignments. The identification of an $E1$ transition is illustrated in Fig. 6 which shows the γ-ray spectrum (upper part) and the conversion electron spectrum (lower part) obtained by gating on the same γ-ray transition (250 keV). One can then make a 1:1 comparison between the corresponding lines in each spectrum to obtain relative conversion coefficients. The extremely weak (nearly unobservable) conversion electron line associated with the $E1$ character for the 199-keV transition is obvious. In the singles measurements in this case, other electron lines appear at this energy and obscure the dramatic effect observed in the gated spectrum shown in Fig. 6. It can be emphasized here that the identification of $E1$ transitions in a decay scheme is critical. They provide the connection between the positive- and negative-parity states. Because they are severely retarded in heavy nuclei, they are very rare (often only one or two are seen). If they are missed or misinterpreted (as $E2$ or $M1$), the scheme may have a number of levels with the wrong parity, which is a disaster for theory. This was precisely

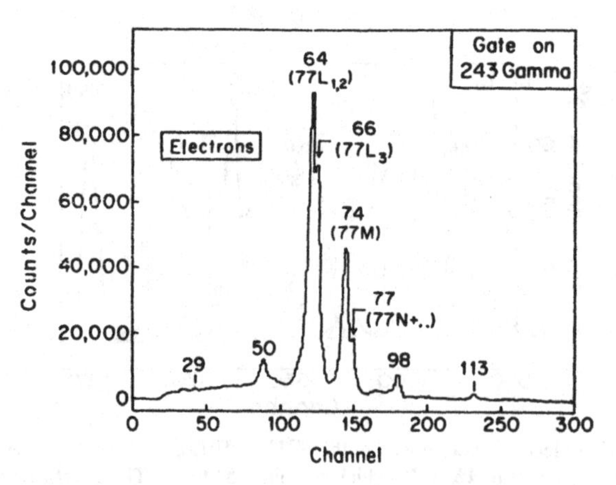

Figure 5. Low-energy conversion electrons in coincidence with the 243-KeV γ ray in $^{185}Au \rightarrow {}^{185}Pt$ decay. The L_{12}, L_3, M, and N electron lines of a 78-keV transition are evident (the L_1 and L_2 subshells and the M and N subshells are unresolved). The data are from Ref. 11.

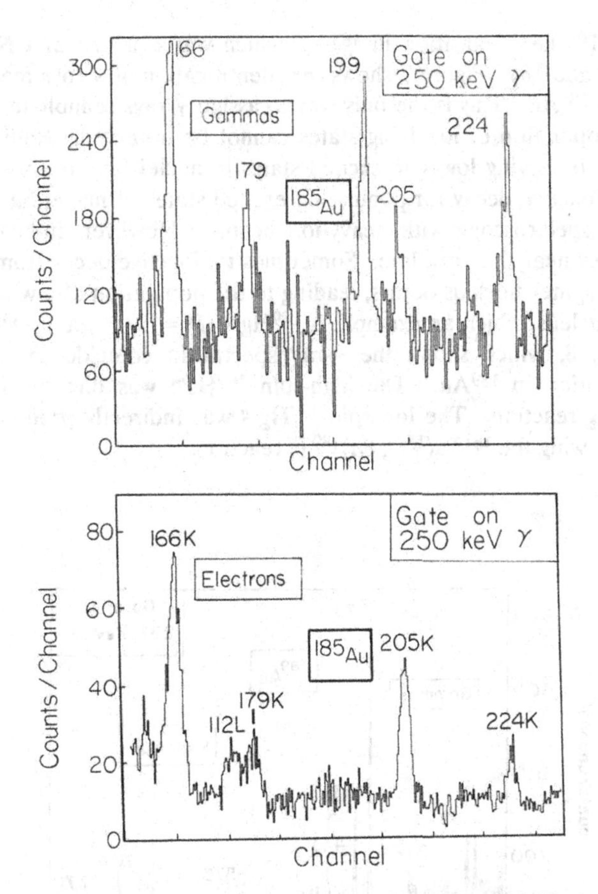

Figure 6. γ rays *a*) and conversion electrons *b*) in coincidence with the 250-keV γ ray in the ^{185}Hg^{g,m} → ^{185}Au decay. The 199 keV transition has $E1$ multipolarity. This is evident from its very weak (nearly unobservable) conversion-electron intensity. The 205-keV transition has $E0+M1+E2$ multipolarity. The figure is taken from Ref. 17.

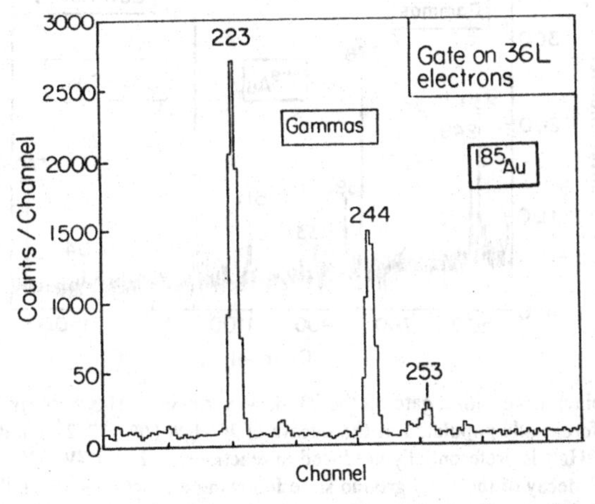

Figure 7. γ rays in coincidence with L-conversion electrons corresponding to the 35-keV transition in the ^{185}Hg^{g,m} → ^{185}Au decay. The figure is taken from Ref. 17.

the case for the 199-keV transition in ^{185}Au, which was clarified at UNISOR (see, for example, Refs. 17 and 18). Figure 7 shows the identification of γ rays feeding a low-lying level at 36 keV in ^{185}Au. This is the only way to assign γ rays reliably to a decay scheme when photon depopulation of low-lying states cannot be observed. Radioactive decay is uniquely suited to observing low-spin excited states in nuclei far from stability. The only alternative to radioactive decay for populating excited states of nuclei far from stability is in-beam reaction spectroscopy with heavy-ion beams. However, in-beam spectroscopy observes only states near the yrast line. Sometimes radioactive decay from high- and low-spin states to a daughter nucleus occurs, leading to the population of a wide range of spins in the daughter nucleus. Such an example is ^{189}Hg^{m} ($I^{\pi} = 13/2^{+}$) and ^{189}Hg^{g} ($I^{\pi} = 3/2^{-}$), illustrated in Fig. 8, which shows the γ-ray spectra in coincidence with a 237-keV, $7/2^{-} \rightarrow 11/2^{-}$ transition in ^{189}Au. The high-spin ^{189}Hg^{m} was directly produced by the ^{182}W(^{12}C,5n)^{189}Hg reaction. The low-spin ^{189}Hg ^{g} was indirectly produced from the β^{+} decay of ^{189}Tl following the ^{181}Ta(^{16}O, 8n)^{189}Tl reaction.

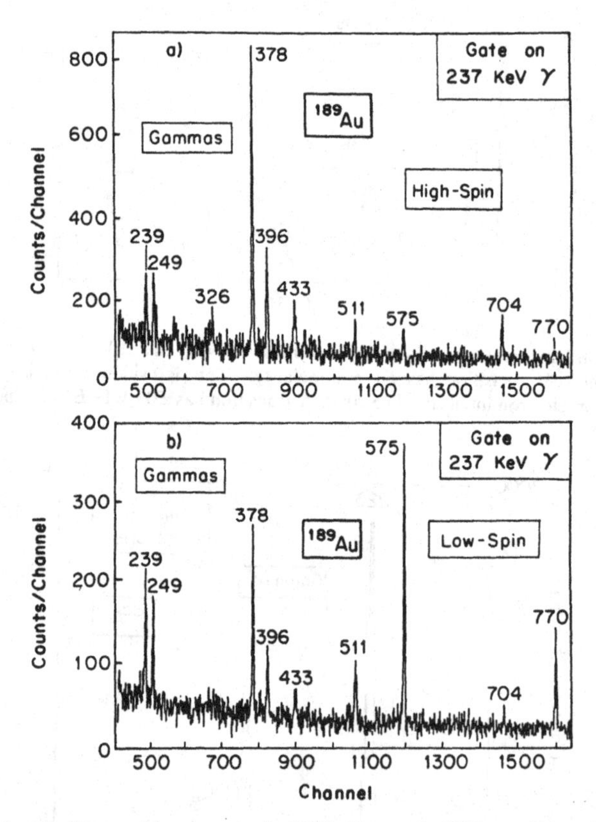

Figure 8. γ rays in coincidence with a gate on the 237-keV γ ray in ^{185}Hg^{g,m} decay. The high- and low-spin states in ^{189}Au are preferentially populated in the decays of ^{189}Hg^{m} ($I^{\pi} = 13/2^{+}$) and ^{189}Hg^{g} ($I^{\pi} = 3/2^{-}$), respectively. High-spin ^{189}Hg^{m} is preferentially produced in reactions such as ^{182}W(^{12}C, 5n)^{189}Hg, while low-spin ^{189}Hg^{g} is produced by β decay of the ^{189}Tl ground state following a reaction such as ^{181}Ta(^{16}O, 8n)^{189}Tl. The $9/2^{-} \rightarrow 7/2^{-}$, 378-keV and $11/2^{-} \rightarrow 7/2^{-}$, 704-keV transitions are clearly enhanced in *a*), while the $3/2^{-} \rightarrow 7/2^{-}$, 575-keV and $5/2^{-} \rightarrow 7/2^{-}$, 770-keV transitions are clearly enhanced in *b*).

4.3 SHAPE COEXISTENCE AND $E0$ TRANSITIONS IN THE $183 \leq A \leq 195$ REGION

Shape coexistence in the $183 \leq A \leq 195$ region is by now widely established (see, for example, Ref. 19). A clear signature of shape coexistence that is emerging[19,20] is the association of $E0$ transitions between coexisting bands. The origin of the $E0$ strength can be understood[21] if the coexisting bands mix. The experimental technique developed and extensively employed at UNISOR to determine accurate α_K values of weak transitions or of transitions involved in complex multiplets is easily demonstrated using Figs. 9 and 10 for ^{190}Hg. By gating on the 416-keV γ ray, for example, and observing both the γ-ray spectrum and the conversion-electron spectrum, one can determine to high accuracy the 1142/1155-keV K-conversion-coefficient ratio. If α_K for the 1142-keV transition can be measured directly and the transition is determined to be a pure multipole, then one immediately has a precise measure of the 1155-keV K-conversion coefficient. For the cases shown in Figs. 9 and 10, the $I^\pi \rightarrow I^\pi$ interband transitions of 737, 933, and 1155 keV are highly converted (contain a sizeable $E0$ component) with $\alpha_K(\text{expt})/\alpha_K(M1, \text{theory})$ values of 2.3 ± 0.2, 3.4 ± 0.4, and 4.6 ± 0.7, respectively. In an odd-mass nucleus, this e^--γ, γ-γ coincidence technique is essential to obtaining accurate conversion coefficients because of the high line density. Consider, for example, the ^{187}Hg $\rightarrow$ ^{187}Au decay shown in Figs. 2 and 4. The line at 330.1 keV in Fig. 4 corresponds to K conversion of a 410.8-keV transition in Au. Note from Fig. 4 that one is not able to resolve such a line from the multiplet structure around 409-411 keV. Clearly one cannot get accurate conversion coefficients for such transitions

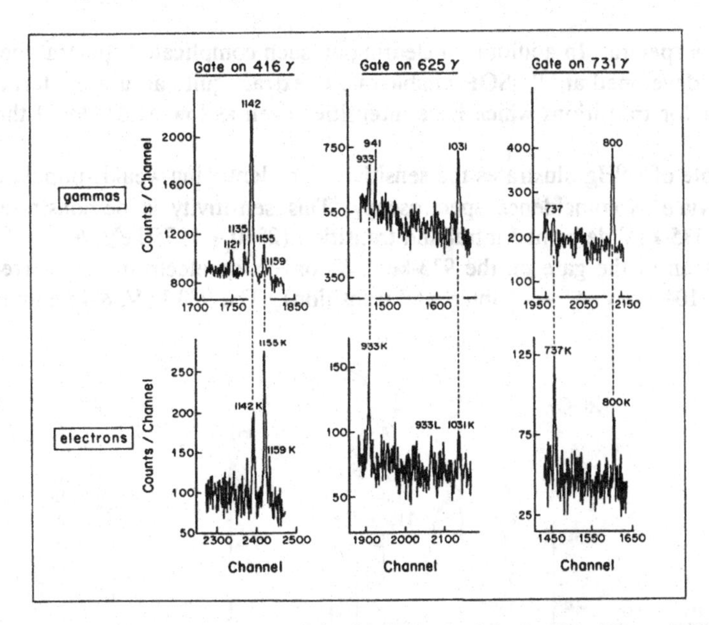

Figure 9. The identification and measurement of sizeable $E0$ strength in the 737-, 933-, and 1155-keV interband transitions in ^{190}Hg, using the e^--γ-coincidence gating technique. (Refer also to the partial ^{190}Hg level scheme shown in Fig. 10.) By gating on the 416-keV γ ray, for example, one can simultaneously determine, to high accuracy, the 1142/1155 keV K-conversion coefficient ratio. If α_K for the 1142-keV transition can be measured directly or be determined to be of pure multipole, then one immediately has a precise measure of the 1155-keV K-conversion coefficient free of interferences from other transitions. That the 1155-, 933-, and 737-keV transitions are highly converted is clear from a comparison of their I_e/I_γ intensity ratios in the gated spectra relative to those for the 1142-, 1031-, and 800-keV transitions, which are either pure $E2$ or mixed $M1+E2$. The data are from Ref. 20.

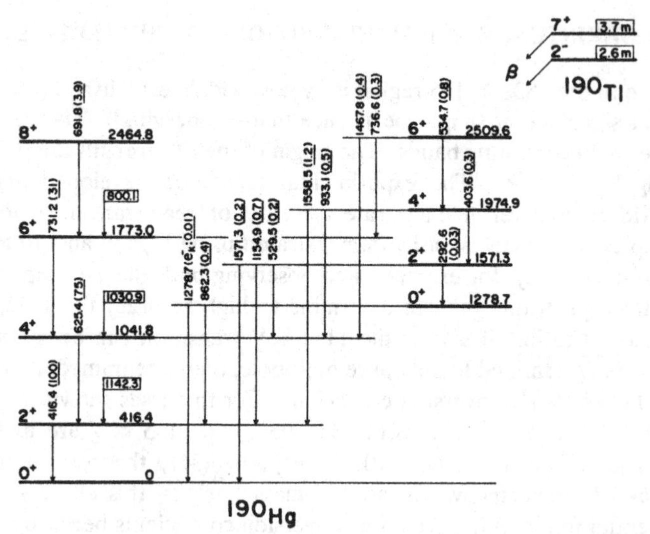

Figure 10. A partial decay scheme of $^{190}Tl^{m,g}$ to ^{190}Hg showing the ^{190}Hg ground state band and the deformed *intruder* band. The 1279-keV $0^+ \to 0^+$ transition is, of course, totally converted, but the $I^\pi \to I^\pi$ interband transitions of 737, 933, and 1155 keV are also highly converted, with $\alpha_K(\text{expt})/\alpha_K$ (*M*1, theory) values of 2.3±0.2, 3.4±0.4, and 4.6±0.7, respectively. The data are from Ref. 20.

from singles spectra. In addition to clearing up such complicated spectra, the coincidence techniques developed at UNISOR enable one to extract quite accurate internal-conversion coefficients for transitions which have intensities even as low as 0.05% of the total decay intensity.

The example of ^{190}Hg illustrates the sensitivity for identifying weakly-populated coexisting structures via e^--γ coincidence spectroscopy. This sensitivity is demonstrated in Fig. 11, where the 535-keV deformed intraband transition (2510 → 1975 keV, $6^+ \to 4^+$; see Fig. 10) is clearly seen in the gate on the 933-keV, *K*-conversion-electron line (correspon-ding to the 1975 → 1042 keV, $4^+ \to 4^+$ interband transition). The 933-keV, *K*-line intensity is 0.027

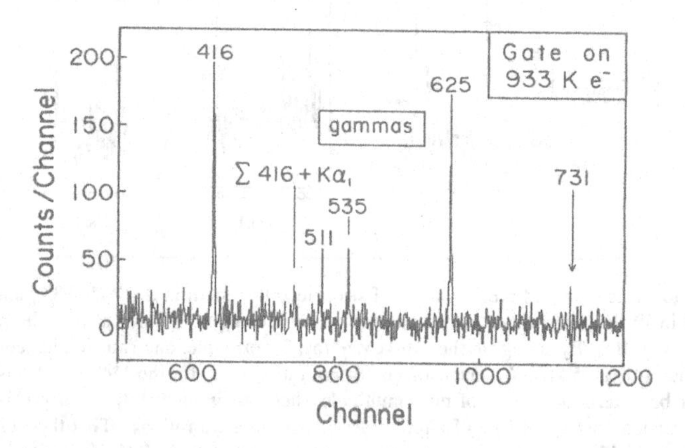

Figure 11. Part of the γ-ray spectrum resulting from a gate on the 933-keV *K*-conversion line in ^{190}Hg. (Refer also to Fig. 10.) This shows that one can get clean, precise determinations using the e^--γ coincidence technique, since in this case the 933-keV *K*-line intensity is only 0.027 units and the 535-keV γ-ray intensity is 0.8 units (relative to 100 units for the strongest line).

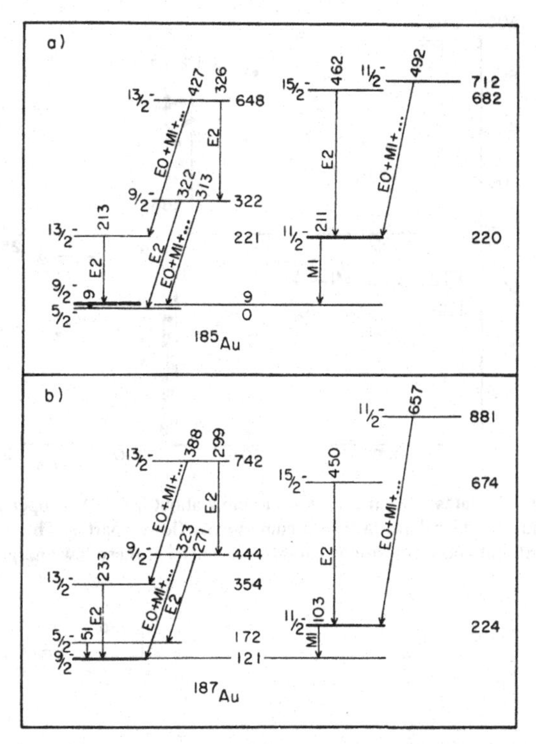

Figure 12. The band structures associated with the $E0$-enhanced transitions of 313, 427, 492 keV (^{185}Au) and 323, 388, 657 keV (^{187}Au). The same γ-γ and e^--γ gating techniques shown in Fig. 9 for ^{190}Hg were used to determine thê internal-conversion coefficients for these $I^\pi \rightarrow I^\pi$ interband transitions.

units, and the 535-keV γ-ray intensity is 0.8 units relative to the 100 unit intensity of the strongest line (416 keV, $2_1^+ \rightarrow 0_1^+$).

In odd-mass nuclei in this region shape coexistence similar to that shown for ^{190}Hg is observed. An example is shown in Fig. 12 for ^{185,187}Au. Using the coincidence-gating techniques described for ^{190}Hg, the 313-, 427-, and 492-keV, $I^\pi \rightarrow I^\pi$ interband transitions in ^{185}Au were determined[22] to have α_K(expt)/α_K($M1$,theory) values of $2.2_{-0.6}^{+1.3}$, $3.0_{-0.8}^{+0.3}$, and $2.7_{-0.8}^{+0.1}$, respectively. The 323-, 388-, and 657-keV, $I^\pi \rightarrow I^\pi$ interband transitions in ^{187}Au were determined[22] to have α_K(expt)/α_K($M1$, theory) values of 2.9±0.7, 5.3±0.8, and 2.8±0.6, respectively. Work at UNISOR on the level schemes of ^{183}Pt (Ref. 23), ^{185}Pt (Ref. 24), ^{187}Pt (Ref. 25), ^{185}Au (Refs. 19, 22, and 26), ^{187}Au (Ref. 22), and ^{195}Pb (Ref. 27) has revealed a total of ≈30 transitions with $E0$ components and energies below 600 keV. A compilation[19] of all known cases of low-energy $E0$ transitions between low-lying levels in odd-mass nuclei is shown in Fig. 13. The occurrence of low-energy $E0$ transitions in neighboring odd and even nuclei suggests that they are related. An example of a highly-converted transition, where the γ ray is too weak to be observed, is shown in Fig. 14 for ^{185}Au → ^{185}Pt decay. Note the two lines at 538 and 623 keV for which no γ-ray intensity is observed. This implies essentially pure $E0$ transitions and is unprecedented in an odd-mass nucleus.

We have suggested[19,22] that these $E0$-enhanced transitions may indicate a new class of nuclear-structure phenomena. Two major qualitative characteristics underlying this

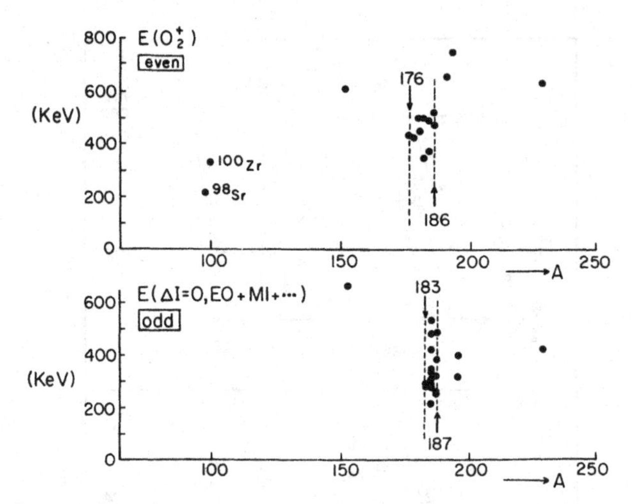

Figure 13. The energy of 0_2^+ states in even-even nuclei below about 800 keV (upper part) and the energy of transitions in odd-mass nuclei which have large $E0$ components (lower part). The data (Ref. 19) represent nearly all known cases of $E0$-enhanced transitions in odd-mass nuclei at these low energies.

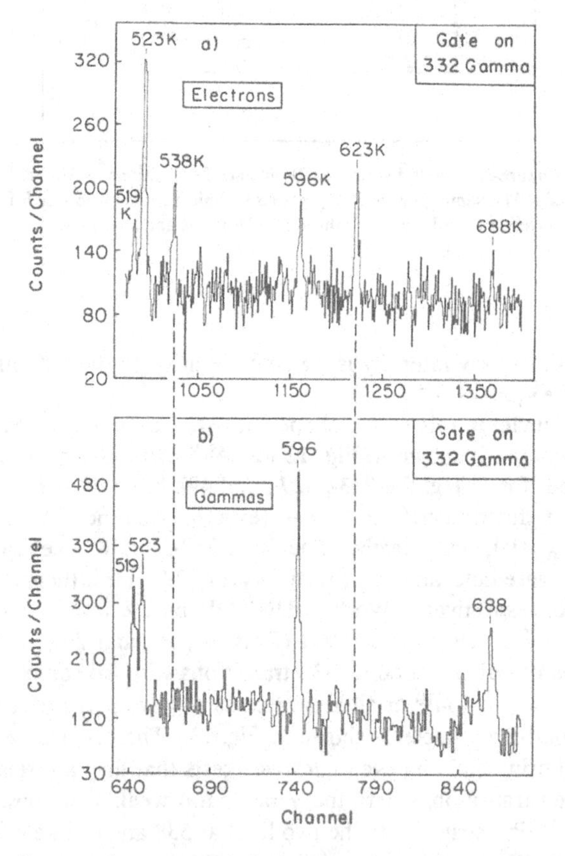

Figure 14. Conversion-electron and γ-ray spectra obtained by gating on the 332-keV γ ray in $^{185}Au \rightarrow ^{185}Pt$ decay. Note the two lines at 538 and 623 keV for which no γ-ray intensity is observed. This implies essentially pure $E0$ transitions and is unprecedented in an odd-mass nucleus (from Ref. 23).

phenomenon have now clearly emerged: Shape-coexisting bands intrude to become low-lying states in nuclei where one kind of nucleon lies at or near a closed shell and the other kind of nucleon near midshell; the intruding bands mix only weakly with neighboring states. An overall parabolic dependence of the intruder band energy is observed with a minimum near mid-shell for the *other* nucleon. This region of the nuclear chart provides the most extensive example of shape coexistence anywhere on the mass surface. It is also the region where the coexisting states are nearly degenerate over the widest range of nucleon numbers. Figure 15 can be used to provide a pictorial view of the intruder-state configurations and how they couple. As drawn, the upper plots (solid circles as proton particles) represent, from left to right, Pt ($Z=78$) and Au ($Z=79$), respectively. The shape-coexisting possibilities for Pt isotopes are then the $\pi(4h)$ and $\pi(2p\text{-}6h)$ configurations. A similar picture for even Hg ($Z=80$) would have the $d_{3/2}$ shell full, with the promoted proton pair coming out of it as sketched. Various couplings of these configurations are possible. For example, proton holes in the Au isotopes can couple to both the spherical and deformed configurations in the corresponding Hg cores, while proton particles in the Au isotopes can couple to the two types of structure in the Pt cores. This is shown in Fig. 12*a* for ^{185}Au, where the couplings $\pi h_{11/2}{}^{-1} \otimes {}^{186}$Hg (spherical) is represented by the band built on the $11/2^-$, 220-keV level; the $\pi h_{11/2}{}^{-1} \otimes {}^{186}$Hg (deformed) coupling by the band built on the $11/2^-$, 712-keV level; the $\pi h_{9/2}{}^{+1} \otimes {}^{184}$Pt (deformed) coupling by the band built on the $9/2^-$, 9-keV level; and the $\pi h_{9/2}{}^{+1} \otimes {}^{184}$Pt (spherical) coupling by the band built on the $9/2^-$, 322-keV

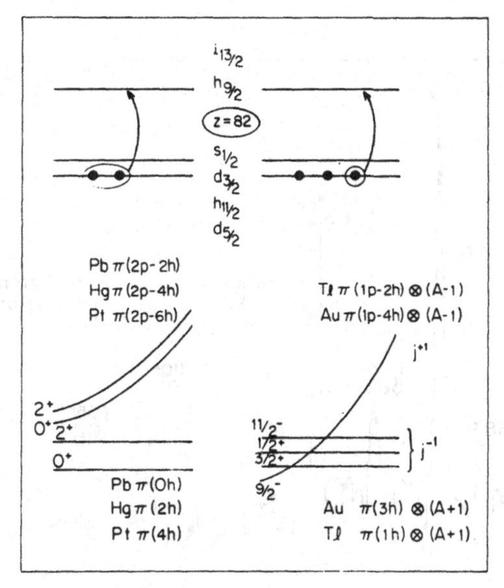

Figure 15. Schematic of the proton-particle and proton-hole configurations for the even-mass Pb, Hg, Pt isotopes and the odd-mass Tl, Au isotopes. As drawn, the upper plots, with the solid circles as proton particles, represent (from left to right) Pt ($Z = 78$) and Au ($Z = 79$), respectively. The lower left shows the normal configurations (0*h*, 2*h*, 4*h*) and the intruding shape coexisting configurations (2*p*-2*h*, 2*p*-4*h*, 2*p*-6*h*) in even-mass Pb, Hg, and Pt. The lower right shows the various particle-hole configurations in odd-mass Au and Tl and their coupling to the several possible Pb, Hg, and Pt core configurations. For Au, the ($A-1$) element is Pt, while the ($A+1$) element is Hg. For Tl, the ($A-1$) element is Hg, while the ($A+1$) element is Pb. The figure is from Ref. 19.

level. Similar couplings can be identified for ^{187}Au as well (part *b* of Fig. 12). The *E*0-enhanced transitions may thus serve, with some caution,[21] as a fingerprint for shape coexistence. Our reliance on e^--γ coincidence spectroscopy and the failure of previous work[18,28] to identify correctly the nature of the phenomenon (because e^--γ coincidences were not measured) illustrates the value of the kind and quality of measurements employed in the research on nuclei far from stability at UNISOR.

4.4 SPECIAL TECHNIQUES AND FUTURE PROSPECTS

Additional techniques employed in conversion-electron spectroscopy involve the use of permanent-magnet, mini-orange spectrometers. These are utilized whenever it becomes necessary to *block* low-energy photons or *filter out* β-decay positrons. A series of these devices with different magnet configurations and strengths is required in order to provide a choice of efficiencies for a corresponding series of electron-momentum intervals. Additional ongoing conversion electron instrumentation work at UNISOR involves the development of a new e^--γ coincidence system based on an 80-mm^2, 500-μm thick, passivated ion-implanted, Si-junction device, which should allow one to extend the e^--γ coincidence capability to include electron energies below 20 keV on a routine basis. Shown in Fig. 16 is a typical low-energy, γ-gated conversion-electron spectrum taken with one of the current spectrometers. The aim of the development of the new spectrometer is to extend the low-energy end of such a γ-gated spectrum to ≈5 keV with similar quality to that displayed in Fig. 16.

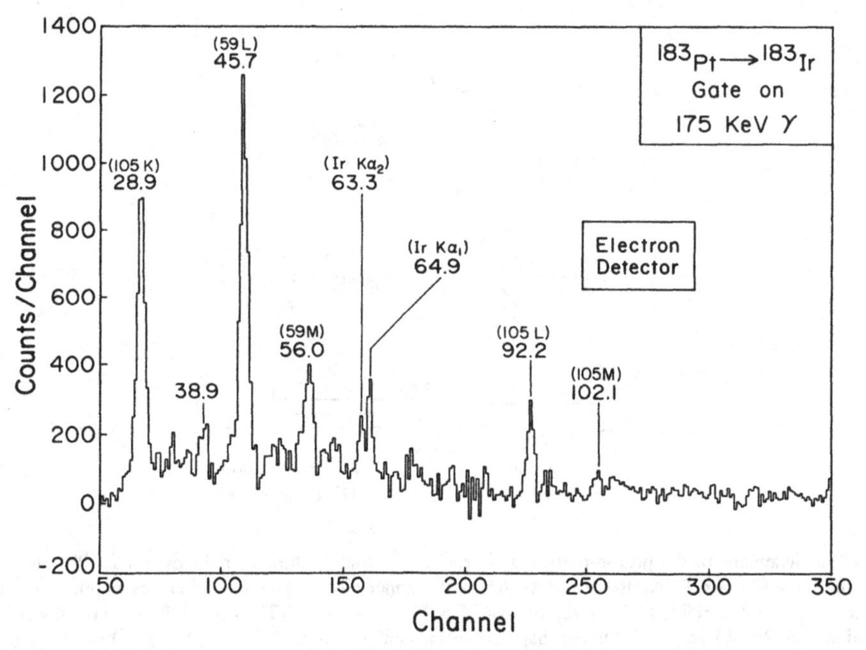

Figure 16. γ-gated electron spectrum of the ^{183}Pt→^{183}Au decay taken with one of the current electron spectrometers. This demonstrates both the quality of the system in terms of low-energy response as well as the enhancement of the quality (background reduction and removal of non-coincident transitions) through the gating technique.

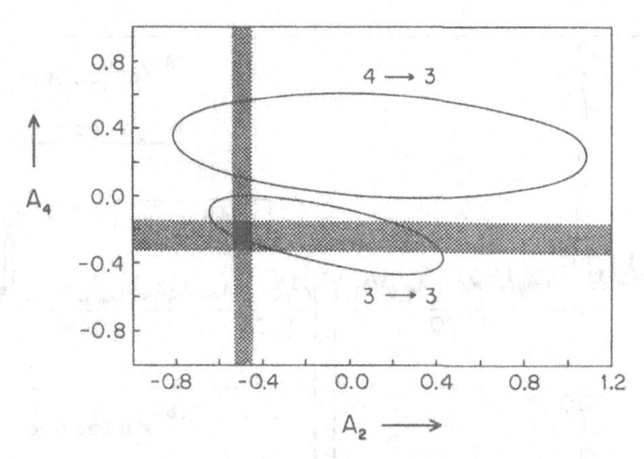

Figure 17. The A_2 and A_4 directional-correlation coefficients plotted as a function of $M1/E2$ mixing ratio δ^2 for 4 → 3 and 3 → 3 transitions. The shaded areas represent $A_2 \pm \Delta A_2$ and $A_4 \pm \Delta A_4$ and show that only by obtaining both coefficients can one obtain a unique value for the mixing ratio (where the shaded regions intersect).

One of the most important recent developments at UNISOR is the commissioning of the Nuclear Orientation Facility (NOF), which has successfully demonstrated a unique design for on-line nuclear orientation systems.[29] Because the radioactive-ion beam enters the ^{4}He/^{3}He-dilution refrigerator from the bottom, one has the option to place detectors at 45° angles around the cold target (catcher). This is a unique feature of the UNISOR NOF and is critical if one intends to obtain both A_2 and A_4 angular-distribution coefficients and thus acquire ***unique*** multipole mixing ratios. That both A_2 and A_4 are required to obtain a unique mixing ratio is clearly shown in Fig. 17. The 45° angle not only enables a unique determination of the multipole mixing ratios, but also the cylindrical geometry about the target allows for the reduction of systematic errors. Results for the first fully-on-line nuclear-orientation experiment ever performed in the U. S. (1989) is presented in Fig. 18 for ^{184}Au → ^{184}Pt decay. Shown are two 5-min collections, one warm and one cold. The 432- and 435-keV transitions are at the 1% intensity level in the decay; and, even with a mere 5 min of collection, one can easily observe a sizeable anisotropy.

Since we contend that $E0$ and $E0$-enhanced transitions may be used as a fingerprint to identify shape coexistence in odd-mass nuclei, it is important to explore fully the applicability of such a fingerprint. One of the major difficulties in this endeavor is that with conventional conversion-electron spectroscopy (conversion-coefficient measurements), the $E0$ admixture to the $I^\pi \rightarrow I^\pi$ transition must be sufficiently large to make the total conversion coefficient (usually a mixture of $M1+E2$) exceed that of a pure $M1$ transition. If it does not, then one simply cannot tell if there is any $E0$ component, and an $E0+E2$ transition can masquerade as $M1+E2$ or pure $M1$. It was fortunate that several $I^\pi \rightarrow I^\pi$ transitions poked their heads above the pure $M1$ values in ^{185,187}Au (Fig. 12). To examine the correspondence fully between shape coexistence and $E0$ enhancement, one needs to obtain the $M1/E2$ mixing ratios in the γ-ray transitions. As shown in Fig. 17, the $M1/E2$ mixing ratios can be determined easily at the UNISOR NOF, and from them and the α_K measurements one can extract the $E0$ strength for the mixed $E0/M1/E2$, $I^\pi \rightarrow I^\pi$ transitions.

With regard to developments for future work on isotopes far from stability taking

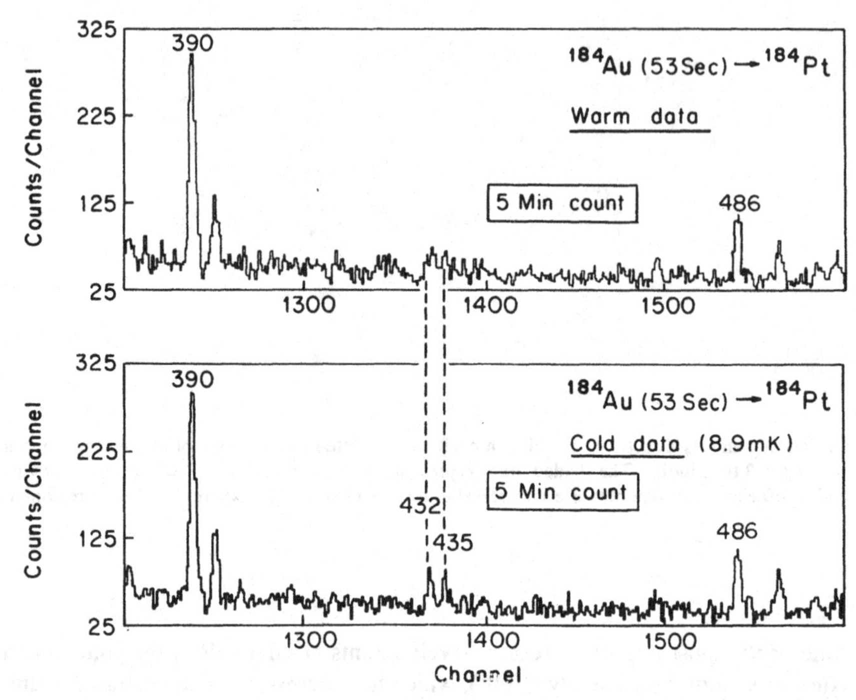

Figure 18. Portions of γ-ray spectra for ^{184}Au → ^{184}Pt decay taken at the UNISOR Nuclear Orientation Facility. The upper spectrum (warm data) was taken at a temperature $T > 100$ mK, while the lower spectrum (cold data) was taken at $T = 8.9$ mK. The collection time in each case was 5 min. The 432- and 435-keV transitions represent only about 1% of the total decay intensity, yet in only 5 min they show a clear anisotropy.

place at UNISOR and the Holifield Heavy-Ion Facility, the two most relevant are a laser ion-source[30] and a recoil mass separator (RMS).[31] A laser ion-source would enable one to study the refractory elements (difficult or impossible with current thermal ion-sources) using the present UNISOR isotope separator. The RMS will be totally revolutionary in far-from-stability work in that it will not only enable one to study refractory elements, but also it will make the very short half-life isotopes accessible to radioactive-decay investigations in a manner similar to those described above.

ACKNOWLEDGMENT

Work supported, in part, by the U. S. DOE under grant/contract DE-FG05-84ER40159 at LSU, DE-FG05-87ER40330 at Georgia Tech., and DE-AC05-76OR00033 at UNISOR.

REFERENCES

1. J. Bonn, G. Huber, H.-J. Kluge, L. Kugler, and E. W. Otten, *Phys. Lett.* **38B**, 308 (1972).
2. K. Heyde, P. van Isacker, M. Waroquier, J. L. Wood, and R. A. Meyer, *Phys. Repts.* **102**, 291 (1983).
3. J. H. Hamilton, P. G. Hansen, and E. F. Zganjar, *Repts. Prog. Phys.* **48**, 631 (1985).

4. K. Heyde et al., *Nucl. Phys.* **A466**, 189 (1987).
5. K. Heyde, J. Ryckebusch, M. Waroquier, and J. L. Wood, *Nucl. Phys.* **A484**, 275 (1988).
6. T. Bjornstad et al., *Nucl. Phys.* **A443**, 283 (1985).
7. S. Hofmann, I. Zychor, F. P. Hessberger, and G. Münzenberg, *Z. Phys.* **A325**, 37 (1986).
8. J. Äystö et al., *Nucl. Instr. Meth.* **B26**, 394 (1987).
9. R. Kirchner, O. Klepper, D. Marx, G.-E. Rathke, and B. Sherrill, *Nucl Instr. Meth.* **A247**, 265 (1986).
10. G. Münzenberg et al., *Nucl. Instr. Meth.* **161**, 65 (1979).
11. J. Schwarzenberg, work in progress.
12. E. F. Zganjar, J. D. Cole, J. L. Wood, and M. A. Grimm, *Fourth Intl. Conf. on Nuclei Far From Stability*, CERN Rept. **81-09**, p. 630 (1981).
13. C. Ekström, S. Ingelman, and G. Wannberg, *Nucl. Instr. Meth.* **148**, 17 (1978).
14. J. Bonn, G. Huber, H.-J. Kluge, and E. W. Otten, *Z. Phys.* **A276**, 203 (1976).
15. D. Vandeplassche, E. Van Walle, C. Nuytten, and L. Vanneste, *Phys. Rev. Lett.* **49**, 1390 (1982); J. Wouters et al., *Nucl. Instr. Meth.* **B26**, 463 (1987).
16. P. Kilcher et al., *Nucl. Instr. Meth.* **A274**, 485 (1989).
17. M. O. Kortelahti et al., *J. Phys.* **G14**, 1361 (1988).
18. C. Bourgeois, P. Kilcher, B. Roussière, and J. Sauvage-Letessier, *Nucl. Phys.* **A386**, 308 (1982).
19. E. F. Zganjar and J. L. Wood, *Hyp. Int.* **43**, 321 (1988); E. F. Zganjar and J. L. Wood, *Nuclear Structure of the Zirconium Region*, p. 88, Springer-Verlag, Berlin, 1988.
20. E. F. Zganjar et al., preprint.
21. K. Heyde and R. A. Meyer, *Phys. Rev.* **C37**, 2170 (1988).
22. C. D. Papanicolopulos et al., *Z. Phys.* **A330**, 371 (1988).
23. J. Schwarzenberg, J. Wood, and E. F. Zganjar, *Bull. Am. Phys. Soc.* **34**, 1825 (1989) and work in progress.
24. J. Schwarzenberg, J. Wood, and E. F. Zganjar, *Bull. Am. Phys. Soc.* **33**, 1603 (1988).
25. B. E. Gnade, R. W. Fink, and J. L. Wood, *Nucl. Phys.* **A406**, 29 (1983).
26. C. D. Papanicolopulos, Ph.D. thesis, Ga. Tech., 1987.
27. J. C. Griffin, Ph.D. thesis, Ga. Tech., 1987.
28. B. Roussière et al., *Nucl. Phys.* **A438**, 93 (1985); ibid., **A485**, 111 (1988).
29. I. C. Girit et al., *Hyperfine Interactions* **43**, 151 (1988).
30. W. M. Fairbank, Jr., and H. K. Carter, *Nucl. Instr. Meth.* **B26**, 357 (1987).
31. J. D. Cole, Chapter 2, This Book.

5. Nuclear Decay Studies Far from Stability

R. B. Firestone, J. M. Nitschke, P. A. Wilmarth,
K. S. Vierinen,* R. M. Chasteler, J. Gilat,†
and A. A. Shihab-Eldin‡
Lawrence Berkeley Laboratory
Berkeley, CA 94720

ABSTRACT

The decays of rare-earth nuclei in the region near $N = 82$ and $Z = 64$ and extending toward the proton drip line have been studied with the OASIS mass-separation facility on-line at the Lawrence Berkeley Laboratory SuperHILAC. Spectra of x rays, γ rays, and β-delayed protons and α particles have been measured in both singles and coincidence modes. Decay schemes have been derived, in many instances, with absolute branching intensities for electron capture, positron emission, and β-delayed particle decay. In favorable cases, $Q(EC)$ values have been determined from the measured electron-capture and positron-decay branchings. Nuclear structure near the closed shells has been qualitatively explained by a simple weak-coupling model, and the disappearance of $Z = 64$ shell effects near $N = 78$ has been explored. The systematics of Gamow-Teller spin-flip transitions have been investigated and will be compared with shell-model predictions.

5.1 INTRODUCTION

Nuclear decay studies far from stability are performed not merely to characterize new isotopes, but also to establish an experimental footing for improving our theoretical understanding of nuclear structure and decay. Although progress has been made in explaining low-lying level structure for a broad range of nuclei, transition probabilities are not yet quantitatively understood. The ability to understand nuclei far from stability is important to astrophysics for extending *r*- and *s*-process calculations to unknown nuclei, and to nuclear engineering for decay-heat calculations. Mass formulas are particularly sensitive to the known Q values for nuclei far from stability. Finally, by studying nuclei far from stability, we probe extremes of both decay energy and proton-neutron ratios, where unforeseen and important new nuclear properties may be exhibited. To understand these nuclei, it is important that complete decay information be determined without resorting to nuclear models, which will bias the interpretation of the results.

The decays of nearly one hundred isotopes and isomers have been studied with the OASIS mass-separation facility on-line at the Lawrence Berkeley Laboratory SuperHILAC. These studies have concentrated on neutron-deficient nuclei with $55 \leqslant Z \leqslant 71$ up to

* Permanent address: University of Helsinki, SF00170, Finland
† Permanent address: Soreq Nuclear Research Center, Yavne 70600, Israel
‡ Permanent address: Kuwait Institute for Scientific Research, Kuwait

Exotic Nuclear Spectroscopy, Edited by W. C. McHarris
Plenum Press, New York, 1990

A = 157 and neutron-rich nuclei with 166 $\leqslant A \leqslant$ 174. An extensive detector array has been constructed to detect x rays, γ rays, protons, α particles, and β particles; also, their coincidences, either β- or particle-delayed or directly emitted by ground-state or isomeric decay. From these data fairly complete decay schemes have been constructed for many isotopes, and decay Q values and EC/β^+ ratios have been determined. Analysis techniques have been developed to infer the electron-capture intensities from x-ray data, positron intensities from 511-keV annihilation intensities, internal-conversion coefficients from coincidence data, and spins from β-delayed-proton final-state feedings.

The results of these experiments are combined in this paper with those from many other laboratories to provide insight into systematic trends of β- and γ-ray transition probabilities near N = 82. It is hoped that the smooth systematic trends in these transition probabilities will provide clues toward interpreting the underlying nuclear structure. Nuclei near N = 82 and Z = 64 are expected to be spherical and should be described by simple shell-model considerations. Away from the shell closures, deformation sets in, which should exhibit itself in the transition probabilities. The Z = 64 shell closure is expected to disappear near N = 78. The qualitative nature of these phenomena will be discussed.

5.2 SOURCE PREPARATION

Most nuclides investigated at OASIS[1] were produced by bombarding various targets with heavy-ion beams of up to 8.5 MeV/u from the Berkeley SuperHILAC. The reactions and beam energies were chosen on the basis of compound-nuclear-reaction cross-sections calculated with the ALICE[2] evaporation code. Targets were mounted near the high-temperature surface-ionization source of the OASIS on-line mass separator facility in a configuration optimized for low-transverse-velocity recoils from compound-nucleus reactions. The ion source is efficient for all isotopes between Cs and Lu; however, elements outside that region had ionization potentials that were too high to allow them to be observed. Mass resolution of about 1 part in 800 was used, and no impurities from adjacent mass chains have been observed. After mass separation of the evaporation residues, a beam of the radioactive reaction products was deflected by an electrostatic mirror to a shielded spectroscopy laboratory ≈4 m above the mass separator. There, the activity was deposited on a programmable moving tape, which positioned it, in a user-selectable time cycle, in the center of an array of β, γ, and charged-particle detectors. Sources could be transported from the collection to detection points in 70 ms, and tape cycles as short as 1.28 s have been used. The ALICE calculations could not be tested quantitatively because of uncertainties in the ionization efficiency and diffusion time; however, the results appear to agree qualitatively with the predictions in most cases. A diagram of the OASIS mass separator is given in Fig. 1.

Neutron-excess nuclei were produced by multinucleon-transfer reactions. Here the targets are located inside the high-temperature region of the ion source and are restricted to refractory materials like W or Ta. To overcome this restriction, heavy-ion rare-earth beams were used to produce projectile-like, neutron-rich rare-earth isotopes. In these experiments natural W targets were bombarded with 8.5-MeV/u ^{170}Er and ^{176}Yb ions to produce isotopes of Dy, Ho, Er, and Tm. Low cross-sections, expected for the desired product nuclei, were partially offset by the scarcity and relatively-long half-lives of the reaction impurities within an isobaric chain. A principal source of contamination came from Lu isotopes produced in these reactions.

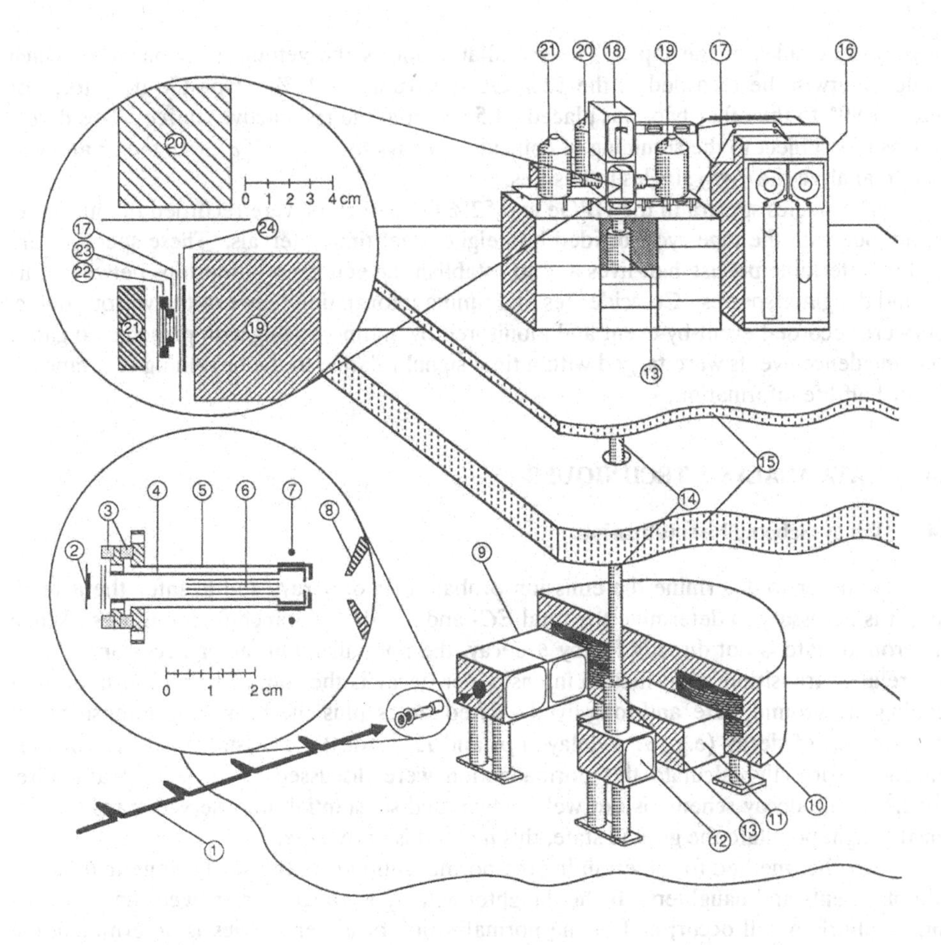

1. SuperHILAC BEAM
2. TARGET
3. INSULATORS (BeO)
4. ION SOURCE ANODE (Ta)
5. ION SOURCE CATHODE (Ta)
6. CAPILLARY TUBES (Ta)
7. EB FILAMENT (Ta)
8. EXTRACTION ELECTRODE
9. EXTRACTION AND FOCUSING
10. ANALYZING MAGNET
11. FOCAL PLANE DETECTOR BOX
12. ELECTROSTATIC MIRROR
13. ELECTROSTATIC QUADRUPOLE
14. TRANSFER LINE
15. CONCRETE SHIELDING
16. TAPE DRIVE (IBM 729)
17. MAGNETIC TAPE
18. DETECTOR BOX
19. N-TYPE Ge DETECTOR (52%)
20. N-TYPE Ge DETECTOR (24%)
21. HPGe DETECTOR
22. 718 μm Si DETECTOR
23. 10.4 μm Si DETECTOR
24. 1mm PILOT F SCINTILLATOR

Figure 1. OASIS experimental apparatus. The exploded drawings circled on the left show the detectors (upper) and ion source (lower) in greater detail.

5.3 DETECTORS

The detector array used for these measurements has evolved significantly since the first OASIS measurements were begun. The present detector geometry is shown in Fig. 1. A Si ΔE-E particle telescope and a hyperpure Ge (HPGe) detector face the front (deposit) side of the collection tape. The telescope records p and α events and identifies β particles which are stopped in the HPGe detector. The HPGe detector also is used to measure x rays and low-energy γ rays. A 1-mm-thick plastic scintillator and a 52% n-type Ge detector face opposite sides of the tape. The scintillator allows the vetoing of β particles which would otherwise be recorded in the 52% Ge spectrum. A 24% n-type Ge detector, oriented at 90° to the other two, was placed ≈4.5 cm from the radioactive source. This detector was less subject to the summing of coincident γ rays than the 52% Ge detector and was used to analyze the γ-ray singles intensities.

The singles spectra in the HPGe and 52% Ge detectors were recorded in multispectrum mode with the tape cycle divided into eight equal time intervals. These spectra were used to determine precise half-lives and to establish the genetic relationships between parent and daughter species. Coincidences and timing information between the various detectors were recorded event-by-event and monitored by on-line sorting with preselected gates; all coincidence events were tagged with a time signal relative to the beginning of a tape cycle for half-life information.

5.4 DATA-ANALYSIS TECHNIQUES

5.4.A Decay-Scheme Normalization

In order to determine the emission probabilities of γ rays and to infer the β feedings, it is necessary to determine the total EC- and β^+-decay branching intensities. When the ground state is not directly fed by β decay, the normalization factor necessary to convert relative transition intensities to intensity per decay is the sum of transition intensities feeding the ground state and long-lived excited states plus the branching intensities for other modes of decay (e.g., α, β-delayed p, and IT modes). The statistical methods required to correctly calculate this normalization were discussed by Browne[3,4] and Firestone.[5] If the decay scheme is not well known, and substantial, unobserved transition intensity might populate the ground state, this method is unreliable.

Another method for determining the normalization is to follow the genetic relationship of parents and daughters. If the daughter activity is much shorter-lived than the parent, equilibrium will occur; and, if the normalization of either species is determined, the other normalization can be directly calculated. It is important to remember that at equilibrium the daughter activity R_d is greater than the parent activity R_p and related to the parent and daughter half-lives by the equation,

$$R_d = R_p \frac{t_{1/2}(p) - t_{1/2}(d)}{t_{1/2}(p)} . \tag{1}$$

When equilibrium is not possible, the relative parent/daughter normalizations can be determined by following the growth and decay.

In the OASIS studies, it was often possible to determine the decay-scheme normalization from the measured EC- and β^+-decay intensities. The methods used to investigate these intensities are discussed below.

5.4.A.1 *EC* Intensity

A signature of electron-capture decay is K-x-ray emission. For the nuclei discussed here, the resolution of the HPGe detector was sufficient to resolve the $K_{\alpha 1,2}$ and $K_{\beta 1,2}$ x rays of adjacent elements. This analysis was performed using a modified version of the

SAMPO code.[6] The relative x-ray intensities for each element are well known,[7] and our measured intensity ratios normally agreed well with the expected values. Table I shows a representative comparison of our measured values with the known branchings. To determine the relative intensity of *EC* decay, the *K*-x-ray intensities were corrected for contributions from internal conversion, although in some cases internal conversion was negligible and could be ignored. Otherwise, the *K*-conversion coefficients must be estimated and a correction applied. Often this correction was large with respect to the *EC* intensity, and the *EC*- feeding could not be reliably determined. If the β feeding were known (Sec. 5.4.A.2), it was sometimes possible to estimate the *EC*- branching intensity from the decay scheme and expected EC/β^+ ratios. After the *K*-x-ray intensity associated with *EC* decay was determined, additional corrections for fluorescence yield[8] and $I_{EC(K)}/I_{EC(tot)}$ were applied.[9]

5.4.A.2 β^+ Intensity

A signature of β^+ decay is the emission of 511-keV annihilation radiation, symbolized by $\gamma^\pm$. In most experiments several β^+ emitters were produced simultaneously. Distinction among the various β^+ emitters could only be made on the basis of half-life and was particularly difficult when complex relationships between parent and daughter activities occurred. To solve this problem, a multilinear analysis method was developed. The key to this method was to follow the decay of a prominent γ ray associated with each β^+-emitting isotope. The relative $\gamma^\pm$ intensity associated with each γ ray is constant at all times. Thus, the total observed $\gamma^\pm$ intensity in any spectrum from *i* positron-emitting species is given by

$$I_\gamma^\pm(tot) = C_\gamma^\pm(1)I_\gamma(1) + C_\gamma^\pm(2)I_\gamma(2) + \ldots + C_\gamma^\pm(i)I_\gamma \,, \tag{2}$$

where $C_\gamma^\pm(i)$ is the proportionality constant relating the γ-ray intensity to the $\gamma^\pm$ intensity associated with that decay. An advantage of this method is that data taken at various dwell times with differing reactions could be analyzed simultaneously. The calculation was performed using the computer code BANAL[10] with the IMSL multilinear-analysis subroutine RLMUL.[11] An example calculation is shown in Table II.

Table I. Comparison of Experimental and Theoretical K-x-ray Intensities

	K_{α_1}		K_{α_1}		K_{β_1}'		K_{β_1}'	
	Exp	Th	Exp	Th	Exp	Th	Exp	Th
Nd	100(2)	100	50(4)	54.9	24(8)	30.0		8.3
Pm	100.0(12)	100	53.6(8)	55.1	30.6(12)	30.1		8.4
Sm	100.0(13)	100	54.6(7)	55.2	31.0(3)	30.2	8.8(3)	8.6
Eu	100.0(10)	100	55.7(8)	55.4	30.9(6)	30.5	8.5(2)	8.7
Gd	100.0(11)	100	56.4(9)	55.6		30.8		8.9
Tb	100(3)	100	50(2)	55.8	34(4)	31.0	10(3)	8.9
Dy	100(2)	100	55.2(13)	56.0		31.2	10(2)	8.9
Ho	100(2)	100	56.6(17)	56.2	29(6)	31.5	8.7(9)	8.8

Table IIa. Multilinear Analysis of A = 142 $\gamma^{\pm}$ Intensities Using the Equation, $I_{\gamma}{}^{\pm}(tot) = C_{\gamma}{}^{\pm}(1)I_{\gamma}(1) + C_{\gamma}{}^{\pm}(2)I_{\gamma}(2) + \ldots + C_{\gamma}{}^{\pm}(i)I_{\gamma}(i)$

Input Relative γ-Ray Intensities $I_{\gamma}(i)$				Input $I_{\gamma}{}^{\pm}(tot)$		Fitted $I_{\gamma}{}^{\pm}(tot)$	
Eu(768)	Gd(179)	Tb(515)	Dy(182)	I(511)	ΔI(511)	I(511)	χ^2
426	100	678	0	6839	280	6830	0.001
531	100	56	7	5330	280	5482	0.298
60	100	0	0	944	90	958	0.024
1051	100	289	32	11476	574	11507	0.003
850	100	157	16	8835	442	8882	0.011
693	100	88	13	7342	367	7172	0.215
492	100	50	9.3	5230	262	5162	0.066
383	100	38	8.4	4148	207	4136	0.003
301	100	25	8.1	3356	168	3354	0.0001
251	100	21	6.3	2811	140	2859	0.120
201	100	19	3.4	2380	119	2347	0.075

Table IIb. Parameters $C_{\gamma}{}^{\pm}(i)$ Calculated with the Computer Code BANAL

Isotope	Parameter	Lower-Bound	Upper-Bound	Std. Error
^{142}Eu	8.83	8.37	9.30	0.16
^{142}Gd	4.28	3.13	5.42	0.40
^{142}Tb	3.89	3.57	4.21	0.11
^{142}Dy	20.9	6.6	35.3	5.0

The resulting $\gamma^{\pm}$ intensity for each isotope must be corrected for annihilation-in-flight[12] and source geometry. These corrections were not entirely straightforward because a 4π positron annihilator was not used. The correction factor was determined by comparison of results using this method with other techniques discussed above, and with β-delayed p data, where the β^+ and EC intensities in coincidence with protons must equal the total p intensity. The uncertainty in this correction was estimated as 10% by comparison with results obtained by the equilibrium method. This comparison for A = 142 nuclei is shown in Table III. The multilinear analysis method described here can also be used to resolve γ-ray multiplets and unusually complex x-ray spectra.

5.4.B Q_{EC} Determination

Two methods have been employed in these experiments to determine Q_{EC} values. The EC/β^+ ratio for the decay to a level can be inferred from the K-x-ray and $\gamma^{\pm}$ intensities in coincidence with transitions deexciting that level. If the level is not significantly populated from above and the deexciting transition is not in coincidence with K x rays from internal conversion, then this method is straightforward. For the 4.7-MeV level in ^{149}Ho, populated by ^{149}Er^m decay, coincidence data for the intense 4.7-MeV γ ray gives EC/β^+ = 0.68±0.34, which corresponds to $Q_{EC} = 9.1_{-0.4}^{+0.9}$. The excitation energy of ^{149}Er^m is 0.74 MeV, so for the ^{149}Er ground state $Q_{EC} = 8.4_{-0.4}^{+0.9}$. Similarly, ^{149}Er^g populates four narrow resonances in ^{149}Ho which decay by β-delayed p emission. From the ratio of

Table III. Summary of Decay Branchings for A = 142

Isotope	$t_{1/2}$[a]	Branching Intensity β^+	EC	p	E_γ	I_γ[b]	I_γ[c]
^{142}Pm	40.5(5) s	0.771(27)	0.229(27		1576.1	0.0196(11)	
^{142}Sm	72.49(5) min	<0.05	>0.95				
^{142}Eu	2.34(12) s	0.899(16)	0.101(16)		768.0	0.102(7)	≡0.102(3)
^{142}Gd	70.2(6) s	0.48(5)	0.52(5)		178.9	0.112(12)	0.113(5)
^{142}Tb	597(17) ms	0.968(4)	0.032(4)	2.4(10)x10^{-5}	515.3	0.249(17)	≡0.249(13)
^{142}Dy	2.3(3) s	0.90(4)	0.10(4)	8(3)x10^{-3}	181.3	0.043(8)	0.51(5)

a Values for ^{142}Pm and ^{142}Sm are from L. K. Peker, Nucl. Data Sheets **43**, 579 (1984). Other values from this work.
b Normalized to measured $EC+\beta^+$ intensity.
c Equilibrium intensity normalized to indicated transitions.

intensities in the p singles and β^+-coincident proton spectra, $(EC+\beta^+)/\beta^+$ can be determined. The value of $(Q_{EC}-B_p)$ can be determined by minimizing the differences between the experimental and theoretical values as a function of Q_{EC}. The result of this minimization for ^{149}Er is shown in Fig. 2, where $(Q_{EC}-B_p) = 7.0_{-0.4}^{+0.5}$ was determined. Wapstra et al.[13] report a systematic p binding energy of $B_p = 1.2\pm0.2$, which gives $Q_{EC} = 8.2\pm0.5$, in good agreement with the value from the 4.7-MeV level analysis.

A second method for determine Q_{EC} was to vary the decay energy so that the measured EC- and β^+-decay branching intensities [see Sec. 5.4.A.1 and Sec. 5.4.A.2] matched those predicted from the established decay scheme. This method is most effective when the decay scheme is well known. Decays in the A = 140 and A = 142 mass chains are particularly suited to this method because their decays are dominated by intense ground-state β feedings. These high-energy, low-log*ft* transitions are so dominant that missing higher-level feedings are not expected to be significant for any but the most neutron-deficient isotopes. A summary of these results is given in Table IV, where they are compared with the evaluated values of Wapstra et al.[14] and the calculated values of Liran and Zeldes.[15] Agreement is excellent, particularly for the previously known values, lending confidence to this method.

5.4.C Internal-Conversion Coefficients

Internal-conversion data are necessary to determine the spins and parities of levels populated in these decays. Unfortunately, conversion electrons were not measured at OASIS, so other methods have been employed to obtain some internal-conversion information. One method was to determine the K-conversion coefficient from the ratio of K-x-ray to γ-ray intensities in coincidence with transitions feeding a level. Another method was to utilize the intensity balance in singles or coincidence data to infer total conversion coefficients. For example, these methods were used to determine the multipolarities of transitions following ^{142}Tb^m IT decay. This decay scheme is shown in Fig. 3, and the coincidence spectra gating on the 182- and 212-keV transitions are shown in Fig. 4. From the 212-keV gate we determined that $\alpha_K(68.5) = 61 \pm 5$, which is consistent with the theoretical $\alpha_K(M2) = 64$.[16] From the relative intensities of the 29.7-, and 68.5-keV γ-ray

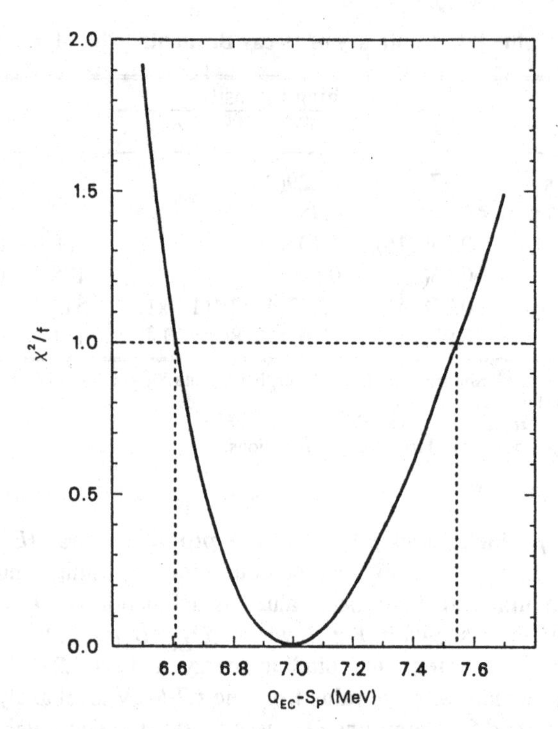

Figure 2. Fit to Q_{EC}-B_p for $^{149}Er^g$ decay to levels associated with structured *p* decay. The χ^2/f from comparison of the experimental $\beta+/(\beta++EC)$ values with the theoretical ratios is plotted on the ordinate for various Q_{EC}-B_p values indicated on the abscissa. The uncertainty is chosen to encompass all values with $\chi^2/f < 1$.

transitions and the Tb *K* x rays in the 182-keV gate, we determined that $\alpha_{tot}(29.7)$ = 44±14, which is consistent with the multipolarity, $M1$ + (6±3)% $E2$. A summary of the conversion coefficients determined in these experiments is given in Table V.

5.4.D Spin Values from Final-State β-Delayed *p* Feedings

The spins of isotopes decaying by β-delayed *p* emission are reflected in the distribution of final-state feedings to the *p*-decay daughter. Unlike α-particle decay, where the formation of the α particle in the nucleus is complex, *p*-decay is a simple process, which is well understood. Thus, if the β-strength distribution associated with the decay to the region of *p* emission is known or the decay can be treated within the framework of a statistical model, the final-state feedings can be calculated and compared with experiment. A detailed discussion of this method for spin assignment has been discussed by Wilmarth.[17] The experimental final-state feedings were determined from *p*-γ-coincidence information, where the ground-state feeding was determined by the difference in the coincidence *p* intensity and the total *p* intensity. Results of this method for odd-*A* precursors are shown in Table VI, and a comparison of results using various β-strength models for ^{153}Yb decay is given in Table VII.

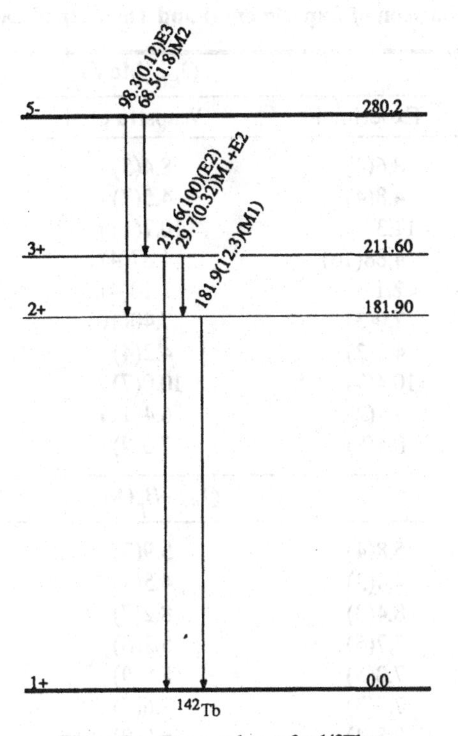

Figure 3. Decay scheme for $^{142}Tb^m$.

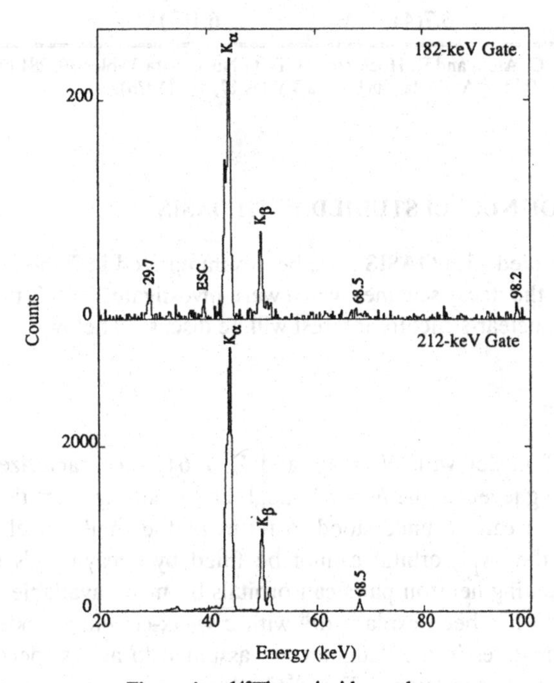

Figure 4. $^{142}Tb^m$ coincidence data.

Table IV. Comparison of Experimental and Theoretical Decay Energies

	Q_{EC}(MeV)		
Isotope	Experiment	Wapstra et al.[a]	Liran-Zeldes[b]
^{140}Eu	8.6(4)	8.4(5)	8.3
^{140}Gd	4.8(4)	4.5(7)	5.5
^{140}Tb	>11.3	10.7(11)	10.9
^{142}Pm	4.88(16)	4.87(4)	5.1
^{142}Sm	<2.1	2.10(4)	2.2
^{142}Eu	7.0(3)	7.40(10)	7.5
^{142}Gd	4.2(3)	4.2(4)	4.6
^{142}Tb	10.4(7)	10.0(7)	9.9
^{142}Dy	7.1(2)	6.4(11)	7.1
^{149}Er	8.4(5)	7.0(9)	8.65
	$Q_{EC}-B_p$(MeV)		
^{145}Dy	5.8(4)	5.9(7)	6.1
^{147}Dy	4.4(3)	4.5(1)	4.8
^{147}Er	8.4(3)	8.2(7)	8.6
^{148}Ho	5.7(5)	5.2(3)	5.9
^{149}Er	7.0(5)	5.8(9)	7.3
^{150}Tm	7.5(3)	7.6(7)	8.4
^{151}Yb	8.8(4)	8.9(9)	9.7
^{152}Lu	9.6(9)	10 (1)	10.7
^{153}Yb	5.7(4)	6.1(5)	6.0

[a] A. H. Wapstra, G. Audi, and R. Hoekstra, At. Data Nucl. Data Tables **39**, 281 (1988).
[b] S. Liran and N. Zeldes, At. Data Nucl. Data Tables **17**, 431 (1976).

5.5 SUMMARY OF NUCLEI STUDIED WITH OASIS

The nuclei studied with OASIS have been summarized in Table VIII. It is not possible to discuss all of the decay schemes which were investigated in this paper, so only a few decays of particular nuclear-structure interest will be discussed below.

5.5.A N = 81 Nuclei

The decay of nuclei with $N = 81$ and $Z \geqslant 64$ is characterized by hindered β transitions to low-lying levels in the $N = 82$ daughter and strong transitions to levels above 4 MeV. These nuclei can be understood in terms of the shell model, where the single neutron vacancy in the $\nu s_{1/2}$ orbital cannot be filled by decaying p's in $\pi d_{5/2}$ or $\pi g_{7/2}$ orbitals. Only by exciting neutron pairs can orbitals be made available for β decay. The decay of ^{145}Gd($1/2^+$) has been explained[18] with a weak-coupling model. In this model, single-particle proton states in the daughter are assumed to act as spectators only weakly interacting with the core excitations. For ^{145}Gd decay, two levels near the 2^+ and two

Table V. Experimental and Theoretical Conversion Coefficients from OASIS

Parent	E_γ	α_K(expt)	α_K(theory)[a] $E1$	$E2$	$E3$	$M1$	$M2$	$M3$	$M4$	Adopted Multipolarity
$^{140}Eu^m$	185.3	0.19(4)	0.046	0.19	0.70	0.28	1.55	6.83	29.5	$E2$
$^{142}Tb^m$	29.7	44(15)[b]	1.50	521		12.5	1130			$M1+E2$
	68.5	61(5)	0.68	2.48	5.93	5.69	63.8	278	911	$M2$
$^{144}Tb^m$	113.7	4.59(20)	0.18	0.81	3.00	1.32	10.1	51.0	241	$E3+M4$
		28(2)[b]	0.21	1.65	22.6	1.57	13.4	96.1	855	
^{145}Ho	66.3	6.5(10)	0.24	2.40	4.59	6.83	77.3	298	788	$M1$
$^{149}Er^{g+m}$	111.3	1.82(11)	0.20	0.82	2.71	1.82	13.6	63.1	268	$M1$
	171.5	0.57(7)	0.064	0.25	0.87	0.49	2.75	11.8	48.8	$M1$
	343.9	0.14(3)	0.011	0.034	0.097	0.074	0.27	0.84	2.56	$M1$
	436.7		0.006	0.018	0.048	0.040	0.13	0.36	0.96	$M1$
	630.5	0.27(3)	0.003	0.008	0.018	0.017	0.047	0.11	0.25	$M4$
^{154}Yb	133.2	0.11(4)	0.13	0.51	1.69	1.19	7.71	34.0	14	$E1$
^{154}Lu	96.6	1.3(3)	0.31	1.09	2.94	3.24	26.2	109	394	$E2$

[a] F. Rosel, H. M. Fries, K. Alder, and H. C. Pauli, At. Data and Nucl. Data Tables 21, 91 (1978).
[b] α_{tot}(exp)

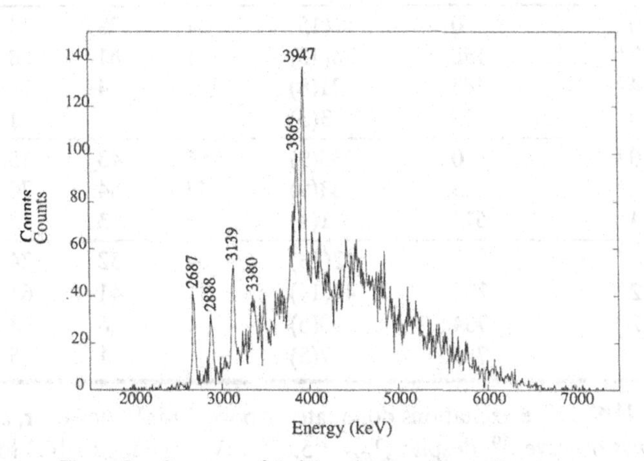

Figure 5. Spectrum of β-delayed protons from ^{149}Er decay.

Table VI. Comparison of Experimental and Calculated p Final-State Branches

Precursor (Adopted J^π)	J^π Daughter	Energy (keV)	Branching Intensity (%) Exp.	1/2+	3/2+	5/2+	7/2+
^{119}Ba	0^+	0	71(10)	60	46	21	
(1/2+)	2^+	337	29(5)	33	45	60	
	4^+	810	1(1)	1	1	9	
^{123}Ce	0^+	0	23(6)		37	14	9
(5/2+)	2^+	197	66(6)		55	66	54
	4^+	570	9(3)		3	14	32
	6^+	1083	2(1)				1
^{125}Ce	0^+	0	36(4)		49	20	14
(5/2+)	2^+	230	53(4)		50	72	65
	4^+	651	9(3)		1	8	21
	6^+	1228	1(1)				
^{127}Nd	0^+	0	60(15)	50	37	14	
(1/2+)	2^+	170	35(13)	48	60	70	
	4^+	520	5(5)	2	3	16	
^{129}Nd	0^+	0	23(7)		44	17	12
(5/2+)	2^+	207	68(7)		54	71	61
	4^+	607	9(3)		2	12	27
^{131}Nd	0^+	0	32(7)		57	26	20
(5/2+)	2^+	254	67(7)		41	68	67
	4^+	710	1(1)		1	4	11
^{131}Sm	0^+	0	41(15)	47	35	13	7
(5/2+)	2^+	158	36(15)	51	61	68	52
	4^+	483	21(8)	2	4	19	39
	6^+	938	3(3)			1	2
^{133}Sm	0^+	0	35(9)	56	43	18	
(3/2+)	2^+	213	63(9)	43	54	70	
	4^+	611	1(1)	1	3	12	
^{135}Sm	0^+	0	42(13)	64	52	24	
(5/2+)	2^+	294	41(14)	31	41	61	
	2^+	754	10(6)	4	6	8	
	4^+	789	7(5)		1	5	

levels near the 4^+ ^{144}Sm core excitations dominate. Above 4 MeV, however, about 12.8% of the β intensity was observed[19] despite Q_{EC} = 5.07 MeV. The log*ft* = 4.3 to this region is similar to that of nearby $\pi h_{11/2} \rightarrow \nu h_{9/2}$ spin-flip transitions. This strong transition presumably populates three-quasiparticle levels with the structure, $(\pi h_{11/2})(\nu h_{9/2})(\nu s_{1/2})$.

Similar decay systematics have been observed for decay of the $1/2^+$ and $11/2^-$ isomer pairs in ^{147}Dy,[20,21] ^{149}Er,[22] and ^{151}Yb.[23,24] Remarkably, the spin-flip decays of the isomer pairs are nearly identical, confirming the weak-coupling assumption that the odd $\nu s_{1/2}$ or $\nu h_{11/2}$ neutron is a spectator. A comparison of the log*ft* values for the dominant

Table VII. Experimental and Calculated β-Delayed p Branches from ^{153}Yb to Levels in ^{152}Er

Levels in ^{152}Er		Final-State Branches (%)			
J^{π}	Energy (keV)	Experiment	Gross Theory $7/2^-$	*QRPA* $7/2^-$	Constant $7/2^-$
0^+	0	57(17)	50	66	49
2^+	808	40(12)	44	32	45
4^+	1481	3(3)	4	2	4

Table VIII. Isotopes Studied at OASIS

Isotope	$t_{1/2}$(s)	Decay	Isotope	$t_{1/2}$(s)	Decay	Isotope	$t_{1/2}$(s)	Decay
^{119}Ba	6.0(3)	βp	^{141}Eu(5/2+)	a	$\beta\gamma$	^{147}Er	2.6(2)	βp
^{120}La	2.8(4)	βp	^{142}Eu	2.34(12)	$\beta\gamma$	^{148}Er	4.4(2)	$\beta\gamma p$
^{122}La	8.7(7)	βp	^{137}Gd	7(3)	βp	^{149}Er(11/2$^-$)	8.9(2)	$\beta\gamma p$
^{123}Ce	3.8(2)	βp	^{139}Gd	5(1)	βp	^{149}Er(1/2+)	4(2)	$\beta\gamma p$
^{125}Ce	9.8(8)	$\beta\gamma p$	^{140}Gd	15.8(4)	$\beta\gamma p$	^{151}Er	a	$\beta\gamma$
^{124}Pr	1.2(2)	βp	^{141}Gd(11/2$^-$)	24.5(5)	$\beta\gamma$	^{174}Er	198(12)	$\beta\gamma$
^{126}Pr	3.2(6)	$\beta\gamma p$	^{141}Gd(1/2+)	14(4)	$\beta\gamma p$	^{149}Tm	0.9(2)	$\beta\gamma p$
^{128}Pr	4(1)	βp	^{142}Gd	70.2(6)	$\beta\gamma$	^{150}Tm	2.2(2)	$\beta\gamma p$
^{127}Nd	1.8(4)	βp	^{140}Tb	2.4(2)	$\beta\gamma p$	^{151}Tm	a	$\beta\gamma$
^{129}Nd	4.9(3)	βp	^{141}Tb	3.5(2)	$\beta\gamma$	^{153}Tm(11/2$^-$)	1.7(2)	$\alpha\beta\gamma$
^{131}Nd	25(5)	βp	^{142}Tb(1+)	0.597(17)	$\beta\gamma p$	^{153}Tm(1/2+)	≈0.5-2.5	$\alpha\beta\gamma$
^{130}Pm	2(1)	βp	^{142}Tb(5$^-$)	0.303(17)	*IT*	^{174}Tm	a	$\beta\gamma$
^{132}Pm	5(1)	$\beta\gamma p$	^{144}Tb(1+)	a	$\beta\gamma$	^{151}Yb(11/2$^-$)	1.6(1)	βp
^{134}Pm(2+)	≈3-20	$\beta\gamma$	^{144}Tb(6$^-$)	4.1(1)	$\beta\gamma$	^{151}Yb(1/2+)	1.6(1)	$\beta\gamma p$
^{134}Pm(5+)	22.6(5)	$\beta\gamma$	^{145}Tb	a	$\beta\gamma$	^{153}Yb	3.9(1)	$\beta\gamma p$
^{135}Pm(5/2+)	≈45	$\beta\gamma$	^{141}Dy	0.9(2)	$\beta\gamma p$	^{154}Yb	0.42(5)	$\alpha\beta\gamma$
^{135}Pm(11/2$^-$)	49(3)	$\beta\gamma$	^{142}Dy	2.3(3)	$\beta\gamma p$	^{155}Yb	1.75(5)	$\alpha\beta\gamma$
^{136}Pm(2+)	≈30-150	$\beta\gamma$	^{143}Dy	3.1(3)	βp	^{152}Lu	0.7(1)	βp
^{136}Pm(5+)	a	$\beta\gamma$	^{144}Dy	9.1(5)	$\beta\gamma p$	^{153}Lu	0.9(2)	$\beta\gamma$
^{142}Pm	a	$\beta\gamma$	^{145}Dy	8(1)	$\beta\gamma p$	^{154}Lu	1.2(1)	$\beta\gamma p\alpha$
^{131}Sm	1.2(2)	βp	^{147}Dy(11/2$^-$)	a	$\beta\gamma$	^{155}Lu(3/2+,1/2+)	0.140(14)	α
^{133}Sm	2.8(5)	βp	^{147}Dy(1/2+)	a	$\beta\gamma p$	^{155}Lu(11/2$^-$)	0.066(7)	α
^{134}Sm	10(3)	$\beta\gamma$	^{149}Dy	a	$\beta\gamma$	^{157}Lu(1/2+,3/2+)	5.7(6)	$\alpha\beta\gamma$
^{135}Sm	10.3(5)	$\beta\gamma p$	^{168}Dy	a	$\beta\gamma$	^{157}Lu(11/2$^-$)	4.8(5)	$\alpha\beta\gamma$
^{136}Sm	a	$\beta\gamma$	^{169}Dy	55(3)	$\beta\gamma$	^{166}Lu(0$^-$)	a	$\beta\gamma$
^{141}Sm	a	$\beta\gamma$	^{144}Ho	0.7(2)	βp	^{166}Lu(3$^-$)	a	$\beta\gamma$
^{142}Sm	a	β	^{145}Ho	2.4(1)	$\beta\gamma$	^{166}Lu(6$^-$)	a	$\beta\gamma$
^{134}Eu	0.5(2)	βp	^{146}Ho	3.1(5)	$\beta\gamma p$	^{168}Lu(3+)	a	$\beta\gamma$
^{135}Eu	1.5(2)	$\beta\gamma$	^{148}Ho	9.7(3)	$\beta\gamma p$	^{168}Lu(6$^-$)	a	$\beta\gamma$
^{136}Eu(3+)	3.7(3)	$\beta\gamma p$	^{149}Ho(11/2$^-$)	21.4(3)	$\beta\gamma$	^{169}Lu(1/2$^-$)	a	*IT*
^{136}Eu(6+,7+)	≈3	$\beta\gamma p$	^{149}Ho(1/2+)	54(2)	$\beta\gamma$	^{171}Lu(1/2$^-$)	a	*IT*
^{140}Eu(1+)	1.51(2)	$\beta\gamma$	^{169}Ho	a	$\beta\gamma$	^{172}Lu(1$^-$)	a	*IT*
^{140}Eu(5$^-$)	0.125(2)	*IT*	^{171}Ho	54(3)	$\beta\gamma$	a = Half-life not determined		
^{141}Eu(11/2$^-$)	3.3(3)	$\beta\gamma$	^{145}Er	0.9(3)	βp	in these experiments.		

configurations contributing to the $N = 81$ decays is given in Table IX. The odd-odd decays display similar decay patterns, further confirming the weak-coupling assumptions.

The spectator neutrons become important when the three-quasiparticle configurations deexcite to the ground configurations. The $\nu h_{9/2}$ neutron can decay by a fast $M1$ spin-flip to fill the $\nu h_{11/2}$ vacancy, an analogous transition to the β decay, but it is very hindered, filling the $\nu s_{1/2}$ vacancy. This effect has been observed in several ways: In ^{145}Gd(1/2$^+$) decay, the three-quasiparticle levels deexcite preferentially to levels above 1.7 MeV. Transitions to low-lying single-particle states are weak, and both $M1$ and $E2$ transitions are of comparable intensity. For 3-MeV transitions, this corresponds to a factor of 60 greater hindrance for $M1$ transitions. In ^{149}Er(11/2$^-$) decay, the three-quasiparticle levels deexcite primarily by $M1$ spin-flip transitions to the $\nu h_{11/2}$ ground state and by weaker $E1$ transitions to the $\nu g_{7/2}$ level. No evidence for low-energy γ-ray transitions is observed. Additional evidence is obtained from β-delayed p data. The p spectrum from ^{149}Er^{g+m} decay is shown in Fig. 5. The structured part of the spectrum is known to belong to the low-spin decay, and the 7(±2)% branching intensity is much larger than the 0.4% predicted from Gross Theory.[25,26] Similarly, the low-spin β-delayed p-branching intensity for ^{147}Dy decay is enhanced by a factor of 20, and Schardt et al.[27] have measured the very-weak γ rays deexciting the levels corresponding to the structured part of the p spectrum. The γ-ray transitions appear to be hindered, enhancing the p intensity. The nature of this hindrance has been discussed by Nitschke et al.[28]

5.5.B Onset of Deformation at $Z = 64$

Unlike the $Z = 50$ and $N = 82$ shells, the $Z = 64$ subshell gap is small, about 1.8 MeV, compared to the ≈4.5-MeV gap observed at $N = 82$. The $Z = 64$ subshell rapidly disappears as we move off the $N = 82$ shell closure. Above $Z = 64$ we observe the emer-

Table IX. Systematics of $N = 81$ β Decays

	Dominant Core Configuration[a]					
	2+,3⁻		4+,5⁻,6+,7⁻		($\nu h_{9/2}$)	
Transition	E_x[b]	log*ft*	E_x[b]	log*ft*	E_x[b]	log*ft*
^{149}Er^m→^{149}Ho	1523	5.2	2498	5.0	4530	4.4
^{149}Er^g→^{149}Ho	1797	>4.2			4700	4.2
^{148}Ho^{g+m}→^{148}Dy	1682	≈5.5	2653	≈4.9	4300	<5.1
^{147}Dy^m→^{147}Tb	1482	5.2	2260	4.9	4800	3.9
^{147}Dy^g→^{147}Tb	1763	5.0			4100	≈3.7
^{146}Tb^{g+m}→^{146}Gd	1971	≈5.4	2841	4.6	4730	4.5
^{145}Gd^g→^{145}Eu	1819	5.4			4500	4.4
^{144}Eu^{g+m}→^{144}Sm	1660	4.9	2450	5.1		

[a]Core configuration that is coupled to the $\pi s_{1/2}$ or the $\pi h_{11/2}$ odd proton in the daughter nucleus.
[b]Intensity-weighted average excitation energy of the core-coupled configurations populated by β decay.

gence of nearly degenerate $\pi s_{1/2}$ and $\pi h_{11/2}$ isomeric states in odd-Z even-nuclei. For odd-A Tb isotopes this pattern holds for $80 \leqslant N \leqslant 86$. At $N = 78$, Redon et al.[29] have established a $5/2^+$ ground state, and no evidence exists for the β decay of the $11/2^-$ isomer. In Fig. 6 the decay scheme for ^{141}Tb, measured at OASIS, is presented. The decay data support a $5/2^-$ spin assignment for ^{141}Tb. Moller[30] has performed a Nilsson/*QRPA* calculation of the excited states in ^{141}Tb which is presented in Fig. 7. For deformations in the ϵ_2 = 0.1–0.2 range, the $5/2^-[532]$ orbital is predicted for the ground state, in agreement with experiment. The ^{143}Tb ground-state spin is less obvious from the *QRPA* calculation; however, the energy of the $5/2^+[402]$ state increases rapidly with deformation and at $\epsilon_2 \approx 0.1$ its energy is near to four or five other states, so it may be the ground state.

5.6 SYSTEMATICS OF SHELL-MODEL TRANSITIONS NEAR N = 82

The available p and n shell-model states important for nuclei near N = 82 are listed in Fig. 8. Single-particle levels corresponding to these states are well known and observed at low excitations in the odd-A nuclei in this region. Additional levels occur in the odd-A nuclei near energies close to the energies of the even-even core excitations (2^+, 3^-, 4^+,

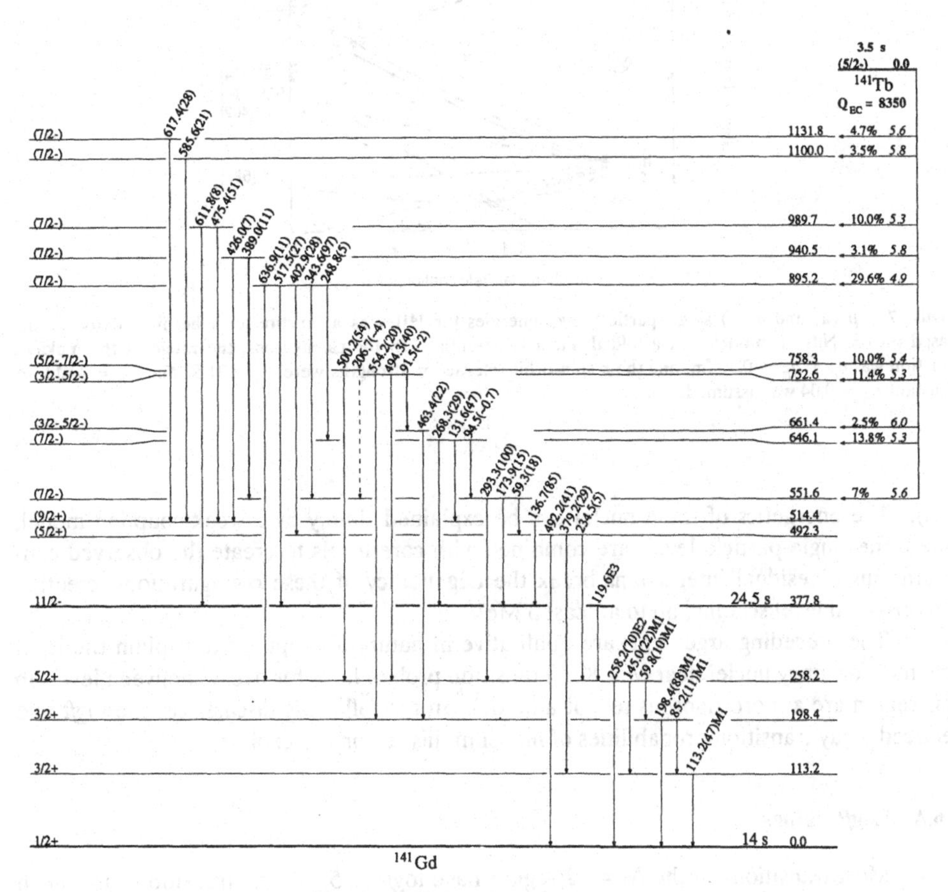

Figure 6. Decay scheme for ^{141}Tb.

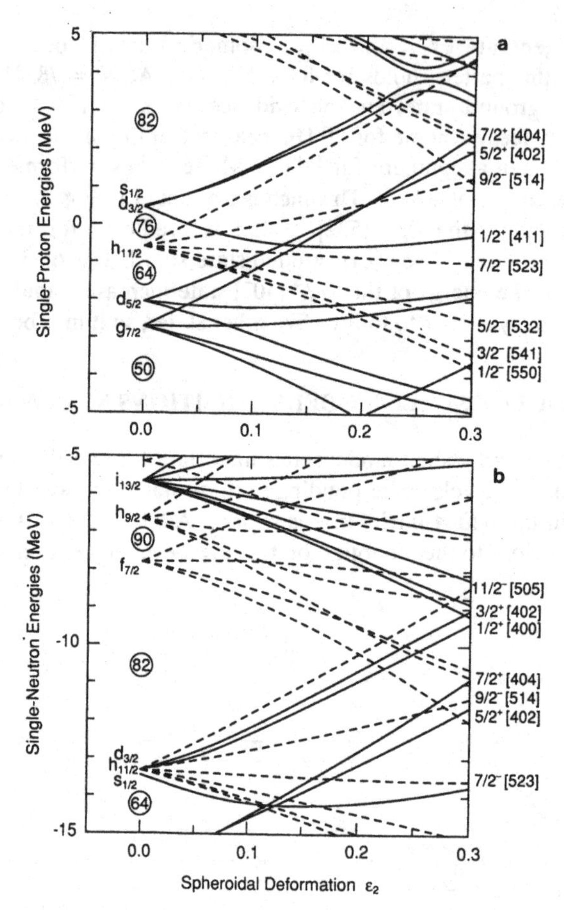

Figure 7. *p* (a) and *n* (b) single-particle level energies for ^{141}Dy as a function of spheroidal deformation, based on the Nilsson model with a folded Yukawa potential. In the calculations, the range of the Yukawa function was $a_p = a_n = 0.80$ fm, and the *p* spin-orbit interaction strengths were $\lambda_p = 31.52$ and $\lambda_n = 34.14$. A constant $\varepsilon_4 = 0.04$ was assumed.

etc.). The energetics of these nuclei can be explained simply by a weak-coupling model, where the single-particle levels are combined with core levels to create the observed configurations. Residual interactions break the degeneracy of these configurations, creating the observed level scheme, up to at least 3 MeV.

The preceding arguments are qualitative in nature and appear to explain the level schemes for many nuclei near $N = 82$. Transition probabilities for decays between levels in this region are a more rigorous test of a model. In the following discussions, the log*ft* and reduced γ-ray transition probabilities of nuclei in this region are explored.

5.6.A Log*ft* Values

Most transitions in the $N = 82$ region have log*ft* > 5. A few transitions are much faster and can dominate a particular decay scheme. Inspection of the shell-model states in

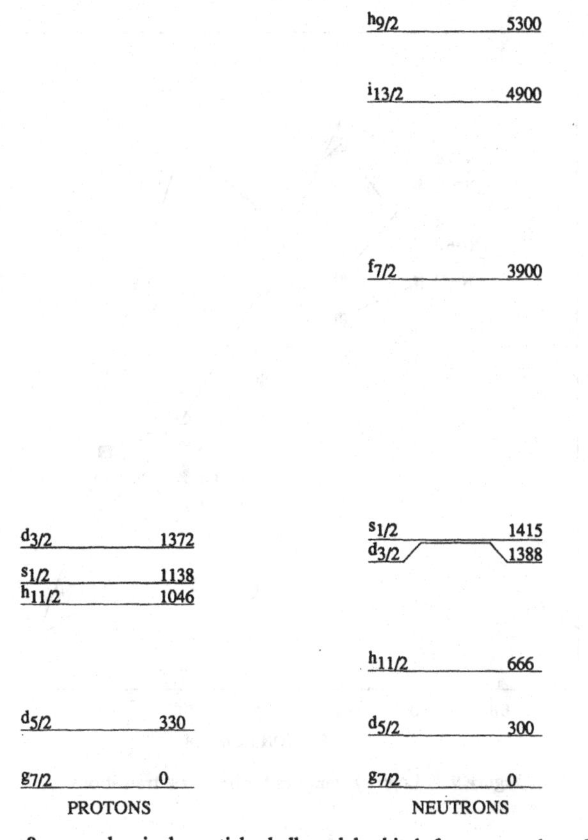

Figure 8. *p* and *n* single-particle shell-model orbitals for rare-earth nuclei.

Fig. 8 reveals two important spin-flip transitions which may be expected. Below $N = 82$, the $\pi d_{5/2} \rightarrow \nu d_{3/2}$ transition is important, and above $Z = 64$ the $\pi h_{11/2} \rightarrow \nu h_{9/2}$ transition is significant. Evidence for both transitions is found in a series of $0^+ \rightarrow 1^+$ transitions. In these transitions either a $(\pi d_{5/2})^2$ or a $(\pi h_{11/2})^2$ pair can be assumed to decay by the spin-flip transition.

The experimental log*ft* values for these transitions are plotted in Fig. 9. For $N < 80$ the transitions should involved the $\pi d_{5/2}$ protons, and for transitions with $N > 82$ the $\pi h_{11/2}$ protons are involved. In Fig. 10 the prediction from the simple shell model for each transition is included for comparison. From the shell model,

$$ft_{\rm SM} = \frac{6160}{g_a^2 B(GT)} \tag{3}$$

where $g_a = 1.263$, $B(GT) = n4l/2l+1$, and n is the number of valence protons. The shell-model predictions are nearly an order of magnitude faster than experiment, a phenomenon commented on previously by Nolte et al.[31] for the $N = 82$ region and by Barden et al.[32] for the $Z = 50$ region ($\pi g_{9/2} \rightarrow \nu g_{7/2}$ transitions). Towner[33] has argued that these discrepan-

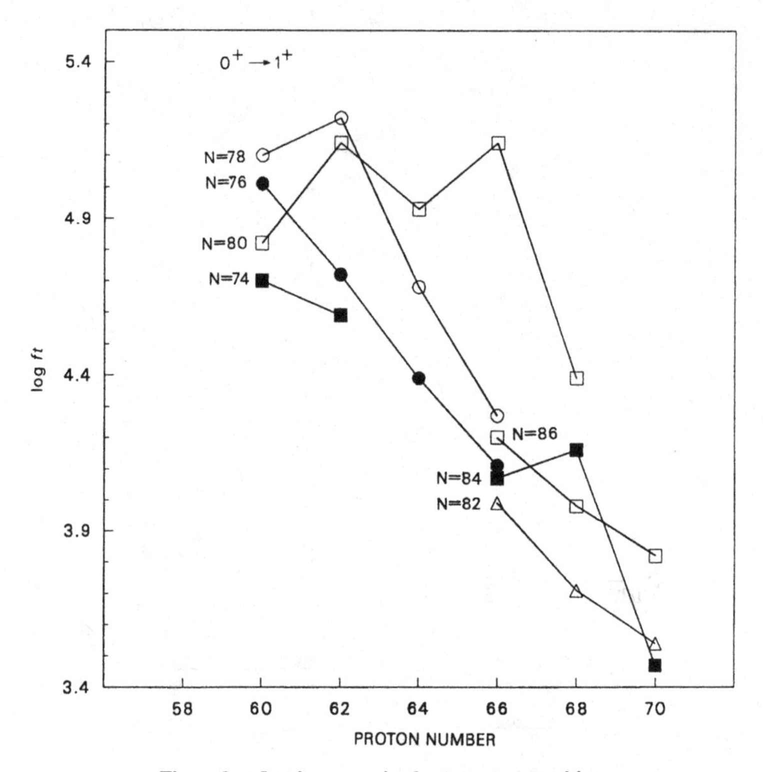

Figure 9. Log*ft* systematics for 0+ → 1+ transitions.

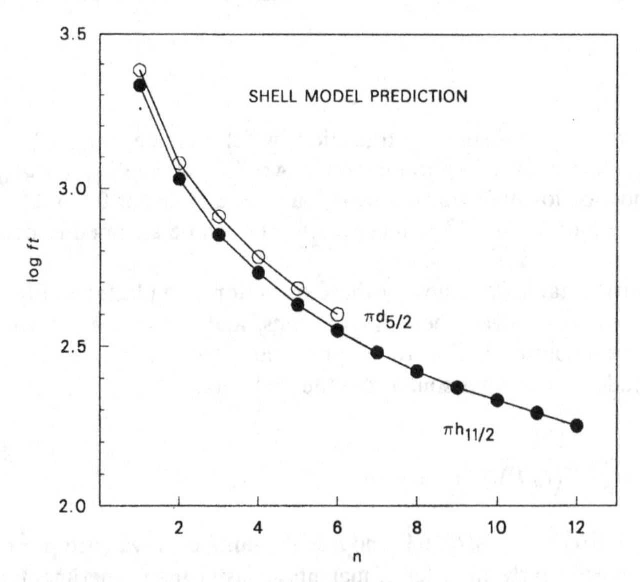

Figure 10. Single-particle shell-model predictions for log*ft* values of $\pi d_{5/2} \rightarrow \nu d_{3/2}$ and $\pi h_{11/2} \rightarrow \pi h_{9/2}$ spin-flip transitions as a function of *n* particles in the valence-*p* orbitals.

cies result from pairing correlations, core polarization, and higher-order phenomena. Nevertheless, several curiosities remain to be explained.

The n dependence appears to be reproduced in the spin-flip transitions, where the average log*ft* value changes between $n = 2$ and 6 by 0.47(9) for $N = 76,78$ and 82,84. This agrees closely with the shell-model expectation of 0.48. The $N = 80$ transitions have been excluded from the average because the shell model predicts that the $\nu d_{3/2}$ orbital is filled, blocking the spin-flip transition. This can be demonstrated by comparing the $N = 76$ and 78 log*ft* values. They differ on average by 0.22(12), which is consistent with 0.3 predicted if the $\nu d_{3/2}$ orbital were half-filled at $N = 78$. It is remarkable that these shell-model trends are preserved, although calculations by Towner[33] do indicate that the hindrance should be constant for $l = 5$ orbitals.

Additional trends in the log*ft* values can be seen in Fig. 9. A constant log*ft* = 5.0 is consistent with all values where $Z = 60$ or $N = 80$. The $\nu s_{1/2}$ and $\nu d_{3/2}$ orbitals are nearly degenerate in this region, perhaps explaining the residual β strength at $N = 80$. Another intriguing trend in Fig. 9 is the low log*ft* values for ^{142,144}Dy. The valence protons in these isotopes should be $(\pi h_{11/2})^2$, yet the log*ft* follows the $\pi d_{5/2}$ spin-flip trend. It is possible that the measured values are somewhat low because of missed β strength to levels near 2 MeV in the daughter, where the $\nu h_{9/2}$ orbitals are expected. This transition has been observed in ^{146}Dy decay[34] with log*ft* = 3.8(2), consistent with the heavier Dy decays. Assuming that this transition occurs with the same log*ft* in the lighter Dy isotopes, about 35% of the decay would populate that resonance in ^{142}Dy decay, effectively raising the log*ft* for the ground state transition to the value from ^{140}Gd decay. A ≈7% affect would occur for ^{144}Dy decay, which is not sufficient to bring the ground-state log*ft* to the value for ^{142}Gd.

Other spin-flip transitions have been observed in odd-Z decays. The measured log*ft* values for $\pi d_{5/2} \rightarrow \pi d_{3/2}$ are summarized in Fig. 11, and the $\pi h_{11/2} \rightarrow \pi h_{9/2}$ transitions are summarized in Fig. 12. These transitions are substantially more retarded than the even-even decays and show little n dependence. The calculations of Towner[33] support the disappearance of n dependence, predicting a rapidly-decreasing hindrance with increasing n. The $1^+ \rightarrow 0^+$ decays are shown in Fig. 13. They are similar to the nearby $0^+ \rightarrow 1^+$ transitions and their log*ft* values increase by 0.50(11) from $n = 1$ to 5, nearly the 0.70 value predicted by the shell model. These transitions are comparable with the $0^+ \rightarrow 1^+$ transitions because they are transitions between the identical levels. Finally, there is a sequence of $9^+ \rightarrow 8^+$ transitions in odd-odd nuclei with $N \geq 83$. The log*ft* systematics for these transitions are shown in Fig. 14. These decays have been described as $(\pi h_{11/2})(\nu f_{7/2}) \rightarrow (\nu h_{9/2})(\nu f_{7/2})$ transitions. Their log*ft* values fall intermediate between the single-particle and even-even decays.

5.6.B Reduced γ-Ray Transition Probabilities

Several families of reduced transition probabilities for single-particle shell model transitions have been studied for nuclei near $N = 82$. The $M4$ Weisskopf reduced transition probabilities for the $\nu h_{11/2} \rightarrow \nu d_{3/2}$ transition are given in Fig. 15. The $B(M4)$ values exceed the Weisskopf estimates and slowly decline with increasing p number. The large strength is consistent with the near closure of the $N = 82$ shell but inconsistent with the expected blocking of the $\nu d_{3/2}$ orbital at $N = 81$. The near degeneracy of the $\nu s_{1/2}$ and $\nu d_{3/2}$ orbitals apparently contributes to the large quasiparticle strength in $\nu d_{3/2}$. At $N = 79$ the $B(M4)$ values are similar to those at $N = 81$ and slightly exceed them for $Z \geq 60$. This

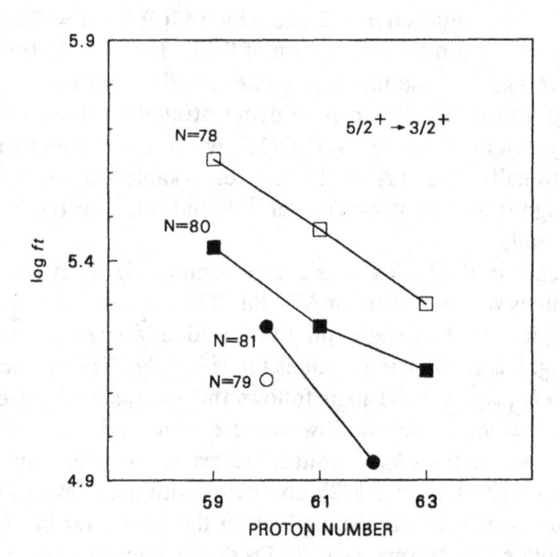

Figure 11. ***Logft*** **systematics for $5/2^{+} \rightarrow 3/2^{+}$ transitions.**

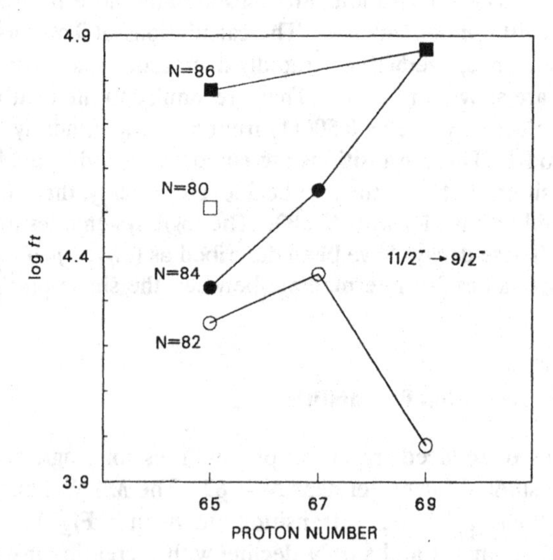

Figure 12. ***Logft*** **systematics for $11/2^{-} \rightarrow 9/2^{-}$ transitions.**

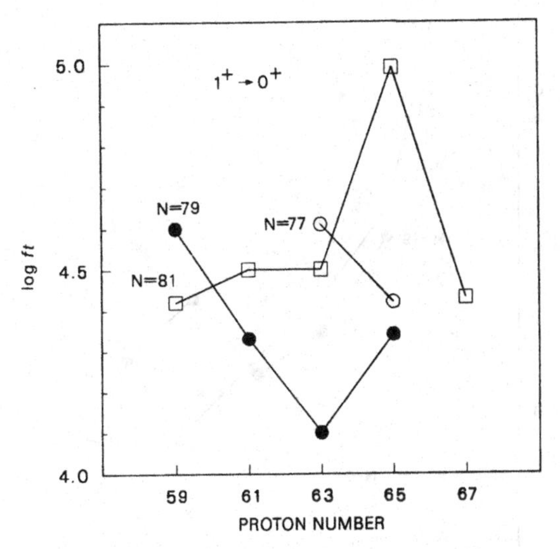

Figure 13. *Logft* systematics for 1+ → 0+ transitions.

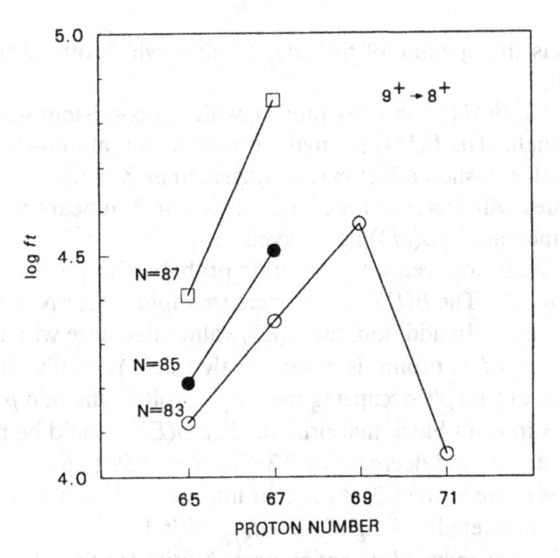

Figure 14. *Logft* systematics for 9+ → 8+ transitions.

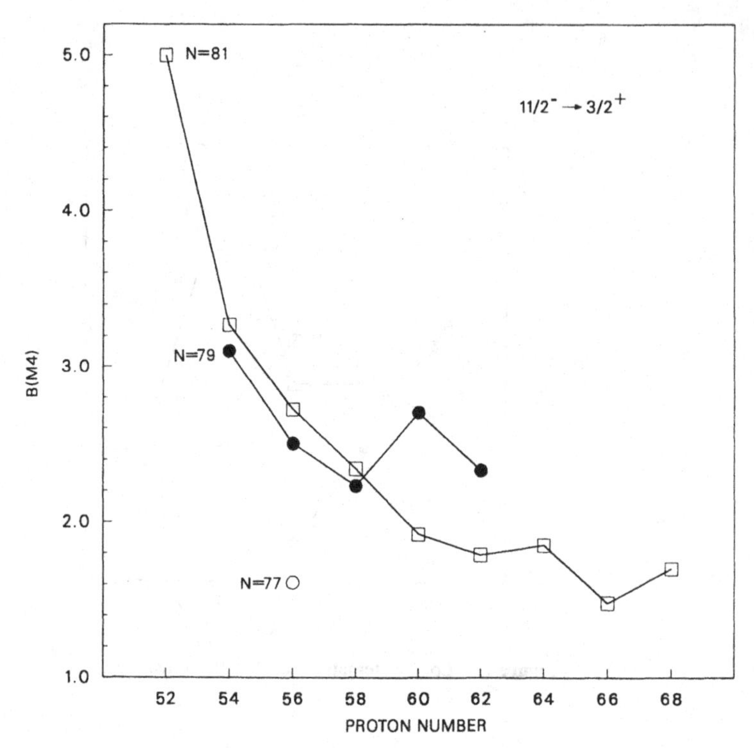

Figure 15. Systematics of Weisskopf reduced $B(M4)$ γ-ray transition probabilities.

probably reflects the opening of the $\nu d_{3/2}$ orbital, which offsets the greater distance from the closed shell.

At $Z = 52$, $B(M4)$ increases rapidly, which is consistent with the proximity to the $Z = 50$ closed p shell. The $B(M4)$ strength decreases until at mid-shell, $Z = 66$, a minimum is reached. A small sub-shell effect may be observed at $Z = 64$. It is tempting to predict that the $B(M4)$ values will increase for $Z \geqslant 68$. This effect appears to be observed at $Z = 68$, where a small increase in $B(M4)$ is observed.

The $E3$ Weisskopf reduced transition probabilities for the $\nu h_{11/2} \rightarrow \pi d_{5/2}$ transition are shown in Fig. 16. The $B(E3)$ values increase rapidly near $N = 82$, greatly exceeding the Weisskopf estimates. In addition, the $B(E3)$ values decrease with increasing p number. As p's are added, the $\pi d_{5/2}$ orbital is progressively filled, partially blocking the $E3$ transition. At $Z = 59$ there are no p's occupying the $\pi d_{5/2}$ orbital (the odd p is in the $\pi h_{11/2}$ orbital), while at $Z = 63$ four p's block that orbital. The $B(E3)$ should be proportional to the number if $\pi d_{5/2}$ vacancies and decrease by 0.33 from $Z = 59$ to 63. At $N = 82$, the experimental $B(E3)$ values decrease by 0.37(6) in that interval. Also, at $Z = 65$, $B(E3)$ is nearly zero, consistent with complete blocking of the $\pi d_{5/2}$ orbital.

$M2$ Weisskopf reduced transition probabilities for the $\pi h_{11/2} \rightarrow \pi g_{7/2}$ transitions are summarized in Fig. 17. These transitions are significantly hindered, except for ^{133}La which is nearly a full Weiskopf unit. The hindrance is not surprising because, except for ^{133}La, the $\pi g_{7/2}$ orbital is fully blocked. The modest strength at $Z = 57$ is consistent with the $\pi g_{7/2}$ orbital being two-thirds full.

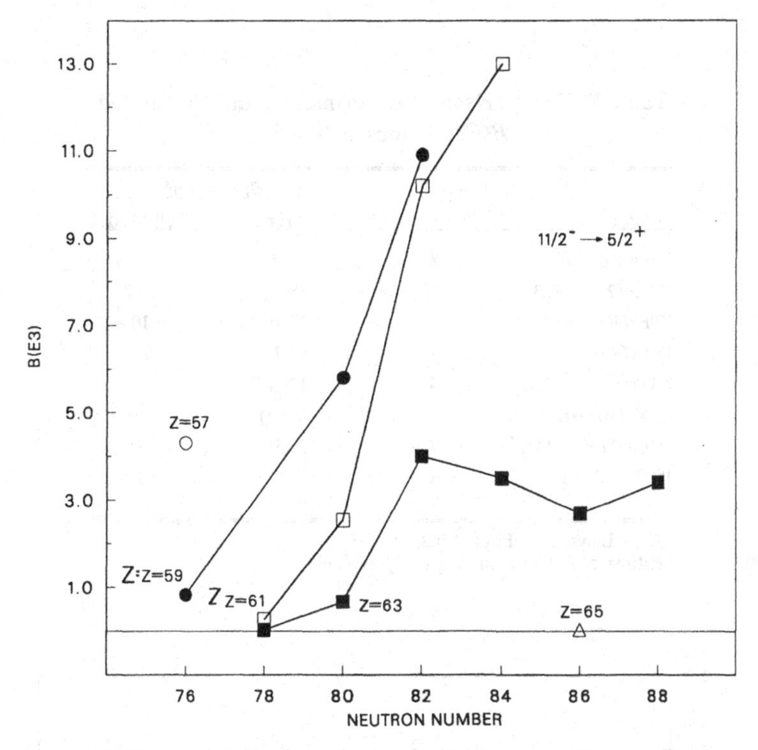

Figure 16. Systematics of Weisskopf reduced $B(E3)$ γ-ray transition probabilities.

A considerable success for the shell model near $N = 82$ has been its ability to explain numerous yrast $E2$ transition energies and transition probabilities. Lawson[35] has shown that seniority ν is a good quantum number and the decays between $(\pi h_{11/2})^n$ configurations give $B(E2)$ values proportional to $[(6-n)/(6-\nu)]^2$. Several experiments[31,36-39] have confirmed the predictions, which are summarized in Table X.

5.7 CONCLUSIONS

A considerable body of data for the neutron-deficient rare-earth nuclei has been measured at OASIS and other facilities. Improved experimental techniques and methods of data analysis have increased the amount of useful information in this region. Smooth trends in the log*ft* data near $N = 82$ and qualitative consistency with the shell model for β and γ transitions give hope that a consistent, simple explanation of these phenomena may be obtained. Towner has shown that pairing effects and core polarization may explain the discrepancies with the shell model; however, these effects are difficult to calculate. Conversely, Lawson has found success in explaining shell-model transitions in several nuclei with very few parameters. Similar explanations of the β transitions may be found by the correct parameterization of the problem. It is remarkable that, while the absolute transition probabilities vary from shell model transitions, the $0^+ \rightarrow 1^+$ β transitions, $E3$ γ-ray transitions, and $N = 82$ even-Z $E2$ transitions all scale with p number exactly as predicted in the shell model.

Table X. Comparison of Experimental and Calculated $B(E2)$ Values at N = 82

Transition	n	$B(E2)e^2fm^4$ Experiment	Shell Model[a]
$^{148}Dy(10^+ \rightarrow 8^+)$	2	44(3)	≡44
$^{149}Ho(27/2^- \rightarrow 23/2^-)$	3	88(6)	92
$^{150}Er(10^+ \rightarrow 8^+)$	4	11.4(14)	10.8
$^{150}Er(8^+ \rightarrow 6^+)$	4	≈37	27.3
$^{151}Tm(27/2^- \rightarrow 23/2^-)$	5	12.1(7)	9.2
$^{152}Yb(10^+ \rightarrow 8^+)$	6	0.9(1)	0
$^{153}Lu(27/2^- \rightarrow 23/2^-)$	7	0.45(9)	9[b]
$^{154}Hf(10^+ \rightarrow 8^+)$	8	2.9(14)	11[b]

[a]R. D. Lawson, Z. Phys. A303, 51 (1981).
[b]Estimated values assuming $(6\text{-}n)^2$ scaling.

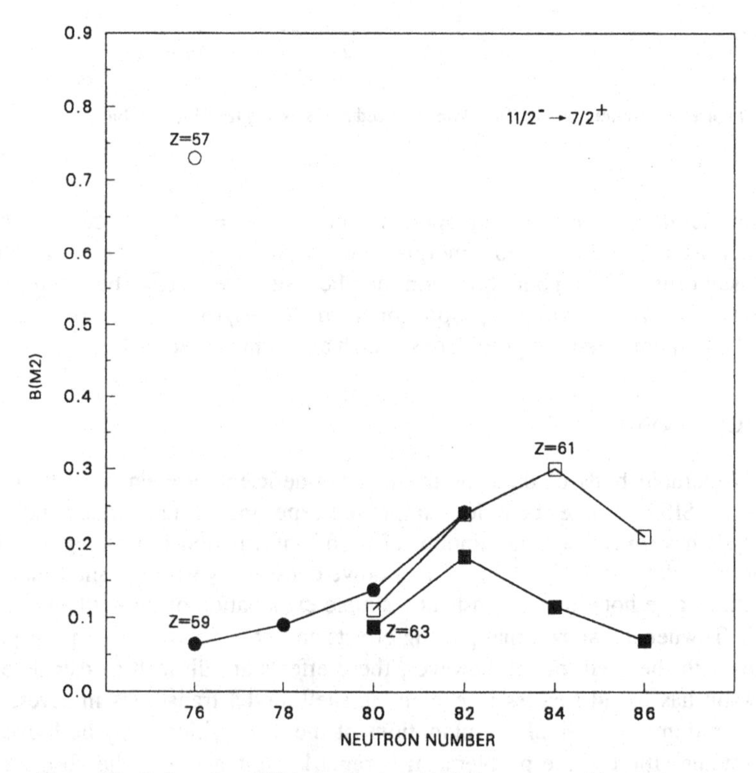

Figure 17. Systematics of Weisskopf reduced $B(M2)$ γ-ray transition probabilities.

The need for more-precise measurements and additional data is apparent. Complete decay-scheme information, including β- and γ-strength measurements, absolute delayed-particle emission probabilities, mass measurements, and level lifetimes are important for understanding nuclear decay and improving our ability to predict unmeasured decay properties. In particular, the neutron-deficient rare-earth nuclei above $Z = 64$ need more investigation because of the importance of the $\pi h_{11/2}$ orbital in fast spin-flip β- and γ-ray transitions. Additional data are also required for neutron-deficient nuclei with $Z \geqslant 72$, where virtually no β-decay studies have been done. These nuclei offer experimental difficulties because they cannot be easily obtained at most mass separators. The low logft = 2.3 for Z = 74, predicted by the shell model, suggests that this region will provide some of the fastest β transitions observed at any mass.

ACKNOWLEDGMENTS

This work was supported by the Director, Office of Energy Research, Division of Nuclear Physics of the Office of High Energy and Nuclear Physics of the U. S. Department of Energy under contract DE-AC03-76SF00098.

REFERENCES

1. J. M. Nitschke, *Nucl. Instr. Meth.* **206**, 341 (1983).
2. W. G. Winn, H. H. Gutbrot, and M. Blann, *Nucl. Phys.* **A188**, 423 (1972); M. Blann and J. Bisplinghoff, U. S. DOE Report UCID-19614 (1982).
3. E. Browne, *Nucl. Instr. Meth.* **A249**, 461 (1986).
4. E. Browne, *Nucl. Instr. Meth.* **A265**, 541 (1988).
5. R. B. Firestone, *Phys. Res.* **A286**, 584 (1990).
6. J. T. Routti and S. G. Prussin, *Nucl. Instr. Meth.*, **72**, 125 (1969).
7. *Table of Isotopes* (7th Ed.), Ed. by C. M. Lederer and V. S. Shirley (Wiley, New York, 1978).
8. W. Bambynek, B. Crasemann, R. W. Fink, H.-U. Freund, H. Mark, C. D. Swift, R. E. Price, and P. V. Rao, *Rev. Mod. Phys.* **44**, 716 (1972).
9. N. B. Gove and M. J. Martin, *At. Nucl. Data Tables* **A10**, 205 (1971).
10. Computer code BANAL for multilinear analysis, R. B. Firestone, unpublished.
11. Least-squares multilinear-analysis subroutine RLMUL, International Mathematical and Statistical Library (IMSL) Version 9.2.
12. G. Azuelos and J. E. Kitching, *At. Data Nucl. Data Tables* **17**, 103 (1976).
13. A. H. Wapstra and G. Audi, *Nucl. Phys.* **A432**, 140 (1985).
14. A. H. Wapstra, G. Audi, and R. Hoekstra, *At. Data Nucl. Data Tables* **39**, 281 (1988).
15. S. Liran and N. Zeldes, *At. Nucl. Data Tables* **17**, 431 (1976).
16. F. Rösel, H. M. Fries, K. Alder, and H. C. Paul, *At. Nucl. Data Tables* **21**, 91 (1978).
17. P. A. Wilmarth, Ph.D. Thesis, University of California, Berkeley, 1988, Lawrence Berkeley Laboratory Report LBL-26101.
18. R. B. Firestone, R. C. Pardo, R. A. Warner, Wm. C. McHarris, and W. H. Kelly, *Phys. Rev.* C **25**, 527 (1982).
19. G. D. Alkhazov, A. A. Bykov, V. D. Wittmann, S. Yu. Orlov, and V. K. Tarasov, *Phys. Lett.* **157B**, 350 (1985).

20. G. D. Alkhazov, A. A. Bykov, V. D. Wittmann, V. E. Starodubsky, S. Yu. Orlov, V. N. Panteleyev, A. G. Polyakov, and V. K. Tarasov, *Nucl. Phys.* **A438**, 482 (1985).
21. D. Schardt, P. O. Larsson, R. Kirchner, O. Klepper, V. T. Koslowsky, E. Roeckl, K. Rykaczewski, K. Zuber, N. Roy, P. Kleinheinz, and J. Blomqvist, *Proceedings of the 7th International Conference on Atomic Masses and Fundamental Constants (AMCO7), Darmstadt-Seeheim, 1984*, Ed. by O. Klepper, p. 222.
22. R. B. Firestone, J. M. Nitschke, P. A. Wilmarth, K. Vierinen, J. Gilat, K. S. Toth, and Y. A. Akovali, Phys. Rev. *C* **39**, 219 (1989).
23. K. S. Toth, Y. A. Ellis-Akovali, J. M. Nitschke, P. A. Wilmarth, P. K. Lemmertz, D. M. Moltz, and F. T. Avignone III, *Phys. Lett.* **178B**, 150 (1986).
24. P. Kleinheinz, B. Rubio, M. Ogawa, M. Piiparinen, A. Plochocki, D. Schardt, R. Barden, O. Klepper, R. Kirchner, and E. Roeckl, *Z. Phys.* **A323**, 705 (1985).
25. K. Takahashi, M. Yamada, and T. Kondoh, *At. Nucl. Data Tables* **12**, 101 (1973).
26. K. Takahashi, private communication (1988).
27. D. Schardt, R. Barden, R. Kirchner, O. Klepper, A. Plochocki, E. Roeckl, P. Kleinheinz, M. Piiparinen, B. Rubio, K. Zuber, C. F. Liang, P. Paris, A. Huck, G. Walter, G. Marguier, H. Gabelmann, and J. Blomqvist, *Proceedings of the Fifth International Conference on Nuclei Far From Stability, Rousseau Lake, Ontario, Canada*, Ed. by I. S. Towner (AIP Conference Proceedings, New York, 1988), p 477.
28. J. M. Nitschke, P. A. Wilmarth, R. B. Firestone, P. Möller, K. S. Toth, and J. Gilat, *Phys. Rev. Lett.* **62**, 2805 (1989).
29. N. Redon, T. Ollivier, R. Beraud, A. Charvet, R. Duffait, A. Emsallem, J. Honkanen, M. Meyer, J. Genevey, A. Gizon, and N. Idrissi, *Z. Phys.* **A325**, 127 (1986).
30. P. Möller, private communication; P. Moller and J. R. Nix, *Nucl. Phys.* **A361**, 117 (1981).
31. E. Nolte, G. Korschinek, and Ch. Setzensack, *Z. Phys.* **A309**, 33 (1982).
32. R. Barden, R. Kirchner, O. Klepper, A. Plochocki, G.-E. Rathke, E. Roeckl, K. Rykaczewski, D. Schardt, and J. Zylicz, *Z. Phys.* **A329**, 319 (1988).
33. I. S. Towner, *Nucl. Phys.* **A444**, 402 (1985).
34. K. Zuber, C. F. Liang, P. Paris, J. Styczen, J. Zuber, P. Kleinheinz, B. Rubio, G. de Angelis, H. Gabelmann, A. Huck, J. Blomqvist, and the ISOLDE collaboration, CERN, *Z. Phys.* **A327**, 357 (1987).
35. R. D. Lawson, *Z. Phys.* **A303**, 51 (1981).
36. P. J. Daly, P. Kleinheinz, R. Broda, A. M. Stefanini, S. Lunardi, H. Backe, L. Richter, R. Willwater, and F. Weik, *Z. Phys.* **A288**, 103 (1978).
37. H. Helppi, Y. H. Chung, P. J. Daly, S. R. Faber, A. Pakkanen, I. Ahmad, P. Chowdhury, Z. W. Grabowski, T. L. Khoo, R. D. Lawson, and J. Blomqvist, *Phys. Lett.* **115B**, 11 (1982).
38. E. Nolte, G. Colombo, S. Z. Gui, G. Korschinek, W. Schollmeier, P. Kubik, S. Gustavsson, R. Geier, and H. Morinaga, *Z. Phys.* **A306**, 211 (1982).
39. J. H. McNeill, J. Blomqvist, A. A. Chishti, P. J. Daly, W. Gelletly, M. A. C. Hotchkis, M. Piiparinen, B. J. Varley, and P. J. Woods, *Phys. Rev. Lett.* **63**, 860 (1989).

6. Accelerated Radioactive Nuclear Beams (Low Energy)

John M. D'Auria
Departments of Chemistry and Physics
Simon Fraser University
Burnaby, British Columbia
Canada V5A 1S6

ABSTRACT

The possibility of producing and accelerating intense beams of short-lived radioactive heavy ions, both for studies of nuclides themselves and for use as projectiles in reactions of considerable interest to the field of nuclear astrophysics, is now considered quite feasible. Two main approaches to producing these Radioactive Nuclear Beams (RNB) are projectile fragmentation of very-high-energy heavy ions and the use of an on-line isotope separator (ISOL) with intermediate-energy protons. The first method optimally produces beams of high energy (50-500 MeV/u) while the second approach leads to intense low-energy RNB (less than 10 MeV/u). A review of the various facilities, proposed, under construction, and/or operational will be presented, including the possibilities of a facility in North America. A report on the TISOL facility, an operational ISOL located at the high-flux, intermediate-energy proton facility, TRIUMF, will also be given, along with the status of the installation of the novel ECR ion source for gaseous low-Z elements.

6.1 INTRODUCTION

The possibility of producing and accelerating intense beams of short-lived radioactive heavy ions, both for studies of the species themselves and to be used as projectiles in reactions of interest, is now considered quite imminent. While several approaches have been used or are proposed for the production of such beams, the two main methods are projectile fragmentation, which will result in high-energy beams ($\geq$50 MeV/u) through the recoiling of the product nuclei, and spallation/fragmentation using intermediate-energy protons followed by a post-accelerator stage, resulting in low-energy (LE) beams ($\leq$10 MeV/u). A review of these, along with related approaches which are presently in place producing such beams but which are more specific in the type of projectile produces, will be outlined in Section 6.2. A review of facilities either operating or proposed will also be given in Section 6.2.

A number of interesting studies can be performed with such a wide variety of projectiles. Of particular interest here are the kinds of studies that are possible with projectiles classified as low-energy (LE), and some of these are presented in Section 6.3. The proposed ISAC (isotope separator/accelerator) facility TRIUMF will be reviewed and a status of the planning presented. In preparation of such a facility, a small test on-line isotope separator was constructed at TRIUMF to initiate a program of developing targets and ion sources of particular usefulness to ISAC. This (TISOL) facility is now operational, and

Exotic Nuclear Spectroscopy, Edited by W. C. McHarris
Plenum Press, New York, 1990

a progress report on some of the results relevant to the proposed ISAC facility, as well as the possible upgrading of TISOL to a production facility for a full scientific program, will be given. Included will be a report on the successful operation of an ECR (electron-cyclotron-resonance) ion source coupled on-line with an ISOL device for the first time.

6.2 METHODS OF PRODUCTION

The main approaches to the production of radioactive nuclear beams (RNB), in general, fall into two categories: namely, those that take advantage of the product nuclei recoiling out of the target at an energy close to that of the energy of the projectile used to produce it, and those that produce the desired projectile initially in some thick target followed by its diffusion into subsequent acceleration stages to reach the desired energy of interest. Each of these has certain advantages and disadvantages, and a thorough exploration of this topic can be found in a recent review by Tanihata.[1]

6.2.A Recoil Techniques

There are two main types of recoil techniques considered optimal for the purpose of producing radioactive nuclear beams. These are projectile fragmentation, which takes advantage of the fragmentation of very-high-energy projectiles in their interaction with a thin target, and transfer reactions of rather low-energy heavy ions. In the first instance, heavy ions of energies higher than a few hundred MeV/u and whose nuclear composition is close to the desired radioactive projectile are intercepted by an appropriate target, and the recoiling fragments emittted in essentially the forward direction are separated by electromagnetic devices in a recoil separator. Such an approach is best illustrated by the present facilities at Berkeley,[2] GANIL,[3] and the facility being constructed at GSI/SIS.[4] In the second approach, low-energy (30–50-MeV) heavy ions undergo, in general, a single-nucleon-transfer reaction, and the recoiling product of interest is again captured by electromagnetic devices, such as a superconducting solenoid lens, to produce a radioactive beam of interest. An example of this, leading to the production of a ^{8}Li beam using a ^{9}Be (^{7}Li, ^{8}Li)^{8}Be reaction,[5] is being implemented at the University of Notre Dame facility[6]; a similar approach is under development at Livermore.[7] Table I provides a comparison of the projected yields of these two techniques, and Table II gives some of the expected or proposed properties of these types of facilities.

6.2.B Diffusion Techniques

The second main approach to the production of a wide range of radioactive nuclear beams takes advantage of diffusion of product nuclei from a target generally operated at high temperatures, followed by some form of electromagnetic on-line mass/isotope separation, and then a booster acceleration stage to attain a final beam energy. In general, two approaches have been used or are proposed by which the desired radioactive species are produced. In specific instances the desired species can be made through the use of an intense beam of low-energy protons with the reaction of interest of the (p,n) type. An example of this is best illustrated by the production of a beam of ^{13}N at the Cyclone facility (Louvain-la-Neuve) which has just started operations.[8]

Table I. Comparison of Methods of Producing Intense Radioactive Beams

	Recoil			Diffusion	
Primary Beam		**Heavy Ion**		**Proton**	
Energy	50 MeV	100 MeV/A	1 Gev/A	20 MeV	≥500 MeV
Intensity (particles/s)	10^{12}	10^{11}	10^{12}	2×10^{15}	3×10^{14}
Target thickness (g/cm^2)	10^{-3}	0.5	1	10^{-1}	100
Cross-Section (mb)	50	1	1	50	1
Production Rate (particles/s)	3×10^7	3×10^7	6×10^8	6×10^{11}	2×10^{12}
Secondary Beam					
Transmission	10^{-2}	10^{-1}	10^{-1}	10^{-2}	10^{-1}
Expected RB Intensity (particles/s)	3×10^5	3×10^6	6×10^7	6×10^9	2×10^{11}

A more versatile approach in terms of the availability of a wide range of possible radioactive beams is the use of intermediate-energy protons (≥500 MeV); through spallation/fragmentation types of reactions almost all radioactive species can be produced. The use of such energetic protons allows the possibility of using thick targets, which in turn leads to rather high intensities of a wide range of final beams. This approach is best illustrated by the proposed ISAC facility at TRIUMF[9] and the proposed facilities at ISOLDE located at CERN,[10] and JHP (Japanese Hadron Project) at KEK in Japan.[11] This latter approach takes advantage of the well-known technology for the production of radioactive beams at on-line isotope facilities, particularly at ISOLDE.[12] Again, Table I provides a summary of the expected production rates at such a facility, while Table II indicates where some of these facilities exist or are proposed.

A key feature of this approach is the choice of an appropriate booster stage or post-accelerator. While the Cyclone facility takes advantage of the existence of a small cyclotron[8] to provide the acceleration, most of the larger proposed facilities suggest that

Table II. Accelerated Radioactive Beams Facilities*

Project (Location)	Production Method	Mass/Energy Range	Status
BEVALAC (LBL, Berkeley)	Fragmentation, Separator	($A < 100$) 30–500 MeV/A	Operating
CYCLONE (Louvain)	30-MeV p + (ISOL) + Cyclotron	$A \leqslant 40$ (?) $\leqslant 110\, q^2/A$	(Operating)
DUBNA (USSR)	ISOL + Post-Acceleration	$A < 230$	Proposal in prep.
GANIL (Caen)	Fragmentation, Separators	$A < 100$ $\leqslant$ 50 MeV/u	Operating
JHP (KEK, Tokyo)	1–GeV p + ISOL + LINAC	$A < 60$ 1–6.5 MeV/A	Proposed
ISOLDE (CERN/Geneva)	600-MeV p + ISOL + LINAC	$A \leqslant 40$ 0.3–2.0 MeV/A	Proposed
QSBTS (LLNL, Livermore)	Transfer Rxs., Quadrupoles	$A \leqslant 20$ (12 MeV)	(Operating)
RPMS/A1200 (MSU)	Fragmentation, Separators	$A < 60$ (?) $1200\, q^2/A$	Operating Planned
RUTHERFORD LAB (UK)	ISOL + Post-Acceleration	$A < 250$ $\leqslant$ 6.5 MeV/A	Proposal in prep.
SC SOLENOID (UND, South Bend)	Transfer Rxs. Separator	$A < 20$ (20 MeV)	Operating
SIS 18/ESR (GSI, Darmstadt)	Fragmentation, Storage Ring, Sep.	$A < 238$ 5–500 MeV/A	1990
TRIUMF (Vancouver)	500-MeV p + ISOL + RFQ + LINAC	$A \leqslant 60$ 0.3–1.5 MeV/A	Proposed

* As of January 1, 1990

the use of a LINAC, similar to that proposed at TRIUMF,[9] is of greater advantage, given it has lower injection and extraction losses. This is of some importance, given the need to minimize transmission losses of the already relatively low-intensity radioactive beams. These LINAC systems generally consist of a first stage, RFQ (RadioFrequency Quadrupole) LINAC which takes the singly-charged, approximately 1-keV/u ISOL ion beam up to an energy of the order of 60 keV/u. This can be followed then by some combination of stripping process and DT (drift tube) LINAC sections to reach the desired, selectable final projectile energy. In general, the diffusion approach to the production of radioactive beams leads to beams of relatively low-energy, i.e., below 10 MeV/u. This is in contrast to the recoil approach using high-energy incident beams, which leads to projectiles with energies of the order of 100 MeV/u. Thus, these two dominant approaches are rather complementary, given that it is expensive to accelerate or decelerate either significantly.

6.3 SCIENTIFIC PROGRAM WITH LE BEAMS

The scientific program is strongly dependent upon the projectile energy and intensity, as expected, although, regardless of energy, there is clearly a very wide range of new and novel types of studies which become possible for the first time. Of interest here are studies possible with LE beams. A wide variety of such studies can be found in some of the recent proposals for such ISOL/post-accelerator systems at the proposed ISAC at TRIUMF,[9,13] at the proposed JHP at KEK,[11] and at the proposed PRIMA facility at ISOLDE in CERN.[10] A brief summary of only some of these are included here.

6.3.A Nuclear Astrophysics

As is well known, stellar evolution and the subsequent formation of elements are dependent upon a sequence of simple nuclear reactions. Such reactions produce radioactive species which decay, given a slow or long time scale. In explosive or high-temperature phenomena such as supernovae, the time scale is sufficiently short that these reactions involve the radioactive species themselves. While there has been success in developing nuclear-reaction networks to help explain stellar processes occurring at low temperatures, it is virtually impossible to make accurate predictions for reactions involving radioactive species. This results from the fact that such simple reactions, particularly in the low-Z region, involve capture into narrow resonances, and the related cross-sections of these are based upon the nuclear structure of the nuclide involved and not on an averaging standard statistical-reaction theory. Thus, measurements of the reaction rates themselves are required, as is used successfully for stable species. Such reactions must be studied at very-low energies below the coulomb barrier and may exhibit cross-sections that are quite small, e.g., in the nb region. Nevertheless, estimates of such reactions involving radioactive nuclear beams incident upon He or H gaseous targets have been made[13,14] and indicate that they are feasible. The forwardly-recoiling product nuclei would be detected with standard reaction-product technology now in place for similar studies with stable projectiles. A brief list of some of the possible reactions of interest is given in Table III. (A more thorough discussion of the importance of these is given elsewhere).[10,15] The energy range of interest for these studies lies from about 0.2 MeV/u up to about 1.5 MeV/u, and beam intensities as low as 10^7 atoms/s can be useful for certain types of reactions.

Table III. Some Reactions of Astrophysical Interest

Reaction	$t_{1/2}$(s) of Reactant	Astrophysical Interest
$^{8}Li(\alpha,n)^{11}B$	0.84	Inhomogeneous Big Bang
$^{13}N(p,\gamma)^{14}O$	598	Hot CN Cycle
$^{14,15}O(\alpha,\gamma)^{18,19}Ne$	71, 122	RP Process, Hot CNO
$^{17,18}F(p,\alpha)^{14,15}O$	65, 6582	Hot CNO Cycle
$^{17,18}F(p,\gamma)^{18,19}Ne$	65, 6582	Hot CNO Cycle
$^{18,19}Ne(p,\gamma)^{19,20}Na$	1.7, 17.2	Hot CNO/RP Process
$^{21}Na(p,\gamma)^{22}Mg$	22.5	NeNaMgAl Cycles, RP Process
$^{26}Al^{m}(p,\gamma)^{27}Si$	6.3	NeNaMgAl Cycles RP Process

6.3.B Nuclear Physics

The most obvious fact when considering the use of radioactive nuclear beams for nuclear physics is the enormous number of projectiles which become available. In essence, any radioisotope could be used as a projectile to perform nuclear-reaction studies. The term "systematic studies" becomes a gross understatement as one contemplates, for example, the availability of perhaps thirty-five isotopes of Cs or forty-eight species of Cs when isomeric states are considered with which a sequence of reactions can be studied. Again, the projectile energy and intensity must be considered, but clearly such studies as coulomb excitation, nucleon-transfer reactions, inelastic scattering, in-beam γ-ray spectroscopy, and production of very-heavy nuclei can be pursued with new vigor into regions previously inaccessible, using very-exotic nuclei as projectiles. Some examples of these are given below, although more complete descriptions can be found elsewhere.[1,10]

The use of the process of coulomb excitation will provide a wealth of new information on the nuclear excitation of a wide range of exotic nuclei used as projectiles. Bombarding a heavy target with a RNB projectile with energies of a few MeV/u will lead to the direct population of members of the ground-state band of the projectile itself. Such processes have reasonably-high cross-sections, and RNB's of the order of 10^6 atoms/s should be sufficient. Aside from the standard study of levels populated by *E2* transitions in, for example, neutron-rich nuclei, information not easily obtainable any other way, the possibilities exist for searching also for other interesting excitations, such as fast $E1$ transitions in nuclei such as ^{11}Be. (The latter possesses the fastest $E1$ transition known.[16])

Nucleon-transfer and inelastic-scattering reactions, again with exotic nuclei as projectiles, also will be a new way of obtaining information on the relevant excitations within these projectiles. For example, the use of ^{8}He as a projectile will allow the possibility of the transfer of up to four neutrons, not only a new way of making very neutron-rich nuclei, but also a means of providing spectroscopic information from the transfer itself. Furthermore, the levels in the projectile itself can be studied using inelastic-scattering techniques. Beam intensities of the order of 10^{10} atoms/s are probably needed to perform some of the desired studies, although low intensities are still quite useful.

In addition, the use of some of these exotic projectiles for the production of nuclei far from stability must be mentioned. Depending upon the available energy and beam intensity, it is not inconceivable to consider such strange beam/target combinations as a ^{63}Ga beam with a ^{40}Ca target to attempt to make isotopes in the ^{100}Sn region. Even with present technology, ^{63}Ga beam intensities of the order of 10^5 atoms are available. Alternatively, the study of very-heavy elements can be reinvigorated by using the wider selection of very-neutron-rich projectiles at energies just over the coulomb barrier.

Of course, the possibility will now exist to provide a wide range of polarized nuclei for use in various types of studies. With the advent of post-acceleration, the possibility of using tilted-foil polarization technique becomes a reality for a wide variety of studies. This is a simple technique and can be used for beams of energies of the order of 0.5 to 1.0 MeV/u. As indicated in the PRIMA proposal, such beams can be used for β-decay asymmetry studies and measurements of nuclear moments.[10]

It should not be ignored that, with the presence of an ISOL device on the front end, a scientific program similar to that presently pursued at such major facilities as ISOLDE in CERN is still possible. Further, also possible are exciting scientific programs in such important areas as condensed-matter physics, radioisotope production for nuclear medicine applications, and ion implantation, which are not mentioned here but should not be overlooked.

6.4 THE ISAC FACILITY

One of the first facilities proposed of the ISOL/post-accelerator type was the ISAC facility to be located at TRIUMF in Vancouver, Canada. Taking advantage of the intermediate energy (500 MeV), high-intensity ($\leq 150\ \mu A$) proton beam, radioactive beams with fluxes approaching 100 nA were predicted. These resultant beams would have had variable energies from 0.2 to 1.5 MeV/u, but with the use of a LINAC as the post-accelerator this higher limit could be incensed to 10 MeV/u relatively easily, using well-known technology. A schematic representation of the proposed facility is presented in Fig. 1, while a more complete description of the facility can be found elsewhere.[9,13] The project was estimated to cost of the order of $20M and would be completed in a period of about four years. Some of the problem areas identified as requiring additional development included: whether or not cryogenic pumping could be used, given the rather hostile environment of the thermally-hot targets of an ISOL; whether or not an ion source could be developed that would exhibit long-term but efficient operation and could produce a wide range of elements of interest to the astrophysics program; whether or not target systems could be

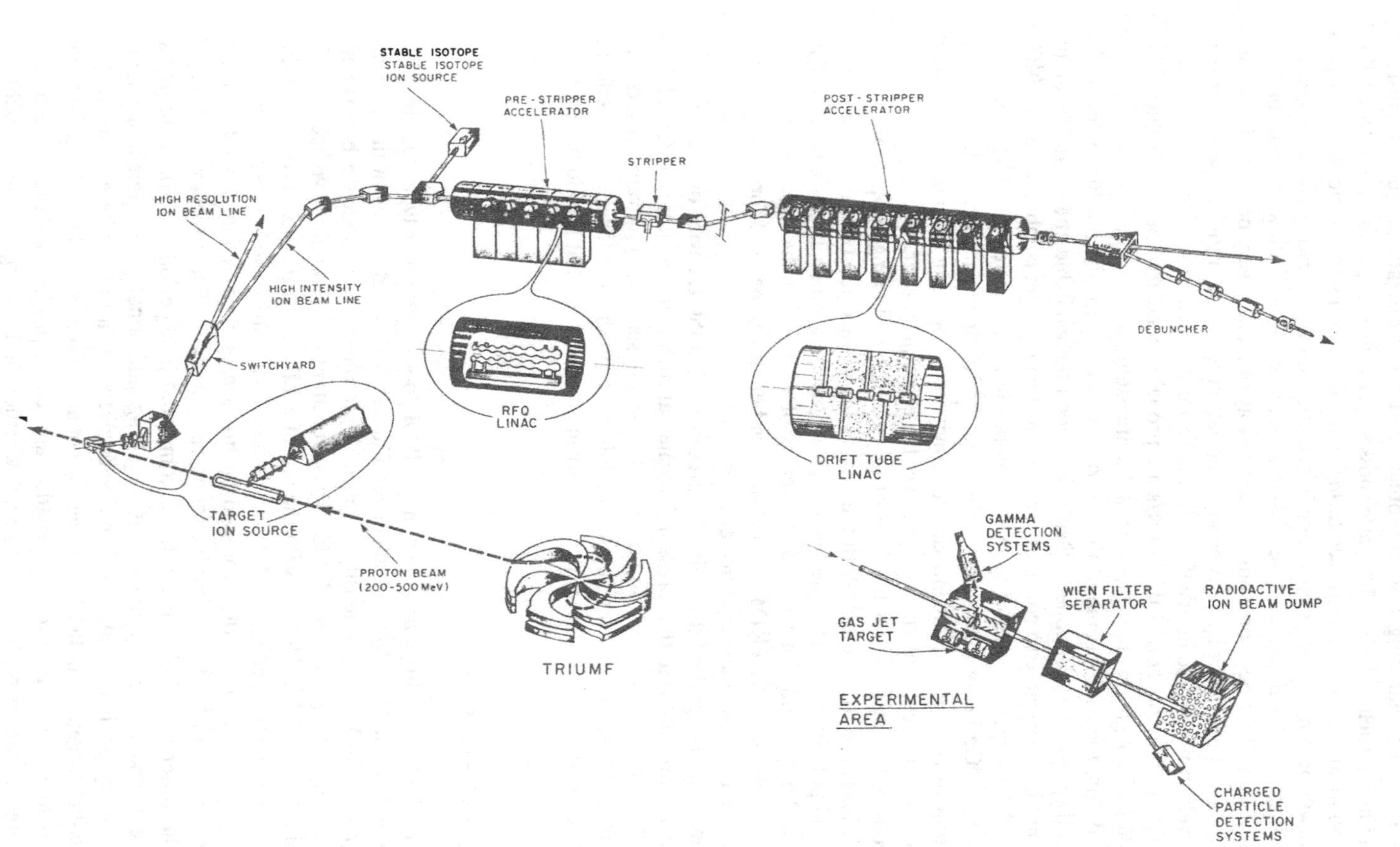

Figure 1. A schematic representation of the proposed ISAC (Isotope Separator and Accelerator) facility at the intermediate-energy (500-MeV) proton facility, TRIUMF. Included is a possible experimental system that could be used to measure the reaction products arising from the interaction of a low-energy radioactive beam, such as ^{13}N with a H_2 gas target.

developed, which could handle the prospect of proton beam intensities as high as 100 μA; whether or not these complex systems could easily and efficiently be manipulated remotely, given such high radiation fields; whether or not the acceleration system could be developed to transmit such low-intensity, low-charge-state beams with acceptable losses and for reasonable costs; and whether or not detection systems could be devised to be used in the expected high-radiation fields of the projectiles themselves. While a decision to proceed on the construction on the ISAC facility has not yet been finalized, a small test on-line isotope separator (TISOL) was installed at TRIUMF to study some of these areas related to the final facility.

6.5 THE TISOL FACILITY

6.5.A General Features

As a test system, it was considered useful that TISOL incorporate some of the features of the proposed ISOL on the front end of ISAC. Thus, TISOL is a vertically-mounted, thick-target, on-line isotope separator, installed at the end of the medium-intensity ($\leqslant$ 10 μA) beam line at TRIUMF. Since installation, two different configurations have been devel-

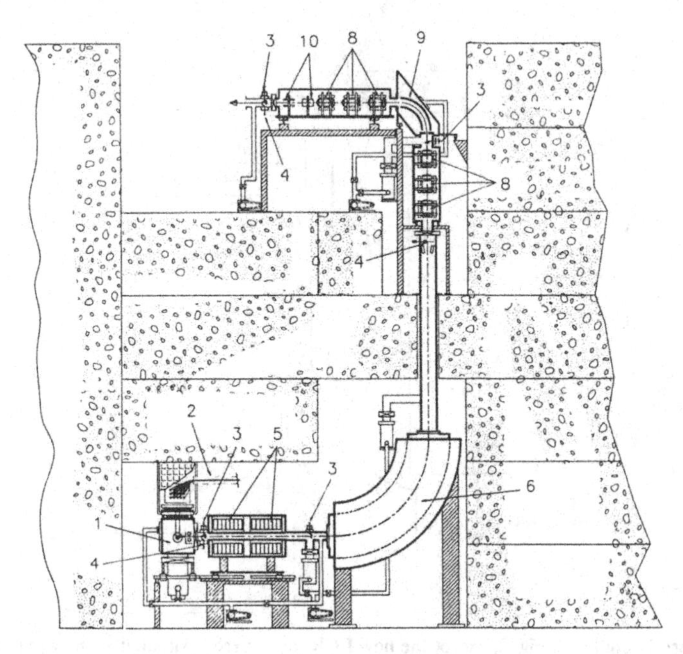

Figure 2. A schematic representation of the elevation view of the TISOL facility located on beam line 4A at TRIUMF. The front end of TISOL shows the configuration for a heated-surface ion source situated in the target/ion source (TIS) vacuum chamber.

oped for the facility, depending upon the type of ion source used. Figure 2 displays a schematic representation of an elevation view of the facility with the heated-surface ion source and associated target chamber in place. This system has been described in some detail elsewhere.[17,18]

Figure 3 shows a similar configuration, only now with the newly-installed and operational Electron Cyclotron Resonance (ECR) ion source in place, along with other modifications to the pumping system and ion-transport system. The ECR system consists of a small water-cooled-target vacuum chamber in which is located the resistance heated Ta oven and a graphite chamber containing the target material itself. A gas-transfer line, connected both to an external gas system and the inlet of the ECR source, is also connected to a small opening at the upper part of the oven/target chamber. Nuclear reaction products resulting from the interaction of the proton beam with the heated target material can diffuse into this opening, but the transfer line is not heated at present, so only gaseous species are expected to reach the ion source. Ions produced in the source are extracted through a 2-mm hole, using a specially-shaped extraction electrode (EE) with a 3-mm opening. As before, the accelerating potential at which the target operates is 20 kV. The EE is movable in the transverse direction and is followed by an einzel lens which can operate at voltages up to 15 kV. This is followed by two magnetic quadrupoles (10 cm in diameter), a magnetic steerer, and then the mass analyzing, magnetic 90° dipole. In addition to the use of

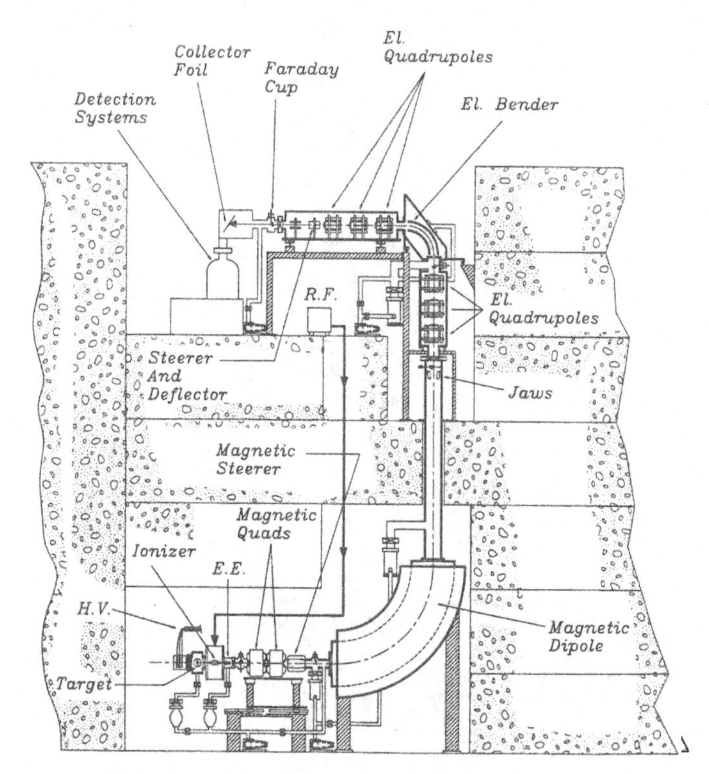

Figure 3. This figure is similar to Fig. 2, except the new ECR ion source is situated at the front end. In this configuration the target chamber is separated from the ion source.

450–L/s turbo pumps common to both configurations, the pumping of the surface target/ion-source system is with a CTI-10 cryopump, while (15-cm) diffusion pumps are used with the ECR system. The latter were used for economic reasons only. Other aspects of the system, including the control system, beam diagnostic elements, etc., are as described in previous reports.[17,18]

6.5.B Target Materials

Since the last report, a series of new target materials has been used successfully to generate a very-wide variety of radioisotopic ion beams from both the heated-surface and ECR ion sources. In general, these targets have been either metal foils or powders in the form of uniquely-prepared pellets. These are placed inside the cylindrically-shaped graphite container, which in turn fits inside the target oven. The targets used have been foils of Ti, Zr, and Nb, and pellets of UO_2/graphite, ZrC, and SiC. These pellets have been prepared by using an organic adhesive to "glue" the powders, plus a very slight pressure to form the powder into a cylindrical shape. Pellets of the order of a few mm in thickness are then sliced. These are heated up to about 2000° C in the target holder under vacuum.

6.5.C Ion Sources

6.5.C.1 Heated Surface. The first ion source to be used successfully and the one for which the greatest number of isotopes have been observed is the well-known heated-surface type. The ionizer itself is simply a material with a relatively-high work function, and the elements ionized are those with relatively-low ionization potentials. Such sources are well known, and the TISOL source has been described in more detail earlier.[18] The TISOL source was primarily Re, also W inside a Ta cylinder. The elements ionized were Li, Na, Al, K, Ga, Rb, Sr, In, Cs, Ba, Yb, and Fr.

6.5.C.2 ECR. Recently an ECR ion source was used successfully with TISOL, leading to the ionization and extraction of radioisotopic beams of Ne, Ar, and Cl. This is the first instance of the coupling of an ISOL device with such a source. General aspects of the design of this source have been described earlier,[19,20] while the specifics of this particular installation and its operation are to be published elsewhere.[21] Figure 3 displays an elevation view of the TISOL-ECR configuration, while Fig. 4 presents a detailed side view of the ECR source itself with the target chamber at its front end. In general, RF waves of 10 GHz at an injected power level of 160 W are used to provide the power to ionize the gaseous material, which reaches the plasma chamber of the source. Since a cooled transfer line was used between the target and the source, only noble gases and other gaseous species are expected to be observed. Using a Ti-foil target heated up to about 1500° C, ion beams of Ne, Ar, and Cl (as Cl^+ and HCl^+) were observed. An extraction potential of 12 kV was used, and in this first run a He carrier gas was used, although the pressure in the target chamber was only of the order of 10^{-5} torr.

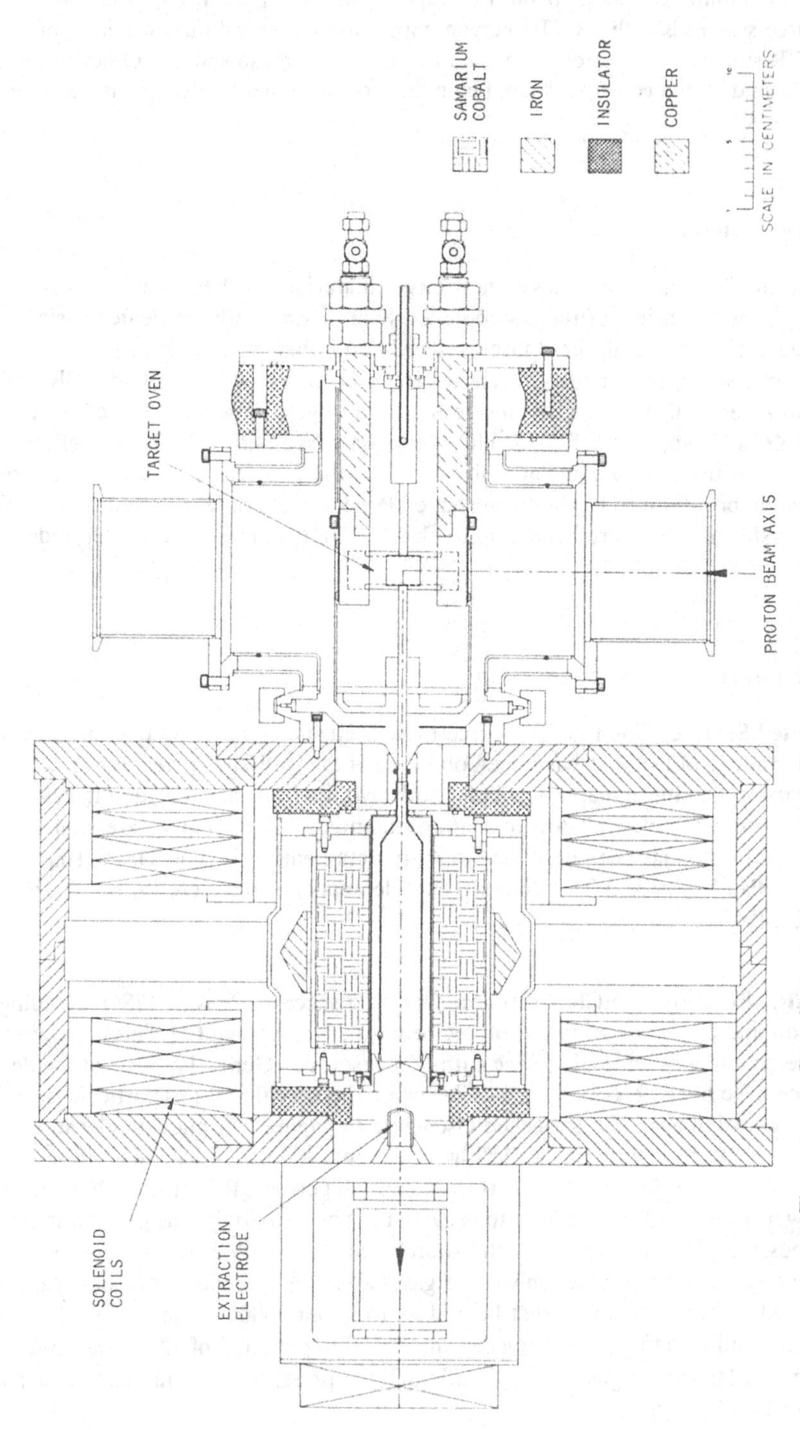

Figure 4. A schematic, detailed representation of the newly installed ECR systems at TISOL. The ion source is the large chamber on the left, while the target chamber is on the right.

Table IVa. Production Yields (particles/s · μA of p) at TISOL (Heated-Surface Ion Source)

Iso-Tope	Half-Life (s)	Target[a] SiC	Nb	Zr	ZrC	UO/C
^{8}Li	0.844	6 x 10^{6}	4 x 10^{4}		3.1 x 10^{6}	2.0 x 10^{5}
^{9}Li	0.177	1.2 x 10^{4}			1.4 x 10^{5}	7.5 x 10^{3}
^{11}Li	87 ms				2 x 10^{3}	
^{20}Na	0.446	1.2 x 10^{6}			2.3 x 10^{3}	
^{21}Na	2.25	5 x 10^{7}		9.0 x 10^{3}	5.5 x 10^{4}	2.6 x 10^{4}
^{24}Na^{m}	20.2 ms	1.2 x 10^{7}				
^{25}Na	59.6	1.5 x 10^{7}	1.6 x 10^{4}	4.4 x 10^{4}	2.3 x 10^{5}	1.3 x 10^{5}
^{26}Na	1.07	2.9 x 10^{5}	3.5 x 10^{2}	1.4 x 10^{3}	1.2 x 10^{4}	1.3 x 10^{5}
^{27}Na	0.304	1.7 x 10^{4}			9.8 x 10^{2}	3.5 x 10^{3}
^{28}Al	134.4	1.6 x 10^{4}			1.1 x 10^{3}	1.3 x 10^{3}
^{29}Al	393.6	8.3 x 10^{3}				5.9 x 10^{2}
^{36}K	0.34			2.1 x 10^{6}		
^{37}K	1.23		5.1 x 10^{3}	7.5 x 10^{5}	4.7 x 10^{4}	
^{38}K	458		2.4 x 10^{5}	1.3 x 10^{6}	4.3 x 10^{6}	
^{42}K	12.36 h		1.0 x 10^{5}	3.5 x 10^{5}		
^{43}K	22.3 h		5.4 x 10^{5}	2.0 x 10^{5}		
^{44}K	1328		2.4 x 10^{4}	4.3 x 10^{4}	4.4 x 10^{4}	
^{45}K	1038		3.0 x 10^{3}	2.1 x 10^{4}	2.2 x 10^{4}	
^{46}K	108		6.5 x 10^{2}	3.4 x 10^{3}	2.4 x 10^{3}	
^{47}K	17.5		2.2 x 10^{2}	1.0 x 10^{3}		

[a]Target thicknesses - SiC (6.06 g · cm^{-2}), Nb (11.57 g · cm^{-2}), Zr (6.32 g · cm^{-2}), ZrC (14.9 g · cm^{-2}), UO/C (1 g · cm^{-2})

Table IVb. Production Yields (particles/s · μA of *p*) at TISOL (Heated-Surface Ion Source)

Iso-Tope	Half-Life (s)	Target[a] Nb	Zr	ZrC
^{75}Rb	21.4	2.1×10^3	3.1×10^3	1.3×10^3
^{76}Rb	37.0	6.0×10^4	8.7×10^4	9.5×10^4
^{77}Rb	222	2.0×10^6	3.5×10^6	1.5×10^6
^{78}Rb^m	345.6	1.2×10^7	1.7×10^6	1.0×10^7
^{78}Rb^g	1060	5.4×10^6	2.8×10^6	7.1×10^6
^{79}Rb	1374	7.0×10^7	5.0×10^7	7.7×10^7
^{80}Rb	34	6.0×10^6	2.0×10^7	9.0×10^6
^{81}Rb^g	4.58h	5.2×10^8	2.6×10^8	1.2×10^9
^{82}Rb^m	6.47h	2.1×10^8	2.1×10^8	4.2×10^8
^{82}Rb^g	76.2	7.0×10^6	1.9×10^7	1.4×10^7
^{83}Rb	86.2d	1.4×10^8	1.1×10^8	1.0×10^9
^{84}Rb^m	1216	8.0×10^6	2.5×10^7	3.7×10^8
^{86}Rb^m	61.0	2.5×10^5	3.6×10^6	6.3×10^6
^{88}Rb	1068	2.8×10^4	2.0×10^6	2.7×10^6
^{89}Rb	912		8.0×10^5	1.2×10^6
^{90}Rb^m	258		1.5×10^5	1.5×10^5
^{90}Rb^g	156		5.0×10^4	3.0×10^4
^{91}Rb	58.4		3.6×10^4	1.1×10^4
^{77}Sr	9			9.0×10^3
^{83}Sr^m	4.95	8.9×10^3	6.1×10^3	
^{83}Sr^g	32.4h	1.7×10^7	8.5×10^6	2.5×10^7
^{85}Sr^m	4060	3.3×10^6	1.9×10^6	
^{87}Sr^m	67.6m	7.0×10^4	4.3×10^5	5.2×10^5
^{91}Sr	9.5h			1.2×10^5

[a]Target thicknesses - SiC ($6.06\ g \cdot cm^{-2}$), Nb ($11.57\ g \cdot cm^{-2}$), Zr ($6.32\ g \cdot cm^{-2}$), ZrC ($14.9\ g \cdot cm^{-2}$), UO/C ($1\ g \cdot cm^{-2}$)

Table IVc. Production Yields (particles/s • μA of *p*) at TISOL (Heated-Surface Ion Source)

Isotope	Half-Life (s)	Target[a] (UO/C)
^{79}Rb	1.374E+03	2.90E+04
^{80}Rb	3.400E+01	3.70E+04
^{86}Rb^{m} (6^{-})	6.102E+01	7.40E+06
^{89}Rb	9.120E+02	1.90E+07
^{90}Rb^{m} (4^{-})	2.580E+02	1.50E+07
^{91}Rb	5.840E+01	1.40E+07
^{92}Rb	4.500E+00	6.80E+05
^{93}Rb	5.860E+00	6.20E+05
^{94}Rb	2.780E+00	3.40E+05
^{95}Rb	3.840E - 01	2.80E+04
^{96}Rb	1.990E - 01	7.60E+03
^{97}Rb	1720E - 01	2.00E+03
^{98}Rb	1140E - 01	1.00E+02
^{122}In^{b}	1.030E+01	3.30E+04
^{122}In^{c}	1.080E+01	1.60E+05
^{124}In^{g}	3.210E+00	2.80E+04
^{124}In^{m}	2.400E+00	5.60E+04
^{126}In^{g}	1.500E+00	9.60E+03
^{126}In^{m}	1.450E+00	9.30E+03
^{128}In^{g}	9.000E - 01	9.80E+02
^{128}In^{m}	9.000E - 01	1.30E+04
^{124}Cs^{g}	2.650E+01	6.00E+04
^{126}Cs	9.840E+01	5.7 E+05
^{128}Cs	2.34 E+02	1.99E+06
^{130}Cs	1.500E+03	4.57E+06
^{139}Cs	5.562E+02	1.33E+07
^{140}Cs	6.370E+01	2.51E+06
^{141}Cs	2.494E+01	3.15E+06
^{142}Cs	1.780E+00	6.02E+04
^{143}Cs	1.780E+00	6.69E+04
^{144}Cs	1.020E+00	1.41E+04
^{145}Cs	5.900E - 01	5.61E+03

[a]Target thickness - 1 g • cm^{-2}

Table V. Production Yields with ECR at TISOL
(Ti Target - 5.45 g/cm^{-2})

Isotope	Half-Life (s)	Production Yield (particles/s/μA of protons)	Comments[a]
^{23}Ne	37.24	1.1×10^5	
^{24}Ne	202.8	2.7×10^4	
^{34}Cl	1920	2.7×10^4	as Cl^+ ion
^{34}Cl	1920	9.6×10^4	as HCl^+ ion
^{38}Cl	2234	8×10^4	as Cl^+ ion
^{35}Ar	1.77	7.2×10^4	
^{41}Ar	1.83 h	4.5×10^6	
^{43}Ar	110.1	1.2×10^4	as Ar^{2+} ion

[a]Unless indicated, all observed species are singly charged.

6.5.D Yields

6.5.D.1 Experimental Setup. Radioactive isotopes produced in thick targets with the 500-MeV proton beam at TRIUMF and extracted as isotopically-separated ion beams using TISOL were measured using standard radiation-detection systems positioned at a fixed collector. Gamma-ray spectra were obtained with a Ge(Li) γ-ray spectrometer, while α spectra were detected with a charged-particle surface-barrier detector. In cases where the half-lives of the species of interest were quite short and a half-life determination was desired, the ion beam could be deflected electrostatically and this process cycled, synchronized with the data acquisition system. All spectra were collected using a PCA-8000 board (manufactured by The Nucleus) in a Multiterm AT (IBM clone). Further details of these measurements can be found elsewhere,[21] and a summary of some of these yield data obtained using both the surface source and the newly-installed ECR source can be found in Table IV and V. Typical target thickness used were in the range of 5-10 g/cm^2 with proton beam currents of 200-300 nA.

In addition to the yields themselves from some target materials and the successful operation of the ECR ion source, not used before with ISOL devices, some of the other new features were the observation of an isotope possessing a half-life of the order of 20 ms, namely $^{24}Na^m$, and isotopes of Al for future possible studies at TISOL.

6.6 SUMMARY AND CONCLUSIONS

It is now apparent that the field of producing and using accelerated radioactive beams for a wide range of new and exciting studies is beginning, and a number of facilities are being planned in several countries around the world. These will take different forms, and all will complement one another to provide different approaches to answer new kinds of questions that can be asked and answered. The proposed ISAC facility proposed to be built at the TRIUMF facility is still awaiting final funding, but the small test facility TISOL is now operational, including a new kind of ion source, an ECR system, and a program of testing and development of systems needed in any ISAC facility. In addition, a standard physics program of the type in progress at the ISOLDE facility at CERN is now possible at TISOL.

REFERENCES

1. I. Tanihata, *Treatise on Heavy Ion Science*, Vol. 8, Ed. by D. A. Bromley (Plenum Press, New York) in press; RIKEN-AF-IVP-58 (1987).
2. I. Tanihata, *Phys. Lett.* **160B**, 380 (1985); *Phys. Rev. Lett.* **55**, 2676 (1985).
3. C. Detraz, *Proc. of Int. Symposium in Heavy Ion Physics and Nuclear Astrophysical Problems* (1988), *Tokyo*, Ed. by S. Kubono, M. Ishihara, and T. Nomura (World Scientific, Singapore, 1989), p. 151.
4. P. Armbruster, H. Geissel, and P. Kienle, ibid, p. 247.
5. F. D. Becchetti et al. *Phys. Rev. C*, in press.
6. F. D. Becchetti, W. F. Liu, D. A. Roberts, J. W. Janeche, J. G. Kolata, A. Morsad, X. J. Kong, and R. E. Warner, ibid. p. 277.
7. R. N. Boyd, K. E. Sale, and G. J. Mathews, in *Proc. of Int. Symposium in Heavy Ion Physics and Nuclear Astrophysical Problems, (1988) Tokyo*, Ed. by S. Kubono, M. Ishihara, and T. Nomura (World Scientific, Singapore, 1989), p. 39.

8. M. Arnould et al., *Nucl. Instr. Meth.* **B40/41**, 498 (1989).

9. J. Crawford, J. K. P. Lee, R. B. Moore, H. Dautet, F. Buchinger, K. Oxorn, L. Nikkinen, J. D'Auria, L. Buchmann, J. Vincent, and J. King, *Nucl. Instr. Meth.*, **B26**, 128 (1987) .

10. C. Rolfs, B. Jonson, B. W. Allardyce, H. Hass, and H. Ravn, CERN Report/EP/HH/tdn/postccc (1989).

11. T. Nomura, *Proc. of Int. Symposium in Heavy Ion Physics and Nuclear Astrophysical Problems, (1988) Tokyo*, Ed. by S. Kubono, M. Ishihara, and T. Nomura (World Scientific, Singapore, 1989), p. 295.

12. "The ISOLDE Users Guide," Ed. by H. J. Kluge, CERN Yellow Report 1986-05 and updates (1988, 1989).

13. L. Buchmann, J. M. D'Auria, J. D. King, G. Machenzie, H. Schneider, R. B. Moore, and C. Rolfs, *Nucl. Instr. Meth.* **B26**, 151 (1987).

14. L. Buchmann and J. M. D'Auria, spokesmen, Proposal 311, Experiment Evaluation Comm., TRIUMF, Vancouver, B. C., Canada (Dec. 1984).

15. *Proceedings of the Accelerated Radioactive Beams Workshop, Parksville, Canada (1985)* Ed. by L. Buchmann and J. M. D'Auria, TRIUMF Report TR1-*85-1*.

16. D. J. Millener, J. W. Olner, E. V. Warburton, and S. S. Hanna, *Phys. Rev. C* **28**, 497 (1983).

17. K. Oxorn et al., *Nucl. Instr. Meth.* **B26**, 143 (1987).

18. J. M. D'Auria, M. Dombsky, L. Buchmann, J. D. Vincent, and J. D. King, *Nucl. Instr. Meth.* **B40/41**, 418 (1989).

19. L. Buchmann, T. Mattmann, J. M. D'Auria, E. DeVita, H. Schneider, A. Otter, H. Sprenger, and P. W. Schmor, *Nucl. Instr. Meth.* **B26**, 253 (1987).

20. P. McNeely, G. Roy, J. Soukup, J. M. D'Auria, L. Buchmann, M. McDonald, P. W. Schmor, H. Sprenger, and J. Vincent, *Int. Conf. on Ion Sources, Berkeley, CA (1989)*; P. McNeeley et al., *Jour. Phys.* **50**, C1-807 (1989).

21. M. Dombsky, J. M. D'Auria, L. Buchmann, J. Vincent, H. Sprenger, P. McNeely, P. W. Schmor, M. MacDonald, and G. Roy, to be submitted to *Nucl. Instr. Meth.*

7. A Technique for Proton Drip-Line Studies via Fusion-Evaporation Reactions

M. F. Mohar, W. Benenson, D. J. Morrissey, R. M. Ronningen,
B. Sherrill, J. Stevenson, J. S. Winfield, and J. Yurkon
National Superconducting Cyclotron Laboratory
Michigan State University
East Lansing, MI 48824

J. Görres
Department of Physics
University of Notre Dame
Notre Dame, IN 46556

K. Subotic
Physics Dpt.-010
Boris Kidrich Institute
POB 533
11001 Beograd, Yugoslavia

ABSTRACT

A survey of the evaporation-residue cross-sections from fusion reactions in the mass region $56 < A < 100$ was begun to investigate the utility of this process for producing proton-rich nuclei of astrophysical importance for future decay studies. The initial study was the fusion of an $E/A = 8$ MeV ^{36}Ar beam with a ^{40}Ca target using the *K*500 cyclotron at the National Superconducting Cyclotron Laboratory, Michigan State University. The details of the new *S*320 spectrograph focal-plane detection system developed to identify the evaporation products are presented, along with preliminary experimental results.

7.1 INTRODUCTION

The study of proton-rich nuclei has provided astrophysicists with important information for the modeling of stellar phenomena. In particular, the *rp* process, or *rapid-proton* capture process,[1] in certain environments is thought to be an important means of stellar energy evolution[2] which results from the synthesis of nuclei along the proton-rich side of the valley of stability. Calculations have been presented for a particular case in which proton capture and β^+ decay occur along the proton-drip line beyond the nickel region (Fig. 1). The extended *rp*-process calculation was carried out by Wallace and Woosely[3] for the particular case of an astrophysical thermonuclear explosion known as a Type-I x-ray burst. In this model, a neutron star accretes matter from a hydrogen-rich companion in a binary star system. The result is a thermonuclear explosion on the surface of the neutron star at very high temperature and density (approx. 10^9 K, hydrogen densities about 10^6 g/cm^3). Hydrogen and helium burning in the synthesis chain provide the energy necessary to sustain

Exotic Nuclear Spectroscopy, Edited by W. C. McHarris
Plenum Press, New York, 1990

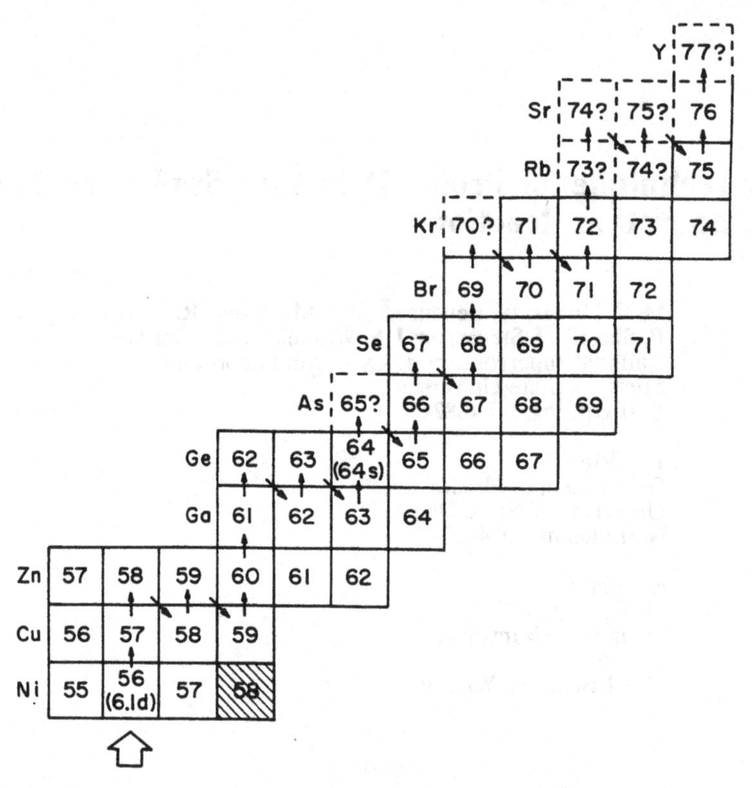

Figure 1. Nuclear flows characterizing the *rp* process above Ni at T ≈ 10^9 K and a hydrogen density ≈ 10^6 g/cm^3, appropriate to an x-ray flash on a neutron star. (Figure from Ref. 2.)

the reaction, and the *p*- and α-capture reactions will continue into the nickel region where the α capture (αp process) ceases because of increasing coulomb barriers. The synthesis is thought to continue up to the mass-100 region via the extended-*rp* process. Information is needed about the structure, proton-capture rates, and β^+- and *p*-decay lifetimes of these proton-rich nuclei in order to calculate the energy evolution of the system and provide a comparison for the various Type-I x-ray burst sites known to exist in our galaxy.[4]

Figure 1 illustrates some of the problems associated with calculating the extended-*rp* process along the proton-drip line. The black arrows indicate the flow of the process, and the dashed boxes represent nuclei with little or no known nuclear-structure information. For these nuclei, semi-empirical mass models are used to estimate the binding energies. However, because of uncertainties in the models, various critical nuclei present in the calculation may be either proton-unbound or so loosely bound that they photo-dissociate at high stellar temperatures. For example, the nucleus ^{65}As occurs at one of these critical points. It has never been observed, and, since it is near the proton-drip line, there is some question as to whether it is proton-bound. There are predictions by Wapstra and Audi[5] that indicate it is bound by 200±500 keV, and the half-life predictions for the bound nucleus range from hundreds of nanoseconds for proton decay to about 100 ms for β^+ decay. A specific problem for the extended-*rp*-process calculation is that if ^{65}As is particle-unbound, then in order for the process to produce nuclei larger in *A* and *Z*, ^{64}Ge must decay to ^{64}Ga and subsequently capture a proton. However, the half-life of ^{64}Ge is comparable to the time scale of the *rp* process. Therefore, in the situation where ^{65}As is unbound, the process would cease with a buildup of ^{64}Ge. Thus, simply showing ^{65}As to be particle-bound has important implications on the extent of the proton burning in the *rp* process. One of our main goals in studying nuclei in this region is to identify and study the structure of as many of these nuclei as possible. Of course, any structure studies done in this region are also important in determining the single-particle and collective properties of the proton-rich nuclei.

A secondary goal of our work is to be able to predict fusion/evaporation product cross-sections for proton-rich nuclei. Computer codes such as ALICE[6] and CASCADE[7] are able to predict the production cross-sections of stable and near-stable nuclei quite well. However, these Hauser-Feshbach-based codes tend to overpredict cross-sections far from stability by as much as three orders of magnitude.[8] Therefore, the feasibility of a given experiment becomes difficult to judge on the basis of these uncertain predictions. This indicates a need to study further the fusion/evaporation reaction mechanism in order to understand the discrepancy between the predictions and experimental results.

Such a study was begun using a spectrograph and detection system to measure the Z, A, and 0° momentum distributions of the fusion/evaporation products. Complete isotope identification and momentum information over the dynamic range of the reaction products allows the reaction mechanism to be fully investigated.

7.2 EXPERIMENTAL TECHNIQUE

The first experiment chosen for this study was the fusion of E/A = 8 MeV ^{36}Ar with a ^{40}Ca target. The beam was provided by the K500 cyclotron at NSCL, and the products were identified and their momentum distributions measured in the S320 spectrograph. The reaction was chosen based on the predictions of the codes CASCADE and ALICE. It was felt that the codes could give a good representation of the dependence of the relative cross-sections on beam and target combinations, as well as on beam energy.

There were some important experimental considerations for this project. Unit Z resolution from Z = 20–40 was required, and thus a detector with a wide dynamical range and an energy resolution of three to four percent was needed. Position resolution at the focal plane of better than 1 mm full-width at half-maximum (FWHM) was necessary in order to obtain a 1:600 momentum resolution. The time-of-flight (TOF) had to be measured to better than 0.8 ns FWHM to obtain 1:600 velocity resolution. This combination would give the slightly better than 1:400 overall resolution necessary to resolve the m/q values (TOF vs position). Also, since the flight times were long compared to the cyclotron *RF* period, a start detector was needed at the acceptance aperture of the spectrograph and a stop detector at the focal plane. It was also important, since the 0° momentum distribution was to be measured, to separate the beam and small-angle elastically-scattered ions from the reaction products before the start detector. Finally, the low energy of the evaporation products—E/A < 2MeV—required that the detector and detector-window thicknesses be minimized in order to reduce energy-loss straggling and multiple scattering. Obviously, this requirement is related to the first three considerations—straggling will affect the energy resolution, multiple scattering will affect the spectrograph resolution, and the need for a start detector must be considered in light of the extra material it will place in the particle's path.

With these considerations in mind, the following detection scheme was designed for the *S*320 spectrograph. Figure 2*a* shows a schematic of the experimental area. A closer

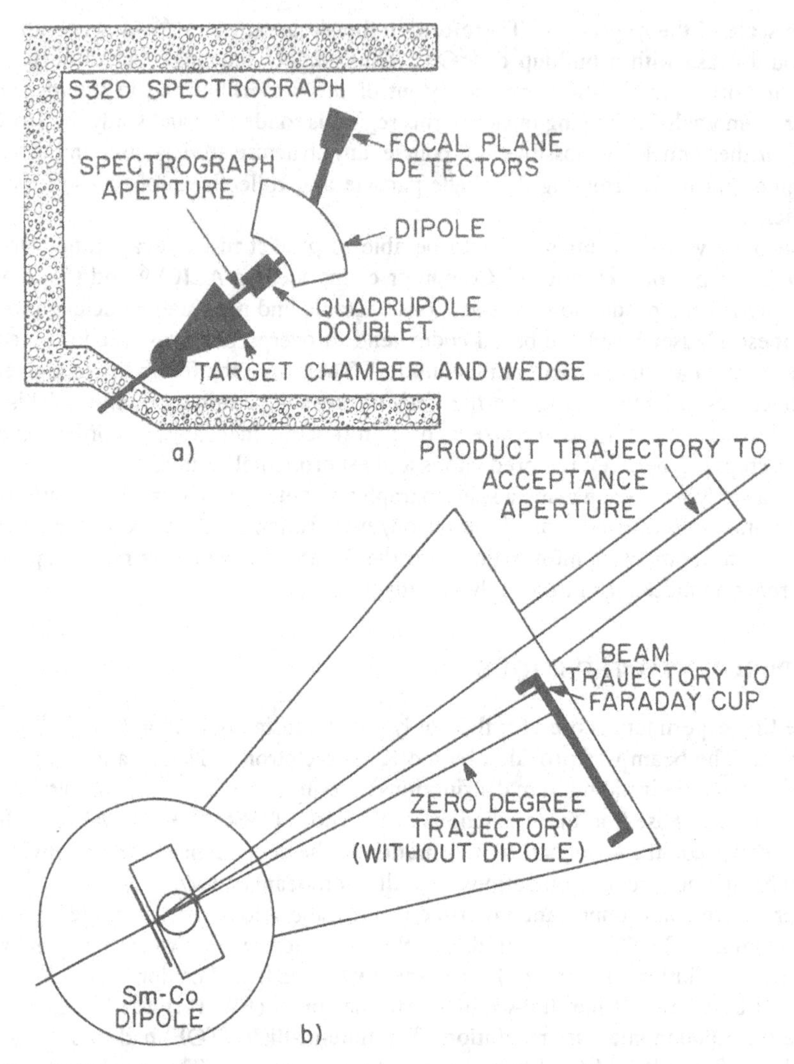

Figure 2. *a*) Schematic of *S*320 spectrograph showing major components. *b*) Close-up view of target and wedge vacuum chamber. Note Sm-Co dipole magnet used to give angular separation between evaporation products and beam-velocity particles.

look at the target chamber and wedge-shaped vacuum chamber (Fig. 2*b*) shows how the products were separated from the beam and small-angle elastics. A Sm-Co dipole magnet was placed immediately after the target position. Since the fusion reaction products had about half the rigidity of the beam, this magnet produced an angular separation between the evaporation products and beam-velocity ions in the wedge vacuum chamber. Then, with the spectrograph set at 7° in the lab, the 0° evaporation products were transported to the spectrograph, while the beam and small-angle elastics, which ranged from 4° to 5°, were collected in the faraday cup. The 0° reaction products thus traveled to the start detector at the acceptance aperture of the spectrograph without the high-rate background of the beam-like particles.

The start detector was a thin-foil scintillator attached to a photomultiplier tube (Fig. 3). The foil used during the experiment was 35 $\mu g/cm^2$ of Bicron BC400 plastic scintillator, made by dissolving the plastic scintillator in xylene and applying it to a glass plate mounted on a turntable. The plate was spun to form a thin uniform layer, and the foil was floated off of the glass plate in water.[9] The foil was mounted on a UVT plastic frame that was glued to the face of a photomultiplier tube. The energy loss in the detector (about 4–5 MeV) was enough to provide an adequate start signal; yet the detector was so thin that it did not significantly degrade the energy or momentum resolution.

The design of the focal plane detector (Fig. 4) was based on the idea that the ions should be stopped in the detector volume. That way, maximum amount of energy loss (ΔE) and total energy (E) information could be used to resolve properly the Z's and to minimize the effects of ΔE straggling. This immediately implies that another position and timing detector is needed before the ΔE detector. The Breskin-type,[10] low-pressure, multiwire proportional counter (LP-MWPC) was chosen, as it provides both a position and a timing signal from the same detector. The position resolution was 0.7 mm, and the rise time of the TOF signal was 2 ns.

The evaporation products were stopped in a gas-filled ion chamber with a segmented anode. The ten 1-in. segments gave a good dynamic range—the number of segments used to create the ΔE signal could be adjusted, depending on the range of the ions in the detector. A plastic scintillator in back gave the E signal for the ions that were not stopped in the ion chamber.

In addition, the angle of incidence of the products at the focal plane had to be measured in order to correct the TOF and ΔE for path-length differences. Two of the twelve anode positions in the ion chamber were occupied by resistive-wire proportional counters to give this information. The two windowless resistive wires, shown in Fig. 4, had two func-

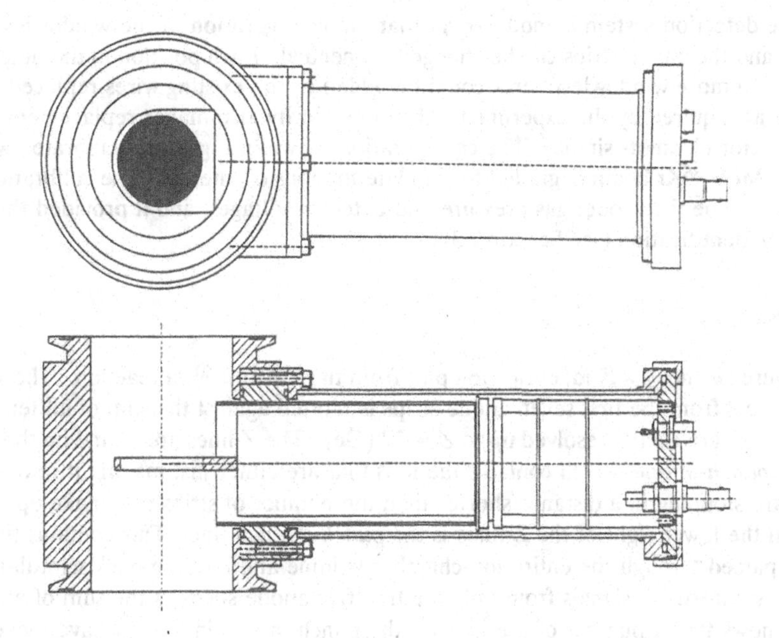

Figure 3. Start detector located at spectrograph aperture. Particles pass through the 35$\mu g/cm^2$ scintillator foil mounted on a UVT frame glued to the photomultiplier face.

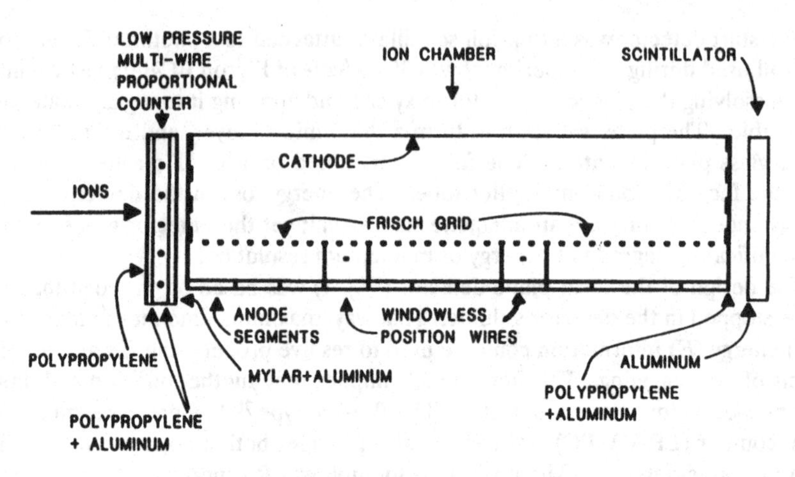

Figure 4. *S320* spectrograph focal plane detector.

tions: first, they allowed the rejection of the ions that were not confined to the detector volume; and, second, they provided the data necessary to calculate the angle of incidence that is needed to correct both the TOF for the angle acceptance of the spectrograph and ΔE for the ion path in the ion chamber. The unusual feature of these segments is that the higher electric field of the proportional wires is shielded from the neighboring anode strips by a grounded box. This way, no additional field-shaping windows are needed in the detector volume. The measured position resolution of the windowless wires was 0.8 mm FWHM.

The detection system is modular, so that the configuration of the windowless position wires and the anode strips can be changed, as needed. Each position in the detector is equivalent, so more windowless wires could be added or the existing wires replaced by anode strips, as required by the experiment. The modularity also makes replacement of damaged detector elements simple. The configuration shown in Fig. 4 was calibrated with an E/A = 2.5 MeV ^{86}Kr beam degraded to simulate the ions of interest. The calibration was used to determine the proper gas pressure and detector voltages, and it provided the scale for the m/q identification (see Section 7.3).

7.3 RESULTS

Figure 5*a* shows a Z-identification plot from the ^{36}Ar on ^{40}Ca reaction. The sum of the energy loss from the first seven anode strips is plotted against the sum of all ten strips, and the Z lines are clearly resolved up to Z = 32 (Ge). The Z lines are bound on the upper left by the *punch-in* line, which contains the ions that are either just making it into the detector or are stopping in a distance shorter than the number of strips that make up the ΔE signal. On the lower right of the Z lines is the *punch-through* line. This contains the ions that have passed through the entire ion-chamber volume and into the back scintillator. A plot of the summed ΔE signals from only the first five anode strips vs the sum of all strips (Fig. 5*b*) shows that a number of the ions on the punch-in line in Fig. 5*a* have moved out along the new Z lines. However, the new Z lines are also broadened a bit, because of the effects of ΔE straggling. This effect is more clearly illustrated in Fig. 5*c*, in which the ΔE in

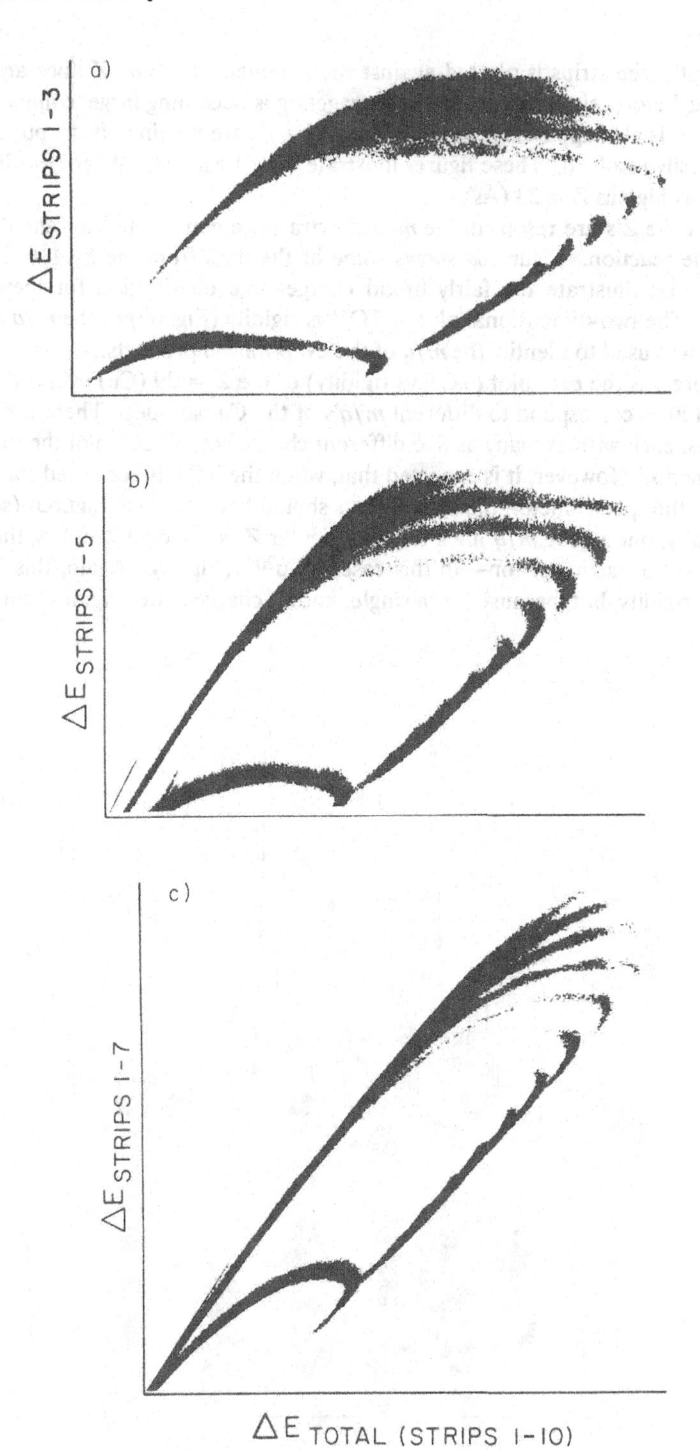

Figure 5. *Z*-identification plots from the segmented anode ion chamber. *a*) Energy loss in the first seven segments vs the total energy loss. *b*) Energy loss in the first five segments vs the total energy loss. *c*) Energy loss in the first three segments vs the total energy loss.

only the first three strips is plotted against the summed ΔE. The Z lines are not so well separated as before, since the energy-loss straggling is becoming large, compared to the ΔE signal. But, it is also apparent that the higher-Z bands are moving off the punch-in line and are more easily resolved. These figures illustrate how the ΔE signal can be adjusted to best resolve Z's as high as $Z = 33$ (As).

After the Z's are resolved, the m/q spectra were constructed for each element produced in the reaction. Figure 6*a* shows some of the data from the $E/A = 2.5$ MeV ^{86}Kr calibration that illustrate the fairly broad charge-state distribution for these low-energy heavy ions. The two-dimensional plot of TOF vs rigidity (Fig. 6*b*) of the m/q's from the Kr calibration was used to identify the m/q of the evaporation products.

Figure 7 is the m/q plot (TOF vs rigidity) of the $Z = 29$ (Cu) evaporation products. The sloped lines correspond to different m/q's of the Cu isotopes. There are seven different isotopes, each with as many as five different charge states. Some of the lines have double assignments. However, it is expected that, when the TOF is corrected for the angle acceptance of the spectrometer, the assignments should become unambiguous (see below).

Finally, one of the m/q lines for a particular Z is selected, yielding the momentum distribution of a particular ion—in this case, $^{64}Cu^{20+}$ (Fig. 8). Again, this is shown as a function of rigidity, but, because it is a single, known charge state, rigidity corresponds di-

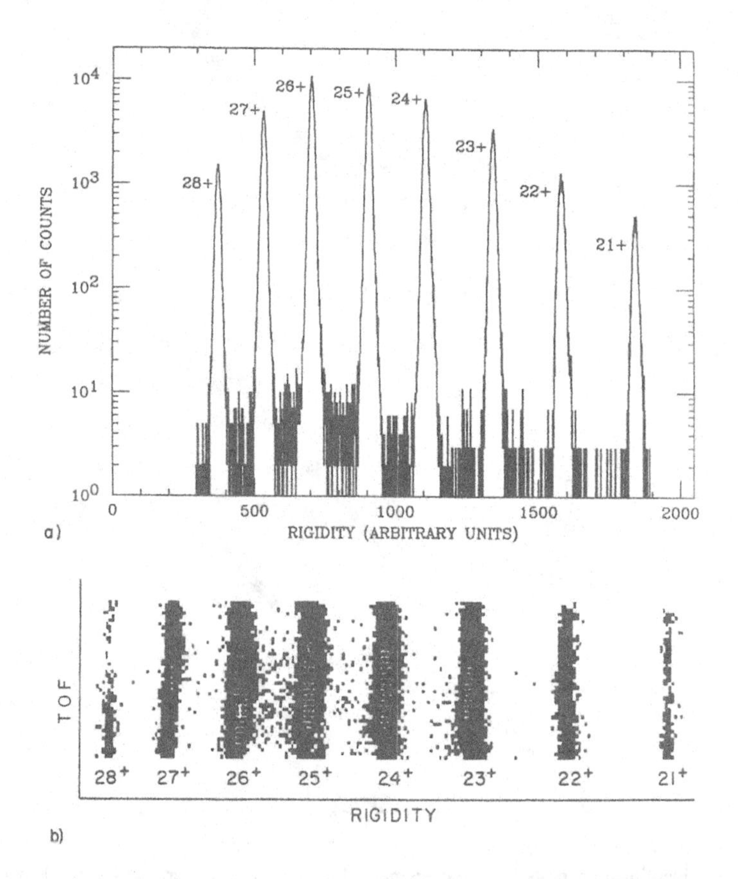

Figure 6. *a*) Charge state distribution of elastically-scattered $E/A = 2.5$ MeV ^{86}Kr on an Au foil. *b*) An m/q plot of the same data (TOF vs rigidity, defined as $B\rho = mv/q$).

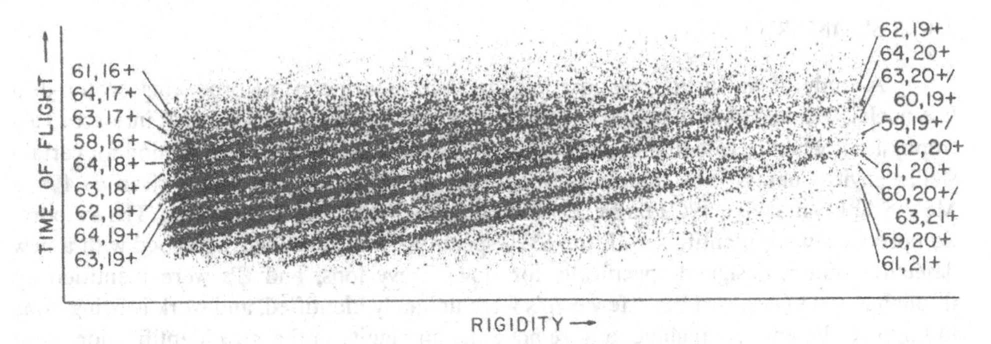

Figure 7. An m/q plot of Cu ions ($Z = 29$).

rectly to momentum. The momentum distribution for each of the isotopes can be integrated and compared to the cross-sections from the Hauser-Feshbach-based codes to look for trends and deviations from the predicted values. Also, the cross-sections at the peak of the momentum distributions for any exotic nuclei of interest will be known, which will enable the accurate prediction of count rates in decay studies. The final task will be to do a computer simulation of the experiment to be sure that the experimental results are reproducible when all of the cross-sections, detector thicknesses, and spectrometer acceptances are included.

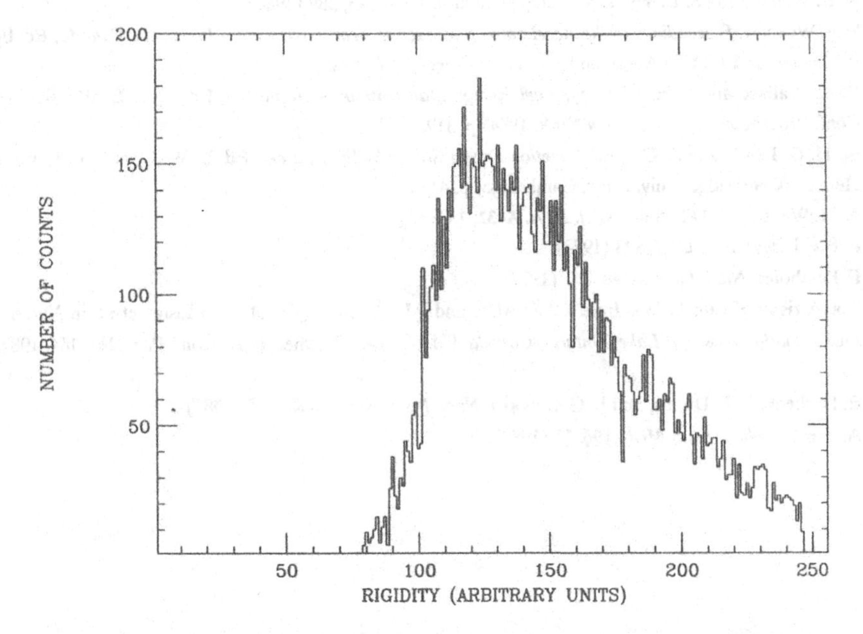

Figure 8. Rigidity plot of $^{64}Cu^{20+}$. Since only one charge state of ^{64}Cu is plotted, the rigidity distribution corresponds directly to a momentum distribution.

7.4 SUMMARY

A study of the fusion/evaporation reaction, as a means of producing proton-drip line nuclei, was begun in the mass region $56 < A < 100$. These nuclei are of interest to astrophysicists who are calculating the *rp* process of synthesis and energy evolution for certain stellar environments. The initial experiment in this study was the fusion of an $E/A = 8$ MeV ^{36}Ar beam with a ^{40}Ca target, carried out at the NSCL K500 cyclotron. The evaporation products were identified with the S320 spectrograph, which was equipped with a new detection system designed specifically for slow, heavy ions, and Z's were identified up through $Z = 33$ (As). All but a few m/q's were uniquely identified, and work is being done to improve the analysis technique to remove the ambiguity in the m/q identification. The momentum distributions for each isotope will be integrated and compared with the cross-sections predicted by various Hauser-Feshbach calculations to look for variations and trends in the production cross-sections. A computer simulation will be performed to model the experiment and be sure that the results are completely understood. Finally, it will be determined if any isotopes of interest were produced in this reaction that may be studied further.

ACKNOWLEDGMENT

The authors gratefully acknowledge the support of National Science Foundation grant PHY 82-15585.

REFERENCES

1. R. K. Wallace and S. E. Woosley, *Astrophys. Journ. Suppl.* 45, 389 (1981) .
2. S. E. Woosley, *Proceedings of the Accelerated Radioactive Beams Workshop, Parksville, Canada*, Ed. by L. Buchmann and J. M. D'Auria, and references therein, p.4 (1985).
3. R. K. Wallace and S. E. Woosley, *High Energy Transients in Astrophysics*, Ed. by S. E. Woosley, (AIP Conf. Proceedings No. 115, New York, 1984) p. 319.
4. W. H. G. Lewin and P. C. Joss, *Accretion Driven Stellar X-Ray Sources*, Ed. by W. Lewin and E. van den Heuvel, (Cambridge Univ. Press, Cambridge) p. 41.
5. A. H. Wapstra and G. Audi, *Nucl. Phys.* **A432**, 1 (1985).
6. F. Plasil, *Phys. Rev. C* **17**, 823 (1978).
7. F. Puhlhofer, *Nucl. Phys.* A280, 267 (1977).
8. Comparison of calculations from CASCADE and ALICE to results of C. J. Lister et al. in *Nuclei Far From Stability, Rosseau Lake, Ontario, Canada*, Ed. by Ian S. Towner, (AIP Conf. Proc. No. 164, 1987) p. 354.
9. E. Norbeck, T. P. Dubbs, and L. G. Sobotka, *Nucl. Instr. Meth.* **A262**, 546 (1987) .
10. A. Breskin, *Nucl. Instr. Meth.* **196**, 11 (1982).

8. Nuclear Structure in the Neutron-Rich Doubly-Magic ^{78}Ni Region

John C. Hill and F. K. Wohn
Ames Laboratory
Iowa State University
Ames, IA 50011

J. A. Winger*
Ames Laboratory
Iowa State University
Ames, IA 50011
and
Brookhaven National Laboratory
Upton, NY 11973

E. K. Warburton and R. L. Gill
Brookhaven National Laboratory
Upton, NY 11973

R. B. Schuhmann
Clark University
Worchester, MA 01610
and
Brookhaven National Laboratory
Upton, NY 11973

ABSTRACT

The magic numbers $Z=28$ and $N=50$ imply that very neutron-rich ^{78}Ni, which has not yet been observed, is doubly magic. The ^{78}Ni region was investigated by studying the $N = 50$ isotones and neutron-rich Zn isotopes. Results on the level structure of ^{83}As, ^{74}Zn, and ^{76}Zn populated in the decays of ^{83}Ge, ^{74}Cu, and ^{76}Cu are presented. The parent nuclides were produced and mass-separated using the TRISTAN facility on-line to the High-Flux-Beam Reactor at Brookhaven. The systematics of the $N = 50$ isotones and even-A Zn isotopes are discussed and compared with shell-model calculations involving active nucleons outside of a ^{78}Ni and ^{66}Ni core, respectively. The extent to which the ^{78}Ni region can be considered doubly magic is assessed.

8.1 INTRODUCTION

The central problem in nuclear physics is the determination of the extent to which the proton and neutron magic numbers known from studies of nuclei near stability remain

*Present Address: National Superconducting Cyclotron Laboratory, Michigan State University, East Lansing, MI 48824

Exotic Nuclear Spectroscopy, Edited by W. C. McHarris
Plenum Press, New York, 1990

valid for nuclei far away from the stable valley. Knowledge of magic numbers far from stability will in turn enable us to predict where regions of spherical and deformed nuclei far from stability can be expected. Numbers that are magic near stability cannot always be extrapolated to regions far from stability. A prime example[1] of this is ^{32}Mg, which has a magic number (N = 20) of neutrons but is quite deformed. This comes about because of the collapse of the gap between the s-d shell and the $f_{7/2}$ shell for very neutron-rich nuclei.

Doubly-magic regions centered about ^{78}Ni and ^{132}Sn will exist if the nucleon numbers Z = 28 and 50 and N = 50 and 82 are magic on the very neutron-rich side of stability. Unfortunately, at the present time it is only possible to study these regions by observing the decay of short-lived neutron-rich nuclei produced in fission or by multinucleon transfer reactions.

The region around ^{132}Sn has been extensively studied—in particular, the neutron-rich Sn nuclei and the N = 82 isotones. It has been shown that the ^{132}Sn region is doubly magic with an especially good shell closure.[2,3] This is confirmed by level structure information on ^{132}Sn, ^{131}Sn (core + one neutron hole),[4] and ^{133}Sb (core + one proton).[5]

The doubly-magic character of a region can most effectively be tested by studying the structure of nuclei in which either the proton or neutron number is magic. This is more difficult for nuclei in the region around ^{78}Ni than for the ^{132}Sn region, since ^{78}Ni is further away from stability on the neutron-rich side. No structure information now exists for Ni (Z = 28) nuclei heavier than ^{68}Ni (see, e.g., Ref. 6). Consequently, the most practical method of exploring the ^{78}Ni region is to study the decay of fission products to levels of the N = 50 isotones above ^{78}Ni. No information is available on the structure of the lighter N = 50 isotones, with the exception of ^{82}Ge (^{78}Ni + four protons), which is populated in the decay[7] of the fission product ^{82}Ga.

The properties of nuclei near ^{78}Ni are also of interest for astrophysical calculations since they lie at the beginning of the r-process path. Also, some of the lighter neutron-rich nuclei in this region provide "seed" nuclei for initiating the r process. Of special astrophysical interest is the N = 50 nucleus ^{80}Zn (^{78}Ni + two protons), which is one of three "waiting-point" nuclei where the r-process path crosses a magic neutron line. The decay of this nucleus has been recently characterized by our group[8,9] and others.[10,11] Also of particular interest are β-decay half-lives and neutron binding energies of nuclei near the r-process path.

All experimental studies presented here were carried out using the TRISTAN separator. Details of the experiments are discussed in Section 8.2. In order to explore the extent to which the ^{78}Ni region is doubly magic, we have carried out a detailed study of the decay of ^{83}Ge to levels in ^{83}As (^{78}Ni + five protons). The results are presented and compared with shell-model calculations in Section 8.3. In Section 8.4 we present information on the level structure of neutron-rich ^{74}Zn and ^{76}Zn, fed by the decay of ^{74}Cu and ^{76}Cu, respectively. Systematics of the even-even Zn isotopes are presented and compared with a restricted shell-model calculation. Our conclusions are presented in Section 8.5.

8.2 EXPERIMENTAL METHODS

All sources were obtained by fission of ^{235}U at the TRISTAN isotope-separator facility operating on-line to the High-Flux Beam Reactor at Brookhaven National Laboratory. The layout for the TRISTAN system is shown in Fig. 1. In this work, a high-temperature plasma ion source[12] containing about 4 g of enriched ^{235}U was exposed to a neutron flux of 3 x 10^{10} $n/\mathrm{cm}^2\cdot\mathrm{s}$. Beams of singly-charged ions were mass-separated and deposited

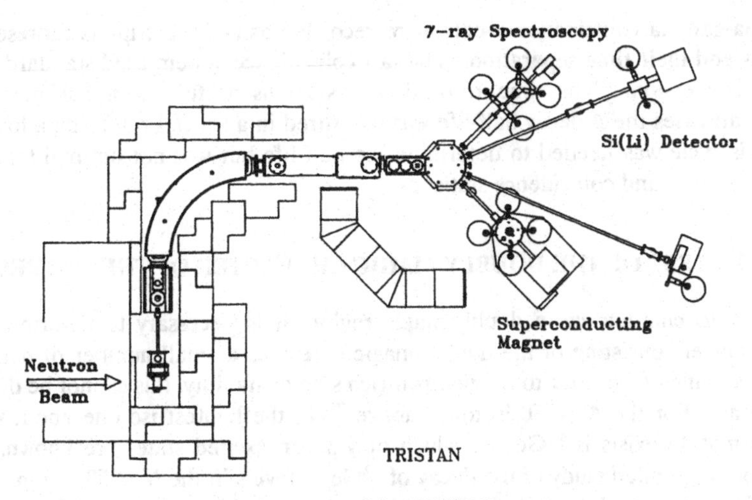

Figure 1. Layout of the TRISTAN separator system.

on a movable Al-coated Mylar tape. No evidence was observed for cross contamination from adjacent masses. In the study of decay of ^{83}Ge, the ion beam consisted primarily of Ge and As. However, in the studies of the decays of the Cu isotopes, most of the activity consisted of Zn and Ga daughters, with γ rays from Cu decays resting on a large background. In addition, nuclides with A = 148 were present in the A = 74 beam because of the formation of doubly-charged ions in the source.

For all three decay studies the experimental procedures were similar. Two Ge γ-ray detectors placed in 180° geometry viewed the source at the point where it was deposited on the tape. A thin plastic scintillator with approximately a 2π-sr acceptance solid angle viewed the source, and the resulting signals served both as a beam monitor and a β gate for the singles spectra. This background-suppression technique was necessary because of low fission yields, particularly for ^{74}Cu and ^{76}Cu. The primary decay data were obtained in runs lasting several days in which the beam was collected and data accumulated for a short time, followed by movement of the tape and repetition of the cycle in order to minimize contributions from long-lived activities. Several other runs involving enhancement of the longer-lived daughter activities were also carried out. By comparison of the relative intensity of γ rays in the various spectra, it was possible to identify γ transitions coming from the decay of interest. In some cases, spectra were also taken at the "daughter" port to enhance the daughter activities.

Singles and coincidence γ-ray measurements were carried out simultaneously, typically covering energy ranges of 20 keV to 5 MeV. The γ energies were calibrated using standard calibration sources and in some cases secondary standards observed in the primary spectra. The efficiency of each detector was determined using standard sources in the same geometry as for the data-taking runs. The uncertainties associated with the energies result from statistical uncertainties in determining peak centroids and system nonlinearities, while the uncertainties associated with the relative intensities reflect uncertainties in the determination of peak areas and detector efficiencies. Corrections for coincidence summing were made to correct the relative instensities for the close geometry and high efficiencies of the detectors. There is also the possiblity for a significant intensity of γ rays below the present detection limits, especially for the Cu decays.

Gamma-gamma coincidence events were recorded as address triplets representing γ-ray energies and their time separation. The fast coincidence system used standard time-to-amplitude conversion. The time resolution was 20 ns at full-width half-maximum (FWHM). In all cases the β-decay half-life was measured in a special run using a long decay cycle. This cycle was needed to determine the half-life but was not optimal for accumulating good singles and coincidence statistics.

8.3 ^{83}As AS A TEST OF THE DOUBLY-MAGIC CHARACTER OF THE ^{78}Ni REGION

In order to characterize a doubly-magic region, it is necessary to determine the properties of nuclei consisting of the doubly-magic core plus a small number of neutrons and/or protons. Since ^{78}Ni is far to the neutron-rich side of stability, this cannot be done at the present time. For the N = 50 isotones above ^{78}Ni, the lightest isotone about which structure information exists is ^{82}Ge, for which only seven excited states are known. We have carried out a detailed study of the decay of ^{83}Ge to levels in the N = 50 isotone ^{83}As. The details of this study and the techniques used have recently been published.[13]

8.3.A Structure of ^{83}As from the Decay of ^{83}Ge

The level scheme for the N = 50 isotone ^{83}As populated in ^{83}Ge decay is shown in Fig. 2. The scheme is based on our γ-ray singles and coincidence measurements. The ^{83}Ge

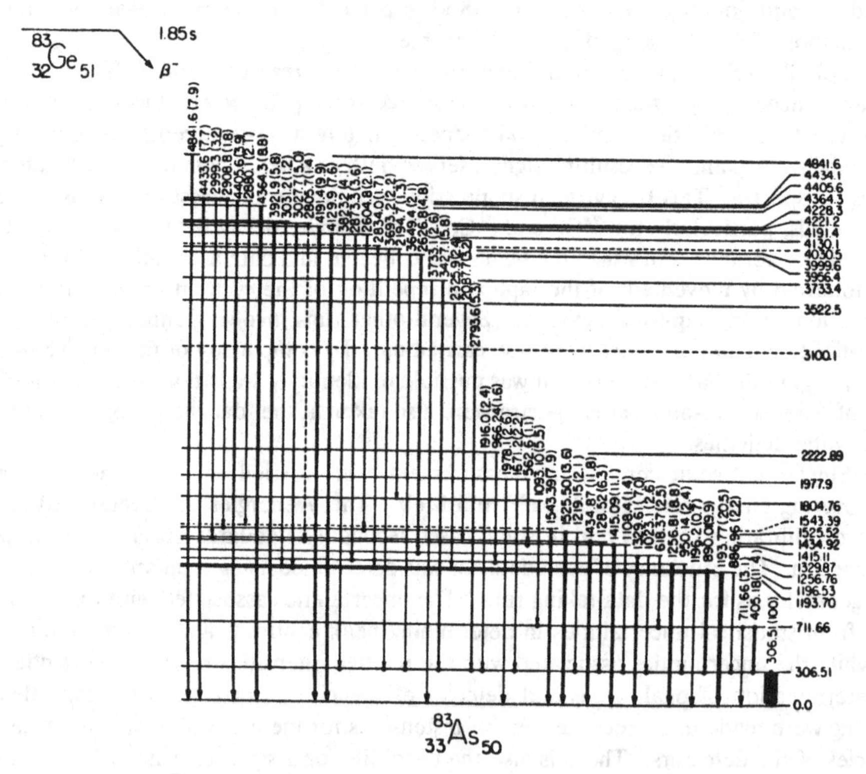

Figure 2. Decay Scheme for ^{83}Ge. All energies are in keV.

half-life was determined to be 1.85 ± 0.06 s by following the decay of the strong 306-keV γ ray. A total of fifty-one γ rays were placed in a level scheme for ^{83}As, corresponding to twenty-eight levels in a range of excitation energy up to 4.84 MeV. Reference 13 contains a complete list of γ energies and intensities and their coincidence relationships. In the decay of ^{83}Ge there is a strong concentration of β intensity to a group of states above 3.5 MeV. Approximately 28% of the β intensity to excited states goes to seven states in ^{83}As lying between 4.13 and 4.43 MeV. In Ref. 13 we argue that the strong β feeding is to a series of neutron-particle–neutron-hole states. These states can be formed when a neutron in one of the filled subshells between $N = 28$ and 50 in ^{83}Ge decays to an available proton orbital in ^{83}As by an allowed β transition. The resulting states would consist of a neutron-particle–neutron-hole configuration consisting of a $d_{5/2}$ particle and a hole in any one of the orbitals ($p_{1/2}, p_{3/2}, f_{5/2}$) in the shell between $N = 28$ and 50. These states could then couple with the five valence protons to give a variety of positive-parity states. The above arguments are consistent with such particle-hole excitations at around 4 MeV relative to the $N = 50$ neutron core.

In this work we can concentrate on the levels in ^{83}As below 2 MeV as a test of the shell model and doubly-magic character of the ^{78}Ni region. It is well established that the first two orbitals in the proton shell between 28 and 50 are $f_{5/2}$ and $p_{3/2}$. Since $Z = 33$ for ^{83}As, J^π for the ground state would be $5/2^-$ for either ordering of the above two orbitals. Furthermore, studies of the ^{86}Kr (d, ^{3}He) ^{85}Br reaction[14] have established that the ground state of the $N = 50$ isotone ^{85}Br is $3/2^-$; thus, our assumption that the $f_{5/2}$ orbital is filled first is reasonable. This is in contrast to the value of $3/2^-$ suggested by As isotope systematics. The first excited state at 306 keV is well established by the coincidence data and the fact that the 306-keV γ ray is over four times more intense than any other γ ray. We postulate that $J^\pi = 3/2^-$ for this level, based on systematics. The second excited state of ^{83}As at 711 keV, which lies lower in energy than the corresponding states[15,16] in ^{85}Br (955 keV) and ^{87}Rb (845 keV), is well established by coincidence data and its J^π value could be $1/2^-$, $3/2^\pm$, $5/2^\pm$, or $7/2^-$, since the level decays both to the $5/2^-$ ground state and the $3/2^-$ level at 306 keV. Shell-model arguments would favor a negative parity for this state.

A major feature of the ^{83}As level scheme is a group of eight levels between 1.1 and 1.6 MeV that, with the exception of the 1543-keV level, are well established by coincidence information and the fact that they feed both the ground and 306-keV states. Similar groupings of levels have been observed[6,16] in the $N = 50$ isotones ^{85}Br and ^{87}Rb, but the level density is somewhat higher for ^{83}As. All of the above levels have significant β feeding, but first-forbidden unique transitions cannot be ruled out and the J^π limitations are the same as for the 711-keV level. The log ft values (see Ref. 13) and energies of these levels are consistent with their interpretation as being mostly negative-parity, three-quasiparticle states outside the ^{78}Ni core. It is striking that between 1.6 and 3.5 MeV only four levels were observed. A similar "gap" is observed in the level structure of ^{85}Br, where only four levels are observed bewteen 2.0 and 3.5 MeV in decay studies.[15]

8.3.B Shell-Model Calculations for the $N = 50$ Isotones

One of the most general predictions of the nuclear shell model is that systems with twenty-eight protons or fifty neutrons are unusually tightly bound or "magic." A major motivation of this work is to test the extent to which the above prediction holds for the region around neutron-rich ^{78}Ni. A basic question about the structure of the proton-deficient

$N = 50$ nuclei, as exemplified by ^{83}As, concerns the degree to which the assumption of an $N = 50$ shell closure leads to an accounting of the low-lying (< 2 MeV excitation) energy levels.

Many model calculations have been made for the $N = 50$ isotones above the ($f_{5/2}$, $p_{3/2}$) subshell closure at $Z = 38$ by assuming an ^{88}Sr closed core with the extra-core protons filling the $p_{1/2}$ and $g_{9/2}$ orbitals. Recently Blomqvist and Rydstrom[17] have presented such calculations for nuclei up to ^{99}In. Calculations using similar assumptions were carried out for ^{89}Y through ^{94}Ru by Ball, McGrory, and Larsen.[18] Gloeckner and Serduck[19] calculated both energy levels and $E2$ and $M4$ transition probabilities in the five $N = 50$ isotones from ^{90}Zr to ^{94}Ru. They concluded that seniority breaking in the $p_{1/2}$-$g_{9/2}$ space is minimal. The conclusion from these studies[17-19] is that the subshell closure at ^{88}Sr is valid and that a good description of the $N = 50$ isotones between ^{88}Sr and ^{100}Sn is obtained by consideration of only the $p_{1/2}$–$g_{9/2}$ proton space.

Until very recently, no calculations had been attempted for the $N = 50$ isotones with $A < 86$. Elementary shell-model considerations suggest that the proton orbitals dominating the low-lying states of the light ($Z = 28$–38) $N = 50$ isotones are $f_{5/2}$ and $p_{3/2}$, with the $p_{1/2}$ and $g_{9/2}$ orbitals lying 1 to 2 MeV higher in excitation energy. ^{78}Ni is assumed to be an inert, doubly-magic core. The above assumptions concerning the proton orbitals are consistent with information from proton-transfer reactions[14,20,21] on ^{86}Kr and ^{88}Sr targets. Such experiments indicate that the $f_{5/2}$ and $p_{3/2}$ orbitals are effectively filled at ^{88}Sr. The above theoretical ideas and empirical facts suggest that a model space restricted to four proton orbitals ($f_{5/2}, p_{3/2}, p_{1/2}, g_{9/2}$) might explain the low-lying features of the $N = 50$ isotones, with the $f_{5/2}$ and $p_{3/2}$ orbitals dominating below ^{88}Sr.

Two separate calculations were performed, utilizing different methods for establishing the two-body matrix elements (TBME) and the single particle energies (SPE). The first calculation used the surface-delta interaction (SDI) to parametrize the TBME of the hamiltonian in terms of one scaling parameter. This parameter, along with the three energy differences of the four SPE, was adjusted to reproduce best the well-established single-particle energy centroids and the well-determined low-lying 0^+, 2^+ and 4^+ states in the $A = 82$–86, $N = 50$ region.

The second calculation[13,22] adopted the opposite extreme of characterizing the TBME. In this calculation all sixty-five TBME of the space, along with the four SPE, were used as free parameters (FP) in an iterative least-squares search to find the optimum effective model interaction. A total of 175 ground-state binding energies and excitation-state energies in the $A = 82$–96 range of the $N = 50$ isotones were used as the data set. From these, thirty-five linear combinations of the sixty-nine TBME and SPE were determined. The $p_{1/2}$ and $g_{9/2}$ aspects of the FP hamiltonian were very well determined in this procedure because the data set was dominated by data for nuclei with $A > 88$. On the other hand, the $f_{5/2}$ and $p_{3/2}$ components, which are the dominating factors for the $A = 82$–87 region, are not so well determined because firm experimental information on energy levels in this region is much less abundant. Recently Ji and Wildenthal[23] have updated the above FP calculations and predicted the low-lying level structure for all the $N = 50$ isotones from ^{80}Zn to ^{87}Rb.

The results of the above shell-model calculations are compared in Fig. 3 with our experimental results for ^{83}As. The calculations yield a ground state of $5/2^-$ and a first-excited state of $3/2^-$. The calculated energy separation between the above $5/2^-$ and $3/2^-$

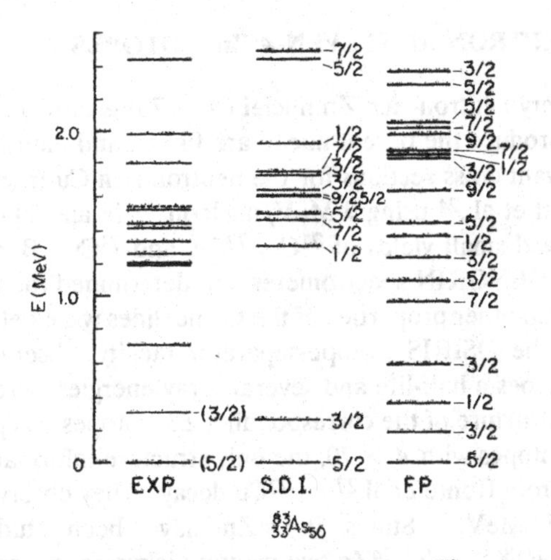

Figure 3. Comparison of experimentally determined excited states in ^{83}As with the results of two shell-model calculations. SDI indicates a calculation using the surface-delta interation and FP indicates the free-parameter calculation. Only negative-parity states are shown.

states is in reasonable agreement with experiment, but the FP calculation significantly underestimates the separation. Both calculations indicate that the $5/2^-$ ground state contains most of the $f_{5/2}$ single-particle strength and that the first-excited $3/2^-$ state contains most of the $p_{3/2}$ single-particle strength. The predominant $p_{1/2}$ single-particle strength is predicted by the FP calculation to lie in the second and third $1/2^-$ states, which have calculated excitation energies between 1.3 and 1.9 MeV. The SDI calculation concentrates most of the $p_{1/2}$ strength in a single state at 1.9 MeV. The first $1/2^-$ state is calculated to have a three-quasiparticle structure with very little $p_{1/2}$ strength. The $g_{9/2}$ single-particle strength is concentrated in a single state above 2.7-MeV excitation energy.

A dominant feature of the experimental level scheme is the cluster of eight states between 1.1 and 1.6 MeV. The experimental density of states between 1 and 2 MeV is fairly well matched by both calculations (excluding the second and third $1/2^-$ states). They each characterize these states as having a three-quasiparticle structure, with the $f_{5/2}$ and $p_{3/2}$ orbitals dominating the orbital occupation.

A striking feature of the experimental spectrum is a well-established low-lying state at 711 keV. This state could not be reproduced in the SDI calculation. The FP calculation yields two states near this energy, with J^π values of $1/2^-$ and $3/2^-$. Both of these states are three-quasiparticle in character. The existence of low-lying three-quasiparticle states in the spectra of nuclei below ^{88}Sr results from a combination of the small values of j for the dominant orbitals and few-valence-particle effects.

Between 1.6 and 3.5 MeV excitation energy, only four states are observed. This results from the fact that the higher states are three- and five-quasiparticle states. Thus, compared to lower-lying states, there is a greater mismatch with the β-decaying state in ^{83}Ge. Also, the smaller Q_β further reduces the transition probability, so few of these states are observed. Reasonable agreement is found between the experimentally-observed levels up to 2 MeV and shell-model calculations based on a doubly-magic ^{78}Ni plus five extra-core protons, restricted to the $f_{5/2}, p_{3/2}, p_{1/2}$, and $g_{9/2}$ orbitals.

8.4 STRUCTURE OF NEUTRON-RICH EVEN-*A* Zn ISOTOPES

Information on the very neutron-rich Zn nuclei ($A > 70$) is difficult to obtain, since the processes employed to produce the parent nuclei are fission and multinucleon-transfer reactions for which the relevant cross sections for the neutron-rich Cu fragments are quite small. Recently, Schmidt-Ott et al.,[24] using 9 MeV/nucleon ^{76}Ge and 11.4 MeV/nucleon ^{82}Se on ^{nat}W targets, observed small yields of ^{71}Cu, ^{72}Cu, and ^{73}Cu. Bernas et al.[25] observed $^{74\text{-}77}$Cu using the LOHENGRIN spectrometer and determined the relative yield of the various Cu nuclides, but no other properties of the Cu nuclides were determined.

Lund et al.,[26] using the OSIRIS isotope-separator facility, observed $^{74\text{-}76}$Cu and ^{78}Cu. For each of these isotopes a half-life and several γ-ray energies were measured, but no information on the level structure of the corresponding Zn isotopes was presented.

For the even-*A* Zn isotopes with $A > 70$, the only structure information from decay studies prior to this work is from Runte et al.[27] on ^{72}Cu decay. They observed nine excited states in ^{72}Zn up to 3.7 MeV. States in ^{74}Zn have been studied using the ^{76}Ge(^{14}C,^{16}O) ^{74}Zn and ^{76}Ge(^{18}O, ^{20}Ne) ^{74}Zn two-proton-pickup reactions.[28,29] Levels in ^{74}Zn at 0.67 and 1.84 MeV were observed.[28] This comprised the only information on excited states in ^{74}Zn prior to this work.

Here we present information on excited states in ^{74}Zn and ^{76}Zn populated in the β^- decay of ^{74}Cu and ^{76}Cu. This is the first information available on the structure of ^{76}Zn. Our information on the decay of ^{74}Cu has been recently published,[30] and a manuscript on ^{76}Cu decay is being prepared.

8.4.A Decay of ^{74}Cu and ^{76}Cu

The details of our ^{74}Cu decay study have recently been published,[30] but a few pertinent facts will be repeated here. A total of nineteen γ rays were placed in a level scheme for ^{74}Zn with ten excited states up to 3 MeV. The half-life of ^{74}Cu was determined (in a special run cycle in which the beam was collected on the tape for 6 s and then deflected for 6 s) to be 1.59 ± 0.05 s, in excellent agreement with the value of 1.60 ± 0.15 s measured by Lund et al.[26] The primary data for the decay of short-lived ^{74}Zn was obtained in a run lasting 4.5 d in which the beam was collected and data accumulated for 3 s, followed by movement of the tape and repetition of the cycle in order to minimize contributions from long-lived activities.

The level scheme for ^{74}Zn populated in ^{74}Cu decay, shown in Fig. 4, is based on the γ-ray singles and coincidence measurements. Since Cu ions comprised a small fraction of the $A = 74$ beam, it was not possible to measure the ground-state β feeding. However, log*ft* calculations have been made assuming zero ground-state β feeding. This would not be true if $J^\pi = 1^+$ for the ^{74}Cu ground state.

The first-excited state at 605 keV is postulated to be 2^+ based on systematics. The energy fits well with the systematics of the neutron-rich even-*A* Zn isotopes. The level is well established by numerous coincidences and the fact that the depopulating γ ray at 605 keV is by far the strongest observed. This state probably corresponds to the one observed in the (^{14}C, ^{16}O) reaction[28] at 0.67 ± 0.03 MeV.

A triplet of states is observed at 1164, 1418, and 1670 keV, all states being well established by energy sums and coincidences. The 1670-keV level is the best candidate for the 2_2^+ level, since it is the only one of the triplet decaying to the ground state. The level

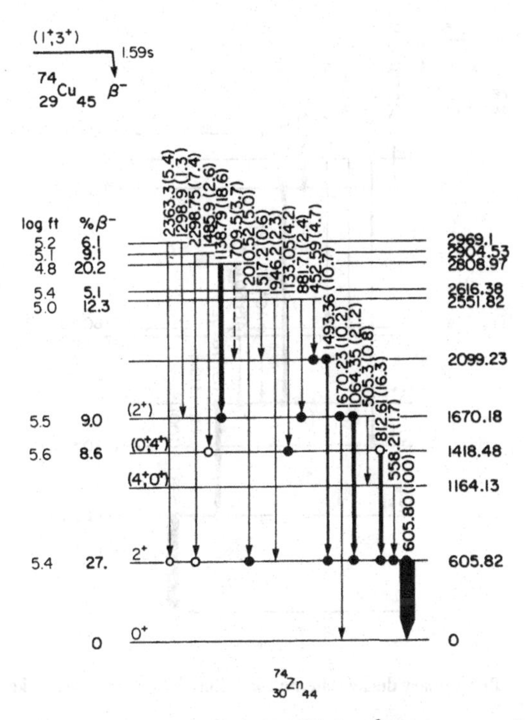

Figure 4. Decay scheme for ^{74}Cu. All energies in keV. The log *ft* values were calculated assuming no β^- feeding to the ^{74}Zn ground state. If the ^{74}Cu ground state is 1^+, then the 1164- and 1418-keV levels would be 4^+ and 0^+, but if the ^{74}Cu ground state is 3^+, then the 1164- and 1418-keV levels would be 0^+ and 4^+, respectively.

at 1418 keV is strongly fed in β decay and is thus the best candidate for $4_1{}^+$ if the ^{74}Cu ground state is assumed to be 3^+. On the same basis, the level at 1164 keV has very little β feeding and is thus the best candidate for the $0_2{}^+$ state, since the β feeding could easily become zero because of possible γ feeding from high-lying unobserved levels, which might populate the 1164-keV level in ^{74}Zn. If a J^π of 1^+ is assumed for the ^{74}Cu ground state, the roles of the 1164- and 1418-keV levels would be reversed. The above assignments are based on systematics and decay patterns.

A cluster of five levels is postulated between 2.5 and 3 MeV. Of the β feeding observed in this work, 52% goes to these levels. The concentration of β strength is striking. A model calculation for the ^{74}Cu ground state[30] indicates the unpaired neutron to be in the $p_{1/2}$ orbital almost 100% of the time, and the exited states in ^{74}Zn have considerable $p_{3/2}$ and $p_{1/2}$ components. Thus the $\nu p_{1/2} \rightarrow \pi p_{1/2}$ and $\nu p_{1/2} \rightarrow \pi p_{3/2}$ allowed β transitions are probably responsible for most of the concentration of β strength observed between 2.5 and 3.0 MeV.

The study of the decay of ^{76}Cu to levels in ^{76}Zn has not been completed, so the information given below should be considered as preliminary. The ^{76}Cu half-life was measured for the 697-keV γ ray to be 0.57 ± 0.07 s, but that measured for the 598-keV γ ray was 0.84 ± 0.07 s. (The 598-keV γ ray was the most intense observed in ^{76}Cu decay). This was surprising in view of the fact that the above two γ rays were strongly in coincidence. We concluded that two isomers of ^{76}Cu were being produced, with one having a half-life of 0.57 s. Statistics were not sufficient to measure the half-life of the other isomer directly. These values are consistent with a half-life of 0.61 s measured for ^{76}Cu by Reeder et al.[31]

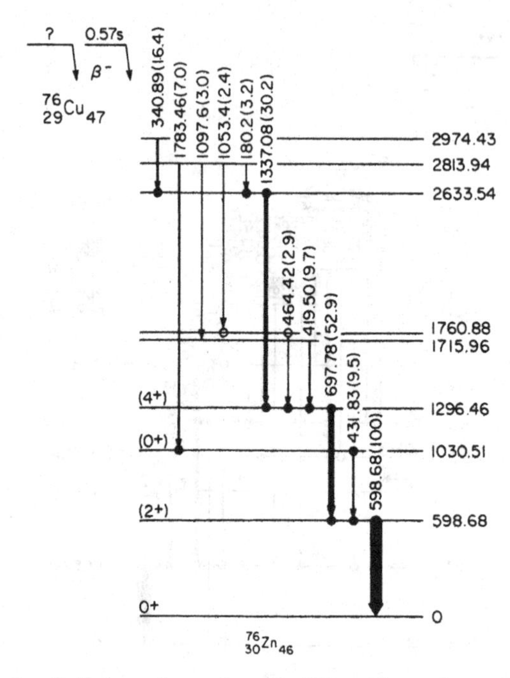

Figure 5. Preliminary decay scheme for ^{76}Cu. All energies are in keV.

from delayed neutron emission but are greater than the value of 0.39 s from measurements[26] at the OSIRIS separator.

A preliminary level scheme for ^{76}Zn populated in ^{76}Cu decay is shown in Fig. 5 based on γ-ray singles and coincidence measurements. We postulate two isomers for ^{76}Cu. Isomerism is not unusual for odd-odd Cu nuclei and has been observed in ^{68}Cu and ^{70}Cu. It usually results in β population of two different sets of levels because of the large J-difference of the isomers.

The first excited state at 598 keV is postulated on the strength of the 598-keV γ ray and given a J^{π} of 2^+ based on systematics. This is the lowest $2_1{}^+$ energy for an even-A Zn nucleus. The second excited state is at 1030 keV and is fed by the low-spin isomer. It is a good candidate for the $0_2{}^+$ state. The third excited state at 1296 keV is probably fed by the high-spin isomer and is a good candidate for the $4_1{}^+$ state. Levels at 1715 and 1760 keV feed only the $4_1{}^+$ state and may be candidates for high-spin states. We also observe a cluster of three states between 2.6 and 3 MeV. Approximately 43% of the observed β^- feeding to excited states in ^{76}Zn goes to these three states. This is similar to the situation in ^{74}Zn, where 52% goes to a cluster of five levels between 2.5 and 3 MeV.

8.4.B Systematics and Calculations for Even-*A* Zn Isotopes

In Fig. 6 the known systematics for even-A Zn isotopes from 0 to 3 MeV are shown from Nuclear Data Sheets[32-38] (A = 60 to 72) and this work (A = 74 to 76). The systematics of the $2_1{}^+$ states are especially interesting. The $2_1{}^+$ energy is essentially constant at around 1 MeV from N = 30 to 38, which corresponds in the simple shell-model picture to

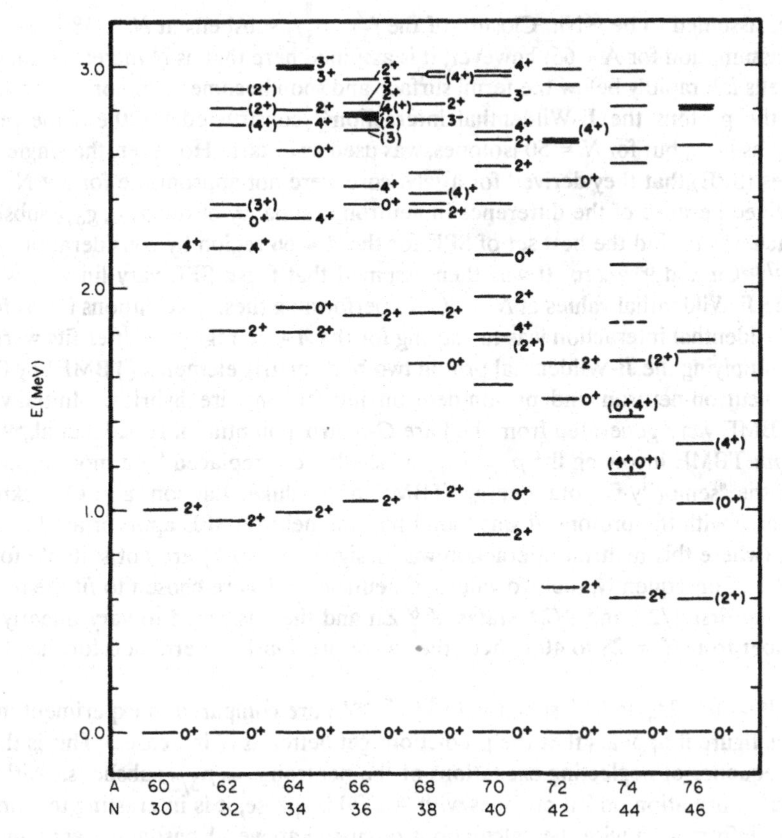

Figure 6. Systematics for even-A Zn isotopes from A = 60 to 76.

the filling of the $f_{5/2}$ and $p_{3/2}$ subshells. An interesting parallel is observed for the Fe isotopes (two proton holes) with N = 30 to 36, where the energy of the 2_1^+ state is roughly constant at about 0.85 MeV.

A striking feature of the Zn systematics is that the energy of the 2_1^+ state drops from 1077 keV at ^{68}Zn, where the $f_{5/2}, p_{3/2}$ neutron subshells are filled, to 598 keV at ^{76}Zn as the $p_{1/2}, g_{9/2}$ neutron subshells are filling. The 0_2^+ and 4_1^+ states appear to follow a similar pattern. These trends are discussed in more detail below.

8.4.C Shell-Model Calculation for $^{68\text{-}74}$Zn

A calculation was made using the code OXBASH[39] as a first step toward the ultimate goal of obtaining an effective shell-model interaction for this region, using least-squares fits to energy spectra. The results are discussed in detail in Ref. 30. The proton model space was taken to be $f_{5/2}, p_{3/2}, p_{1/2}, g_{9/2}$, outside a closed $f_{7/2}$ Z = 28 core. The neutron space was taken to be $p_{1/2}, g_{9/2}$, outside a closed $f_{7/2}, f_{5/2}, p_{3/2}$ N = 38 core. Thus, the core was assumed to be ^{66}Ni. Closure of the $f_{5/2}$, $p_{3/2}$ subshells at N = 38 is certainly not a good assumption for A ≈ 66; however, it is assumed here that as N increases above 38 these subshells fall rapidly below the fermi surface and soon become of minor importance.

For the protons the Ji-Wildenthal interaction[22] constructed for the same proton model space as here, but for N = 50 isotones, was used as a start. However, the single-particle energies (SPE) that they derived for a ^{78}Ni core were not appropriate for a ^{66}Ni core. This is expected because of the difference in neutron occupancy of the $p_{1/2}$, $g_{9/2}$ subshells. Our approach was to find the best set of SPE for the A ≈ 66 region by consideration of the spectra of $^{67\text{-}69}$Cu and $^{68\text{-}69}$Zn. It was then assumed that these SPE vary linearly with N ending at the Ji Wildenthal values at N = 50. In performing these calculations it was found that the Ji-Wildenthal interaction was too strong for the A ≈ 66 region. Better fits were obtained by multiplying the Ji-Wildenthal proton two-body matrix elements (TBME) by 0.7.

The neutron-neutron and proton-neutron interactions are hybrids. Initially, the necessary TBME were generated from the bare G-matrix potential of Hosaka et al.,[40] then the pn and nn TBME involving the $p_{1/2}$, $g_{9/2}$ subshells were replaced by a modification by Brown[41] of the "seniority-fit total-energy" TBME of Serduke, Lawson, and Gloeckner.[42] As was the case with the protons, it was found that the neutron SPE appropriate for the A ≈ 90 region (where this neutron interaction was designed to work) are not suitable for the A ≈ 66 region. Consequently, the two values of neutron SPE were chosen to fit the binding energies of the first $1/2^-$ and $9/2^+$ states of ^{69}Zn and then assumed to vary linearly with proton number from Z = 28 to 40 (where they were previously determined for the A ≈ 90 region).

The J^π = 0^+, 2^+, and 4^+ spectra of ^{68,70,72,74}Zn are compared to experiment in Fig. 7. From this figure it appears that the predictions get better as N increases. This is the expected consequence of neglecting excitations of the neutron $f_{5/2}$, $p_{3/2}$ subshells. Although experimental information on Zn nuclides with A ≥ 70 is sparse, it is interesting to compare the available information with the calculations discussed above. Reasonable agreement is observed for 0^+, 2^+ and 4^+ states up to about 2 MeV. However, the trend of decreasing energy with increasing A for the $2_1{}^+$ states starting at A = 70 is not adequately reproduced.

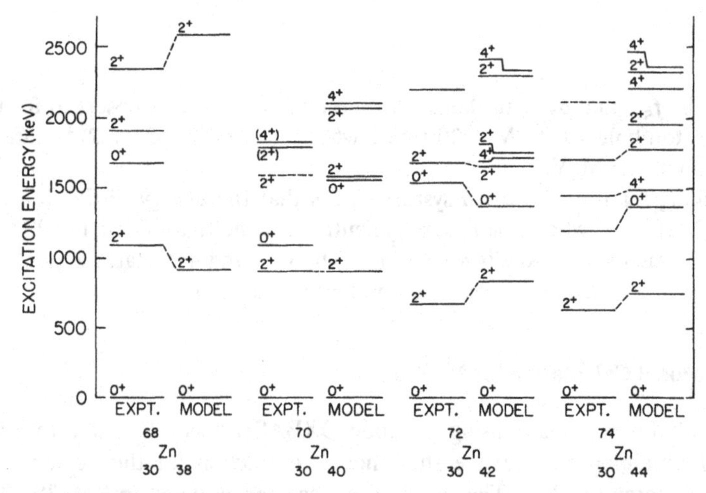

Figure 7. Comparison of predicted and experimental J^π = 0^+, 2^+, and 4^+ levels of ^{68,70,72,74}Zn. All predicted 0^+, 2^+, and 4^+ levels below 2.4-MeV excitation are shown. Possible correspondences between predicted and experimental levels are indicated by dashed lines.

The major occupancy of the calculated wave functions for the 0_1^+, 2_1^+, 0_2^+, 2_2^+, and 4_1^+ states are $f_{5/2}$ and $p_{3/2}$ for the protons and $p_{1/2}$ and $g_{9/2}$ for the neutrons (because of the restricted model space for neutrons).

Presently, an attempt is being made to extend the shell-model calculations to all even-*A* Zn nuclei from ^{60}Zn to ^{76}Zn. For these calculations, the shell-model space includes all neutron subshells for *N* between 28 and 50 and the proton subshells for *Z* between 28 and 40, so only the $\pi g_{9/2}$ subshell is excluded. In these calculations ^{56}Ni is taken to be a doubly-magic core about which both neutrons and protons are coupled. This assumption should be valid, since the shell closure at *N*, *Z* = 28 is expected to be quite strong. In the initial calculations, an SDI interaction is being used to determine the TBME while the SPE come from analysis of nearby odd-*A* nuclei. (These calculations follow closely those of Koops and Glaudemans[43] where the $g_{9/2}$ orbital was excluded). The basic goal of these calculations is an attempt to reproduce the striking behavior of the 2_1^+, 0_2^+, and 4_1^+ energies seen in systematics. Especially striking is the behavior of the 2_1^+ energies, which remain constant at ≈ 0.6 MeV for larger *N*. Calculations in this region are quite complex because of the large model space used, the large number of active particles, and the "dynamic" behavior of the interaction.[22,30,43] This dynamic behavior manifests itself in the way in which the TBME and SPE must be adjusted to describe different portions of the shell. Specifically, it is hard to determine the proper way to vary the interactions as the shell fills. Neither a simple linear variation of SPE between the appropriate values at the ends of the shell nor a simple scaling of TBME have been found to be satisfactory. In addition, basic shell-model considerations indicate that the excitation energy of the 2_1^+ state should increase rather than decrease as the occupancy of the $\pi f_{5/2}$ and $\nu g_{9/2}$ subshells increases. Consequently, no calculation involving manipulation of SPE or SDI strengths has yet been able to reproduce the observed 2_1^+ energy systematics. Future calculations will need to deal with a more realistic interaction and a better understanding of the dynamic behavior of the parameters.

8.5 CONCLUSIONS

The structure of the *N* = 50 isotone ^{83}As has been used to test the supposition that ^{78}Ni is doubly magic. Assuming that ^{83}As can be modeled as a doubly-magic ^{78}Ni core plus five protons, restricted to the $f_{5/2}$, $p_{3/2}$, $p_{1/2}$, and $g_{9/2}$ orbitals, reasonable agreement is found between the observed and calculated levels up to 2 MeV. The calculations indicate the $f_{5/2}$, $p_{3/2}$, and $g_{9/2}$ single-particle strengths to be highly concentrated in single levels, while the $p_{1/2}$ strength is slightly more fragmented. The level density of states between 1 and 2 MeV is adequately mirrored by the calculations, which indicate that these states are predominantly three-quasiparticle in character. We conclude that there is no need to introduce intruder configurations to describe the low-lying level structure observed in ^{83}As.

Shell-model calculations of the even-*A*, neutron-rich Zn isotopes were carried out with active protons in the $f_{5/2}$, $p_{3/2}$, $p_{1/2}$, and $g_{9/2}$ subshells and with the neutrons confined to the $p_{1/2}$ and $g_{9/2}$ subshells. Reasonable agreement is obtained between experiment and the calculation for excited states below 2.0 MeV, for which the calculated proton strength is almost entirely in the $f_{5/2}$ and $p_{3/2}$ orbitals. However, calculations involving the full *N*,*Z* = 28 to 50 shell do not reproduce the observed decrease in the energy of the 2_1^+ states for the even-*A* Zn isotopes beyond ^{68}Zn. These states drop to an essentially constant value of ≈ 0.6 MeV, which cannot yet be explained in terms of simple shell-model considerations. It would be of great interest to study the structure of ^{78}Zn, about which no information presently exists. Even if a study of ^{78}Zn were to reveal only the energy of the $2_1{}^+$ state, this alone would provide significant new information on the behavior of the very neutron-rich Zn isotopes.

In conclusion, calculations indicate that the shell closures at $Z = 28$ and $N = 50$ are probably still valid for very neutron-rich nuclei. Improved calculations are necessary for the nuclei just above $Z = 28$ in the $N = 28$ to 50 shell to provide further justification for this belief. However, it is still reasonable to assume that the ^{78}Ni region is doubly magic.

ACKNOWLEDGMENTS

The authors wish to thank a large number of people for their help in this work. B. H. Wildenthal kindly supplied the shell-model code used for the surface-delta calculations and Wildenthal and X. Ji carried out the FP calculation for ^{83}As. D. S. Brenner helped in the measurements of the Cu decays. We are especially indebted to A. Piotrowski, who designed the plasma ion source that allowed us to study Cu activities. We acknowledge fruitful conversations with R. F. Casten concerning the systematics of the even-A Zn isotopes. This work was supported by the U. S. Department of Energy under contracts W-7405-ENG-82 and DE-AC02-76CH00016.

REFERENCES

1. B. H. Wildenthal and W. Chung, *Phys. Rev. C* **22**, 2260 (1980).
2. T. Bjornstad, M. J. G. Borge, J. Blomqvist, R. D. Von Dincklage, G. T. Ewan, P. Hoff, B. Jonson, K. Kawade, A. Kerek, O. Klepper, G. Lovhoiden, S. Mattson, G. Nyman, H. L. Ravn, G. Rudstam, K. Sistemich, O. Tengblad, and ISOLDE Collaboration, *Nucl. Phys.* **A453**, 463 (1986).
3. K. Kawade, K. Sistemich, G. Battistuzzi, H. Lawin, K. Shizuma, and J. Blomqvist, *Z. Phys.* **A308**, 33 (1982).
4. B. Fogelberg and J. Blomqvist, *Nucl. Phys.* **A429**, 205 (1984).
5. J. Blomqvist, A. Kerek, and B. Fogelberg, *Z. Phys.* **A314**, 199 (1983).
6. M. Bernas, P. Dessagne, M. Langevin, G. Parrot, F. Pougheon, E. Quiniou, and P. Roussel, *J. Phys. Lett. (Paris)* **45**, 851 (1984).
7. P. Hoff and B. Fogelberg, *Nucl. Phys.* **A368**, 210 (1981).
8. R. L. Gill, R. F. Casten, D. D. Warner, A. Piotrowski, H. Mach, J. C. Hill, F. K. Wohn, J. A. Winger, and R. Moreh, *Phys. Rev. Lett.* **56**, 1874 (1986).
9. J. A. Winger, J. C. Hill, F. K. Wohn, R. Moreh, R. L. Gill, R. F. Casten, D. D. Warner, A. Piotrowski, and H. Mach, *Phys. Rev. C* **36**, 758 (1987).
10. G. Rudstam, P. Aagaard, P. Hoff, S. Johansson, and H.-U. Swicky, *Nucl. Instr. Meth.* **186**, 365 (1981).
11. B. Ekström, B. Fogelberg, P. Hoff, E. Lund, and A. Sangariyavanish, *Phys. Scr.* **34**, 614 (1986); a report of this work was given in E. Lund, K. Aleklett, B. Fogelberg, and A. Sangariyavanish, in *Atomic Masses and Fundamental Constants*, Vol. 26 of *THD Schriftenreihe Wissenschaft and Technik*, Ed. by O. Klepper (GSI, Darmstadt, 1984), p. 102.
12. A. Piotrowski, R. L. Gill, and D. C. McDonald, *Nucl. Instr. Meth.* **B26**, 249 (1987).
13. J. A. Winger, J. C. Hill, F. K. Wohn, R. L. Gill, X. Ji, and B. H. Wildenthal, *Phys. Rev. C* **38**, 285 (1988).
14. E. C. May and S. A. Lewis, *Phys. Rev. C* **5**, 117 (1972); A. Pfeiffer, G. Mairle, K. T. Knopfle, T. Kihm, G. Seegert, P. Grabmayr, G. J. Wagner, V. Bechtold, and L. Friedrich, *Nucl. Phys.* **A455**, 381 (1986).
15. M. Zendel, N. Trautmann, and G. Herrmann, *J. Inorg. Nucl. Chem* **42**, 1387 (1980).

16. P. Luksch and J. W. Tepel, *Nucl. Data Sheets* **27**, 389 (1979).
17. J. Blomqvist and L. Rydstrom, *Phys. Sci* **31**, 31 (1985).
18. J. B. Ball, J. B. McGrory, and J. S. Larson, *Phys. Lett.* **41B**, 581 (1972).
19. D. H. Gloeckner and F. J. D. Serduke, *Nucl. Phys.* **A220**, 477 (1974).
20. L. R. Medsker, H. T. Fortune, S. C. Headley, and J. N. Bishop, *Phys. Rev. C* **11**, 474 (1975).
21. J. F. Harrison and J. C. Hiebert, *Nucl. Phys.* **A185**, 385 (1972).
22. X. Ji and B. H. Wildenthal, *Phys. Rev. C* **37**, 1256 (1988).
23. X. Ji and B. H. Wildenthal, *Phys. Rev. C* **40**, 389 (1989).
24. W.-D. Schmidt-Ott, P. Koschel, F. Meissner, U. Bosch, E. Runte, R. Kirchner, H. Folger, O. Klepper, E. Roeckl, D. Schardt, and K. Rykaczewski, in *Proceedings of the Fifth International Conference on Nuclei Far from Stability* AIP Conf. Proc. No. 164, Ed. by I.S. Towner (AIP, New York, 1988), p. 365.
25. M. Bernas, P. Armbruster, J. P. Bocquet, R. Brissot, H. R. Faust, and P. Roussel, in Ref. 24, p. 768.
26. E. Lund, B. Ekström, B. Fogelberg, and G. Rudstam, in Ref. 24, p. 578.
27. E. Runte, W.-D. Schmidt-Ott, P. Tidemond-Petersson, R. Kirchner, O. Klepper, W. Kurcewicz, E. Roeckl, N. Kaffrell, P. Peuser, K. Rykaczewski, M. Bernas, P. Dessagne, and M. Langevin, *Nucl. Phys.* **A399**, 163 (1983).
28. M. Bernas, P. Dessagne, M. Langevin, J. Payet, F. Pougheon, R. Roussel, W.-D. Schmidt-Ott, P. Tidemand-Petersson, and M. Girod, *Nucl. Phys.* **A413**, 363 (1984).
29. R. Haupt, C.-A. Wiedner, G. J. Wagner, K. Wannebo, T. S. Bhatia, H. Hafner, R. Maschum, W. Saathoff, and S. T. Thornton, *Z. Phys. A* **317**, 193 (1984).
30. J. A. Winger, J. C. Hill, F. K. Wohn, E. K. Warburton, R. L. Gill, A. Piotrowski, and D. S. Brenner, *Phys. Rev. C* **39**, 1976 (1989).
31. P. L. Reeder, private communication.
32. P. Andersson, L. P. Ekstrom, and J. Lyttkens, *Nucl. Data Sheets* **48**,251 (1986).
33. M. L. Halbert, *Nucl. Data Sheets* **26**, 5 91979).
34. M. L. Halbert, *Nucl. Data Sheets* **28**, 179 (1979).
35. N. J. Ward and F. Kerns, *Nucl. Data Sheets* **39**, 1 (183).
36. M. R. Bhat, *Nucl. Data Sheets* **55**, 1 (1988).
37. M. R. Bhat, *Nucl. Data Sheets* **51**, 95 (1987).
38. M. M. King, *Nucl. Data Sheets* **56**, 1 (1989).
39. B. A. Brown, A. Etcheogoyen, W. D. M. Rae, and N. S. Godwin, OXBASH, 1984 (unpublished).
40. A. Hosaka, K.-I. Kubo, and H. Toki, *Nucl. Phys.* **A444**, 76 (1985).
41. B. A. Brown, private communication.
42. F. J. D. Serduke, R. D. Lawson, and D. H. Gloeckner, *Nucl. Phys.* **A256**, 45 (1976).
43. J. E. Koops and P. W. M. Glaudemans, *Z. Phys.* **AZ80**, 181 (1977).

9. Nuclear Spectroscopy in the Rare-Earth Region Near the Proton Drip Line

K. S. Toth
Oak Ridge National Laboratory
Oak Ridge, TN 37831, USA

J. M. Nitschke, P. A. Wilmarth, and K. S. Vierinen*
Lawrence Berkeley Laboratory
Berkeley, CA 94720, USA

ABSTRACT

We have used the isotope separator facility OASIS, on-line at the Lawrence Berkeley Laboratory SuperHILAC, to investigate rare-earth nuclei close to the proton drip line. Single-particle states near the 82-neutron shell have been examined and their excitation energies determined. Numerous new isotopes, isomers, and β-delayed proton emitters have been discovered. In addition, the α-decay properties of nuclides with $N \geqslant 84$ have been reexamined; this has led to the discovery of several previously unobserved α transitions. The overall experimental program is summarized and some recently obtained results on ^{145}Dy, ^{147}Er, ^{147}Tm, ^{153}Lu, ^{155}Lu, and ^{157}Lu are discussed.

9.1 INTRODUCTION AND EXPERIMENTAL METHOD

With the use of the OASIS separator facility,[1] on-line at the Lawrence Berkeley Laboratory SuperHILAC, we have investigated the decay properties of numerous short-lived neutron-deficient rare-earth nuclei with $65 \leqslant Z \leqslant 71$. Figure 1 shows a portion of the nuclidic chart which encompasses the mass region where these radioactivities are located. Some of the nuclei, on the lighter side of the 82-neutron shell, lie on or close to the proton drip line. For many of them, β-delayed proton (and in a few instances direct-proton) emission becomes a probable mode of decay. Note that most of the nuclei in Fig. 1 with $N \geqslant 84$ are α-particle emitters. This is due to the extra enhancement in decay energies that comes about as a result of the $N = 82$ shell. Because of the shell's stability α-decay energies reach a maximum for $N = 84$ nuclei, and no α emission has been observed in rare-earth isotopes with $N < 84$. Table I lists the investigated radioactivities together with their half-lives and spin assignments. We should add that, in a separate experimental program, the OASIS separator has also been used to study other proton-rich rare earth nuclei, primarily those with $Z \leqslant 66$ (see Ref. 2).

The isotopes in Table I were produced in fusion reactions induced by bombardments of ^{96}Ru, ^{92}Mo, ^{94}Mo, ^{95}Mo, ^{96}Mo, and ^{93}Nb with ^{64}Zn and ^{58}Ni projectiles.

* On leave from the University of Helsinki, SF- 00170, Finland

Exotic Nuclear Spectroscopy, Edited by W. C. McHarris
Plenum Press, New York, 1990

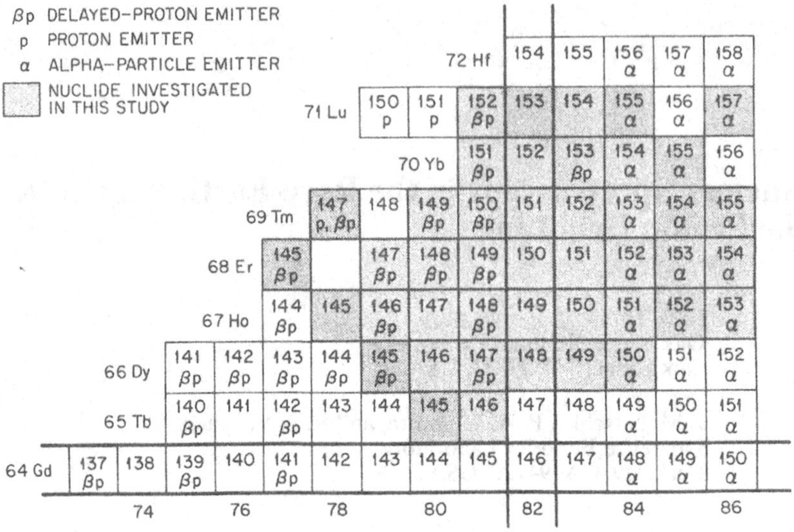

Figure 1. Portion of nuclidic chart where isotopes investigated in this experimental program are indicated by shaded squares.

Reaction products recoiling out of the (2-3)-mg/cm^2 targets were stopped in a Ta catcher foil located inside a surface ionization source. After diffusing out of the catcher and being ionized, the recoils were accelerated to 50 keV and mass separated. Products of a single mass were then selected by an analyzing slit in the focal plane of the separator, transported ionoptically to a fast-cycling, computer-controlled, tape system, and positioned between an array of detectors (Fig. 2). This consisted of a Si particle ΔE-E telescope and a hyperpure Ge detector facing the radioactive layer, with a 1-mm thick plastic scintillator and an n-type Ge detector (relative efficiency of 52%) located on the other side of the tape. In addition, a 24% n-type Ge detector, oriented at 90° with respect to the other two Ge detectors, was placed $\approx$4.5 cm from the radioactive source. Coincidences between particles, γ rays, x rays, and positrons were recorded in an event-by-event mode, with all events tagged with a time signal for half-life information. Singles γ-ray data were acquired with all three Ge detectors. While counting was in progress, a fresh source of radioactivities was collected and then transported to the detector station when the counting interval had terminated. Time intervals were selected on the basis of known or expected half-lives of the evaporation residues of interest.

9.2 RESULTS AND DISCUSSION

9.2.A Delayed Proton Emission for N = 81 Precursors

As a result of high-level densities, β-delayed-proton decay originating from nuclei with $A \gtrsim 80$ yields proton spectra that are basically featureless. Near major closed shells, however, pronounced peaks have been observed in some delayed-proton spectra. Are these peaks due to nuclear structure effects or to a fluctuation phenomenon which can be explained by a statistical model approach?

Table I. Isotopes Studied.

Isotope	J^{π}	$t_{1/2}$(s)	Delayed Protons
^{145}Er[a]		0.9(3)	yes[b]
^{145}Ho[a]	$11/2^-$	2.4(1)	
^{145}Dy	$11/2^-$	15(2)	yes
^{145}Dy[c]	$1/2^+$	8(2)	yes[b]
^{146}Ho		3.6(3)	yes[b]
^{146}Dy	0^+	29(3)	
^{146}Tb	1^+	≈8	
^{147}Tm[d]	$11/2^-$	≈0.5	
^{147}Er	$11/2^-$	2.5(2)	yes
^{147}Er[c]	$1/2^+$		yes[b]
^{147}Ho	$11/2^-$	5.8(2)	
^{147}Ho[c]	$(1/2^+)$		
^{148}Er	0^+	4.4(2)	yes[b]
^{148}Ho	6^-	9.7(3)	yes[b]
^{148}Ho	1^+	2.2(11)	yes[b]
^{149}Tm[a]	$(11/2^-)$	0.9(2)	yes[b]
^{149}Er	$11/2^-$	9(1)	yes
^{149}Er	$1/2^+$		yes
^{149}Ho	$11/2^-$	21(1)	
^{149}Ho	$1/2^+$	54(5)[e]	
^{150}Tm	(6^-)	2.2(2)[e]	yes[b]
^{150}Tm	(1^+)		yes[b]
^{150}Er	0^+	20(1)	
^{151}Yb[c]	$11/2^-$	1.6(1)	yes[b]
^{151}Yb[a]	$1/2^+$		yes[b]
^{151}Tm	$11/2^-$	4.3(2)	
^{151}Tm[c]	$1/2^+$	8	
^{152}Lu[a]	(6^-)	0.7(1)	yes[b]
^{152}Yb	0^+	3.1(2)	
^{152}Tm	(2^-)		
^{153}Lu	$11/2^-$	0.9(2)	
^{153}Yb	$7/2^-$	3.9(1)	yes[b]
^{153}Tm	$11/2^-$	1.7(2)	
^{153}Tm	$1/2^+$	(2.5)	
^{154}Lu	(7^+)	1.2(1)	yes[b]
^{154}Yb[d]	0^+	≈0.4	
^{155}Lu	$11/2^-$	0.066(7)	
^{155}Lu	$(1/2^+,3/2^+)$	0.14(2)	
^{155}Yb[d]	$7/2^-$	1.7(1)	
^{155}Tm	$11/2^-$	21.6(2)	
^{155}Tm[c]	$1/2^+$	44.(4)	
^{157}Lu[d]	$11/2^-$	≈5	
^{157}Lu[c,d]	$(1/2^+,3/2^+)$	≈6	

[a]Isotope discovered in this study.

[b]Delayed-proton emission observed for the first time.

[c]Isomer discovered in this study.

[d]Beta-decay branch of nuclide identified for the first time.

[e]New half-life measured in this investigation.

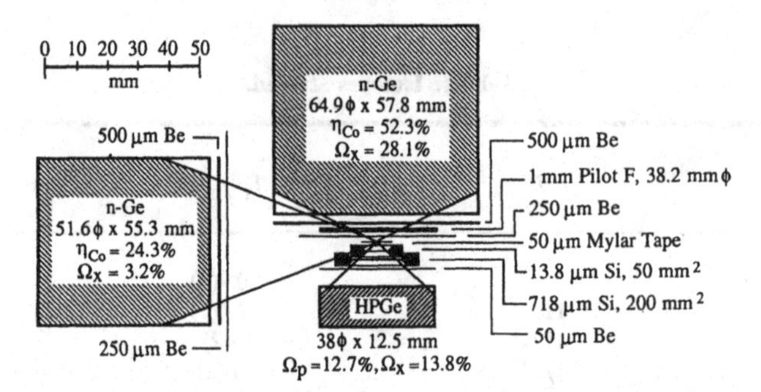

Figure 2. Detector arrangement used in nuclear spectroscopic studies at the OASIS on-line facility.

To shed light on this question we investigated spectra arising from an isotonic sequence of six β-delayed-proton precursors with $N = 81$ and $66 \leqslant Z \leqslant 71$. They are shown in Fig. 3, where one notes that the mean energy values of the spectra from the odd-odd isotopes, ^{148}Ho, ^{150}Tm, and ^{152}Lu, are higher than those of the spectra from the even-Z nuclei, ^{147}Dy, ^{149}Er, and ^{151}Yb. This is explained by the fact that for the odd-odd precursors, Gamow-Teller β decay occurs without breaking a proton pair, so that the proton-emitting nucleus is left in a state of high excitation energy (and high-level density); the resultant spectra are adequately described[3] by the statistical model. In contrast, the proton spectra

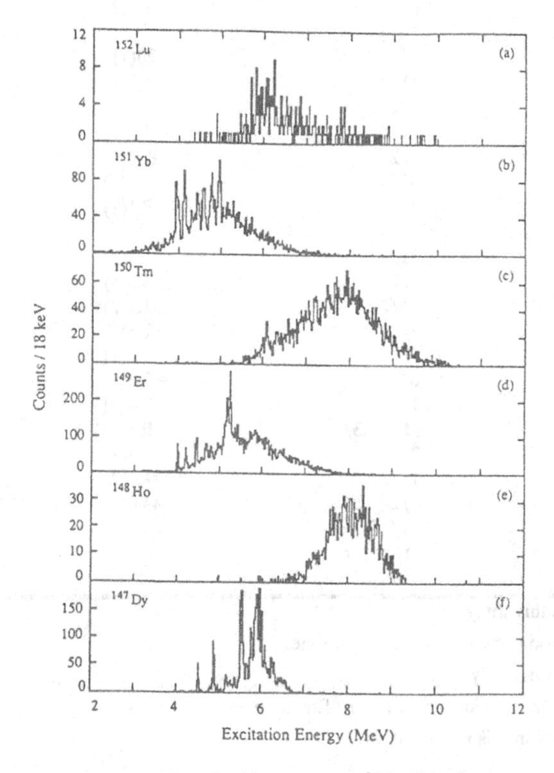

Figure 3. Beta-delayed proton spectra of $N = 81$ precursors.

of the even-Z precursors exhibit pronounced structure which has been shown[4] to be associated with the decay of the $1/2^+$ ground states. The structureless components of these spectra are associated primarily with the $11/2^-$ isomeric states in the same nuclei (because of angular-momentum considerations, these protons originate[4] from levels at high excitation energies). Gamma-ray decay studies, β-strength-function measurements, and calculations of state densities of Gamow-Teller strength distributions have led us to suggest[5] that some of the structure near $N = 82$ arises from the preequilibrium decay of doorway states populated in β decay and that it can be understood within the framework of a model based on the work by Bloch and Feshbach.[6]
These peaks disappear from observed spectra almost as soon as the β-decay daughter no longer has a major closed-shell configuration, though there may be weak structures appearing in the spectra of the $N = 79$ precursors, ^{145}Dy and ^{147}Er. This and other points are addressed in a recent thesis[7] where the results obtained at the OASIS facility for delayed-proton precursors, ranging from ^{119}Ba to ^{154}Lu, are described and discussed.

9.2.B Single-Particle States Near $N = 82$

The study of low-lying states in nuclei near closed shells is of interest because these levels are relatively pure shell-model states and they provide an opportunity for comparison with calculations based on various models. Nuclei near $N = 82$ and $Z = 64$ are particularly interesting because they should provide us with information about the underlying spherical proton structure in isotopes located halfway between the major proton shells of 50 and 82. Much of our experimental effort has been focused on obtaining structure information for nuclei in this mass region. For example, in Ref. 8 we presented experimental systematics of proton states in odd-Z, $N = 82$ nuclei and of neutron states in even-Z, $N = 81$ nuclei, together with results of spherical Hartree-Fock-Bogoliubov calculations. Here we discuss briefly some information that we have garnered from the β decays of ^{147}Tm, ^{145}Dy, ^{147}Er, and ^{153}Lu.

The $h_{11/2}$ ground state of ^{147}Tm has been known[9] for several years to be a direct proton emitter. Cross-section measurements and barrier-penetration calculations indicate that this level should decay mainly by β decay. In 1987 we reported[10] the identification of this β-decay branch through the observation of Er K x rays whose intensity decreased in a manner consistent with the ^{147}Tm half-life of $\approx$0.5 s. Further analysis of the data has revealed, in addition to the Er K x rays, an 80.9-keV γ ray that follows the β decay of ^{147}Tm. This transition, together with the Er $K_{\alpha 1}$ peak, can be seen in Fig. 4*a*, where γ rays in coincidence with annihiliation radiation are displayed. The spectrum was measured during 1.28-s counting cycles designed to emphasize the production of ^{147}Tm *vis-a-vis* isobars with longer half-lives. Indeed, the 80.9-keV γ ray is scarcely visible in Fig. 4*b*, where data, once again in coincidence with annihilation radiation but accumulated during 4-s cycles, are displayed.

We believe that the 80.9-keV γ ray is the transition that proceeds from the first-excited, $d_{3/2}$ neutron-hole level in ^{147}Er to the $s_{1/2}$ ground state. Our assignment is based on energy systematics of $s_{1/2}$, $d_{3/2}$, and $h_{11/2}$ neutron-hole states in this mass region that were recently updated[11] following our study of the β decay of ^{145}Ho ($h_{11/2}$), the $N = 78$ isotone just below ^{147}Tm. Figure 5 shows these systematics. One notes that the 80.9-keV energy for the $d_{3/2}$ level in ^{147}Er fits well into the overall picture, i.e., this level lies at 1.58 keV in ^{141}Sm, then increases in excitation energy to 45.1 and 66.3 keV in ^{143}Gd and ^{145}Dy, respectively. A discussion of these single-neutron level systematics is presented in Ref. 11.

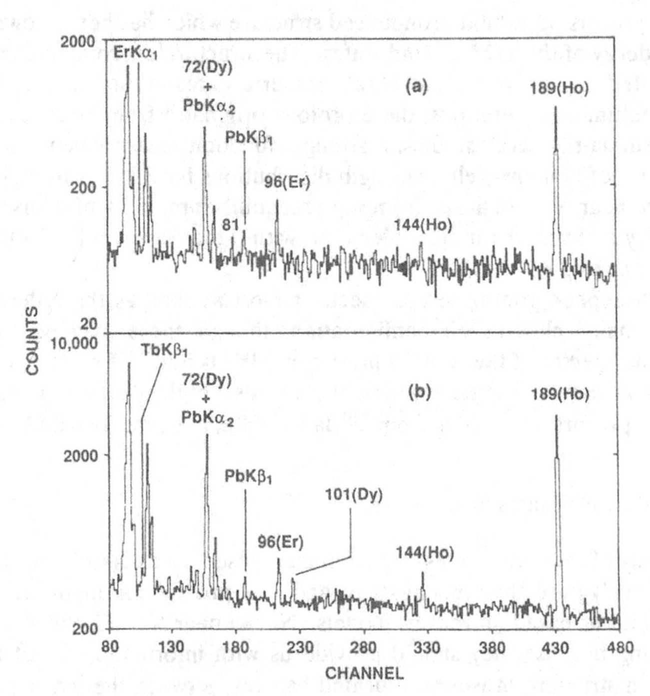

Figure 4. Gamma-ray spectra observed in coincidence with annihilation radiation for $A = 147$ nuclides during 1.28-s a) and 4-s b) counting cycles. The 81-keV γ ray in part a) is assigned to ^{147}Tm β decay. Transitions assigned to ^{147}Dy, ^{147}Ho, and ^{147}Er are labeled by their elemental symbols.

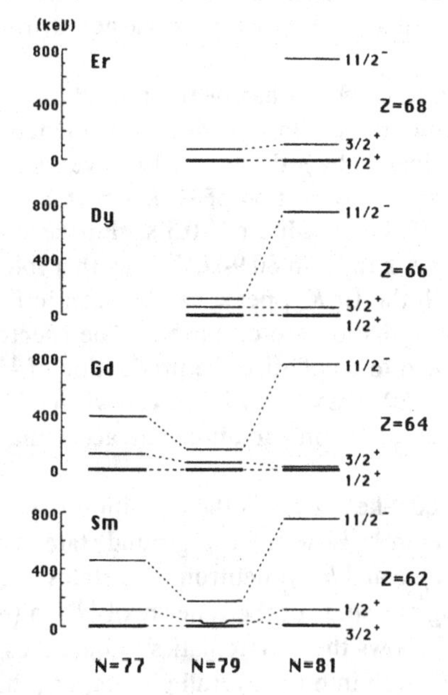

Figure 5. Systematics of $s_{1/2}$, $d_{3/2}$, and $h_{11/2}$ neutron-hole states in Sm, Gd, Dy, and Er nuclei with $N = 77$, 79, and 81.

The radionuclides ^{145}Dy and ^{147}Er are known[12] β-delayed-proton precursors. However, little information is available concerning levels in their β-decay daughters, ^{145}Tb and ^{147}Ho. No γ rays have been assigned to ^{147}Er decay, while a total of only four transitions have been observed[13] to follow ^{145}Dy decay. We have investigated the β decay of the $s_{1/2}$ ground and the $h_{11/2}$ isomeric states in ^{145}Dy and ^{147}Er. Here we report on the identification of low-lying single-proton levels in ^{145}Tb and ^{147}Ho based on the observation of three γ rays in cascade, characteristic of odd-*Z*, even-*N* nuclei in this region, that connect and deexcite the $g_{7/2} \rightarrow d_{5/2} \rightarrow d_{3/2} \rightarrow s_{1/2}$ states.

Figures 6*a* and 6*b* show γ-ray spectra in coincidence with the $g_{7/2} \rightarrow d_{5/2}$ transitions in ^{145}Tb (184.5 keV) and ^{147}Ho (616.7 keV). The energies of the $d_{5/2} \rightarrow d_{3/2}$ and $d_{3/2} \rightarrow s_{1/2}$ transitions are 145.1 and 108.1 keV in ^{145}Tb [Fig. 6*a*] and 292.0 and 96.1 keV in ^{147}Ho. There is also a crossover 253.1-keV transition, $d_{5/2} \rightarrow s_{1/2}$, in ^{145}Tb. We show updated single-proton level systematics for N = 80, 82, and 84 (Tb, Ho, and Tm) nuclei in Fig. 7. One sees that the new states in ^{145}Tb and ^{147}Ho follow the trend established by the previously available information, i.e., the gap between the $d_{5/2}$ and the $d_{3/2}$ orbitals increases as the atomic number increases. Also, we indicate the $s_{1/2}$ orbital as the ground state in Tb nuclei; the $h_{11/2}$ state, which is metastable for $Z \leqslant 65$, becomes the ground state for the Ho and Tm isotopes (see Ref. 14).

Part of our investigation included a search for the β decay of ^{153}Lu, which should populate single-neutron states in the N = 83 nucleus ^{153}Yb. A low production yield was anticipated for the ^{92}Mo(^{64}Zn, *p*2*n*) channel, and indeed there was no obvious indication of Yb *K* x rays. Observation of their presence was severely hampered by the much more intense K_{α} and K_{β} x rays of elements with $Z \leqslant 69$. However, ^{153}Yb levels have recently been studied[15] via in-beam γ ray spectroscopy. In addition to high-lying, high-spin states, the authors located the $h_{9/2}$ and $i_{13/2}$ neutron levels at 567 and 1202 keV, respectively. The

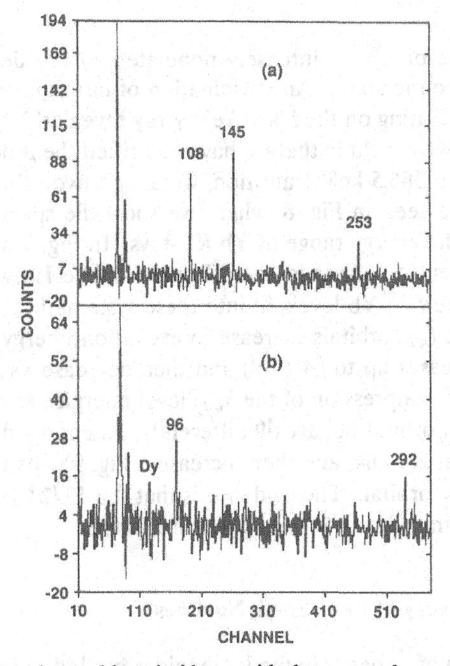

Figure 6. Gamma-ray spectra observed in coincidence with the $g_{7/2} \rightarrow d_{5/2}$ transitions in ^{145}Tb and ^{147}Ho following the β decays of ^{145}Dy a) and ^{147}Er b).

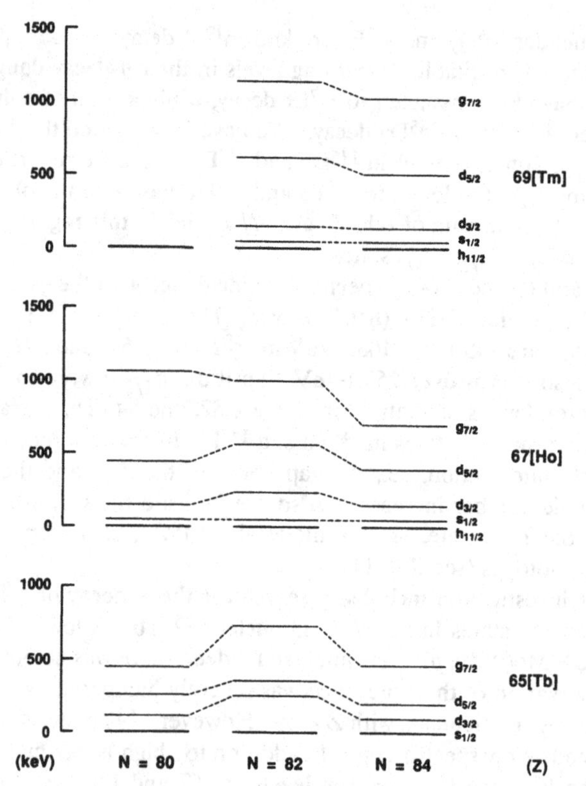

Figure 7. Systematics of $s_{1/2}$, $d_{3/2}$, $d_{5/2}$, and $g_{7/2}$ proton states in Tb, Ho, and Tm nuclei with N = 80, 82, and 84.

$h_{9/2}$ state should be the one most intensely populated by the decay of the $11/2^-$ ($h_{11/2}$ proton orbital) ^{153}Lu ground state. An examination of our data showed the presence of a weak 566.6-keV γ ray. Gating on the 566.5-keV γ ray revealed Yb K x rays in coincidence. Based on this evidence we conclude that we have identified the β decay of ^{153}Lu.

In addition to the 566.5-keV transition, we assign two other γ rays to the decay of ^{153}Lu. All three can be seen in Fig. 8, where we show the spectrum observed in coincidence with photons in the energy range of Yb K x-rays. In Fig. 9 we have plotted the energies of $h_{9/2}$ and $i_{13/2}$ neutron states in even-Z nuclei above Eu with N of 83, 85, and 87. The 566.5- and 1202.0-keV ^{153}Yb levels fit into these systematics. It has been noted[16] that the $h_{9/2}$, $p_{1/2}$, $p_{3/2}$, and $f_{5/2}$ orbitals increase in excitation energy with respect to the $f_{7/2}$ ground states as Z increases up to 64 (Gd) and then decrease as Z increases further (see Fig. 9). Also, we note a compression of the $h_{9/2}$ level energies as the neutron number becomes larger. The $i_{13/2}$ orbital behaves[16] differently; its energy decreases with increasing Z, reaches a minimum at Z = 64, and then increases (Fig. 9). Its behavior with N also differs from that of the $h_{9/2}$ orbital. The evidence is that the $13/2^+$ levels are not pure single-particle states; they are mixed with 3^- octupole core states.

9.2.C Alpha-Particle Decay of Rare-Earth Nuclides

The investigation of α decay in the lanthanides has led to the discovery of many isotopes, the determination of nuclear masses and isomeric excitation energies, and the first

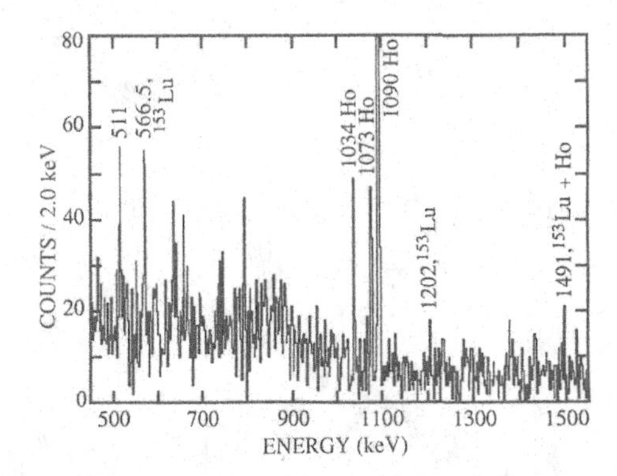

Figure 8. Gamma-ray spectrum measured at A = 153 in coincidence with photons in the energy range of K_{α_1} x rays from Yb.

observation of the subshell closure at Z = 64. We have reinvestigated some of the α emitters and have determined more precisely a number of α-branching ratios. Also, we have recently observed[14] fine structure in the α spectrum of ^{153}Tm. These new transitions populate the $d_{3/2}$ (220.4-keV) and $d_{5/2}$ (564.4-keV) single-proton levels[17] in ^{149}Ho. Herein we review briefly our study of the ^{155}Lu and ^{157}Lu α-decays.

Figure 10 shows the α spectrum that we recorded for A = 155 during collection and assay cycles of 1.28 s. Above the intense 5.194-MeV α peak, which belongs to ^{155}Yb decay, there are two weak α groups with energies of 5.579 ± 0.005 and 5.648 ± 0.005 MeV. The higher energy group, presumed to follow the α decay of the $h_{11/2}$ ground state ($t_{1/2}$ = 66 ms) of ^{155}Lu, has been known[18] since 1965. Recently, Hofmann et al.[19] observed α decay (E_α = 5.575 ± 0.010 MeV) from a second low-lying level in ^{155}Lu. Our data shown in Fig. 10 therefore confirm the existence of this new α-emitting level, which we believe is either the $s_{1/2}$ or $d_{3/2}$ single-proton state. No half-life for it was reported in Ref. 19; we measure a $t_{1/2}$ of 140 ± 20 ms. Because of the low cross-section for the ^{94}Mo(^{64}Zn, $p2n$) reaction and the fact that ^{155}Lu decays mainly by α emission, we saw no evidence for the isotope's β-decay branch.

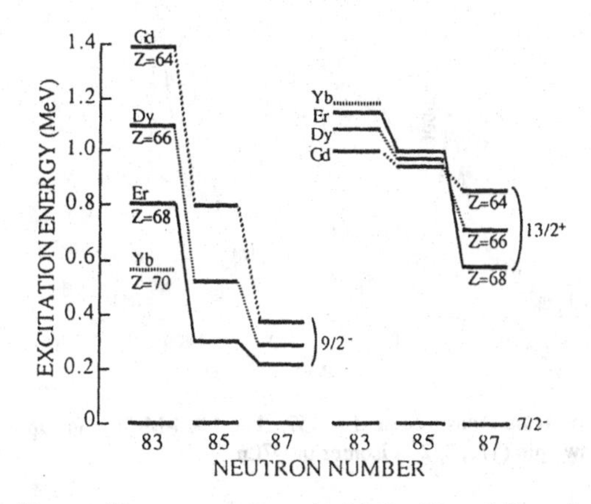

Figure 9. Systematics of $h_{9/2}$ and $i_{13/2}$ neutron states in Gd, Dy, Er, and Yb nuclei with N = 83, 85, and 87.

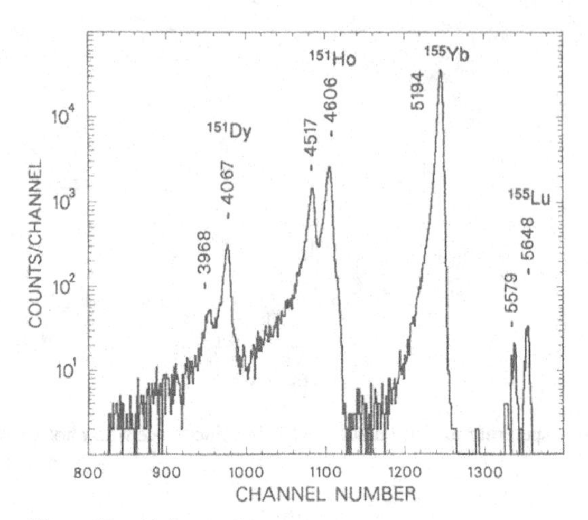

Figure 10. Alpha-particle spectrum measured at A = 155.

Figure 11 shows the α spectrum that we measured for A = 157. In addition to α particles emitted by ^{149}Tb, ^{153}Ho, ^{157}Yb, ^{153}Er, ^{153}Tm, and the known[20] ^{157}Lu ground state (presumably $h_{11/2}$), we observe a new α group, 4.924 ± 0.020 MeV, which we assign to the decay of the previously unreported low-spin (1/2, 3/2$^+$) isomer in ^{157}Lu. Preliminary analysis of our data indicates that the half-life of this isomer is ≈6 s rather than the 5.4 ± 0.2 s value adopted[20] for the ground state. Information on the β decay properties of the two ^{157}Lu levels has hitherto not been available.[20] We observe numerous γ rays following ^{157}Lu β decay and are in the process of constructing a scheme for the ^{157}Yb levels that are populated. Based on this decay scheme, the α branches and α-decay rates for these two α emitters will be determined. It may then be possible to say if the ^{157}Lu isomer is a $d_{3/2}$ or $s_{1/2}$ proton level.

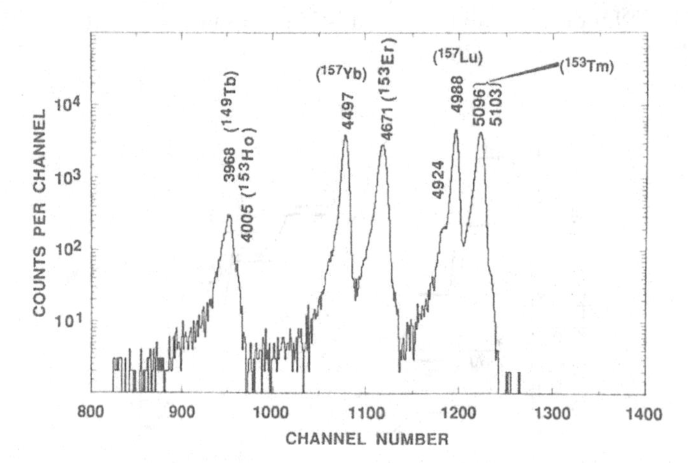

Figure 11. Alpha-particle spectrum measured at A = 157. The 4.924-MeV α group is assigned to the decay of the previously unobserved low-spin (1/2, 3/2$^+$) isomer in ^{157}Lu.

ACKNOWLEDGMENTS

We wish to thank Y.A. Akovali, L.F. Archambault, F.T. Avignone III, R.M. Chasteler, R.B. Firestone, J. Gilat, M.O. Kortelahti, A.A. Shihab-Eldin, D.C. Sousa, and A.A. Wydler for their contributions to this experimental program. We also express thanks to the staff of the SuperHILAC for providing smooth operation of the accelerator. Oak Ridge National Laboratory is operated by Martin Marietta Energy Systems, Inc., for the U.S. Department of Energy under Contract No. DE-AC05-840R21400. Work at the Lawrence Berkeley Laboratory is supported by the Director, Office of Energy Research, Division of Nuclear Physics of the Office of High Energy and Nuclear Physics of the U.S. Department of Energy under Contract DE-AC03-76SF00098.

REFERENCES

1. J.M. Nitschke, *Nucl. Instr. Meth.* **206**, 341 (1983).
2. J.M. Nitschke et al., in *Nuclei Far From Stability* - 1987, Ed. by I.S. Towner, AIP Conference Proceedings No. 164 (American Institute of Physics, New York, 1988), p. 697.
3. J.M. Nitschke et al., *Phys. Rev. C* **37**, 2694 (1988).
4. K.S. Toth et al., *Phys. Lett.* B178, **150** (1986).
5. J.M. Nitschke et al., *Phys. Rev. Lett.* **62**, 2805 (1989).
6. B. Bloch and H. Feshbach, *Ann. Phys.* **23**, 47 (1963).
7. P.A. Wilmarth, Ph.D. thesis, University of California, Berkeley, 1988, Lawrence Berkeley Laboratory Report LBL-26101.
8. K.S. Toth et al., *Phys. Rev. C* **32**, 342 (1985).
9. P.O. Larsson et al., *Z. Phys.* **A314**, 9 (1983).
10. K.S. Toth et al., in *Nuclei Far From Stability* - 1987, Ed. by I.S. Towner, AIP Conference Proceedings No. 164 (American Institute of Physics, New York 1988), p. 718.
11. K.S. Vierinen et al. *Phys. Rev. C* **39**, 1972 (1989).
12. D. Schardt et al., in *Proceedings of the 7th International Conference on Atomic Masses and Fundamental Constants,* Darmstadt-Seeheim, 1984, ed. by O. Klepper (Technische Hochshule, Darmstadt, 1984), p. 229.
13. L.K. Peker, *Nucl. Data Sheets* **49**, 1 (1986).
14. K.S. Toth et al., *Phys. Rev. C* **38** 1932 (1988).
15. J.H. McNeill et al., *Z. Phys.* **A332**, 105 (1989).
16. K.S. Toth et al., *Phys. Rev. C* **19**, 482 (1979).
17. R.B. Firestone et al., *Phys. Rev. C* **39**, 219 (1989).
18. R.D. Macfarlane, *Phys. Rev.* **137** B1448 (1965).
19. S. Hofmann et al., *Z. Phys.* **A333**, 107 (1989).
20. R.G. Helmer, *Nucl. Data Sheets* **55**, 71 (1988).

10. Shapes and Moments of Odd-A Pt Nuclei ($177 \leq A \leq 195$)

J. Rikovska
Clarendon Laboratory
Parks Road, OX1 3PU
Oxford, U. K.

ABSTRACT

The shape of transitional Pt nuclei has been of interest for a long time. Taking advantage of recent new experimental data on isotope shift and electromagnetic moments of odd-A Pt isotopes, together with more spectroscopic information, this paper presents the first comprehensive microscopic-model calculation of these isotopes. The Lund particle-triaxial-rotor model has been applied to both positive- and negative-parity energy levels, transition probabilities, and ground- and isomeric-state magnetic-dipole and electric-quadrupole moments in terms of two shape parameters, ε_2 and γ. The calculation indicates a *gradual* development in the shape of odd-A Pt nuclei from ^{177}Pt ($\gamma = 16°$) via triaxial $^{187\text{-}193}$Pt ($28.5° \leqslant \gamma \leqslant 35°$) to close-to-oblate ^{195}Pt ($\gamma = 52°$) on the grounds of overall agreement between calculated and experimental energy levels and electromagnetic properties. The previously suggested *sharp* prolate-oblate or prolate-triaxial shape change between ^{185}Pt and ^{187}Pt is thus not supported. On the other hand, an important role of the odd neutron, most likely stabilizing the γ-soft even-even Pt core at a particular value of γ, dependent on the neutron orbital, and creating in this way rigid non-axial shapes of odd-A Pt nuclei, is demonstrated.

10.1 INTRODUCTION

Data on structure of the ground and excited states of odd-A Pt isotopes ($99 \leqslant N \leqslant 115$) has been reported in numerous experimental studies. These involve energy levels and electromagnetic transitions from α decay,[1] β decay,[2-13] nuclear-orientation data,[14-22] lifetimes of isomeric states,[1,23,24] laser-spectroscopy measurements,[25-29] and nuclear-reaction data.[24,30-37]

There is extensive experimental (see, e. g., Refs. 38-45) and theoretical[46-49] information on even-even Pt nuclei, used as cores in the particle–triaxial-rotor calculation. Total potential-energy-surface calculations have predicted a complex shape development from light prolate (with ^{176}Pt possibly triaxial) to heavier oblate via γ-soft nuclei around $N = 108$. Experimental ground-state band E_4/E_2 ratios, summarized in Table I, are mostly close to 2.5, suggesting a γ-soft shape. Also, in all light even-even Pt nuclei ($N \leqslant 108$), the moment of inertia dependence on $(\hbar\omega)^2$ shows deviations from a straight line at lowest frequencies, which may indicate considerable softness of these nuclei.[39]

The particle–axially-symmetrical-rotor model has been applied to negative-parity bands in ^{185}Pt, yielding only a modest agreement with energy levels[6,7] and failing to ex-

Exotic Nuclear Spectroscopy, Edited by W. C. McHarris
Plenum Press, New York, 1990

Table I. Comparison of Deformation Parameters Extracted from Different Experiments and Model Calculations

A	ε_2^a	ε_2^b	$\lvert\beta_2\rvert^c$	$\lvert\beta_2\rvert^d$	$\lvert\beta_2\rvert^e$	$\lvert\langle\beta_2^2\rangle^{1/2}\rvert^f$	E_2 exp^g (MeV)	E_2 cal (MeV)	E_4/E_2 exp	E_4/E_2 cal
176		0.151	0.20(1)	0.22			0.264	0.093	2.14	3.22
177	0.270	0.232								
178		0.171	0.22(1)	0.24			0.170		2.51	
179		0.239								
180		0.225		0.24			0.153	0.109	2.68	3.27
181	0.240	0.232								
182		0.212		0.23			0.155	0.116	2.71	3.11
183	0.240	0.225				0.24(1)				
184		0.205	0.23(2)	0.22		0.21(1)	0.163	0.129	2.67	3.00
185	0.230	0.212				0.23(1)				
186		0.192	0.18(1)	0.20	0.188	0.20(1)	0.192	0.263	2.56	2.66
187	0.169	0.192				0.19(1)				
188		0.158	0.17(2)	0.18	0.170	0.18(1)	0.266	0.301	2.53	2.70
189	0.155	0.151				0.17(1)				
190		−0.151	0.17(2)	0.15	0.155	0.16(1)	0.296	0.315	2.49	2.67
191	0.150	−0.151				0.15(1)				
192		−0.144	0.16(1)	0.14	0.144	0.15(1)	0.317	0.272	2.48	2.78
193	0.150	−0.144				0.14(1)				
194		−0.138	0.14(1)	0.13	0.134	0.1434(26)	0.328	0.327	2.47	2.80
195	0.120	−0.144				0.14(1)				
196		−0.131	0.13(1)	0.12	0.125	0.13(1)	0.356		2.46	

a ε_2 values fitted in the present PTR calculation for ground states of $^{177-195}$Pt ($\gamma \neq 0,60°$, see Table III.
b ε_2 values calculated by Moller and Nix[66] for axially-symmetrical shape and $\varepsilon_4 \neq 0$ and $\varepsilon_6 \neq 0$.
c β_2 values extracted by Bengtsson et al.[48] from experimental $B(E2)$ values. To deduce $\lvert\beta_2\rvert$, for $^{176-188}$Pt, prolate shape and $B(E2, 8^+ \to 6^+)$; for $^{190-196}$Pt, oblate shape and $B(E2, 2^+ \to 0^+)$ were used.
d β_2 obtained as minima in the potential energy surfaces calculated with the Woods-Saxon potential. See Ref. 48 for details.
e Results of standard self-consistent HFB using Kumar-Baranger many-body wave-function[72] calculated without hexadecapole deformation with $\gamma = 0°, 20.2°$, and $60°$ for A = 186, 188, and 190-196, respectively. See Refs. 46 and 47 for details.
f β_2 values extracted from known $\delta \langle r^2 \rangle$ using weighted least-squares fit.[29]
g Experimental data taken from Refs. 38 (A = 176, 178), 39 (A = 180), 40 (A = 184), 43 ($182 \leq A \leq 192$), 41 (A = 186), 44 ($186 \leq A \leq 194$). Where not specified, current *Nucl. Data Sheets* were used.

plain ground-state magnetic moments of $^{185-191}$Pt (Ref. 27; see also Table II). Gnade et al.[8] used the particle–triaxial-rotor model for ^{187}Pt but did not obtain satisfactory results. On the other hand, successful calculations of positive-parity bands in $^{189-193}$Pt (see, e.g., Refs. 36, 37) have been reported using the Meyer-ter-Vehn particle–triaxial-rotor model. Hecht and Satchler used their particle–triaxial-rotor model to calculate ^{195}Pt.[50] They obtained a very good fit to lowest-lying energy levels but failed to reproduce transition probabilities. Yamazaki and Sheline[51] applied the Nilsson model and the Faessler-Greiner model to ^{195}Pt in order to calculate energy levels and spectroscopic factors. Dracoulis et al.[30] performed band-mixing calculations using a particle-axially-symmetrical-rotor model with coriolis coupling and variable moment of inertia. They obtained a reasonable fit to aligned angular momenta vs rotational frequency for the $1/2^-[521]$, $5/2^-[512]$, and $7/2^+[633]$ bands in ^{177}Pt. The only disagreement with experiment is that their calculations

Table II. Electromagnetic Moments of $^{177-193}$Pt

A	I^{π}	Q exp[a] (eb)	Q cal (eb)	μ exp (μ_N)	μ cal[h] (μ_N)	μ cal[i] (μ_N)	μ cal[j] (μ_N)	μ cal[k] (μ_N)
177	$5/2^-$		+3.00		−0.31 −0.52[m]			
177m	$1/2^-$				+0.46 +0.47[m]			
181	$1/2^-$				+0.45 +0.47[m]			
183	$1/2^-$			+0.521(27)[b]	+0.45 +0.47[m]			
183m	$7/2^-$		+3.35 +3.68[m]		+0.86 +0.91[m]			
185	$9/2+$	+4.3(5)[c] +2.1(9)[d]	+3.46 +3.93[m]	−0.83(1)[c] ±0.774(14)[d]	−0.93 −0.90[m]	−0.45	−0.73	−0.79
185m	$1/2^-$			+0.540(9)[b]	+0.46 +0.47[m]			
187	$3/2^-$	−1.10(8)[b] −1.13(5)[c] −0.85(40)[d]	−0.91 −0.89[p]	−0.425(19)[b] −0.397(5)[c] ±0.408(8)[d]	−0.36 −0.13[p]	+0.39	+0.53	+0.39
189	$3/2^-$	−1.21(21)[b] −1.03(5)[c] −0.65(26)[e]	−0.79 −0.80[p]	−0.441(8)[b] −0.421(5)[c] ±0.434(9)[e] ±0.427(9)[f]	−0.44 −0.24[p]	+0.39	+0.39	+0.31
191	$3/2^-$	−0.86(11)[b] −0.98(5)[c] −0.64(26)[e]	−0.74 −0.76[p]	−0.494(7)[b] −0.501(4)[c] ±0.500(10)[e] ±0.492(10)[f] ±0.491(11)[g]	−0.51 −0.31[p]	+0.39	+0.31	+0.19
193	$1/2^-$			+0.603(8)[b]	+0.51 +0.47[p]			

Data taken from Refs. 28(b), 27(c), 17(d), 21(e), 22(f), 18(g).

a Q_{exp} in Refs. 27 and 28 do not include the Sternheimer correction.

h Present work, g_{eff} = 0.7; for model parameters, see Table III.

i Calculation of Duong et al.[27], using Nilsson model and g_{eff} = 0.6.

j-k Calculation of Duong et al.[27], using particle-axially symmetrical rotor model with single-particle wavefunctions from HFB calculation; g_{eff} = 0.6, the core nucleus is $^{A-1}$Pt for (j) and $^{A+1}$Pt for (k). For more details, see Ref. 27.

m Present work, γ = 0.

p Present work, γ = 60°.

overestimated signature splitting as a function of spin. However, this fit is dependent on too many parameters, as Dracoulis et al. had to use a different value for the unperturbed moment of inertia for each band. Also, a coriolis attenuation factor of 0.85 and 0.6 was required for the $5/2^-[512]$ and $7/2^+[633]$ band fits, respectively.

The present calculation aims for a systematic investigation of shape and its evolution in the ground state and excited state configurations in $^{177\text{-}195}$Pt, using the Lund particle–triaxial-rotor-model.[52] We benefit from the extensive range of experimental data available at present except in ^{179}Pt, which for that reason was not included in the present study. The main interest of this work is concentrated on whether or not there is a sharp shape change between ^{185}Pt and ^{187}Pt, similar to the one in Hg (see, e. g., Ref. 53) and Au isotopes,[54] as has been suggested by several experimental groups.[16,17,25,27]

We stress that we are interested in the shape of odd-*A* Pt *ground states* and properties of *low-lying* excitations (E_{ex} < 1.5 MeV, $I \leqslant 21/2$ for yrast bands, E_{ex} < 0.5 MeV for non-yrast states). We do not attempt to understand the behaviour of higher-spin members of known rotational bands in this region. It has become apparent that the cranked shell model is much more appropriate for this purpose.[30-32,34] However, at low frequencies, considering just the lowest few members of a rotational band, we are more concerned with the relative position of bands heads close to the fermi level and with electromagnetic properties of the lowest-lying excited states.

This work supersedes our preliminary results for negative-parity states.[55] Analysis of both negative- *and* positive-parity states shows, especially for $^{189\text{-}193}$Pt where relatively few unambiguously assigned negative-parity states are available, that the value of γ close to that obtained from the best fit of positive-parity states can also be used successfully to reproduce the main features of the negative-parity states. Also, a slightly different constant in the moment of inertia, more suitable for the Pt region, has been used in the present work (see Section 10.2).

The paper starts in Section 10.2 with an outline of the main features of the model, together with a discussion of the physical constraints upon its parameters and their chosen values. In Section 10.3, we summarise the results of the calculation for negative-parity states, divided into yrast bands in $^{177\text{-}185}$Pt (Section 10.3.*A*), non-yrast negative-parity states in ^{183,185}Pt (Section 10.3.*B*), and low-lying negative-parity excitations in $^{187\text{-}195}$Pt (Section 10.3.*C*). Positive-parity bands based on the $i_{13/2}$ spherical state in $^{177\text{-}193}$Pt are discussed in Section 10.4, and the results for electromagnetic moments are covered in Section 10.5. Properties of the even-even cores used in the calculation are considered in Section 10.6. The final section gives summary remarks and conclusions.

10.2 CALCULATION PROCEDURE AND PARAMETERS OF THE MODEL

The model has been described in detail in several papers (see, e. g., Refs. 52, 56, 57). The single-particle hamiltonian used has the form,[52]

$$H_{sp} = H_{ho}(\varepsilon_2,\gamma) - \kappa(N)\hbar\omega_0\{2\mathbf{l}\cdot\mathbf{s} + \mu(N)(\mathbf{l}^2 - <\mathbf{l}^2>N)\}, \qquad (1)$$

where H_{ho} is the deformed harmonic-oscillator hamiltonian, and all the other symbols have their usual meaning. Diagonalization of the hamiltonian in the $|Nlj\Omega>$ basis and inclusion of pairing effects gives adiabatic quasi-particle energies and wave-functions written as an expansion,

$$\chi_\nu = \sum_{Nlj\Omega} C^{(\nu)}_{Nlj\Omega} \phi_{Nlj\Omega}, \tag{2}$$

and their conjugate states,

$$\tilde{\chi}_\nu = \sum_{Nlj\Omega} (-1)^{(j-\Omega)} C^{(\nu)}_{Nlj\Omega} \phi_{Nlj-\Omega}, \tag{3}$$

where ν is the sequential number of a wave-function χ_ν. The core hamiltonian is defined as

$$H_{core} = \sum_{i=1}^{3} (\hbar^2/2\Im_i)(\mathbf{I}_i - \mathbf{j}_i)^2, \quad (i = 1,2,3 \text{ for } x,y,z), \tag{4}$$

where **I** is the total nuclear spin and **j** is the spin of the odd particle. $_i$ is the hydrodynamical moment of inertia, calculated as

$$\Im_i(K) = 4B\epsilon^2 \sin^2(\gamma + 2K\pi/3), \tag{5}$$

using Grodzin's empirical relation,

$$\hbar^2/B = 1000/A^{7/3}(\text{MeV}). \tag{6}$$

H_{core} is exactly diagonalized in the strong-coupling basis,

$$\Psi^{I(\nu)}_{MK} = [(2I+1)/16\pi^2]^{1/2} \sum_{Nlj\Omega} C^{(\nu)}_{Nlj\Omega} \{D^I_{MK}\phi_{Nlj\Omega} + (-1)^{(I-j)} D^I_{M-K}\phi_{Nlj-\Omega}\}, \tag{7}$$

yielding energy levels and total wave-functions,

$$\Phi^I_M = \sum_{K,\nu} a^{I(\nu)}_K \Psi^{I(\nu)}_{MK}. \tag{8}$$

In all cases, five single-particle orbitals above the fermi level and five below were included in the diagonalization. The positive- and negative-parity systems were treated separately.

We comment that this choice of hamiltonian implies that the core is approximated as a rigid body, in principle not including vibrations or effects like γ-softness, etc. Nevertheless, Larsson et al.[56] pointed out that there is some shape-vibration amplitude present in the model within uncertainties of the core quadrupole components Q_{20} and Q_{22}, which can contribute a sizeable fraction of the total deformation in transitional nuclei.

All parameters varied during the present calculation are summarized in Table III. We see that our calculation is dependent on up to five free parameters, which were changed from isotope to isotope. However, well-defined constraints on almost all of the free parameters exist, which must be satisfied to ensure the fitted parameters have a physical meaning. Let us discuss these constraints in more detail.

As can be seen in Table I, there is full experimental information about the value of quadrupole deformation β_2 as extracted from both $B(E2)$ values and ground- and isomeric-

Table III. Summary of the PTR Model Calculation Parameters

A	N	π	ε_2	γ (deg)	λ_{fit} ($\hbar\omega_0$)	λ_{BCS} ($\hbar\omega_0$)	Δ_{fit} (MeV)	Δ_{BCS} (MeV)	ξ
177	99	−	0.270	17.0	6.57	6.57	0.807	0.791	0.80
		+	0.270	16.0					0.77
181	103	−	0.240	16.0	6.67	6.68	0.798	0.831	1.00
		+	0.240	17.5					0.86
183	105	−	0.240	20.0	6.72	6.72	0.810	0.810	1.00
		+	0.230	21.0					0.87
185	107	−	0.227	23.0	6.74	6.77	0.787	0.847	1.00
		+	0.230	22.0					0.73
187	109	−	0.169	30.0	6.79	6.79	1.060	0.970	1.00
		+	0.180	28.5					0.85
189	111	−	0.155	32.0	6.81	6.83	0.982	0.937	1.00
		+	0.176	32.0					0.86
191	113	−	0.150	31.7	6.83	6.86	0.955	0.877	1.00
		+	0.168	33.0					0.83
193	115	−	0.150	35.0	6.92	6.90	0.860	0.780	1.00
		+	0.168	31.0					0.95
195	117	−	0.120	52.0	6.92	6.92	0.773	0.773	1.00

state mean-square radii. However, one should take these values with caution, as they have been extracted from experiment assuming rigid axially-symmetrical nuclei. Furthermore, there is no exact analytical relation between ε_2 and β_2 for non-axially symmetrical shapes, and we can only estimate $\varepsilon_2 \approx 0.85$–$0.95\ \beta_2$ at deformations around 0.2.[58] Despite these uncertainties, data shown in Table I serve as a useful guide to the dependence of deformation upon neutron number.

On the other hand, there is very little guidance concerning γ deformation, and this parameter is not open to direct measurement. The only experimental information available, in our case, is signature splitting in rotational bands, but this may not be directly relevant to the shape of the nuclear ground state. Theoretical calculations of Pt deformation parameters, including γ, have been made only for even-even nuclei.[46-49,59-61] We will discuss later the relation between the parameters of the even-even cores of odd-A Pt nuclei given by the present calculation and the characteristics of their even-even neighbours. Here we only note that this relation is not by any means straightforward, and theoretical predictions of shapes of even-even Pt isotopes may mislead if it is assumed that addition of an odd neutron does not affect the shape.

Standard BCS procedure was used for calculating pairing parameters, with G_{n0} = 19.2 MeV and G_{n1} = 7.4 MeV.[58] λ and Δ were varied within a narrow interval around the BCS values, as is seen in Table III. Deviation of λ and Δ from the BCS values led to improved fits for all isotopes except ^{183}Pt and ^{195}Pt. Whilst the differences between λ_{fit} and λ_{BCS} were very small, deviations from BCS values were found for Δ_{fit} for neutron numbers $107 \leqslant N \leqslant 111$. The behaviour of Δ_{fit} can be understood as caused by the existence of a subshell closure at $N = 108$. This closure has been identified before in Pt isotopes[60,61] and its effect on the pairing energy discussed theoretically. We have investigated the single-particle level spectrum in the γ plane ($0° \leqslant \gamma \leqslant 60°$) for $0.15 \leqslant \varepsilon_2 \leqslant 0.25$, and the presence of the subshell gap at $N = 108$ is clearly seen.

It is well known that agreement between particle-rotor calculations and experiment is generally improved if a coriolis attenuation parameter is introduced. There are two ways of doing this, i) using a factor ξ, multiplying all single-particle matrix elements[56] and/or ii) using a factor of the form $(u_1 u_2 + v_1 v_2)^{\eta}$ to be attached to each single-particle matrix element.[62]

In our calculation of negative-parity states, attenuation of the coriolis-mixing matrix elements ξ and η was not required ($\xi = \eta = 1.0$) apart from ^{177}Pt (see Table III). An attenuation of up to 30% was needed in calculation of positive-parity bands based on a high-j, $i_{13/2}$ orbital ($\xi > 0.7$, $\eta = 1$). Coriolis attenuations of this magnitude are commonly used in particle-rotor calculations.

For each nucleus, the number of experimental data fitted was never less than twenty, including energy levels and electromagnetic properties. Unless there was a clear basis in the physics, smooth variation of all parameters with neutron number was required.

There are three more parameters of the model which were set and kept constant for most of the nuclei. These are the strengths of the $\mathbf{l} \cdot \mathbf{s}$ and $\mathbf{l}^2$ terms in the modified-oscillator potential, κ and μ, and the constant B in expression (6) for the moment of inertia.

Parameters κ and μ are generally dependent on the main oscillator quantum number N. We initially took $\kappa = 0.062$ (for both $N = 5$ and 6) and $\mu = 0.43$ ($N = 5$) and $\mu = 0.34$ ($N = 6$) following Bengtsson and Ragnarsson in their calculation of high-spin states in the rare-earth region. However, we found that a better overall fit was achieved for $\kappa = 0.0636$ ($N = 5$ and 6), $\mu = 0.40$ ($N = 5, A = 177$–185, and $N = 6$, all A), and $\mu = 0.39$ ($N = 5, A = 187$–193). This choice changed the relative energies of the negative-parity spherical $f_{7/2}, p_{3/2}, f_{5/2}$, and $p_{1/2}$ states and the positive-parity $i_{13/2}$ state, resulting in much-improved agreement with these energies as found in ^{207}Pb, the nearest close-to-spherical nucleus.

We note that Nilsson et al.[58] used neutron parameters $\kappa = 0.0636$ and $\mu = 0.393$ in the mass region around $A = 187$, whilst $\kappa = 0.062$ and $\mu = 0.40$ were adopted by Bacelar et al.[64] for $N = 6$ neutron states in their calculation of ^{167,169}Yb; both these sets are well in line with those used here.

The constant B has been calculated from Grodzin's formula (6). In our previous calculation,[55] the same formula was used but with the value of 1225 instead of 1000, which gave a less-satisfactory fit to experiment and higher values of ε_2.

Finally, we note that we did not use any of the other options included in the present model—namely, variable moment of inertia or taking the $E_{2^+(1)}$ and $E_{2^+(2)}$ experimental energies—as input data for calculation of the moment of inertia. Also, we treated the recoil term in the core hamiltonian (4) as a one-body operator.

The calculation has been done for each nucleus, starting from an axially-symmetrical shape (prolate for $^{177\text{-}185}$Pt, oblate for $^{187\text{-}195}$Pt). The fit was then improved by varying the deformation parameter γ.

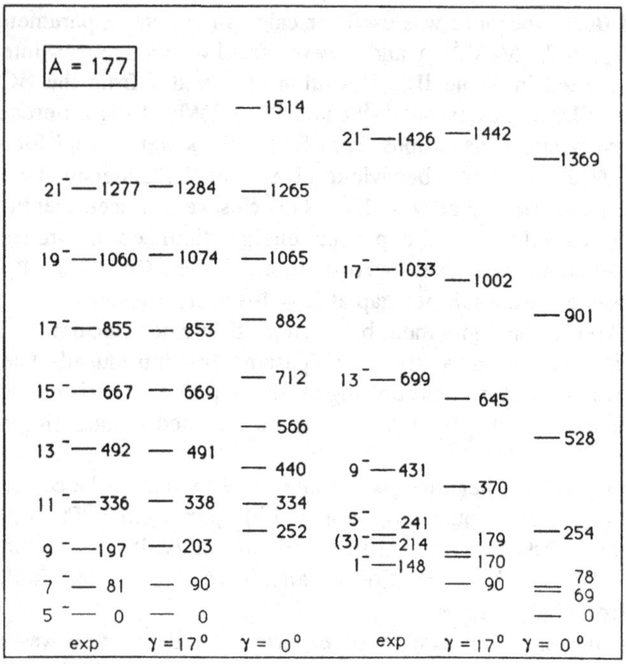

Figure 1. Negative-parity bands in ^{177}Pt. Results of the PTR calculation for states with $I \leqslant 21/2$ are shown for axially-symmetrical and non-axial shapes. All spins given in the figure are twice the experimental value. Experimental data are taken from Refs. 30 and 1.

10.3 RESULTS AND DISCUSSION

10.3.A Negative-Parity Yrast Bands in $^{177-185}Pt$

Results of our fit to energy levels with $I^{\pi} \leqslant 21/2^{-}$ in these nuclei are summarized in Figs. 1-4. All known bands in each of the nuclei were calculated using the same model parameters. Transition rates $R = T_{\gamma}\ (\Delta I = 2)/T_{\gamma}\ (\Delta I = 1)$ were calculated and $B(M1)/B(E2)$ ratios were extracted from experimental data according to the formula,[30]

$$\Lambda = \frac{B(M1; I \to I-1)}{B(E2; I \to I-2)} = 0.0693(16\pi/5)\frac{[E_{\gamma}(I \to I-2)]^{5}}{[E_{\gamma}(I \to I-1)]^{3}R(1+\delta^{2})}, \quad (9)$$

where δ is the $E2/M1$ multipole mixing ratio, taken in the Krane-Steffen sign convention. Where available, these are presented in Tables IV-VII. In all cases in this paper, *experimental* energies were used to calculate R, Λ, and δ. For $I \to I$–1 transitions with unknown δ, a pure $M1$ multipolarity was assumed.

10.3.A.1 ^{177}Pt

Dracoulis et al.[30] identified two bands in ^{177}Pt based on $1/2^{-}$ and $5/2^{-}$ states and suggested $1/2^{-}[521]$ and $5/2^{-}[512]$ Nilsson assignments, mainly on the basis of previous re-

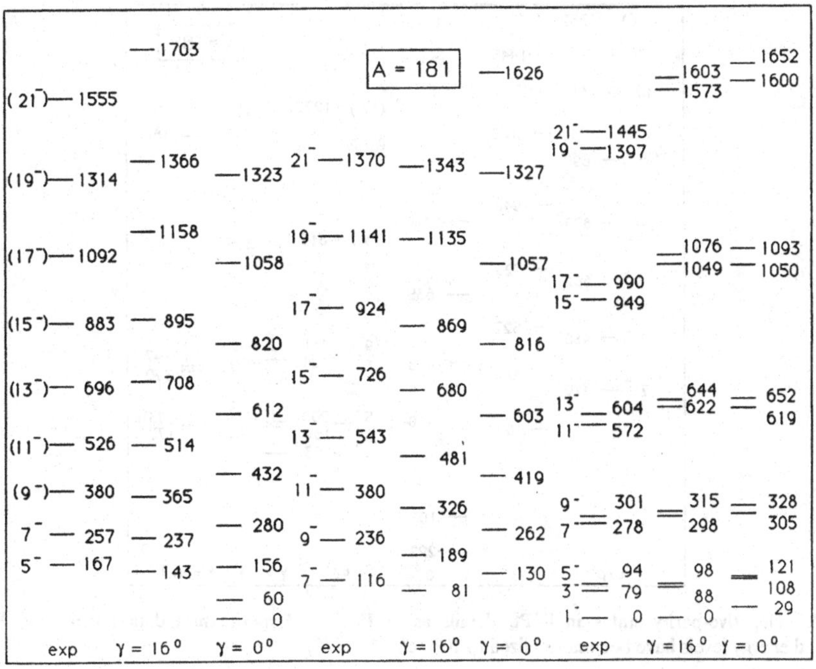

Figure 2. Negative-parity bands in ^{181}Pt, details as Fig. 1. Experimental data taken from Ref. 31.

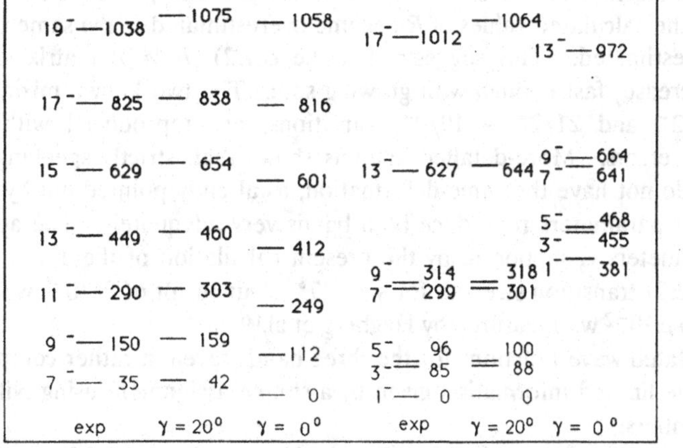

Figure 3. Negative-parity states in ^{183}Pt, details as Fig. 1. Experimental data taken from Refs. 5 and 32.

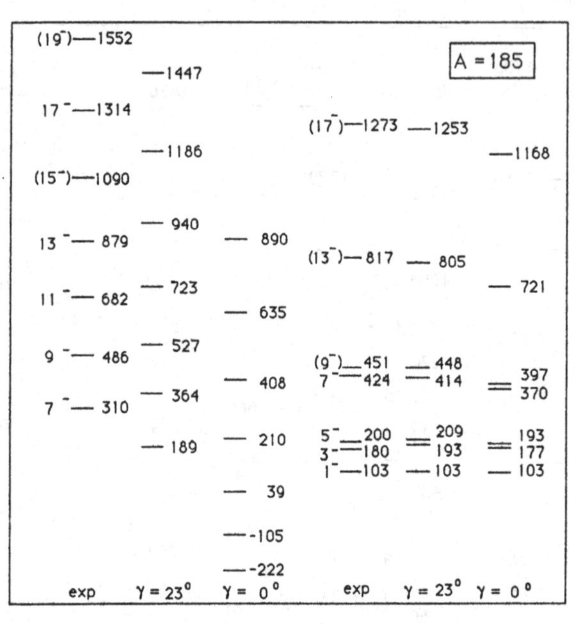

Figure 4. Negative-parity states in ^{185}Pt, details as in Fig. 1. Experimental data taken from Ref. 34. Calculated energy levels have been normalized to 103 keV (I = 1/2).

sults by Hagberg et al.[1] and their interpretation of the α decay of ^{181}Hg to levels in ^{177}Pt. As a result of their particle–axial-rotor calculation, they obtained deformation parameters, $\varepsilon_2 = 0.20, \gamma = 0$. The values, $\varepsilon_2 = 0.27$ ($\gamma = 17^0, \gamma = 0$), extracted for this nucleus from our fit to only lower-spin parts of the bands are somewhat higher. We tried to obtain a fit of the same quality as that shown in Fig. 1 (for $\varepsilon_2 = 0.27$ and $\gamma = 17°$) with the deformation of Dracoulis et al. without success. We were unable to reproduce either the $5/2^-$ state as the ground state or the level spacing within the bands. These authors do not report comparison between experimental values of R and λ and those calculated using their particle-axial-rotor wave-functions, only λ values obtained using the formalism of Donau.[65] In our calculation, we obtain very satisfactory agreement with experimental R and λ for the $5/2^-$ band for states with $I^\pi \leq 15/2^-$, taking both $\gamma = 0$ and $\gamma = 17°$. For higher members of the band, the calculated values of R become overestimated at the same time as values of λ are underestimated. This suggests that the $B(E2)$ [$B(M1)$] matrix elements do not decrease [increase] fast enough with growing spin. The two known mixing ratios, for the $11/2^- \rightarrow 9/2^-$ and $21/2^- \rightarrow 19/2^-$ transitions, are reproduced within rather large experimental errors. More-detailed analysis shows that, strictly speaking, the $5/2^-$ and $1/2^-$ bands do not have the same deformation, as already pointed out by Dracoulis et al. However, our parameters reproduce both bands very adequately. The adoption of these average parameters is supported by the present calculation of the $1/2^- \rightarrow 5/2^-$ 148-keV interband $B(E2)$ transition strength for $\gamma = 17°$. Our result of 9×10^{-3} wu agrees with the value of $3.0(4)\times10^{-2}$ wu measured by Hagberg et al.[66]

Calculated wave-functions for the three bands reveal a rather complex structure, in contrast to the limited information given by a simple assignment using Nilsson asymptotic quantum numbers.

The $1/2^-$ band contains (in the notation of Eqs. 2,3,7,8) a single K = 1/2 component with $|a_{1/2}{}^{1/2\nu}|^2 = 0.97(1.00)$ for $\gamma = 17°$ (0), made up from four roughly-

Table IV. Transition Probability Ratios in ^{177}Pt

Band-Head I^{π} = 5/2$^-$ Ground State

I^{π}	E^{exp}_{level} (MeV)	$E^{exp}_{\gamma(\Delta I=2)}$ (MeV)	$E^{exp}_{\gamma(\Delta I=1)}$ (MeV)	$B(E2)^{cal}_{\Delta I=2}$ (e^2b^2)	$B(E2)^{cal}_{\Delta I=1}$ (e^2b^2)	$B(M1)^{cal}_{\Delta I=1}$ (μ_N^2)	R^{a}_{exp}	R_{cal}	δ_{exp}	δ_{cal}	Λ^{b}_{exp} (μ_N^2/e^2b^2)	Λ_{cal} (μ_N^2/e^2b^2)
9/2$^-$	0.197	0.197	0.116	c) 0.70	2.21	0.21	0.57(13)	0.40		−0.31	0.21(5)	0.30
				d) 0.74	2.03	0.18		0.49		−0.32		0.24
11/2$^-$	0.336	0.255	0.139	c) 1.19	1.73	0.23	1.57(16)	1.31	$-0.35^{+0.14}_{-0.25}$	−0.32	$0.16^{+0.01}_{-0.03}$	0.19
				d) 1.26	1.51	0.20		1.59		−0.32		0.16
13/2$^-$	0.492	0.294	0.156	c) 1.52	1.36	0.23	2.23(17)	2.41		−0.32	0.17(1)	0.15
				d) 1.60	1.12	0.18		3.23		−0.32		0.11
15/2$^-$	0.667	0.330	0.175	c) 1.75	1.09	0.23	2.7(6)	3.50		−0.32	0.18(4)	0.13
				d) 1.88	0.85	0.20		4.36		−0.30		0.11
17/2$^-$	0.855	0.364	0.189	c) 1.91	0.89	0.21	4.1(7)	6.51		−0.26	0.15(3)	0.07
				d) 2.06	0.65	0.15		8.14		−0.33		0.07
19/2$^-$	1.060	0.394	0.205	c) 2.04	0.74	0.20	3.9(6)	7.03		−0.33	0.19(3)	0.10
				d) 2.24	0.51	0.18		8.78		−0.29		0.08
21/2$^-$	1.277	0.422	0.217	c) 2.13	0.63	0.18	6.3(11)	9.64	$-0.25^{+0.19}_{-0.38}$	−0.34	$0.14^{+0.01}_{-0.04}$	0.08
				d) 2.36	0.40	0.12		16.0		−0.33		0.05

Table IV. (continued)

Band-Head I^π = 7/2⁺ 95 keV

I^π	E^{exp}_{level} (MeV)	$E^{exp}_{\gamma}(\Delta I=2)$ (MeV)	$E^{exp}_{\gamma}(\Delta I=1)$ (MeV)	$B(E2)^{cal}_{\Delta I=2}$ (e^2b^2)	$B(E2)^{cal}_{\Delta I=1}$ (e^2b^2)	$B(M1)^{cal}_{\Delta I=1}$ (μ_N^2)	R^{a}_{exp}	R_{cal}	δ_{cal}	Λ^{b}_{exp} (μ_N^2/e^2b^2)	Λ_{cal} (μ_N^2/e^2b^2)
11/2⁺	0.210	0.114	0.070	c) 0.40	2.41	0.05	0.28(8)	0.21	−0.40	0.14(4)	0.11
				d) 0.56	2.25	0.05		0.38	−0.39		0.089
13/2⁺	0.265	0.124	0.055	c) 0.87	2.17	0.11	1.03(25)	0.93	−0.20	0.12(3)	0.13
				d) 1.05	1.86	0.10		1.23	−0.20		0.095
15/2⁺	0.441	0.231	0.176	c) 1.22	1.85	0.16	1.02(9)	0.51	−0.50	0.082(7)	0.13
				d) 1.40	1.45	0.11		0.83	−0.53		0.078
17/2⁺	0.532	0.267	0.091	c)1.49	1.55	0.21	15.9(33)	8.50	−0.21	0.079(16)	0.14
				d)1.72	1.15	0.19		11.0	−0.19		0.11
19/2⁺	0.778	0.337	0.246	c) 1.69	1.30	0.25	2.1(3)	1.12	−0.47	0.097(14)	0.15
				d) 1.89	0.87	0.12		2.44	−0.55		0.063
21/2⁺	0.902	0.370	0.124	c) 1.85	1.10	0.29	13.2(57)	15.5	−0.20	0.19(8)	0.16
				d) 2.15	0.74	0.27		19.6	−0.17		0.13

[a] $R = T\gamma(I \to I\text{-}2)/T\gamma(I \to I\text{-}1)$.

[b] $\Lambda = B(M1, I \to I\text{-}1)/B(E2, I \to I\text{-}2)$ calculated from Eq. 9. Where experimental value δ_{exp} is not known, pure $M1$ multipolarity of the $I \to I\text{-}1$ transition is assumed.

[c] $\gamma = 0$.

[d] $\gamma = 17°$ (5/2⁻ band) and $\gamma = 16°$ (7/2⁺ band).

Experimental values taken from Ref. 30.

Table V. Transition Probability Ratios in ^{181}Pt

Band-Head I^{π} = $7/2^{-}$ 116 keV

I^{π}	E^{exp}_{level} (MeV)	$E^{exp}_{\gamma(\Delta I=2)}$ (MeV)	$E^{exp}_{\gamma(\Delta I=1)}$ (MeV)	$B(E2)^{cal}_{\Delta I=2}$ (e^2b^2)	$B(E2)^{cal}_{\Delta I=1}$ (e^2b^2)	$B(M1)^{cal}_{\Delta I=1}$ (μ_N^2)	R^{a}_{exp}	R_{cal}	δ_{cal}	Λ^{b}_{exp} (μ_N^2/e^2b^2)	Λ_{cal} (μ_N^2/e^2b^2)
$13/2^-$	0.543	0.307	0.162	c) 1.04	1.48	0.15	6.0(1)	2.61	−0.42	0.074(1)	0.14
				d) 0.88	1.33	0.016		9.73	−1.23		0.018
$15/2^-$	0.726	0.346	0.184	c) 1.31	1.20	0.18	2.7(4)	3.47	−0.40	0.21(3)	0.14
				d) 1.16	1.03	0.018		15.2	−1.16		0.016
$17/2^-$	0.924	0.382	0.199	c) 1.49	0.98	0.19	6.4(9)	4.92	−0.38	0.11(2)	0.13
				d) 1.43	0.78	0.017		26.6	−1.12		0.012
$19/2^-$	1.141	0.415	0.216	c) 1.63	0.81	0.22	20(12)	5.61	−0.34	0.042(26)	0.13
				d) 1.55	0.60	0.019		34.2	−1.01		0.012
$21/2^-$	1.370	0.445	0.229	c) 1.73	0.67	0.21	9(3)	7.40	−0.34	0.11(4)	0.12
				d) 1.75	0.49	0.018		49.2	−0.99		0.010

Table V. (continued)
Band-Head I^{π} = 5/2⁻ 167 keV

I^{π}	E^{exp}_{level} (MeV)	$E^{exp}_{\gamma(\Delta I=2)}$ (MeV)	$E^{exp}_{\gamma(\Delta I=1)}$ (MeV)	$B(E2)^{cal}_{\Delta I=2}$ (e^2b^2)	$B(E2)^{cal}_{\Delta I=1}$ (e^2b^2)	$B(M1)^{cal}_{\Delta I=1}$ (μ_N^2)	R^{a}_{exp}	R_{cal}	δ_{cal}	Λ^{b}_{exp} (μ_N^2/e^2b^2)	Λ_{cal} (μ_N^2/e^2b^2)
$9/2^-$	0.380	0.214	0.124	c) 0.36	1.96	0.07	0.36(18)	0.65	−0.55	0.46(23)	0.19
				d) 0.49	1.59	0.13		0.54	−0.36		0.26
$11/2^-$	0.526	0.269	0.145	c) 0.78	1.71	0.08	0.12(4)	2.38	−0.56	2.7(9)	0.10
				d) 0.90	1.23	0.19		1.39	−0.31		0.21
$13/2^-$	0.696	0.315	0.170	c) 1.12	1.39	0.07	0.92(33)	5.01	−0.63	0.48(17)	0.06
				d) 1.16	0.94	0.18		2.55	−0.32		0.16
$15/2^-$	0.883	0.357	0.187	c) 1.35	1.12	0.07	1.5(6)	8.54	−0.62	0.41(16)	0.05
				d)1.44	0.72	0.26		3.19	−0.26		0.18
$17/2^-$	1.092	0.396	0.209	c) 1.54	0.91	0.06	2.4(1.4)	13.0	−0.68	0.31(18)	0.04
				d) 1.53	0.54	0.18		5.76	−0.30		0.12
$19/2^-$	1.314	0.431	0.223	c) 1.66	0.75	0.06	3.0(1.5)	18.0	−0.66	0.31(16)	0.04
				d) 1.77	0.45	0.30		5.22	−0.23		0.17
$21/2^-$	1.555	0.463	0.241	c) 1.77	0.63	0.06	8(3)	21.8	−0.65	0.13(5)	0.03
				d) 1.77	0.34	0.16		10.7	−0.29		0.10

Table V. (continued)
Band-Head $I^\pi = 9/2^+$ 276 keV

I^π	E^{exp}_{level} (MeV)	$E^{exp}_{\gamma(\Delta I=2)}$ (MeV)	$E^{exp}_{\gamma(\Delta I=1)}$ (MeV)	$B(E2)^{cal}_{\Delta I=2}$ (e^2b^2)	$B(E2)^{cal}_{\Delta I=1}$ (e^2b^2)	$B(M1)^{cal}_{\Delta I=1}$ (μ_N^2)	R^{a}_{exp}	R_{cal}	δ_{cal}	Λ^{b}_{exp} (μ_N^2/e^2b^2)	Λ_{cal} (μ_N^2/e^2b^2)
$15/2^+$	0.542	0.222	0.160	c) 0.94	1.53	0.12	0.38(11)	0.58	−0.48	0.24(7)	0.14
				d) 0.90	1.32	0.09		0.72	−0.51		0.10
$17/2^+$	0.643	0.262	0.101	c) 1.17	1.28	0.18	1.5(5)	5.39	−0.23	0.56(19)	0.15
				d) 1.18	1.05	0.17		5.52	−0.21		0.14
$19/2^+$	0.887	0.345	0.244	c) 1.35	1.08	0.21	1.7(2)	1.23	−0.46	0.14(2)	0.16
				d) 1.34	0.81	0.14		1.80	−0.49		0.10
$21/2^+$	1.013	0.370	0.126	c) 1.49	0.90	0.27	4.5(5)	13.2	−0.20	0.53(6)	0.17
				d) 1.59	0.68	0.26		14.1	−0.17		0.16

[a-c] As footnotes Table IV.
[d] $\gamma = 17.5°$ (positive parity) and $\gamma = 16°$ (negative parity).
Experimental data taken from Ref. 31.

Table VI. Transition Probability Ratios in ^{183}Pt

Band-Head I^π = 7/2⁻ 35 keV

I^π	E^{exp}_{level} (MeV)	$E^{exp}_{\gamma}(\Delta I=2)$ (MeV)	$E^{exp}_{\gamma}(\Delta I=1)$ (MeV)	$B(E2)^{cal}_{\Delta I=2}$ (e^2b^2)	$B(E2)^{cal}_{\Delta I=1}$ (e^2b^2)	$B(M1)^{cal}_{\Delta I=1}$ (μ_N^2)	R^a_{exp}	R_{cal}	δ_{exp}	δ_{cal}	Λ^b_{exp} (μ_N^2/e^2b^2)	Λ_{cal} (μ_N^2/e^2b^2)
$11/2^-$	0.290	0.255	0.139	c) 0.46	2.13	0.020	5.8(7)	2.60		−1.19	0.048(6)	0.043
				d) 0.44	1.62	0.010		3.90		−1.47		0.023
$13/2^-$	0.449	0.299	0.160	c) 0.86	1.81	0.022	18(4)	6.4	−2.5(2)	−1.21	0.0031(7)	0.025
				d) 0.84	1.26	0.010		10.5		−1.50	0.022(3)[e]	0.012
$15/2^-$	0.629	0.339	0.180	c) 1.16	1.49	0.023		10.9		−1.21		0.020
				d) 1.13	0.796	0.011		3.19		−1.41		0.010

Table VI. (continued)

Band-Head $I^{\pi} = 9/2^{+}$ 196 keV

I^{π}	E^{exp}_{level} (MeV)	$E^{exp}_{\gamma(\Delta I=2)}$ (MeV)	$E^{exp}_{\gamma(\Delta I=1)}$ (MeV)	$B(E2)^{cal}_{\Delta I=2}$ (e^2b^2)	$B(E2)^{cal}_{\Delta I=1}$ (e^2b^2)	$B(M1)^{cal}_{\Delta I=1}$ (μ_N^2)	R^{a}_{exp}	R_{cal}	δ_{exp}	δ_{cal}	Λ^{b}_{exp} (μ_N^2/e^2b^2)	Λ_{cal} (μ_N^2/e^2b^2)
$15/2^{+}$	0.478	0.234	0.161	c) 0.70	1.73	0.17	0.70(7)	0.41	−1.7(5)	−0.43	0.043	0.24
				d) 0.67	1.21	0.096		0.66		−0.48	0.17(2)[e]	0.14
$17/2^{+}$	0.590	0.274	0.112	c) 0.95	1.48	0.23	37(7)	3.0	$-0.6^{+3.0}_{-0.4}$	−0.24	0.017(14)	0.24
				d) 0.97	0.93	0.18		4.0		−0.21	0.023(4)[e]	0.18
$19/2^{+}$	0.834	0.356	0.244	c) 1.16	1.26	0.27	7.6(1.3)	0.98	$-1.1^{+0.3}_{-0.5}$	−0.44	0.014(5)	0.23
				d) 1.13	0.72	0.13		1.93		−0.48	0.031(5)[e]	0.12
$21/2^{+}$	0.996	0.376	0.132	c) 1.31	1.06	0.32	20(2)	9.0		−0.20	0.12(1)	0.24
				d) 1.40	0.66	0.25		12		−0.18		0.18

[a-c] As footnotes Table IV.
[d] $\gamma = 21^{\circ}$ (positive parity) and $\gamma = 20^{\circ}$ (negative parity).
[e] Calculated assuming pure $M1$ $I \rightarrow I$-1 transition.
Experimental data taken from Ref. 32.

Table VII. Transition Probability Ratios in ^{185}Pt

Band-Head I^{π} = 7/2$^-$ 310 keV

I^{π}	E^{exp}_{level} (MeV)	$E^{exp}_{\gamma(\Delta I=2)}$ (MeV)	$E^{exp}_{\gamma(\Delta I=1)}$ (MeV)	$B(E2)^{cal}_{\Delta I=2}$ (e^2b^2)	$B(E2)^{cal}_{\Delta I=1}$ (e^2b^2)	$B(M1)^{cal}_{\Delta I=1}$ (μ_N^2)	R^{a}_{exp}	R_{cal}	δ_{cal}	Λ^{b}_{exp} (μ_N^2/e^2b^2)	Λ_{cal} (μ_N^2/e^2b^2)
11/2$^-$	0.682	0.371	0.195	c) 0.40	1.89	0.019	1.2(2)	3.82	−1.62	0.55(9)	0.047
				d) 0.51	0.70	0.26		1.20	−0.27		0.51
13/2$^-$	0.879	0.393	0.198	c) 0.76	1.61	0.019	2.8(4)	10.1	−0.33	0.30(4)	0.025
				d) 0.74	0.94	0.35		1.63	−0.27		0.47
15/2$^-$	1.090	0.409	0.211	c) 1.02	1.32	0.019		14.4	−0.29		0.019
				d) 0.95	0.84	0.40		1.88	−0.25		0.42
17/2$^-$	1.314	0.435	0.223	c) 1.22	1.08	0.018		21.5	−0.27		0.015
				d) 1.18	0.68	0.44		1.25	−0.23		0.37

Table VII. (continued)

Band-Head I^π = 9/2⁺ Ground State

I^π	E^{exp}_{level} (MeV)	$E^{exp}_{\gamma}(\Delta I=2)$ (MeV)	$E^{exp}_{\gamma}(\Delta I=1)$ (MeV)	$B(E2)^{cal}_{\Delta I=2}$ (e^2b^2)	$B(E2)^{cal}_{\Delta I=1}$ (e^2b^2)	$B(M1)^{cal}_{\Delta I=1}$ (μ_N^2)	R^{a}_{exp}	R_{cal}	δ_{exp}	δ_{cal}	Λ^{b}_{exp} (μ_N^2/e^2b^2)	Λ_{cal} (μ_N^2/e^2b^2)
13/2⁺	0.212	0.212	0.117	c) 0.35	1.98	0.13	0.56(5)	0.43	−0.30(7)	−0.38	0.31(8)	0.37
				d) 0.34	1.45	0.08		0.67		−0.41		0.24
15/2⁺	0.373	0.278	0.161	c) 0.67	1.80	0.19	1.0(2)	0.83	−0.30(7)	−0.41	0.25(8)	0.28
				d) 0.69	1.07	0.12		1.37		−0.40		0.17
17/2⁺	0.531	0.318	0.157	c) 0.94	1.56	0.25	2.85(12)	1.98	−0.28(6)	−0.32	0.19(4)	0.26
				d) 0.96	0.95	0.17		3.0		−0.31		0.18
19/2⁺	0.753	0.380	0.222	c) 1.14	1.33	0.30	2.39(8)	1.66	−0.33(8)	−0.38	0.19(5)	0.26
				d) 1.13	0.82	0.15		3.19		−0.43		0.18
21/2⁺	0.939	0.407	0.185	c) 1.31	1.13	0.34	6.5(4)	4.35	−0.26(8)	−0.28	0.18(6)	0.26
				d) 1.36	0.75	0.25		6.18		−0.27		0.18

[a-c] As footnotes Table IV.
[d] γ = 22° (positive parity) and γ = 23° (negative parity).
Experimental data taken from Refs. 33 and 34.

equal ($|C_{Nlj\Omega}{}^{\nu}|^2$ = 0.18 – 0.26) contributions from $|5h_{9/2,\ 1/2}\rangle$, $|5f_{7/2,\ 1/2}\rangle$, $|5f_{5/2,\ 1/2}\rangle$, and $|5p_{3/2,\ 1/2}\rangle$ for both values of γ. This suggests strong mixing and makes it impossible to find a dominant component with a definite spherical parentage. We note here that our calculation predicts that practically the same structure of the lowest-lying $1/2^-$ state remains almost the same throughout the whole region of $^{177\text{-}191}$Pt, showing very little sensitivity to γ. This finding is supported by almost identical experimental low-lying level spacings in the $1/2^-$ bands in these nuclei (see Figs. 1–4) and very similar values of measured magnetic moments (Table II).

The $5/2^-$ band has a single K = 5/2 component with $|a_{5/2}{}^{5/2\nu}|^2$ = 0.97 (0.85) for γ = 17° (0), with a dominant contribution from $|5f_{7/2,\ 5/2}\rangle$ with $|C_{Nlj\Omega}{}^{\nu}|^2$ = 0.55 (0.70). We note that no admixture of a $7/2^-$ state has been found in the $5/2^-$ band at low spin for either value of γ.

The two different sets of deformation parameters and different structure of wave-functions obtained by us and by Dracoulis et al. indicate considerable model dependence for these results, at least for the lower-spin data. The high deformation of ^{177}Pt is somewhat unexpected, as N = 99 is well below the middle of the neutron shell where deformation is broadly on the decrease. However, Moller and Nix[66] predict a rather sudden increase in deformation for ^{177}Pt as compared with ^{176}Pt and ^{178}Pt (see Table I). The higher deformation may be connected with shape coexistence between the 4h and 6h-2p states, which are thought to change position between ^{176}Pt and ^{178}Pt, with the more deformed 6h-2p state contributing mostly to the ground state of $^{178\text{-}186}$Pt.[38,40,67] Although this is a very intriguing topic, it is beyond the scope of this paper and will be discussed in future publications.

10.3.A.2 ^{181}Pt

Three negative-parity bands have been identified in ^{181}Pt (Ref. 31) based on $1/2^-$, $5/2^-$, and $7/2^-$ band heads with Nilsson-model assignments $1/2^-$[521], $5/2^-$[512], and $7/2^-$[514], respectively. We show, in Fig. 2, the results of energy level calculations for γ = 0 and γ = 16°, both at ϵ_2 = 0.24. It is clear that axial symmetry does not yield the correct relative position of the band heads; instead, it predicts the $5/2^-$ level as the ground state.

However, comparison of experimental and calculated transition-probability ratios R, summarized in Table V for the $5/2^-$ and $7/2^-$ bands, shows interesting features. Whilst the γ = 0 results are reasonable for the $7/2^-$ band, they are too high and vary too much in the $5/2^-$ band. The γ = 16° results, by contrast, exceed experiment in the $7/2^-$ band but have the correct trends in the $5/2^-$ band. Thus, our calculation of ^{181}Pt seems to be suggesting different shapes for the two bands.

The wave-functions for the $5/2^-$ band show very similar structure to the $5/2^-$ band in ^{177}Pt for both values of γ. Relative amplitudes of the main contributions in the total wave-function do not change with increasing spin. Thus, we have no explanation for the reported irregular behavior of R and λ in this band.

On the other hand, the calculation predicts very different structures for the $7/2^-$ band for axial and non-axial shapes. For γ = 16°, there is only a single K = 7/2 component with $|a_{7/2}{}^{7/2\nu}|^2$ = 0.88 with a dominant contribution from $|5h_{9/2,7/2}\rangle$ with $|C_{Nlj\Omega}{}^{\nu}|^2$ = 0.75. For γ = 0, we calculate two significant components in the total wave-function, namely K = 7/2 with $|a_{7/2}{}^{7/2\nu} = \nu 1|^2$ = 0.38 and K = 5/2 with $|a_{5/2}{}^{7/2\nu} = \nu 2|^2$ = 0.60. Thus, the fact that our calculation of R and λ is closer to experiment for γ = 0 may relate to the strong mixing of the K = 5/2 and 7/2 components of the band. This supports the suggestion of de Voigt et al.[31] that the irregularities in transition-probability ratios, even in the lower part of the $7/2^-$ band, cannot be explained without some mixing.

10.3.A.3 ^{183}Pt

Two negative-parity bands, based on the $7/2^-$ isomeric state and the $1/2^-$ ground state, have been established in ^{183}Pt (Ref. 32) and interpreted as $7/2^-[514]$ and $1/2^-[521]$ Nilsson configurations, respectively. In our calculations, the best overall fit is obtained for $\varepsilon_2 = 0.24$. As in ^{177}Pt and ^{181}Pt, the relative position of the band-heads is well reproduced for a non-axial shape with $\gamma = 20°$, $\gamma = 0$ giving the wrong ground state. There is only a limited amount of experimental information on R and Λ available. Although the calculated values of R are underestimated by about a factor of 2, values of Λ for the $7/2^-$ band are in good agreement with experiment for both $\gamma = 0$ and $20°$. The same is true for $\delta(E2/M1)$ for the 180-keV transition and the magnetic moment of the $1/2^-$ ground state (see Table II). The structure of the $7/2^-$ state wave-function is rather simple, and for both values of γ is of spherical $h_{9/2}$ parentage.

10.3.A.4 ^{185}Pt

Pilotte et al.[34] recently identified two negative-parity bands in ^{185}Pt with band-heads $1/2^-$ and $7/2^-$, associated with Nilsson configurations $1/2$–$[521]$ and $7/2^-[503]$, respectively. Results of our fit to energy levels of the bands are shown in Fig. 4. It can be seen that, whilst the $1/2^-$ band, as in lighter Pt isotopes, changes very little with γ, both the band-head energy and electromagnetic properties of the $7/2^-$ band depend critically on γ.

In ^{185}Pt there are two $7/2^-$ orbitals calculated close to the fermi surface, originating from the spherical $h_{9/2}$ and $f_{7/2}$ states. For $\gamma = 0$, the lower one has a very similar wave-function to the $7/2^-$ band head in ^{183}Pt but gives poorer agreement with experiment. For $\gamma = 23°$, the wave-function of the lower $7/2^-$ band has two significant components, both with $K = 7/2$, with $|a^{7/2\nu} = {}^{\nu}1_{7/2}|^2 = 0.65$ and $|a^{7/2\nu} = {}^{\nu}2_{7/2}|^2 = 0.25$. It can be seen in Fig. 4 and in Table VII that full agreement with experiment can be achieved for $\gamma = 23°$, which brings a different dominant contribution of the $7/2^-$ orbital lying closest to the fermi surface. In this way, the most striking feature of the $7/2^-$ band, i.e., the order-of-magnitude difference between its measured $B(M1)/B(E2)$ ratios for intraband transitions (and the same ratios for the $7/2^-$ band in ^{183}Pt) is well reproduced and explained as a consequence of different spherical origin of the two $7/2^-$ bands in question.

10.3.B Non-Yrast Negative-Parity States in ^{183,185}Pt

In contrast to $^{177\text{-}181}$Pt, where very little is known about non-yrast negative-parity states,[1,2,3] considerable spectroscopic information is available on low-lying negative-parity excited states in ^{183}Pt[5] and ^{185}Pt.[6,7] Roussiere et al. consistently interpret their data in terms of nuclear models, valid for axially-symmetrical rigid, well-deformed nuclei only, seeking to identify configurations predicted by the models to lie close to the fermi surface in these two nuclei. In their analysis, they rely upon similarities to decay properties of states identified in neighbouring nuclei (e. g., Os) in order to deduce spin-parity assignments of experimental energy levels in ^{183,185}Pt. We show, in Figs. 5 and 6, comparison of calculated energy levels as a function of γ with the quasi-band structure suggested by

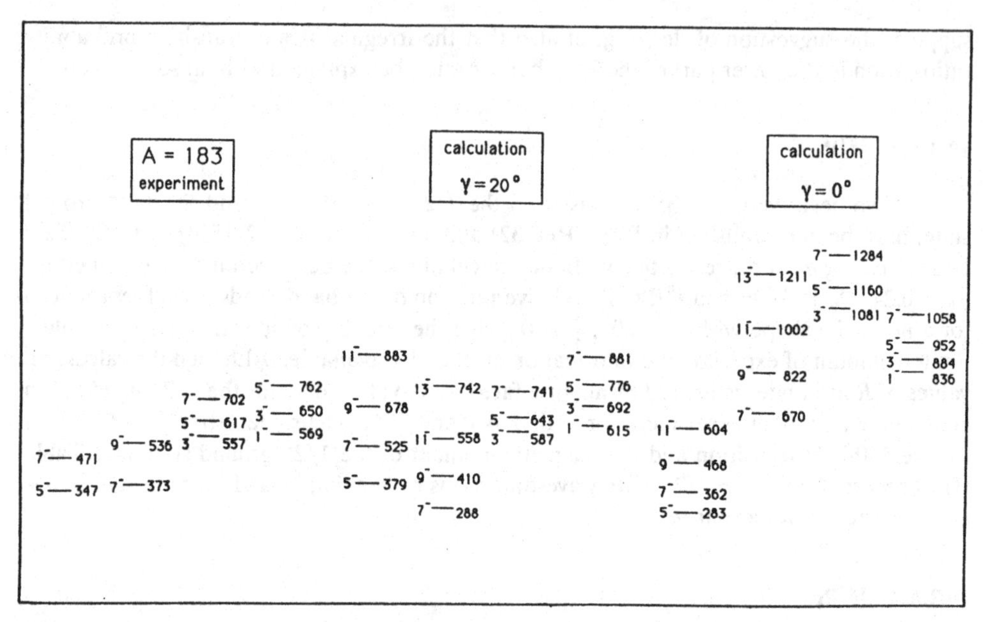

Figure 5. PTR calculation of non-yrast negative-parity excited states in ^{183}Pt. Experimental levels are grouped into quasi-bands suggested by Roussiere et al.[5] See, also, Fig. 1 caption and the text for explanation.

Roussiere et al. It is immediately clear that, in both nuclei, the experimental data are reproduced for non-zero γ in both band-head position and band spacing to within 50 keV (^{183}Pt) –200 keV (^{185}Pt), whilst the assumption of axial symmetry leads to discrepancies of several hundred keV in the quasi-band-head energies.

However, there are two comments to be made:

First, the quasi-band structure proposed by Roussiere et al. in ^{183,185}Pt is directed by their model expectations. The "bands" sometimes have as few as two members and may not even be supported by the existence of interband transitions. In accepting and fitting their proposed band structure, we have been guided by the selection of series of levels with common major-K components. However, our calculation provides more states than shown in Figs. 5 and 6; and there are different possible ways of interpreting the full experimental data of Roussiere et al., some of which have the virtue of incorporating some observed states that were not included in their proposed bands. More detailed analysis of these two nuclei, using the present model, is under consideration and will be published later.

Second, in ^{185}Pt, the calculation for $\gamma = 23°$, despite reproducing all the other states to better than 200 keV, does not reproduce the low-lying $1/2^-$ state at 388 keV (originally assigned $3/2^-$, $5/2^-$ in Ref. 6) and the quasi-band that Roussiere et al. associate with this state in their latest paper, based on analogy with ^{183}Os.[7] The lowest calculated $1/2^-$ band, with band head at 662 keV, has different level spacings and cannot be associated with their proposed $1/2^-$ band. In the calculation for $\gamma = 0$, we find a band with similar properties to the experimental "band" based at 388 keV, but all other calculated states for axial symmetry are in marked disagreement with experiment. We note, however, that the calculated $1/2^-$ band for $\gamma = 23°$ fits very well with a $1/2^-$ band, formed of 645-,

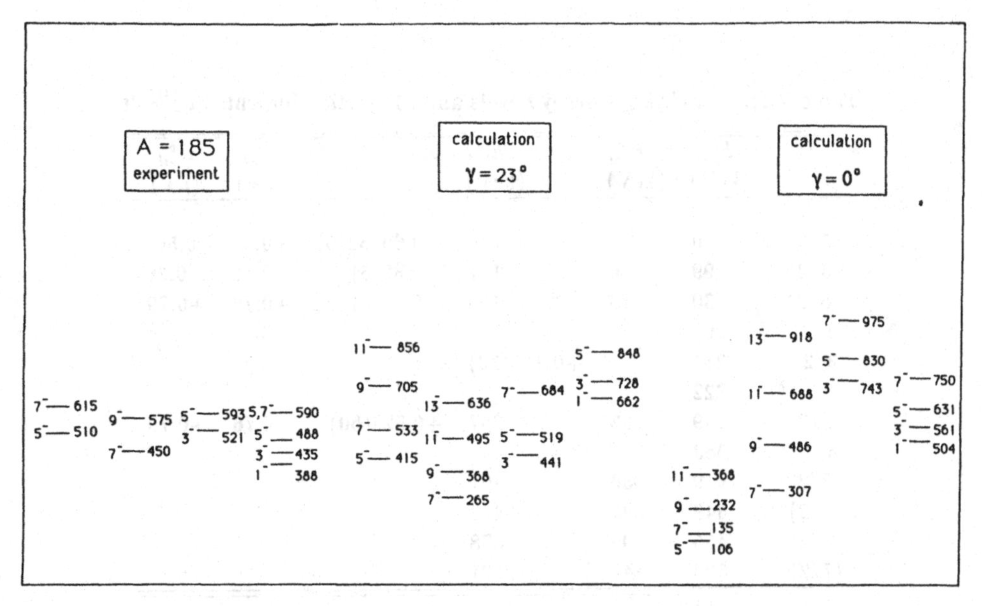

Figure 6. PTR calculation of non-yrast states in ^{185}Pt, details as in Fig. 4. Experimental data taken from Refs. 7 and 71 (spin assignment of the level at 590 keV). Calculated energy levels have been normalized to 103 keV (I = 1/2).

728-, and 816-keV states, proposed by Roussiere et al. in their first publication.[6]

We have also investigated the wave-functions of the calculated non-yrast bands at higher energy. The structure of these bands increases in complexity with increasing energy, and any assignment of asymptotic Nilsson quantum numbers to such non-yrast band heads conveys little information. Also, the "band" character (similar single-particle make-up for members with increasing spin) deteriorates, leading to a loss in meaning of the band concept.

10.3.C Low-Lying Negative-Parity Excitations in $^{187\text{-}195}$Pt

All known low-lying negative parity states with E_{exc}<500keV and $I^{\pi} \leqslant 7/2^{-}$ in $^{187\text{-}193}$Pt are summarized in Figs. 7–10 and Table VIII (^{195}Pt). Detailed experimental information is limited in that, apart from ^{187}Pt,[8] very few excited states have unambiguous spin-parity assignment. In contrast with the lighter Pt nuclei, little clear band structure has been established below 500 keV. In ^{189}Pt, Daly and co-workers[35] have singled out two sequences of negative-parity levels based on the 173-keV $9/2^{-}$ state and the 273-keV $7/2^{-}$ state, respectively, which they group as bands associated with the $h_{9/2}$ neutron orbital. The argument for this assignment (for the band based on the $9/2^{-}$ level) is that both the $9/2^{-}$ state and its feeding $11/2^{-}$ 493-keV state are populated in the allowed β decay of ^{189}Au^{m} characterized by $h_{11/2}$ proton orbital.[68] Both proposed bands have irregular level spacing which may be a sign of strong mixing. Taking the high-spin bands in ^{189}Pt as our starting point, a very reasonable fit to their level energies is obtained with deformation parameters ϵ_2 = 0.18, γ = 26° and ξ = 0.8, the calculation showing strong mixing between the two bands. Analysis of the wave-functions identifies the $9/2^{-}$ band as of $h_{9/2}$ spherical origin and the $7/2^{-}$ band of mixed $h_{9/2}$ and $f_{7/2}$ parentage.

Table VIII. Low-Lying Energy Levels and Magnetic Moments in ^{195}Pt

I^{π}	E_{exp} (keV)	E^{a}_{cal} (keV)	E^{b}_{cal} (keV)	μ_{exp} (μ_N)	μ^{a}_{cal} (μ_N)	μ^{b}_{cal} (μ_N)
$1/2^-$	0	0	0	+0.60952(6)	+0.51	0.50
$3/2^-$	99	56	112	−0.62(5)	−0.49	−0.28
$5/2^-$	130	78	140	+0.90(6)	+0.78	+0.79
$3/2^{-\,c}$	200					
$3/2^{-\,c}$	211		+0.156(32)			
$1/2^{-\,c}$	222					
$5/2^-$	239	252	267	+0.523(50)	+0.76	+0.79
$5/2^{-\,c}$	389					
$3/2^-$	419	356	407			
$(7/2)^-$	449	392	452			
$5/2^-$	455	500	578			
$(7/2)^-$	508	447	521			

[a] $\varepsilon_2 = 0.12, \gamma = 52°$.
[b] $\varepsilon_2 = 0.12, \gamma = 60°$.
[c] The level is not calculated for given deformation parameters.
Experimental data taken from compilations 12 and 74.

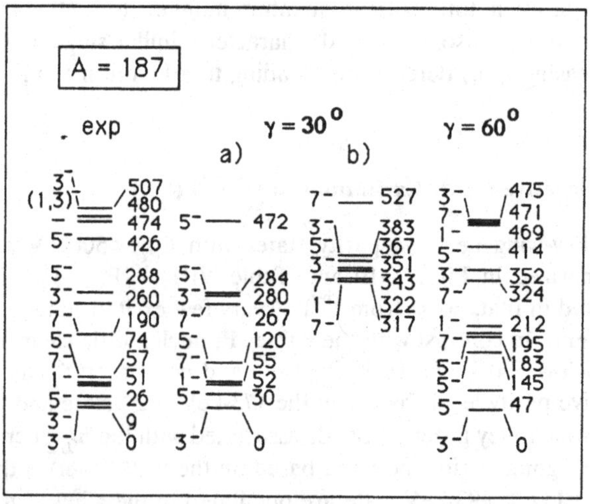

Figure 7. Results of the PTR calculation for low-lying negative-parity excited states in ^{187}Pt with $I^{\pi} \leq 7/2^-$ $E < 0.5$ MeV for both triaxial and oblate shapes. All calculated states are shown in this energy region. For the triaxial shape, they are divided into group *a*) - most likely corresponding to observed levels, and group *b*) - calculated states with no apparent experimental equivalent. Experimental data are taken from Ref. 8.

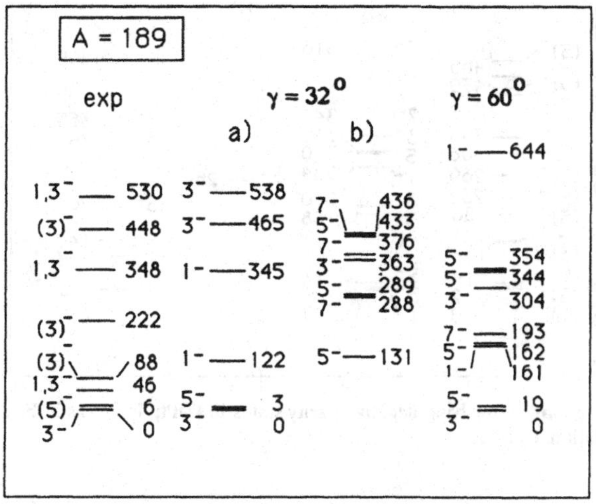

Figure 8. PTR calculation for low-lying negative-parity states in ^{189}Pt, details as in Fig. 7. Experimental data taken from Refs. 9 and 75.

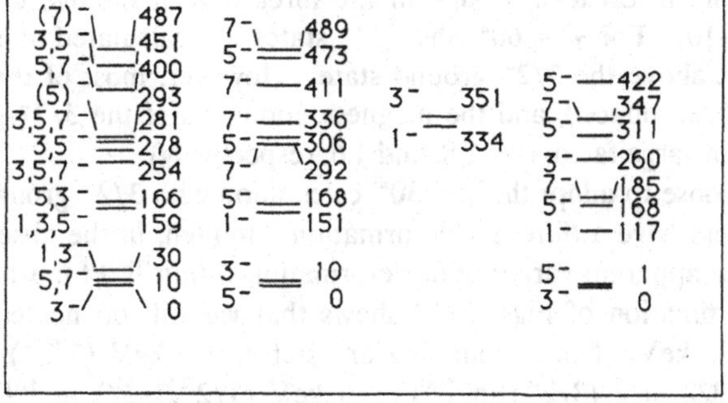

Figure 9. PTR calculation for low-lying negative-parity states in ^{191}Pt, details as in Fig. 7. Experimental data taken from the compilation, Ref. 10.

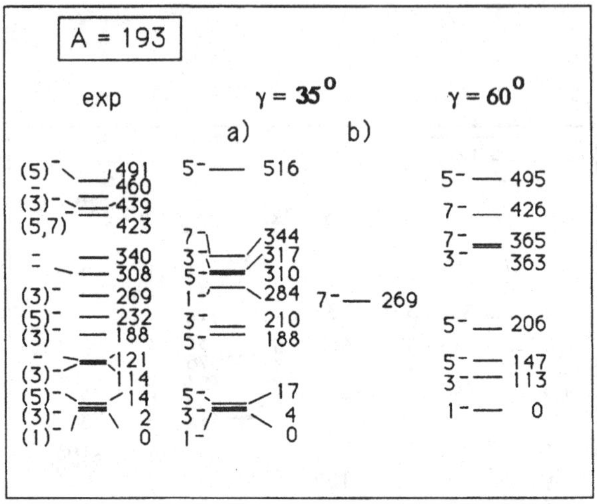

Figure 10. PTR calculation for low-lying negative-parity states in ^{193}Pt, details as in Fig. 7. Experimental data taken from the compilation, Ref. 11.

The other lower-spin $I \leqslant 7/2^-$ states, however, are poorly fitted with these deformation parameters, and for these states a much improved fit is achieved with somewhat lower deformation, $\varepsilon_2 = 0.169, 0.155, 0.150$, and 0.150, respectively, in ^{187}Pt, ^{189}Pt, ^{191}Pt, and ^{193}Pt. We show, in Figs. 7–10, all excited states having $I \leqslant 7/2^-$ in this energy region calculated for $\gamma \approx 30°$ and $\gamma = 60°$. For the case $\gamma \approx 30°$, the calculation predicts the lowest $3/2^-$ state in ^{187,189,191}Pt at 95, 91, 84 keV, respectively, above a predicted $9/2^-$ ground state. Experimentally, $9/2^-$ isomers are found at 86, 173, and 101 keV[24] in these nuclei with $3/2^-$ ground state. The calculated lowest $3/2^-$ states are positively identifiable with the experimental ground states through their magnetic dipole and electric quadrupole moments which are calculated in excellent agreement with experiment (see Table II). Moreover, the great majority of excited states with $I \leqslant 7/2$ are reproduced to within a few tens of keV in the three nuclei, relative to the lowest $3/2^-$ states (see Figs. 7–10). For $\gamma = 60°$, the $9/2^-$ states are calculated at 133, 243, and 282 keV in ^{187,189,191}Pt above the $3/2^-$ ground state. However, most of the excited $I \leqslant 7/2$ states are not well reproduced, and the magnetic moments of the $3/2^-$ ground states fall below experiment by large factors (3, 1.8, and 1.6, respectively).

Thus, we choose to adopt the $\gamma \approx 30°$ calculation with $3/2^-$ ground state and infer that the $h_{9/2}$ isomers have a different deformation although, in the absence of measured transition rates, this apparent variation of deformation within ^{189}Pt must remain tentative. Also, detailed examination of Figs. 7–10 shows that we still do not calculate all states observed below 500 keV. Notably missing are states at 9 keV ($3/2^-$) in ^{187}Pt, 46 keV ($1/2^-$,$3/2^-$), and 222 keV ($3/2^-$) in ^{189}Pt, 30 keV ($1/2^-$,$3/2^-$) in ^{191}Pt, and 114 keV ($3/2^-$) in ^{193}Pt.

It is of interest to examine the wave-functions of the $3/2^-$ ground states of $^{187\text{-}191}$Pt, as they have known magnetic-dipole and electric-quadrupole moments which are fitted by calculations for non-axial shapes only. The structure of all three ground states for the optimum fit shape is very similar and consists of significant components of $f_{5/2}$, $p_{3/2}$, and $p_{1/2}$ spherical parentage.

Strong mixing present amongst the low-lying excited states, close in energy and with similar values of spin, hinders more detailed interpretation of the wave-functions. No clear band structure has been found in $^{187\text{-}193}$Pt below 500 keV. This contrasts with the lighter Pt nuclei and may well be a consequence of lower deformation and stronger coriolis mixing present in the heavier Pt nuclei.

The spectrum of energy levels in ^{195}Pt differs from the pattern known in $^{187\text{-}193}$Pt in two respects. The level density becomes lower, and the lowest $3/2^-$ and $5/2^-$ states increase in energy by about 100 keV with respect to $^{187\text{-}193}$Pt. ^{195}Pt is a marginal case for application of the present model, as its deformation is rather low, and this should be kept in mind in assessing its success. ^{195}Pt is the only nucleus considered in this work which has both known magnetic moments and absolute transition probabilities for low-lying excited states. This type of data is a very sensitive test of the structure of calculated wave-functions. We performed our calculations as a function of γ starting at $\gamma = 35°$ (the value fitted to ^{193}Pt but were unable to achieve any kind of sensible fit to energy levels, nor were we able to reproduce the $1/2^-$ ground state for any γ value less than 50°. Our results are shown in Tables VIII-IX. All fitted properties of the ground, 99-, 130-, and 239-keV states are reproduced surprisingly well, with no obvious preference for $\gamma = 52°$ or 60°. On the other hand, the states with energies 200, 211, 222, and 389 keV are missing in the calculated spectrum. Either these states have different deformation or are of a nature which is outside the model.

As was mentioned in the introduction, several other model calculations have been applied to ^{195}Pt (see, e. g., Refs. 50, 51). In particular, Yamazaki and Sheline[51] applied the Nilsson model, assuming an oblate nuclear shape and the Faessler-Greiner rotation-vibration model to calculate energy levels and spectroscopic factors obtained from (d,p) and (d,t) reactions with considerable success. They predicted a quasi-band structure of low-lying energy levels. Our calculation does not agree with their proposed bands. For example, the ground-state band suggested in Ref. 51 consists of $1/2^-$, $3/2^-$, and $5/2^-$ levels at 0, 211, and 239 keV, respectively. We find very good agreement with the properties of the ground and 239-keV $5/2^-$ state, but the 211-keV state is not present in our calculation.

We consider ^{195}Pt to be the first nucleus in the sequence $^{177\text{-}195}$Pt to have a close-to-oblate shape.

To summarize our discussion of the properties of negative-parity states in $^{177\text{-}195}$Pt, there are several points to stress:

First of all, in the region of lighter Pt nuclei, where level spacing within rotational bands and transition-probability ratios have somewhat limited sensitivity to deviation from axially-symmetrical shape (apart from the $7/2^-$ band in ^{185}Pt), the most pronounced effect of non-zero γ is the relative position of band heads and selection of the ground state in agreement with experiment. This may be understood as reflecting the correct level ordering close to the fermi level in the ε_2, γ plane. It might be thought that changing γ may be an artificial effect, substituting for inadequate treatment of pairing or of the parameters of the deformed harmonic-oscillator potential. We have examined both possibilities and did not find a fit to experiment of the same quality as presented here for $\gamma \neq 0$, with γ constrained to 0 and smooth variation of κ, μ, Δ, and λ within small ranges. We also tried the effect of introducing a deformation parameter ε_4 with $\gamma = 0$ without success.

Second, for heavier Pt nuclei, $^{187\text{-}193}$Pt, the main argument for accepting a triaxial shape, besides the apparent success of our calculation of the ground-state magnetic-dipole moments, is the density of experimental levels. We investigated this problem by calculating energy levels of all the isotopes in question as a function of ε_2 (within limits suggested by

Table IX. Transition Probabilities in ^{195}Pt

E_{level} (keV)	$I^{\pi}(i)$	$I^{\pi}(f)$	E_{γ}^{exp} (keV)	$B(E2)_{exp}$ (Wu)	$B(E2)_{cal}^{a}$ (Wu)	$B(E2)_{cal}^{b}$ (Wu)	$B(M1)_{exp}$ (mWu)	$B(M1)_{cal}^{a}$ (mWu)	$B(M1)_{cal}^{b}$ (mWu)	δ_{exp}	δ_{cal}^{a}	δ_{cal}^{b}
99	$3/2^-$	$1/2^-$	99	11.2(8)	21.9	26.0	16.9(5)	48.7	44.4	−0.130(4)	−0.11	−0.14
130	$5/2^-$	$3/2^-$	31	4.6(20)	0.43	1.0	26(3)	4.8	6.1	−0.021(4)	−0.015	−0.020
		$1/2^-$	100	7.5(5)	4.2	6.2						
239	$5/2^-$	$3/2^-$	140	8(4)	6.3	6.8	15(2)	13.7	16.9	−0.17(4)	−0.15	−0.15
		$1/2^-$	239	8.4(8)	22.7	23.7						

[a-b] As footnotes Table VIII.
Experimental data taken from compilation, Ref. 12.

experiment) and $0 \leq \gamma \leq 60°$. The experimental level density in the low-energy region (where no states outside this model are expected), can be achieved only for non-axial shapes. The additional levels originate naturally from additional degrees of freedom connected with departure from axial symmetry.

Third, our calculation yields a smooth variation of γ as a function of neutron number, which may be seen as a sign of gradual development of the nuclear shape. Had γ been just a fitting parameter without physical content, more irregular behaviour would have been expected.

10.3.D Positive-Parity $i_{13/2}$ Bands in $^{177\text{-}193}$Pt

Odd-A Pt nuclei show a family of $i_{13/2}$ neutron hole states developing from strongly-coupled level ordering in $^{177\text{-}185}$Pt (although with certain staggering between signature partners) to decoupled bands in $^{187\text{-}193}$Pt. There is a striking internal similarity between the bands in lighter Pt isotopes, $A \leq 185$ (see Fig. 11). The heavier, $A \leq 187$ bands form a second close group with very different character (Figs. 12–15).

Detailed calculations for $^{177\text{-}185}$Pt show that in order to reproduce the signature splitting at low spins ($I \leq 21/2$), slowly increasing $16° \leq \gamma \leq 22°$ values are needed; $\gamma = 0$ gives clearly worse results. The calculated ratios R and Λ, on the other hand, are less sensitive to γ and, within experimental errors, no decisive preference emerges for $\gamma = 0$ or $\gamma \neq 0$ for ^{177}Pt and ^{185}Pt. As was found for their negative-parity bands, positive-parity bands in ^{181}Pt and ^{183}Pt show more-pronounced staggering in both R and λ than the calculation predicts, although the general trend is well reproduced.

The wave-functions of the $i_{13/2}$ bands in $^{177\text{-}185}$Pt show only two significant components, $K = 7/2$ and $9/2$, gradually changing from a pure $7/2^+[633]$ orbital in ^{177}Pt to pure $9/2^+[624]$ in ^{185}Pt. Both orbitals contribute to the wave-function of the $9/2^+$ bands in ^{181}Pt [$|a_{7/2}{}^{9/2(\nu=\nu 1)}|^2 = 0.59$ (0.67), $|a_{9/2}{}^{9/2(\nu=\nu 2)}|^2 = 0.33$ (0.27) for $\gamma = 17.5°$ (0)], and in ^{183}Pt [$|a_{7/2}{}^{9/2(\nu=\nu 1)}|^2 = 0.23$ (0.13), $|a_{9/2}{}^{9/2(\nu=\nu 2)}|^2 = 0.74$ (0.85) for $\gamma = 21°$ (0)], and in these nuclei, R and Λ are very sensitive to details of the mixing. No evidence for mixing of $K = 7/2^+$ and $9/2^+$ components, suggested by Dracoulis and al.,[30] was found in the $I \leq 21/2$ members of the $7/2^+$ band in ^{177}Pt.

The $9/2^+$ band in ^{185}Pt is of special interest, as its band head becomes the ground state in this nucleus. This state is often referred to as being of prolate shape.[6,7,34] Indeed, the signature splitting in this band is the least-pronounced in the $^{177\text{-}185}$ Pt region. Results for all experimentally-known electromagnetic properties of this band, including ratios R, Λ and mixing ratios δ, as well as the magnetic-dipole and electric-quadrupole moments (with their present errors) of the $9/2^+$ ground state, show excellent agreement with experiment for both $\gamma = 0$ and $\gamma \neq 0$ with possibly a slight preference to non-axial shape at higher spins (see Table VII). Thus, the spectroscopic data have very limited sensitivity to the shape of the $9/2^+$ state, in agreement with detailed calculation of the $9/2^+[624]$ orbital across the ε_2, γ plane which reveals practically no dependence on γ, as expected for an orbital in the middle of the $i_{13/2}$ shell.

Nevertheless, our calculation of the energy levels of the band members with $I \leq 21/2$ agreed with experiment within less than 10 keV for $\gamma = 22°$, reproducing well small deviations from the typical strongly-coupled, rigid-axial-rotor level spacing. In Fig. 16, we show the sensitivity to γ of the calculation of the moment of inertia parameter as a function of $I(I+1)$. The agreement in Fig. 16 clearly supports a non-axial shape for the $9/2^+$ band.

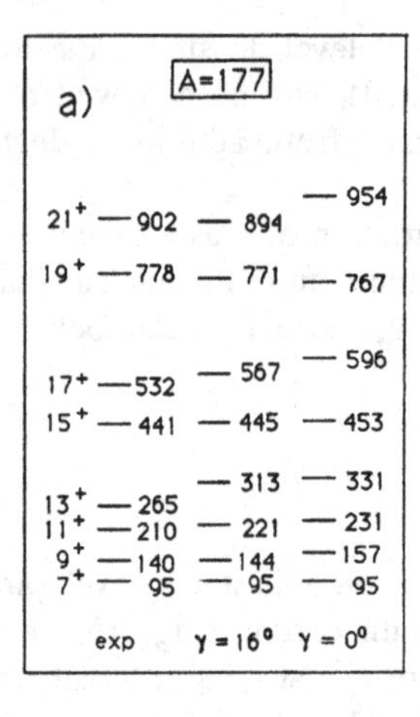

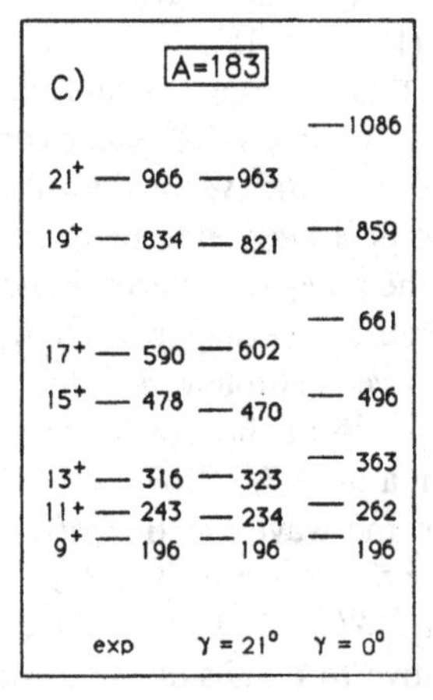

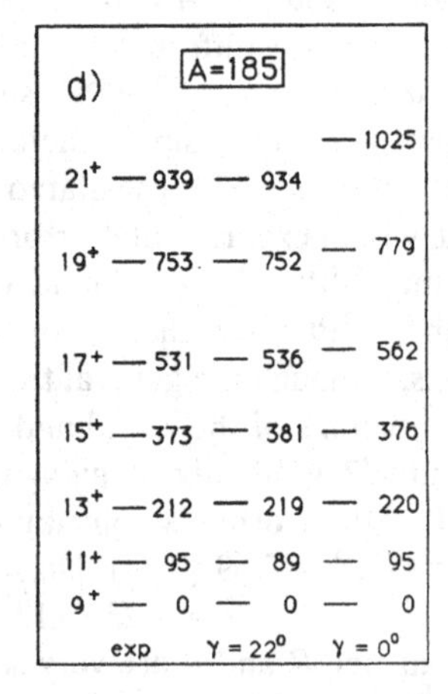

Figure 11. Positive-parity states in $^{177\text{-}185}$Pt with $I \leqslant 21/2$. Results of PTR calculation for both axially-symmetrical and non-axial shapes are compared with experimental data taken from Refs. 30*a*), 31*b*), 32*c*), and 34*d*).

This is in line with the systematic variation in γ found throughout the Pt isotopes. $\gamma = 0$ for the positive-parity states in ^{185}Pt would be highly anomalous.

In the heavier isotopes, a pattern typical of a decoupled band with sizeable energy staggering between states with different signature is seen (Figs. 12-15). Comparison of $\gamma \approx 30°$ and $\gamma = 60°$ calculations, as shown for ^{187}Pt and ^{193}Pt, reveals the inadequacy of the oblate axially-symmetric shape in describing the positive-parity bands in these nuclei. Much the same situation arises in ^{189,191}Pt. Existence of triaxial shapes in these nuclei has been firmly established in numerous papers, using the Meyer-ter-Vehn particle–triaxial-rotor model.[24,36,37] Our calculation shows that the present model gives very similar results and very reasonable agreement with experiment for both energy levels and electromagnetic properties (see Tables X-XII). We note that, in our calculation, only a small coriolis attenuation was needed ($\leqslant 17\%$) as compared to the previous calculation which required $\eta = 5$.[36]

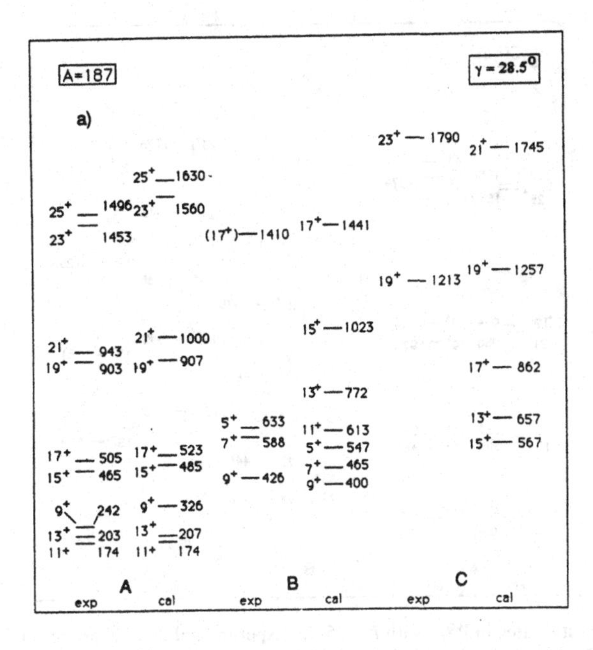

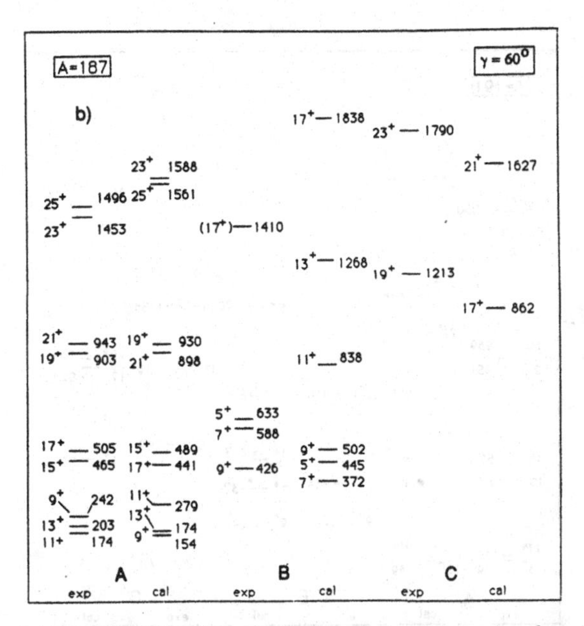

Figure 12. Positive-parity states in ^{187}Pt with $I \leqslant 25/2$. Experimental data taken from Refs. 8 (and 73 I^{π} = $5/2^+, 7/2^+$, and $9/2^+$). Results of the PTR calculation are shown for *a*) $\gamma = 28.5°$ (triaxial) and *b*) $\gamma = 60°$ (oblate) shape. Calculated energy levels have been normalized to 174 keV ($I = 11/2$).

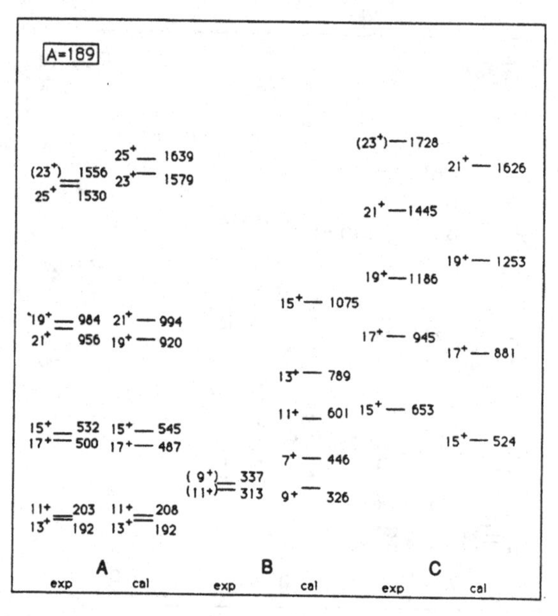

Figure 13. Positive-parity states in ^{189}Pt with $I \leqslant 25/2$. Experimental data taken from the compilation, Ref. 9, are compared with PTR calculation for $\gamma = 32.0°$. Calculated energy levels have been normalized to 192 keV ($I = 13/2$).

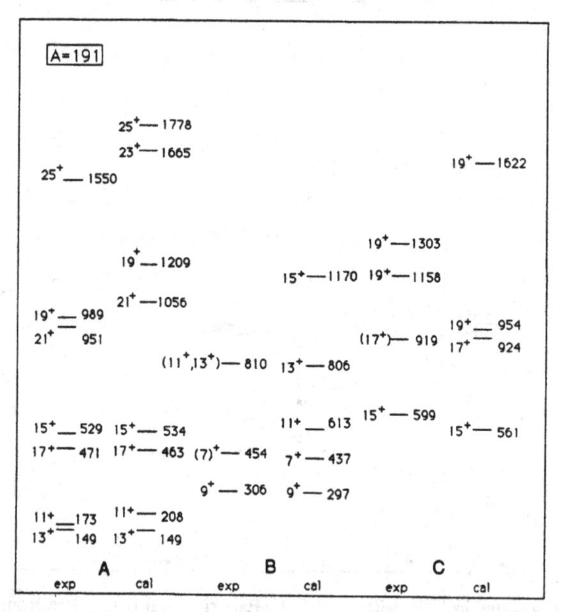

Figure 14. Positive-parity states in ^{191}Pt with $I \leqslant 25/2$. Experimental data taken from Refs. 10 and 36 are compared with PTR calculation for $\gamma = 33.0°$. Calculated energy levels have been normalized to 149 keV ($I = 13/2$). Two possible candidates for the $19/2^+$ member of the band C are shown.

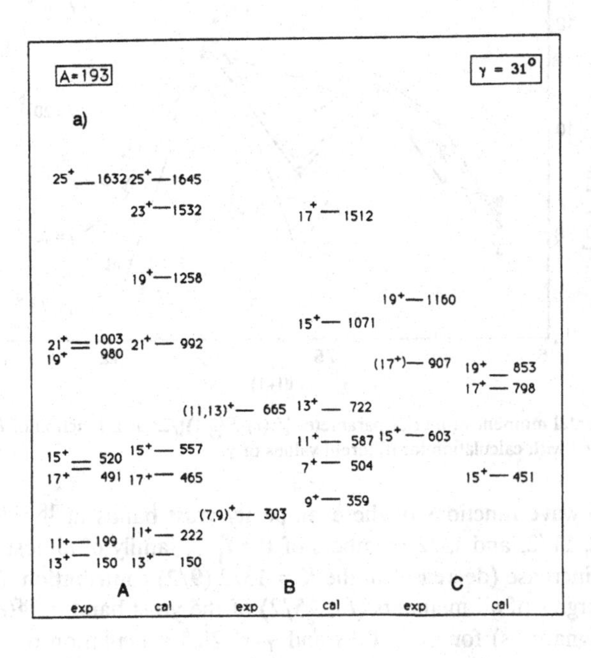

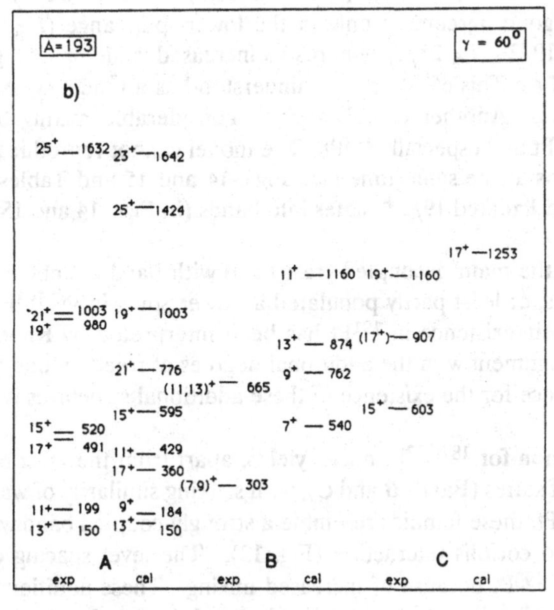

Figure 15. Positive-parity states in ^{193}Pt with $I \leqslant 25/2$. Experimental data taken from Refs. 36 and 76. Results of the PTR calculation are shown for *a*) $\gamma = 31°$ (triaxial) and *b*) $\gamma = 60°$ (oblate). Calculated energy levels have been normalized to 150 keV ($I = 13/2$). Experimental error of level energies at 303 and 665 keV is ± 3 keV.[76]

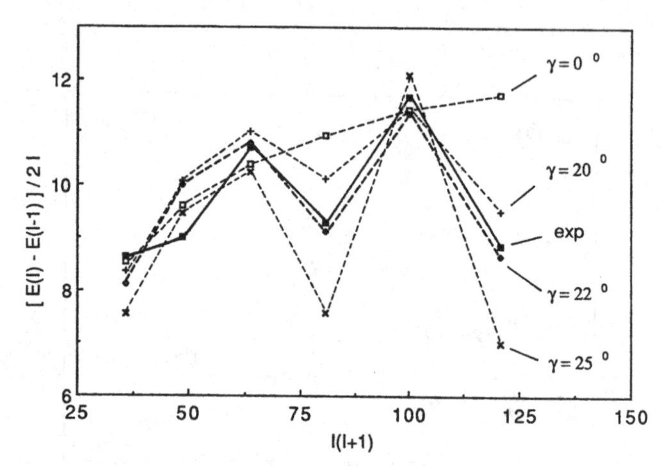

Figure 16. Experimental moment-of-inertia parameter $[E(I)\text{-}E(I\text{-}1)]/2I$ as a function of $I(I + 1)$ for the $9/2^+$ band in ^{185}Pt compared with calculation for different values of γ.

Calculated wave-functions of the even-parity yrast bands in $^{187\text{-}193}$Pt show components of $K = 9/2$, $11/2$, and $13/2$ members of the $i_{13/2}$ family of almost equal magnitude with only a slight increase (decrease) of the $K = 13/2$ $(9/2)$ contribution towards ^{193}Pt.

Whilst energies of all members ($I \leqslant 25/2$) of the yrast band in ^{187}Pt are well reproduced (in both signatures) for $\varepsilon_2 = 0.18$ and $\gamma = 28.5$, calculation of the same band in $^{189\text{-}193}$Pt, using $\varepsilon_2 = 0.176$, $\gamma = 32°$, $\varepsilon_2 = 0.168$, $\gamma = 33°$, and $\varepsilon_2 = 0.168$, $\gamma = 31°$, respectively, gives good agreement only in the lower-spin range ($I \leqslant 17/2$) of the band. Fitting the region, $19/2 \leqslant I \leqslant 25/2$, requires an increased value of γ (e. g., to 36°) for ^{189}Pt at the same ε_2 and ξ. This effect may be understood as a tendency towards oblate shape with increasing spin. Another complication is considerable mixing between close-lying $19/2^+$ states in ^{191}Pt and especially ^{193}Pt. The model cannot reproduce both energy levels and branching ratios at the same time (see Figs. 14 and 15 and Tables XI and XII). We chose to place the calculated $19/2^+$ states into bands (in Figs. 14 and 15) according to their branching ratios.

Apart from the main decoupled yrast band with band members of both signatures, two more structures, at least partly populated at lower spins in $^{187\text{-}193}$Pt, can be identified at low energy. Their existence in ^{191}Pt has been interpreted by Khoo et al.[37] as arising from incomplete alignment with the additional degrees of freedom due to triaxiality. More experimental evidence for the existence of these additional structures is available in ^{187,189,} and ^{193}Pt.

Our calculation for $^{187\text{-}193}$Pt nuclei yields, apart from the yrast band (Band A), two separate families of states (Bands B and C), with striking similarity of wave-functions within each family. In ^{187}Pt, these families resemble a strongly-coupled band with inversion of the lowest levels due to coriolis interaction (Fig. 12). The level spacing changes to a more irregular one in $^{189\text{-}193}$Pt because of increased mixing. These additional calculated bands agree reasonably well with experimentally-observed energy levels and branching ratios where available. The wave-functions of Bands B and C are quite striking. The main K components of Band B are $K = 5/2$, $7/2$, and $9/2$, each of comparable amplitude and close to the amplitudes of their respective $K = 9/2$, $11/2$, and $13/2$ components in the yrast band. In Band C, the major components have $K = 13/2$, $15/2$, (for members with $I = 13/2$, $15/2$), and (for members with $I > 15/2$) $17/2$. Thus, we have $\overline{K_B} + 2 = \overline{K_A} = \overline{K_C} - 2$,

Table X. Transition Probability Ratios in ^{187}Pt (Positive Parity)

$I^{\pi}(i)$	E^{exp}_{level} (MeV)	$I^{\pi}(f_1)$	$E^{exp}_{\gamma}(f_1)$ (MeV)	$I^{\pi}(f_2)$	$E^{exp}_{\gamma}(f_2)$ (MeV)	$B(E2)^{cal}_{\Delta I=2}$ (e^2b^2)	$B(E2)^{cal}_{\Delta I=1}$ (e^2b^2)	$B(M1)^{cal}_{\Delta I=1}$ $(\mu_N)^2$	R^{a}_{exp}	R_{cal}	δ_{exp}	δ_{cal}
$9/2^{+}$ [b]	0.426	$13/2^+$	0.223	$11/2^+$	0.252	0.12	0.38	0.079	0.13(2)	0.030		+0.46
$7/2^{+}$ [b]	0.588	$11/2^+$	0.414	$9/2^+$	0.346	0.23	0.075	0.0027	1.26(14)	5.2		+1.5
$15/2^+$	0.465	$11/2^+$	0.291	$13/2^+$	0.262	0.31	0.41	0.030	0.45(3)	0.50		−0.81
$19/2^+$	0.903	$15/2^+$	0.438	$17/2^+$	0.398	0.58	0.32	0.023	11.7(7)	1.8	$-0.94^{+0.06}_{-0.05}$	−1.2
$19/2^+$	1.213	$15/2^+$	0.748[c]	$17/2^+$	0.708	0.0073	0.039	0.0051		0.18	$-1.82^{+0.30}_{-0.23}$	−1.6

[a] As footnote Table IV.
[b] Experimental data taken from Refs. 8(b) and 73.
[c] Not observed in experiment.

Table XI. Transition Probability Ratios in ^{191}Pt (Positive Parity)

$I^{\pi}(i)$	E^{exp}_{level} (MeV)	$I^{\pi}(f_1)$	$E^{exp}_{\gamma}(f_1)$ (MeV)	$I^{\pi}(f_2)$	$E^{exp}_{\gamma}(f_2)$ (MeV)	$B(E2)^{cal}_{\Delta I=2}$ (e^2b^2)	$B(E2)^{cal}_{\Delta I=1}$ (e^2b^2)	$B(M1)^{cal}_{\Delta I=1}$ $(\mu_N)^2$	R^{a}_{exp}	R_{cal}	δ_{exp}	δ_{cal}
$9/2^+$	0.306	$13/2^+$	0.157	$11/2^+$	0.133	0.46	0.046	0.063	0.56(8)	0.23	|0.26(5)|	−0.10
$7/2^+$	0.454	$11/2^+$	0.281	$9/2^+$	0.148	0.28	0.51	0.017	3.3(3)	6.3		−0.94
$15/2^+$	0.529	$11/2^+$	0.356	$13/2^+$	0.380	0.23	0.077	0.0186	1.2(1)	1.3		−1.2
$19/2^+$	0.989	$15/2^+$	0.461	$17/2^+$	0.519	0.33	0.022	0.0010	2.0(2)	6.9		−2.3
$19/2^+$	1.158	$15/2^+$	0.559	$17/2^+$	0.687	0.48	0.33	0.049	0.66(2)	0.37		−1.5
$19/2^+$	1.303	$15/2^+$	0.703	$17/2^+$	0.832	0.044	0.0040	0.0035	2.4(6)	1.7		+0.74

[a] As footnote Table IV.
Experimental data taken from Refs. 10 and 36.

Table XII. Transition Probability Ratios in ^{193}Pt (Positive Parity)

$I^\pi(i)$	E^{exp}_{level} (MeV)	$I^\pi(f_1)$	$E^{exp}_\gamma(f_1)$ (MeV)	$I^\pi(f_2)$	$E^{exp}_\gamma(f_2)$ (MeV)	$B(E2)^{cal}_{\Delta I=2}$ (e^2b^2)	$B(E2)^{cal}_{\Delta I=1}$ (e^2b^2)	$B(M1)^{cal}_{\Delta I=1}$ $(\mu_N)^2$	R^a_{exp}	R_{cal}	δ_{cal}
$15/2^+$	0.520	$11/2^+$	0.321	$13/2^+$	0.370	0.29	0.0010	0.0048	3.6(4)	2.8	+0.14
$15/2^+$	0.603	$11/2^+$	0.404[c]	$13/2^+$	0.454	0.021	0.60	0.065	<0.004	0.011	−1.1
$19/2^+$	0.980	$15/2^+$	0.461	$17/2^+$	0.490	0.19	0.0161	0.0010	1.45(7)	5.0	−1.6
$19/2^+$ [b]	0.980	$15/2^+$	0.461	$15/2^+$	0.377	0.19	0.15		8.3(1.4)	3.5	

[a] As footnote Table IV.
[b] Ratio of $B(E2)$'s only.
[c] Not observed in experiment.
Experimental data taken from Ref. 36.

where $\bar{K}$ is an average value for each band. It is tempting to recall the prediction of the unified model of Faessler,[69] experimentally confirmed in ^{165}Ho and ^{159}Tb,[70] of the existence of two γ bands in odd-A nuclei as a consequence of the dynamical departure of these nuclei from axial symmetry. States of this nature could be calculated by the present model, if the core-vibration amplitude were within the allowed limits (see Section 10.2 and Ref. 56).

We note that our calculations produce more energy levels of unrelated character, not shown in the figures for simplicity. Several experimental positive-parity states, in particular with non-unique spin assignment,[8,71] have also been omitted.

To summarize, the systematic study of positive-parity states in $^{177\text{-}193}$Pt, all of the same $i_{13/2}$ spherical origin, shows evidence for a smooth departure from axial symmetry in the lighter Pt nuclei (despite limited sensitivity of the $7/2^+[633]$ and $9/2^+[624]$ orbitals to γ), resulting in quite pronounced triaxial shapes in $^{187\text{-}193}$Pt. The lack of experimental data available on positive-parity states in ^{195}Pt means that we have not carried this study to an oblate limit. The main argument for these conclusions is the presence of signature energy splitting in the yrast positive-parity bands in $^{177\text{-}185}$Pt, which is larger than expected for an axially-symmetric $i_{13/2}$ orbital. No admixtures of low-K members of the $i_{13/2}$ family, which would be responsible for increased signature splitting due to increased coriolis interaction, have been found in these bands. In the heavier $^{187\text{-}193}$ Pt isotopes, the signature splitting becomes so pronounced that no other mechanism but the existence of a non-axial shape of the nucleus could give a satisfactory description of this effect.

10.3.E Calculation of Electromagnetic Moments

Results of the calculation of electromagnetic moments are given in Table II ($^{177\text{-}193}$Pt) and Table VIII (^{195}Pt). Previous calculations of Duong et al.[27] and results for axially-symmetrical shapes in the present calculation are shown for comparison. It can be seen that wave-functions obtained from fits previously described yield, in all cases, magnetic moments with correct sign and magnitude, in contrast to the calculation of Duong et al. The excellent agreement with experiment is strong evidence in support of the fitted model parameters. Turning to electric-quadrupole moments, we reproduce the change in sign (and magnitude) of the quadrupole moment of ^{185}Pt, as compared to $^{187\text{-}191}$Pt, without invoking a sudden prolate-oblate shape change, as suggested frequently in the past.[16,17,25,27] To make more use of the limited sensitivity of the nuclear-charge moments to deviation from axial symmetry is not possible at present, given the large experimental errors in published quadrupole moments deriving from Sternheimer corrections and normalisation problems.

As mentioned already in Section 10.3.*C* above, interesting results have been obtained in ^{195}Pt, where magnetic moments of excited states are known (see Table VIII). Together with the fit to energy levels, the close agreement with the magnetic moments can be regarded as strong supporting evidence for at least some states in ^{195}Pt approaching the oblate shape.

10.3.F Calculated Properties of the Even-Even Cores

Comparison of experimental energies of the first 2^+ state and the ground-state band E_4/E_2 ratios in even-even Pt nuclei with the same parameters obtained for the cores used in the present calculation, is given in Table I. We do not fix any core parameters in our calculation and obtain the core first 2^+-state energy and the E_4/E_2 ratios only as a result of the overall fit to a wide range of experimental data. It is clear, from Table I, that the core required in our calculation of an odd-A nucleus differs considerably from the even-even nucleus with $A-1$ for $176 \leqslant A \leqslant 188$ and the E_4/E_2 ratios do not agree in the whole region. As a common feature, calculated cores are more deformed and show less evidence for γ softness. Without going into details, it seems feasible to connect this phenomenon with a polarizing effect of the odd neutron, coupled to otherwise very soft, even-even nuclei. As the result, the core achieves a certain stable deformed shape, most likely dependent upon the particular odd-neutron orbital. If this is true, we should find shape coexistence for different single-particle excitations within the same nucleus. Our results of the fit to negative-parity states in $^{185-195}$Pt (see Sections 10.3.*A-C*), together with a small but persistent difference in the deformation parameters between positive- and negative-parity systems and, in particular, a very different shape for the $h_{9/2}$ band in ^{189}Pt as compared to low-I adjacent states, seems to give evidence for such coexistence. We note that a similar conclusion has been drawn by Ragnarsson and Semmes[52] for the case of ^{185}Au (an odd proton coupled to the even-even Pt core). Their total-energy-surface calculation gives clearly different minima for different configurations in the same nucleus. However, not surprisingly, it is difficult to pursue this point in great detail, except in special cases such as the $h_{9/2}$ band. In its present form, the model cannot include self consistently different configurations with individual deformations in one nucleus. Difficulties in such a calculation involve the absolute energy scale and the treatment of mixing, which would be necessary to understand transition probabilities between coexisting states with different deformations.

10.4 CONCLUSIONS

The present particle–triaxial-rotor calculation of both positive- and negative-parity energy levels and electromagnetic properties in $^{177-195}$Pt gives very good broad agreement with experimental data. The main conclusions of this calculation can be summarized as follows:

1) Fitted values of the deformation parameters ε_2 and γ clearly indicate a gradual change of shape from $\gamma = 16°$ (^{177}Pt) to close-to oblate (^{195}Pt).

2) Our results indicate shape coexistence for different excitations within the same nucleus. This phenomenon can be seen as a consequence of the polarizing effect of the odd neutron on the soft core, driving the odd-A system to different but stable deformations according to the orbital of the odd particle (hole). The proposition that the odd neutron stabilizes deformation of the γ-soft core at certain ε_2 and γ values, leading to the above-mentioned shape coexistence, makes good physical sense and, a postiori, justifies the use of a model that assumes a rigid triaxial core.

3) The calculation offers an abundance of predicted values for comparison with future experiments, in the hope that detailed testing and refinement of this interpretation may result.

ACKNOWLEDGMENTS

It is a pleasure to thank I. Ragnarsson and P. Semmes for many helpful and elucidating discussions and comments. Fruitful, informative talks with P. von Brentano, D.

Brink, V. Janzen, H.-J. Kluge, E. Otten, P. Quentin, B. Roussiere, J. Sauvage, and E. Zganjar are also gratefully acknowledged. Special thanks are given to N. J. Stone for many discussions and continued interest, support, and encouragement during the course of this work. This work has been supported by the Nuclear Structure Committee of the UK SERC.

REFERENCES

1. E. Hagberg, P. G. Hansen, P. Hornsoj, B. Jonson, S. Mattson, and P. Tidemand-Petersson, *Nucl. Phys.* **A318**, 29 (1979).
2. F. Braganca-Gil et al., *Portgal. Phys.* **15**, 59 (1984).
3. B. Roussiere, C. Bourgeois, P. Kilcher, J. Sauvage, M. G. Porquet, A. Wojtasiewicz, M. I. Macias-Marques, and F. Braganca-Gil, *Int. Conf. on Nuclear Shapes (Crete, 1987)*, Vol. 1, p. 79.
4. J. W. Gruter, B. Jonson, O. B. Nielsen, CERN-76-13, 1976.
5. B. Roussiere, C. Bourgeois, P. Kilcher, J. Sauvage, and M. G. Porquet, *Nucl. Phys.* **A504**, 511 (1989).
6. B. Roussiere, C. Bourgeois, P. Kilcher, J. Sauvage, and M. G. Porquet, *Nucl. Phys.* **A438**, 93 (1985).
7. B. Roussiere, C. Bourgeois, P. Kilcher, J. Sauvage, M. G. Porquet, and A. Wojtasiewicz, *Nucl. Phys.* **A485**, 111 (1988).
8. B. E. Gnade, R. W. Fink, and J. L. Wood, *Nucl. Phys.* **A406**, 29 (1983).
9. R. B. Firestone, *Nuclear Data Sheets* **34**, 537 (1981).
10. E. Brownde, *Nuclear Data Sheets* **56**, 593 (1989).
11. V. S. Shirley, *Nuclear Data Sheets* **32**, 593 (1981).
12. Zhou Chunmei, *Nuclear Data Sheets* **57**, 1 (1989).
13. J. F. W. Jansen, A. Faas, and W. J. B. Winter, *Z. Phys.* **261**, 95 (1973).
14. E. van Walle, J. Wouters, D. Vanderplassche, N. Severijns, and L. Vanneste, *Hyp. Int.* **22**, 507 (1985).
15. E. van Walle, Ph.D. Thesis, Leuven, 1987.
16. R. Eder et al., *Hyp. Int.* **43**, 469 (1988).
17. R. Eder et al., *Proceedings of the Eighth International Conference on Hyperfine Interactions, Prague, 1989*, in print in *Hyp. Int.*
18. W. M. Lattimer, K. S. Krane, N. J. Stone, and G. Eska, *J. Phys.* **G7**, 1713 (1981).
19. S. Berkes, R. Haroutunian, and G. Marest, *Journ. Phys.* **G6**, 775 (1980).
20. I. Berkes, G. Marest, and H. Sayouty, *Hyp. Int.* **15/16**, 983 (1983).
21. R. Eder, E. Hagn, and E. Zech, *Phys. Lett.* **158B**, 371 (1985).
22. S. Ohya, K. Nishimura, N. Okabe, and N. Mutsuro, *Hyp. Int.* **22**, 585 (1985).
23. A. Visvanathan, E. F. Zganjar, J. L. Wood, R. W. Fink, L. L. Riedinger, and F. E. Turner, *Phys. Rev. C* **19**, 282 (1979).
24. M. Piiparinen, S. K. Saha, P. J. Daly, C. L. Dors, F. M. Bernthal, and T. L. Khoo, *Phys. Rev. C* **13**, 2208 (1976).
25. B. Roussiere et al., *Hyp. Int.* **43**, 473 (1988).
26. J. K. P. Lee et al., *Phys. Rev. C* **38**, 2985 (1988).
27. H. T. Duong et al., *Phys. Lett.* **B217**, 401 (1989).
28. T. Hilberath, St. Becker, G. Bollen, M. Gerber, H.-J. Kluge, U. Kronert, G. Passler, *Z. Phys.* **A332**, 107 (1989).
29. T. Hilberath, Ph.D. Thesis, Mainz, 1990.
30. G. D. Dracoulis, B. Fabricius, R. A. Bark, A. E. Stuchbery, D. G. Popescu, and T. Kibedi, *Nucl. Phys.* **A510**, 533 (1990).
31. M. J. A. de Voigt et al., *Nucl. Phys.* **A507**, 447 (1990).
32. J. Nyberg et al., *Nucl Phys.* **A511**, 92 (1990).

33. V. P. Janzen et al., *Phys. Rev. Lett.* **61**, 2073 (1988).
34. S. Pilotte et al., *Phys. Rev.* C **40**, 610 (1989).
35. P. J. Daly, C. L. Dors, H. Helppi, M. Piiparinen, S. K. Saha, F. M. Bernthal, and T. L. Khoo, Michigan State Cyclotron Laboratory, *Annual Report 1978*, p.50.
36. S. K. Saha, M. Piiparinen, J. C. Cunnane, P. J. Daly, C. L. Dors, T. L. Khoo, and F. M. Bernthal, *Phys. Rev.* C **15**, 94 (1977).
37. T. L. Khoo, F. M. Bernthal, C. L. Dors, M. Piiparinen, S. K. Saha, P. J. Daly, and J. Meyer-ter-Vehn, *Phys. Lett.* **60B**, 341 (1976).
38. G. D. Draculis, A. E. Stuchbery, A. P. Byrne, A. R. Poletti, S. J. Poletti, J. Gerl, and R. A. Bark, *J. Phys.* **G12**, L97 (1986).
39. M. J. A. de Voigt et al., *Nucl. Phys.* **A507**, 472 (1990).
40. U. Garg, A. Chaudhury, M. W. Drigert, E. G. Funk, J. H. Michelich, D. C. Radford, H. Helppi, R. Holzmann, R. V. F. Janssens, T. L. Khoo, and J. L. Wood, *Phys. Lett.* **180B**, 319 (1986).
41. G. Hebbinghaus, T. Kutsarova, W. Gast, A. Kramer-Flecken, R. M. Lieder, and W. Urban, *Nucl. Phys.* **A514**, 225 (1990).
42. G. Kajrys et al., private communication, 1988, in *Nuclear Data Sheets* **59**, 133 (1990).
43. M. Finger et al., *Nucl. Phys.* **A188**, 369 (1972).
44. M. Piiparinen, J. C. Cunnane, P. J. Daly, C. L. Dors, F. M. Bernthal, and T. L. Khoo, *Phys. Rev. Let.* **34**, 1110 (1975).
45. J. C. Cunnane, M. Piiparinen, P. J. Daly, C. L. Doors, T. L. Khoo, and F. M. Bernthal, *Phys. Rev. C* **13**, 2197 (1976).
46. A. Ansari, *Phys. Rev. C* **33**, 321 (1986).
47. A. Ansari, *Phys. Rev. C* **38**, 953 (1988).
48. R. Bengtsson, T. Bengtsson, J. Dudek, G. Leander, W. Nazarewicz, and Jing-ye Zhang, *Phys. Lett.* **B183**, 1 (1987).
49. S. J. Krieger, P. Quentin, M. S. Weiss, J. Meyer, M. Meyer, N. Redon, H. Flocard, and P.-H. Heenen, *Nucl. Phys.* **A500**, 308 (1989).
50. K. T. Hecht and G. R. Satchler, *Nucl. Phys.* **32**, 286 (1962).
51. Y. Yamazaki and R. K. Sheline, *Phys. Rev. C* **21**, 531 (1976).
52. I. Ragnarsson and P. B. Semmes, *Hyp. Int.* **43**, 425 (1988) and private communication, 1989, 1990.
53. J. Bonn, G. Huber, H.-J. Kluge, L. Kugler, and E. W. Otten, *Phys. Lett.* **38B**, 308 (1972).
54. K. Wallmeroth et al., *Nucl Phys.* **A493**, 224 (1989).
55. J. Rikovska, D. E. Brown, and N. J. Stone, in the *Proceedings of XII Workshop on Nuclear Physics, Iguazu Falls, Argentina*, Ed. by M. C. Cambiaggio, A. J. Kreiner, and E. Ventura (World Scientific, Singapore, 1990), p. 170.
56. S. E. Larsson, G. Leander, and I. Ragnarsson, *Nucl. Phys.* **A307**, 189 (1978).
57. Ch. Vieu, S. E. Larsson, G. Leander, I. Ragnarsson, W. de Wieclawik, and J. S. Dionisio, *J. Phys.* **G4**, 531 (1978) and *Z. Phys.* **A290**, 301 (1979).
58. S. G. Nilsson, Chin Fu Tsang, A. Sobieczewski, Z. Szymanski, S. Wycech, Ch. Gastafson, I.-L. Lamm, P. Moller, and B. Nilsson, *Nucl. Phys.* **A131**, 1 (1969).
59. J. Savage-Letessier et al., *Nucl. Phys.* **370**, 231 (1981).
60. I. Ragnarsson et al., *Nucl. Phys.* **A233**, 329 (1974).
61. R. Bengtsson, P. Moller, J. R. Nix, and Jing-ye Zhang, *Physica Scripta* **29**, 402 (1984).
62. F. S. Stephens, P. Kleinheinz, R. K. Sheline, and R. S. Simon, *Nucl. Phys.* **A222**, 235 (1974).
63. T. Bengtsson and I. Ragnarsson, *Nucl. Phys.* **A436**, 14 (1985).
64. J. C. Bacelar, M. Diebel, C. Ellegaard, J. D. Garrett, G. B. Hagemann, R. Herskind, A. Holm, C.-X. Yang, J.-Y. Zhang, P. O. Tjom, and J. C. Lisle, *Nucl. Phys.* **A442**, 509 (1985).
65. F. Donau, *Nucl. Phys.* **A471**, 469 (1987).

66. P. Moller and J. R. Nix, Los Alamos Preprint LA-UR-86-3983, 1986.
67. C. D. Papanicolopulos, M. A. Grimm, J. L. Wood, E. F. Zganjar, M. O. Kortelahti, J. D. Cole, and H. K. Carter, *Z. Phys.* **A330**, 371 (1988).
68. J. Jastrebski et al., *J. de Phys.* **34**, 755 (1973).
69. J. M. Eisberg and W. Greiner, *Nuclear Models*, North Holland, 1975, **Vol. 1**, p. 262.
70. R. M. Diamond, B. Elbeck, and F. S. Stephens, *Nucl. Phys.* **43**, 560, (1963).
71. E. Browne, *Nuclear Data Sheets* **58**, 441 (1989).
72. K. Kumar and M. Baranger, *Nucl. Phys.* **A122**, 273 (1968) and references therein.
73. G. Kajrys, private communication, 1989.
74. P. Raghavan, *Atomic Data and Nuclear Data Tables* **42**, 189 (1989).
75. J. Kalifa, G. Berrier-Ronsin, M. Vergnes, G. Rotbard, J. Vernotte, Y. Deschamps, and R. Seltz, *Phys. Rev.* **C22**, 997 (1980).
76. G. R. Smith, N. J. di Giacomo, M. L. Munger, and R. J. Peterson, *Nucl. Phys.* **A290**, 72 (1977).

11. Towards Superheavy Nuclei—Status and Prospects for the Production and Investigation of Heavy Elements

G. Münzenberg and P. Armbruster
Gesellschaft für Schwerionenforschung
D-6100 Darmstadt, West Germany

ABSTRACT

The recent experiments on the synthesis of the trans-actinide elements 104 to 109 led to the discovery of a region of nuclei with an unexpected high stability against spontaneous fission, which could be explained in the frame of our experimental results by large ground-state shell-effects. These nuclei are located already in a region of macroscopic instability and owe their existence predominantly to microscopic effects, as has been predicted for the superheavy nuclei. The newly discovered heavy elements are stabilized by a gap in the Nilsson levels coupled to a strong hexadecapole deformation. In our experiments we learned that the chance to proceed to still heavier elements beyond 109 by complete fusion of heavy ions is determined by the possibility of their production. The cross-sections are limited by the strongly reduced fusability of massive nuclear systems at the coulomb barrier, as well as by the small survival probability of the excited compound nucleus on its way to the ground state. In this paper we discuss our experimental results on the decay properties of the heaviest nuclei, as well as the results on their production. We consider this within the framework of recent theories on nuclear ground-state stability of the heaviest elements and the fusability of massive systems.

11.1 INTRODUCTION

The investigation of the heaviest elements is one of the most difficult and challenging experimental tasks, as the very limits of nuclear stability are touched. The competition between the cohesive nuclear forces and—towards the upper end of the periodic table—the rapidly-increasing disruptive coulomb force pushes nuclear half-lives down to milliseconds for the known isotopes of the heaviest elements 107 to 109. In addition, it also cuts down the production cross-sections to below the nanobarn region. Thus, investigation of the elements at the upper end of the periodic table has to cope with nuclei having very short half-lives and production rates of less than a few atoms per day.

11.2 THE PRODUCTION OF HEAVY ELEMENTS

11.2.A Overview

Production of trans-actinide elements beyond Lr was possible only by heavy-ion fusion. The complete amalgamation of target and heavy-projectile nuclei creates new isotopes far from the region of available target materials and ion beams in one step. By irridating targets from the artificial trans-uranium elements (typical Cm and Cf isotopes) with beams

Exotic Nuclear Spectroscopy, Edited by W. C. McHarris
Plenum Press, New York, 1990

of C, B, N, and O, it was possible to reach element 106.[1]

More symmetric target-projectile combinations permit the production of weakly-excited compound systems, an idea firstly published by Oganessian.[2] The use of targets of the strongly-bound doubly-magic nucleus ^{208}Pb or close to it and beams of ^{50}Ti, ^{54}Cr, or ^{58}Fe leads to compound nuclei with excitation energies of about 20 MeV, compared to 40–50 MeV for reactions using actinide targets. This is advantageous for the production of very-heavy and highly-fissile elements. With this method of "cold fusion," which was the type of reaction investigated thoroughly in the last two years, it has been possible to synthesize the heaviest elements known at present, 107 to 109.[3,4]

Figure 1 shows the excitation functions for the production of elements with even proton numbers, Fm to 108, measured at the Separator for Heavy-Ion Reaction Products (SHIP) in the complete-fusion reactions of ^{40}Ar, ^{48}Ca, ^{50}Ti, ^{54}Cr, and ^{58}Fe with ^{208}Pb. The onset of cold fusion is observed at an excitation energy of about 30 MeV, with the $2n$ channel close to 20 MeV. This drop in the excitation energy causes an increase of the evaporation-residue cross-section.

For the elements heavier than No the excitation functions show two remarkable features: The cross-sections are reduced by about one to two orders of magnitude per two elements, while the excitation energy at the barrier (calculated for instance by Bass,[5] arrows in Fig. 1) is almost constant but decreases for the heaviest elements.

The production cross-section for a fusion-evaporation product is given by the product of the fusion cross-section and the survival probability of the compound nucleus:

$$\sigma(E) = \pi \lambda\!\!\!^{-2} \sum (2\ell + 1) p(E,\ell) \omega(E+Q, \ell) \quad (1)$$

Here λ is the de-Broglie wavelength of the entrance channel, Q the reaction Q-value, p the fusion probability, and ω the survival probability, the latter two depending on the angular momentum ℓ of the system. The experimentally-observed decrease in cross-section might therefore have two reasons: A reduced fusability for heavy systems caused by dynamic effects, or a decrease in survival probability due to the increasing fissility of the heavy systems.

To study entrance-channel effects in the fusion of massive systems, a systematic investigation of evaporation-residue production has been carried out at SHIP in recent

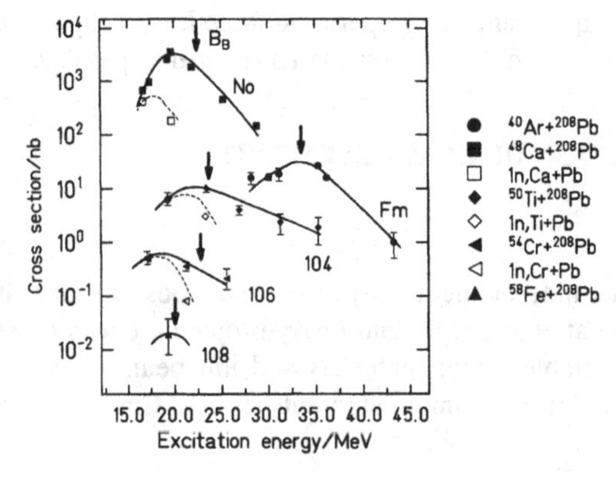

Figure 1. Excitation functions for the production of the heaviest known elements with Pb-targets. The Bass barrier[5] is indicated with arrows.

years.[6-13] Figure 2 shows the production cross-sections for evaporation residues in the region from Er ($Z = 68$) to element 109, produced by the fusion of heavy ions and measured at the Bass barrier, plotted versus the fissility of the fusing system in the entrance channel as given by Blocki et al.[14] We find a rapid decrease in the evaporation-residue cross-sections close to $x_m = 0.70$, which corresponds[15] to the reaction, Zr + Zr → Hg. The cross-sections drop from the order of the geometrical cross-section for Hg to about 10 pb, more than 10 orders of magnitude, for the production of elements 108 and 109. For Hg and heavier systems symmetric reactions were chosen, and for the heavy elements Pb- and Bi-based reactions were chosen for Fig. 2; thus, automatically, cold compound systems have been selected, which deexcitate by evaporation of one or two neutrons.

A first answer to the question as to whether the strong decrease in the production cross-section results from a reduced fusability, caused by the large coulomb forces in the entrance channel, or from a strongly decreasing survival probability of the compound nucleus, due to an increasing fissility for the heavier elements, can be given if we look into other nuclear-reaction channels. Simultaneous with the decrease in the production cross-section, the transfer cross-sections increase (Fig. 2)[16] and for $x_m \approx 0.75$ a new reaction channel is opened, fast fission,[17,18] which is the link between nuclear transfer and complete fusion. This points towards a hindrance in fusion.

The reduced fusability of massive nuclei can be explained in terms of dissipative losses of the two amalgamating nuclei on their way to fusion. These losses raise the fusion barrier above the conventional barrier located near the contact point of target and projectile, which leads to a decrease of the fusion probability at the conventional barrier: As the projectile cannot pass inside the barrier, the two colliding nuclei reseparate after interaction, and the missing cross-section shows up in the transfer products.[16]

For very-heavy systems even this condition is not sufficient for complete fusion. Very-heavy nuclei undergo fission at small ground-state deformations. For those systems, the saddle point for fission from the ground-state (unconditional saddle) is located inside the saddle for the binary system (conditional saddle). To achieve complete fusion, the system has to pass inside this second saddle. Those systems which have passed the first, but not the second saddle, decay into fission-like products; this fast fission becomes important the region, $x_m \approx 0.75$.

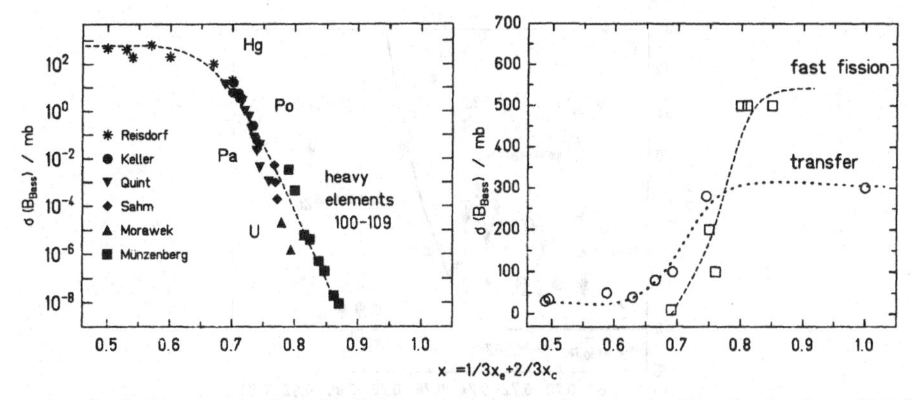

Figure 2. Production cross-sections for fusion-evaporation products at the Bass barrier for cold-fusion reactins, e.g., symmetric systems and heavy elements with Pb on Bi targets (full symbols) left panel, and transfer at the coulomb barrier (circles), and fast fission (squares) right panel, plotted vs the fissility parameter of the entrance channel.

11.2.B The Extra Push in the Complete Fusion of Massive Nuclei

The barrier shift in the fusion of massive systems has been studied at SHIP. From experimental excitation functions for the production of evaporation residues, the extra-extra push referred to the Bass barrier has been extracted by fusion-evaporation calculations, where the barrier height (and of course barrier fluctuations) have been chosen to match the experimental cross-sections. Mainly symmetric target-projectile combinations, leading to compound nuclei in the region from Hg to U, were investigated.[6-10] Figure 3 shows the experimental extra-extra push data plotted versus the entrance-fissility parameter x_m, which is averaged from the entrance fissility of the fusing system for equilibrated charge x_e and the fissility parameter of the compound nucleus x_c.[14] Figure 3 also contains the cold-fusion reactions, ^{40}Ar, ^{48}Ca, ^{50}Ti + Pb, leading to the compound nuclei, Fm, No, and 104; also, ^{48}Ca + ^{209}Bi, leading to Lr (the data points of Fig. 3 contain a subset of the reactions displayed in as in Fig. 2, using the same symbols). We find a strong onset of the extra-extra push close to x_m = 0.70, which rises up to more than 30 MeV for x_m = 0.75, reflecting the cross-section curve of Fig. 2 discussed above.

The general trend of a fast increase of the extra push within a very narrow window of the entrance fissility parameter is well explained by the calculations of Blocki et al., which come from a further development of Swiatecki's extra-push model.[20] In this theory, the rate of energy flow is described by one-body dissipation. The dissipative losses occur through wall-and-window friction: The nucleons, which can move freely inside the nuclei because of the pauli principle, interact with the walls of the deforming body of the nuclei on their way to fusion and also exchange energy through the window, which is opened by the neck of the amalgamating system (Fig. 4).

A similar model developed by Nix et al.[19] uses the surface-plus-window mechanism. In contrast to Swiatecki's model, where the nucleons move randomly, in this model the velocity distribution is constrained by the many-body wave-function to be a slater determinant of single-particle wave-functions. Also, the total strength of the nuclear interaction is a free parameter, whereas the Blocki model does not contain any free parameters. In the calculations of Nix et al., the one-body dissipation is about only one-tenth of that in the Blocki theory. As a consequence, the onset of the extra-extra push starts later.

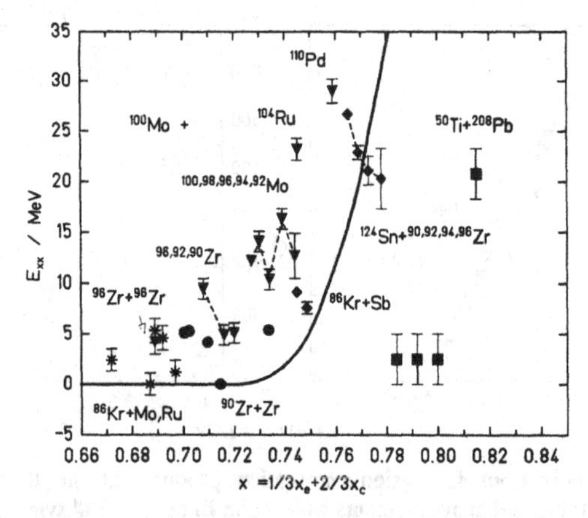

Figure 3. The experimental extra-extra push data compared to the calculations of Blocki (solid line)[14].

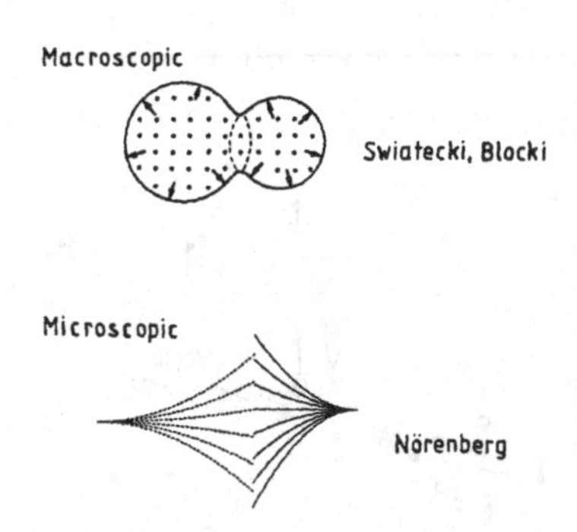

Figure 4. The macroscopic and microscopic explanations of energy dissipation in the fusion process.

Of course, these macroscopic theories can predict only the general trend of an extra push in the fusion of heavy systems, but not the effects of nuclear structure, which go opposite to the macroscopic expectations (as the data do for the target-projectile combinations along the isotopic chains of ^{100,98,96,94,92}Mo and ^{96,92,90}Zr).

The extra push for combinations with closed-shell nuclei such as ^{92}Mo or ^{90}Zr in the entrance channel is significantly lower than that with soft mid-shell nuclei such as ^{100}Mo or ^{90}Zr. These structure effects were explained within the framework of the dissipative-adiabatic model.[21] In this model the energy dissipation is linked to the filling of nuclear shells. In the phase of approach, the degeneracy of the nuclear levels is removed and the levels split, while the nucleons stay on their orbits and can be raised to higher energy levels. This effect cancels out if all levels are filled.

Figure 5 shows a comparison of the experimental data and the calculations. The calculations nicely reproduce the nuclear-structure effects; they tend, however, to be too small for the large extra-push values. The reason is that the theory takes into account only the early stages of the fusion process. Possibly a macroscopic-microscopic approach to account for the later stages of the nuclear system on its way to fusion (where structure effects are destroyed) might be included in the sense of a macroscopic-microscopic fusion theory. Such structure effects will also explain the extremely small extra-push in cold-fusion reactions, where targets near the doubly-magic nucleus ^{208}Pb are used.[15] It is noteworthy to mention that the surface-friction model of Fröbrich, which, similar to the dissipative adiabatic approach,[22] takes into account only the touching phase of fusion, rides straight through the points from this calculation.

An interesting question would be how much ground-state properties of the compound system for cold fusion would affect the extra-extra push; for instance, ground-states with saddle-point configurations located at large deformations.

Let us now briefly discuss the exit-channel effects, e.g., the influence of the survival probability on the production cross-section. The heaviest element for the production of which an excitation function has been measured is element 107. Figure 6 shows the excitation function for the reaction, ^{209}Bi(^{54}Cr, xn)$^{273-x}$107, which is rather narrow with a half-width of about 6 MeV, indicating that the competition of penetrating the fusion barrier and survival of the compound nucleus is extremely strong.[23] Remarkably, the maximum forma-

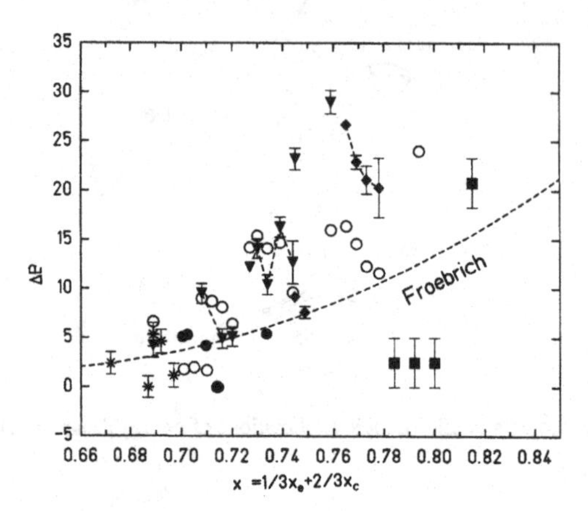

Figure 5. Comparison of the experimental extra-extra push data (dots) to the calculations of Berdichevsky, Nörenberg et al.[21] (circles) and Fröbrich (dotted line), respectively.[22]

tion cross-section is located close to the Bass barrier at 21 ± 2 MeV and not shifted to higher energies by the extra-extra push when proceeding to the heavier elements, where the production cross-sections peak near 20 MeV excitation energy.[4] The maximum cross-section for the 1-*n* deexcitation channel is 163 ± 34 pb and exceeds that of the 2-*n* channel (36_{-16}^{+25} pb) of Pb by a factor of about four. This is the first time that we observed the 2-*n* channel to be weaker than the 1-*n* deexcitation channel. Elements 108 and 109 were only observed in heavy-ion-fusion, γ-1*n* reactions.

The decrease towards the heavier elements of the peak cross-section for the 2-*n* channels with respect to those of the 1-*n* channels can be explained by a decrease in the stability of excited systems, e.g., the survival probability of predominantly shell-stabilized compound nuclei. The increase of the fusion cross-section caused by an enhanced barrier penetrability at higher energies is over-compensated for by a decrease of the survival probability of the more highly-excited compound nucleus. This is an effect which may be rather pronounced for the shell-stabilized nuclei, as shell effects start to be damped out with a damping constant of 18 MeV^{-1},[24] which becomes important above excitation energies of 20 MeV.[25] This assumption is supported by the experimentally-observed fact that the maximum production cross-section of all trans-actinide elements produced by cold fusion is located near 20 MeV excitation energy.

Extrapolations on the basis of our measured cold-fusion reactions predict a produc-

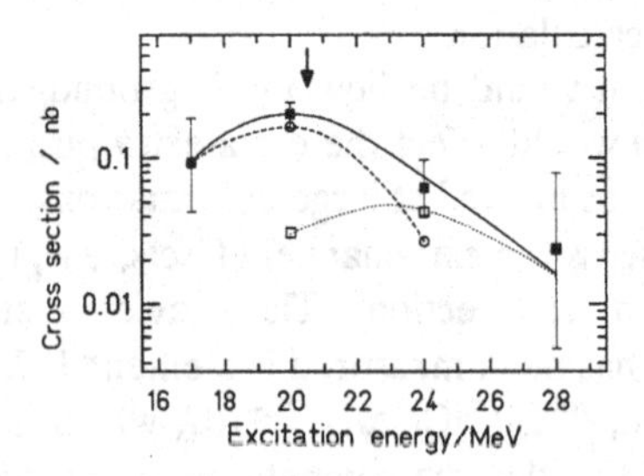

Figure 6. Excitation function for the production of element 107 in irradiations of ^{209}Bi with ^{54}Cr.[23] The arrow indicates the Bass-barrier.

tion cross-section of about 1 pb for element 110, or irradiation times of about three weeks for the production of one single atom of this element, if we use the reaction $^{208}Pb(^{62}Ni,n)$. One experiment to produce element 110 in an irradiation of ^{208}Pb with ^{64}Ni failed already; the limiting cross-section of <8 pb was certainly too high. This path to the heavier elements is closed by the fusion-hindrance of massive nuclear systems.

The use of more-asymmetric target-projectile combinations where the fusion hindrance is smaller, on the other hand, is limited by the survival probability of heated, shell-stabilized nuclei, since the Bass barriers for those systems are located near 50 MeV excitation energy. Here we might profit from the fact that we produce more neutron-rich systems in which the neutrons are less bound, which helps a lot for the nuclei to survive the initial stages of deexcitation. Element 106, for instance, was produced in the reaction, $^{249}Cf(^{18}O,4n)^{263}106$, at an excitation energy of 40 MeV,[1] with a cross-section of about 300 pb, which is comparable to the cold-fusion cross-section of 500 pb for the reaction, $^{208}Pb(^{54}Cr,2n)$.[24]

11.3 THE INVESTIGATION OF HEAVY ELEMENTS AT SHIP

11.3.A Overview

The main experimental work at SHIP concentrated on the investigation of the trans-actinides with Z = 104 to 109, which were produced in irradiations of Pb or Bi with ^{50}Ti, ^{54}Cr, or ^{58}Fe.[3,4] In this chapter, we present our recent experimental results on the ground-state properties of elements 107 and 109 and their interpretation. Our present knowledge about the decay of the heaviest elements is shown in Fig. 7.

Figure 7 shows that all known isotopes of the heaviest elements beyond element 105 have half-lives of less than one second and range down into the millisecond region. Therefore, the appropriate method for their investigation is separation in flight followed by implantation into Si surface-barrier detectors to observe their decay. In fact, this method is

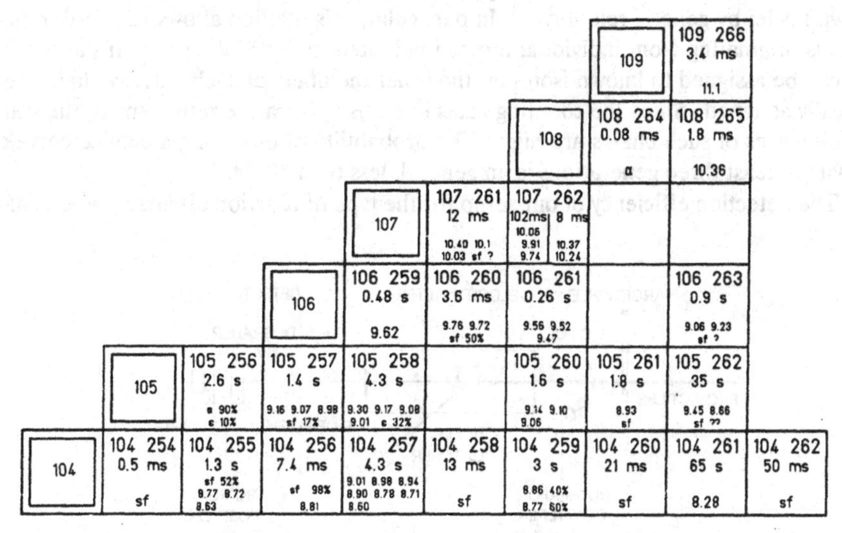

Figure 7. The upper end of the periodic table with the presently-known isotopes.

already very valuable for the investigation of the isotopic chain of element 104, where millisecond fission of the even-even isotopes and few second half-lives of the odd-mass isotopes alternate; in-flight separation for the first time allowed the investigation of isotopic chains of this element in a single experiment.[12]

11.3.B Experimental Method

The in-flight separation at SHIP takes advantages of the reaction kinematics of complete fusion. The evaporation residues produced in the complete-fusion reaction receive full momentum transfer from the projectile and recoil from the target with the center-of-mass velocity. All other products from binary interactions move with different velocities, for the coulomb forces drive them apart after interaction; hence, they can be separated in flight with velocity as the appropriate parameter for separation. To achieve assignment, the separated nuclei are identified by genetic relationships of α decay, a method introduced for the identification of new elements by Ghiorso et al.[1]

A schematic drawing of our setup is given in Fig. 8. The heavy-ion beam from the UNILAC heavy-ion accelerator with energies of about 5 MeV/u, close to the coulomb barrier, has typical intensities of $2x10^{12}$ particles/s and is produced from isotopically-enriched source material. The thin foil targets of about 400-$\mu g/cm^2$ thickness are evaporated onto a C backing and covered with a thin C layer to avoid sputtering and to improve radiative cooling. In order to endure the high beam intensities, they are mounted on a rotating target wheel.[26]

The heavy recoils are separated in-flight with SHIP, a velocity filter similar to the well-known wien filter but with spatially-separated electric and magnetic deflection fields. To obtain high background suppression, SHIP has two separation stages.[27] The detector system consists of a secondary-electron-emission time-of-flight detector with thin C foils and an array of seven position-sensitive Si surface-barrier detectors. Combined time-of-flight and energy measurements of α decay or spontaneous fission of a heavy nucleus at its position of implantation, plus the corresponding time intervals, permit the identification of unknown nuclei by genetic relations.[28] In particular, this method allows us to follow α-decay chains originating from individual atoms implanted in the Si detector. If parts of such chains can be assigned to known isotopes, the other members of such a decay chain are automatically assigned. Since the counting rates in our α spectra are rather small, the statistical significances of such chains are high. The probability of observing a chance-correlated chain with at least three generations is, in general, less than 10^{-14}.

The detection efficiency of our setup for the type of reaction discussed here is of the

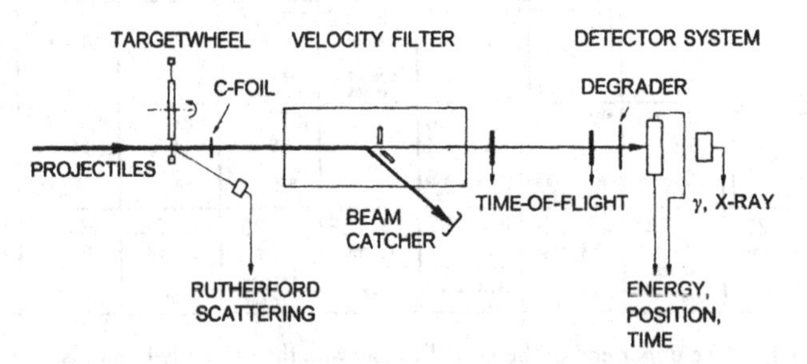

Figure 8. Experimental setup for the detection of heavy elements with SHIP.

order of 25% for evaporation residues created by neutron evaporation from the compound nucleus and one order of magnitude less for those from α,xn channels. The probability for detecting the full α energy from the decay of an implanted nucleus is about 60%; for the detection of the total kinetic energy of fission, 50%. This results form the fact that the implantation depth of the recoils is smaller than the range of the decay products in the detector.

11.3.C Experimental Results for Elements 107 and 109

Our experiments in the search for new elements started with the investigation of elements with odd proton numbers, 107 and 109. The reason was that at that time fission decay was believed to become the predominant decay mode for the trans-actinides. Our aim, however, was to produce α-emitting isotopes, in order to apply the method of genetic relationships for their safe identification. Thus, we identified the isotopes, $^{262}107$ (Ref. 29) and $^{266}109$ (Ref. 30), produced in bombardments of ^{209}Bi with ^{54}Cr and ^{58}Fe, respectively.

It turned out that those experiments were extremely difficult: The isotope $^{262}107$ has a complex α-decay scheme. In our first experiment, where we observed only six atoms of element 107, we already found two different half-lives for this isotope, indicating ground-state and metastable-state decays. For element 109 we were only able to produce one single atom in our first experiment[30] because of the small cross-section—of the order of 10 pb.

11.3.D Element 107

In our experiments we observed two isotopes of element 107, having masses 262 and 261 (Ref. 23). Both of them were produced in irradiations of ^{209}Bi with ^{54}Cr. Figure 9 shows the α spectra from the irradiation at 4.88 MeV/u. The integral projectile dose of $2.8x10^{17}$ needed for these spectra was achieved within 70 h. Spectrum 9*a*) shows all α decays in the energy range of 7.5 to 10.5 MeV, accumulated during the intervals between

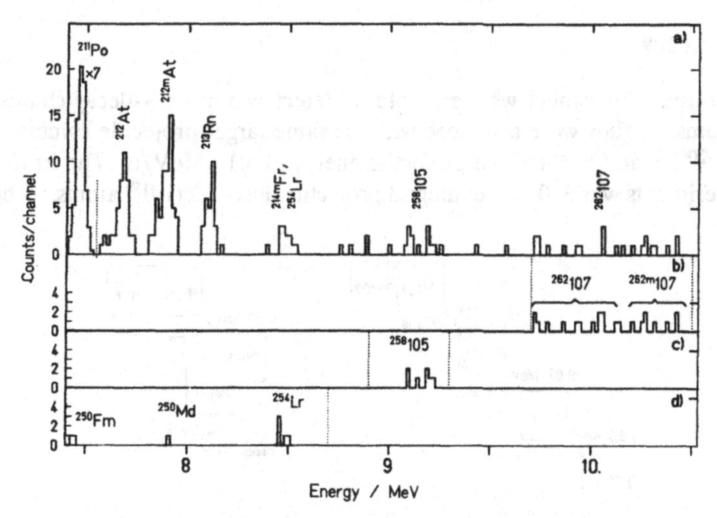

Figure 9. α spectra of element 107 and following generations.[23]

the UNILAC beam bursts. In *9b*), we accumulated all decays which followed the implantation of a heavy ion in the Si detector within a position window of ±0.4 mm and a time window of 1 s. The two bottom spectra show the genetic relationships of α decay: *9c*) the daughter and *9d*) the granddaughter and grand-granddaughter generations. This analysis proves that in this irradiation the α-emitting isotope $^{262}107$ was produced. The losses in the course of the subsequent generations are caused by the 50% escape probability for γ rays and by a 30% *EC* branch in $^{258}105$, leading to $^{258}104$, which decays by spontaneous fission.

A detailed analysis of the decay chains and their assignments was performed by analysing the single-atom decay sequences. Two examples for $^{262}107$ are shown in Fig. 10. These chains additionally contain the time intervals measured between single α decays.

For $^{261}107$ we measured a half-life of 11.8 ms and three α-decay energies. We did not observe spontaneous fission for this isotope. The fission branch is $b_f \leqslant 10\%$ for $^{261}107$. According to results from Dubna, the isotope $^{261}107$ produced in the same reaction should undergo spontaneous fission with a 1–2 ms half-life.[32] Our first experiments on elements 107 and 105[13,29] showed that at least their arguments leading to this assignment were incorrect.

For the analysis of the time distances in order to evaluate the half-lives of the observed isotopes in our case of poor statistics, we plotted the measured time intervals in logarithmic bins,[33] resulting in a distribution which peaks at the average lifetime. Figure 11 shows our analysis for the decays of $^{262}107^{g,m}$ as obtained from the experiments discussed here; the curve for $^{258}105$ was taken from our experiments on the synthesis of element 105.[13]

Table I shows a compilation of our results on the decay properties of both isotopes, including the numbers of decays observed from the following generations; fission of element 107 was not observed. For $^{262}107$, we observed two transitions with 102-ms and 8-ms half-lives. The assignment of one or both transitions to the ground state and isomeric state of $^{262}107$ is based only on α-decay energy systematics and has been made for simplicity. The fission branch for this isotope is $b_f < 3\%$ (at a confidence level of 68%). The upper limit for a fission branch of $^{261}107$ is $b_f < 10\%$.

11.3.E Element 109

In a recent experiment we were able to detect two more α-decay chains from element-109 atoms.[31] They were produced with the same target-projectile combination as before, $^{58}Fe + ^{209}Bi$, and at the same projectile energy of 5.17 MeV/u. The irradiation time for both experiments was 330 h, the integral projectile dose 1.2×10^{18} atoms. The produc-

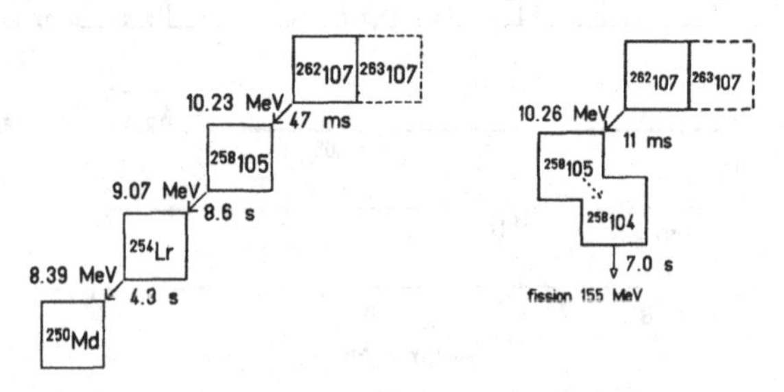

Figure 10. Decay chains from two atoms of $^{262}107^m$.

Table I. Measured Spectroscopic Properties of the Transitions Assigned to Element 107 and the Number of Correlated Decays from Later Generations

Assignment	E_α (MeV)	No. of Events	No. of Correlations with 105	with Lr	with Fission	with Md,Fm	$t_{1/2}$ (ms)
$^{261}107$	10.40	4	2	2		1	$11.8^{+5.3}_{-2.8}$
	10.10	4	1	1	2	1	
	10.13	2	1	1	0	1	
$^{262}107$	10.06	7	2	2	3		102±26
	9.91	4	1	1	2	2	
	9.74	4	2	2	1	1	
$^{262}107^m$	10.37	8	3	3	1	3	8.0±2.1
	10.24	6	3	1	2	1	

tion cross-section is 10_{-6}^{+10} pb. Figure 12 shows the old and the new chains of element 109; they fit nicely to the two chains from $^{262}107$ in Fig. 10. The new chains show that the observed atom of $^{266}109$ decays via the "isomeric" state of $^{262}107$, indicating that $^{266}109$ also might have a complicated α-decay scheme. The half-life of $^{266}109$ is $3.4_{-1.3}^{+6.1}$ ms, and its α-decay energy is 11.0 ±0.03 MeV. The time intervals from these chains for the decays of $^{262}107$ and $^{258}105$, are shown in Fig. 11 as hatched histograms.

Experiments on the synthesis of the heaviest elements also have been carried out at Dubna, where the same reactions were investigated with nuclear-chemical methods.[36] In order to circumvent the half-life problem, the long-lived isotopes at the end of the α-decay chains of the heavy elements were detected. At the end of the $^{262}109$ chain ^{246}Cf appears, the daughter of ^{250}Fm (Fig. 12). ^{246}Cf is an α emitter with a 36-h half-life. In the Dubna

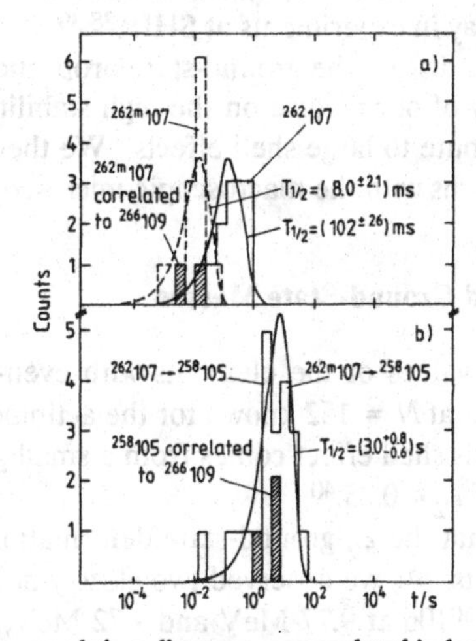

Figure 11. Distribution of the measured time distances, accumulated in logarithmic bins. Histograms indicate the decay chains of $^{262}107$; hatched histograms, the decay chains of $^{266}109$.

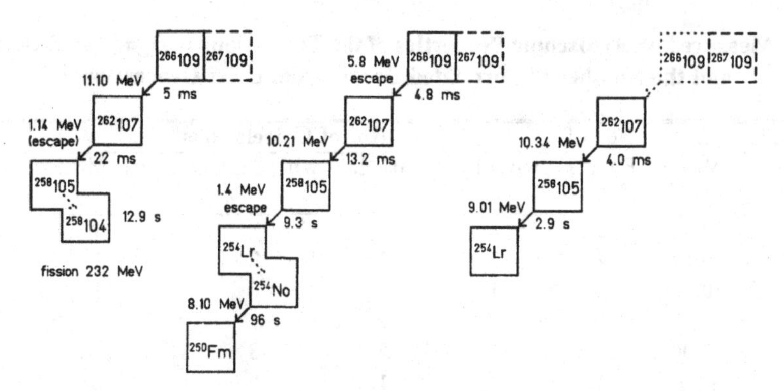

Figure 12. The three decay chains of element 109 observed in the irradiations of ^{209}Bi with ^{58}Fe.

experiments, this isotope was observed in irradiations of ^{209}Bi with ^{54}Cr, as well as with ^{58}Fe, from which the production of ^{262}107 and ^{266}109, respectively, was concluded.

In general, the experimental results obtained at Dubna lead to the same conclusions as those from our experiments, though they can, strictly speaking, serve only as indirect confirmations which do not contradict our results.

11.4 GROUND-STATE PROPERTIES OF THE HEAVIEST ELEMENTS

11.4.A Introduction

The most exciting result of heavy-element research in recent years—besides the discovery of the new elements 107 to 109—was the observation that the role of fission decay decreases compared with α decay when we pass beyond element 104 (see Fig. 7). The first evidence for this unexpected nuclear stability at the very upper end of the nuclear table was obtained conclusively in experiments on the synthesis of element 106 at Dubna[37] and later by direct observation of α decay in experiments at SHIP.[38,39]

In the following we will discuss the ground-state properties of the heaviest elements and try to test the consistency of our picture on the high stability of the heaviest elements against fission, which we attribute to large shell effects. We then discuss briefly the results of recently improved calculations with the macroscopic-microscopic approach.

11.4.B α-Decay Properties and Ground-State Masses

The experimental Q_α values of the elements with even-proton number in Fig. 13 show that the small shell effect at $N = 152$ known for the actinide elements is still observed for the heaviest elements. This shell effect comes from a small gap in the Nilsson levels at a ground-state deformation of $\varepsilon_2 \approx 0.25$.[40]

This is an indication that the ε_2 ground-state deformation is rather constant in this region of the heaviest elements. As we observed two closely-neighboring α-decay energies for the doubly-even nucleus ^{260}106 at 9.77 MeV and 9.72 MeV, respectively, (these we attribute to transitions in the lowest rotational levels 0^+, 2^+ of the daughter nucleus ^{256}104) we can make a direct comparison of ground-state deformations. The nuclear moment of

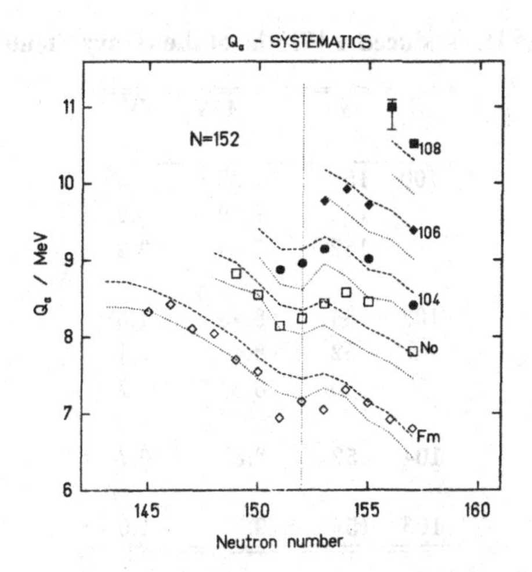

Figure 13. Q_α values of the heaviest elements with even-proton number. Symbols: Experiment; dashed line = Møller/Nix[34] and dotted line = Møller/Myers et al. 88.[35]

inertia is $2J/\hbar^2 = (120 \pm 15)$ MeV^{-1}, which is also in agreement with shell-model calculations.[40]

Of course, this result indicates that no influence of the doubly-closed shell predicted for $Z = 114$ and $N = 184$ shows up. A similar conclusion can be made from the reduced α width of the doubly-even isotopes of the heaviest elements, shown in Table II. The table shows the reduced α widths close to 1, a value observed for nuclei far from β stability.[41] They were calculated from the Rasmussen formula[42] and referred to ^{212}Po.

11.4.C Ground-State Masses

When we compare our experimental Q_α values with two recent calculations of Møller and Nix[34] and Møller et al.,[35] we find that the Møller/Nix calculations predict the Q_α values nicely within an error of 0.1 MeV to 0.5 MeV for even the heaviest known isotopes, whereas the Møller/Myers formula fails in the actinide region by up to more than 0.5 MeV, with an increasing deviation towards the upper end of the periodic table (Fig. 14). It predicts a higher nuclear stability than observed experimentally. As both calculations use the same microscopic correction, obtained with the folded Yukawa potential[44], this effect is caused by deviations in the macroscopic surfaces used in both mass formulae. The macroscopic part of the Møller, Nix mass formula is obtained from the liquid-drop model with the Yukawa-plus-exponential potential, further developed from the Krappe, Nix, and Sierk formula.[45] The macroscopic part of the Møller, Myers calculation is a further development of the droplet model from Myers[46], now including the finite range of the nuclear force.

Figure 15 shows a comparison of both mass formulae with the experimental masses. Here we took experimental masses from literature or from our $N - Z = 48$ isotopes, and, for isotopes with unknown ground-state masses, from the systematics of Wapstra[43] (no symbols). While the experimental masses deviate from the Møller, Nix predictions by −1 MeV to +0.5 MeV for Pu and heavier elements. The experimental masses agree nicely with those calculated from Møller, Myers, et al.; however, they deviate rapidly towards the heaviest elements by up to 2 MeV for the heaviest elements. This deviation may be caused

Table II. Reduced α Widths of the Heavy Elements

Z	N	E_α (MeV)	W_α
100	156	6.97	1.5
	154	7.19	0.9
	152	7.04	2.2
102	154	8.43	1.5
	152	8.40	2.1
	150	8.41	0.7
104	152	8.81	0.7
106	154	9.77	1.0

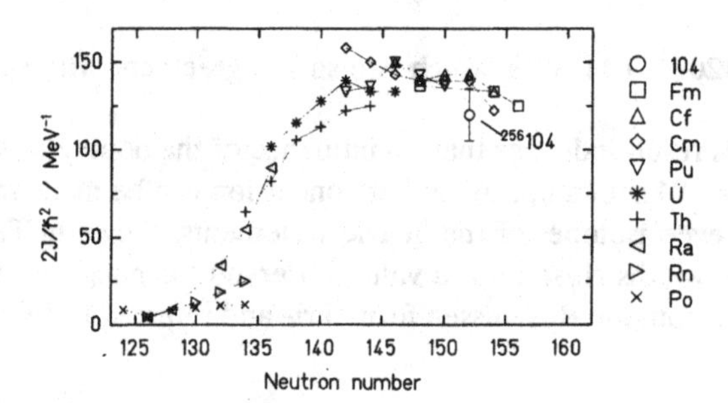

Figure 14. Moments of inertia for the $Z > 82$ nuclei.

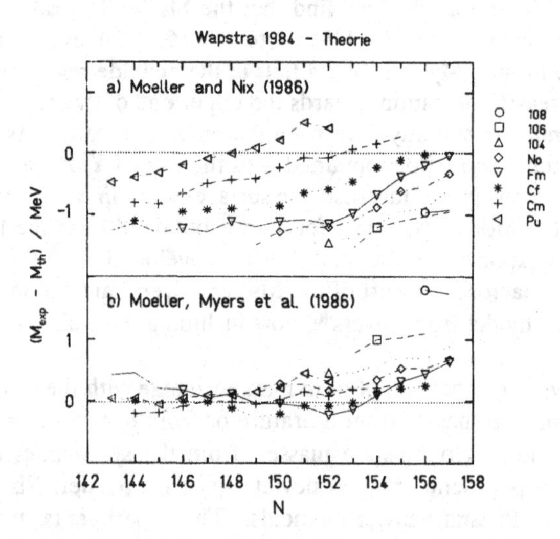

Figure 15. Differences between the experimental masses (resp. masses from Wapstra systematics[43]) to the calculations of Møller and Nix[34] and Møller, Myers, et al.[35]

by the redistribution energy, a term describing the charge distribution within the nuclear drop, which is not used in the liquid-drop formula.[47]

11.4.D Shell Effects and Fission Barriers

The fission half-life systematics of the doubly-even isotopes show a change in the half-lives beyond No (Fig. 16). Up to No we find, as expected in the frame of the liquid-drop model, a general decrease of fission half-lives with increasing fissility parameter x. For elements 104 to 108, the partial-fission half-lives become independent of this macroscopic parameter, suggesting that microscopic effects determine fission half-lives, i.e., nuclear stability, in this region. This observation is corroborated by the observation that the half-life fluctuations in the isotopic chains of the elements Cm to No increase towards the heavier elements. They are caused by the small shell effect at 152 neutrons.

For the direct evaluation of shell effects from the experimental masses, an appropriate macroscopic surface is needed, the choice of which is rather problematic after we have seen in the previous section that the two most recent theories deviate from each other by about 2.8 MeV for the heaviest elements. Here we are confronted with an inherent problem of the macroscopic-microscopic approach. The only experimental quantity to compare with theory is the nuclear mass, so we have no criteria to apply in choosing which of the macroscopic models we should prefer. Nevertheless, an experimental test of the theoretical models, especially in the region of the heavy elements, is crucial for testing their predictive power. The microscopic part, for instance, is important for learning about stability against nuclear fission and thus about the predictive power of the nuclear-mass formulae for heavy and superheavy nuclei.

Therefore, we will try to evaluate the shell effects obtained with various models. Here we will take the Lysekil formula,[48] since as it is the simplest and still is frequently used and the two recent liquid-drop[34] and droplet formulae[35] to extract shell effects from the experimental masses and discuss them in the frame of our present knowledge about heavy elements, including the other important part of our experimental results, the fission half-lives. We then examined which of the surfaces gives the most consistent picture.

Figure 17 shows the shell effects obtained with the macroscopic surfaces [Møller and Nix: Dots; Møller et al.: Triangles; Myers and Swiatecki (Lysekil): Squares]. The microscopic part of the Møller and Nix mass formula is also indicated (circles). For all

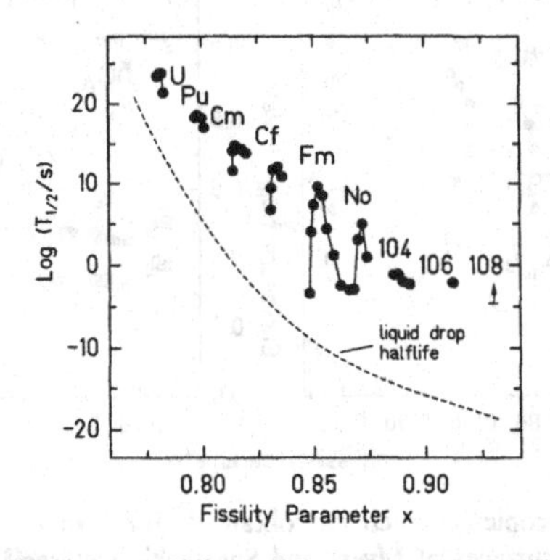

Figure 16. Logarithm of the partial-fission half-lives of the even-even nuclei plotted vs the fissility parameter. The half-lives expected, if the shell stabilization were to be absent, are indicated as a dashed line.

four groups we find continuously-increasing shell effects towards the heavier elements. The smallest "experimental" shell effects are obtained with the Lysekil formula, the largest ones with the macroscopic surface of Møller and Nix. The differences between both start from about 2 MeV at U and reach to as much as 4 MeV for element 108. For U we extract "experimental" shell effects close to zero from the Lysekil formula, and around Pu and Cm, where most of the fission isomers are observed, they are below 2 MeV, which is rather a small value when we remember that the fission isomers live from the strong wiggles in the shell effects of deformed nuclei.

With the droplet surface of Møller and Myers, we obtain "experimental" shell effects close to those with the liquid drop surface of Møller and Nix. In the region above Cf they start to differ with increasing element number. Which set of shell effects appropriately explain our experimental data can be decided using the experimental fission half-lives. The fission half-life is given by

$$t_{1/2} = \ln 2/np\,, \tag{2}$$

where n is the number of barrier assaults, which is $2.5\mathrm{x}10^{20}$/s, and p the probability of the nucleus for penetrating the fission barrier. If we treat the fission barrier in the parabolic approximation, the penetrability is

$$p = [1 + \exp(2\pi B_F/\hbar\omega_F)]^{-1}\,. \tag{3}$$

Here, B_F represents the height of the fission barrier and $\hbar\omega_F$ the barrier curvature. The height of the fission barrier is in good approximation,

$$B_F \approx B_{LD} - E_{\text{shell}}\,, \tag{4}$$

where B_{LD} is the liquid drop part of the barrier, which we will take from semi-empirical formula.[52] In Eq. (4) the shell correction at the fission saddle point is neglected, a correction which should be small (< 1 MeV) for the isotopes discussed here.

With our three sets of "experimental" shell corrections we obtain the barrier-curvature energies displayed in Fig. 17. For U to Pu we find rather a good agreement for all

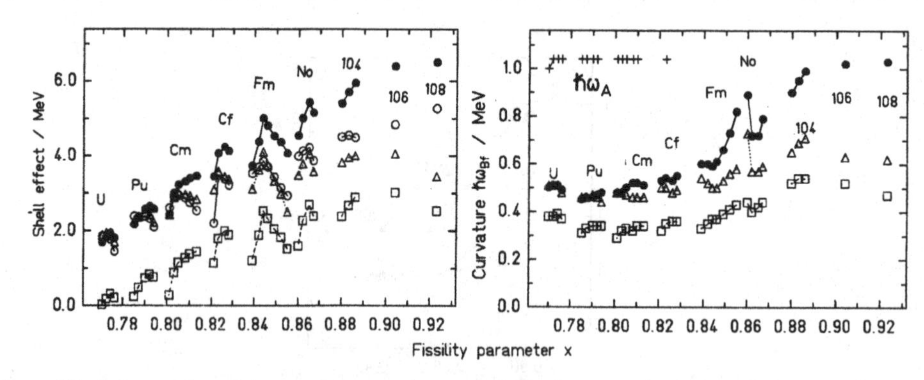

Figure 17. *Left panel*: Microscopic corrections obtained for experimental masses (resp. Wapstra systematics[43]) and: Macroscopic surfaces of Myers and Swiatecki (Squares[48]), Møller and Nix (dots[34]), and Møller and Myers (triangles[35])—the circles refer to the microscopic calculation of Møller. *Right panel*: Barrier curvature parameters obtained with the microscopic corrections. *Crosses*: Curvatures of the inner barriers.[49]

three data sets, as the barriers are dominated by the liquid-drop part (B_{LD} = 4–6 MeV). The liquid-drop barrier itself would have a curvature of $\hbar\omega$ = 0.6 MeV for x = 0.775,[53] a value not too far from the average of $\hbar\omega$ = 0.45 MeV observed here. With the increasing influence of the shell effects, however, the barrier curvatures obtained with the three sets of shell effects begin to differ more and more, especially in the trans-actinide region, where the liquid-drop barrier falls below 0.5 MeV and does not contribute significantly any more to nuclear stability.

The small shell corrections obtained with the Lysekil macroscopic surface have to be compensated by large barrier thicknesses, i.e., respectively small barrier curvatures, in order to achieve the experimental half-life. With the set from the droplet surface, the barrier curvature increases to about 0.7 MeV. With the shell effects from the Møller and Nix liquid drop surface, the trans-actinides have barrier curvatures close to 1 MeV. This value is close to the barrier curvature of the inner barrier, measured for the heavy actinides to be 0.8 MeV to 1 MeV.[49] Here, the outer hump of the doubly-humped fission barrier, located on top of the liquid-drop barrier at large deformations, drops down as the liquid-drop barrier remains effective. Thus, the fission barrier of the trans-actinides is thin and single-humped, determined only by the evaluation of the ground-state shell effect vs nuclear deformation. This picture is consistent with calculations, e.g. Ref. 54.

11.5 WHAT DID WE LEARN ABOUT THE STABILITY OF HEAVY AND SUPER-HEAVY NUCLEI?

The unexpected exciting result from the investigation of the trans-actinide nuclei is the observation that strong shell effects stabilize these nuclei against fission. These nuclei have a liquid-drop barrier of about 0.5 MeV[52,54] and are consequently macroscopically instable. In that sense, a new region of superheavy nuclei has been discovered. The nuclei in this region, however, owe their stability to a strong ground-state hexadecapole deformation and not to a double shell closure, like the superheavy nuclei predicted in the classic calculations.[55,58] That the unexpected high stability is coupled to a hexadecapole deformation has been calculated first by Cwiok et al.[59] and recently by Møller et al. and Sobiczewski et al. with the folded Yukawa[44] and Woods-Saxon potentials,[54,60] respectively.

The maximum of that stabilization is predicted for N = 162 and Z = 108. This new region of hexadecapole-stabilized nuclei interconnects the upper end of the periodic table and the region of the "classic" spherical superheavies near N = 184 and Z = 114. In our new picture, these no longer form an island which is separated from the upper end of the periodic table.[54,60]

Half-life calculations including fission are rather difficult and hence tend to have large errors, as they have to include the evaluation of the fission barrier and the inertial mass in the multidimensional deformation space.

Figure 18 shows the comparison of the experimental partial-fission half-lives for the trans-actinide elements 104 and 106 with recent calculations of Møller et al.[50] and Böning et al.[51] The calculation of Møller et al. shows a drop of the fission half-life of about six orders of magnitude when N = 164 is approached. This is because of the expectation that these nuclei split into two magic N = 82 nuclei, which opens a new fission path leading to the compact scission shape of a nucleus splitting into two spherical fragments. In fact, this new fission valley explained, for example, the symmetric fission of ^{258}Fm and the short half-life (0.38 ms) of this nucleus.

The calculation of Böning et al. does not show this decrease; in contrast, the fission

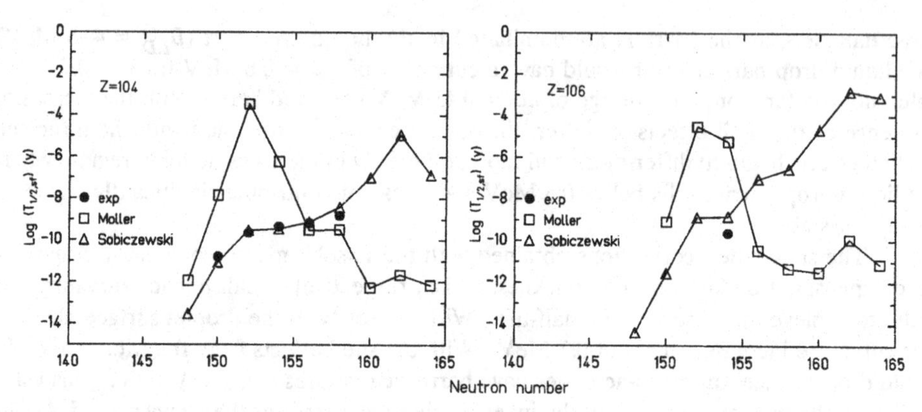

Figure 18. Experimental (dots) and calculated fission half-lives from Møller et al.[50] (squares) and Böning et al.[51] (triangles).

half-lives profit from the increasing stabilization towards $N = 162$ isotopes with the maximum predicted hexadecapole deformation. They can also explain the Fm half-lives.[61] The main difference between the two calculations is the increased deformation space.[54,59,60] As these calculations seem to reproduce the presently known trans-actinide half-lives better, we will use them for the prediction of the fission half-lives of the heavy and superheavy nuclei, discussed in the following.

Figure 19 shows the evaluation of the half-lives of the heavy and superheavy elements Fm to 120 as a contour plot plotted vs the neutron number.[15,51,60,62] The figure shows the weak stabilizing influence of the $N = 152$ neutron gap and the maximum stability of the hexadecapole deformed region near $N = 162$, as well as the enhanced stability at the $N = 182$ shell closure. As all decay modes are considered in the contour plot, the maximum in the half-lives is shifted down to $Z = 110$. The figure also shows a band of nuclei with half-lives ≈1 s, extending from the neutron rich side of element 105 to the double shell closure. Those nuclei should be detectable with fast on-line chemistry. The 10^{-6} s limit refers to in-flight separation.

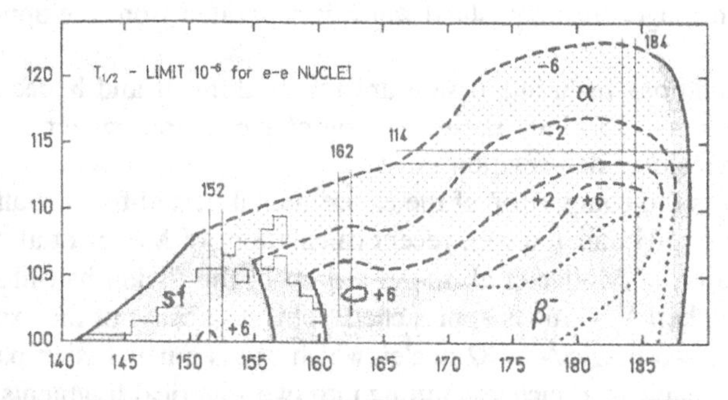

Figure 19. The upper end of the Chart of Nuclides limited by a half-life of 10^{-6} s for even-even isotopes.[52,64] The lines indicate half-lives in seconds (log $T_{1/2}$ given as number). The known isotopes are shown in the lower-left corner. Full line: spontaneous fission; hatched line: α decay; dotted line: β^- decay. The picture has been taken from Ref. [15].

11.6 CONCLUSION AND OUTLOOK

Heavy-element research of recent years has led to the discovery of a region of shell-stabilized nuclei in the trans-actinides. They are superheavy nuclei in the sense that they are already located in a region of macroscopic instability. In contrast to the classic superheavy nuclei, they are not stabilized by a double shell closure but by a large hexadecapole deformation. This newly-discovered region of α-emitting nuclei is predicted to interconnect the known nuclei of the upper end of the periodic table and those around the double shell closure at $N = 182$ and $Z = 114$.

The heaviest elements identified at present, 107, 108, and 109 are produced with cross-sections of 100–10 pb in cold-fusion reactions with Pb or Bi targets. The small production cross-sections result from a hindrance in the fusion of massive systems. Thus, prospects for going far beyond element 109 are poor. For elements 110 and 111, production cross-sections of the order of 1 pb are expected. With our present techniques, this corresponds to production rates of 1 atom/3 weeks. Certainly, we will try to produce element 110 in the reaction, $^{208}Pb + {}^{62}Ni$. The question as to whether the more asymmetric combinations with actinide targets, which are the only ones to come closer to $N = 162$, would be more favorable is not yet fully answered. From our present knowledge, the high excitation energies at the barrier are unfavorable, so that one could only hope for sub-barrier fusion.

To meet the experimental problems of increasingly smaller cross-sections, we have completed SHIP by a *DQQ*-separator which can be operated in combination with SHIP as a recoil mass-separator with moderate resolution ($M/\Delta M \approx 130$) but high transmission,[39,63] Fig. 20. This separator would reduce background of scattered particles, target-like recoils, and transfer products, however, at the expense of a slightly smaller transmission. The advantage of this setup will be fully available when more powerful ion sources with intensities of more than 10^{13}/s becomes available.

Of course, we also discuss the use of radioactive neutron-rich secondary beams; with the intensities available at SIS (Table III), we will not able to produce heavy elements.

The scientific goal of heavy-element research in the future besides the production of new elements should be to understand the fusion of massive systems better, especially applied to the production of heavy elements and, of course, to produce and investigate more transactinides, especially in the neutron-rich region $N \approx 156$, and to learn more about the shell-stabilized superheavy elements at the upper end of the periodic table.

Table III. Calculated Production Rates for Neutron-Rich Isotopes using ^{86}Kr Beams of 5 x 10^{11}/s and 10 g/cm^2 C Targets[64]

Isotope	Intensity (s^{-1})
^{50}Ca	4×10^5
^{52}Ca	10^5
^{52}Ti	3×10^7
^{56}Cr	7×10^7
^{60}Fe	2×10^8

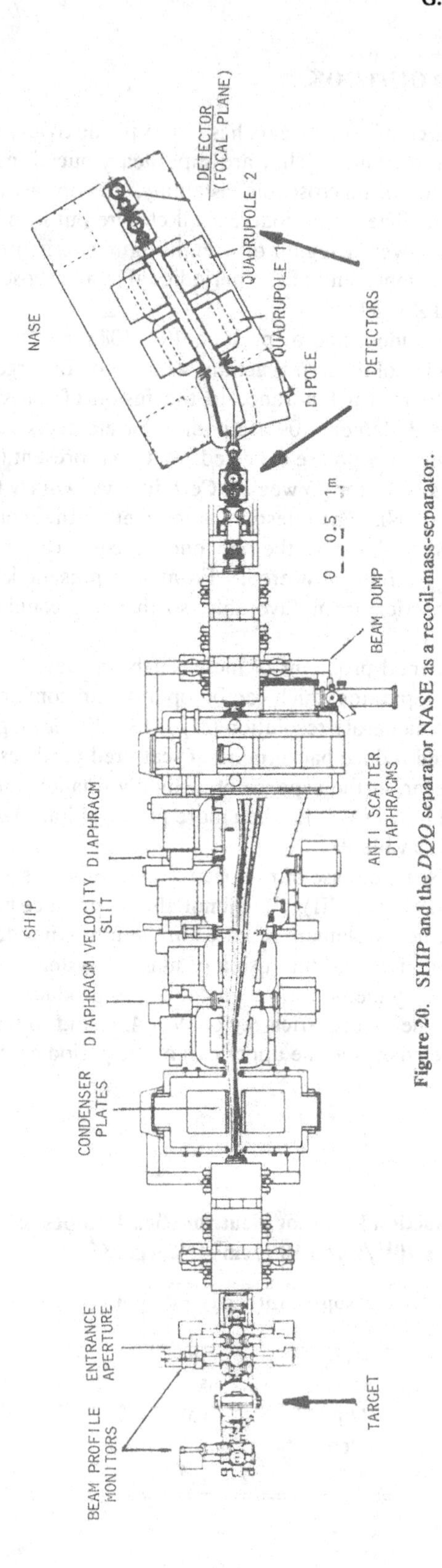

Figure 20. SHIP and the *DQQ* separator NASE as a recoil-mass-separator.

ACKNOWLEDGMENTS

We would like to thank S. Hofmann, F. P. Hessberger, M. E. Leino, W. Reisdorf, and K. H. Schmidt, as the main co-workers in these experiments; and A. Sobiczewski for many useful discussions.

REFERENCES

1. A. Ghiorso, J. M. Nitschke, J. R. Alonso, C. T. Alonso, M. Nurmia, and G. T. Seaborg, *Phys. Rev. Lett.* **33**, 1490 (1974).
2. Y. Ts. Oganessian, *Lecture Notes in Physics* **33**, 221 (1974).
3. P. Armbruster, *Ann. Rev. Nucl. Part. Sci.* **35**, 135 (1985).
4. G. Münzenberg, *Reports on Progress in Physics* **51**, 57 (1988).
5. R. Bass, Proc. Symp. "Deep Inelastic and Fusion Reactions with Heavy Ions", *Lecture Notes in Physics* **117**, 420 (1980).
6. C. C. Sahm, H. G. Clerc, K. H. Schmidt, W. Reisdorf, P. Armbuster, F. P. Hessberger, J. G. Keller, G. Münzenberg, and D. Vermeulen, *Nucl. Phys.* **A441**, 316 (1985).
7. W. Reisdorf, F. P. Hessberger, K. D. Hildenbrand, S. Hofmann, G. Münzenberg, K. H. Schmidt, W. F. W. Schneider, K. Sümmerer, G. Wirth, J. V. Kratz, K. Schlitt, and C. C. Sahm, *Nucl. Phys.* **A444**, 154 (1985).
8. J. G. Keller, K. H. Schmidt, F. P. Hessberger, G. Münzenberg, W. Reisdorf, H. G. Clerc, and C. C. Sahm, *Nucl. Phys.* **A452**, 173 (1986).
9. U. Morawek, D. Ackermann, T. Brohm, H. G. Clerc, U. Gollerthan, E. Hanelt, M. Horz, W. Schwab, B. Voss, K. H. Schmidt, J. J. Gaimard, and F. P. Hessberger, *GSI Ann. Rep.* GSI **89-1**, 38 (1989).
10. A. B. Quint, Thesis, Technische Hochschule Darmstadt (1988).
11. K. H. Schmidt, *Nucl. Phys.* **A488**, 47 (1988).
12. F. P. Hessberger, G. Münzenberg, S. Hofmann, W. Reisdorf, K. H. Schmidt, H.-J. Schött, P. Armbruster, R. Hingmann, B. Thuma, and D. Vermeulen, *Z. Phys.* **A321**, 317 (1985).
13. F. P. Hessberger, G. Münzenberg, S. Hoffman, Y. K. Agarwal, K. Poppensieker, W. Reisdorf, K. H. Schmidt, J. R. H. Schneider, W. F. W. Schneider, H.-J. Schött, P. Armbruster, B. Thuma, C. C. Sahm, and D. Vermeulen, *Z. Phys.* **A322**, 557 (1985).
14. J. P. Blocki, H. Feldmeier, and W. J. Swiatecki, *Nucl. Phys.* **A459**, 145 (1986).
15. P. Armbruster, Int. Nuclear Physics Conf., São Paulo, Brazil (1989).
16. R. Bellwied, J. V. Kratz, W. Reisdorf, D. Schüll, B. Kohlmeyer, and R. Kunkel, *Lect. Notes in Physics* **317**, 125 (1988).
17. B. Heusch, C. Volant, H. Freisleben, R. P. Chestnut, K. D. Hildenbrand, F. Pühlhofer, W. F. W. Schneider, B. Kohlmeyer, and W. Pfeffer, *Z. Phys.* **A288**, 391 (1988).
18. W. Q. Shen, J. Albinski, A. Gobbi, S. Gralla, K. D. Hildenbrand, N. Herrmann, J. Kuzminski, W. I. J. Muller, H. Stelzer, J. Töke, B. B. Back, J. Bjørnholm, and S. P. Sørensen, *Phys. Rev.* **C36**, 115 (1987).
19. J. R. Nix and A. J. Sierk, Preprint LA-UR-87-133, Los Alamos National Laboratory (1987).
20. W. J. Swiatecki, *Phys. Rev.* **24**, 113 (1981).
21. D. Berdichevski, A. Lukasiak, W. Nörenberg, and A. Rozmej, *Nucl. Phys.* **A499**, 609 (1989).
22. H. Fröbrich, Proc. XXVI Int. Winter Meeting on Nucl. Phys., Bormio (1986).
23. G. Münzenberg, P. Armbruster, S. Hofmann, F. P. Hessberger, H. Folger, J. G. Keller, V. Ninov, K. Poppensieker, A. B. Quint, W. Reisdorf, K.-H. Schmidt, J. R. H. Schneider, H.-J. Schött, K. Sümmerer, I. Zychor, M. E. Leino, D. Ackermann, U. Gollerthan, E. Hanelt, W. Morawek, D. Vermeulen, Y. Fujita, T. Schwab, Preprint GSI-88-59 (1988) and *Z. Phys.* **A333**, 163 (1989).
24. A. V. Ignatyuk, M. G. Itkis, V. N. Orlovich, G. N. Smirenski, and A. S. Tishin, *Sov. J. Nucl. Phys.* **21**, 612 (1975).

25. A. S. Iljinov and E. A. Cherepanov, Int. Rep. JINR P7-84-86, Dubna (1984).
26. G. Münzenberg, F. P. Hessberger, W. Faust, H. Folger, S. Hofmann, K. H. Schmidt, H.-J. Schött, and P. Armbruster, *Nucl. Instr. and Methods*, **A282**, 28 (1989).
27. G. Münzenberg, W. Faust, S. Hofmann, P. Armbruster, K. Güttner, and H. Ewald, *Nucl. Instr. Meth.* **161**, 65 (1979).
28. S. Hofmann, G. Münzenberg, F. P. Hessberger, and H.-J. Schött, *Nucl. Instr. Meth.* **223**, 312 (1984).
29. G. Münzenberg, S. Hofmann, F. P. Hessberger, W. Reisdorf, K. H. Schmidt, J. H. R. Schneider, P. Armbruster, C. C. Sahm, and B. Thuma, *Z. Phys.* **A300**, 107 (1981).
30. G. Münzenberg, W. Reisdorf, S. Hofmann, Y. K. Agarwal, T. P. Hessberger, K. Poppensieker, J. R. H. Schneider, W. T. W. Schneider, K.-H. Schmidt, H.-J. Schött, P. Armbruster, C. C. Sahm, and D. Vermeulen, *Z. Phys.* **A315**, 145 (1985).
31. G. Münzenberg, S. Hofmann, F. P. Hessberger, H. Folger, V. Ninov, K. Poppensieker, A. B. Quint, W. Reisdorf, H.-J. Schött, K. Sümmerer, P. Armbruster, M. E. Leino, D. Ackermann, U. Gollerthan, E. Hanelt, W. Morawek, Fujita, T. Schwab, and A. Turler, *Z. Phys.* **A330**, 435 (1988).
32. Yu. Ts. Oganessian, A. G. Demin, N. A. Danilov, G. N. Flerov, M. P. Ivanov, A. S. Iljinov, M. N. Kolesnikov, B. N. Markov, V. M. Plotko, and S. P. Tretyakova, *Nucl. Phys.* **A273**, 505 (1976).
33. K. H. Schmidt, C. C. Sahm, K. Pielenz, and H. G. Clerc, *Z. Phys.* **A316**, 19 (1984).
34. P. Møller and J. R. Nix, *At. Nucl. Data Tables* **39**, 213 (1988).
35. P. Møller, W. D. Myers, W. J. Swiatecki, and J. Treiner, *Atomic Nuclear Data Tables* **39**, 225 (1988).
36. Yu. Ts. Oganessian, M. Hussenois, A. G. Demin, Yu. P. Kharitanov, H. Bruchertseifer, O. Constantinescu, Yu. S. Korotkin, S. P. Tretyakova, U. K. Utsonkov, I. V. Shirokovsky, and J. Estevez, Int. Conf. on Nucl. and Radiochemistry, Lindau (1984); published in *Radiochimica Acta* **37**, 113 (1984).
37. A. G. Demin, S. P. Tretyakova, V. K. Utyonkov, and I. V. Shirokovski, *Z. Phys.* **A315**, 197 (1984).
38. G. Münzenberg, S. Hofmann, H. Folger, F. P. Hessberger, J. G. Keller, K. Poppensieker, A. B. Quint, W. Reisdorf, K. H. Schmidt, H.-J. Schött, P. Armbruster, M. E. Leino, and R. Hingmann, *Z. Phys.* **322**, 227 (1985).
39. G. Münzenberg, P. Armbruster, G. Berthes, F. P. Hessberger, S. Hofmann, W. Reisdorf, K. H. Schmidt, and H.-J. Schött, *Nucl. Instr. Meth* **B26**, 294 (1987).
40. C. Gustafson, J. L. Lamm, B. Nilsson, and S. G. Nilsson, *Ark. Fysik* **36**, 613 (67).
41. E. Roeckl, D. Schardt, in *Particle Emission from Nuclei II*, Ed. by D. N. Peonaru and M. S. Ivascu (CRC Press, Boca Raton, USA, 1989), p. 1.
42. J. O. Rasmussen, *Phys. Rev.* **113**, 1593 (1959).
43. A. H. Wapstra, G. Audi, and R. Hoekstra, *Atomic Nuclear Data Tables* **39**, 281 (1988).
44. G. A. Leander, P. Møller, and J. R. Nix, *Z. Phys.* **A323**, 41 (1986).
45. H. J. Krappe, J. R. Nix, and A. J. Sierk, *Phys. Rev.* **C20**, 992 (1979).
46. W. D. Myers, *Droplet Model of Atomic Nuclei* (IFI Plenum, New York, 1977).
47. W. J. Swiatecki, private communication (1987).
48. W. D. Myers and W. J. Swiatecki, *Ark. Fys.* **36**, 343 (1967).
49. B. Bjørnholm and J. E. Lynn, *Rev. Mod. Phys.* **52**, 725 (1980).
50. P. Møller, J. R. Nix, and W. J. Swiatecki, *Nucl. Phys.* **A492**, 349 (1989).
51. K. Böning, Z. Patyk, A. Sobiczewski, and S. Cwiok, *Z. Phys.* **325**, 479 (1986).
52. M. Dahlinger, D. Vermeulen, and K. H. Schmidt, *Nucl. Phys.* **A376**, 94 (1982).
53. J. R. Nix, *Ann. Phys.* **41**, 52 (1967).
54. Z. Patyk, J. Skalski, A. Sobiczewski, and S. Cwiok, *Nucl. Phys.* **A502**, 591 (1989).
55. V. M. Strutinsky, *Nucl. Phys.* **A95**, 420 (1967).
56. E. O. Fiset and J. R. Nix, *Nucl. Phys.* **A193**, 647 (1972).
57. S. G. Nilsson, J. R. Nix, A. Sobiczewski, Z. Szymanski, S. Wycech, C. Gustafson, P. Møller, *Nucl. Phys.* **A115**, 545, (1968).

58. U. Mosel, W. Greiner, *Z. Phys.* **222**, 261, (1969).

59. S. Cwiok, V. V. Pashkevich, J. Dudek, and W. Nazarewicz, *Nucl. Phys.* **A410**, 254 (1987).

60. A. Sobiczewski, Z. Patyk, and S. Cwiok, *Phys. Lett.* **B186**, 6 (1989).

61. S. Cwiok, P. Rozmej, A. Sobiczewski, and Z. Patyk, *Nucl. Phys.* **A491**, 281 (1989).

62. A. Sobiczewski, Z. Patyk, and S. Cwiok, *Phys. Lett.* **224**, 1 (1989).

63. G. Berthes, Thesis, Int. Report GSI 87-12 (1987).

64. K. Sümmerer, D. J. Morrissey, *First Conf. on Radioactive Beams*, Berkeley 1989, World Scientific, 122 (1990).

12. Reflection-Asymmetric Shapes in Nuclei

I. Ahmad, M. P. Carpenter, H. Emling,*
R. Holzmann,* R. V. F. Janssens,
T. L. Khoo, E. F. Moore, and L. R. Morss
Argonne National Laboratory
Argonne, IL 60439

J. L. Durell, J. B. Fitzgerald, A. S. Mowbray,
M. A. C. Hotchkis, and W. R. Phillips
Department of Physics
University of Manchester
Manchester, M13 9PL, England

M. W. Drigert
INEL, EG&G Idaho, Inc.
Idaho Falls, ID 83415

D. Ye
University of Notre Dame
Notre Dame, IN 46556

Ph. Benet
Purdue University
West Lafayette, IN 47907

ABSTRACT

In the last few years, theoretical calculations and experimental results have established that nuclei in the mass 224 and $Z = 56$, $N = 88$ regions possess reflection-asymmetric (octupole) shapes. In both regions it is found that even-even nuclei are reflection symmetric in their ground states. However, these nuclei develop octupole deformation with rotation. Recently we have studied such phenomena in neutron-rich Ba and Ce nuclei produced in spontaneous fission. Detailed properties of these nuclei will be discussed.

12.1 INTRODUCTION

Soon after the collective nuclear model[1,2] was established, low-lying negative-parity states were observed[3] in even-even Ra and Th nuclei from α-particle spectroscopic studies. These negative-parity levels formed a rotational band with spin sequence, 1, 3, 5, and K quantum number 0. Since these states had energies much lower than the expected two-quasi-particle states, these were interpreted as octupole vibrations about a spheroidal

*Present address: GSI, Planckstrasse 1, D-6100, Darmstadt, West Germany.

equilibrium shape. Ever since the discovery of the octupole vibrations, scientists have been thinking about the possibility of octupole deformations in nuclei, i.e., nuclei with octupole equilibrium shape. These octupole shapes are symmetric about the Z axis but reflection asymmetric about the XY plane and resemble a pear.

The octupole deformation or octupole vibration is produced by the long-range octupole-octupole correlations. These correlations depend on the $r^3Y_3 \cdot r^3Y_3$ matrix elements between single-particle states with $\Delta \ell = \Delta j = 3$ and the energy difference between them. A look at the shell states (Fig. 1) indicates that valence neutrons in the $N \approx 134$ nuclei occupy the $j_{15/2}$ and $g_{9/2}$ orbitals, and valence protons in the $Z \approx 88$ nuclei occupy the $i_{13/2}$ and $f_{7/2}$ orbitals. Thus, in nuclei with $N \approx 134$ and $Z \approx 88$, both neutrons and protons will have maximum octupole deformation. Experimentally, one observes that the energies of the $K,I^\pi = 0,1^-$ states in the mass-224 nuclei decrease[4] as one goes from ^{228}Th to ^{222}Th, indicating that octupole correlations are increasing. Alpha-decay hindrance factors show a similar behavior.

Interest in octupole shape was revived in the early eighties because of two observations. First, the mass formula[5], which reproduced the experimental masses well in all other regions, predicted less binding in the mass-224 region. Inclusion of a small octupole deformation (β_3) in the potential improved the fit. This pointed out large octupole-octupole correlations or deformation in the mass-224 region. Another observation was the presence of the low-lying 0^+ state in ^{234}U, observed in the (p,t) reaction,[6] which could not be explained on the basis of the vibrational model. Chasman[7] found that lowering of the 0^+ state in ^{234}U could be accounted for, if one includes octupole-octupole correlations in the potential. When this calculational technique was extended to odd-mass nuclei in the mass-224 region, it predicted[8] parity doublets with large $B(E3)$ values between its members, a characteristic signature of octupole deformation. The calculations[9] for the actinides indicate that the additional binding energy due to the β_3 deformation is only ≈ 1 MeV, compared with the binding energy of ≈ 10 MeV due to the quadrupole deformation (Fig. 2).

It is well known that the rotation of symmetric and asymmetric molecules generate different rotational bands. In the case of the former, the band has a $0^+, 2^+, 4^+, \ldots$, level

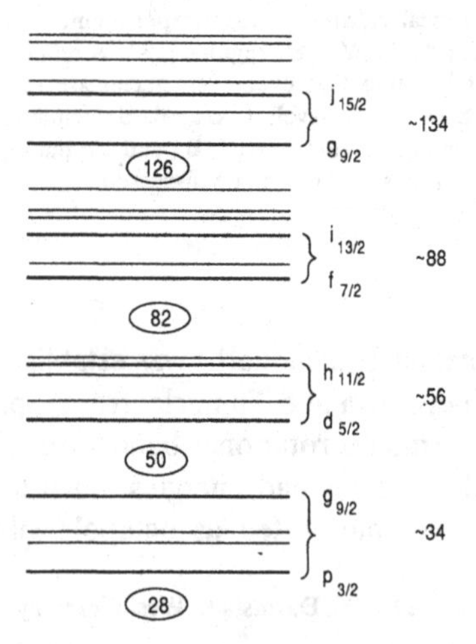

Figure 1. Shell-model states for neutrons, showing the locations where states with $\Delta j = 3$ occur in nuclei.

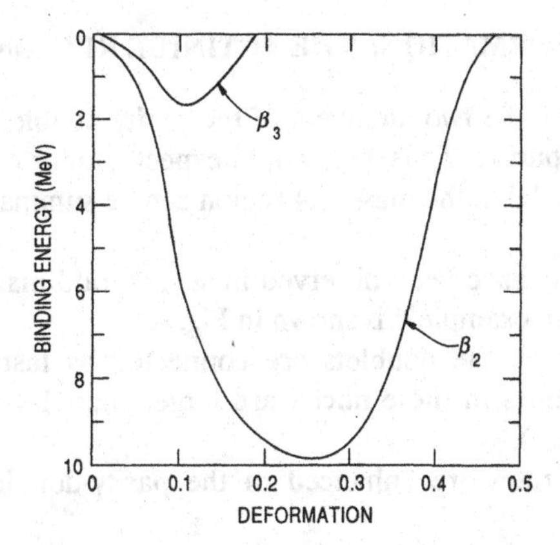

Figure 2. **Calculated increase in binding energy as a function of quadrupole deformation β_2 and octupole deformation β_3 for an actinide nucleus.**

sequence, while the latter has a 0^+, 1^-, 2^+, 3^-, . . ., sequence. Thus, rotation of reflection-symmetric and reflection-asymmetric nuclei will produce different spectra, and these can be used as a signature of octupole shape. Many spheroidal nuclei are known to show a 0^+, 2^+, 4^+, . . ., level sequence in their ground states, but no one has observed the 0^+, 1^-, 2^+, 3^-, . . ., level sequence in any nucleus. Thus, it can be safely said that there is no even-even reflection-asymmetric nucleus in the ground state. It turns out that the octupole-octupole correlations in nuclei can be enhanced either by rotating the nucleus or by placing an unpaired nucleon in the system. The signature of reflection asymmetry in an odd-mass nucleus is the presence of a parity doublet (Fig. 3)—a pair of almost degenerate states with the same spin, opposite parity, and connected by a large $B(E3)$ value. Octupole deformations were soon observed in odd-mass nuclei,[10] as well as in even-even nuclei at moderate spins.[11]

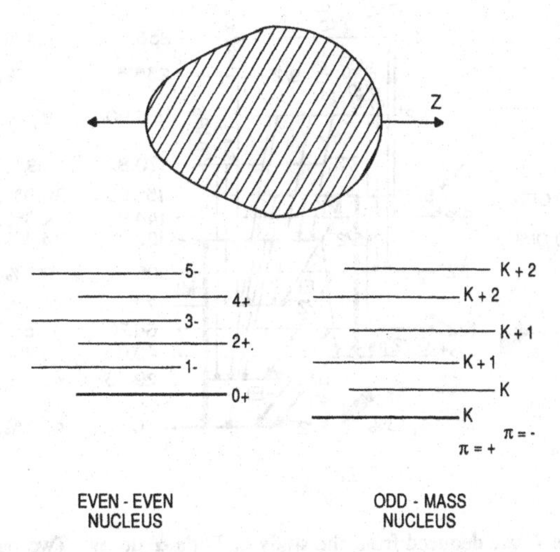

Figure 3. **Signatures of reflection asymmetry in even-even and odd-mass nuclei.**

12.2 OCTUPOLE DEFORMATION: THE ACTINIDE REGION

In odd-mass nuclei, the two members of the parity doublet have very similar wave-functions except for the parity. Thus, one would expect similar properties for both bands. The data on odd-mass nuclei in the mass-224 region can be summarized as follows:

1) Parity doublets have been observed in several odd-mass Ra, Ac, and Pa nuclei.[12-15] An example[14] is shown in Fig. 4.
2) The members of the doublets are connected by fast $E1$ transitions.[15] The $B(E1)$ values in these nuclei are larger than 1.0 x 10^{-3} Weisskopf units (see Fig. 5).
3) Alpha decay rates are enhanced to the parity-doublet partner of the favored band.
4) Coriolis matrix elements and $M1$ tansition rates are attenuated.
5) Octupole deformation has also been found at moderate spins[16] in ^{221}Th and ^{223}Th.

In even-even nuclei, levels with alternating parity have been observed in several Ra and Th nuclei. In general, these are connecting by fast $E1$ transitions.

12.3 OCTUPOLE DEFORMATION IN $N \sim 88$, $Z \sim 56$ NUCLEI

As shown in Fig. 1, octupole correlations are also expected in the $Z \sim 56$, $N \sim 88$ nuclei. However, in these nuclei, spacings for the $\Delta\ell = 3$, $\Delta j = 3$ levels are larger than in

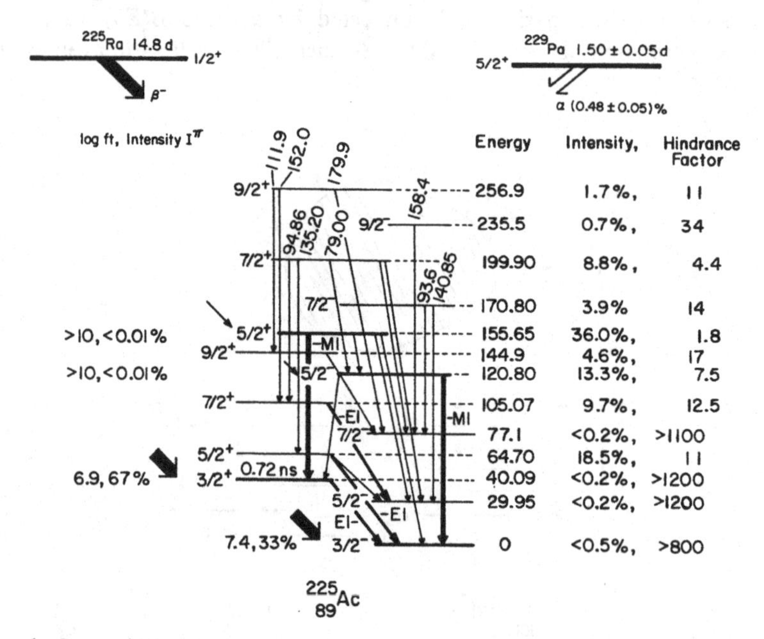

Figure 4. Level scheme of ^{225}Ac deduced from the study of ^{229}Pa α decay. Two parity doublets, 5/2± and 3/2±, were identified.

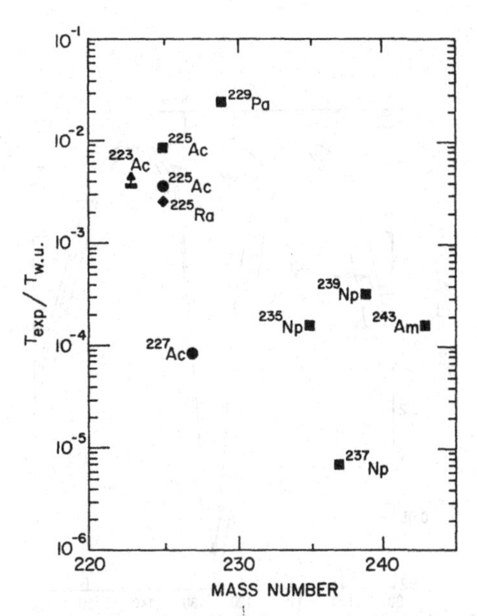

Figure 5. *E*1 rates in odd-mass actinides. Plotted are the ratios of experimental $B(E1)$ values and Weisskopf estimates. The Weisskopf estimate (T_{wu}) was calculated from the formula $T_{wu} = 1.0 \times 10^{14} \cdot A^{2/3} \cdot E_\gamma$, where A is the mass number and $E\gamma$ is the γ-ray energy in MeV.

the mass-224 region. Hence, one expects smaller octupole-octupole correlations than in the mass-224 region. Calculations by Nazarewicz et al.[17] predicted octupole minima for ^{144}Ba and ^{146}Ba ground states. Later mean-field calculations[18] show that the octupole shape stabilizes with rotation.

The nuclei with $Z \approx 56$, $N \approx 88$ are neutron rich and are difficult to access for usual in-beam γ-ray spectroscopy. These nuclei are copiously produced by spontaneous fission. There are only two conveniently available spontaneous-fission sources, ^{252}Cf ($t_{1/2}$ = 2.646 y, fission branching = 3.09%) and ^{248}Cm (3.4 x 10^5 y, = 8.26%). The mass distributions[19,20] of these nuclei along with that for the 60.5-d ^{254}Cf (fission = 99.7%) [Ref. 21] are shown in Fig. 6.

12.4 EXPERIMENTAL PROCEDURE

Level structures of neutron-rich nuclei produced in the spontaneous fission of ^{252}Cf and ^{248}Cm were studied in two experiments carried out with the Argonne–Notre-Dame BGO γ-ray facilty. In the first experiment, a 60-μCi source (which corresponds to 2 μCi of fission) embedded in a Be cylinder was used, and the detector system consisted of seven compton-suppressed Ge detectors, one low-energy-photon spectrometer (LEPS), and fourteen BGO hexagons. Data were collected for a period of 3 d. In the experiment performed this year in May, a 5-mg ^{248}Cm source was used and the γ-ray facility consisted of ten compton-suppressed Ge detectors, two LEPS, and fifty BGO hexagons (see Fig. 7). The Cm source was prepared by mixing Cm_2O_3 with 150 mg of KCl and pressing it under a

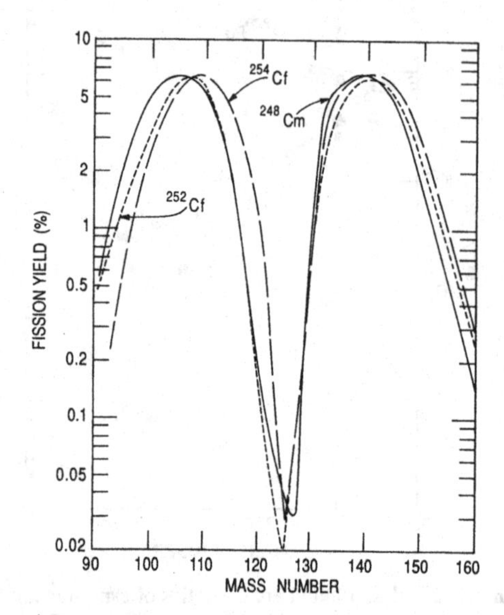

Figure 6. Fission yields for ^{248}Cm, ^{252}Cf, and ^{254}Cf determined by radiochemical methods.

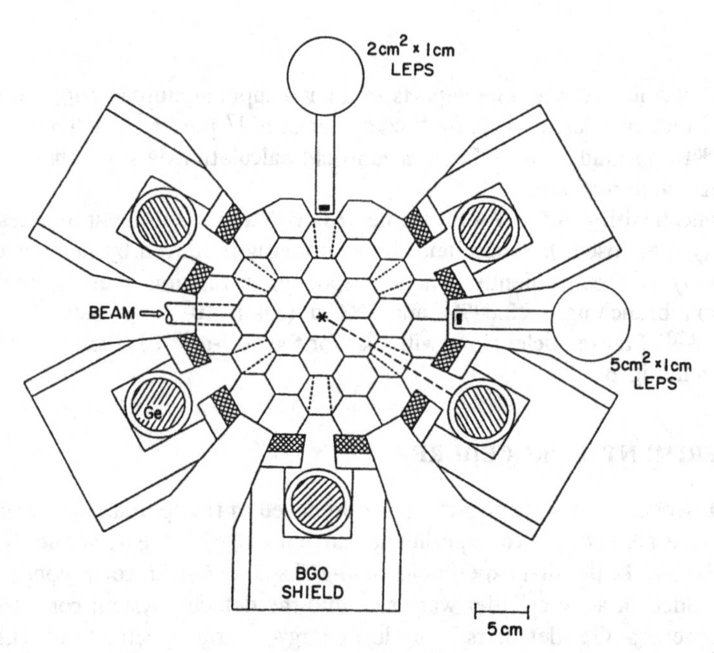

Figure 7. Argonne–Notre-Dame γ-Ray Facility used in the experiments described here. For the present experiment, two 90° Ge detectors were removed. A 2 cm^2 x 1-cm LEPS detector was placed at 0°

pressure of 600 MPa to make a pellet. The Cm was chemically separated from fission products and other actinides just before the preparation of the source. This kind of source was required in order to measure low-energy γ rays and K x rays, and to stop the fission fragments as soon as they were emitted to avoid doppler broadening of γ-ray peaks. Spectra were collected for a period of 10 d. Events were recorded when two Ge's (LEPS) and at least three of the hexagons fired. These quintuple coincidence events were collected in the event-by-event mode on magnetic tape and were later sorted with appropriate gates.

12.5 RESULTS

Gamma-ray transitions observed in gated spectra were assigned to nuclides on the following basis:

1) In many nuclei, the $2^+ \rightarrow 0^+$ transitions are known from the early Berkeley work.[22] Thus, the spectrum gated by the $2^+ \rightarrow 0^+$ γ rays in ^{144}Ba will contain all ^{144}Ba γ rays plus those in the complementary Zr isotopes. Next, gating on the Zr transitions, one would be able to distinguish which γ rays belong to ^{144}Ba.

2) One can sometimes identify a transition in a nucleus about which nothing was previously known by measuring, through an isotopic sequence, γ-ray yields in coincidence with complementary fragments. This has been done for known transitions in $^{138\text{-}140}$Xe and for candidate lines in ^{141}Xe and ^{142}Xe. The ratio of the intensities of the lines to that of the ^{140}Xe line, observed in coincidence with different mass complementary fragments, is shown in Fig. 8. The variations show that the lines proposed for ^{141}Xe and ^{142}Xe do indeed correspond to transitions in those nuclei.

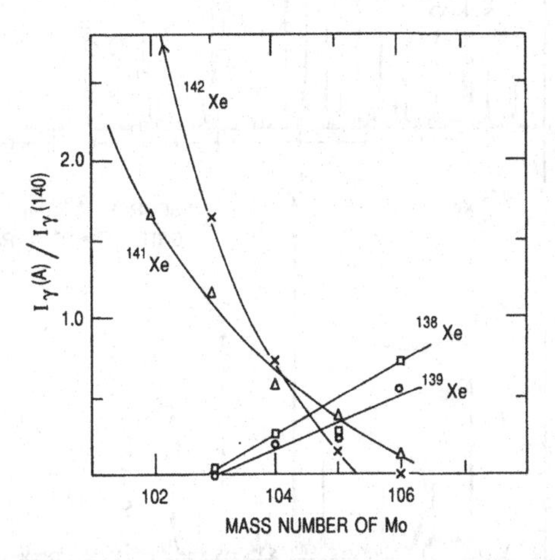

Figure 8. Yields of lines in Xe isotopes (expressed as ratios to that for a ^{140}Xe line) observed in coincidence with different mass complementary Mo fragments.

3) The multipolarities of strong transitions were deduced from the measured angular correlations. In some cases, the conversion coefficients of the low-energy transitions were deduced from the measured *K* x-ray intensities. The Ge detectors were sensitive only to γ rays above 150 keV because the constant-fraction discriminator was used in the slow-rise-time reject mode. For energies below ~ 150 keV, the two LEPS' were useful.

Gamma-ray spectra were generated for many nuclides by placing appropriate gates. Because of the compton suppression, the spectra were extremely clean. The sensitivity of the measurement was good enough to observe γ rays with intensities greater than 1.0% for high-yield fission products. Examples of γ-ray spectra are shown in Fig. 9.

12.6 DISCUSSION

From the analysis of the present data, level schemes of many nuclei around mass 100 and mass 144 were deduced. Level schemes of ^{144}Ba and ^{142}Xe are displayed in Fig. 10. As can be seen, the levels in ^{144}Ba above 7 $\hbar$ have alternating parities, indicative of a single rotational band. On the other hand no negative-parity band has been observed in ^{142}Xe, suggesting lack of strong octupole correlation in this nucleus. Our results indicate

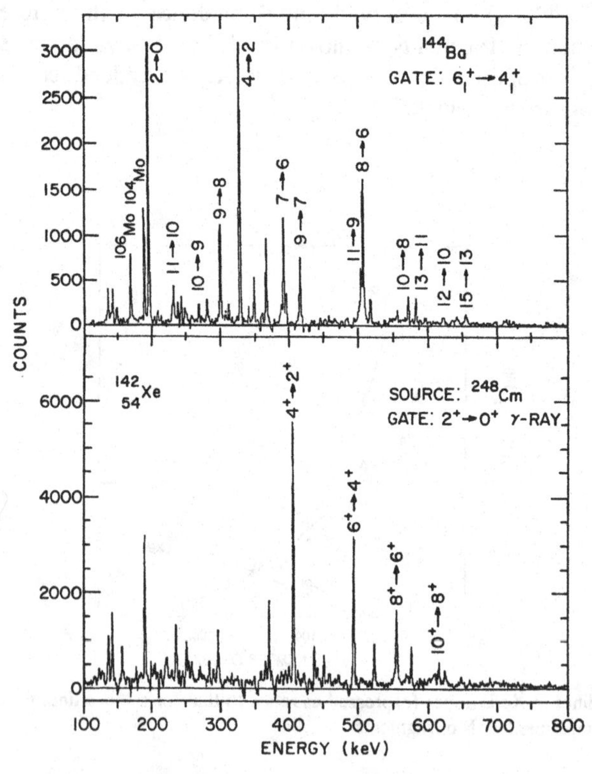

Figure 9. Samples of coincidence spectra for ^{144}Ba and ^{142}Xe.

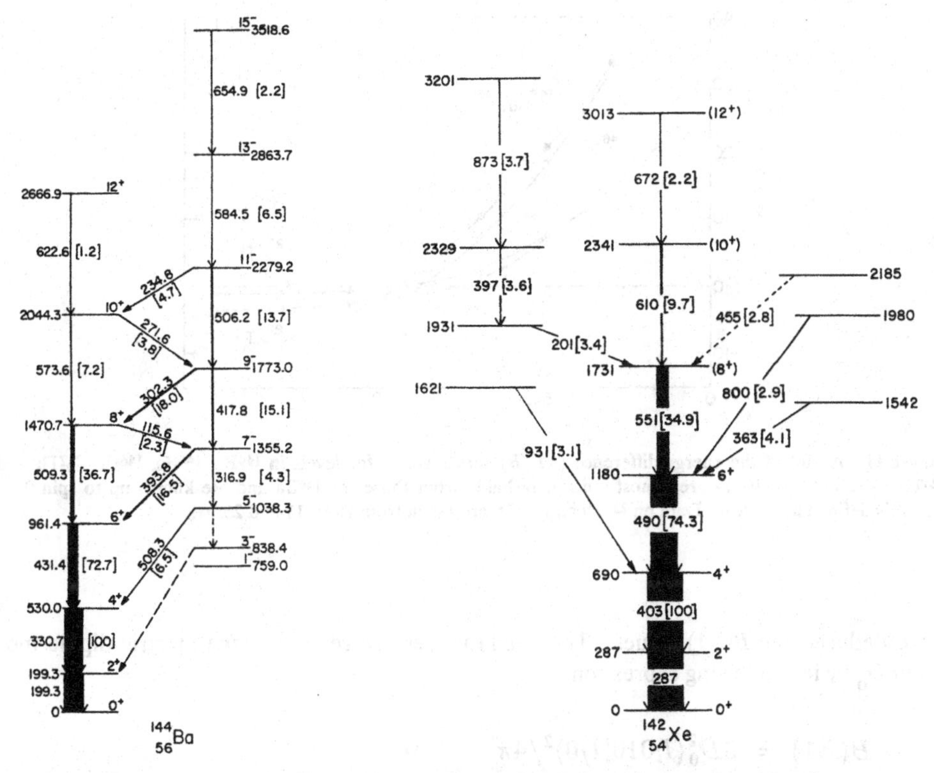

Figure 10. Partial level schemes for ^{144}Ba and ^{142}Xe deduced from the present study.

presence of octupole deformation[23,24] in ^{144}Ba, ^{146}Ba, and ^{146}Ce above spin 7 $\hbar$; ^{142}Ba, ^{148}Ce, ^{150}Ce, and ^{142}Xe do not exhibit the intertwined negative and positive-parity levels. In the case of the octupole limit, the positive- and negative-parity levels constitute a single rotational band with one value of moment inertia. How well the levels merge to the octupole limit can be judged from a plot of δE against the spin, where δE is the difference between the observed energy of a negative parity state and the energy calculated from the energies of the neighboring positive parity states. The quantity δE can be calculated with the expression,

$$\delta E = E_{I^-} - \frac{(I+1)\cdot E_{(I-1)^+} + I\cdot E_{(I+1)^+}}{2I+1}$$

For an ideal octupole nucleus, δE (or $\delta E/B$, where B is the rotational constant) should be zero. In Fig. 11, the values of $\delta E/B$ are plotted against spin I for the nuclei studied in the present work and for ^{222}Th and ^{224}Th.[11,25] As can be seen in the figure, δE is quite high at low spin, but it decreases as the spin increases and is near zero at 8-9 $\hbar$, indicating increase of octupole correlations with rotation and stabilization of octupole deformation above spin 8 $\hbar$.

Another characteristic feature of the octupole shape is the enhancement in the $E1$ transition rates. Using the experimentally measured values of the quadrupole moment from the lifetime of $2^+ \rightarrow 0^+$ states[26] and assuming that they do not change with spin, we

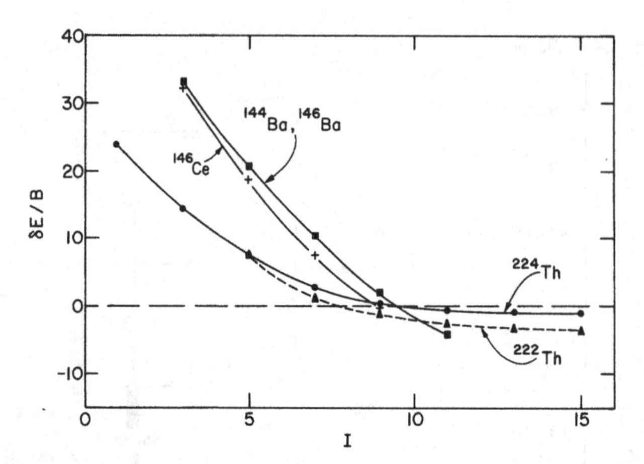

Figure 11. A plot of the energy difference ($\delta E/B$) versus spin I for levels in ^{144}Ba, ^{146}Ba, ^{146}Ce, ^{222}Th and ^{224}Th. The data for ^{146}Ba are almost indistinguishable from those for ^{144}Ba and are known up to spin 9 $\hbar$. $\delta E/B$ is defined in the text. Data on ^{222}Th and ^{224}Th are taken from Refs. 11 and 25.

have deduced the $B(E1)$ values. The $B(E1)$ values are related to the electric dipole moment D_0 by the following expression:

$$B(E1) = 3D_0^2\langle I_i 010|I_f 0\rangle^2/4\pi$$

The dipole moments thus determined are given in Table I. Also included are experimental values for ^{146,148}Nd[27] and ^{150}Sm.[28]

Electric dipole moments have been calculated by Leander et al.[29] for the nuclei in the mass-224 region. It has been pointed out by Bohr and Mottelson[30], also Strutinski,[31] that a nucleus with reflection-asymmetric shape will have an electric dipole moment in the intrinsic reference frame. Only liquid-drop contributions were considered by these authors. Leander et al.[29] have calculated the dipole moment by adding the contributions from the liquid-drop term as well as shell-correction terms. Thus the dipole moment can be written as:

$$D_0 = D_{LD} + D_{SC}$$

The shell-correction term contains a term due to neutrons and a term due to protons. These terms were determined by considering the orbits occupied by the nucleons and calculating the displacement of the center of charge from the average position. Leander et al. normalized their calculations to the experimental dipole moment of ^{222}Th and found a good agreement for other nuclei in the mass-224 region. Soon afterward, Dorso et al.[32] showed that the inclusion of the neutron-skin effect (which was ignored by Leander et al.) makes the liquid-drop term much smaller and hence the agreement worse.

The calculational technique used by Leander et al., was used by Nazarewicz[18] to calculate the dipole moment of nuclei in the mass-144 region. In his calculation, the liquid drop term was not included. The dipole moments calculated only on the basis of the shell-

Table I. Experimental and Theoretical Intrinsic Electric Dipole Moments

Nucleus	D_0 (exp) e fm	D_0 (theory) e fm
^{142}Xe		0.02
^{144}Ba	0.13 ± 0.01	0.09
^{146}Ba	0.04 ± 0.01	0.03
^{144}Ce	0.17 ± 0.05	0.16
^{146}Ce	0.16 ± 0.04	0.18
^{146}Nd[a]	0.18 ± 0.01	0.20
^{148}Nd[a]	0.23 ± 0.03	0.22
^{150}Sm[a]	0.20 ± 0.01	0.25

[a]Values for these nuclei are taken from Refs. 27 and 28.

shell-correction term are included in Table I, and these are in fair agreement with the experimental data.

12.7 CONCLUSION

Experimental data show that there is no even-even nucleus with a reflection-asymmetric shape in its ground state. Maximum octupole-octupole correlations occur in nuclei in the mass-224 ($N \approx 134$, $Z \approx 88$) region. Parity doublets, which are the characteristic signature of octupole deformation, have been observed in several odd mass Ra, Ac, and Pa nuclei. Intertwined negative- and positive-parity levels have been observed in several even-even Ra and Th nuclei above spin ≈8 $\hbar$. In both cases, the opposite-parity states are connected by fast $E1$ transitions. In some medium-mass nuclei, intertwined negative and positive parity levels have also been observed above spin ≈7 $\hbar$. The nuclei which exhibit octupole deformation in this mass region are ^{144}Ba, ^{146}Ba and ^{146}Ce; ^{142}Ba, ^{148}Ce, ^{150}Ce, and ^{142}Xe do not show these characteristics.

ACKNOWLEDGMENT

This work was supported by the U. S. Department of Energy, Nuclear Physics Division, under contracts No. W-31-109-ENG-38 and No. DE-FG02-87ER40346, by the Science and Engineering Research Council of the U. K. under Grant No. GR.E.6535, and by the National Science Foundation Grant No. PHY-88-02279. The authors are also indebted for the use of ^{248}Cm to the Office of Basic Energy Sciences, U. S. Department of Energy, through the transplutonium-element production facilities at the Oak Ridge National Laboratory. Helpful discussions with R. R. Chasman are also acknowledged.

REFERENCES

1. A. Bohr, *K. Danske Vidensk. Selsk. Mat.-Fys. Medd.* **26**, No. 14 (1952).
2. A. Bohr and B. R. Mottelson, *K. Danske Selsk. Mat.-Fys. Medd.* **27**, No. 16 (1953).
3. F. S. Stephens, F. Asaro, and I. Perlman, *Phys. Rev.* **96**, 1568 (1954); *Phys. Rev.* **100**, 1543 (1955).
4. C. M. Ledere and V. S. Shirley (John Wiley, Inc., New York, 1978) 7th ed.
5. P. Möller and J. R. Nix, *Nucl. Phys.* **A361**, 117 (1981).
6. J. A. Maher, J. R. Erskine, A. M. Friedman, R. H. Siemssen, and J. P. Schiffer, *Phys. Rev. C* **5**, 1380 (1972).
7. R. R. Chasman, *Phys. Rev. Lett.* **42**, 630 (1979).
8. R. R. Chasman, *Phys. Lett.* **96B**, 7 (1980).
9. R. R. Chasman, *Nuclear Structure, Reactions and Symmetries*, Ed. by R. A. Meyer and V. Paar (World Scientific Co., Singapore, 1986).
10. I. Ahmad, J. E. Gindler, R. R. Betts, R. R. Chasman, and A. M. Friedman, *Phys. Rev. Lett.* **49**, 1758 (1982).
11. D. Ward, G. D. Dracoulis, J. R. Leigh, R. J. Charity, D. J. Hinde, and J. O. Newton, *Nucl. Phys.* **A406**, 591 (1983).
12. R. K. Sheline and G. A. Leander, *Phys. Rev. Lett.* **51**, 359 (1983).
13. G. A. Leander and Y. S. Chen, *Phys. Rev. C* **37**, 2744 (1988).
14. I. Ahmad, J. E. Gindler, A. M. Friedman, R. R. Chasman, and T. Ishii, *Nucl. Phys.* **A472**, 285 (1987).
15. I. Ahmad, R. Holzmann, R. V. F. Janssens, P. Dendooven, M. Huyse, G. Reusen, J. Wauters, and P. Van Duppen, *Nucl. Phys.*, **A505**, 257 (1989).
16. M. Dahlinger, E. Kankeleit, D. Habs, D. Schwalm, B. Schwartz, R. S. Simon, J. D. Burrows, and P. A. Butler, *Nucl. Phys.* **A484**, 337 (1988).
17. W. Nazarewicz, P. Olanders, I. Ragnarsson, J. Dudek, G. A. Leander, P. Moller, and E. Ruchowska, *Nucl. Phys.* **A429**, 269 (1984).
18. W. Nazarewicz, *Int. Conf. on Nuclear Structure through Static and Dynamic Moments* (Melbourne, Australia), 180 (1987).
19. K. F. Flynn, J. E. Gindler, and L. E. Glendenin, *J. Inorg. Nucl. Chem.* **37**, 881 (1975).
20. K. F. Flynn, J. E. Gindler, and L. E. Glendenin, *J. Inorg. Nucl. Chem.* **39**, 759 (1977).
21. J. E. Gindler, K. F. Flynn, L. E. Glendenin, and R. K. Sjoblom, *Phys. Rev. C* **16**, 1483 (1977).
22. E. Cheifetz, J. B. Wilhelmy, R. C. Jared, and S. G. Thompson, *Phys. Rev. C* **4**, 1913 (1971).
23. W. R. Phillips, I. Ahmad, H. Emling, R. Holzman, R. V. F. Janssens, T. L. Khoo, and M. W. Drigert, *Phys. Rev. Lett.* **57**, 3257 (1986).
24. W. R. Phillips, R. V. F. Janssens, I. Ahmad, H. Emling, R. Holzmann, T. L. Khoo, and M. W. Drigert, *Phys. Lett.* **212**, 402 (1988).
25. P. Shuler et al., *Phys. Lett.* **174B**, 241 (1986).
26. G. Mamane, E. Cheifetz, E. Dafni, A. Zemel, and J. B. Wilhelmy, *Nucl. Phys.* **A454**, 213 (1986).
27. W. Urban, R. M. Lieder, W. Gast, G. Hebbinghaus, A. Kramer-Flecken, T. Morek, T. Rzaca-Urban, W. Nazarewicz, and S. L. Tabor, *Phys. Lett.* **200B**, 424 (1988).
28. W. Urban, R. M. Lieder, W. Gast, G. Hebbinghaus, A. Kramer-Flecken, K. P. Blume, and H. Hubel, *Phys. Lett.* **185B**, 331 (1987).
29. G. A. Leander, W. Nazarewicz, G. F. Bertsch, and J. Dudek, *Nucl. Phys.* **A453**, 58 (1986).
30. A. Bohr and B. R. Mottelson, *Nucl. Phys.* **4**, 529 (1957).
31. V. M. Strutinskii, *Atomnaya Energiya* **4**, 150 (1956); Atomic Energy (USSR) **4**, 164 (1956).
32. C. O. Dorso, W. D. Myers, and W. J. Swiatecki, *Nucl. Phys.* **A451**, 189 (1986).

13. Reflection Asymmetry and Fission Yields in the Mass-145 Region and a Possible Route to Superheavy-Element Synthesis

J. David Robertson
Department of Chemistry
University of Kentucky
Lexington, KY 40506

and

William B. Walters
Department of Chemistry and Biochemistry
University of Maryland
College Park, MD 20742

ABSTRACT

New data from the study of the level structures of ^{143}Ba, ^{147}La, and other nuclides in the mass-145 region will be reported. The evidence for reflection asymmetric contributions to the low-energy structure of the nuclides in this mass region will be presented. The relationship between the high fission yields that have been determined for these nuclides and their possible reflection asymmetric shapes will be discussed.

In this paper, we wish to explore the notion that there is a relationship between the structure of the nuclides in the $Z = 56, A = 88$ region and the strong fission yields observed for these nuclides in both spontaneous and neutron-induced fission of a wide range of fissile nuclides. To explore this possible connection, we must first establish the nature of the fission yields in the heavy-mass region of the double-humped asymmetric-fission-mass yield distribution and then explore the structure of the nuclides in that mass region. Then we will discuss the reflection-asymmetric shapes of the fragments at the scission point and note calculations indicating that these shapes increase the probability of sub-barrier fusion.

In a strictly liquid-drop approach, fission would be symmetric into equal mass fragments. Therefore, it has long been accepted that nuclear structure must play a role in driving asymmetric fission, a role that is dominant at low excitation energies. At the same time, there is the symmetric liquid-drop fission that begins to emerge when the energy of the compound nucleus increases, with the result that the yields of the symmetric nuclides in the fission valley rise. Just changing from thermal neutron to fission spectrum neutrons brings about a three-fold rise in the yields of the symmetric fission products. The most commonly accepted approach to including the effects of the structure of the possible fission

Exotic Nuclear Spectroscopy, Edited by W. C. McHarris
Plenum Press, New York, 1990

fragments was devised by Strutinsky.[1] The subsequent discovery of large, odd-even effects on the independent yields of individual fission nuclides served as a powerful reminder of just how strong the role of the structure and mass of the fragment plays in directing the division of the fissioning nucleus. The enormously reduced yields of the odd-odd nuclides in Th and U fission demonstrate the effect of just 2 MeV in the mass of the product nuclide.

In this paper, we wish to explore the implications to be drawn from the data shown in Fig. 1. There we show the relationship between the average mass in the high- and low-mass regions of the double-humped asymmetric-fission mass-yield distribution and the mass of the fissioning nuclide.[2] We draw attention to the fact that, over a very wide range of fissioning nuclides in both neutron-induced and spontaneous fission, the average mass of the heavy nuclides is near 140, while the mass of the light nuclide follows a direct linear relationship with the mass of the fissile nuclide. *We take these data to indicate that there is some feature of the structure of the nuclides in the heavy-mass group that leads to these results.* That is, the nucleus fissions into some nuclide in the heavy-mass group, and whatever is left over makes up the light-mass fragment.

What is remarkable about these data is the wide range of nuclear structural features present in the light-mass nuclides that appear to have no effect whatsoever on these yields. For the very-light fissile nuclides, the low-mass yields peak near mass 90, and this peak moves steadily through mass 100 up to mass 110. Kratz et al.[3] have discussed the onset of deformation in the Kr nuclides and in their paper showed the large range of 2^+ energies in

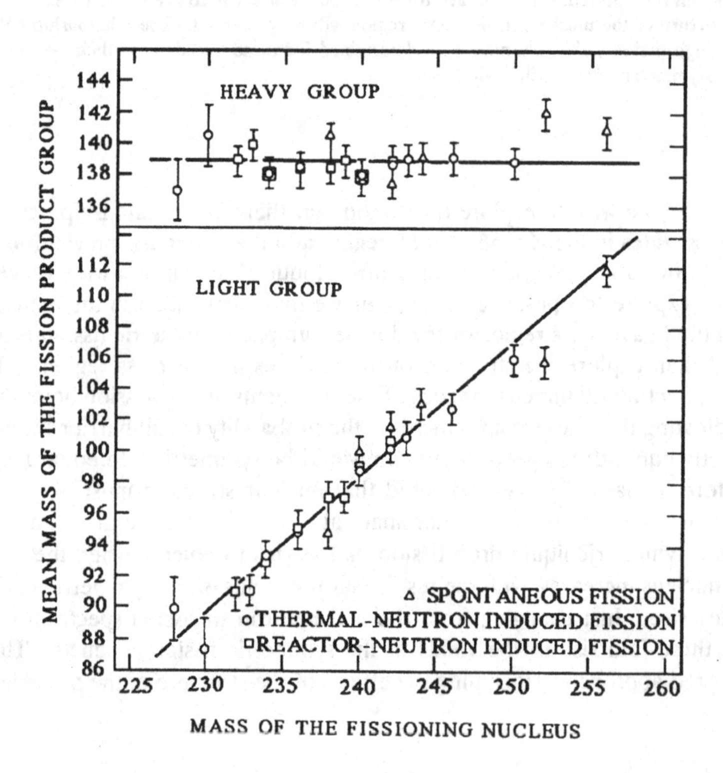

Figure 1. Average fission yields in the light and heavy-mass fragment as a function of the mass of the fissioning nuclide.

the neutron-rich fission products in the low-mass peak (see Fig. 3 in Chap. 28). Whereas the nuclides around mass 90 are near the $N = 50$ closed neutron shell and show only moderate departure from sphericity, the nuclides near mass 100 show some of the most prolate ground-state deformation values that are known.[4] For the higher-mass nuclides near mass 100, limited deformation and shape coexistence have been proposed.[5] In spite of all of the various shapes and orbitals available to the nuclides in the low-mass peak, none of them appears to be able to deter the fissioning nuclide from forming the fragments in the heavy-mass peak.

Now, we must examine the heavy-mass peak in detail. The heavy-mass nuclides are seen to break into two groups, one narrowly peaked around the doubly-closed-shell nuclide, ^{132}Sn, and a second broad group peaked around mass 143. Since we first drew attention to the possible relationship between the structure of these nuclides and fission yields, a considerable body of new experimental data and new calculations have become available.[6] For example, the special nature of nuclides near ^{132}Sn in fission has been shown in recent studies by Hulet et al. of the fission of nuclides with $Z \approx 100$ and $N \geq 156$, which fission symmetrically into two Sn fragments with relatively high kinetic energies.[7] In those studies, high-kinetic-energy fission to Sn fragments with neutron numbers less than 82 has been observed, along with low-kinetic-energy fission into a somewhat broader range of final masses for nuclides with $Z > 100$. Hulet et al. conclude that the high-kinetic-energy mode arises from fission into two nearly-spherical fragments with Sn structures that leaves little energy left over for internal nuclear excitation. These data are of considerable significance. From them, the conclusion can be drawn that asymmetric fission is facilitated by something other than energy. That is, symmetric fission into two ^{132}Sn fragments does occur with the release of 250 MeV in kinetic energy. For Fm nuclides with exactly 100 protons, symmetric fission into two Sn nuclides simply overwhelms the asymmetric channels. Yet, as soon as additional protons are present, even one, and certainly with four in ^{260}(Rf-Ku), asymmetric fission returns, yielding the structurally and kinetically-favored nuclides in the mass-140 range, with the leftovers in the mass-120 range but releasing only 220 MeV, fully 30 MeV less than the symmetric path.

Wang and Hu have identified several principal modes of fission.[8] They specifically recognize the symmetric mode as contrasted with asymmetric fission that has two modes, one leading to nuclides near ^{132}Sn and one leading to nuclides in the mass-140 region. They parameterize the fragments, however, only as a function of β_2 and do not include β_3 as shown in Fig. 2.

We suggest that the important feature of the structure of the other group of heavy-mass nuclides that is centered around mass 144 is the presence of important reflection-asymmetric or octupole configurations near the ground state. The full nature of these configurations is still under consideration: whether what is observed represents merely a tendency toward octupole vibrations or represents actual ground-state octupole structures. We shall use the term "octupole structures" to cover both of these possibilities and also the possibility of nuclear cluster structures as proposed by Iachello and Jackson in their nuclear-vibron model.[9]

To that end, let us review the evidence that has accumulated for such structures in this mass range. The first item was a paper by Leander et al.,[10] in which a difference of nearly 1 MeV was shown to exist between the calculated and observed mass of ^{145}Ba. Lesser values of this difference were shown to exist for a relatively small group of nearby nuclides. We subsequently investigated the structure of ^{145}Ba and could not identify parity-doublet structure comparable to that found in the Rn and Ra nuclides near mass 224.[11]

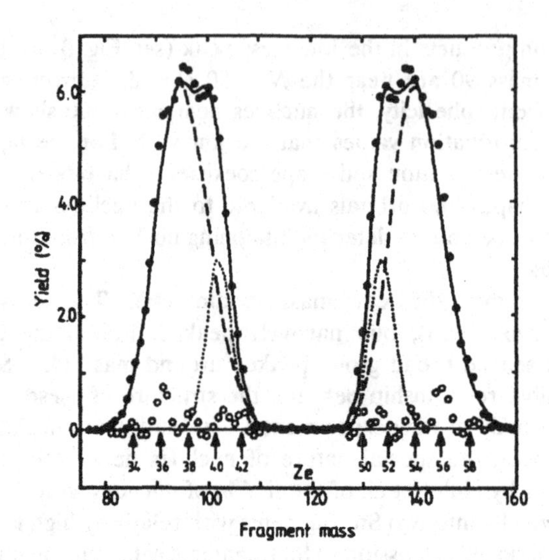

Figure 2. Breakup of the upper mass-yield curve into components with peaks around mass 132 and 145 (from Ref. 8).

We have also investigated the structures of isotopic ^{143}Ba so as to be able to observe the development of the structure of ^{145}Ba from the closed neutron shell at $N = 82$ outward. More important has been the ability to follow the changes in the $N = 87$ isotones moving down in charge from $Z = 64$ to 56. And, we studied the structure of isotonic ^{147}Ce in order to be able to follow the changes in the $N = 89$ isotones from $Z = 64$, where the structures are well known and quite deformed, down to $Z = 56$. In that study, we noted the close similarity in the level structures of ^{145}Ba and ^{147}Ce with considerable dissimilarity between these two nuclides and the higher-Z nuclides.[12]

But, a considerable body of other data do support octupole deformation in the mass-145 region but to a lesser extent than found in the mass-224 region. Strong support for the idea of octupole configurations in a limited mass region around ^{145}Ba has come from the discovery of bands with alternating parity with strong $E1$ transitions competing with $E2$ crossover transitions.[13,14] In these two papers, these bands have been identified in ^{144,146}Ba, and ^{146}Ce, and they were not found in ^{144,148}Ce. Similar structures have been identified at higher energies in Nd and Sm nuclides.[15,16] More recently, excited bands have also been identified in ^{142}Xe, but evidence for octupole deformation at low energy is not strong.[17] The narrow range of these bands corresponds rather well to the initial proposal for their range by Leander et al. and also corresponds to the region of highest yields in fission.

Another important series of data arise from the measurements of spins, parities, and electric-quadrupole and magnetic-dipole moments for the neutron-rich Xe, Cs, and Ba nuclides.[18-20] The spins, parities, and moments for these nuclides and some Nd nuclides for comparison are given in Table I. The spin- and parity-values of $3/2^+$ found for ^{143,145}Cs indicate that these nuclides are deformed to some degree. We have also included the data for two other $3/2^+$ Cs nuclides from the middle of the $N = 50$ to 82 shell. As can be seen, the magnetic moments are comparable, but the qudrupole moments are somewhat lower. The spins and parities of the Ba and Xe nuclides with neutron numbers 83, 85, 87, and 89 are seen to be identical, and the quadrupole moments are seen to be somewhat smaller for

Table I. Spins, Parities, and Electric-Quadrupole and Magnetic-Dipole Moments

Nuclide	N	Z	J^π	μ (nm)	Q (b)
^{137}Cs	82	55	7/2+	+2.8413	+0.05
^{139}Cs	84	55	7/2+	+2.696	−0.075
^{141}Cs	86	55	7/2+	+2.438	−0.4
^{143}Cs	88	55	3/2+	+0.870	+0.47
^{145}Cs	90	55	3/2+	+0.784	+0.62
^{119}Cs	64	55	3/2+	+0.836	+0.9
^{121}Cs	66	55	3/2+	+0.770	+0.838
^{137}Xe	83	54	7/2−	−0.968	−0.48
^{139}Xe	85	54	3/2−	−0.304	+0.40
^{141}Xe	87	54	5/2−	+0.10	−0.58
^{143}Xe	89	54	5/2−	−0.46	+0.93
^{139}Ba	83	56	7/2−	−0.975	−0.50
^{141}Ba	85	56	3/2−	−0.346	+0.43
^{143}Ba	87	56	5/2−	+0.454	−0.81
^{145}Ba	89	56	5/2−	−0.272	+1.15
^{143}Nd	83	60	7/2−	−1.065	−0.48
^{145}Nd	85	60	7/2−	−0.656	−0.33
^{147}Nd	87	60	5/2−	+0.578	+0.9
^{149}Nd	89	60	5/2−	+0.351	+1.3
^{145}Sm	83	62	7/2−	−1.1	−0.59
^{147}Sm	85	62	7/2−	−0.815	−0.259
^{149}Sm	87	62	5/2−	−0.6238	+1.014
^{151}Sm	89	62	5/2−	−0.363	+0.66

the Xe nuclides than for the Ba nuclides, indicating somewhat less deformation.

The mere existence of any significant deformation at in ^{141}Xe, which has only nine particles beyond the doubly-magic nuclide, ^{132}Sn, is remarkable. Casten, Brenner, and Haustein[21] have examined proton-neutron interactions as a source of deformation. They have defined a factor, $P = N_\pi N_\nu/(N_\pi + N_\nu)$, which represents an average of proton-neutron interactions, and observed that a value of $P \approx 4$ is required for the onset of deformation. Such a value is achieved with eight protons and eight neutrons or slight variations from those numbers. For this mass region, such a nuclide would be ^{148}Ce. We must point out at this juncture that, while ^{146}Ce shows octupole deformation at low energy, ^{148}Ce does not. That is, once proton-neutron-induced deformation sets in, that structure is sufficiently lowered in energy to make the apparent energy of any octupole band appear to be at a higher energy. Casten, Frank, and von Brentano also examined deformation as a function of the product $N_\pi N_\nu$ and found two separate curves for this mass region—one for neutron numbers less than 90 and one for neutron numbers greater than 90.[22] We cite these data to support our contention that the deformation found around mass 145 arises from a different source than that found beyond $N = 90$ and $Z = 58$, which can be reasonably attributed to proton-neutron interactions arising from occupancy of high-j, low-K orbitals for both protons and neutrons which may or may not be in spin-orbit partner orbitals.[23,24]

The detailed structure of the odd-Z Cs, La, and Pr nuclides has been studied, and the low-energy $3/2^+$ level that becomes the ground state in ^{143}Cs is found to evolve smoothly downward in energy as neutrons are added to the closed-shell ^{137}Cs.[25,26] Among the $N = 86$ odd-Z isotones, the $3/2^+$ level is even at a lower position in ^{143}La, and it is likely that the ground-state spins and parities of ^{145,147}La are also $3/2^+$. Leander et al. demonstrated how the addition of additional octupole deformation could account for this lowering of the $3/2^+$ level. The addition of additional degrees of deformation has been discussed by Nazarewicz[27] and Cwiok and Nazarewicz.[28] Also, diagrams of possible shapes have been shown in Cwiok, Rozmej, Boning, and Sobiczewski for nuclides in the Ra region.[29]

The structures of the odd-neutron Xe and Ba isotones are marked by their similar quadrupole and dipole moments, as well as similar spins and parities. The $7/2^-$ values for the $N = 83$ neutron nuclides reprsent the single $f_{7/2}$ neutron beyond the closed shell. The $3/2^-$ spin and parity values for the $N = 85$ isotones is well described in the cluster-vibration model as arising from the coupling of a three-neutron cluster to the $N = 82$ core.[30] While the $5/2^-$ level in the $N = 87$ nuclides could also arise from a cluster mechanism, as it surely does in the heavier $N = 87$ isotones, the negative quadrupole moment suggests, rather, that this level is a member of a $K = 1/2$ band. The inversion of the signs of the dipole and quadrupole moments for the $5/2^-$ ground states for both Xe and Ba between $N = 87$ and 89 suggests further deformation such that the last neutron jumps any intervening $K = 3/2$ band.

It is our contention that the nuclides in the region around ^{145}Ba are characterized by deformation not readily described in the simple Nilsson model, which takes into account only quadrupole and/or hexadecapole deformation, and that additional degrees of freedom are required. Better fits to the data have been obtained by including odd orders of deformation, including β_5 and β_7.[31] One degree of freedom generally not included in many calculations is that represented clustering as described by the nuclear-vibron model.[32,33] Even going from $N = 83$ to 85, the effect of the small three-neutron cluster in the Ba nuclides results in a reduction of 1400 keV in the energy of the first $5/2^+$ level. In Fig. 3, we show the similarity of the cluster and octupole deformations by showing the nucleus as broken into two spherical components, one a large ^{132}Sn fragment and the other with the additional nucleons beyond mass 132, thirteen in the case of ^{145}Ba. The advantage of this configuration is that it carries with it a dipole moment and, consequently, a mechanism for enhanced $E1$ transitions. Alternately, Kusnezov and Iachello have described the even-even Ba nuclides in an IBM calculation including both p and f bosons.[34] Some support for this notion of clustering has been found in the discovery of the emission of ^{13}C and heavier fragments in decay of heavy elements, leading to nuclides near the doubly-closed shell,

OCTUPOLE DEFORMATION CLUSTER DEFORMATION

Figure 3. Schematic drawing of the octupole and cluster shapes.

^{208}Pb. Stronger support has been indicated by Lonnorth from a study of $E1$ transitions in nuclides in the region around ^{208}Pb.[35] Shriner et al.[36] have discussed the relative merits of the two approaches and noted that the IBM without the p boson cannot fit the data but that, with the p boson, the results are comparable with the fits obtained by the vibron approach, which explicitly includes clustering at the α level.

We suggest that it is the larger cluster that is important for the nuclides near ^{145}Ba and for fission.[37] That small clusters play a role has already been shown for the $N = 85$ isotones, where the lowering of the position of the j–1, 5/2 and j–2, 3/2 level can be well described in the cluster-vibration model. For this paper, it is the shape of the nucleus during the scission process that is important. *We argue that it is the ability of these fragments near ^{145}Ba to assume the octupole-distorted or cluster configurations that makes their formation the major channel for fission to these particular nuclides.* If nuclear deformation is to play any role in fission, the inability of the strong prolate deformation in the mass-100 region suggests that prolate deformation is not an important factor in determining fission yields. While the fissile nucleus is quite prolate at the saddle point, unless the nucleus is to be expected to fission lengthwise, the deformability in the prolate direction should not be a factor. Rather, we argue that it is the ability to be drawn to a point that is important in the fragments to be formed.

The recognition that the octupole or pear shape could play a role in the structure of nuclides near the scission point is not a new one. Nix noted, in his Fig. 3, a series of shapes in which the one proceeding scission involved two pear-shaped fragments.[38] A number of authors have recently investigated the various shapes and shape changes undergone by nuclides in the fission process. For example, Mignen, Royer, and Sebille have examined the shapes that can evolve in fission leading to both spherical (^{132}Sn) and deformed (^{145}Ba) shapes, as shown in Fig. 4,[39] which distinctly shows octupole-shaped products as the result of fission from an elongated saddle point and spherical products as a result of fission from a more compact configuration. Cwiok, Rozmeij, Sobiczewski, and Patyk show a series of shapes corresponding to two different paths of fission, one leading toward a compact shape

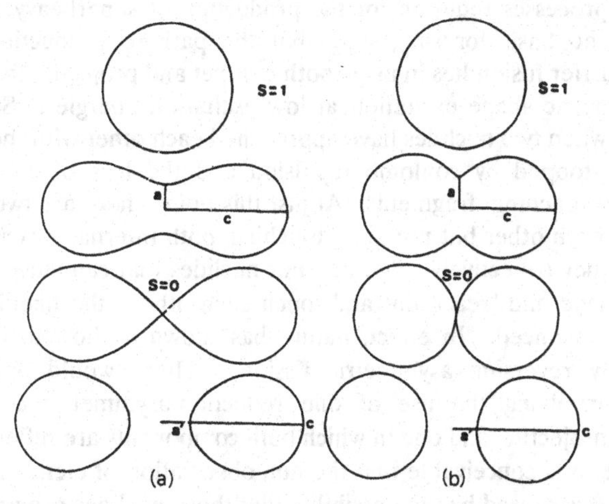

Figure 4. Evolution of the cassinian ovals (a) and of the elliptic lemniscates (b) as a function of the parameter s. The shapes vary continuously from a sphere ($s = 1$) to two separated fragments. A neck appears when $s < 1/\sqrt{3}$ in the first valley and when $s < 1/\sqrt{2}$ in the second one. The configuration at the separation ($s = 0$) is the Bernoulli lemniscate for the cassinian ovals and two touching spheres for the elliptic lemniscates. The bodies have a symmetry plane and an axis of revolution. a is the neck radius or the half distance between the tips of the fragments, while c is the half-elongation of the system (from Ref. 39).

and one to a distended shape.[40] Sperber et al. also examined neck formation and exhibited pear-shaped fission products arising from scission of an elongated nuclide.[41]

A framework in which to test these ideas in calculations has been presented by Wilkins, Steinberg, and Chasman.[42] They approached the scission point by considering the effects of the deformed shell corrections in the product nuclides and postulated that the extremely-prolate (now considered superdeformed) shapes present along the $N = 88$ isotone chain would influence the distribution of fission products. Their results were favorable in many ways but failed to hold the heavy-mass products' average mass at 140; the heavy-mass products increased in mass so as to follow the $N = 88$ isotonic chain rather than staying centered, not only at $N = 88$, but also at $Z = 56$. Because they included the effects of the strong shell closure at ^{132}Sn, they did reproduce these yields as nearly constant and did account for the symmetric fission of elements with $Z > 100$ and $N > 156$, which, of course, produce product nuclides near ^{132}Sn. As we have noted earlier, if prolate deformation were important at the scission point, then higher-than-average yields should not only be associated with the $N = 88$ isotones, but also with a number of nuclides in the mass-100 region. Inasmuch as that part of the assumption indicating increasing mass in the heavy-product group with increasing mass of the fissile nuclide was not borne out, we would argue against the superdeformed shapes associated with $N = 88$ isotones playing a strong role in low-energy nuclear fission. On the other hand, if it is possible to incorporate the shell corrections already outlined by Leander et al. into the Wilkins, Steinberg, and Chasman framework, then these ideas could be put to a severe test.

Recent calculations by Catara, Dasso, and Vitturi have provided support for the idea that these reflection-asymmetric structures can also significantly increase reaction cross-sections for subbarrier fusion, as shown in Fig. 5.[43] The same shapes that are necessary for a large nucleus to pass through in the scission process would appear to be equally important to the fusion process, particularly near the coulomb barrier. Consequently, the use of projectiles and targets in nuclear reactions that have low-energy or ground-state reflection-asymmetric deformations could enhance the cross-section for sub-barrier fusion processes required for the production of superheavy elements. These calculations form the basis for suggesting that the path to production of superheavy elements via sub-barrier fusion lies in using both a target and projectile that are susceptible to reflection-asymmetric shape excitation at low excitation energies. Sub-barrier fusion occurs at the point when two nuclides have approached each other with the approach being slowed down and stopped by coulomb repulsion and the loss of energy via coulomb excitation of the approaching fragments. At just this point, there are two nuclides, sitting for an instant near each other but not quite touching, both internally excited by the strong coulomb field that they collectively generate. For nuclides that can move to the reflection-asymmetric pear shape and "reach out and touch each other," the likelihood of fusion is potentially greatly enhanced. In effect, nature has shown us how to make superheavy elements by simply reversing asymmetric fission. There would appear to be two approaches—one involving the use of one reflection-asymmetric component of the reaction, target, or projectile, and one in which both components are reflection asymmetric. On the reverse side, it is conceivable that the non-observation of elements in the $Z > 108$ region with $N > 170$ is caused by the possibility that those nuclides can and do fission very rapidly into two reflection asymmetric fragments in the ^{144}Ba region or one ^{144}Ba-like fragment and one ^{132}Sn-like fragment.

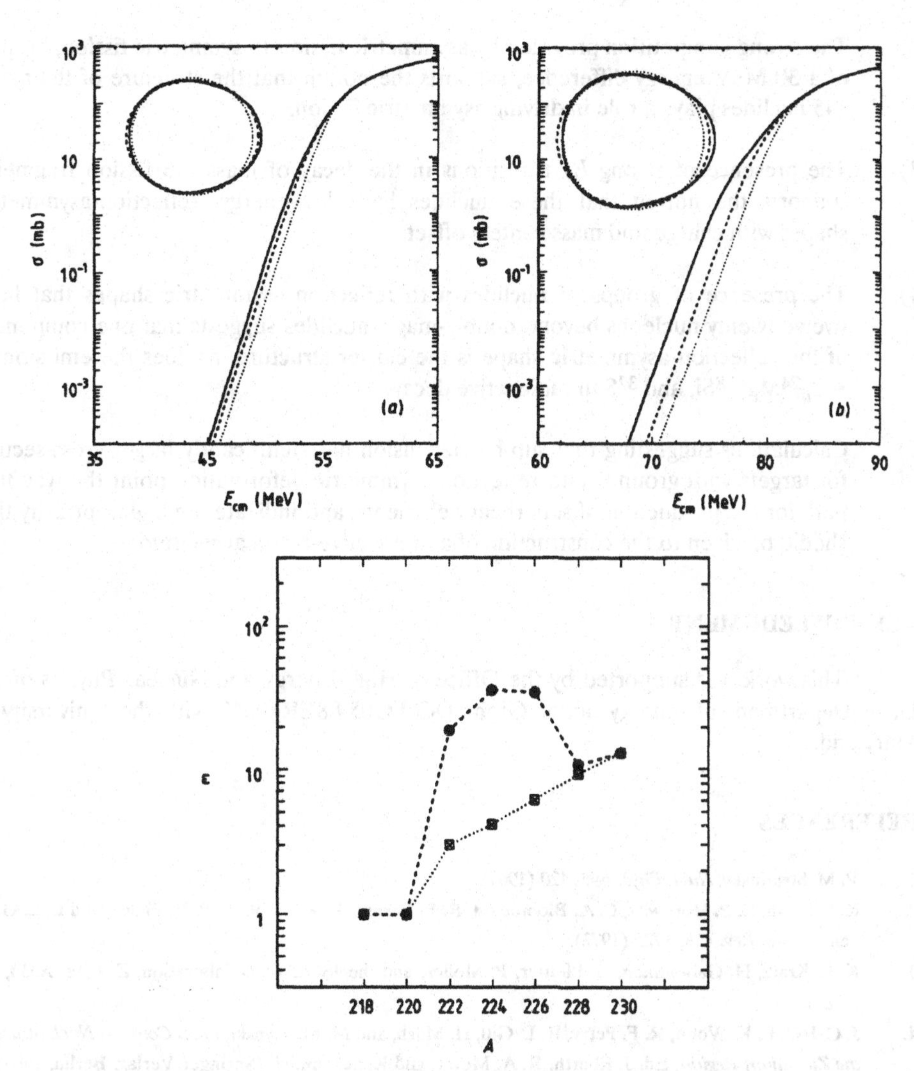

Figure 5. Top: Sub-barrier fusion cross-sections calculated with no deformation (solid lines), only quadrupole deformation (dashed curve) and quadrupole plus octupole deformation (dotted curves) for ^{16}O incident upon ^{144}Ba (*a*) and ^{224}Ra (*b*). Bottom: Cross-section difference between calculated sub-barrier fusion with (circles) and without (squares) octupole deformation for Ra nuclides (from Ref. 43).

We summarize our points as follows:

1) The production of nuclides near mass 145 in the fission of a wide range of heavy elements indicates that the structure of these nuclides in the mass-145 region is a driving element in the asymmetric-fission process.

2) The strong competition provided by asymmetric fission to symmetric fission, in spite of a 30 MeV energy difference, supports the notion that the structure of the mass-145 nuclides plays a role in driving asymmetric fission.

3) The presence of strong $E1$ transitions in the decay of mass-145 fission fragments supports the notion that these nuclides have low-energy, reflection-asymmetric shapes with charge and mass centers offset.

4) The presence of groups of nuclides with reflection asymmetric shapes that have twelve-twenty nucleons beyond doubly-magic nuclides suggests that one component of the reflection-asymmetric shape is the cluster structure, as does the emission of ^{13}C, ^{24}Mg, ^{28}Si, and ^{32}S in radioactive decay.

5) Calculations suggesting that sub-barrier fusion has significantly larger cross-section for targets with ground-state reflection-asymmetric deformation point the way to a path for the production of superheavy elements, and indicate the higher priority that should be given to the construction of a radioactive-beam accelerator.

ACKNOWLEDGMENT

This work was supported by the Office of High Energy and Nuclear Physics of the U. S. Department of Energy under Grant DG-FG05-88ER40418 with the University of Maryland.

REFERENCES

1. V. M. Strutinsky, *Nucl. Phys.* **A95**, 420 (1967).
2. K. F. Flynn, E. P. Horowitz, C. A. Bloomqvist, R. F. Barnes, R. K. Sjoblom, P. R. Fields, and L. E. Glendenin, *Phys. Rev. C* **5**, 1725 (1972).
3. K.-L. Kratz, H. Gabelmann, B. Pfeiffer, P. Moller, and the ISOLDE collaboration, *Z. Phys.* **A330**, 229 (1988).
4. J. C. Hill, F. K. Wohn, R. F. Petry, R. L. Gill, H. Mach, and M. Moszynski, *Proc. Conf. on Nucl. Struct. of the Ziconijum Region*, Ed. J. Eberth, R. A. Meyer, and K. Sistenmich (Springer-Verlag, Berlin, 1988), p. 64.
5. R. A. Meyer, D. F. Kusnezov, M. A. Stoyer, and R. P. Yaffe, in *Proceedings of the Conference on Nuclear Structure of the Ziconium Region*, Ed. by J. Eberth, R. A. Meyer, and K. Sistemich (Springer-Verlag, Berlin, 1988), p. 82.
6. J. D. Robertson and W. B. Walters, *Proc. Conf. on High-Spin Nuclear Structure and Novel Nuclear Shapes*, Argonne National Laboratory, April 13, 1988, ANL-PHY-88-2, p. 106.
7. E. K. Hulet et al., *Phys. Rev. C* **40**, 770 (1989).
8. F.-C. Wang and J. M. Hu, *J. Phys. G* **15**, 829 (1989).
9. F. Iachello and A. D. Jackson, Phys. Lett. **108B**, 15q (1982).
10. G. A. Leander, W. Nazarewicz, P. Olanders, I. Ragnarsson, and J. Dudek, Phys. Lett. 152B, 284 (1985).
11. J. D. Robertson, S. H. Faller, W. B. Walters, R. L. Gill, H. Mach, A. Piotrowski, E. F. Zganjar, H. Dejbakhsh, and R. F. Petry, *Phys. Rev. C* **34**, 1012 (1986).
12. J. D. Robertson, P. F. Mantica, Jr., C. A. Stone, S. H. Faller, E. M. Baum, and W. B. Walters, *Phys. Rev. C* **40**, 2804 (1989).

13. W. R. Phillips, I. Ahmad, H. Emling, R. Holzmann, R. V. F. Janssens, T. L. Khoo, and M. W. Drigert, *Phys. Rev. Lett.* 57, 3257 (1986).
14. W. R. Phillips, R. V. F. Janssens, I. Ahmad, H. Emling, R. Holzmann, T.. Khoo, and M. W. Drigert, *Phys. Lett.* **212B**, 402 (1988).
15. W. Urban, R. M. Lieder, W. Gast, G. Hebbinghaus, A. Kramer-Flecken, T. Morek, T. Rzaca-Urban, W. Nazarewicz, and S. L. Tabor, *Phys. Lett.* **185B**, 331 (1987).
16. W. Urban, R. M. Lieder, W. Gast, G. Hebbinghaus, A. Kramer-Flecken, T. Morek, T. Rzaca-Urban, W. Nazarewicz, and S. L. Tabor, *Phys. Lett.* **200B**, 424 (1988).
17. A. S. Mowbray et al., *Phys. Rev. C* **42**, 1126 (1990).
18. A. C. Mueller, F. Buchinger, W. Klempt, E. W. Otten, R. Neugart, C. Ekstrom, and J. Heinemeier, *Nucl. Phys.* **A403**, 234 (1983).
19. W. Borchers, E. Arnold, W. New, R. Nuegart, K. Wendt, G. Ulm, and the ISOLDE Collaboration, *Phys. Lett.* **B216**, 7 (1989).
20. C. Thibault et al., *Nucl. Phys.* **A367**, 1 (1981).
21. R. F. Casten, D. S. Brenner, and P. E. Haustein, *Phys. Rev. Lett.* **58**, 658 (1987).
22. R. Casten, W. Frank, and P. von Brentano, *Nucl. Phys.* **A444**, 133 (1985).
23. B. Pfeiffer and K.-L. Kratz, in *Proceedings of the Conference on Nuclear Structure of the Zirconium Region*, Ed. by J. Eberth, R. A. Meyer, and K. Sistemich, (Springer-Verlag, Berlin, 1988), p. 368.
24. W. Nazarewicz and T. Werner in *Proceedings of the Conference on Nuclear Structure of the Zirconium Region*, Ed. by J. Eberth, R. A. Meyer, and K. Sitemich, (Springer-Verlag, Berlin, 1988), p. 277.
25. S. H. Faller, J. D. Robertson, E. M. Baum, C. Chung, C. A. Stone, and W. B. Walters, *Phys. Rev. C* **38**, 307 (1988).
26. E. M. Baum, J. D. Robertson, P. F. Mantica, S. H. Faller, C. A. Stone, W. B. Walters, R. A. Meyer, and D. F. Kusnezov, *Phys. Rev. C* **39**, 1514 (1989).
27. W. Nazarewicz, in *Proc. Intern. Conf. on Nuclear Structure Through Static and Dynamic Moments,* Ed. by H. H. Bolotin (Conf. Proc. Press, Melbourne, 1987) Vol II, p. 180.
28. S. Cwiok and W. Nazarewicz, *Nucl. Phys.* **A469**, 367 (1989).
29. S. Cwiok, P. Rozmej, K. Bőning, and A. Sobiczewski, in *Proc. Intern. Conf. on Nuclear Structure Through Static and Dynamic Moments*, Ed. by H. H. Bolotin (Conf. Proc. Press, Melbourne, 1987), Vol. I, p. 18.
30 G. Vanden Berghe and V. Paar, *Z. Phys.* **A294**, 183 (1980).
31. S. Cwiok and W. Nazarewicz, *Nucl. Phys.* **A469**, 367 (1989).
32. F. Iachello, in *Proceedings of the Conference on Nuclear Structure with Heavy Ion*, Ed. by R. A. Ricci and C. Villi (Italian Physical Society, Bologna, 1985) Vol 2.
33. F. Iachello, *Nucl. Phys.* **A396**, 233c (1983).
34. D. Kusnezov and F. Iachello, *Phys. Lett* **B209**, 420 (1988).
35. T. Lonnorth, *Z. Phys.* **A331**, 11 (1988).
36. J. F. Shriner, Jr., P. D. Cottle, J. F. Ennis, M. Gai, D. A. Bromley, J. W. Olness, E. K. Warburton, L. Hildingsson, M. A. Quader, and D. B. Fossan, *Phys. Rev. C* **32**, 1888 (1985).
37. D. A. Bromley, in *Proc. of the 4th Int. Conf. on Clustering Aspects of Nucl. Struct. & React.*, Ed. by J. Lilley and M. Ngarajan (D. Reidel, Dord, 1984).
38. J. R. Nix in *Proc. of the Int. Conf. on the Properties of Nuclides Far From the Region of Beta Stability*, (CERN 70-30) Vol 2, p. 605 (1970).
39. J. Mignen, G. Royer, and F. Sebille, *Nucl. Phys.* **A489**, 461 (1988).
40. S. Cwiok, P. Rozmej, A. Sobiczewski, and Z. Patyk, *Nucl. Phys.* **A491**, 281 (1989).
41. D. Sperber, J. Stryjewski,, and M. Zielinska-Pfabe, *Proc. Conf. Theoretical and Experimental Methods of Heavy-Ion Physics*. Ed. by Z. Wilheli and G. Szeflinska, (Harwood Academic Publishers, Chur, 1988) Vol. **15**, p. 81.

42. B. Wilkins, E. Steinberg, and R. Chasman, *Phys. Rev. C* **14**, 1832 (1976).
43. F. Catara, C. H. Dasso, and A. Vitturi, *J. Phys. G: Nucl. Phys.* **15**, L191 (1989).

14. Quadrupole and Octupole Shapes in Nuclei

Douglas Cline
Nuclear Structure Research Laboratory
University of Rochester
Rochester, NY 14627

ABSTRACT

Heavy-ion coulomb excitation measurements can determine almost the complete set of ⩾100 *E*2 matrix elements involving the lowest ⩽30 collective states in nuclei, as well as *E*3 matrix elements. The completeness of the *E*2 data is sufficient to determine directly the centroids and fluctuation widths of the *E*2 properties in the principal axis frame, providing a direct measure of the correlation of the *E*2 data for low-lying states. This adds a powerful new dimension to the study of quadrupole collectivity in nuclei. The *E*3 matrix elements provide a sensitive test of octupole deformation in nuclei. The results and model implications of recent coulomb excitation measurements of the quadrupole shapes in odd- and even-*A* nuclei will be discussed, including examples of shape transitions, shape coexistence, and coriolis effects. The results of a study of octupole deformation in ^{148}Nd also will be presented.

14.1 INTRODUCTION

Nuclear shapes are a sensitive probe of nuclear structure. Collective rotational- and vibrational-shape degrees of freedom have been shown to be a dominant feature of low-lying spectra in nuclei. Strongly-deformed quadrupole collective bands occur in many nuclei, and evidence for octupole deformation is manifest in some nuclei.

The $E\lambda$ properties are the most direct and unambiguous measure of λ-multipole shape degrees of freedom. Coulomb excitation is the preeminent probe of collective $E\lambda$ matrix elements in the yrast domain; that is, it selectively excites collective bands in the yrast domain with cross-sections that are directly related to the $E\lambda$ matrix elements involved in the excitation. Moreover, coulomb excitation is well understood, allowing for quantitative studies of nuclear structure unimpeded by uncertainties in our knowledge of the interaction and reaction mechanism that plague many other spectroscopic probes employed in nuclear science. Recent advances[1] in the field of coulomb excitation make it feasible to measure essentially complete sets of $E\lambda$ matrix elements for low-lying states of stable nuclei, providing a powerful new probe of nuclear shapes. The completeness and extent of such information for $\Delta I = 2$ provides a direct measure of the centroids and widths of the fluctuating-shape degrees of freedom for low-lying states.[1] This paper presents results of recent coulomb-excitation studies of both quadrupole and octupole deformation in nuclei.

Exotic Nuclear Spectroscopy, Edited by W. C. McHarris
Plenum Press, New York, 1990

14.2 EXPERIMENTAL TECHNIQUES

A renaissance in the field of coulomb excitation has resulted from the ability of the latest generation of heavy-ion accelerators to provide copious beams of all stable nuclei with energies up to 5 MeV per nucleon. Now it is feasible using high-Z projectiles, e.g., ^{208}Pb, to coulomb excite the lowest state of each spin, the yrast sequence, to spin 30 $\hbar$ in strongly-deformed actinide nuclei and to excite lower-spin collective states lying within 2 MeV of the yrast sequence.

Gamma-ray spectroscopy using Ge γ-ray detectors is the only viable experimental technique for resolving the many states coulomb excited when heavy ions are utilized. Unfortunately, the recoil velocities of the excited nuclei can be large, $v/c \approx 0.1$, which leads to considerable doppler shift and doppler broadening of detected γ rays. The doppler broadening can be reduced by using thin targets so that the excited nuclei recoil into vacuum and by observing the deexcitation γ-rays in coincidence with scattered ions detected at known scattering angles, in order to specify the recoil direction and velocity. Thus, the individual γ-ray signals can be corrected for the doppler shift on an event-by-event basis.

Most of the heavy-ion-induced coulomb excitation experiments have employed variants of the large solid-angle position-sensitive, parallel-plate, avalanche detector array[2] illustrated in Fig. 1. The scattered projectile and ejectile are detected in kinematic coincidence as well as in coincidence with an array of compton-suppressed Ge detectors observing the deexcitation γ rays. The optimum geometry depends on the projectile-to-target mass ratio. Figure 2 illustrates a typical doppler-corrected coincident γ-ray spectrum for an even-A nucleus.[3] The energy resolution is 0.5% FWHM and the peak-to-background ratio is 100 for the intense peaks. The high selectivity of coulomb excitation is reflected by the simplicity of such spectra. Spectra for odd-A nuclei[4] are more complicated, as illustrated in Fig. 3, because there are twice as many states and both cascade and cross-over deexcitation transitions occur.

The proposed GAMMASPHERE γ-ray detector system (see Chapter 15 in This Book) will provide about a hundred-fold increase in statistics for particle γ-γ coincident coulomb-excitation data. This will make it viable, for the first time, to separate clearly the interesting sideband transitions from the dominant ground-band transitions and provide about an order of magnitude increase in sensitivity of the measurements. This new capability should dramatically advance the study of nuclear shapes and possibly lead to the discovery of new phenomena.

A unique determination of the many electromagnetic properties of an investigated nucleus requires coulomb-excitation data covering a wide range of coulomb-interaction strength. Over determining the problem requires data recorded using both a wide range of projectile Z values and a wide range of scattering angles. The experimental geometry shown in Fig. 1 provides coulomb excitation data simultaneously over a wide range of scattering angles. Several closed-shell nuclei, such as ^{16}O, ^{40}Ca, ^{58}Ni, ^{136}Xe, and ^{208}Pb, are chosen to excite the nuclei of interest, since such closed-shell nuclei are not appreciably coulomb excited, simplifying the γ-ray spectra and the analysis.

In several cases, we have also measured lifetimes of coulomb-excited states by the recoil-distance technique. These both simplify and serve as an excellent check on the results of the complicated analysis of the coulomb-excitation yield data.

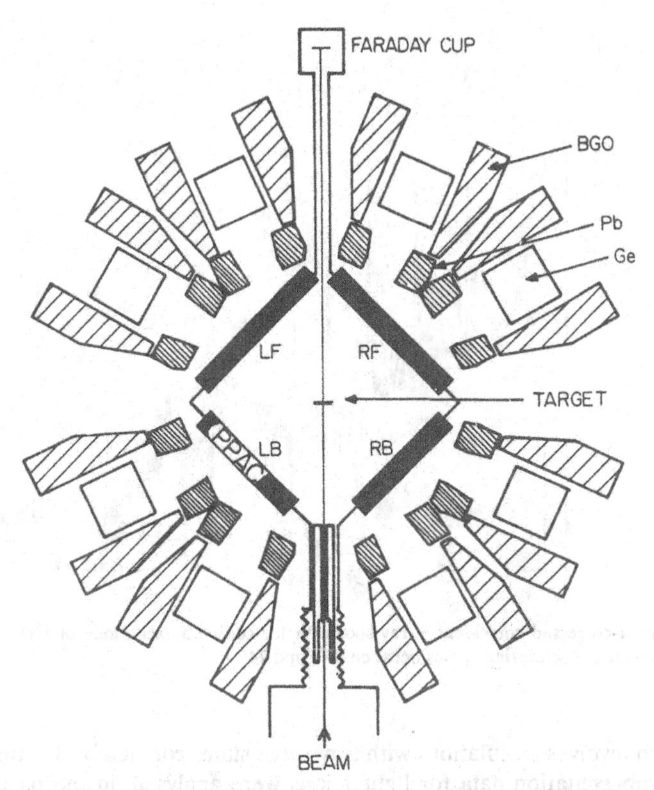

Figure 1. Rochester coulomb-excitation apparatus. A position-sensitive, parallel-plate, avalanche detector array detects both the scattered projectile and ejectile in kinematic coincidence, as well as in coincidence with an array of compton-suppressed Ge detectors observing the deexcitation γ rays. The particle detectors measure both θ and ϕ in the range $8° < \theta < 76°$, $104° < \theta < 164°$, and $-35° < \phi < 35°$. The 0.8° θ resolution and 0.6 ns time resolution of the particle detectors are sufficient to resolve the various kinematic solutions by both the correlation of the scattering angles and the time of flight difference.

14.3 EXTRACTION OF ELECTROMAGNETIC MATRIX ELEMENTS FROM COULOMB-EXCITATION DATA

The semiclassical approximation provides an excellent description for heavy-ion-induced coulomb excitation. Using the semiclassical description, the time-dependent excitation amplitudes for the coulomb-excited nucleus are governed by a set of coupled differential equations that are solved for motion along the classical hyperbolic trajectory in the coulomb field, with the condition that both nuclei initially are in their ground states. The subsequent deexcitation of the coulomb-excited nuclei can be treated separately, because the lifetimes of the excited states are several orders of magnitude larger than the effective collision time.

The analysis of these coupled equations for heavy-ion-induced coulomb excitation data is difficult because of the dramatic increase in the number of matrix elements involved when many states are excited. The cross-sections depend, in a complicated, non-linear way, on both the signs and magnitudes of many $E\lambda$ matrix elements, while $M1$ matrix elements are needed to account for the deexcitation γ decay. For example, coulomb excitation of

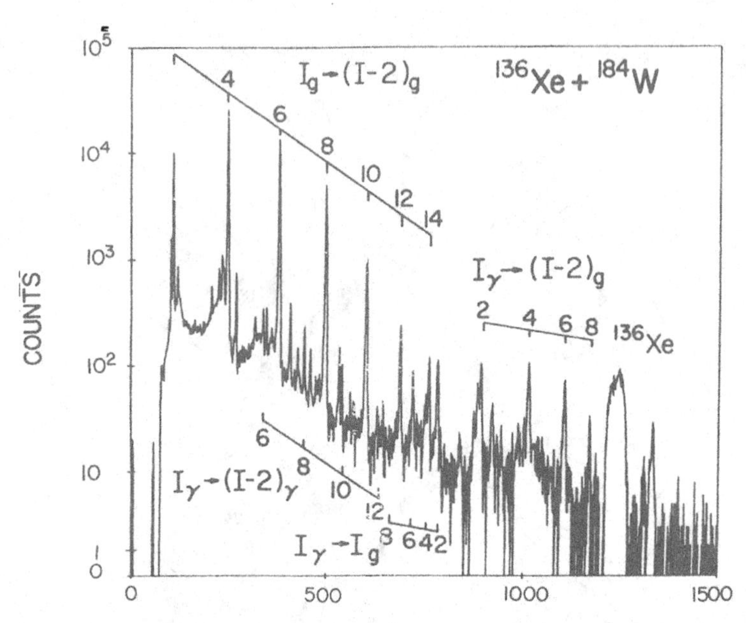

Figure 2. Doppler-corrected coincident γ-ray spectrum for coulomb excitation[3] of ^{184}W by 4.125-MeV per nucleon ^{136}Xe detected at scattering angles between 54° and 74.

^{165}Ho by ^{208}Pb involves calculations with fifty-three states coupled by 393 matrix elements.

Coulomb-excitation data for lighter ions were analysed, in the past, by comparing data with semiclassical calculations using matrix elements predicted by models. The semiclassical coulomb excitation code COULEX[5] played a pivotal role in such model-dependent analyses for nearly two decades. A full, model-independent analysis requires that the inverse problem be solved, namely, a least-squares fit of the matrix elements to the observed data. Unfortunately, it is impractical to perform such least-squares searches utilizing the full coupled-channel procedure, because of the enormous computational load involved in numerically solving the many coupled equations. A fast and sufficiently-accurate approximation is necessary for solving the coupled equations during the least-squares search procedure. The Rochester coulomb-excitation code GOSIA, developed by Czosnyka et al.,[6] replaces the time-dependent collision functions with a mean time-independent effective interaction strength. This simplification results in an analytical solution to the coupled-channels problem, which is used to evaluate the derivatives needed to search for a minimum in the least-squares statistic built from the coulomb-excitation data plus other available data, such as lifetimes, mixing, and branching ratios. The approximations used in GOSIA reproduce the matrix-element dependence for the excitation quite well and give excellent agreement by introducing correction factors that are computed periodically by comparison of approximate with exact calculations. Gradient minimization, modified to handle better the expected shapes of the least-squares statistic, is used to locate the minimum. This is exploited by using several sets of random numbers as starting values in the fitting procedure to eliminate any bias and to check the uniqueness of the solution.

The largest version of the code GOSIA can make a least-squares fit of ≤ 400 matrix elements ($E1$, $E2$, $E3$, $E4$, $E5$, $E6$, $M1$, $M2$) to several thousand data from up to fifty independent coulomb-excitation experiments, as well as other lifetime, branching ratio, and

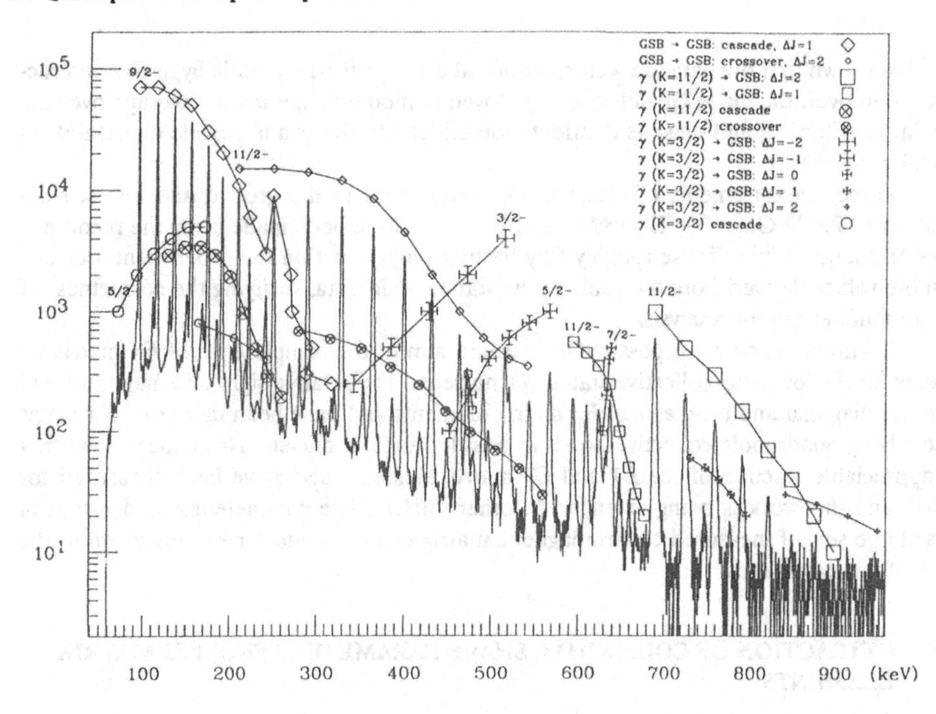

Figure 3. Doppler-corrected coincident γ-ray spectrum for coulomb excitation[4] of ^{165}Ho by 215-MeV ^{58}Ni ions.

$E2/M1$ mixing ratio data. The code allows integration over solid angles of the detectors and energy loss in the target.

The estimation of the errors of the fitted matrix elements is difficult because of the dominant cross-correlation effects for this strongly-coupled nonlinear system. Several techniques have been tested to estimate the errors. The curvature matrix method and the unity increase in the least-squares statistic both have severe problems. The quadratic approximation often is not applicable, and the unavoidable presence of so-called "nuisance parameters", i.e., matrix elements which do not influence the observations, cause the error matrix to be ill-defined. The only viable method is to construct an approximate probability distribution in the space of the fitted parameters and to request that the total integrated probability be equal to the confidence limit chosen, i.e., 68.3%.

The computer time required to make a complete, model-independent, least-squares analysis of coulomb-excitation data depends on the number of states involved, the number of unknowns, the ground-state spin, and the number of experiments. A complete analysis of a typical thirty-level system with 150 matrix elements takes about 30 CPU-hours on an ETA10 supercomputer. Now it is viable to analyse data even for odd-A nuclei.

The primary assumption of coulomb excitation is that the interaction is purely electromagnetic. This is ensured by using a conservative safe-energy criterion,[1] and this has been checked by performing quantal calculations that include both the nuclear and coulomb interactions.[7] Quantal corrections to the semiclassical approximation are unimportant for heavy-ion-induced coulomb excitation. The deorientation effect is the only unknown in computing the γ-ray angular distributions. A recent study of the deorientation effect[7] has shown that the data are well reproduced using primarily a static hyperfine interaction. Moreover, the many Ge detectors employed in modern experiments average over the angular distribution, and thus the deorientation effect introduces a negligible uncertainty in typical experiments.

Direct measurements of lifetimes for states, both in the ground and lowest side-bands in ^{110}Pd, ^{156}Gd, ^{166,168}Er, ^{148}Nd, and ^{182,184}W, have been made using the recoil-distance technique.[3, 8-10] These roughly fifty lifetimes agree, within the experimental errors, with the values derived from the coulomb-excitation yield data, verifying the correctness of the coulomb-excitation analysis.

In summary, now it is possible to measure almost the complete set of $E\lambda$ matrix elements for the low-lying collective states in a nucleus. The relative signs and magnitudes of both the diagonal and off-diagonal $E2$ matrix elements and have been measured for states of low-lying quadrupole collective bands in a wide range of nuclei. This paper shows that an appreciable fraction of the $E1$ and $E3$ matrix elements also have been measured for ^{148}Nd, and this work is being extended to other nuclei. The completeness and extent of these large sets of measured electromagnetic matrix elements add a new dimension to the study of nuclear shapes.

14.4 EXTRACTION OF COLLECTIVE SHAPE PARAMETERS FROM E2 MATRIX ELEMENTS

The use of rotationally-invariant products[1,11-13] of $E2$ operators to directly relate properties in the principal axis frame to those in the laboratory frame has proven to be a powerful method for interpreting the wealth of data produced by coulomb excitation. It provides considerable insight into the underlying collective degrees of freedom.

The electric-quadrupole tensor can be rotated into an instantaneous principal axis frame that has only two non-zero quadrupole moments. By analogy with Bohr's parameters (β,γ), we can express the principal axis-frame electric moments in terms of two parameters, Q, δ, where $E(2,0) = Q\cos\delta$, $E(2,+2) = E(2,-2) = Q\sin\delta \lambda\sqrt{2}$, and $E(2,1) = E(2,-1) = 0$. Zero-coupled products of $E2$ operators can be evaluated in the principal axis frame, e.g.:

$$\{E2 \times E2\}^0 = (1/\sqrt{5})Q^2 \tag{1}$$

$$\{[E2 \times E2]^2 \times E2\}^0 = -(\sqrt{2/35})Q^3 \cos 3\delta \tag{2}$$

For any state s, expectation values of the $E2$ rotational invariants in the laboratory frame can be evaluated by summation over the products of the experimental $E2$ matrix elements formed by making intermediate state expansions of the expectation values of the rotational invariants; that is, the rotational invariants can be evaluated in the laboratory frame, using experimental data, if the relative signs and magnitudes of the $E2$ matrix elements are available. Higher-order products of $E2$ operators can be formed, using various intermediate couplings which, when evaluated, involve different subsets of matrix elements. This provides a check of the completeness and self-consistency of the evaluation of the invariants. Thus, for any state s, the expectation values of all the rotationally-invariant products of the $E2$ operator can be evaluated directly, determining the distribution of the intrinsic-frame electric-quadrupole moments, parametrized by Q and δ. Knowing the values of the various invariants for any state determines directly the centroids, fluctuation widths, skewnesses, cross-correlation coefficients, etc., describing the distribution in the Q-δ plane of the expectation value of the $E2$ moments for that state. The invariants for different states will be correlated if quadrupole collectivity is manifest.

Although this technique has been discussed in the context of its application to the collective model, it is completely model-independent and is applicable to any spherical tensor operator where the strength distribution is sufficiently localized to be evaluated reliably. The invariants can be evaluated exactly for models and approximately for experimental data; thus, they are equivalent to observables. The significance and usefulness of presenting the experimental data in the form of these model-independent invariants depends on the degree to which the nuclear properties are correlated by shape degrees of freedom. The recent advances in the field of coulomb excitation make it possible to determine the set of $E2$ matrix elements required to apply this model-independent method and thereby express a wealth of data in a form that exhibits clearly the extent to which the data are correlated by quadrupole collectivity. The rotational invariants provide the most insight into the underlying collective correlations at the expense of some loss in precision due to incomplete summation. Model assumptions can be used to relate the Q-δ $E2$ distributions to the β-γ shape distribution. A more detailed description of this method is given in Ref. 1.

14.5 QUADRUPOLE COLLECTIVITY IN STRONGLY-DEFORMED NUCLEI

Coulomb excitation has been used extensively to populate states up to spins as high as 30 $\hbar$ in the ground bands in many strongly-deformed, rare-earth and actinide nuclei. As summarized in Ref. 1, in-band $B(E2)$ values and static quadrupole moments have been measured for the ground bands in several nuclei. A notable feature of these results is that, for strongly-deformed nuclei, the measured ground band $B(E2)$ values and static $E2$ moments obey the simple rigid prolate-spheroidal-rotor relation with remarkable accuracy except at band intersections or for nuclei adjacent to shape transitions.

A nice example is the case of ^{168}Er. Neutron capture γ-decay work[14] located thirty-six low-lying rotational bands in this nucleus, providing one of the most complete examples of rotational structure in nuclei. A complementary measurement of a similarly complete set of $E2$ matrix elements for this nucleus will provide a remarkably complete set of data that should be a stringent test of collective models. Coulomb excitation of ^{168}Er by ^{40}Ca, ^{58}Ni, and ^{208}Pb ions was used[9] to measure fifty-four $E2$ matrix elements involving the ground and γ bands from 925 pieces of data. Lifetimes were also measured using the recoil-distance method.
Figure 4 summarizes the transition and diagonal matrix elements for the ground and γ bands of ^{168}Er. The predictions of the rigid-spheroidal-rotor model, the asymmetric-rigid-rotor model (not shown), and the interacting-boson model (IBA) all are in good agreement with the data shown in Fig. 4. The interband $E2$ matrix elements between the ground and γ bands are more sensitive to the model predictions, as shown in Fig. 5. The asymmetric rigid rotor, assuming $\gamma = 8°$, reproduces the data well, although these data are relatively insensitive to γ softness about such a small centroid of the asymmetry. The IBA predictions deviate somewhat from these data. The only noticeable deviation from the rotor model is a possible reduction in the collectivity for the 6^+ and 8^+ states in the γ band. Unfortunately, higher-lying sidebands were not observed. It should be possible to study these interesting sidebands when GAMMASPHERE is operational. The agreement with the rotor model also can be interpreted as a measure of the reliability of the $E2$ matrix elements derived from coulomb-excitation work.

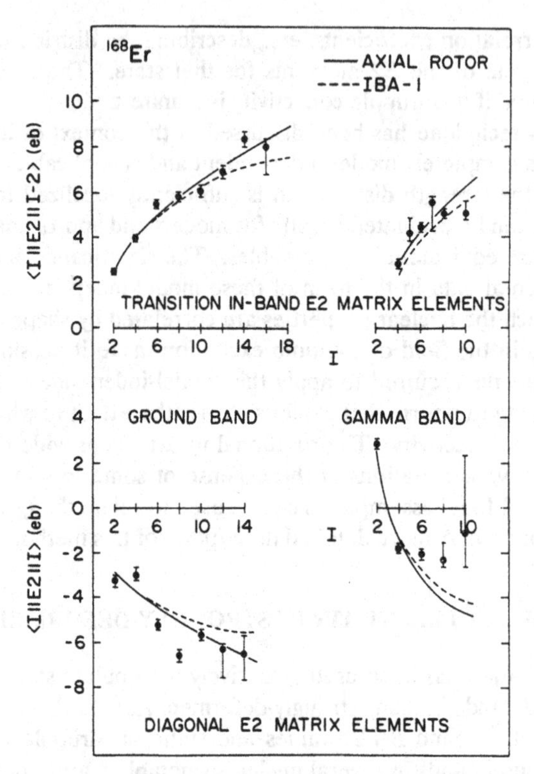

Figure 4. The in-band $E2$ matrix elements for the ground and γ bands in ^{168}Er. The lines represent calculations using the axial rotor model (solid line) and the IBA-1 model (dashed line).

The rotational invariants,[1] evaluated using the measured $E2$ properties of ^{168}Er and shown in Fig. 6, exhibit the expected behavior for a good rotor. The centroids of $\langle Q^2 \rangle$ and $\langle \cos 3\delta \rangle$ are approximately spin-independent to spin 16 of the ground band. The asymmetry $\langle \cos 3\delta \rangle$ for the ground-band states corresponds to $\delta \approx 9°$, i.e., an almost pure prolate shape. The widths $\sigma(Q^2)$ and $\sigma(\cos 3\delta)$ suggest a relative "stiffness" in both the magnitude Q^2 and asymmetry δ for ground-band states in ^{168}Er. The γ band again shows a possible reduction in collectivity for the highest spin states. This work confirms that quadrupole collectivity is the dominating feature for the low-lying states in ^{168}Er.

14.6 SHAPE TRANSITION AND BAND MIXING IN W, Os, AND Pt NUCLEI

Quadrupole collectivity is a prominent feature of shape-transitional nuclei, such as the even-A Pt, Os, and W nuclei. Coulomb excitation of ^{182,184}W, ^{186,188,190,192}Os, and ^{194}Pt by ^{40}Ca, ^{58}Ni, ^{136}Xe, and ^{208}Pb ions[3,15] populated levels up to spin 12^+ of the ground band, 10^+ of the γ band, and 6^+ of the 4^+ band in most of the nuclei studied. Figure 7 shows a typical level scheme resulting from this work. The magnitudes and observable signs of about forty transition and diagonal matrix elements were determined, with reasonably small errors, from up to 1900 pieces of data for each of the seven nuclei studied. These matrix elements include, in most nuclei, static electric-quadrupole moments of members of the ground band to spin 10^+, the γ band to spin 8^+, and the 4^+ band head, in

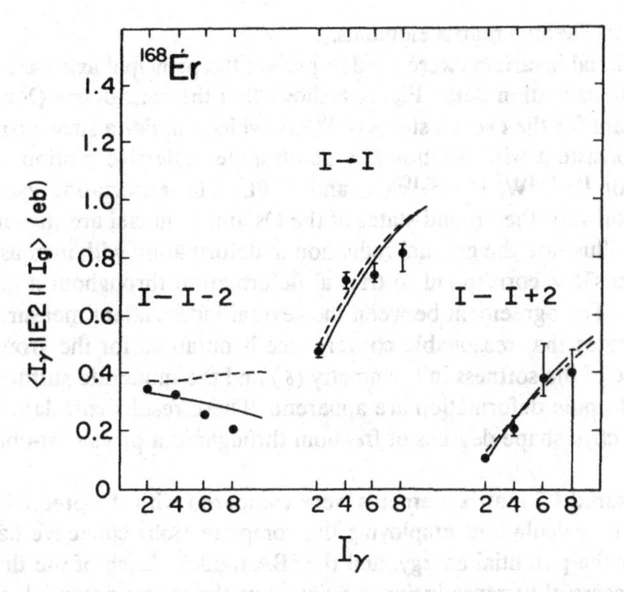

Figure 5. The inter-band $E2$ matrix elements between the ground and γ bands of ^{168}Er. The solid line represents calculations using the asymmetric rotor model with $\gamma = 8°$. The dashed line shows the predictions of the IBA-1 model.

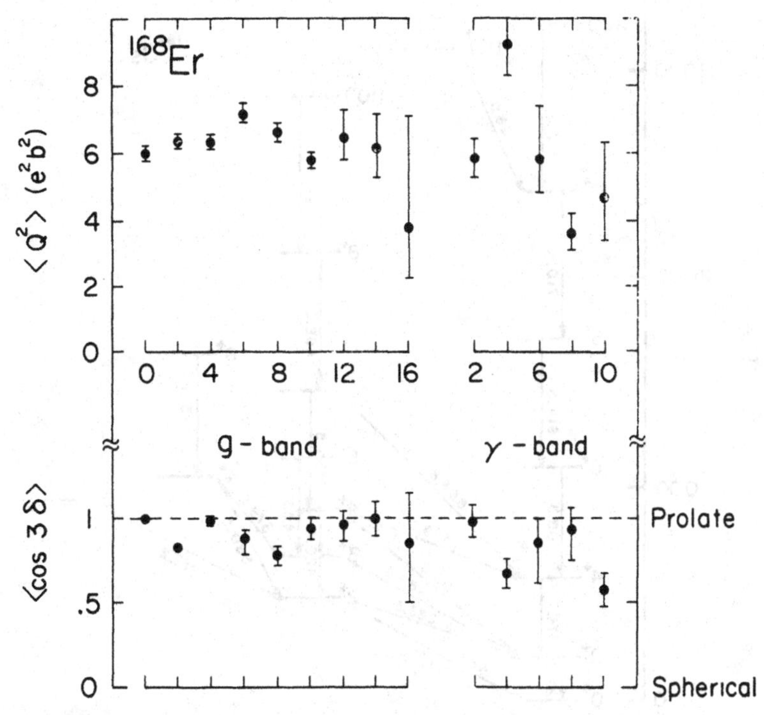

Figure 6. Centroids for the magnitude and asymmetry of the intrinsic-frame $E2$ properties of the states in the ground and γ bands of ^{168}Er.

addition to the off-diagonal matrix elements.

The rotational invariants were used to project the principal axis-frame $E2$ properties from the coulomb excitation data. Figure 8 shows that the centroids $\langle Q^2 \rangle$ and $\langle \cos 3\delta \rangle$ are almost constant for the excited states of ^{192}Os, which implies a strong correlation of the $E2$ properties consistent with rotation-like quadrupole collective motion. Similar results were obtained for ^{182,184}W, ^{186,188,190}Os, and ^{194}Pt. The expectation values of both the centroids and widths for the ground-states of the Os and Pt nuclei are shown in Fig. 9. The centroids $\langle Q^2 \rangle$ illustrate the gradual reduction in deformation with increasing mass, while the centroids $\langle \cos 3\delta \rangle$ correspond to triaxial deformation throughout a prolate-to-oblate shape transition. The agreement between the several independent measures of $\cos 3\delta$ and the widths illustrates that reasonable convergence is obtained for the ground states. The mass dependence of the softness in asymmetry (δ) and the moderate stiffness in magnitude (Q^2) of the quadrupole deformation are apparent. These results correlate well, using only quadrupole collective shape degrees of freedom throughout a prolate-to-oblate shape transition.

The measured $E2$ matrix elements were compared with the predictions of the rigid asymmetric model, calculations employing the complete Bohr collective hamiltonian with γ-soft shapes for the potential energy, and the IBA model. Each of the three models was only partially successful in reproducing the data, but the γ-soft potentials were more successful than γ-stiff model.

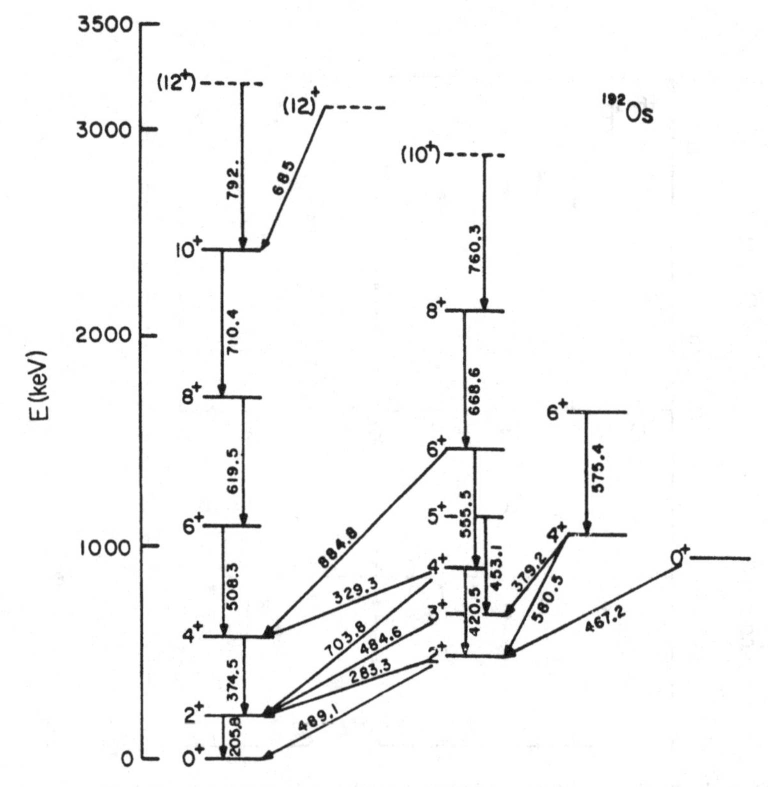

Figure 7. Level scheme of ^{192}Os derived from coulomb excitation.

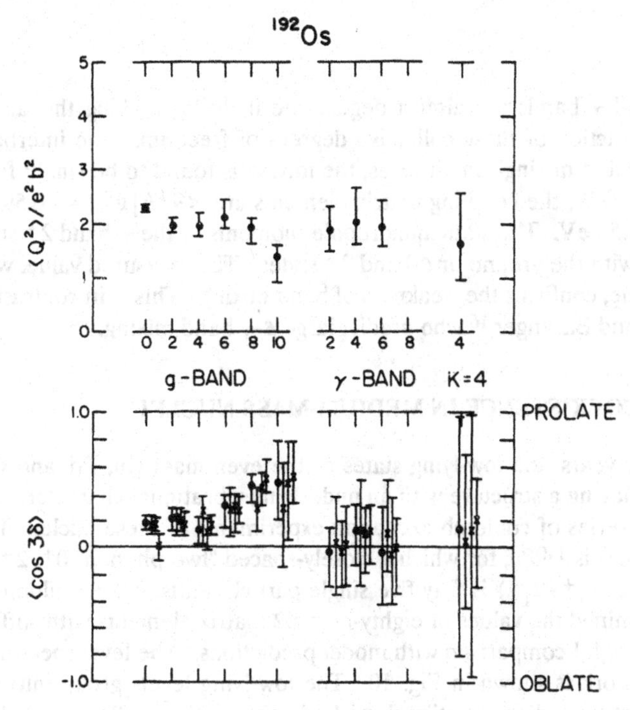

Figure 8. Centroids Q^2 and cos3δ versus spin for the ground band, γ band, and 4+ band head in ^{192}Os.

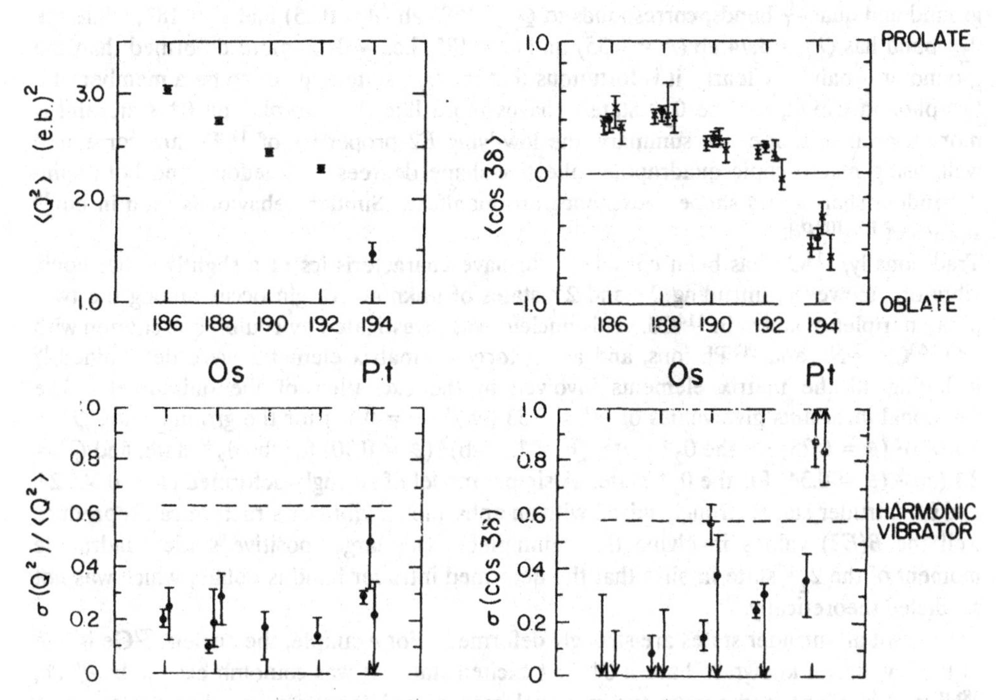

Figure 9. Centroids (upper) and fluctuation widths (lower) for the magnitude and asymmetry of the intrinsic frame $E2$ properties of the ground states of ^{186}Os, ^{188}Os, ^{190}Os, ^{192}Os, and ^{194}Pt.

The β and γ bands are almost degenerate in ^{182}W, making this an ideal case for probing the interaction of these collective degrees of freedom. The interband transitions are sensitive to band mixing. In all cases, the mixing is found to be small; for example, for the 2^+ states in ^{182}W, the coupling matrix elements are $<\gamma|h|g> = -0.59\pm0.27$ keV and $<\beta|h|\gamma> = -1.5$ keV. The static quadrupole moments of the γ-band 2^+ state is very sensitive to mixing with the ground or β-band 2^+ states. The measured value, which is 80% of the unmixed value, confirms the weakness of band mixing. This is in contrast to the predictions of Kumar and Baranger,[16] who predict large β-γ band mixing.

14.7 SHAPE COEXISTENCE IN MEDIUM-MASS NUCLEI

For many years, the low-lying states of the even-mass Ru, Pd, and Cd nuclei have been viewed as having a structure with an underlying vibrational character. This stimulated us to perform a series of coulomb-excitation experiments of these nuclei. The earliest example we studied[17] is ^{110}Pd, for which a closely-spaced "two-phonon" 0^+, 2^+, 4^+ triplet exists and the $B(E2;\ 0_1^+\to2_1^+)$ is fifty-five single-particle units. The coulomb-excitation experiments determined the values of eighty-nine $E2$ matrix elements with sufficient accuracy to make a meaningful comparison with model predictions. The level spectrum of ^{110}Pd derived from this work is shown in Fig. 10. The low-lying levels group into four rotational bands, in contrast to earlier vibrational model interpretations. The in-band transition and diagonal $E2$ matrix elements for these bands have a rotational character that, for the ground and quasi-γ bands, corresponds to $Q_o = 2.85$ eb ($\beta \approx 0.25$) and $\delta = 18°$, while the 0_2^+ band has $Q_o = 3.74$ eb ($\beta = 0.35$) and $\delta = 18°$, i.e., $\approx$ 31% more deformed than the ground or γ bands. Clearly, it is fortuitous that the 0_2^+ state appears to be a member of a two-phonon triplet, for the 0_3^+ state behaves more like the two-phonon 0^+ state and is more triaxial in shape. In summary, the low-lying $E2$ properties of ^{110}Pd are correlated well, using macroscopic quadrupole-collective-shape degrees of freedom, and both spin-dependent shapes and shape coexistence are manifest. Similar behavior is seen in our[18] studies of ^{106,108}Pd.

Traditionally, ^{114}Cd has been considered to have characteristics of a slightly anharmonic vibrator. However, intruding 0^+ and 2^+ states of unknown origin occur among the two-phonon triplet of states in ^{114}Cd. This nucleus was investigated by coulomb excitation with ^{16}O, ^{40}Ca, ^{58}Ni, and ^{208}Pb ions, and about forty $E2$ matrix elements were determined,[19] including all the matrix elements involved in the excitation of the quintuplet. The rotational invariants give values of $Q^2 = 0.53$ $(eb)^2$ ($\beta = 0.17$) for the ground state, $Q^2 = 1.1$ $(eb)^2$ ($\beta = 0.25$) for the 0_2^+ state, $Q^2 = 1.5$ $(eb)^2$ ($\beta = 0.30$) for the 0_3^+ state, and $Q^2 = 2.1$ $(eb)^2$ ($\beta = 0.34$) for the 0_4^+ state. A simple model of strongly-deformed ($\beta = 0.30$) 2^+ and 0^+ intruder states strongly mixed with an anharmonic-vibrator structure can reproduce well the $B(E2)$ values involving the quintuplet. The large, positive static quadrupole moment of the 2_2^+ state implies that the deformed intruder band is oblate, which was not predicted theoretically.

Not all intruder states are strongly deformed. For example, the nucleus ^{72}Ge is one of the few nuclei known to have a 0^+ first excited state. It was coulomb excited by ^{208}Pb, ^{58}Ni, and ^{16}O ions, and twenty-one matrix elements involving the seven lowest states were determined.[20] These results are consistent with a soft triaxial quadrupole structure ($\beta = 0.3$, $\gamma = 28.5$), perturbed by an isolated, presumed spherical, intruder 0^+ state. Similar studies[7] of ^{76,80,82}Se also show collective band behavior characteristic of a soft triaxial rotor with $\delta \approx 26°$. No evidence for shape coexistence was observed.

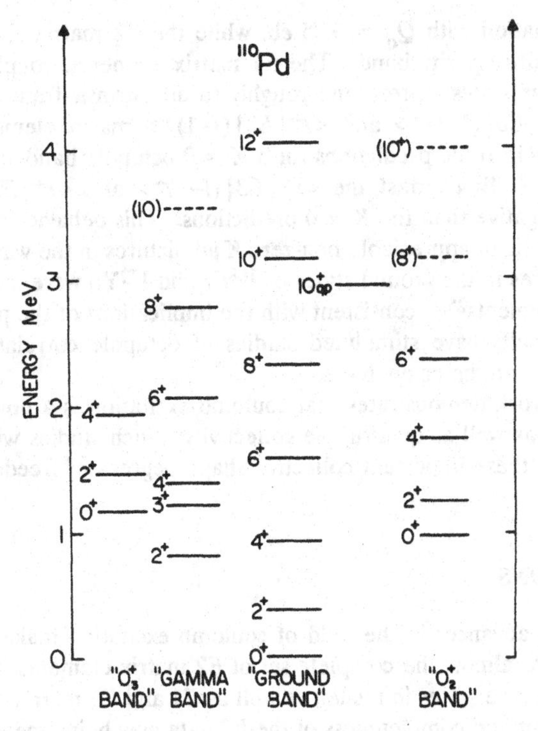

Figure 10. ^{110}Pd level spectrum derived from coulomb excitation.

In summary, the $E2$ data in all the nuclei discussed are correlated well, using collective models that have fluctuating triaxial quadrupole-deformed shapes and, in some cases, include coexistence of differently-shaped configurations.

14.8 OCTUPOLE DEFORMATION IN ^{148}Nd

There is mounting evidence for the importance of reflection-asymmetric shapes in certain nuclei, e.g., the existence of alternating-parity bands, enhanced $E1$ transitions, parity doublets in odd-A nuclei, etc. The apparent reflection asymmetry has been interpreted as arising from octupole correlations, but the evidence for this interpretation is ambiguous. Enhanced $E1$ matrix elements are not a reliable indicator of octupole correlations because they are strongly affected by a fluctuating single-particle contribution.[21,22] The only direct and unambiguous measure of octupole deformation are $E3$ matrix elements. Coulomb excitation is the only viable way of measuring $E3$ matrix elements, which limits study to stable nuclei. Nuclei with Z = 54-60 and N = 88 exhibit characteristic reflection-asymmetric behavior[23], and these are being studied by heavy-ion coulomb excitation.

The nucleus ^{148}Nd has been coulomb excited[24] by ^{58}Ni, ^{92}Mo, and ^{208}Pb, and $E1$, $E2$, and $E3$ matrix elements have been extracted with good precision from the ^{58}Ni and ^{92}Mo yield data plus recoil-distance lifetime measurements.[10] The characteristic alternating-parity sequence is exhibited by the level scheme shown in Fig. 11. The positive-parity ground-band transition $E2$ matrix elements and static $E2$ moments fit well to the prolate spheroidal rotor relation with Q_0 = 3.75 eb, while the $E2$ matrix elements are 10-15% smaller for the negative-parity band. The $E1$ matrix elements, coupling the alternating negative and positive states, correspond roughly to an intrinsic-frame dipole moment of 0.33 $e$fm. The $\langle I^-|E3|(I-3)^+\rangle$ and $\langle I^-|E3|(I-1)^+\rangle$ matrix elements for $I \leq 11$ correlate remarkably well to the predictions for a $K = 0$ octupole band with an intrinsic moment Q_3 = 1.52 eb$^{3/2}$. By contrast, the $\langle I^+|E3|(I-3)^-\rangle$ and $\langle I^+|E3|(I-1)^-\rangle$ are systematically more negative than the $K = 0$ predictions. This behavior of the $E3$ matrix elements is characteristic of appreciable non-zero K admixtures in the wavefunctions. Large, non-zero K admixtures in the ground states of ^{168}Er and ^{172}Yb were measured by inelastic α scattering measurements[25,26] consistent with the implications of the present ^{148}Nd work. These intriguing results have stimulated studies of octupole correlations in ^{150}Nd and ^{150}Sm, and these data are being analysed.

This recent work demonstrates that coulomb excitation is a powerful probe of octupole correlations, as well as quadrupole collectivity. Such studies will elucidate the intriguing interplay of these important collective shape degrees of freedom and the role of triaxiality.

14.9 CONCLUSIONS

Considerable advances in the field of coulomb excitation make it feasible, for the first time, to measure almost the complete set of $E2$ matrix elements involving the lowest twenty to thirty collective states in nuclei, as well as $E1$ and $E3$ matrix elements for five to ten states. The extent and completeness of the $E2$ data now being measured are sufficient to determine directly the centroids and fluctuation widths of the $E2$ expectation values in the principal axis frame for several low- lying collective states, i.e., the shape degrees of

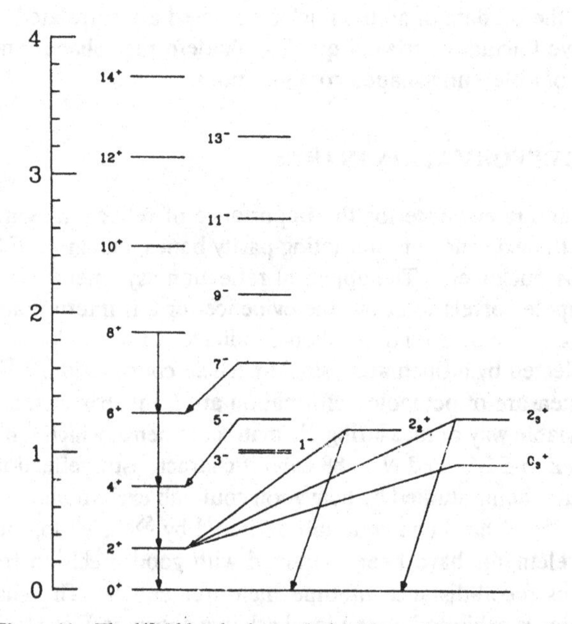

Figure 11. The ^{148}Nd level scheme derived from coulomb excitation.

freedom. These show clearly the extent to which the data are correlated due to quadrupole collectivity, adding a new dimension to the study of nuclear shapes.

These new techniques are being applied to a wide range of nuclei, including odd-A nuclei where measured $M1$ and $E2$ matrix elements probe both the single-particle and collective degrees of freedom. The even-A examples discussed show that the measured $E\lambda$ properties are correlated very well, using collective shape degrees of freedom in both strongly-deformed and shape-transitional nuclei. Collective models employing either geometric or boson degrees of freedom are able to reproduce the general trends of the data for the quadrupole collectivity, but available model calculations fail to give a detailed fit to the large sets of $E2$ data. More refined collective model calculations are needed.

The commissioning of the next generation γ-ray detector arrays, such as GAMMASPHERE, will improve the sensitivity of coulomb excitation measurements by about an order of magnitude greatly expanding the role of coulomb excitation as a probe of nuclear structure. This technique will be exploited to study further the limits of validity of the collective model, the role of symmetries in collective motion, shape transitions, shape coexistence, octupole deformation, and the interplay of single-particle and collective degrees of freedom in nuclear structure. It may be possible to exploit the completeness of the data sets now being measured to determine directly the model content of wavefunctions of states without resorting to large model calculations, which obfuscate model interpretations of nuclear structure.

ACKNOWLEDGMENTS

The author wishes to thank his many collaborators, especially T. Czosnyka, L. Hasselgren, R. Ibbotson, A. E. Kavka, B. Kotlinski, C. White, C. Y. Wu, and E. G. Vogt. This work was supported by the National Science Foundation.

REFERENCES

1. D. Cline, *Ann. Rev. Nucl. Part. Sci.* **36**, 683 (1986).
2. D. Cline, B. Kotlinski, *Univ. Rochester Nucl. Struct. Res. Lab. 1982-1983 Biennial Report*, pp. 363-365 (1984).
3. C.Y. Wu, D. Cline, E.G. Vogt, W.J. Kerman, T. Czosnyka, A.E. Kavka, and R.M. Diamond, *Phys. Rev. C* **40** (1989) R3.
4. E.G. Vogt, D. Cline, T. Czosnyka, B. Kotlinski, C.Y. Wu, A.E. Kavka, W.J. Kernan, R.M. Diamond, F.S. Stephens, J. deBoer, Ch. Lauterback, and C. Fahlander, *Rochester Nuclear Structure Research Laboratory Annual Reports, 1982-1988.*
5. A. Winther and J. deBoer, in *Coulomb Excitation*, Ed. by K. Alder and A. Winther, (Academic Press, New York, 1966, p. 303.
6. T. Czosnyka, D. Cline, and C.Y. Wu, *Bull. Am. Phys. Soc.* **28**, 745 (1983).
7. A.E. Kavka, Ph.D. Thesis, University of Uppsala, 1990.
8. B. Kotlinski, A. Backlin, D. Clark, and D. Cline, *Nucl. Phys.* **A503**, 575 (1989).
9. B. Kotlinski, A. Backlin, D. Cline, C.Y. Wu, R.M. Diamond, A.O. Macchiavelli, and M.A. Delaplanque; submitted to *Nucl. Phys.*
10. R. Ibbotson, B. Kotlinski, D. Cline, K. Helmer, A.E. Kavka, A. Renalds, E.G. Vogt, P.A. Butler, C.A. White, R. Wadsworth, and D.L. Watson; submitted to *Nucl. Phys.*

11. D. Cline, *Proc. of the Orsay Colloquium on Intermediate Nuclei*, Ed. by R. Foucher, (Inst. de Phys. Nucleaire d'Orsay (1971)) pp. 4-34.
12. D. Cline and C. Flaum, *Proc. of the Int. Conf. on Nucl. Struct. Studies using Electron Scattering and Photoreaction, Sendai 1972*, Ed. by K. Shoda and H. Ui, (Sendai: Tohoku Univ. (1972)) 5, 61-82.
13. Kumar, K., *Phys. Lett.* 28, 249 (1972).
14. W.F. Davidson, W.R. Dixon, and R.S. Storey, *Can. J. Phys.* 62, 1538 (1983).
15. C.Y. Wu, Ph.D. Thesis, University of Rochester (1983).
16. K. Kumar and M. Baranger, *Nucl. Phys.* A122, 273 (1968).
17. L. Hasselgren and D. Cline, *Proc. of the Int. Conf. on Interacting Bose-Fermi Systems in Nuclei*, Ed. F. Iachello (Plenum, New York, 59); L. Hasselgren et al., *Nucl. Phys.* (to be published).
18. L.E. Svensson, Ph.D. Thesis, University of Uppsala, 1989.
19. C. Fahlander et al., *Nucl. Phys.* A485, 327 (1988).
20. B. Kotlinski et al., submitted to *Nucl. Phys.*
21. G.A. Leander et al., *Nucl. Phys.* A453, 58 (1986).
22. R.J. Poynter et al., *Phys. Lett.* B232, 447 (1989).
23. W. Urban et al., *Phys. Lett.* 200B, 424 (1988).
24. C.A. White et al., unpublished.
25. I.M. Govil, H.W. Fulbright, D. Cline, et al., *Phys. Rev. C* 33, 793 (1986).
26. I.M. Govil, H.W. Fulbright, and D. Cline, *Phys. Rev. C* 36, 333 (1987).

15. GAMMASPHERE

I-Yang Lee
Oak Ridge National Laboratory*
Oak Ridge, TN 37831-6371

ABSTRACT

Gamma-ray detector systems are important in many areas of science, and the significant advances in such systems open possibilities to address new and interesting questions. A high-resolution γ-ray detector system with completely new capabilities is possible today, and a proposal for its construction is under consideration by the DOE. The new capabilities have to do with the resolving power of γ-ray detectors; that is, their ability to isolate one sequence of γ-rays from a complex spectrum. Higher resolving powers can be achieved by using the additional information contained in the higher-fold coincidence events, and about three orders of magnitude improvement over existing detector systems is anticipated. Much of the frontier work involving γ-ray detection requires the study of weak signals embedded in very large backgrounds, and this enormous gain in resolving power will have a large impact on these studies. The system proposed consists of 110 large compton-suppressed Ge detectors and has about six times higher efficiency for a $\approx$1 MeV γ-ray than any existing detector system. The primary use envisioned for such a detector system would be in studies of the structure of nuclei; however, it will have an impact in many other fields, including atomic physics, electroweak physics, astrophysics, and some "new and undefined" physics.

15.1 INTRODUCTION

Arrays of compton-shielded Ge detectors, due to their high resolution, efficiency, and low background, have become a powerful tool for γ-ray spectroscopy. Many recent advances in high-spin physics, such as the observation of superdeformed states,[1] are the results of using these arrays. Presently, about ten such arrays are operational or near completion. These arrays typically have twenty detectors covering 10-15% of the total solid angle and a total full-energy efficiency of about 2%. Although they provide high efficiency for γ-γ two-fold coincidence experiments and moderate efficiency for three-fold coincidences, they are not adequate for higher-fold coincidences. However, recent developments in high-spin γ-ray spectroscopy have demonstrated the need for higher-fold coincidence experiments. In the meantime, because of advances in detector and computer technology, it is technically possible to build detector arrays with much higher efficiency and to acquire high-fold data at desired rates. GAMMASPHERE is designed to carry out four- and five-fold coincidence experiments with high efficiency. The advantage of high-fold coincidence is the improvement of the resolving power; i.e., the ability to identify weak cascades from a large number of cascades.

---

*Operated by Martin Marietta Energy Systems, Inc. under contract DE-AC05-84OR21400 with the U. S. Department of Energy.

Exotic Nuclear Spectroscopy, Edited by W. C. McHarris
Plenum Press, New York, 1990

15.2 RESOLVING POWER

To demonstrate the concept of resolving power, a typical decay scheme of a collective nucleus formed in a heavy-ion induced reaction is shown in Fig. 1. Eight parallel rotational bands are observed. The γ-decay pathways follow these bands for many transitions and eventually feed into the ground-state band. The energy difference between two consecutive transitions is about 60 keV. Therefore, in a singles (one-dimensional) spectrum, there is one γ-ray peak from each band in a region of 60 keV. A Ge detector has a resolution of about 3 keV; therefore, the 60-keV region can only be divided into twenty resolvable bins. A schematic illustration is shown in Fig. 2*a*. Twenty γ rays, each from a different rotational band, are distributed randomly in this region. It is obvious that, if more than two γ rays occupy the same bin, they cannot be resolved. The probability of this happening is calculated to be 0.26; thus, it is difficult to resolve all twenty bands, as shown in the figure. However, a γ-γ coincidence experiment with two detectors will produce a two-dimensional spectrum; a 60-keV-x-60-keV section is illustrated in Fig. 2*b*. Again, twenty γ-γ coincidence events were randomly distributed in the spectrum. Since there are now 20^2 = 400 bins in this two-dimensional spectrum, the chance of having more than two cascades occupying the same bin is very small (0.0012). Therefore, the resolving power, the number of cascades that can be resolved, is increased by a factor of approximately twenty. Similar arguments can be applied to three- and higher-fold coincidence spectra, so the resolving power improves as the power of the coincidence fold.

15.3 THE IMPACT ON PHYSICS

The high resolving power provided by GAMMASPHERE will open up new scientific opportunities for a broad range of nuclear studies. In the following, we will present a few examples which have been discussed in the GAMMASPHERE proposal[2] to demonstrate the capabilities of GAMMASPHERE.

15.3.A Super- and Hyperdeformed States

High-spin states corresponding to a superdeformed nucleus were found[1] for the first time in 1986. A rotational band having a very large moment of inertia and deformation was discovered in ^{152}Dy in the I = 22 to 60 range. This discovery has opened a new direction in nuclear physics; and, subsequently, many other cases of superdeformation have been observed in the $A \approx 130$, 150, and 190 regions. These nuclei have an axially-symmetric, prolate-spheriodal shape, and the principal axes have a ratio of lengths significantly larger than the "usual" 1.3 to 1. Like usual deformed nuclei, they exist because shell effects provide extra stability for such shapes.

The observation and study of such highly deformed nuclei will give us a chance to explore new phenomena and ideas: 1) The particular grouping of single-particle levels, which leads to the prediction of regions of superdeformed nuclei, is more generally related[3] to special approximate symmetries (called pseudo-spin and pseudo-$SU(3)$ symmetries) of the hamiltonian. These can be studied by testing the location and strength of shell effects at large deformation. 2) We will also have a first look at nuclei where the ratio of coulomb to surface energy is significantly different from those we know. This means that the basic elements determining nuclear structure have a different relationship to one another. 3) A truly new phenomenon which should be accessible with GAMMASPHERE will be the

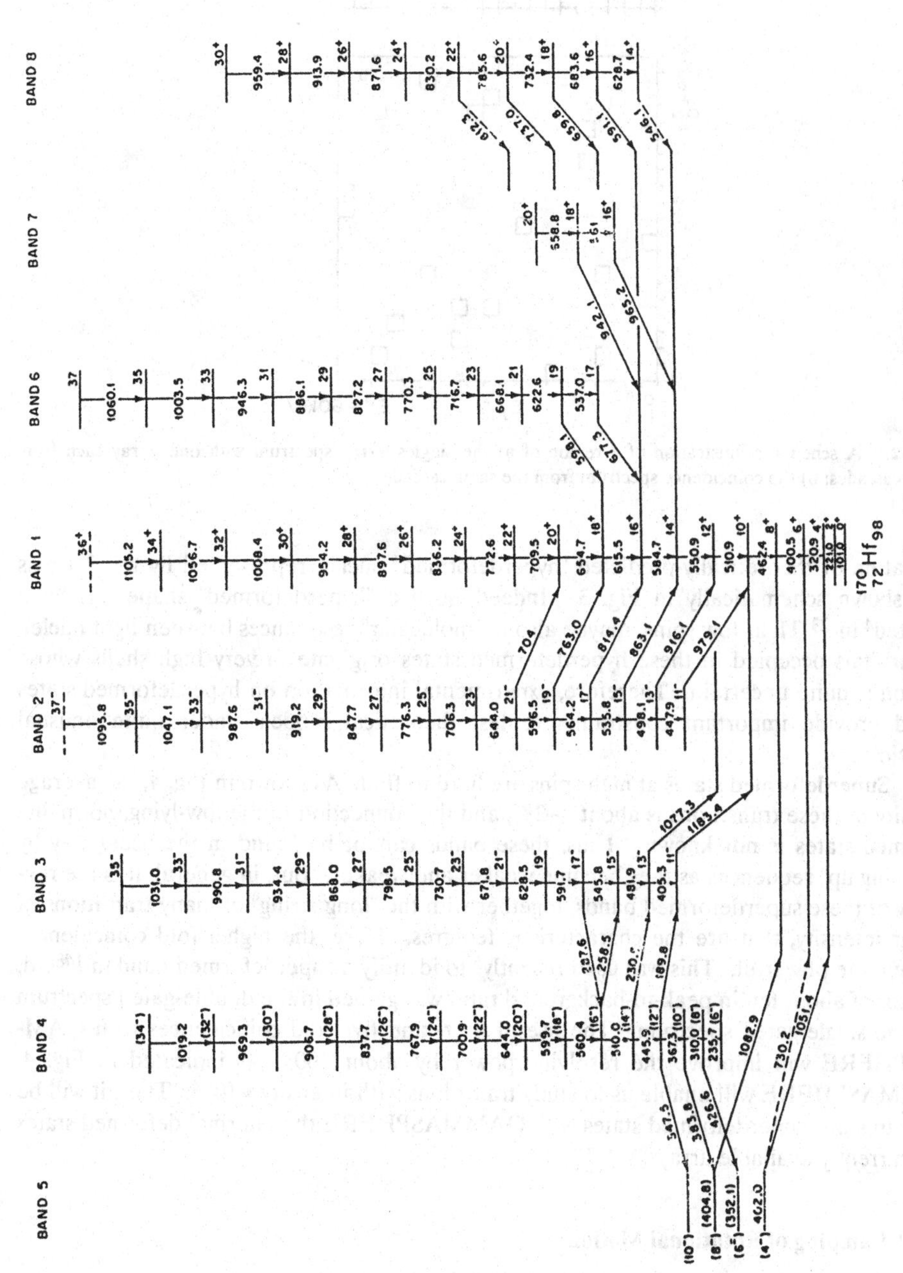

Figure 1. A typical decay scheme for a rotational nucleus produced in a heavy-ion-induced fusion reaction (from C. Baktash et al.).

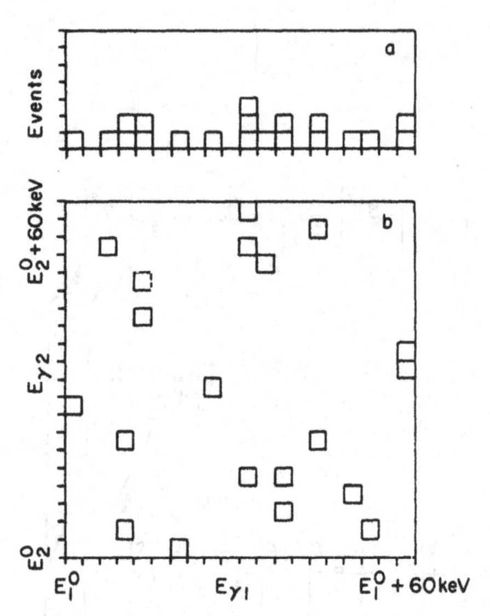

Figure 2. A schematic illustration of a region of a) the singles γ-ray spectrum with one γ ray each from twenty cascades; b) the coincidence spectrum from the same cascades.

population of theoretically-predicted "hyperdeformed" nuclei, represented by the 3:1 axis ratio shown schematically in Fig. 3. Indeed, such a "hyperdeformed" shape has been reported[4] in ^{231}Th at low spins, maybe also in "molecular" resonances between light nuclei. The orbitals occupied in these hyperdeformed states originate in very high shells whose position is quite uncertain. Therefore, experimental information on hyperdeformed states should provide important constraints to test theoretical models under quite unusual conditions.

Superdeformed states at high spins are hard to find. As shown in Fig. 4, the average intensity of these transitions is about 1–2%, and the connection to the low-lying, normally-deformed states is not known. Thus, these bands cannot be found in the usual way by "following up" sequences as they become weaker and weaker. But, in general, it is the regularity of these superdeformed bands, together with the "long string" of many transitions of similar intensity, that are the characteristic features. Here, the higher-fold coincidences will be quite powerful. This was used recently[5] to identify a superdeformed band in ^{148}Gd. A factor of about ten in peak-to-background ratio was gained in the double-gated spectrum over the single-gated spectrum. The use of four- and five-fold coincidences with GAMMASPHERE will improve the resolving power by about 100. As indicated in Fig. 4, GAMMASPHERE will enable us to study transitions with intensity $\approx 10^{-4}$. Thus, it will be easier to study superdeformed states with GAMMASPHERE than normal deformed states with currently-available arrays.

15.3.B Damping of Rotational Motion

The nuclear level density increases exponentially with increasing excitation energy. At excitation energy U about 2 MeV (or temperature $T \approx 0.3$ MeV), where the average separation between states becomes comparable to the residual interaction between those states, the states mix heavily over a spreading width $\Gamma_\mu(U)$, producing complicated eigen-

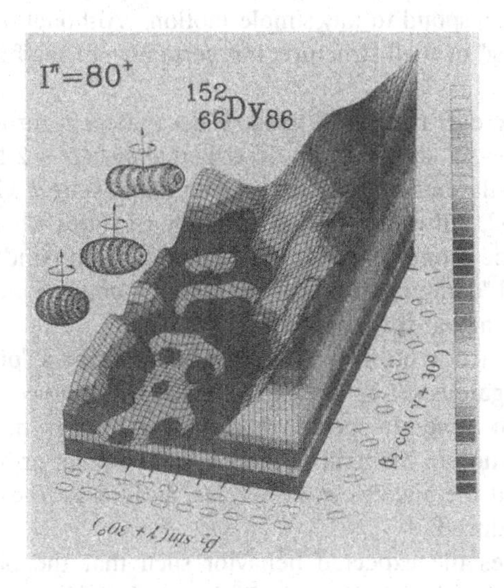

Figure 3. The calculated potential-energy surface for ^{152}Dg at I = 80. The inserts correspond to the shapes of the nucleus at three minima. The left axis is parallel to the γ = 60° line and the right axis to the γ = −30° line.

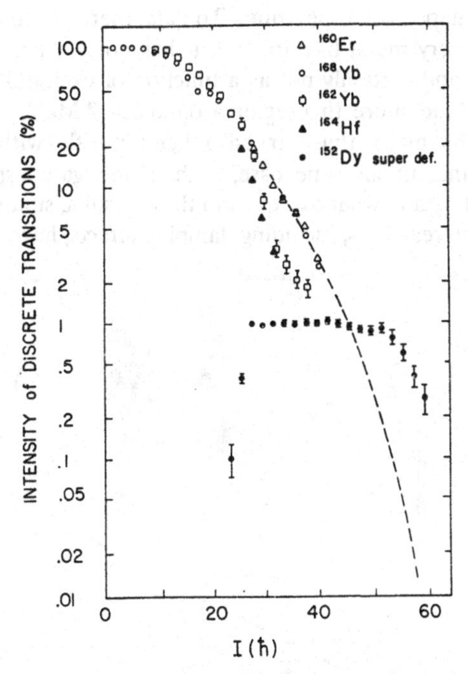

Figure 4. Discrete line intensities vs spin for some well-deformed nuclei and for the superdeformed band in ^{152}Dy. The lines show results from simulation calculations. GAMMASPHERE will push the intensity limit of an observable state from 1% to 0.01%.

states that no longer correspond to any simple motion. Although this change presumably corresponds to a "melting" of shell structure, it is certainly not the full shell structure that is destroyed.

In a normal (discrete) rotational band, γ decay occurs through a particular set of states (a band) I, I–2, I–4,.... At the point described above ($U \approx 2$ MeV), these bands will become mixed, and calculations suggest that a given initial state I will no longer decay to a particular final state, I–2, but rather to a distribution of states whose energy spread is related to the spread in the moments of inertia of the admixed bands. This situation is depicted in Fig. 5, is called "damping of rotational motion", and is contrasted with the normal rotational behavior shown near the yrast line.

The spectrum of a discrete rotational band consists of a "picket fence" of equally-spaced γ rays; and, if a gate is set on one of these, the resulting coincident spectrum has a hole at the gate position. Even in spectra consisting of lines from many rotational bands which have a wide variation in moment of inertia, such a hole, or "dip", should persist. In fact, no such dip exists at the higher γ-ray energies, and it was to explain this absence that damping[6] was first introduced.

Damping modifies the expected behavior such that the observed dip should be wider, eventually approximating the (inverted) shape of the damping width, Γ_{rot}. For Γ_{rot} values above 300 keV, this dip would be a very broad, shallow feature. (The area is conserved, so that, as it becomes wider, it becomes shallower.) Such a feature would be very difficult to observe in a spectrum whose shape is not well known. The shape is not well known because the gate also imposes a strong spin selection (and, perhaps, also other selections) that affect the shape of the spectrum. To date, there is good, general evidence for damping (or something very much like it), but, it has not yet been possible to measure damping widths directly and certainly not as a function of excitation energy, as would be needed to probe in detail the interesting region around $U \approx 2$ MeV.

The measurements made thus far have been mostly with singly-gated spectra (double coincidences) and, in just one case, with doubly-gated spectra[7] (triple coincidences). In order to understand what to expect in these results, simulations of the cascades following heavy-ion fusion reactions, including damping effects, have been made. The sim-

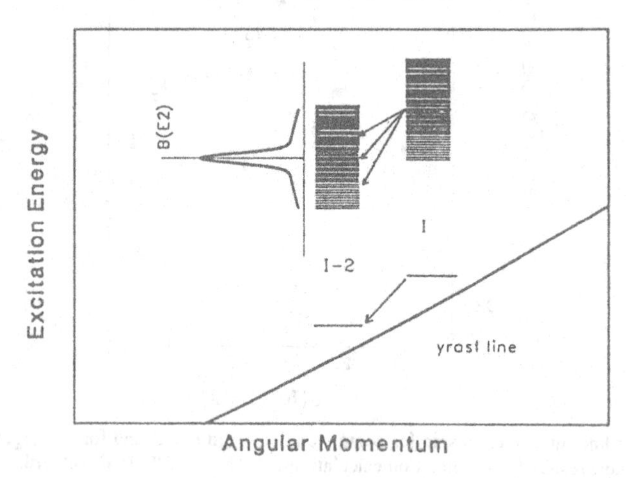

Figure 5. At high energies above the yrast line, a mixed state of spin I will decay to any one of a distribution of states at I-2.

ulation shows that there is a very large difference in the feeding effects between the singly-gated and singles spectra. The doubly-gated spectrum is much more like the singly-gated spectra, but the difference in feeding effects in these spectra is still about as large as that due to the dip associated with the damped rotational behavior. However, when comparing triply-gated (four-fold coincidence) with doubly-gated spectra, the simulations indicate that by far the largest difference between the two spectra is, indeed, because of the dip associated with the damped rotational behavior. Furthermore, the dip has a width that is related to the input damping width. Thus, the simulation strongly suggests that, if we could work with triply-gated spectra, we could measure the damping width directly. Note that its variation with excitation energy can be obtained both from the variation with γ-ray energy (related to excitation energy) and from total-energy and multiplicity gates provided by GAMMASPHERE.

The rates for such experiments with GAMMASPHERE are quite plausible. Considering a 2-d run and gates 20-keV wide (small, compared with both the $\approx$60 keV average separation between rotational energies and the 300 keV damping width), we find about 2.5 x 10^5 (full-energy) events in the triply-gated spectrum, and a dip area of 10^4 counts. By contrast, the best existing arrays today would produce 200 counts in the full spectrum and $\approx$10 in the dip—clearly unusable.

15.3.C Structure of Giant Resonances

Giant resonances can be excited by inelastic scattering. A study of the photon decay of these resonances, while difficult due to its small probability, can provide information different from that coming from nucleon decay. Recently, photon decay of the giant quadrupole resonances in ^{208}Pb has been studied[8] in the excitation energy region from 9 to 15 MeV. The nucleus was excited by a 381-MeV ^{17}O beam. Since this state decays predominantly by neutron emission with only a 10^{-4} branch for γ decay, it was necessary to use the Spin Spectrometer, a 4π solid-angle γ-ray detector array, and particle-γ coincidence techniques to observe the γ rays. The results are shown in Fig. 6. The giant quadrupole resonance was observed to decay by an $E2$ branch to the ground state and an $E1$ branch to several excited 1^- states. Most interestingly, the $E1$ decay to the collective 3^- state at 2.61 MeV is strongly suppressed in comparison to the decay to the non-collective 3^- state at 4.97 MeV. These results confirm the isoscalar character of the 10.6 MeV resonance. Unfortunately, because of the limited resolution of the NaI detectors, it is now possible to study only closed-shell nuclei with low-level densities for the low-lying states.

The strong coupling between the giant resonances and the low-lying, surface-vibrational modes of the nucleus has been studied theoretically since the 1960s. This coupling can be investigated experimentally by study of photon-decay branches from the dipole resonance to low-lying collective states. This area of study is currently of intense interest at electron accelerators where measurements are made of the inelastic scattering of tagged photons. GAMMASPHERE would make possible the study of scattered ion-γ-γ coincidences. In the experiment, the forward Ge detectors would be replaced by BaF_2 detectors, leaving the backward hemisphere for low-energy γ-ray detection. The inelastically-scattered ion would give the excitation energy, a high-energy γ ray detected in the BaF_2 detectors would define the primary decay from the giant resonance, and a low-energy γ ray detected in the Ge detectors would indicate the final state of the primary decay. In this mode, the γ decay of giant resonances of any nuclei can be studied.

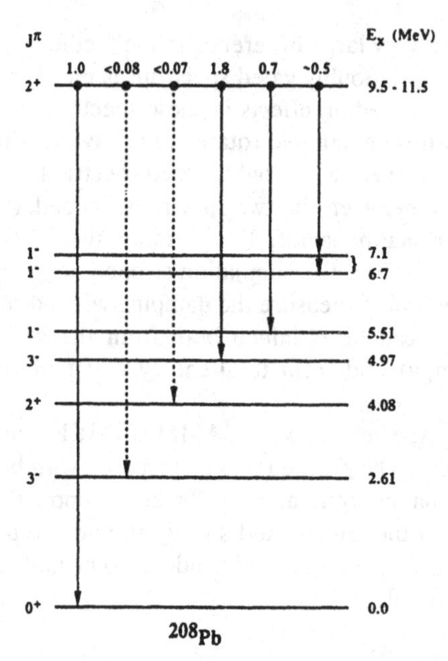

Figure 6. γ decay of the 9.5-11.5-MeV excitation-energy region of ^{208}Pb.

15.3.D Heavy-Ion-Induced Transfer Reactions

One- and two-nucleon transfer reactions induced by heavy ions provide a means of studying transfer between high-spin states populated by coulomb excitation prior to transfer. Thus, such reactions make it possible to study single-particle and pairing effects under the influence of considerable collective angular momentum and provide a selective new population mechanism for studying high-spin states.

Recent studies of heavy-ion transfer reactions, using the Spin Spectrometer,[9,10] have given new insight into the reaction mechanism. The major conclusions are as follows: 1) Heavy-ion transfer reactions comprise a major fraction of the total reaction cross-section near the coulomb barrier. 2) Heavy-ion-induced transfer reactions on deformed nuclei selectively populate "cold" states in the yrast domain; i.e., those close to the yrast "zero-temperature" line, up to spin 30 with large cross-sections. 3) These reactions have been used[10] to populate states to spins of almost 30 $\hbar$ in actinide nuclei. Moreover, the fission channel is unimportant, in contrast to the situation when other reactions are used to populate such states. 4) The most recent study of transfer to strongly-deformed, rare-earth nuclei has shown that two-neutron transfer to the ground band, at large separation distances between the colliding nuclei, exhibits an oscillatory behavior with separation distance and is strongly enhanced.

Heavy-ion-induced transfer reactions require detection of the scattered particles in coincidence with the deexcitation γ rays in order to determine the two-body kinematics necessary to separate transfer from competing reactions, as well as to correct the γ spectra for doppler-shift effects. GAMMASPHERE will have an inner sphere sufficiently large to accommodate the particle-detector arrays needed for this work. It will provide more than an order of magnitude increase in particle-γ coincidence rate relative to what has been used in past transfer experiments, improving the sensitivity of the measurements. However, it is the two orders of magnitude increase in the particle-γ-γ triple coincidence rate and the ability to perform even higher-fold coincidence experiments that is the most significant advance and the one that will open new frontiers. As shown in Fig. 7, particle-γ-γ, or higher-fold coincidences, are needed for study of heavy-ion induced transfer reactions, since at least one of the coincident γ rays is used to identify the reaction product, and it is the remaining coincident γ rays that provide the interesting physics.

In addition, GAMMASPHERE will have sufficient resolution in both total energy and multiplicity to allow study of selective regions of spin and excitation energy in order to

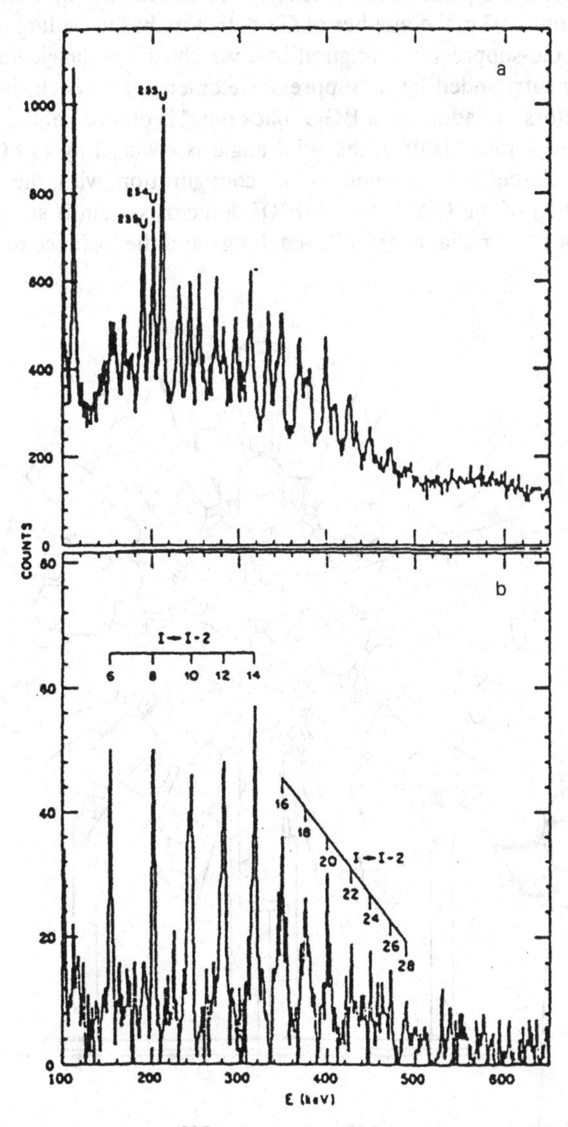

Figure 7. γ-ray spectra for the reaction, ^{235}U + 1394-MeV ^{206}Pb. a) The total particle-γ spectrum. b) particle-γ-γ coincidence spectrum with summed gates on the $8^+ \rightarrow 6^+$, $10^+ \rightarrow 8^+$, and $12^+ \rightarrow 10^+$ transitions of ^{234}U.

separate transfer to the interesting ground state band region from that to the dominant population of complicated multi-quasiparticle states which decay to the states of interest.

15.4 DESIGN AND SCHEDULE OF GAMMASPHERE

The design goal of the GAMMASPHERE detector system is to achieve high energy resolution, high efficiency, a good response function, and resistance to neutron damage. Within the limits of existing technology, the obvious choice of the detectors is the large-volume *n*-type Ge detector with BGO compton suppressors. To minimize pile-up and doppler broadening, for a typical fusion reaction with a γ-ray multiplicity of about 20 and recoil velocity of about 0.03 *c*, the number of Ge detectors has to be larger than 100. After considering several Ge-suppressor configurations, we chose a "honeycomb" design, where each Ge detector is surrounded by six suppressor elements, and each suppressor element serves two Ge detectors. In addition, a BGO "back plug" is placed behind each Ge to catch the forward-scattered γ rays. Half of the solid angle is covered by 110 Ge and the other half by 360 BGO elements in a geometrical configuration with the symmetry of an icosahedron. A sketch of the GAMMASPHERE detector system is shown in Fig. 8. The size of a Ge detector is 7-cm diameter by 7.5-cm long, and the distance to target is 26 cm.

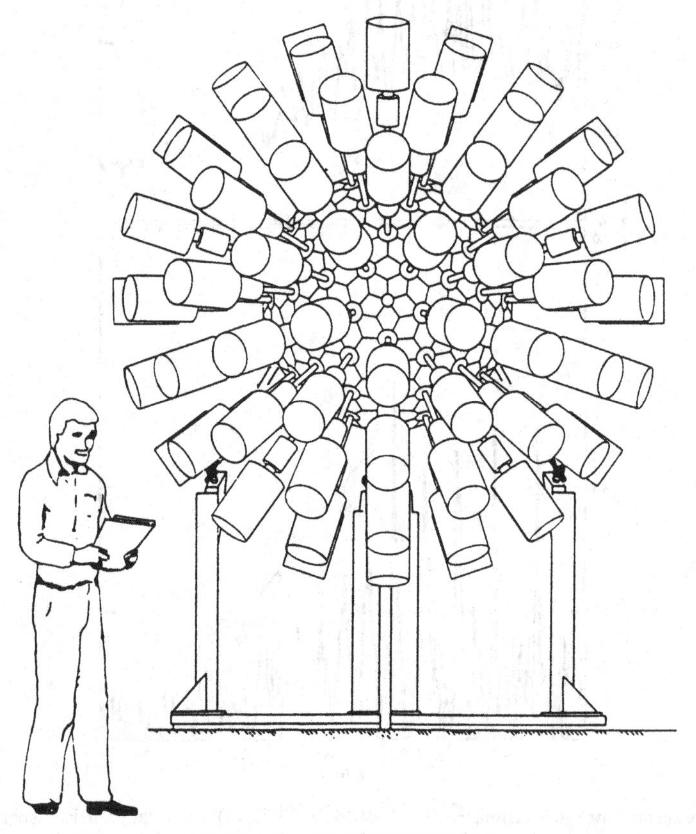

Figure 8. A sketch of the GAMMASPHERE detector system.

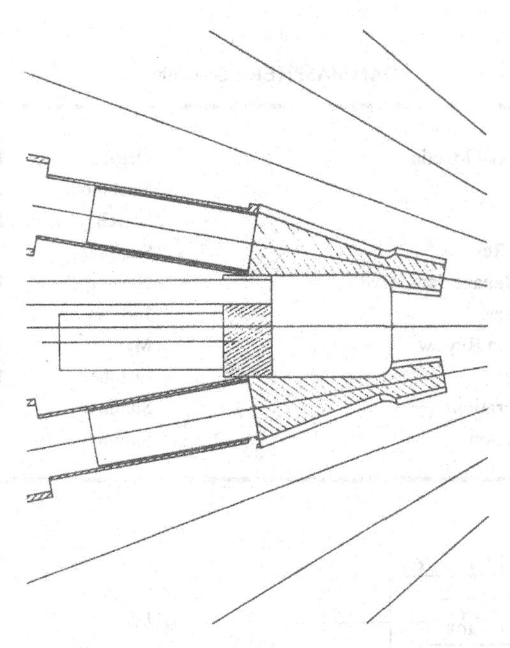

Figure 9. Side view of a Ge detector and its BGO compton suppressors.

A BGO element has a length of 17 cm, and the distance to the target is 22 cm. Fig. 9 shows a typical Ge detector with two of the suppressor elements and a back plug. An inner ball of 240 BaF_2 detectors will be placed in front of the suppressors. It would protect the suppressor elements from direct γ-ray hits, thus preventing false rejection. It also can distinguish γ rays from neutrons and give a better γ-ray multiplicity value. For experiments studying high-energy γ rays (E_γ > 5 MeV), 55 BaF_2 detectors will be available to replace the Ge detectors.

The electronics system, as shown in Fig. 10, is designed to have a system dead-time of 10 μsec. This is achieved by using ADC's and TDC's with 5-μs digitizing time and a data-read-out time of 100 ns/word. At the designed maximum event rate of 50k/s and a typical length of event of 100 bytes, the data rate will be 5 Mbyte/s. It is expected that, in most of the experiments on-line data selection will be carried out, so that only a small fraction of the data has to be stored in event-by-event mode. The on-line data modification, selection, histogramming, and storage require an estimated processing power of 65 Mips. Since the data from each event can be analyzed independently, parallel processing is the most straightforward and economical approach. Currently, a single-board computer can provide about 10 Mips per board. It is planned to use such a computer as the building block of a parallel computer based on VMEbus.

The schedule of the project is shown in Table I. In the summer of 1988, the proposal was submitted to DOE for peer review and received strong endorsement. In November of 1988, the proposal underwent cost, schedule, and management review and, again, was highly endorsed. The project was presented to the DOE/NSF Nuclear Science Advisory Committee (NSAC), which on 5 March 1989 recommended that the project be funded and the site-selection process be begun as soon as possible. Consequently, a DOE panel reviewed siting proposals from ANL, LBL, and ORNL and recommended that ORNL be

Table I
GAMMASPHERE Schedule

Organizational Meeting	October	1987
Workshop	November	1987
Proposal	March	1988
DOE Mail Review	Summer	1988
Cost and Management Review	November	1988
NSAC Review	January	1989
Site-Selection Review	May	1989
(Funding	October	1990)
(Initial Operation	Summer	1992)
(Full Operation	Summer	1993)

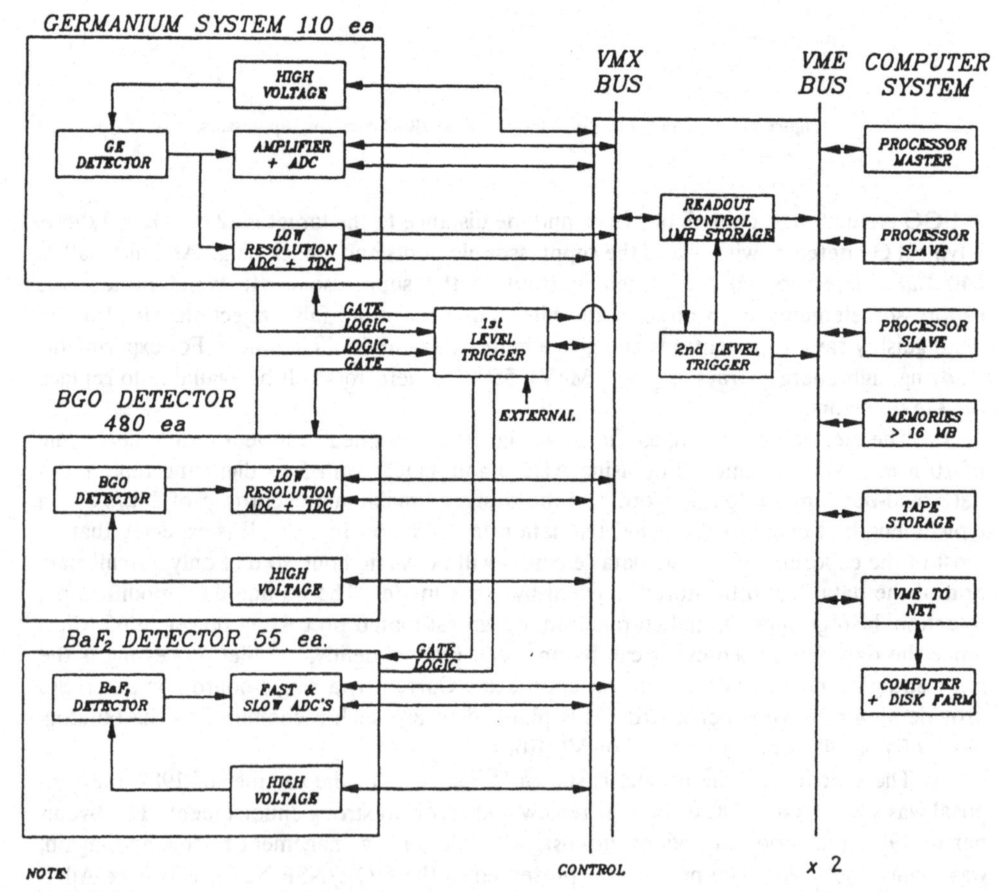

Figure 10. Schematic diagram for the GAMMASPHERE electronics system.

the site of GAMMASPHERE. DOE is expected to include initial funds for GAMMASPHERE construction in the FY 1991 budget, which starts on 1 October 1990.

Currently, the design and prototyping of the components of GAMMASPHERE are being carried out in several laboratories. The designs of the Ge detectors and the BGO suppressors have been completed. They are based on the results of extensive simulation calculations of the detector response. A prototype Ge detector and seven associated BGO elements have been ordered by ANL. The delivery is scheduled in early 1990. The design of the BaF_2 inner ball is being carried out at ORNL. Due to space limitation, a suitable method of reading out the scintillation light has to be found. Among the options being studied are the use of light pipe, small phototube, and photodiode.

The design of building and beamline are being carried out at ORNL. Fig. 11 shows the floor plan of GAMMASPHERE and a Recoil Mass Spectrometer. GAMMASPHERE can be used in a stand-alone beamline, at the target position of the RMS or at the focal position of the RMS.

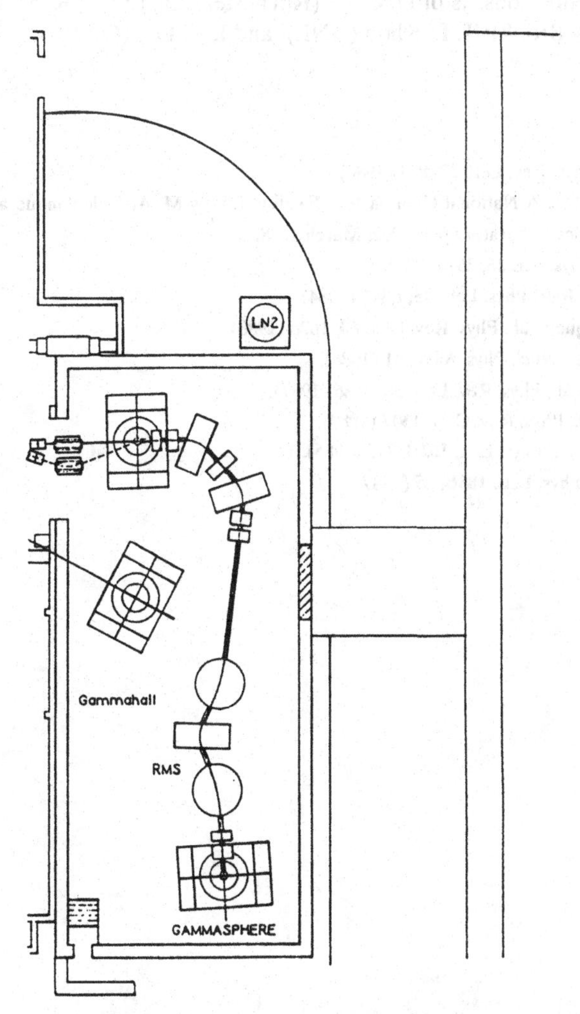

Figure 11. The floor plan for the GAMMASPHERE (at HHIRF of ORNL).

According to the schedule, the project will be complete in three years at a cost of about 18M$. However, experiments can start after two years when about half of the detectors are delivered and the essential part of the data acquisition system is ready. It is expected that about twenty experiments will be carried out yearly for a total running time of 2500 hours. The exciting new opportunities provided by such an array will make GAMMASPHERE, together with a similar array being developed in Europe, the Eurogam, the premier facilities for nuclear-structure studies for the next decade.

ACKNOWLEDGMENT

This project has made rapid progress since it was first proposed by Frank Stephens during the summer of 1987. The enormous amount of effort of Frank and the wide participation of the community are the main reason for the success of the project at this stage. Most of the development tasks have been organized and carried out by the steering committee, which currently consists of D. Cline (Rochester), chairman; R. M. Diamond (LBL); D. B. Fossan (Stony Brook); T. L. Khoo (ANL); and I. Y. Lee (ORNL).

REFERENCES

1. P. Twin et al., Phys. Rev. Lett. **57**, 811 (1986).
2. GAMMASPHERE, A National Gamma-Ray Facility, Ed. by M. A. Deleplanque and R. M. Diamond, Lawrence Berkeley Laboratory pub-5202, March 1988.
3. A. Bohr et al., Phys. Scr. **26**, 267 (1982).
4. B. Fabbro et al., Jour. Phys. Lett. **45**, L843 (1984).
5. M. A. Deleplanque et al., Phys. Rev. Lett. **60**, 1626 (1988).
6. B. Lauritzen et al., Nucl. Phys. **A456**, 61 (1986).
7. F. S. Stephens et al., Phys. Rev. Lett. **58**, 2186 (1987).
8. J. R. Beene et al., Phys. Rev. C **39**, 1307 (1989).
9. M. W. Guidry et al., Phys. Lett. **163B**, 795 (1985).
10. C. Y. Wu et al., Phys. Lett. **188B**, 25 (1987).

16. Evidence for Octupole-Octupole and Quadrupole-Octupole Excitations in Spherical Nuclei

S. W. Yates, R. A. Gatenby,† E. M. Baum,
and E. L. Johnson
Department of Chemistry
University of Kentucky
Lexington, KY 40506

J. R. Vanhoy*
Department of Physics and Astronomy
University of Kentucky
Lexington, KY 40506

T. Belgya, B. Fazekas, and G. Molnár
Institute of Isotopes
of the Hungarian Academy of Sciences
Budapest H1525, Hungary

ABSTRACT

The nuclei ^{96}Zr and ^{144}Sm have been studied with the $(n,n'\gamma)$ reaction, and lifetimes of many states have been extracted from the observed doppler shifts of the deexciting γ rays. A large number of fast $E1$ transitions have been observed, and the identification of possible members of the octupole-octupole and quadrupole-octupole multiplets in these nuclei has been made.

16.1 INTRODUCTION

The role of vibrational excitations in nuclei has been the subject of study for many years, but our knowledge of these fundamental modes remains incomplete. In the quadrupole case, equally-spaced, degenerate phonon multiplets are expected, and there are many examples in even-even nuclei near closed shells where the E_{4^+}/E_{2^+} ratio is near the harmonic value of 2. However, the anticipated, closely spaced 0^+, 2^+, 4^+ two-phonon triplet is seldom observed. Evidence for three-phonon quadrupole states was sparse until the recent report[1] of a complete quintet of levels in ^{118}Cd, but recent lifetime measurements[2,3] in this and neighboring nuclei seem inconsistent with this interpretation. In closed-shell nuclei, the octupole vibrations often have relatively low excitation energies and compete successfully with the quadrupole mode.[4] Data on multi-phonon states involving octupole excitations are rare.

In two heavy nuclei, ^{146}Gd and ^{208}Pb, the 3^- state actually lies lower than the

† Permanent address: Bettis Atomic Power Laboratory, West Mifflin, PA 15122
* Permanent address: Physics Department, U. S. Naval Academy, Annapolis, MD 21402

Exotic Nuclear Spectroscopy, Edited by W. C. McHarris
Plenum Press, New York, 1990

quadrupole phonon and is the first excited state in each. These states decay with large, similar $E3$ transition probabilities of 37 and 34 Weisskopf units, respectively, suggesting that they are indeed collective octupole excitations. The unusual properties of these two nuclei have led to a number of searches[5-7] for the expected $3^- \otimes 3^-$ quartet of states with spins and parities of 0^+, 2^+, 4^+, and 6^+ at about twice the energy of the 3^- phonon. No clear-cut identification of the members of the two-phonon quartet has emerged in either nucleus. In ^{208}Pb the large level density in the vicinity of the predicted energy of the two-phonon states contributes to the difficulties in identifying these levels. The octupole phonon in ^{146}Gd lies about 1 MeV lower in energy than the 3^- octupole state in ^{208}Pb. The smaller level density expected in the region of the predicted $(3^- \otimes 3^-)_I$ states, has led to ^{146}Gd as the site of searches[7,8] for two-phonon octupole excitations.

Only in recent years has convincing evidence for two-phonon octupole states been reported. The most compelling data are the observations[9,10] of stretch-coupled states of this type in ^{147}Gd and ^{148}Gd. The identifications of these states by their characteristic cascades of two $E3$ transitions were possible because, serendipitously, they occur as yrast states in these nuclei, and lower multipolarity decays do not occur readily. But, since these states involve the coupling of one or two neutrons to the two-phonon octupole excitation ($J^\pi = 19/2^-$, $\nu f_{7/2} \otimes 3^- \otimes 3^-$ in ^{147}Gd and $J^\pi = 12^+$, $\nu^2 \otimes 3^- \otimes 3^-$ in ^{148}Gd), their descriptions are not so straightforward as would be the expected case in a doubly-closed-shell nucleus. Because of these ambiguities, the identification of two-phonon octupole multiplets in even-even nuclei remains an important goal.

Coupling between the quadrupole and octupole vibrational modes should produce[11] a quintet of negative-parity states with angular momenta ranging from 1 to 5. Therefore, low-lying two-phonon octupole vibrations and excitations of the quadrupole-octupole type are expected in the same energy region.

Since strong octupole excitations should persist in nuclei near closed shells, we have attempted to locate two-phonon octupole states in ^{144}Sm, which is only two protons removed from ^{146}Gd, has as its second excited state a 3^- level, and is stable. Recent measurements[12] indicate that the octupole excitation in this nucleus, with $B(E3; 3^- \rightarrow 0^+) = 38 \pm 3$ wu, is very similar to that of ^{146}Gd. With the recognition[13] that the first 3^- state of ^{96}Zr, a nucleus with a double *sub*shell closure, is a strong octupole, $B(E3; 3^- \rightarrow 0^+) = 38_{-15}^{+49}$ wu, this nucleus emerges as another likely candidate for the observation of two-phonon octupole excitations. We review here, also, the status of our efforts to identify these excitations in ^{96}Zr.

16.2 EXPERIMENTAL METHODS

Neutron-induced reactions have been useful weapons in the arsenal of the nuclear spectroscopist for many years; however, the use of neutron-scattering reactions for spectroscopic purposes has not enjoyed the widespread acceptance of some other reactions, e.g., the (n,γ) reaction. The inelastic-neutron-scattering (INS) or $(n,n'\gamma)$ reaction has been successfully employed in nuclear-structure studies at the University of Kentucky for a number of years, and pulsed-beam time-of-flight methods in neutron and γ-ray spectroscopy have been an area of special emphasis. Many of the advantages of the INS reaction arise because of the absence of coulomb effects; thus, the nuclear levels can be excited with low incident particle energies—i.e., the nuclei are produced quite "cold." With monoenergetic, accelerator-produced neutrons, low-lying levels can be studied without the attendant complications associated with the presence of radiation from higher-lying levels. A second advantage of INS is the general non-selectivity of the reaction; the population of a level is predominantly determined by the neutron penetrabilities.

The experimental methods employed in both neutron- and γ-ray-detection measurements at the University of Kentucky have been described[14,15] previously. Rather than repeating these details, we will attempt to indicate the advantages and limitations of these measurements. It should be noted, of course, that INS measurements are only applicable to stable nuclei. While we have successfully employed[16] scattering samples as small as 0.008 mole for these measurements, the best overall combination of sensitivity and accuracy is achieved with samples of 0.2 to 0.5 moles. The availability of large, isotopically-separated samples is thus very important if detailed spectroscopic data are desired. Following INS, the spectroscopic information can be obtained from either the inelastically-scattered neutron or the γ-ray deexciting the residual nucleus.

Neutron time-of-flight spectroscopy has been employed extensively in a variety of studies in our laboratory, and it has been amply demonstrated that, with pulsed, monoenergetic neutrons, TOF spectroscopy of the inelastically scattered neutrons can also yield valuable spectroscopic information about the lowest-lying nuclear levels.[17] In some cases the angular distributions of the scattered neutrons can be interpreted to characterize the final states. On the other hand, the resolution for neutron detection is typically poor, and such measurements would not be applicable to nuclei with high level densities. These difficulties can be alleviated, however, by observing the γ rays which deexcite the excited levels, rather than the inelastically-scattered neutrons.

In the γ-ray detection mode, a Ge detector (at present we employ an n-type spectrometer of 35% relative efficiency surrounded by a 16.5—cmx20.3—cm BGO anti-compton annular shield) is used to detect the deexcitation photons, so the resolution of the experiment is greatly improved. It is this mode which is most fruitful for the study of heavy nuclei, and hereafter we will limit our discussion to γ-ray detection measurements following INS.

16.2.A Gamma-Ray Excitation Functions

The accurate determination of the threshold for a particular γ-ray uniquely places the level from which the transition arises. When necessary, thresholds can be determined with uncertainties of only a few keV, but threshold determinations to within 30 to 50 keV are generally sufficient for transition placements, even in heavy nuclei.

These excitation functions of γ-ray yields can be compared with cross-sections calculated with statistical models and are useful for inferring level spins and parities. Examples of the use of excitation functions to infer level assignments are given in Fig. 1.

16.2.B Gamma-Ray Angular Distributions

The angular distributions of both neutrons and γ rays from INS contain valuable spectroscopic information. Unfortunately, except in special cases like the spin-0 example discussed in Ref. 17, the neutron distributions usually reflect the spin of the excited state only weakly at neutron energies of a few MeV, and the anisotropies are small. On the other hand, the INS reaction aligns the excited nuclei, so the γ-ray angular distributions from the decays of the excited levels exhibit anisotropies reflecting this alignment, the spins

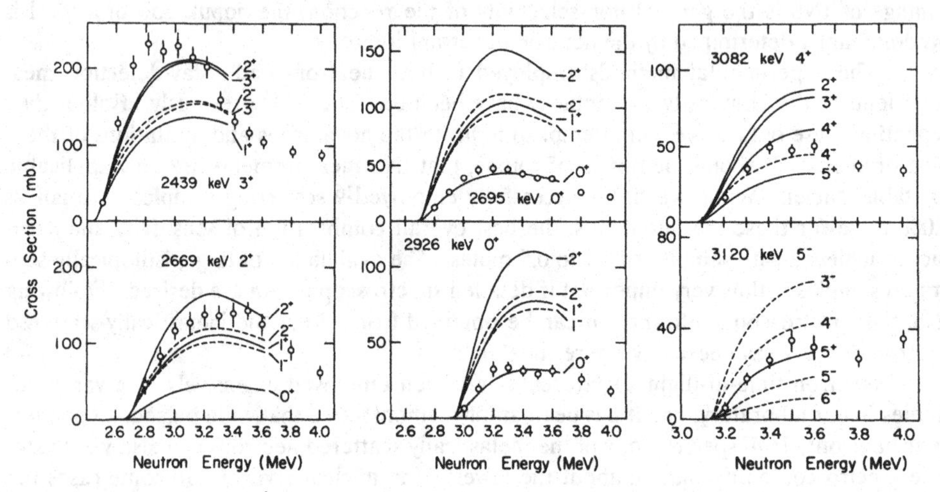

Figure 1. Experimental $(n,n'\gamma)$ reaction cross-sections (Ref. 14) for selected ^{96}Zr levels compared with Hauser-Feshbach-Moldauer cross-sections (lines) for selected ^{96}Zr levels.

of the levels, and the multipolarities of the transitions. In even-even nuclei it is often possible to find levels which decay by more than one transition—one of known pure multipolarity and others with unknown mixtures of multipolarities. In these cases, the pure multipole transition can be used to deduce the excited-state alignment, and the other distributions can then be analyzed to determine multipole mixing ratios. This thus affords a model-independent method for determining mixing ratios. In practice, the alignments calculated with statistical models are generally sufficient, and tests have shown[15] that mixing ratios determined using these calculated alignments agree well with those determined with the model-independent methods. Figure 2 illustrates the quality of the agreement which can be anticipated for experimental and theoretical γ-ray angular distributions.

16.2.C Lifetimes

In our pulsed-beam experiments, it is generally possible to observe level-decay lifetimes greater than 5 ns with conventional timing methods, and we routinely search for lifetimes greater than this. Another time regime can also be investigated in $(n,n'\gamma)$ measurements by employing the doppler-shift attenuation method (DSAM). Prior to our measurements, this method with the INS reaction had been confined[18] to nuclei with $A < 60$. We have shown that, given the proper set of conditions, this method can be extended to much heavier nuclei, and our initial work in this area has recently been published.[19] While the recoil velocity imparted ($v/c \approx 0.001$) in the scattering reactions on even heavy nuclei is sufficient to produce observable doppler shifts, the lifetimes of the low-lying states of these nuclei are usually greater than 1 ps. Thus, this method is practically limited to studies of high-energy (>500 keV) transitions. The formalism for analyzing these data has been described[20] for the case of reactor fast neutrons, and, in general, the analysis for our experiments which employ monoenergetic, accelerator-produced neutrons is even more straightforward, although a number of difficulties, such as assessing the stopping powers of the slowly moving ions, remain.

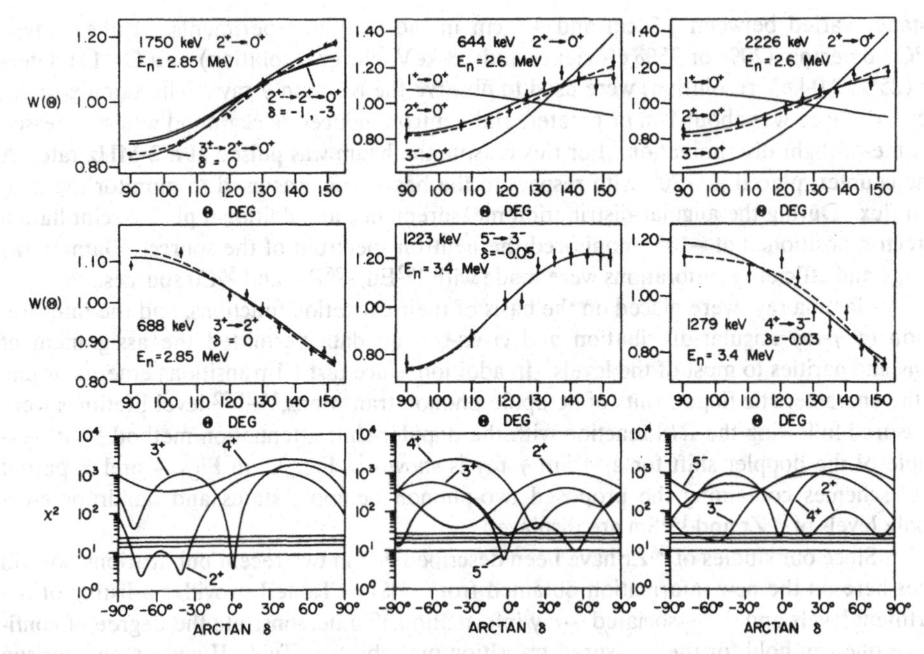

Figure 2. Angular distributions (Ref. 14) of ^{96}Zr γ rays from the $(n,n'\gamma)$ reactions deexciting the 1750-keV and 2226-keV 2^+ states (top), as well as the 2439-keV 3^+, 3120-keV 5^-, and 3176-keV 4^+ levels (middle). Dashed lines are fits; solid lines are angular distributions calculated for the indicated spin sequences and mixing ratios δ. At the bottom, results of χ^2 tests are indicated for the corresponding mixed transitions in the middle row. The horizontal lines indicate 32%, 5%, 1%, and 0.1% confidence limits, respectively.

The doppler-shifted γ-ray energy, $E_\gamma(\theta)$, measured at a detector angle of θ with respect to the incident neutrons can be related to E_γ, the energy of the γ ray emitted by a nucleus at rest, by the expression:

$$E_\gamma(\theta) = E_\gamma\ (1 + F_{exp}(t)\, v_{cm} \cos\theta\ /c), \tag{1}$$

where v_{cm} is the velocity of the center of mass in the inelastic neutron scattering collision with the atom and c is the speed of light. $F_{exp}(\gamma)$ is the experimental attenuation factor determined from the measured doppler shift and must be compared with theoretical attenuation factors to determine the lifetime. In our initial study of level lifetimes in ^{96}Zr, a number of difficulties inherent to our measurements were encountered. These problems have been partially considered in our recent publication.[19]

16.3 EXPERIMENTAL MEASUREMENTS

Inelastic neutron scattering (INS) experiments were performed with the University of Kentucky 7.0-MV Van-de-Graaff accelerator to search for two-phonon states in ^{96}Zr and ^{144}Sm. Cylindrical powder samples in thin polyethylene vials were suspended before a tritium gas cell at an angle of 0° to the incident beam. In both cases the scattering samples were ≈20g of enriched oxides—^{96}Zr (59.6%) and ^{144}Sm (85.6%). The sample-to-cell distance varied between 3.3 cm and 4.5 cm in the various experiments. Either *n*-type HPGe detectors (20% or 35% efficiency and 1.8-keV energy resolution) or a Ge(Li) detector (25%, 2.0-keV resolution) were used to observe the reaction γ rays. The sample-to-detector distance was about 1 m or greater, and neutron-induced background was suppressed by time-of-flight discrimination. For this reason, the beam was pulsed at a 2-MHz rate. A long counter, placed at −20° with respect to the beam axis, was used to monitor the neutron flux. During the angular-distribution measurements, an additional plastic scintillation detector, positioned at −10°, monitored the neutron spectrum of the source. Gamma-ray energy and efficiency calibrations were made with ^{152}Eu, ^{226}Ra, and ^{56}Co sources.

Gamma rays were placed on the basis of their excitation functions, and the interpretation of γ-ray angular-distribution and cross-section data permitted the assignment of spins and parities to most of the levels. In addition, since fast $E1$ transitions emerge as one of the characteristic fingerprints of octupole phonon transitions,[14,21,22] level lifetimes were measured following the INS reaction with the doppler-shift attenuation method.[19] An example of the doppler shift for a ^{144}Sm γ ray is shown in Fig. 3. In Figs. 4 and 5, partial level schemes containing the proposed two-phonon octupole states and quadrupole-octupole levels of ^{96}Zr and ^{144}Sm are displayed.

Since our studies of ^{96}Zr have been described[14,19] in two recent publications, we will focus here on the new information obtained from ^{144}Sm. Table I provides a listing of the pertinent levels and the associated γ-ray information. To demonstrate the degree of confidence one can hold for the measured transition probabilities, Table II gives a comparison of some level lifetimes we have determined using DSAM with those previously deduced in ^{144}Sm from nuclear-resonance-fluorescence measurements.[23]

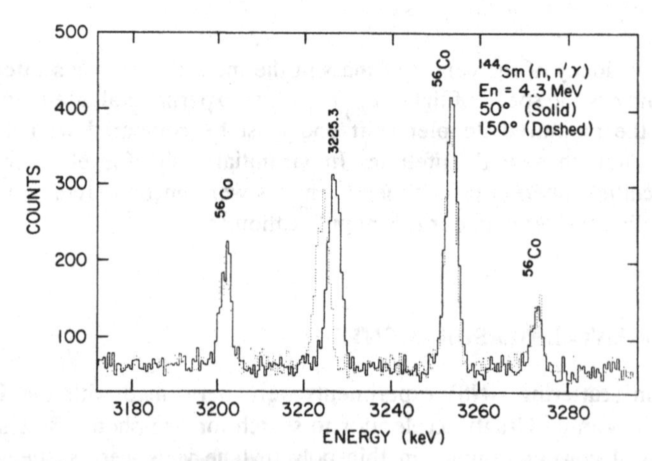

Figure 3. Portion of the γ-ray spectrum measured following the ^{144}Sm $(n,n'\gamma)$ reaction induced by 4.3-MeV neutrons. The shift in energy of the 3225.3-keV γ ray between the two angles of measurement is evident. The additional peaks are γ rays from a ^{56}Co calibration source, which was placed near the detector during the measurements.

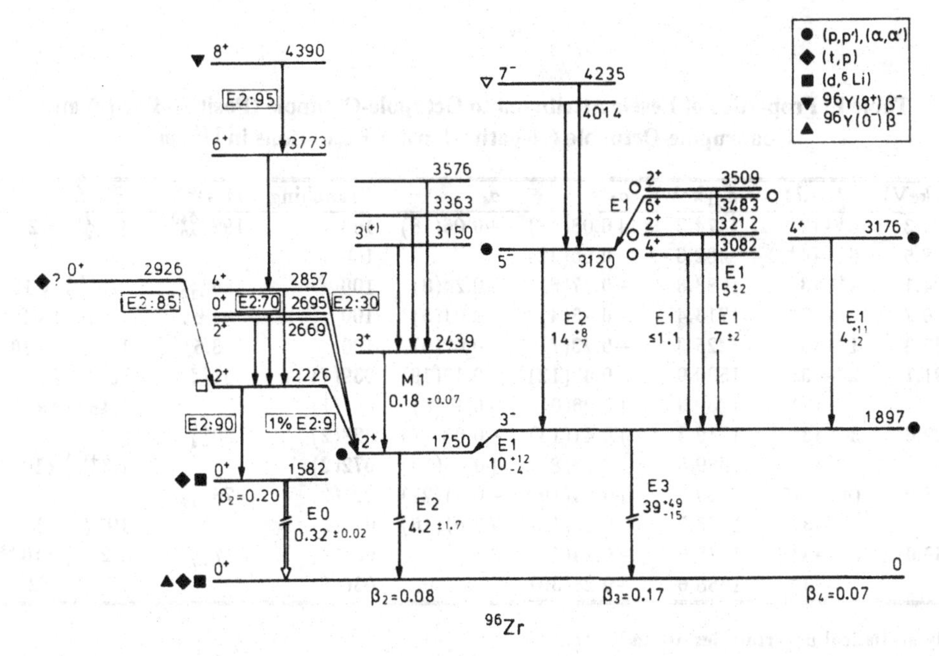

Figure 4. Classification[14] of ^{96}Zr levels according to their γ-decay properties and populations in different reactions. $B(E2)$ percentages are presented in boxes, while reduced transition probabilities are given beside the arrows in Weisskopf units (10^{-4} wu in the case of $E1$ transitions). $\rho^2(E0)$ values are in Bohr-Mottelson units of 0.5 $A^{-2/3}$.

16.4 RESULTS AND DISCUSSION

16.4.1 The Nucleus ^{144}Sm

Perhaps the most characteristic signature of two-phonon octupole structures we can hope to observe is the occurrence of fast $E1$ transitions from 2^+ and 4^+ members of the two-phonon octupole quartet to the 3^- octupole phonon. Additionally, these states are not expected to exhibit extensive γ-ray branching to other levels. Only two levels in ^{144}Sm, the 4^+ level at 3494.1 keV and the 2^+ level at 3523.7 keV, are observed to decay solely by $E1$ transitions to the octupole state. Support for the assertion that these are collective two-phonon octupole levels comes from the observation of $E1$ transitions, which are considerably faster [$B(E1) \sim 10^{-3}$ wu] than the majority of $E1$ transitions in this mass region.[24] The fact that these levels have not been observed previously,[25] particularly in single-nucleon transfer reactions, is also consistent with their being of a more complex, two-phonon origin. However, while the γ-ray transitions from these levels are considerably faster than "normal" $E1$ transitions, there are a number of other fast $E1$ transitions from excited states of ^{144}Sm to the 3^- phonon.

Candidates for the 0^+ and 6^+ members of a two-phonon octupole multiplet are difficult to identify, particularly since clearly distinguishing features, such as direct $E3$ transitions to the 3^- phonon, are not anticipated.[22] Low-lying 0^+ states at 2477.6, 2822.6, and 3134.3 keV have been reported[25] previously and are observed in our INS measurements. Each decays by γ-ray emission to the first excited state. The lowest of these has been firmly established[26] as a $\pi(h_{11/2})_0^2 j_0^{-2}$ excitation of the ^{146}Gd core. The 2822.6-keV level

Table I. Properties of Levels Attributed to Octupole-Octupole (Positive-Parity) and Quadrupole-Octupole (Negative-Parity) Excitations in ^{144}Sm

E_x(keV)	$J_i^\pi \to J_f^\pi$	E_γ(keV)	a_2	a_4	Branching	τ(fs)[a]	$B(\sigma L)$[b]
3134.3	$0^+ \to 2^+$	1474.2	+0.08(13)	+0.06(18)	100	190^{+340}_{-80}	14^{+11}_{-9} $E2$
3308.5	$6^+ \to 4^+$	482.6	−0.65(32)		100		
3494.1	$4^+ \to 3^-$	1683.8	−0.47(6)	+0.24(8)	100	86^{+15}_{-12}	$8.7^{+1.4}_{-1.3} \times 10^{-4}$
3523.7	$2^+ \to 3^-$	1713.4	−0.23(4)	−0.01(5)	100	61±4	$1.2\pm0.1\times10^{-3}$
3225.3	$1^- \to 0^+$	3225.3	−0.23(7)	−0.02(9)	100	$8.5^{+1.2}_{-1.1}$	$1.3\pm0.2\times10^{-3}$
3391.1	$2^- \to 3^-$	1580.9	−0.48(12)	−0.18(19)	039(1)	31^{+5}_{-4}	23^{+4}_{-3} $E2$
	$\to 2^+$	1730.9	+0.06(6)	−0.17(8)	061(1)		$1.4\pm0.2\times10^{-3}$
3529.6	$3^- \to 3^-$	1719.3	+0.41(13)	−0.03(17)	028(2)	30^{+5}_{-4}	
	$\to 2^+$	1869.5	−0.28(8)	+0.05(10)	072(2)		$1.3^{+0.3}_{-0.2}\times10^{-3}$
3597.0	$(4^-) \to 5^-$	770.5	+0.04(19)	−0.21(22)	025(2)	76^{+57}_{-27}	
	$\to 3^-$	1786.7	+0.13(12)	−0.01(16)	075(2)		10^{+6}_{-4} $E2$
3669.0	$5^- \to 4^+$	1478.0	−0.14(11)		064(8)	17^{+10}_{-7}	$4.2^{+4.4}_{-2.1}\times10^{-3}$
	$\to 3^-$	1858.6	+0.28(39)		036(8)		17^{+19}_{-9} $E2$

[a] Only statistical uncertainties are indicated.
[b] $E2$ transitions are indicated. All others for which values are reported are $E1$.

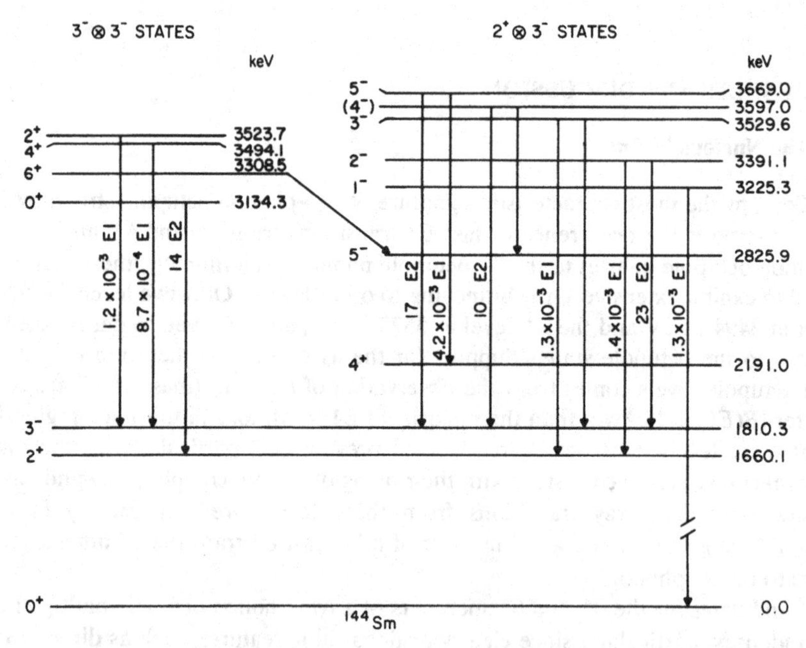

Figure 5. Partial ^{144}Sm level scheme containing proposed octupole-octupole and quardupole-octupole states. Measured transition rates in Weisskopf units are given on the transition arrows. Table I contains additional information about the deexciting γ rays.

Table II. Comparison of Additional ^{144}Sm Lifetimes with those Previously Measured by Nuclear Resonance Fluorescence

E_x(keV)	τ(fs)	
	DSAM	γ,γ'(Ref. 23)
2423.3	37^{+5}_{-4}	43 ± 6
2799.7	82^{+8}_{-7}	140 ± 27
3225.3	$8.5^{+1.2}_{-1.1}$	3.0 ± 0.3
3890.1	$2.0^{+1.9}_{-1.0}$	3.1 ± 0.5

is the excited state most strongly populated in the (t,p) two-neutron transfer reaction[27] and can be identified as the neutron pairing vibration. The 3134.3-keV state is also strongly populated, although with considerably smaller cross-section than for the 2822.6 keV state, in the two-neutron-transfer reaction and is the best candidate for the $(3^- \times 3^-)_0{}^+$ state. The large two-neutron transfer strength to this state can be taken as an indication of substantial interaction between the pairing-vibrational and octupole two-phonon modes. There are no other candidates in this energy region for the 0^+ member of the multiplet, with the possible exception of the 3823.6-keV level, which has not been assigned a definite spin. This level is the only higher-lying state whose decay to the first 2^+ level appears in any way to exhibit the isotropic γ-ray angular distribution expected for a 0^+ state.

The only definite 6^+ level in this region lies at 3308.5 keV. This state deexcites only to the 5^- level at 2925.9 keV by an $E1$ transition that is too low in energy to permit a meaningful lifetime determination by the DSAM method.

The predictions of Vogel and Kocbach[28] indicate that the $2^+ \times 3^-$ states should lie in the 3.0–4.5-MeV region in ^{144}Sm. An additional search was therefore undertaken for the expected quintet of quadrupole-octupole states having spins and parities of 1^-, 2^-, 3^-, 4^-, and 5^-.

Several authors[12,29,30] have suggested the 3225.3-keV level to be the 1^- member of the multiplet. Barfield and coworkers[12] have recently offered support, based primarily on the previously observed[30] fast E1 transition from this level and the $B(E1)$ they measured for the $3_1{}^- \rightarrow 2_1{}^+$ transition, for the interpretation of this state as a two-phonon $(2^+ \times 3^-)_1$-excitation. The $B(E1, 1^- \rightarrow 0^+)$ we deduce from our measured lifetime of the 3225.3-keV level is not so large (see Table II) as that reported by Metzger.[23,30] The slower rate measured in our study could result from feeding from a higher-lying level. Although a branch to the $2_1{}^+$ quadrupole phonon had not been observed,[23,30] Barfield et al.[12] suggest that branching from this 1^- level to the first excited state should occur with a $B(E1, 1^- \rightarrow 2^+)/B(E1, 1^- \rightarrow 0^+)$ ratio of unity. Neither this expected $1^- \rightarrow 2^+$ branch nor γ-ray decay to the 3^- one-phonon level are observed in the present measurements. With Barfield's assumption of a $B(E1)$ ratio of unity, though, the branch to the 1660.1-keV 2^+ level should have been observed. We have determined an upper limit of this ratio to be 0.3.

While the 2^- and 3^- members of the quintet will not likely decay directly to the ground state, transitions to the $2_1{}^+$ and $3_1{}^-$ levels are expected. Despite the fact that it is more than 1 MeV lower than the predicted[28] energy, the best 2^- candidate lies at an energy of 3391.1 keV. This level is the only identified 2^- state that decays to both of the one-phonon states, and the $E1$ transition of the 2^+ quadrupole phonon is fast (see Fig. 5). On similar grounds, the best 3^- candidate is the state at 3529.6 keV. We should point out, however, that there are several other 2^- and 3^- levels that decay soley to the first excited state by fast $E1$ transitions.

The most likely 4^- multiplet member, based on the observed fast $E2$ transition to the 3^- octupole level, is the 3597.0-keV state, although the spin assignment is tentative.

There are several well-established 5^- levels in ^{144}Sm; however, only one is observed to decay to both the 3_1^- and 4_1^+ levels. A decay to the first 4^+ level might be anticipated since this state is of similar structure as the 2_1^+ state.[31] Even in view of the large uncertanties, the fast transitions from the 3669.0-keV level support the interpretation of this level as a quadrupole-octupole two-phonon state.[29]

The observation of related positive and negative parity states in the 3-4-MeV excitation region could also be taken as evidence of α clustering.[32]

16.4.2 The Nucleus ^{96}Zr

As shown in Ref. 14, the low-lying levels of the ^{96}Zr nucleus fall into three distinct categories according to whether their decays favor the first 0^+, 2^+, or 3^- excited states. These relationships are shown in Fig. 4, where only the most prominent decay branches are included for each level. Here we will focus on those that decay preferentially to the 1897-keV first 3^- state which has been established as a strong octupole.[13]

There are three groups of levels associated with the 3^- state, as shown on the right side of Fig. 4. There is a group of states decaying to the first 3^- and 5^- state by $E1$ transitions and exhibiting substantial (p, p') and (α, α') inelastic scattering strength.[33,34] Such a pattern may be the signature of two-phonon octupole vibrational states expected in the Zr region, where the nuclei exhibit octupole softness but are not octupole-unstable because of the large energy separation between the spherical orbitals involved.[35] Although the observed $E1$ transition rates are generally lower than the values obtained in ^{144}Sm or in octupole-deformed actinides,[36,37] they are substantially faster than the average $E1$ rate of about 5×10^{-5} wu extracted for this mass region from recent compilations.[24,38] Since fast $E1$ transitions may indicate strong octupole correlations, these levels have been examined[14,22] as possible members of a two-phonon structure.

The low-lying states of ^{96}Zr have been classified according to their γ-decay properties. From the observation of fast $E1$ transitions to the 3^- octupole state, the 3212-keV and 3509-keV 2^+ states, as well as the 3082-keV 4^+ state, are suggested as possible candidates for two-phonon vibrational octupole states. Whether the 3483-keV 6^+ state can also be considered[21,39] as such is less clear at present. The 3176-keV 4^+ level, sharing its $E1$ decay strength with the 3082-keV state, seems more reasonably described as a hexadecapole vibration. The 3120-keV 5^- level decaying to the 3^- octupole state by a collective $E2$ transition may be interpreted as either a dyotriacontapole or a quadrupole-octupole excitation. Although the observed collectivity of the $E2$ transition connecting the 3120-keV 5^- state with the 3^- octupole level supports the quadrupole-octupole interpretation, identification of other members of such a quintet would be required to test this assumption.

16.5 CONCLUSIONS

We have provided direct evidence, in the form of observed decay patterns and fast $E1$ transitions, for the existence of two-phonon octupole vibrational states in ^{96}Zr and ^{144}Sm. The wealth of fast $E1$ transitions observed in ^{144}Sm may be explained by the shell structure of this mass region. Also presented are possibilities for the 5^- quadrupole-octupole state in ^{96}Zr and the $(2^+ \otimes 3^-)_I$ quintet of states in ^{144}Sm. Further experimental tests, like inelastic scattering with special emphasis on two-step processes, are needed to clarify the viability of the present proposals which are based mainly on the observation of electromagnetic decay properties.

ACKNOWLEDGMENTS

We wish to thank J. Blomqvist, E. A. Henry, F. Iachello, P. Kleinheinz, D. Kusenezov, H. Mach, M. T. McEllistrem, R. A. Meyer, K. Sistemich, D. W. Wang, and J. L. Weil for fruitful discussions and for their interest in this work. The support of the Hungarian Academy of Science and the U. S. National Science Foundation under grants No. INT-8512479 and No. PHY-8702369 are gratefully acknowledged.

REFERENCES

1. A. Aprahamian, D. S. Brenner, R. F. Casten, R. L. Gill, and A. Piotrowski, *Phys. Rev. Lett.* **59**, 535 (1987).
2. H. Mach, M. Moszyński, R. F. Casten, R. L. Gill, D. A. Brenner, J. A. Winger, W. Krips, C. Wesselborg, M. Büscher, F. K. Wohn, A. Aprahamian, D. Alburger, A. Gelberg, and A. Piotrowski, *Phys. Rev. Lett.* **63**, 143 (1989).
3. J. Kumpulainen, R. Julin, J. Kantele, A. Passoja, W. H. Trzaska, E. Verho, and J. Vaaramaki, private communication (1989).
4. A. Bohr and B. Mottelson, *Nuclear Structure*, (W. A. Benjamin, New York, 1975) Vol. 2.
5. M. A. J. Mariscotti, D. R. Bes, S. L. Reich, H. M. Sofia, P. Hungerford, S. A. Kerr, K. Schrechenbach, D. D. Warner, W. F. Davidson, and W. Gelletly, *Nucl. Phys.* **A407**, 98 (1983).
6. R. Julin, J. Kantele, J. Kumpulainen, M. Luontama, A. Passoja, W. Trzaska, E. Verho, and J. Blomqvist, *Phys. Rev. C* **36**, 1129 (1987).
7. S. W. Yates, L. G. Mann, E. A. Henry, D. J. Decman, R. A. Meyer, R. J. Estep, R. Julin, A. Passoja, J. Kantele, and W. Trzaska, *Phys. Rev. C* **36**, 2143 (1987).
8. S. W. Yates, R. Julin, P. Kleinheinz, B. Rubio, L. G. Mann, E. A. Henry, W. Stoeffl, D. J. Decman and J. Blomqvist, *Z. Phys.* **A324**, 417 (1986).
9. P. Kleinheinz, J. Styczen, M. Piiparinen, J. Blomqvist, and M. Kortelahti, *Phys. Rev. Lett.* **53**, 1531 (1984).
10. S. Lunardi, P. Kleinheinz, M. Piiparinen, M. Ogawa, M. Lach, and J. Blomqvist, *Phys. Rev. Lett.* **53**, 1531 (1984).
11. O. Scholten, F. Iachello and A. Arima, *Ann. Phys. (NY)* **115**, 312 (1978).
12. A. F. Barfield, P. von Brentano, A. Dewald, K. O. Zell, N. V. Zamfir, D. Bucurescu, M. Ivascu, and O. Scholten, *Z. Phys.* **A332**, 29 (1989).
13. G. Molnár, H. Ohm, G. Lhersonneau and K. Sistemich, *Z. Phys.* **A331**, 97 (1988).
14. G. Molnár, T. Belgya, B. Fazekas, A. Veres, S. W. Yates, E. W. Kleppinger, R.A. Gatenby, R. Julin, J. Kumpulainen, A. Passoja, and E. Verho, *Nucl. Phys. A500*, 43 (1989), and references therein.
15. S. W. Yates, *Proc. ACS Symposium on Recent Advances in the Study of Nuclei Off the Line of Stability*, Chicago, 1985 ACS Symposium Series **324**, 470 (1986).
16. R. A. Gatenby, E. W. Kleppinger, and S. W. Yates, *Nucl. Phys.* **A492**, 45 (1989).
17. S. W. Yates, A. Khan, M. C. Mirzaa, and M. T. McEllistrem, *Phys. Rev. C* **23**, 1993 (1981).
18. A. K. Georgiev, M. K. Goergieva, D. V. Elenkov, and G. Kh. Tumbev, *Sov. J. Nucl. Phys.* **48**, 207 (1988).

19. T. Belgya, G. Molnár, B. Fazekas, A. Veres, R. A. Gatenby, and S. W. Yates, *Nucl. Phys. A500*, 77 (1989), and references therein.
20. D. Elenkov, D. Lefterov, and G. Toumbev, *Nucl. Instr. Methods* **228**, 62 (1984).
21. E. A. Henry, R. A. Meyer, A. Aprahamian, K. H. Maier, L. G. Mann, and N. Roy, in *Proc. Int. Workshop on Nuclear Structure in the Zirconium Region*, Bad Honnef, FRG, 24-28, April 1988, Ed. by K. Sistemich, J. Eberth, and R. A. Meyer (Springer, Berlin, 1988) p. 223.
22. D. F. Kusnezov, E. A. Henry, and R. A. Meyer, *Phys. Lett.* **228**, 11 (1989).
23. F. R. Metzger, *Phys. Rev. C* **17**, 939 (1978).
24. P. M. Endt, *Atomic Data Nucl. Data Tables*, **26** (1981) 47.
25. J. K. Tuli, *Nucl. Data Sheets* **56**, 607 (1989).
26. R. Julin, M. Luontama, A. Passoja, and W. Trzaska, in *Proc. Int. Symposium on In-Beam Nuclear Spectroscopy*, Debrecen, Hungary, 1984, Ed. by Zs. Dombrádi and T. Fényes (Akadémiai Kiadó, Budapest, 1985) p. 369.
27. E. R. Flynn, J. van der Plicht, J. B. Wilhelmy, L. G. Mann, G. L. Struble, and R. G. Lanier, *Phys. Rev. C* **28**, 97 (1983).
28. P. Vogel and L. Kocbach, *Nucl. Phys.* **A176**, 33 (1971).
29. R. Martin, L. Bimbot, S. Gales, L. Lessard, D. Spalding, W. G. Weitkamp, O. Dietzschand, and J. L. Foster, Jr., *Nucl. Phys.* **A210**, 221 (1973).
30. F. R. Metzger, *Phys. Rev. C* **14**, 543 (1976).
31. M. Waroquier and K. Heyde, *Z. Phys.* **268**, 11 (1974); *Nucl. Phys.* **A164**, 113 (1971).
32. F. Iachello and A. D. Jackson, *Phys. Lett.* **108B**, 151 (1982).
33. M. Fujiwara, Y. Fujita, S. Y. Hayakawa, H. Ikegami, I. Katayama, K. Katori, S. Morinobu, Y. Tokunaga and T. Yamazaki, *RCNP Osaka University Ann. Rep.* (1983), p. 59.
34. M. Lahanas, D. Rychel, P. Singh, R. Gyufko, D. Kolbert, B. van Kruechten, E. Madadakis and C. A. Wiedner, *Nucl. Phys.* **A455**, 399 (1986).
35. W. Nazarewicz and T. Werner, in *Proc. Int. Workshop on Nuclear Structure in the Zirconium Region*, Bad Honnef, FRG, 24-28, April 1988, Ed. by K. Sistemich, J. Eberth and R. A. Meyer (Springer-Verlag, Berlin, 1988), p. 277.
36. I. Ahmad, R. R. Chasman, J. E. Gindler and A. M. Friedman, *Phys. Rev. Lett.* **52**, 503 (1984).
37. L. Lönnroth, *Z. Phys.* **A331**, 11 (1988).
38. P. M. Endt, *Atomic Data Nucl. Data Tables* **23**, 547 (1979).
39. R. A. Meyer, *Proc. Int. Conf. on Nuclear Structure through Static and Dynamic Moments*, Melbourne, Australia, 25-28, August 1981, Ed. by H. H. Bolotin (Conf. Proc. Press, Melbourne, 1987), Vol. II, p. 15.

17. Spin-Stabilized Deformation in Transitional Nuclei

P. Chowdhury, B. Crowell, P. J. Ennis, C. J. Lister, and Ch. Winter
A. W. Wright Nuclear Structure Laboratory
Yale University
New Haven, CT 06511

H. R. Andrews, D. Horn, D. C. Radford, and D. Ward
Atomic Energy of Canada Limited
Chalk River Nuclear Laboratories
Chalk River, Ontario, Canada K0J 1J0

J. K. Johansson and J. C. Waddington
Tandem Accelerator Laboratory
McMaster University, Hamilton, Ontario, Canada L8S 4K1

S. Pilotte
Laboratoire de Physique Nucléaire
Université de Montréal, Quebec, Canada H3C 3J7

ABSTRACT

Nuclei that lie in the transition region between those that display pure single-particle and pure collective behavior at low spins tend to yield to the fictive forces generated by rotation and undergo structural changes. The role of angular momentum in stabilizing exotic shapes in nuclei has been dramatically demonstrated by observation of superdeformation at high spins in the $A \approx 130$, 150, and 190 regions. Current theoretical predictions also point towards the lighter $80 < A < 100$ region as likely terrain for such spin-stabilized deformations. This chapter centers around the high-spin spectroscopy of two transitional nuclei in this mass region, ^{86}Zr and ^{101}Ag, which were studied with the 8π spectrometer at Chalk River and the Gamma-Array at Yale. A status report on the assembly of a unique combination of detection devices for nuclear-structure studies at Yale will also be presented.

17.1 INTRODUCTION

The role of angular momentum in modifying the total potential energy surface in nuclear-shape degrees of freedom is a topic of widespread interest. Shifts of the minima in the total energy surface as a function of the nuclear spin have been predicted by theoretical calculations, and the resulting changes in nuclear structure have been documented by experiments in a variety of nuclear regions. Angular-momentum-induced stabilization of exotic shapes in nuclei has been dramatically demonstrated by the observation of superdeformation at high spins in the $A \approx 130$ and 150 regions,[1] and more recently, in the $A \approx 190$ region.[2] Current theoretical predictions,[3-5] also point towards the lighter $80 < A < 100$ region as likely terrain for such spin-stabilized deformations.

Nuclei that lie in the transition region between those which display pure single-par-

Exotic Nuclear Spectroscopy, Edited by W. C. McHarris
Plenum Press, New York, 1990

ticle and pure collective behavior at low spins tend to yield to the fictive forces generated by rotation and undergo structural changes more easily than nuclei in either limit. Here we describe the high-spin spectroscopy of two transitional nuclei in this mass region, ^{101}Ag and ^{86}Zr, which were studied with Yale Gamma-Array and the 8π spectrometer at Chalk River. We also take this opportunity to provide a status report on the performance of the ESTU accelerator and our efforts towards the assembly of a unique combination of detection devices for nuclear structure studies at the Wright Nuclear Structure Laboratory at Yale.

17.2 ^{101}Ag

The $A = 100$ region is of special interest to shell-model enthusiasts, as it offers the as yet unattained doubly-magic self-conjugate nucleus ^{100}Sn with fifty neutrons and fifty protons. The excited level structures of nuclei neighboring this double shell closure exhibit the characteristic signatures of particle-hole excitations about the $N = 50$, $Z = 50$ core. However, as one moves away from this shell closure, the onset of collectivity at low spins is evident from the observation of well-developed rotational-band structures. Additional interest in this region arises from the recent observation[6] of discrete-line collective structures with apparently large moments of inertia at very high spins. In order to investigate the microscopic basis for collectivity in this region, one needs to study the interplay of single-particle and collective degrees of freedom in generating angular momentum. To this end, the detailed spectroscopy of a transitional nucleus at high spins promises to provide direct clues through changes in structure along the yrast line.

The nucleus ^{101}Ag, with forty-seven protons and fifty-four neutrons, is such a transitional nucleus. The region of interest is shown in Fig. 1, with $Z < 50$ and $N > 50$ nuclei. The odd-A isotopes of Ag display single-particle behavior for $N < 52$ and collective behavior for $N > 56$. Similarly, the isotones of ^{101}Ag switch from single-particle behavior for $Z > 48$ to collective behavior for $Z < 46$. This is amply demonstrated by the low-spin level schemes of ^{97}Ag (Ref. 7), ^{103}Ag (Ref. 8), ^{102}Cd (Ref. 9), and ^{100}Pd (Ref. 10).

The nucleus ^{101}Ag had been previously studied[11] via (p,2n) and (^{12}C,p2n) reactions before the advent of compton-suppressed Ge (CSG) detectors, and spins had been assigned as high as 25/2. We chose to repeat the ^{92}Mo(^{12}C,p2n) reaction as a test run with first beams from our ESTU accelerator and with four CSG detectors. Subsequently, we decided

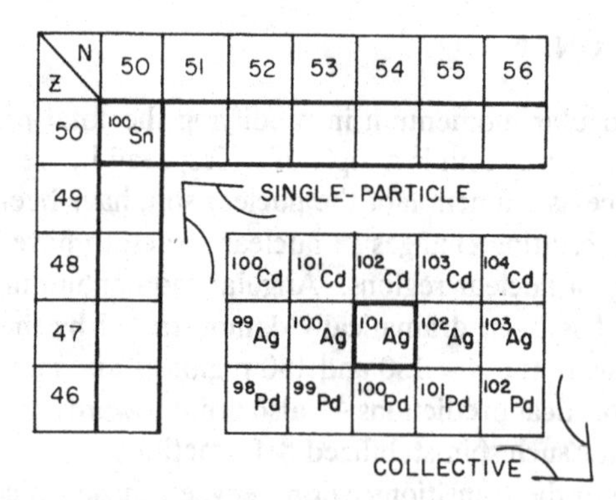

Figure 1. A close-up of the $N > 50$, $Z < 50$ nuclear region, highlighting the transitional nature of the nucleus ^{101}Ag.

to increase the angular momentum input with heavier ions, and chose the ^{70}Ge(^{35}Cl,2p2n)^{101}Ag reaction. By then, our full array of five CSG detectors, shown in Fig. 2, was operational. The target was 800-μg/cm^2 ^{70}Ge evaporated on 30-mg/cm^2 Pb. Excitation functions were measured for ^{35}Cl beam energies from 110 to 182 MeV, the latter corresponding to 16.5 MV on the ESTU terminal. An optimum beam energy of 140 MeV was chosen for the γ-γ coincidence experiment. In addition to the five CSG detectors, four 3"x3" NaI detectors were used as a filter to reject low-multiplicity events. A modest 5 x 10^7 events were recorded in the γ-γ matrix, of which more than half were coincident with at least one NaI detector.

The preliminary level scheme of ^{101}Ag deduced from this data is shown in Fig. 3. The number of γ rays observed in the present work is more than twice that observed previously, and the highest levels observed in this study are also at approximately twice the excitation energy. While the present results corroborate the previously published level scheme[11] at excitation energies below ≈2 MeV, considerable discrepancies exist in the medium spin region between ≈2 MeV and ≈4 MeV, which was as high as the previous work extended. The data above ≈4 MeV excitation energy are essentially comprised of previously-unobserved transitions. Multiple-decay branches provide reasonable confidence in the medium-spin region of the present level scheme, while the ordering of the levels at the highest spins, where no crossover transitions are observed, should be considered as tentative. Since angular distributions are yet to be measured, spins and parities have not been assigned. However, the existence of crossover transitions provides strong evidence for dif-

Figure 2. The five compton-suppressed Ge detectors of the Yale Gamma Array.

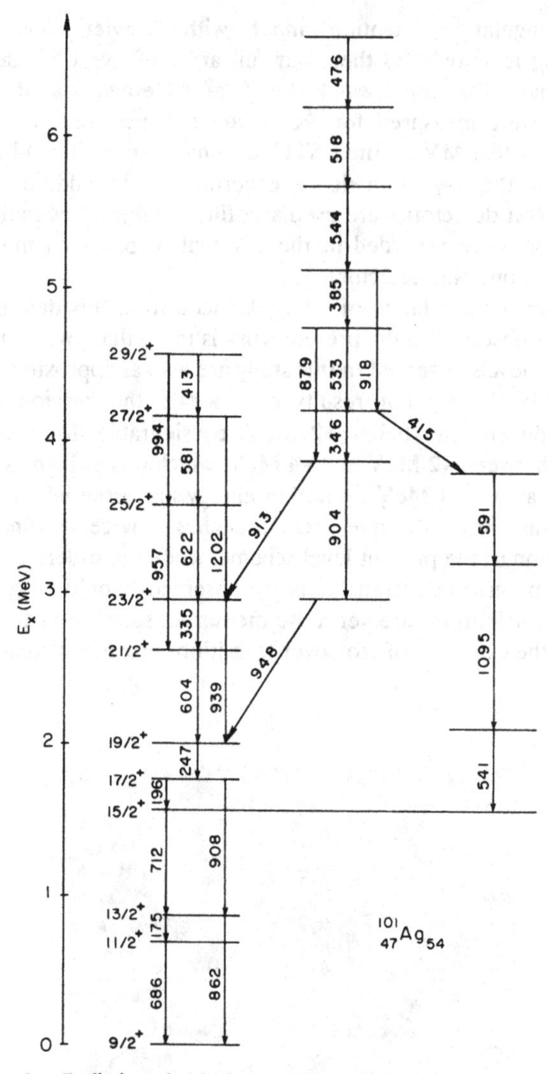

Figure 3. Preliminary level scheme of ^{101}Ag deduced from present work.

ferentiating between dipole and quadrupole transitions.

Without detailed spin assignments, no quantitative discussion on changes in structure can be sustained. However, a few qualitative points can definitely be made. The low-spin region, where the spins and parities are known, is predominantly of single-particle nature, as highlighted by the presence of the strong $19/2^+ \rightarrow 17/2^+$ 247-keV dipole transition in the yrast sequence. Above the $19/2^+$ level, however, the level scheme seems to exhibit signatures of modest collectivity; viz., lower energy dipoles in the 300–650-keV range, with $E2$ crossovers in the 850–1200-keV range. If the relationship between energy and multipolarity extends to the highest spins where no crossover transitions are observed, the 346-533-385-544-518-476-keV sequence would be the dipole transitions of a band of which the lowest two $E2$ crossovers are observed. It is an interesting exercise to compare this band in ^{101}Ag to a similar one observed[8] in the neighboring ^{103}Ag nucleus, as shown in Fig. 4. The energies are remarkably similar, and in both cases the decay out of the bottom of the band is fragmented and complicated. Definitive statements on changes in structure of the ^{101}Ag nucleus with increasing angular momentum will have to await the completion of the level scheme, with spin and parity assignments from future angular distribution measurements. The present data, the first with ESTU beams and the Yale Gamma Array, point towards an evolution to weak collectivity at high spins for the transitional ^{101}Ag, comparable to that observed in the neighboring higher-mass odd-A Ag nuclei.

17.3 ^{86}Zr

The second nucleus that we discuss is ^{86}Zr, with forty protons and forty-six neutrons. The transitional nature of this nucleus is immediately obvious if one plots the first excited 2^+ states in even-even Zr nuclei between $N = 40$ and 50, (see Fig. 5). The $2^+ \rightarrow 0^+$ transition energies go from 2186 keV in the nucleus ^{90}Zr with a $N = 50$ closed shell, to 290 keV in the well-deformed nucleus ^{80}Zr. The $B(E2)$ values also range from 6 wu in spherical ^{90}Zr to 88 wu in deformed ^{82}Zr. The nucleus ^{86}Zr treads a middle ground with a $2^+ \rightarrow 0^+$ transition energy of 752 keV and a $B(E2)$ value of 14 wu. The level schemes of the even-even neighbors ^{84}Zr and ^{88}Zr also highlight the transitional nature of ^{86}Zr. While the level scheme of ^{88}Zr, known[12] up to a spin of 21 and an excitation energy of 11 MeV, exhibits the classic signatures of generating angular momentum by aligning particles, the nucleus ^{84}Zr shows[13] a regular collective stretched $E2$ cascade up to spin 34^+. Previous work[14] on the nucleus ^{86}Zr via a ^{60}Ni(^{29}Si,2pn) reaction had resulted in a level scheme up to spin 15 and an excitation energy of 7.4 MeV, consistent with a non-collective character.

Calculated shell-energy landscapes,[4] as a function of nucleon number and deformation, point towards stabilization of superdeformed shapes for nucleon numbers lying between 40 and 46 and provide a strong motivation for studying ^{86}Zr at high spins. On a less

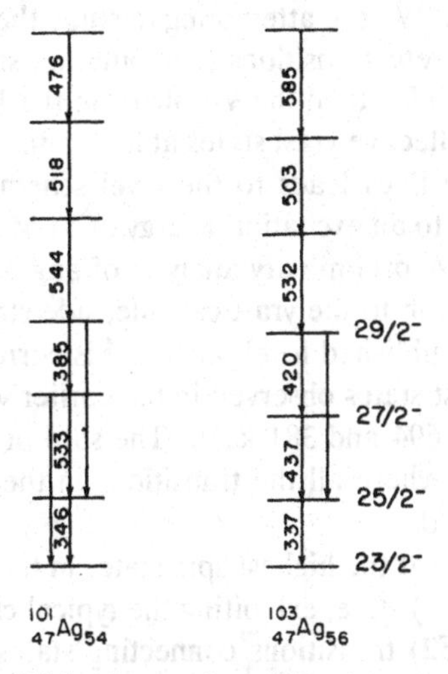

Figure 4. Comparison of high-spin cascades in ^{101}Ag (present work) and ^{103}Ag (Ref. 8).

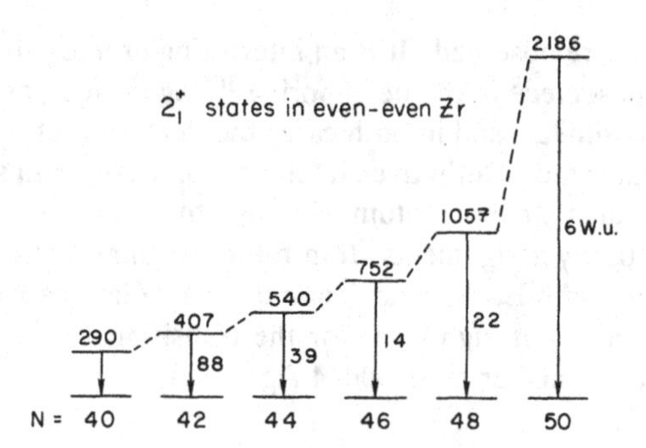

Figure 5. Systematics of 2+ states in even-even Zr isotopes with $40 < N < 50$. The transition energies and the $B(E2)$ strengths for the $2^+ \rightarrow 0^+$ transitions are marked.

exotic level, given the transitional nature of ^{86}Zr, it is of inherent interest to map the onset of collective behavior as a function of angular momentum in this nucleus.

In order to optimize the conditions for trapping the residual nucleus in a superdeformed minimum, we chose a four-particle evaporation channel in order to populate the ^{86}Zr closer to the yrast line. the reaction used was ^{60}Ni(^{30}Si,2p2n)^{86}Zr, with a ^{30}Si beam energy of 135 MeV on a stacked 2x480-μg/cm^2 self-supporting ^{60}Ni target. The experiment was performed at the 8π spectrometer at Chalk River Nuclear Laboratories in collaboration with CRNL, McMaster University, and University of Montreal. The results that we present were obtained from analyzing a high-multiplicity γ-γ matrix containing about 7x10^7 events, which required at least fourteen out of seventy-two elements of the BGO array to fire.

Fig. 6*a* shows a total projection spectrum of the high-multiplicity γ-γ matrix. The immediate feature that stands out in the spectrum is the pronounced broad bump of continuum transitions centered just above 2 MeV. That the bump comprises predominantly of high-spin transitions is evident from the increase in its intensity in the high-fold cut of $K >$ 14 compared to a low-fold cut of $9 < K < 14$, as shown in Fig. 6*b*, where K represents the number of BGO elements that fired. We are in the process of analyzing the features of the bump in a quantitative fashion. We are attempting to study the evolution of the continuum bump in coincidence with discrete transitions depopulating states with increasing spin in the yrast cascade in order to understand the γ-flow from the high-spin collective cascades in the continuum to the non-collective yrast states at low spin.

Analysis of the discrete lines leads to the level scheme for ^{86}Zr shown in Fig. 7. Levels have been identified up to an excitation energy of 14 MeV, which is about twice that observed in previous work.[14] A preliminary analysis of angular correlations allows tentative spin assignments. In addition to the yrast cascade, side structures have been extended considerably. The previously-published level scheme[14] is corroborated except for the spin assignments of the top two yrast states observed in the earlier work, which are depopulated by transitions with energies of 694 and 381 keV. The sum of clean gates in the medium-spin region is shown in Fig. 8, where all the transitions in the yrast cascade placed in the present level scheme are marked.

A feature that stands out at the highest spin states in the level scheme is the coupled band structure built on the (14$^+$) state, exhibiting the typical characteristics of a signature-split band, with quadrupole ($E2$) transitions connecting states of the same signature and "dipole" ($M1/E2$) transitions connecting states of opposite signature.

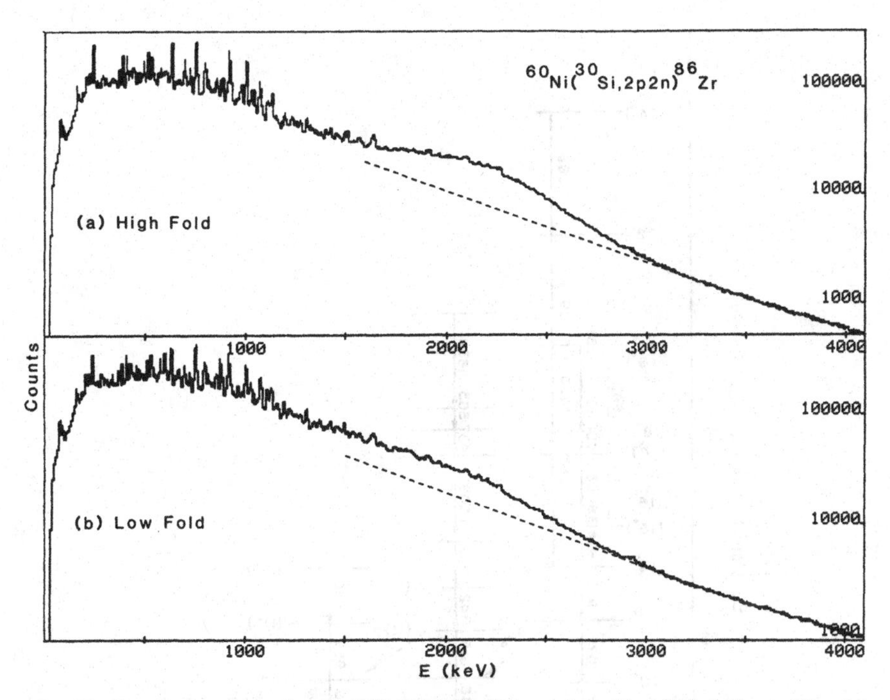

Figure 6. Total-projection spectra for the 135-MeV $^{30}Si + ^{60}Ni$ reaction with *a*) high-fold cut of $K > 14$ and *b*) low-fold cut of $9 < K < 14$.

It is instructive to plot the $\Im^{(1)}$ moment of inertia versus rotational frequency for the yrast cascade in ^{86}Zr, and compare it with the same for the neighboring ^{84}Zr nucleus (Fig. 9). In ^{84}Zr the two upbends at $\hbar\omega$ values of 0.5 MeV and 0.6 MeV had been interpreted,[13] using cranked-shell-model calculations, as alignments of two $g_{9/2}$ quasi-protons followed by two $g_{9/2}$ quasi-neutrons. Beyond these two alignments, the variation of $\Im^{(1)}$ with frequency for ^{84}Zr stabilizes to a remarkably constant value of 25.5 MeV^{-1} over a wide frequency range of almost 0.5 MeV. In comparison, in the low frequency region below 0.6 MeV in ^{86}Zr, the $\Im^{(1)}$ moment of inertia shows two sharp backbends around $\hbar\omega \approx 0.5$ MeV. Considering the generally non-collective nature of this transitional nucleus at low spins, the sharp backbends are perhaps better attributed to uncorrelated single-particle residual interactions rather than to distinct band crossings. On the other hand, given the transitional nature of ^{86}Zr, alignments of two $g_{9/2}$ quasi-protons and two $g_{9/2}$ quasi-neutrons are also possible explanations from a weakly collective point of view. Above spin 14 and $\hbar\omega \approx 0.55$ MeV, however, the variation in $\Im^{(1)}$ seems to decrease and shows a tendency to stabilize at values of $\approx$22.5 MeV^{-1}. Both decay sequences feeding the 14^+ state, which show characteristics of being opposite signature partners of the same quasi-particle orbit, exhibit this tendency towards stabilization of $\Im^{(1)}$ at around the same value.

This behavior is very similar to that observed in ^{84}Zr, especially in terms of the threshold spin and angular frequency. The comparison with ^{84}Zr suggests that following the alignment of two valence quasi-protons and two valence quasi-neutrons in ^{86}Zr, the aligned quasi-particles are able to polarize and deform the soft transitional core sufficiently

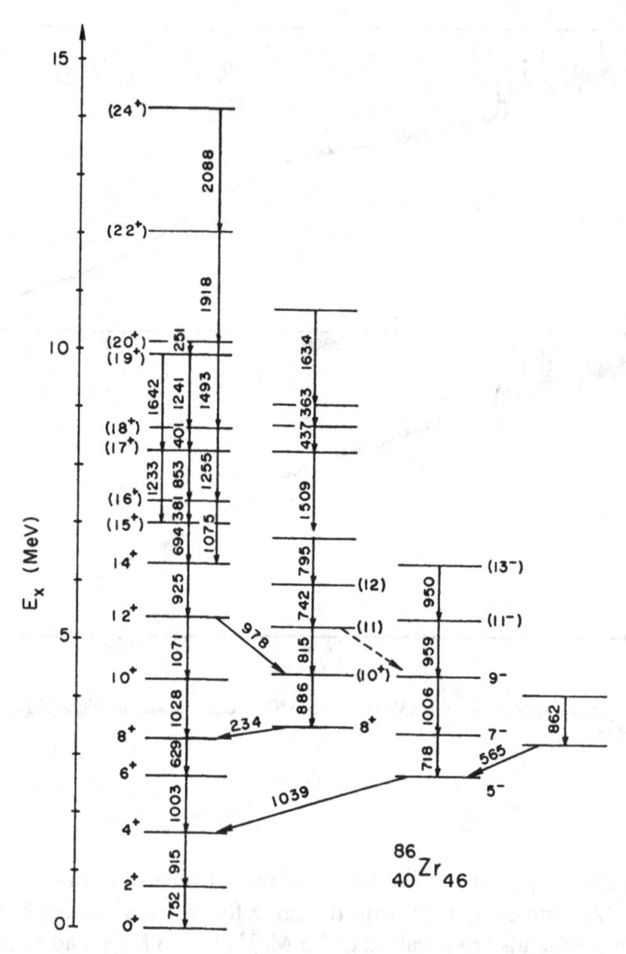

Figure 7. Level scheme of ^{86}Zr deduced from present work.

that it becomes energetically more favorable to generate higher angular momenta by collective rotation rather than by quasi-particle alignment. From earlier lifetime measurements[14] in ^{86}Zr, the $B(E2)$ values for the yrast transitions up to the 14^+ level vary between 3 and 14 wu. We have recently performed a doppler-shift-attenuation measurement at Chalk River for the lifetimes of the highest-spin states, in order to quantify the onset of spin-stabilized deformation along the yrast line in ^{86}Zr for angular frequencies above ≈ 0.55 MeV and spins greater than 14^+.

The search for shape transitions at high spin naturally leads one to an analysis of the two-dimensional E_γ-E_γ correlations. Fig. 10 shows a two-dimensional E_γ–E_γ correlation matrix following subtraction of uncorrelated background and efficiency correction. A ridge structure parallel to the diagonal is immediately noticeable in the 1–2-MeV region. Cuts were made perpendicular to the diagonal, and Fig. 11*a*-*c* shows three 50-keV-wide cuts in the 1250–1400-keV range. These cuts are relatively uncontaminated by strong discrete γ rays and show a sustained ridge-valley structure of constant width, while Fig. 11*d* shows the sum of the three cuts. The ridge separation of 300 keV leads to a dynamic moment of iner-

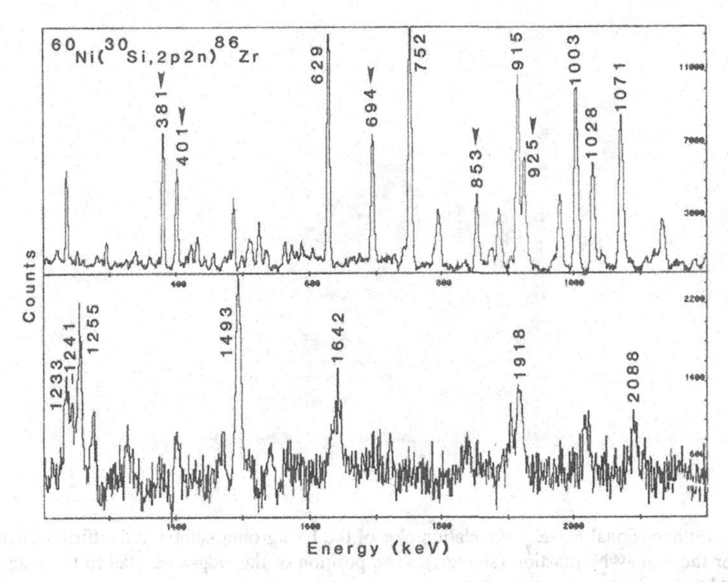

Figure 8. Sum of clean gates in ^{86}Zr in the 12 < I < 19 region. The gating transitions are indicated by arrowheads.

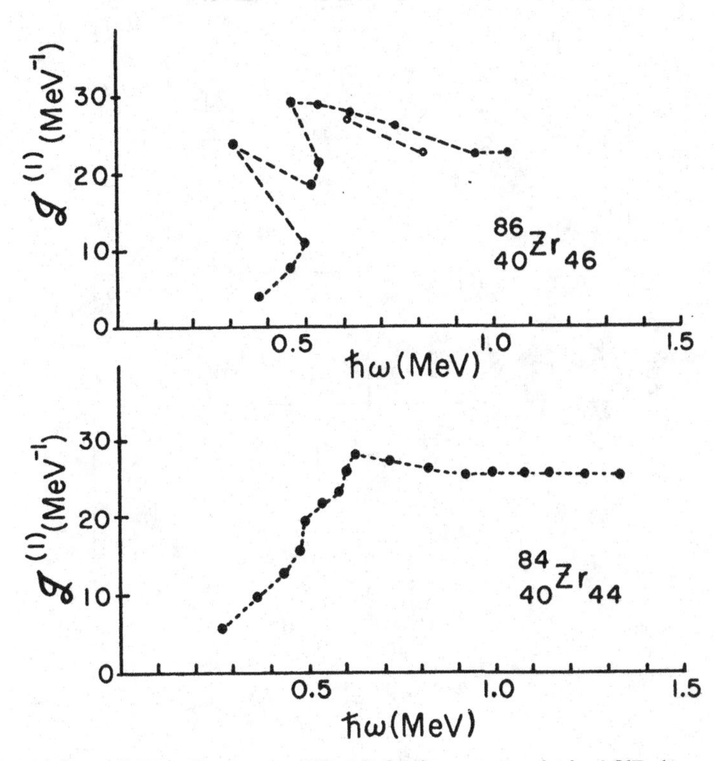

Figure 9. Comparison of kinematic moments of inertia for the yrast cascades in *a*) ^{86}Zr (present work) and *b*) ^{84}Zr (Ref. 13). Filled and open circles represent even- and odd-spin states, respectively.

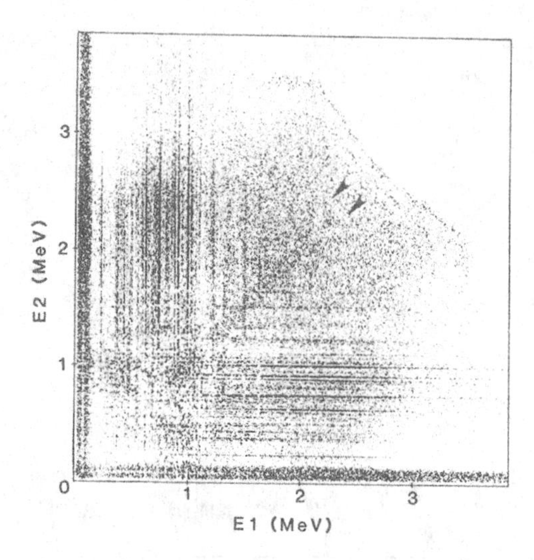

Figure 10. Two-dimensional E_γ-E_γ correlation plot of the background-subtracted, efficiency-corrected high-fold matrix for the ^{30}Si+^{60}Ni reaction (see text). The position of the ridges parallel to the diagonal are indicated by arrowheads.

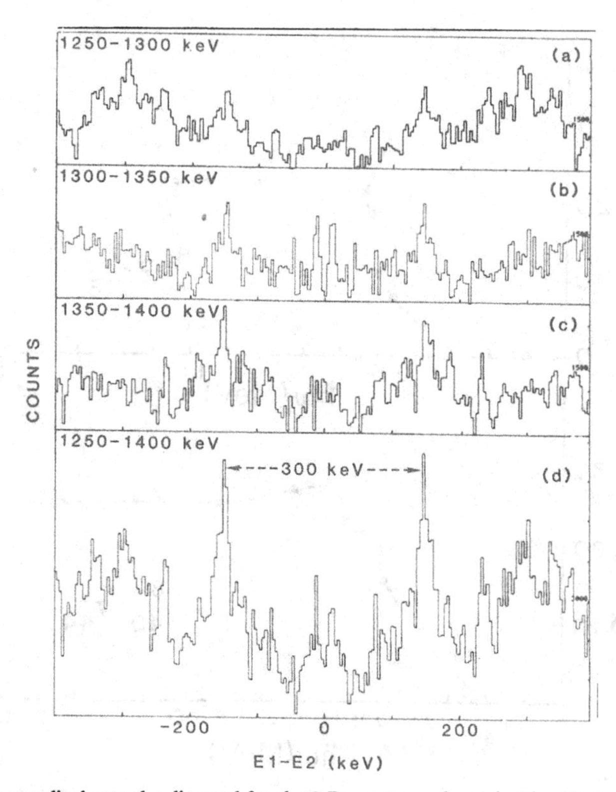

Figure 11. Cuts perpendicular to the diagonal for the 2-D spectrum shown in Fig. 10 at specific energy intervals (see text).

tia $\Im^{(2)}$ of 26.7 MeV^{-1}, corresponding to a value of $\beta \approx 0.48$ for an axially symmetric ellipsoid. These numbers can be compared to those for ^{84}Zr, where Q_0 was measured to be 3.5 *e*b for spins greater than 18^+, leading to values of $\beta = 0.43$ for an axially-symmetric ellipsoid. We are now in the process of trying to identify, if possible, discrete line transitions which contribute to this ridge structure in the E_γ-E_γ correlations. All the various pieces of the puzzle seem very exciting at this point, and whether we will raise the *superdeformation* flag in this nucleus is to be decided in the very near future.

17.4 INSTRUMENTATION

As promised in the introduction, we give a short update on our instrumentation developments. At Yale we have made considerable progress in putting together a unique multi-purpose detection facility for nuclear spectroscopy, which is categorized below. Currently, we have five CSG detectors of ≈25% efficiency, which are our state-of-the-art workhorses for high-resolution γ-ray spectroscopy. We have ordered five planar 36-mm-diameter by 1-mm-thick Ge detectors for detecting x rays and low-energy γ rays with high efficiency, to be used for channel selection. We have already implemented and taken data with a five-segment NE213 liquid-scintillator neutron wall to isolate weak reaction channels. We are awaiting delivery of crystals and electronic components for a thirty-eight element BGO array. The individual detectors of hexagonal cross-section will be close-packed to act as both a multiplicity filter and a high-efficiency calorimeter. Finally, a sixteen-element, large-solid-angle charged-particle calorimeter is under construction for sensitive determination of the nuclear temperature.

On the accelerator front, the ESTU has already delivered beams with 18 MV on the terminal. To date, we have had heavy-ion beams of ^{12,13}C, ^{16}O, ^{28,29}Si, ^{32,34,36}S, ^{35}Cl, ^{40}Ca, and ^{197}Au on target for experiments. The list is rapidly expanding, and the stability of the accelerator during experiments has been very commendable.

17.5 CONCLUSIONS

Definitive conclusions on the physics, from the studies presented here, are yet to be finalized. However, intriguing signatures of structure changes are evident at high spins in transitional nuclei in the $A \approx 80$ and 100 regions.

In the case of ^{101}Ag we have evidence for weakly-collective behavior at high spins, comparable to the heavier neighboring odd-*A* isotopes. Angular distributions and lifetime measurements are planned in the near future in order to quantify this. In ^{86}Zr we find compelling evidence of strong collectivity at high spins. The ridge structure in γ-γ correlation data suggests large deformations. The new data emerging from these studies on spin-stabilized deformation in the lighter-mass transitional nuclei promises to provide exciting physics in the near future in a new region of the nuclear chart.

ACKNOWLEDGMENT

Work supported by U.S.D.O.E. Contract #DE-AC02-76ER03074.

REFERENCES

1. P. J. Nolan and P. J. Twin, *Ann. Rev. Nucl. Part. Sci.* **38**, 533 (1988), and references therein.
2. E. F. Moore et al., *Phys. Rev. Lett.* **63**, 360 (1989).
3. W. Nazarewicz et al., *Nucl. Phys.* **A 435**, 397 (1985).
4. I. Ragnarsson and T. Bengtsson, in *Proc. Int. Workshop on Nuclear Structure of the Zirconium Region*, Bad Honnef, FRG, 1988, p. 193; W. Nazarewicz and T. Werner, ibid., p. 277.
5. J. Dudek et al., *Phys. Rev.* C **35**, 1489 (1987).
6. A. O. Macchiavelli et al., *Phys. Rev.* C **38**, 1088 (1988).
7. W. F. Piel, Jr., et al., *Phys. Rev.* C **37**, 1067 (1988).
8. J. Treherne et al., *Nucl. Phys.* **A342**, 357 (1980).
9. J. Treherne et al., in *Proc. Int. Conf. on Nuclear Far From Stability*, Helsingør, Denmark, 1981, p. 459.
10. W. F. Piel, Jr., et al., *Phys. Rev.* C **23**, 708 (1981).
11. A. W. B. Kalshoven et al., *Nucl. Phys.* **A 346**, 147 (1980).
12. E. K. Warburton et al., *Phys. Rev.* C **31**, 1184 (1985).
13. H. G. Price et al., *Phys. Rev. Lett.* **51**, 1842 (1983).
14. E. K. Warburton et al., *Phys. Rev.* C **31**, 1211 (1985).

18. Shape Evolution Studies in the $A = 70$ Region

A. V. Ramayya and J. H. Hamilton
Department of Physics
Vanderbilt University
Nashville, TN 37235

ABSTRACT

The neutron-deficient nuclei in the mass region of A between 70 and 80 rapidly change their collective properties with proton and neutron numbers, reflecting the influence of competing gaps in the Nilsson single-particle levels. Systematic studies of nuclei in this region across long isotopic and isotonic sequences and across major shells have revealed coexisting different shapes: prolate, oblate, and triaxial at low spin; also, stabilization of the prolate shape at high spin with large deformation ($\beta \geqslant 0.4$). Many of these properties can be understood qualitatively in terms of the shell gaps in the Nilsson single-particle levels at larger deformation. A review of our knowledge of selected nuclei in this region and the recent experimental developments which offer very powerful techniques will be discussed.

18.1 INTRODUCTION

Since the discovery of shape coexistence in ^{72}Se,[1,2] the nuclei in the mass region $A \approx$ 70–80 have been the subject of extensive investigation both experimentally and theoretically. Calculations[3] employing the deformed shell model and the shell-correction procedure explain qualitatively the shape coexistence with two minima in the potential-energy surface (PES), one at oblate and the other at prolate deformation. Recent microscopic calculations for this region with mean-field approaches,[4] which are essentially parameter free, have had some successes. These neutron-deficient transitional nuclei rapidly change their collective properties with proton and neutron numbers through the influence of competing gaps between the Nilsson single-particle levels at different deformation. The Nilsson diagram for this region shows a large gap for $N, Z = 34$ at prolate and oblate deformations; for $N, Z = 38$ at large ($\beta_2 \approx 0.4$) prolate deformation; and for $N, Z = 36$ at large oblate deformation. Unusually large ground state deformations, $\beta \approx 0.4$, observed in this region, beginning with ^{74,75}Kr[5] and then ^{80}Sr, are the result of a reinforcing effect of the shell gaps in the single-particle levels for both $Z, N = 38$.

Studies of these nuclei to high spin and extension to very neutron-deficient nuclei across a long isotopic chain are expected to give new insights into the stability and evolution of nuclear shapes and structure. In this chapter we will present a brief review of what we have learned about selected nuclei in this region and the new facilities which are needed to extend our knowledge.

Exotic Nuclear Spectroscopy, Edited by W. C. McHarris
Plenum Press, New York, 1990

18.2 SHAPE COEXISTENCE AND HIGH-SPIN STUDIES IN ^{70}Se AND ^{72}Se

Until about 1974, the nuclei in the mass region $A \approx$ 70–80 were assumed to be quasi-vibrational and uninteresting, largely because of insufficient data and our simplistic concepts about the properties of the mean fields on which the shell model is founded. Studies[1,2] of the Se isotopes (also, many other nuclei) by the Vanderbilt group, and later by many others, have shown that these nuclei exhibit complex spectra and coexistence of two bands built on quite different deformations.

Recent studies[6] of ^{72}Se at HHIRF (Oak Ridge) revealed discrete γ rays from states up to 24$^+$ (tentatively to 26$^+$) and provided more accurate energies for the higher levels than recently reported.[7] The γ-ray spectrum obtained by summing the coincidence spectra of the yrast gates above 6$^+$ is shown in Fig. 1. Individual and summed gates of the transitions in the yrast cascade above 6$^+$ are strikingly clean with side feeding at or below the 10% level. The yrast bands in ^{72}Se from our work and ^{70}Se (Ref. 7) are shown in Fig. 2.

The kinematical ($\Im^{(1)} = I_x/\omega$) and dynamical moments of inertia ($\Im^{(2)} = dI_x/d\omega$) for ^{72}Se along with ^{70}Se[7] and ^{84}Zr[8] are shown in Fig. 3. Shape coexistence discovered at low spin in ^{72}Se[1,2] is seen in the forward bending of $\Im^{(1)}$ with increase in $\hbar\omega$. Very short lifetimes can be estimated from the doppler-shift properties of the γ rays to be below 0.3 ps, which gives an estimated value for $\beta \approx 0.4$. Recent cranking calculations[7] based on the deformed Woods-Saxon potential indicate large interactions for proton and neutron configurations in ^{72}Se, in agreement with the gradual increase of $\Im^{(1)}$ in Fig. 3. These calculations also show the nuclei to be very γ-soft for the ground state and at low spins ($\hbar\omega \leqslant 0.3$ MeV). Three minima in the Total Routhian Surface (TRS) are seen in Fig. 4: two

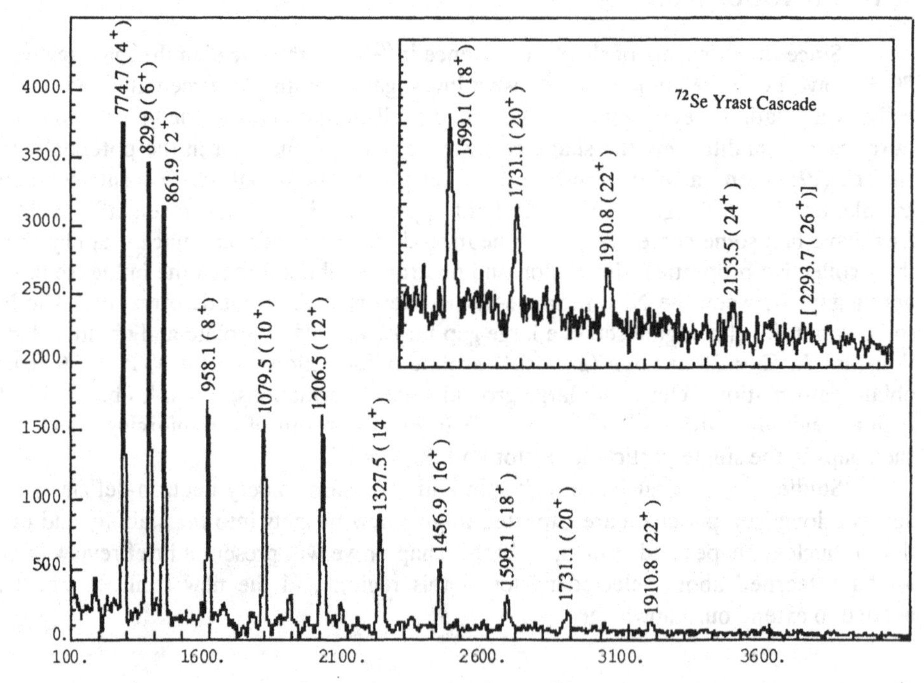

Figure 1. ^{72}Se γ-ray spectrum obtained by summing the coincidence spectra of the yrast gates above 6$^+$.

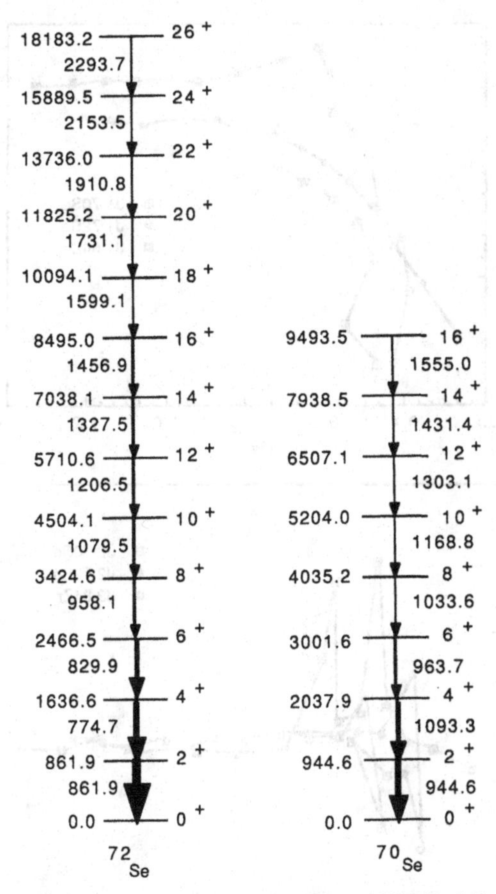

Figure 2. Yrast band in ^{72}Se (Ref. 6) and ^{70}Se (Ref. 7).

oblate minima with $\beta_2 \approx 0.25$ ($\gamma = 60°, -60°$) correspond to the same configuration at $\omega = 0$), and the excited collective prolate minima at $\beta_2 \approx 0.32$ and $\gamma = 0°$. The prolate and oblate minima are well separated by a potential barrier of about 330 keV, reflecting the low-spin coexistence. At higher frequencies for ^{72}Se the prolate minimum stabilizes $\beta_2 \approx 0.4$ and $\gamma = 5°$, and it does not change in the β-γ plane up to $\hbar\omega \geqslant = 1.2$ MeV. For ^{70}Se the calculations indicate a similar deformation pattern. However, for $0 \leqslant \hbar\omega \leqslant 0.4$ MeV, the oblate minimum persists as the dominant structure. At higher frequencies ($\hbar\omega \geqslant 0.5$ MeV) the region above the 8^+ state, the well-deformed configuration, stabilizes at the near prolate shape with $\gamma \leqslant 15°$ ($\beta_2 \approx 0.27$).

At the highest spins in ^{72}Se and ^{84}Zr both $\Im^{(1)}$ and $\Im^{(2)}$ are remarkably constant and approach the rigid-body moment of inertia. For ^{84}Zr calculations of Dudek et al.[9] suggest that neutron and proton pairing correlations should disappear in the highest observed states—the Mottelson-Valatin[10] pairing phase transition. The unusually large and constant moments of inertia in both ^{72}Se and ^{84}Zr are consistent with such a collapse in pairing, as has been proposed earlier to occur in the new islands of strong ground-state deformations in the mass 70 and 100 regions when shell gaps for both N and Z reinforce each other. It is a challenge to established definitively the existence or non-existence of

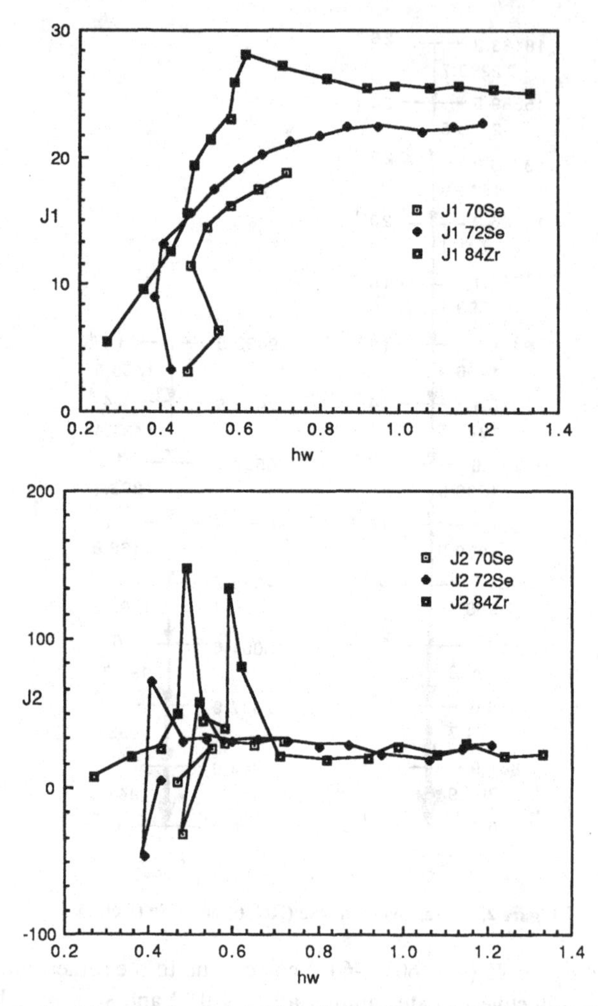

Figure 3. Kinematical and dynamical moments of inertia for ^{72}Se (Ref. 6), ^{70}Se (Ref. 7), and ^{84}Zr (Ref. 8).

such a pairing collapse. (Recent microscopic calculations[4] in a mean-field approach predict several other positive parity bands which are not seen.)

Now, turning our attention to odd-A nuclei, the properties of the levels of ^{69}Se and ^{71}Se provide new insight into the shapes of these nuclei. Detailed calculations of deformed single-particle potentials with the Strutinsky method[11] or even simpler Hartree-Fock calculations with effective interactions[12] predict extra stability for oblate shapes for N, Z - 34,36. Since an odd neutron can provide a sensitive probe of the quadrupole field of the even-even core and the sign of the quadrupole moment can be determined from the sign of the mixing ratio, $\delta = \langle J_i | E2 | J_f \rangle / \langle J_i | M1 | J_f \rangle$ for $\Delta J = 1$ transitions between favored and unfavored bands and the signature splitting of these bands.

Searches for such unfavored bands in ^{71}Se were carried out at the University of Rochester with a Recoil Mass Spectrometer (RMS) and also at the University of Köln. A five-sector neutron detector plus NaI detectors and Ge detectors at target position of the

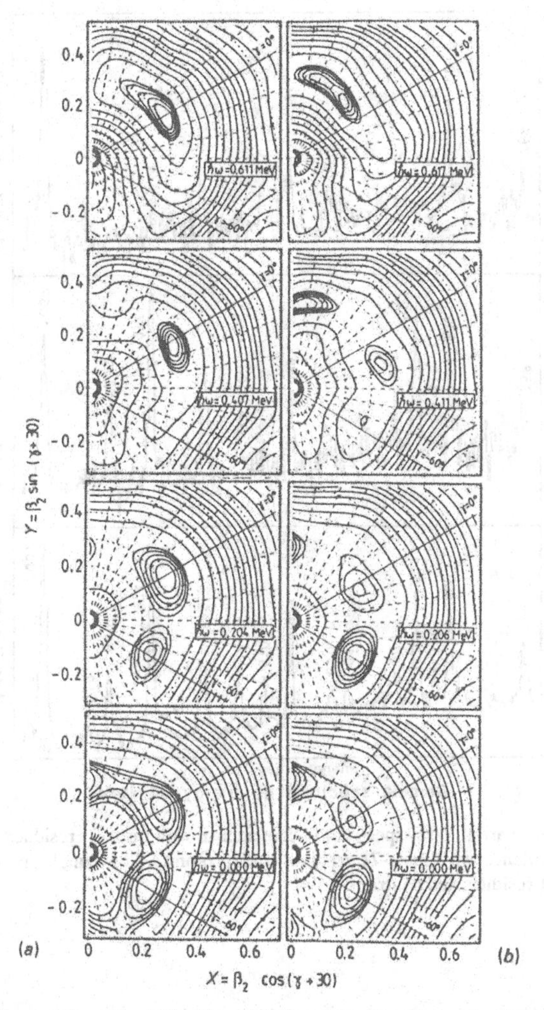

Figure 4. Total routhian surfaces for ^{70}Se b) and ^{72}Se a) in the (β_2 - γ) plane according to the Lund convention.

RMS were used to identify new transitions in ^{71}Se. The following double- or triple-coincidence events were sorted, using ^{16}O on ^{58}Ni: mass-71–γ, mass-71–n-γ and mass-71–p-γ. In Fig. 5 are shown the partial γ-ray spectra in coincidence with the types of events discussed above. One cay clearly see that the 894-keV transition belongs to ^{71}Se, which was not identified in earlier work. The coincidence and $\gamma\gamma(\theta)$ experiments carried out at the University of Köln establish the 894-keV γ ray to be $11/2^+ \rightarrow 9/2^+$ with δ = 1.5 ± 0.3.

The level scheme of ^{71}Se is shown in Fig. 6. The levels to the left in Fig. 6 form a ΔI = 1 sequence based on the $9/2^+$ level. These levels appear to be associated with the oblate 9/2[404] orbital. There is a sizable signature splitting in this ΔI = 1 sequence, which can be reproduced by assuming a triaxial (γ) deformation in the non-adiabatic model. A predominantly oblate deformation is implied by the measured mixing ration of δ = +1.5 ± 0.3 for the $11/2^+ \rightarrow 9/2^+$ transition. For high-Ω $g_{9/2}$ orbitals and a deformation range 0.2 < β < 0.3, $g_K - g_K$ is negative.

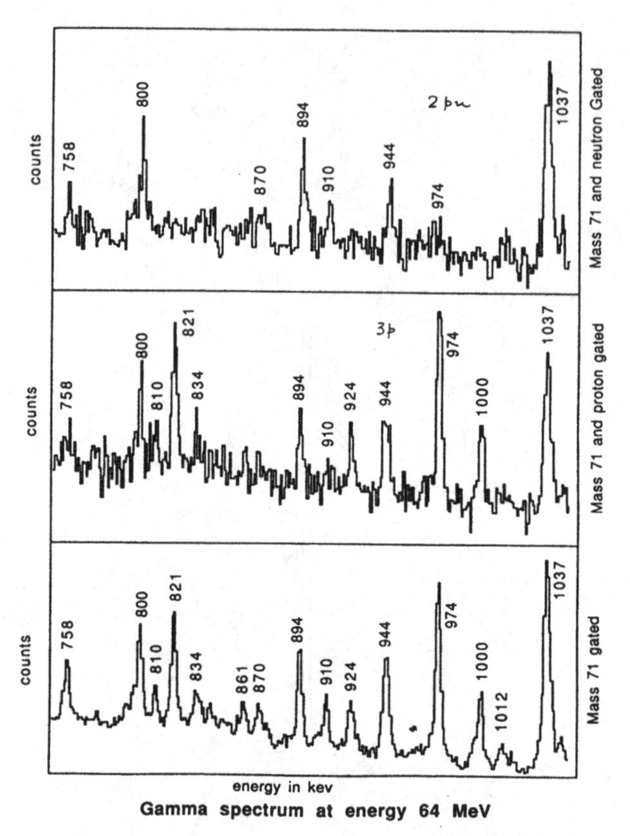

Figure 5. Bottom panel: Partial γ-ray spectrum in coincidence with mass-71 residues. Middle panel: partial γ-ray spectrum in coincidence with mass-71 residues and protons. Top panel: partial γ-ray spectrum in coincidence with mass-71 residues and neutrons.

The relative sign $[(\delta) = \text{sign}\,(g_K - g_K)/Q_0]$ implies a negative value for the intrinsic quadrupole moment Q_0 for a positive δ. Although the mixing ratio indicates an oblate deformation, the energy-level spacings seem to indicate triaxial structure. However, for the case of ^{69}Se, both energy-level spacings and δ for the $11/2^+ \rightarrow 9/2^+$ seem to indicate oblate shape according to the calculations[13] for a deformed-configuration-mixing shell model based on Hartree-Fock states.

18.3 GROUND-STATE DEFORMATIONS AND SHAPE COEXISTENCE IN LIGHT Kr ISOTOPES

Extensive spectroscopy of ^{74,76}Kr nuclei has indicated large ground-state prolate deformation with excited near spherical states.[14] These large ground-state deformations are associated with the shell gaps at N and Z of 38 at $\beta \sim 0.4$. Gamma-ray spectroscopy on ^{75}Kr has indicated[15] that this nucleus has a prolate to somewhat triaxial shape with $\beta \approx 0.37$ and $\gamma \leqslant 15°$. More recent, extensive studies of ^{75}Kr to high spins using the cranked shell model with pairing calculations suggest more precise values of $\beta \approx 0.4$ and $\gamma \leqslant 5°$.

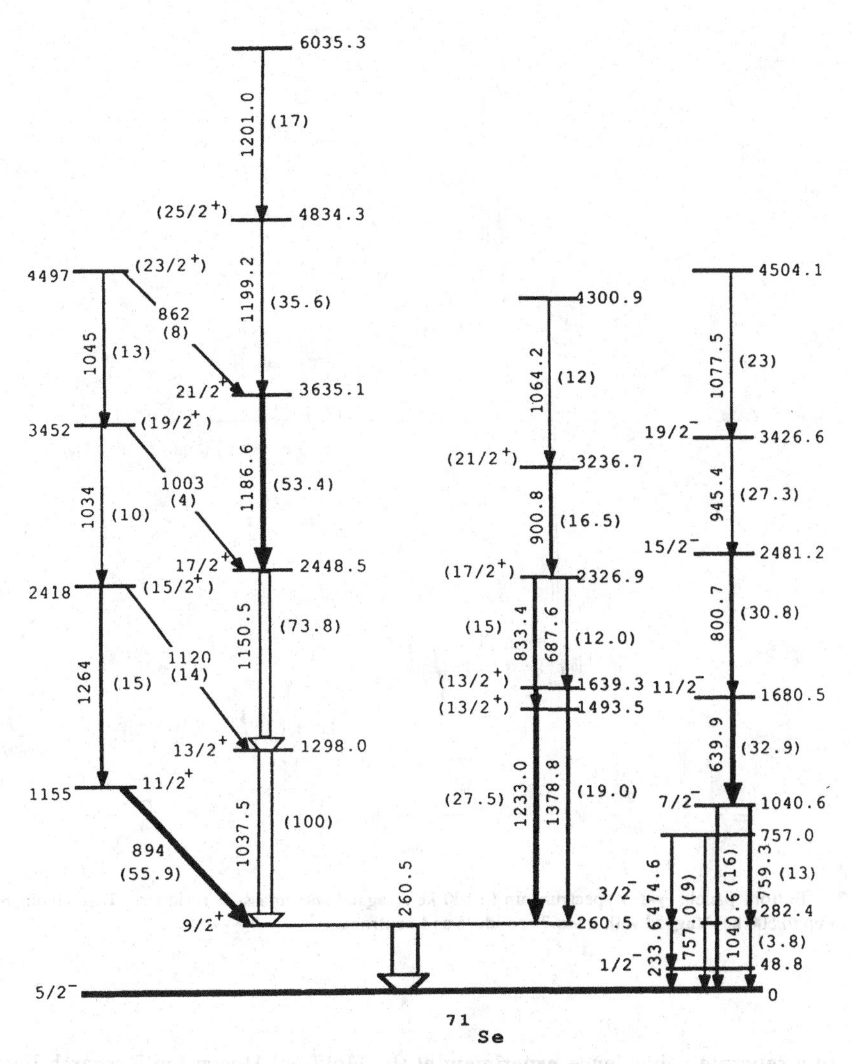

Figure 6. Level scheme of ^{71}Se.

In an effort to gain further insight into the origin and development of deformation, we have sought to identify levels in ^{73}Kr. Furthermore, it is an experimental challenge to identify the transitions in ^{73}Kr because the cross-section is very small. For an unambiguous identification, one has to tag the γ rays (seen with Ge detectors), which are unique to the nucleus of interest. Such a tagging procedure was carried out[16] using the RMS at the University of Rochester. At the target position, in addition to the γ-ray detectors, a neutron detector was used to give additional tagging. For the reaction, 95–MeV ^{35}Cl on ^{40}Ca, the only competing channels for mass 73 ar ^{73}Br (2p) and ^{73}Br (pn). Additional tagging with a neutron detector will unambiguously identify the transitions in 73Br (pn). Additional tagging with a neutron detector will unambiguously identify the transitions in ^{73}Kr. In Fig. 7 are shown the γ-ray spectra up to 590 keV tagged with mass-73 residues and mass-73 residues plus neutrons. Comparison of the top and bottom panels clearly shows that the 125–keV and 142–keV transitions belong to ^{73}Kr.

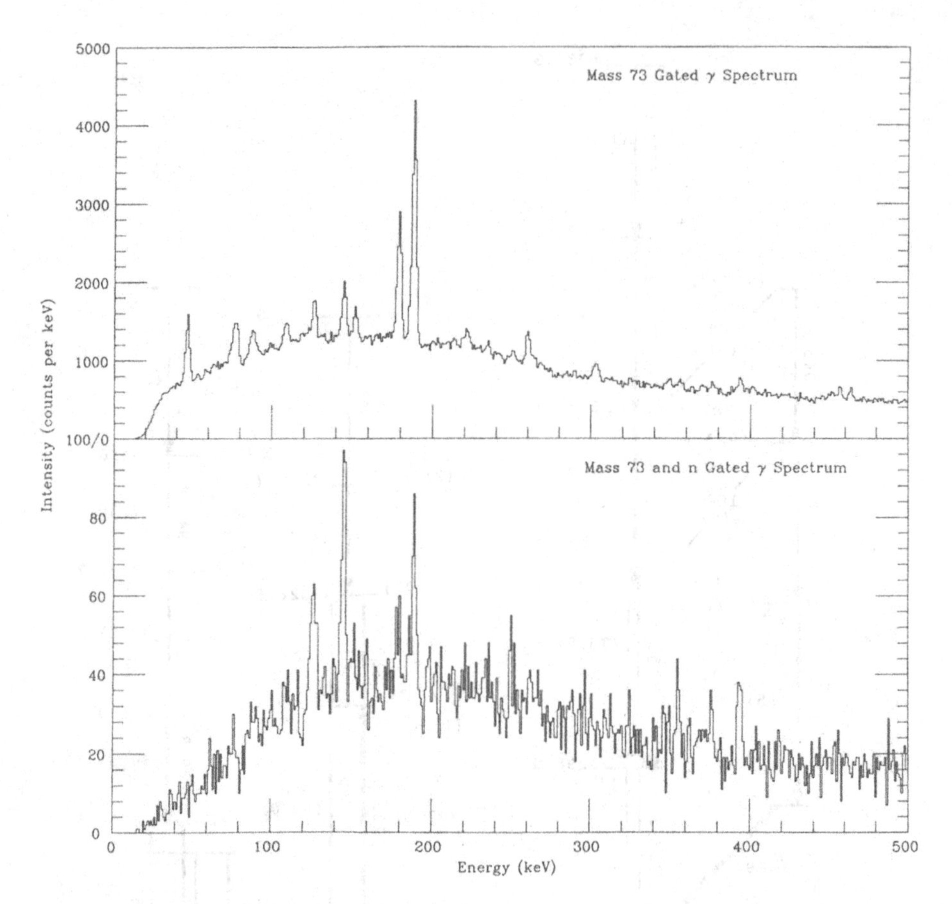

Figure 7. Bottom panel: γ-ray spectrum up to 500 keV tagged with mass-73 residues. Top panel: γ-ray spectrum up to 500 keV tagged with mass-73 residues and neutrons.

In a separate coincidence experiment at the Holifield Heavy Ion Research Facility with twenty compton-suppressed Ge detectors in the close-packed ball, we have been able to establish two bands in ^{73}Kr. These two bands are shown in Fig. 8.

We are in the process of analyzing the data further in order to make spin and parity assignments. In the absence of spin and parity assignments, we can make some comments based on the systematics. Based on the deformed shell model, if the ^{73}Kr nucleus has the same deformation as that of the ^{75}Kr, the ground-state spin should be $3/2^+$ for a prolate shape or $7/2^+$ for an oblate shape (or $3/2^-$ for $\beta = 0.48$). It is very important to measure the ground-state spin of ^{73}Kr. The two bands, one with positive parity and the other with negative parity, are expected to consist of $\Delta J = 1$ transitions, as in ^{75,77}Kr.

18.4 FUTURE EXPERIMENTAL TECHNIQUES

The experiments on ^{71}Se and ^{73}Kr described in this paper are triple-coincidence experiments but with only one γ ray. It is already clear that in the future one has to perform

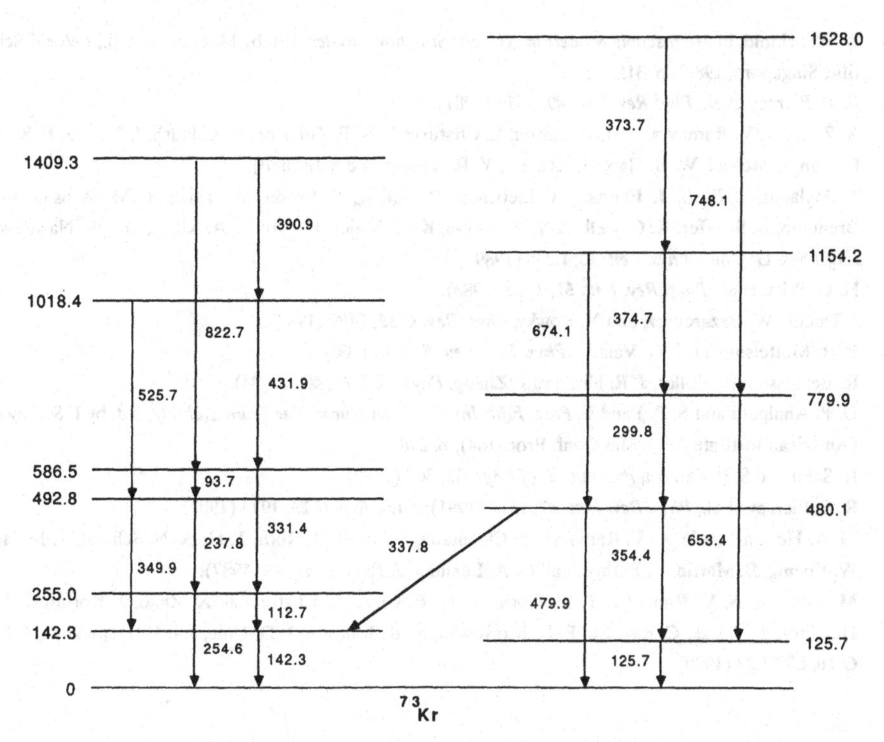

Figure 8. Level scheme of ^{73}Kr.

recoil-mass-γ-γ-coincidence experiments to establish the levels and extract nuclear structure information. The new Fragment Mass Analyzer at Argonne National Laboratory and RMS at Oak Ridge National Laboratory offer large transmissions. In addition, high-rigidity machines will open up inverse reactions. In such reactions kinematic focussing will further increase efficiencies to perform mass-γ-γ-coincidence experiments, and the high energies of the fragment will allow Z identification. Such a high-rigidity RMS is being developed by a Vanderbilt-INEL-ORNL-Tennessee-UNISOR collaboration. With the development of GAMMASPHERE at HHIRF, the combination of such a RMS and *GS* will enable us to trace the structure of nuclei out nearly to the proton drip line.

ACKNOWLEDGMENT

This work was supported in part by U. S. Department of Energy Grant No. DE-FG05-88ER40407.

REFERENCES

1. J. H. Hamilton et al., *Phys. Rev. Lett.* **32**, 239 (1974).
2. A. V. Ramayya et al., *Phys. Rev. C* **12**, 1360 (1975).
3. R. Bengtsson and W. Nazarewicz, *Proc. XIX Winter School on Physics*, Ed. by Z. Stachura, Report IFJ No. 1263 (1984), P. 171.

4. K. W. Schmid, in *Microscopic Models in Nuclear Structure Physics*, Ed. by M. Guidry et al., (World Scientific, Singapore, 1989), p. 315.
5. R. B. Piercey et al., *Phys Rev. Lett.* **47**, 1514 (1981).
6. X. Zhao, A. V. Ramayya, J. H. Hamilton, L. Chaturvedi, N. R. Johnson, C. Baktash, I. Y. Lee, F. K. McGowan, J. McNiel, W. B. Gao, H. Xie, and Y. R. Ying (to be published).
7. T. Mylaeus, J. Bush, J. Eberth, M. Liebchen, R. Sefizig, S. Skoda, W. Teichert, M. Wiosna, P. V. Brentano, K. Schiffer, K. O. Zell, Z. V. Ramayya, K. H. Maier, H. Growe, A. Kluge, and W. Nazarewicz, *J. of Phys. G: Nucl. Phys. Lett.* **15**, L295 (1989).
8. H. G. Price et al., *Phys. Rev. Lett.* **51**, 1842 (1983).
9. J. Dudek, W. Nazarewicz, and N. Rowky, *Phys. Rev. C* **35**, 1489 (1987).
10. B. R. Mottelson and J. G. Valatin, *Phys. Rev. Lett.* **5**, 511 (1960).
11. R. Bengtsson, P. Moller, J. R. Nix, and J. Zhang, *Phys. Scri.* **29**, 402 (1984).
12. D. P. Ahalpara and S. P. Pandya, *Proc. Fifth Int. Conf. on Nuclei Far From Stability*, Ed. by I. S. Towner, (American Institute of Physics Conf. Proc. 164), p. 278.
13. R. Sahu and S. P. Pandya, *Pramana J. of Phys.* **32**, 367 (1989).
14. R. B. Piercey et al., *Phys. Rev. Lett.* **47**, 1514 (1981); *Phys. Rev. C* **25**, 1914 (1982).
15. M. A. Herath-Banda, A. V. Ramayya, L. Cleemann, J. Eberth, J. Roth, T. Heck, N. Schmal, T. Mylaeus, W. Koenig, B. Martin, K. Bethge, and G. A. Leander, *J. Phys. G* **13**, 43 (1987).
16. M. Satteson, A. V. Ramayya, T. M. Cormier, H. E. Gove, L. Chaturvedi, X. Zhao, J. Kormicki, J. H. Hamilton, I. Y. Lee, C. Baktash, F. K. McGowan, N. R. Johnson, J. D. Cole, and H. Dejbakhsh, *J. Phys. G* **16**, L27-L32 (1990).

19. Shape and Structural Changes in Nuclei at High Spin and Temperature

T. L. Khoo, I. Ahmad, R. Holzmann,* R. V. F. Janssens,
E. F. Moore, and F. L. H. Wolfs
Argonne National Laboratory
Argonne, IL 60439

M. W. Drigert
Idaho National Engineering Laboratory
E. G. & G.
Idaho Falls, ID 83415

K. B. Beard, D-Z. Ye, and U. Garg
Department of Physics
University of Notre Dame
Notre Dame, IN 46556

Ph. Benet, P. J. Daly, and Z. Grabowski
Department of Chemistry
Purdue University
West Lafayette, IN 47907

ABSTRACT

Structural changes both along and above the yrast line have been observed in experiments performed with the Argonne-Notre Dame BGO Gamma Facility at ATLAS. Transitions from prolate to oblate shapes have been detected in transitional Dy nuclei. Above the yrast line, quasicontinuum $E2$ transitions reveal clear signatures of structural changes which may be related to phase transitions predicted by theory. Superdeformed bands have been discovered in some of these nuclei, as well as in a new region of superdeformation in ^{191}Hg. The initial population of the superdeformed bands appears to be close to the secondary minimum and may explain why the SD bands, which lie several MeV above the yrast line, do not rapidly tunnel to the normal states.

[The above abstract is the one published in the program booklet for the ACS Symposium. Teng-Lek Khoo presented a paper in Miami Beach on "Superdeformation in the Hg Region," but this paper was not written up for this book.]

*Present address: GSI, Darmstadt, Germany.

Exotic Nuclear Spectroscopy, Edited by W. C. McHarris
Plenum Press, New York, 1990

20. Shell-Model Calculations for Exotic Nuclei

B. A. Brown
National Superconducting Cyclotron Laboratory
and Department of Physics and Astronomy
Michigan State University
East Lansing, MI 48824

E. K. Warburton
Brookhaven National Laboratory
Upton, NY 11973

B. H. Wildenthal
Department of Physics and Astronomy
University of New Mexico
Albuquerque, NM 87131

ABSTRACT

Recent theoretical calculations for the properties (mass, lifetime, size, and decay modes) of light nuclei (A < 50) far from stability will be discussed. The calculations are based on the mixing of many shell-model configurations (up to about 7000) where the hamiltonian has been determined from the binding energies and excitation energies of nuclei near the valley of stability. For the nuclear sizes these wave functions are combined with Hartree-Fock mean-field calculations. The comparison to experiment tests the model assumptions. On the proton-rich side, some of the interesting aspects concern the understanding of the Gamow-Teller strength functions and the isospin forbidden decay modes observed. On the neutron-rich side, we will discuss the large rms radii for the neutron rich Li and Be isotopes, the intruder states seen around A, Z = 32, 12, and the relation of the lifetime and decay modes to the Gamow-Teller strength function in the final nucleus.

20.1 INTRODUCTION

In this paper we review the progress of the shell-model approach to understanding the properties of light exotic nuclei (A < 40). By "shell-model" we mean the consistent and large-scale application of the classic methods discussed, for example, in the book of de-Shalit and Talmi.[1] Modern calculations incorporate as many of the important configurations as possible and make use of realistic effective interactions for the valence nucleons.[2] Properties such as the nuclear densities depend on the mean-field potential, which is usually treated separately from the valence interaction. We will discuss results for radii which are based on a standard Hartree-Fock approach with Skyrme-type interactions.[3]

With the present generation of shell-model codes run on VAX computers, treatment of cases with J-scheme dimensions of up to about 7,000 is routine.[4] At this level we are able to consider the full basis for both protons and neutrons in the $0p$ or $1s0d$ major shells, as well as for protons in the $0p$ shell combined with neutrons in the $1s0d$ shell. For

Exotic Nuclear Spectroscopy, Edited by W. C. McHarris
Plenum Press, New York, 1990

many cases in which protons in the 1*s*0*d* shell are combined with neutrons in the 1*p*0*f* shell, the dimensions are already over 7,000. However, many interesting *sd-pf* cases are possible, and we have considered a few cases in the *sd-pf* model space with dimensions up to about 12,000.[5]

In Table I we summarize the model spaces and interactions used in the discussions. We give the interactions used within the major shells (*p*, *sd*, and *pf*) and the cross-shell interaction for the combined shells (*p-sd* and *sd-pf*). The total interaction in the combined shells may be unified, as in the case of the PG interaction, or it may be composed of three parts representing the components within the two major shell plus the cross terms. The ones of this type discussed here are M-W-MK for the region $2 \leqslant Z \leqslant 8$ and $8 \leqslant N \leqslant 20$, and W-MG-WBMB for $8 \leqslant Z \leqslant 20$ and $20 \leqslant N \leqslant 40$.

All interactions have been adjusted by a least-squares fit to some selection of binding energy and excitation energy data (the fit-data set). This was accomplished by varying the individual two-body matrix elements (TBME), by varying the well-determined linear combinations of two-body matrix elements and keeping the rest fixed at some *G*-matrix values (TBME+G), by varying parameters of some potential (POT), or by varying the Talmi integrals (TI). We note that all of these effective interactions are very close to that expected from microscopic *G*-matrix calculations based on the nucleon-nucleon interaction, but that the small differences from the *G* matrix are very important for detailed spectroscopy.[2,4]

Within the above framework we are able in principle to discuss all observed properties of the exotic nuclei. Many of these properties will now be discussed, roughly in order of increasing complexity. First, the binding energy systematics of neutron rich nuclei are discussed in the framework of $0\hbar\omega$ shell-model configurations (Section 2). "$0\hbar\omega$" means that the nucleons are assumed to fill the lowest available major-shell configurations. The deviation between experiment and theory in this context for nuclei around ^{32}Mg will then be addressed in terms of cross-shell ($n\hbar\omega$) excitations (Section 3). Next we discuss the β decay properties, starting with half-life comparisons and going into some details of the Gamow-Teller strength functions and the subsequent delayed particle emission (Section 4). Finally, we address the question of the large radii and large Coulomb break-up cross-

Table I. Table of Model Spaces and Interactions

Model Space	Interaction	Type	Code	Ref.
0*p* (*p*)	Cohen-Kurath	TBME	CK	6
	Millener	POT	M	7
1*s*0*d* (*sd*)	Wildenthal	TBME+G	W	8
	Brown et al.	POT	SDPOTA	9
1*p*0*f* (*pf*)	McGrory	TBME+G	MG	10
p-sd	Poppelier et al.	TI	PG	11
	Millener-Kurath	POT	MK	12
sd-pf	Warburton et al.	POT	WBMB	13

sections found for the neutron rich Li and Be isotopes Section 5). Summaries will be included in each section.

20.2 BINDING ENERGIES

The first property usually measured for the most exotic nuclei is whether or not they are bound with respect to nucleon decay. When a nucleus is so bound, it can only decay by the weak interaction and will have a half-life on the order of milliseconds or greater. This will be referred to as a "stable" nucleus in the discussion below. A nucleus which does decay by nucleon emission will be referred to as "unstable." If a nucleus is stable, its mass excess can be measured. Even for unstable nuclei, the mass excess of the ground and excited states can be measured by transfer reactions if the appropriate target is available. The only important nucleon-decay channel for the ground states of unstable neutron-rich nuclei is neutron emission. The Q values for these decays can be inferred easily from the binding energy plots shown below.

20.2.A *p* and *p-sd* Model Spaces

The He isotopes were recently studied by the MSU group, where ^{10}He was not observed and thus is most probably unstable.[14] In Fig. 1 we show the experimental binding energies for the ground and excited states of the He isotopes compared to a recent *p*-shell calculation. In this calculation three parameters of a potential model for the interaction and the two single-particle removal energies (SPRE) were fitted to the eight known data, including the binding energy of ^{9}He obtained from a pion double-charge-exchange reaction.[15] The fit is good for the known states and in agreement with the experimental prediction that ^{10}He is unstable. In the calculation, ^{10}He is clearly unbound to two-neutron emission and may or may not be bound to one-neutron emission.

In the extreme *j-j* coupling limit, the He binding energies can be interpreted in the shell model as follows: The $p_{3/2}$ SPRE, as determined by the ^{5}He (ground state) –^{4}He energy difference, and the $p_{1/2}$ SPRE, as determined from the ^{5}He(excited state) –^{4}He energy difference, are both negative (unbound). The ^{6}He–^{4}He binding-energy difference is then two times the $p_{3/2}$ SPRE (which by itself would make ^{6}He unbound) plus the pairing energy in the $(p_{3/2})^2$ configuration. This latter term is positive (attractive) and large enough to make ^{6}He bound. If ^{8}He had a pure $(p_{3/2})^4$ configuration, the effective $p_{3/2}$ SPRE is given by the ^{8}He–^{7}He energy difference. This is seen to be positive as a result of the addition of the residual interaction within the $p_{3/2}$ shell. Similarly, the effective $p_{1/2}$ SPRE given by the ^{9}He–^{8}He energy difference is negative but less so than it was in ^{5}He–^{4}He. Finally, we see that ^{10}He is unstable, because the positive pairing energy for the $(p_{1/2})^2$ configuration is weaker than it was for ^{6}He and not strong enough to make up for the twice the negative $p_{1/2}$ SPRE contribution. The calculations shown in Fig. 1 are more complicated than this extreme *j-j* coupling picture, since they take all *p*-shell configurations into account. However, the dominant components of the wave-functions are those assumed above, and thus the qualitative features of the discussion still hold.

The neutron-rich Li, Be, B, C, and N nuclei have been experimentally studied all the way to the neutron drip line.[16-18] Poppelier et al.[11] have made binding energy comparisons between their *p-sd* PG calculations and experiment. Vieira[19] has made binding energy comparisons between the *p-sd* M-W-MK calculations and experiment. In both cases there are differences of up to 2 MeV between experiment and theory, and because of this there

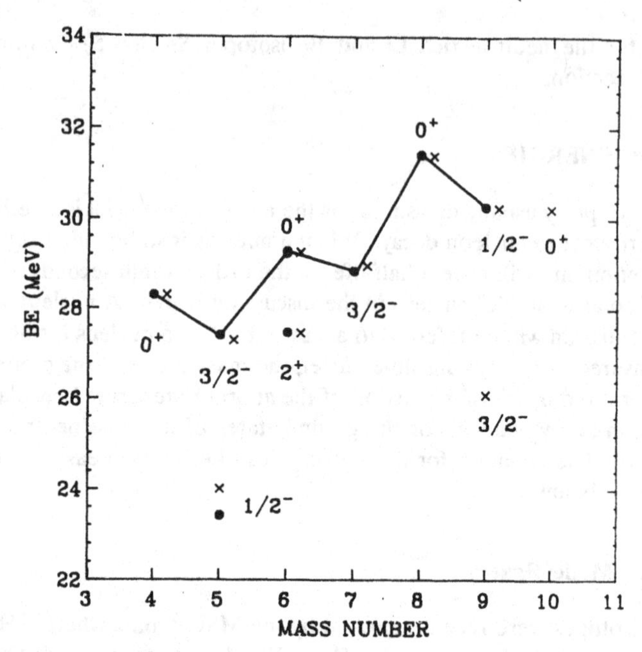

Figure 1. Binding energies of states of the He isotopes.[14] The experimental energies are shown by circles and experimental ground-state binding energies are connected by a line. The calculated energies are shown by the crosses (slightly offset to the right in the case when they are close to the circles).

several cases where the predicted stability property is in disagreement with experiment. The calculations could probably be improved by use of a better effective interaction, but the limit within the model is probably around the 0.5 MeV rms deviation obtained for the well-known p-shell nuclei.[6]

20.2.B *sd* and *sd-pf* Model Spaces

There has been excitement recently about the apparent instability of ^{26}O.[16] We show in Fig. 2 binding energy curves for the O isotopes. The experimental data are compared with three predictions: the global predictions of Moeller and Nix,[20] the W-*sd* calculation up to $N = 20$ plus the W-MG-WBMB-*sd-pf* calculation beyond $N = 20$, and the SD-POTA-*sd* calculation. The shell-model predictions are clearly better than the global-model predictions in this case. The average deviation between the shell-model predictions and experiement is consistent with the average 180-keV rms deviation found for 447 ground and excited states over the entire *sd* shell.[2,8] The global model of Moeller and Nix as well as most other global models[21] predict that ^{26}O is stable, in contradiction to experiment. However, this is not too surprising, given the rather poor agreement for the other O isotopes. It is more surprising that the W-*sd* calculations also predicts that ^{26}O is stable by about 1 MeV. However, the SDPOTA-*sd* calculation predicts ^{26}O to be unstable, but only by 20 keV! This difference is an indication of the rather large model dependence which can exist in the shell-model extrapolations to exotic nuclei. With both W and SDPOTA, ^{27}O and ^{28}O are predicted to be unstable. We now discuss some aspects of the model dependence.

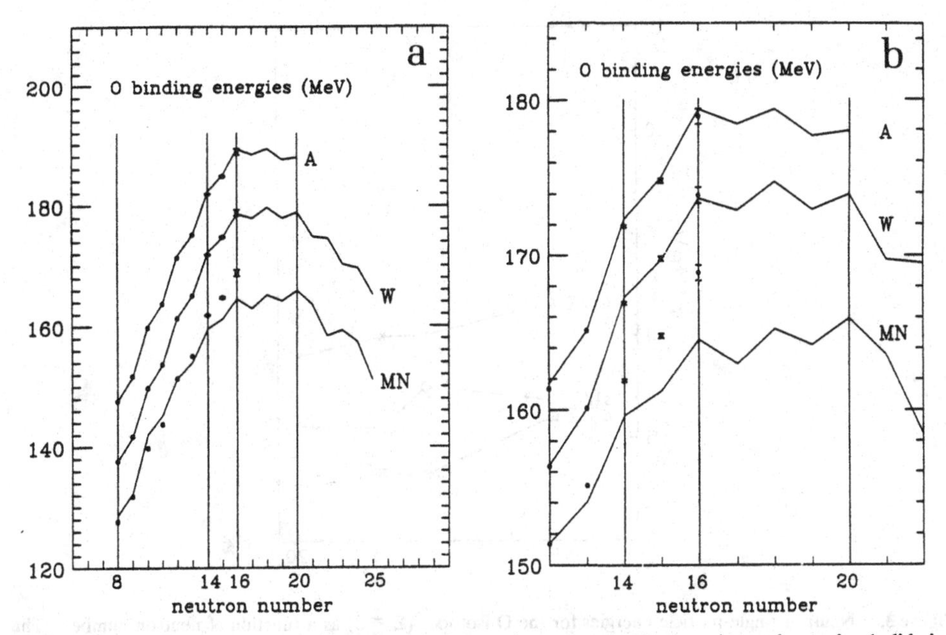

Figure 2. Binding energies of the ground states of the O isotopes. *a*) The experimental energies (solid circles) are compared to three calculations (solid lines) discussed in the text: Moeller-Nix (MN), W-*sd* for $N \leqslant 20$ plus W-MG-WBMB-*sd-pf* for $N \geqslant 20$ (W), and SDPOTA-*sd* (A). For display 10 MeV has been added to the W comparison and 20 MeV has been added to the A comparison. *b*) An expanded version of *a*) showing the region near the stability line. For display 5 MeV has been added to the W comparison and 10 MeV has been added to the A comparison.

The trend of the O binding energies can be understood in the extreme *j-j* coupling limit in much the same way as was discussed for the He isotopes. We show in Fig. 3 the effective neutron single-particle energy (ESPE) as a function of neutron number in this *j-j* coupling limit. (By convention, ESPE = –SPRE, where SPRE is the single-particle removal energy used in the discussion for the He isotopes. The neutron ESPE are seen to be rather constant as a function of neutron number. This leads to a simple qualitative interpretation for the binding energy curve shown in Fig. 1. Between $N = 8$ and 14 the neutrons fill the $d_{5/2}$ orbit, which is bound in ^{16}O by about 4 MeV. This, together with the attactive pairing energy, provides the sharp increase in binding energy observed between $N = 8$ and 14. Between $N = 14$ and 16 the neutrons fill the $s_{1/2}$ orbit, which is less bound than the $d_{5/2}$. Thus one starts to see less increase in the binding energy at this point. Between $N = 16$ and 20 the neutrons fill the $d_{3/2}$ orbit, which has close to zero energy, and the binding energy curve becomes flat in this region. This flatness, of course, makes it difficult to predict exactly which nuclei will be stable, which is what makes this region so interesting from the stability point of view. Beyond $N = 20$ the neutrons must start to go into the *pf*-shell orbits which are unbound. Hence, the binding energy curve decreases beyond this point. This marks the end of where the O isotopes can be studied and also the end of where they need to be understood for astophysical purposes.

At the next level of detail we should take into account the small shifts in the neutron ESPE shown in Fig. 3. These are again due to the interactions within the shells. For exam-

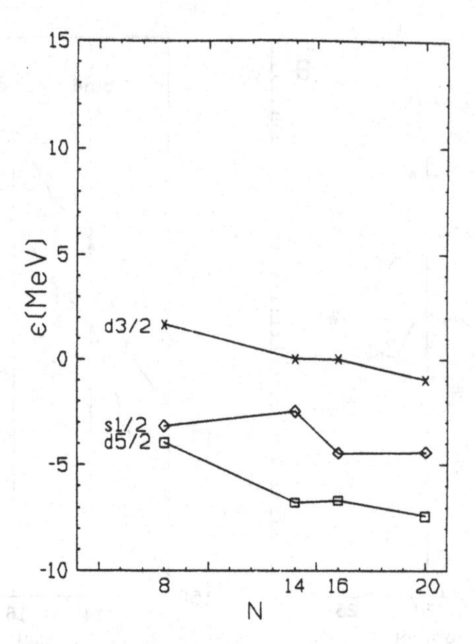

Figure 3. Neutron single-particle energies for the O isotopes ($Z = 8$) as a function of neutron number. The values at $N = 14$ and 16 were obtained in the extreme *j-j* coupling limit.

ple the shift in the $s_{1/2}$ ESPE between ^{16}O and ^{22}O is due the monopole average over two-body matrix elements:

$$\sum_J (2J+1)\langle d_{5/2}, s_{1/2}, J|V|d_{5/2}, s_{1/2}, J\rangle / \sum_J (2J+1)$$

It is important to note that these two-body matrix elements can in principle be obtained from information on excited states in ^{18}O and ^{19}O, etc. Data on excitation energies relevant to these $d_{5/2}$-$s_{1/2}$ two-body matrix elements were included in the 447 fit-data set used to obtain the W and SDPOTA interactions. Hence, the good agreement for the ^{23}O and ^{24}O mass predictions may not be surprising, even though data on these two nuclei were not included in the fit-data set. In contrast, beyond ^{24}O the ESPE depend on the $d_{5/2}$-$d_{3/2}$ two-body matrix elements. Data on excitation energies relevant to these do not exist because they are more highly excited configurations and lie in a large level density of intruder states. Hence, beyond ^{24}O the predictions rely more on assumptions in the calculation, which cannot be tested from previously known data. For the W interaction this is the Kuo-Brown *G* matrix used for the poorly determined linear combinations, and for the SDPOTA interaction this is the particular form of the potential model assumed (a modified surface one-boson-exchange potential).

The actual *sd*-shell calculations shown in Fig. 2 go beyond *j-j* coupling and include all possible *sd*-shell configurations. But again the dominant configurations are those assumed above. These calculations, it should be remembered, are based on zeroth-order perturbation theory together with an assumed constancy of the bare SPE and an assumed

sumed above. These calculations, it should be remembered, are based on zeroth-order perturbation theory together with an assumed constancy of the bare SPE and an assumed simple $(A/18)^{0.3}$ mass dependence of the two-body matrix elements.[2,8] Presumably, nature is more complicated than this. But, given the continued success of the shell-model in correlating essentially all observed data, these complications must get folded into the effective nature of the interactions in a away that may never be fully quantitatively understood.

In contrast to the relative constancy of the neutron ESPE as a function of neutron number, they quickly decrease as a function of proton number, as shown in Fig. 4 for $N = 20$ and $Z = 8$ to 20. By the time one reaches ^{40}Ca, all of the *sd*-shell orbits as well as the *pf*-shell orbits are bound. At $N = 20$ the ESPE of the neutron orbits as a function of N should again be relatively constant, and applying the same qualitative argument as above, we can see that all Ca isotopes out to ^{60}Ca should be stable. The stability in the region between ^{60}Ca and ^{70}Ca depends on exactly what the value of the ESPE for the $g_{9/2}$ orbit is at this point.

Binding-energy comparisons have been made for Z = 9–20, similar to the ones shown in Fig. 2 for O. We show the differences between experiment and the theoretical predictions in Fig. 5. The predictions are based on W-*sd* for $N \leqslant 20$ and W-MG-WBMB-*sd*-*pf* for $N > 20$. New data on the F and Ne isotopes have been included.[22]

We first discuss the stability properties which are not shown in Fig. 5. ^{28}F is predicted to be unstable, and ^{29}F is predicted to be stable by both W and SDPOTA, in agreement with experiement.[16] W-MG-WBMB predicts all F nuclei to be unstable beyond this point. ^{29}Ne is predicted to be unstable by 129 keV with the W interaction and by 4 keV with the SDPOTA interaction. Hence, the disagreement with experiment[16] is not too surprising. ^{30}Ne and ^{32}Ne are predicted to be stable and ^{31}Ne unstable, in agreement with experiment. Beyond these, only ^{34}Ne is predicted to be stable. All remaining nuclei in the *sd*-shell ($N < 20$) for $Z > 10$ are predicted to be stable, in agreement with experiment.

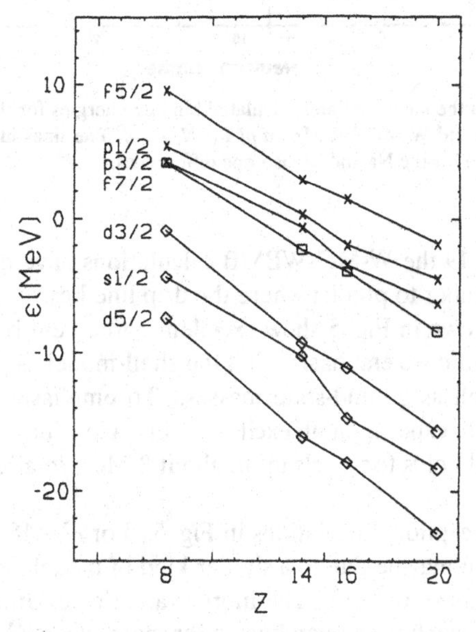

Figure 4. Neutron single-particle energies for the $N = 20$ isotones as a function of proton number. The values at $Z = 14$ and 16 were obtained in the extreme *j-j* coupling limit.

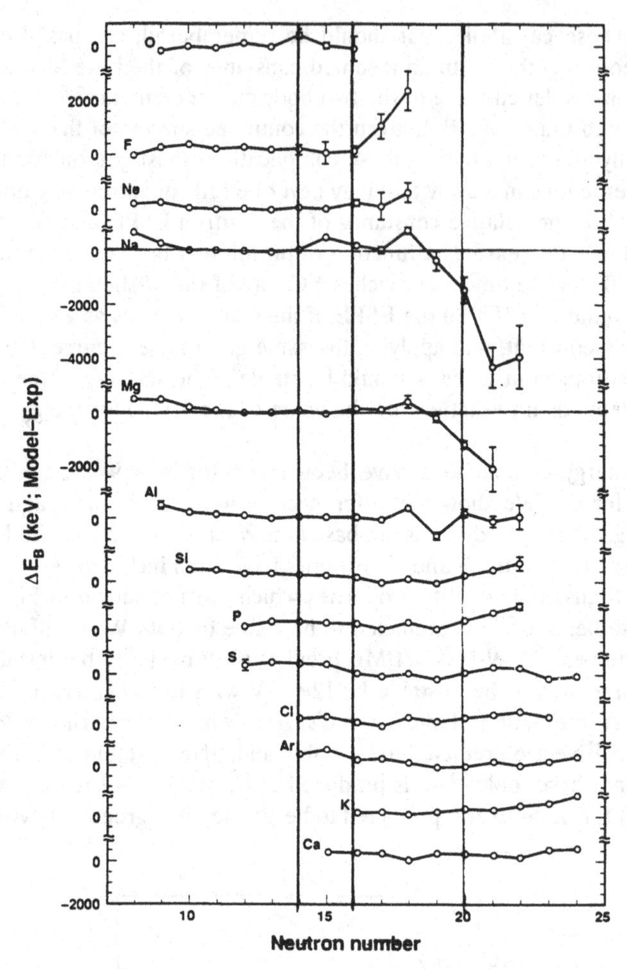

Figure 5. Difference between the measured and calculated binding energies for the *sd* and *sd-pf* regions. The calculation is W-*sd* for $N \leqslant 20$ and W-MG-WBMB-*sd-pf* for $N \geqslant 20$. The lines indicate the semi-magic numbers $N = 14$ and 16 and the zero for the Na and Mg isotope differences.

And for $N > 20$ and $Z > 19$ the W-MG-WBMB calculations have not yet been carried out far enough in neutron number to predict where the drip line lies.

The differences shown in Fig. 5 show excellent agreement between experiment and theory for most cases. Again we emphasize that the shell-model is a successful model for excitation energies, as well as ground-state masses. To emphasize this point we show in Fig. 6, the predicted and experimental excitation energies for ^{27}Al.[23] The quality of agreement shown for ^{27}Al holds for levels up to about 8 MeV in all *sd*-shell nuclei near the valley of stability.

There are some exceptional deviations in Fig. 5. For $N = 18$ and $9 \leqslant Z \leqslant 13$ there is a pronounced glitch, which may be due to a similar kind of model-dependence left in the W interaction, as discussed above for ^{26}O. The more dramatic deviations in the most neutron rich Na and Mg isotopes, which have been known for many years,[24] point to the intruder

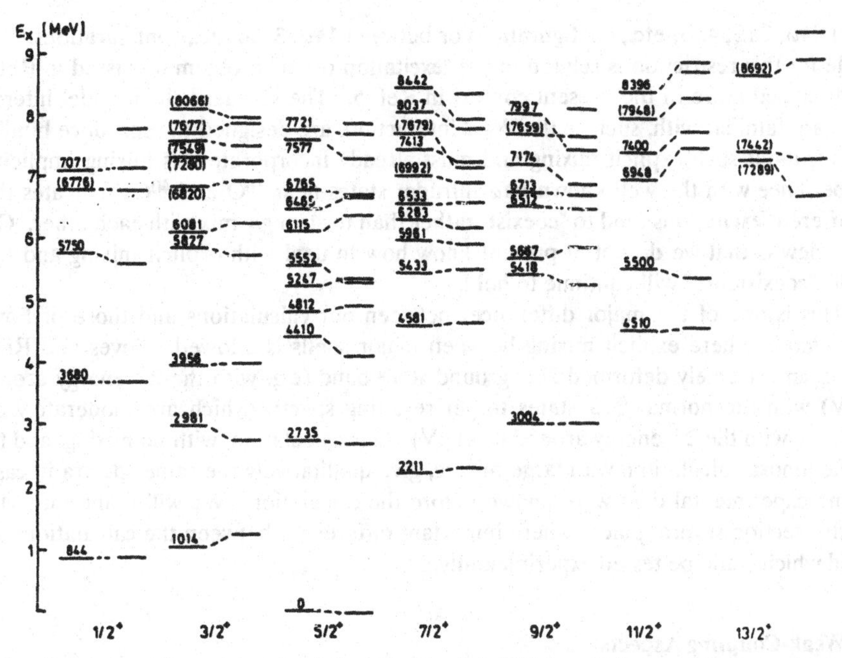

Figure 6. Comparison of experimental and theoretical levels in ^{27}Al from Lickert et al.[23] For each spin, the numbers label the energy in keV for the experimental levels on the left-hand side which are connected by a dashed line to the associated theoretical levels on the right-hand side.

state problem which is discussed in the next section.

The next major step in the *sd*-shell calculations will be to incorporate all of the new data which has appeared since the original fit-data set was put together about ten years ago.

20.3 CROSS-SHELL EXCITATIONS AROUND ^{32}Mg

The anomalies in the binding energies, excitation energies, half-lives, and radii of the most neutron rich Na and Mg isotopes relative to the type of $0\hbar\omega$ calculations described above have been known for a long time.[24] The situation has been referred to as the "collapse of the conventional shell model."[25] Several studies over the past ten years have indicated that the problem is due to low-lying intruder states from the *pf* shell.[26-28] We report here on a new set of systematic calculations with the W-MG-WBMB interactions for these intruder states.[5] These new calculations incorporate all of the *sd* and *pf* orbits and represent the most ambitious calculations to date. An advantage (and disadvantage) of our calculations is that there are no adjustable parameters. The W-*sd* and MG-*pf* parts of the interactions were already well established. The WBMB cross-shell interaction[13] was obtained from a fit of the Millener-Kurath potential-model[12] parameters and some fine tuning of some individual two-body matrix to reproduce the excitation energies of $1p$-$1h$ states in ^{40}Ca and ^{40}K.

An important aspect of our calculations is that we do not allow for explicit mixing between $0\hbar\omega$, $2\hbar\omega$, $4\hbar\omega$, etc., configurations or between $1\hbar\omega$, $3\hbar\omega$, etc., configurations. The rationale for this restriction is related to the "excitation-order" problem discussed in Ref. 2 and is discussed more in the present context in Ref. 5. The standard shell-model interactions we are familiar with, such as the W-*sd* interaction, are designed to reproduce binding energies without such explicit mixing and must already incorporate this mixing implicitly. Our experience with the well-known $n\hbar\omega$ intruder states near ^{16}O and ^{40}Ca indicates that these different excitations tend to "coexist" rather than to strongly mix with each other. Our point of view is that we do not at present know how to deal with explicit mixing and that hopefully "coexistence" will continue to hold.

This is one of the major differences between our calculations and those of Poves and Retamosa, where explicit mixing between major shells is allowed. Poves and Retamosa mix an extremely deformed $2\hbar\omega$ ground state band (e.g. with the 2^+ energy around 200 keV) with the normal $0\hbar\omega$ states to get resulting spectra which are moderately deformed (e.g. with the 2^+ energy around 800 keV). Our calculation, with no mixing, and the Poves-Retamosa calculation, with large mixing, give qualitatively the same spectra in cases where the experimental data were known before the calculation. We will point out at the end of this section several places where important differences between the calculations are expected which could be tested experimentally.

20.3.A Weak-Coupling Aspects

In our study of the ^{32}Mg region, we first calculated all of the $0\hbar\omega$ and $1\hbar\omega$ spectra which had dimensions less than about 12,000. Comparison to experiment for the $Z \geqslant 14$ region was good.[29,30] However, as discussed in the last section, the calculated $0\hbar\omega$ binding energies in the region of ^{32}Mg are markedly smaller than experiment. We next looked at $n\hbar\omega$ excitations for $N = 20$ and $Z < 14$. Full-space $1\hbar\omega$ and $2\hbar\omega$ *sd-pf* calculations are possible for $^{28,29}F$ and ^{30}Ne. From these studies we determined that the proton excitations from *sd* to *pf* could safely be ignored. The lowest state was found to correspond to a $2\hbar\omega$ neutron excitation. This state had an excitation energy relative to the lowest $0\hbar\omega$ state in these three nuclei of 2.96, 1.34, and −0.79 MeV, respectively. The full-space calculations for most of the nuclei of interest around ^{32}Mg have dimensions which are too large to handle. We investigated several truncations which have promise for future calculations. However, the most interesting and useful result found in these investigations is that a weak-coupling model can be used to relate the excitation energy of the $n\hbar\omega$ configurations to the calculated $0\hbar\omega$ binding energies in nuclei with neighboring neutron numbers.

Many examples of the weak-coupling model have been discussed for nuclei around ^{16}O and ^{40}Ca where the residual interaction between particles and holes has an isospin-dependence.[31] In our case the form of the particle-hole interaction is much simpler since only neutrons are presumed to be excited. The motivation for the weak-coupling formula is as follows. Consider, as an example, the $N = 19$ nucleus ^{31}Mg, which has a $0\hbar\omega$ configuration of one neutron hole ($1h$) in the *sd* shell relative to the $N = 20$ neutron closed shell. The $n\hbar\omega$ configuration then has the form, $(\nu pf)^n - (\nu sd)^{-n+1})$. In analogy with the expression for one-particle one-hole states in closed shell nuclei, the energy of this $n\hbar\omega$ configuration relative to the $N = 20$ closed shell is

$$E(19, n\hbar\omega) - E(20) = \varepsilon(n \text{ particles}) - \varepsilon[(n+1) \text{ holes}] - n(n+1)C,$$

where $E(N)$ is the interaction energy of the $0\hbar\omega$ configurations (the negative of the binding energy) for the neutron number N (at a fixed Z), and where the last term takes into account the average particle-hole interaction C between neutrons. The particle energy is given by

$$\varepsilon\,(n \text{ particles}) = E(20+n) - E(20),$$

and the whole energy is given by

$$\varepsilon\,[(n+1) \text{ holes}] = E(20) - E(19-n).$$

By combining these results we obtain

$$\begin{aligned} Ex^{wc}(19, n\hbar\omega) &= E(19, n\hbar\omega) - E(19) \\ &= E(19-n) + E(20+n) - E(20) - E(19) - n(n+1)C. \end{aligned}$$

For $N \leqslant 20$ this generalizes to

$$Ex^{wc}(N, n\hbar\omega) = E(N-n) + E(20+n) - E(20) - E(N) - n(20-N+n)C,$$

and for $N > 20$,

$$Ex^{wc}(N, n\hbar\omega) = E(N+n) + E(20-n) - E(20) - E(N) - n(N-20+n)C.$$

The weak-coupling model should be a valid approximation if the dominant interactions are taken into account by the $E(N)$ and the residual interaction C is weak. This appears to be true in our case. Empirically, we obtain $C = 143$ keV for $Z = 8$ and $C = 240 + 10(Z-9)$ keV otherwise. There is the additional assumption that the energies and structure of the multi-particle and multi-hole configurations do not change when they are coupled together.

The excitation energies for the $2\hbar\omega$ and $1\hbar\omega$ configurations based on the full space and weak-coupling models are compared in Fig. 7 and Fig. 8, respectively. In addition, in these figures we give the excitation energies in the weak-coupling model, where the full basis calculation is not possible, and in Fig. 8 we also give the $3\hbar\omega$ excitation in the weak-coupling model. Examination of Figs. 7 and 8 shows that the region of nuclei shown in Fig. 9 do not have $0\hbar\omega$ ground states, and we will refer to this as the "island of inversion."

20.3.B Mechanisms For Lowering the $n\hbar\omega$ Excitations

We now discuss the mechanisms by which the $n\hbar\omega$ excitations are lowered in energy to create the island of inversion. The first aspect is the single-particle energy gap between the *sd* and *pf* shell. The neutron effective single-particle energies (ESPE) calculated with our interaction are given in Fig. 4. For ^{28}O ($Z = 8$) and ^{40}Ca ($Z = 20$), the results are exact and given by $E(N=21) - E(N=20)$ for the *pf* particle states and by $E(N=20) - E(N = 19)$ for the *sd*-hole state.

For ^{34}Si($Z = 14$) and ^{36}S($Z = 16$) the results shown in Fig. 4 are approximate and obtained under the assumption that the protons have a sub-shell closure of $(d_{5/2})^6$ and $(d_{5/2})^6(s_{1/2})^2$, respectively. Our full calculations for ^{34}Si and ^{36}S include all possible *sd*-shell proton configurations.

The gap between *sd* and *pf* shells shows a moderate decrease from $E_{gap} = 7239$

Z/N	18	19	20	21	22	23
15	^{33}P +4839:W	^{34}P +3781:W	^{35}P +2698:W	^{36}P +2811:W	^{37}P *D(1/2)* **18823*	^{38}P *D(4^-)* *149590*
14	^{32}Si +4134:W	^{33}Si +2716:W	^{34}Si +1816:W +1554:T	^{35}Si +2063:W	^{36}Si +2261:W	^{37}Si *D(7/2)* *169606*
13	^{31}Al +3396:W	^{32}Al +2214:W	^{33}Al +854:W	^{34}Al +786:W	^{35}Al *D(5/2)* **46116*	^{36}Al *D(4^-)* *149590*
12	^{30}Mg +2104:W	^{31}Mg +780:W	^{32}Mg − 926:W −966:T	^{33}Mg −1090:W	^{34}Mg − 685:W	^{35}Mg *D(3/2)* **62787*
11	^{29}Na +2317:W	^{30}Na +776:W	^{31}Na −502:W −764:T	^{32}Na −1295:W	^{33}Na − 427:W	^{34}Na *D(4^-)** **44735*
10	^{28}Ne +1843:W	^{29}Ne +609:W	^{30}Ne − 698:W − 788:F	^{31}Ne − 891:W	^{32}Ne − 128:W	^{33}Ne + 480:W
9	^{27}F +3273:W	^{28}F +2110:W +2251:F	^{29}F +1286:W +1338:F	^{30}F +1444:W	^{31}F +1329:W	^{32}F +1557:W
8	^{26}O +4246:W +4495:F	^{27}O +3550:W +3568:F	^{28}O +3038:W +2956:F	^{29}O +2870:W +2904:F	^{30}O +2945:W +2905:F	^{31}O +2768:W

Figure 7. Neutron-rich nuclei in the $A = 32$ region. The listed numbers are the predicted excitation energies (in keV) of the $2\hbar\omega$ ground state relative to the $0\hbar\omega$ ground state, $Ex(2\hbar\omega)$. The symbols denote weak-coupling, W, the full W-MG-WBMB-*sd-pf* model space, F, and a truncated version of this model space, T (not discussed in the text). We label the excitation energies beyond our computational resources with asterisks and, for these cases, show the J dimension of the $N+2$ isotope necessary to provide the weak-coupling (W) prediction. The magic numbers $Z = 8$ and $N = 20$ are emphasized with double lines.

Z/N	18	19	20	21	22	23
15	^{33}P +5720:W *****:F *****:3	^{34}P +1673:W *****:F *****:3	^{35}P +4637:W +4290:F +8531:3	^{36}P +1025:W *****:F *****:3	^{37}P +4750:W *****:F *****:3	^{38}P *****:W *****:F *****:3
14	^{32}Si +5647:W *****:F *****:3	^{33}Si + 935:W *****:F *****:3	^{34}Si +4747:W +4227:F +7710:3	^{35}Si + 881:W *****:F +4042:3	^{36}Si +4994:W *****:F *****:3	^{37}Si +1079:W *****:F *****:3
13	^{31}Al +4505:W *****:F *****:3	^{32}Al + 959:W *****:F *****:3	^{33}Al +3145:W +3086:F +5291:3	^{34}Al − 105:W *****:F *****:3	^{35}Al +3077:W *****:F *****:3	^{36}Al *****:W *****:F *****:3
12	^{30}Mg +4678:W *****:F +8024:3	^{31}Mg + 8:W *****:F +1948:3	^{32}Mg +2972:W +2484:F +3588:3	^{33}Mg − 934:W − 729:F + 87:3	^{34}Mg +2808:W *****:F *****:3	^{35}Mg − 529:W *****:F *****;3
11	^{29}Na +3682:W *****:F +6142:3	^{30}Na + 306:W − 87:F +1830:3	^{31}Na +2404:W +2083:F +2387:3	^{32}Na − 808:W −1347:F + 43:3	^{33}Na +1611:W *****:F *****:3	^{34}Na + 60:W *****:F *****:3
10	^{28}Ne +4426:W +3939:F +7108:3	^{29}Ne + 211:W + 123:F +1861:3	^{30}Ne +3119:W +2666:F +3535:3	^{31}Ne − 909:W −1192:F + 270:3	^{32}Ne +2926:W *****:F +4713:3	^{33}Ne − 146:W *****:F +2371:3
9	^{27}F +3538:W +3821:F +8567:3	^{28}F + 662:W + 687:F +4093:3	^{29}F +2714:W +2721:F +4982:3	^{30}F + 624:W + 630:F +2777:3	^{31}F +2872:W +2780:F +5253:3	^{32}F + 509:W *****:F +2887:3
8	^{26}O + 5480:W + 4808:F +11488:3	^{27}O +2110:W +2015:F +6188:3	^{28}O +4972:W +4255:F +8354:3	^{29}O + 928:W + 833:F +4383:3	^{30}O +4804:W +4020:F +8084:3	^{31}O +1003:W + 878:F +4408:3

Figure 8. Neutron-rich nuclei in the $A=32$ region. The top two listed numbers are the $Ex(1\hbar\omega)$ in keV for weak-coupling, W, and the full W-MG-WBMB-*sd-pf* model space, F. The bottom number is $Ex(3\hbar\omega)$ for weak coupling. We label the excitation energies beyond our computational resources with asterisks. The magic numbers $Z=8$ and $N=20$ are emphasized with double lines.

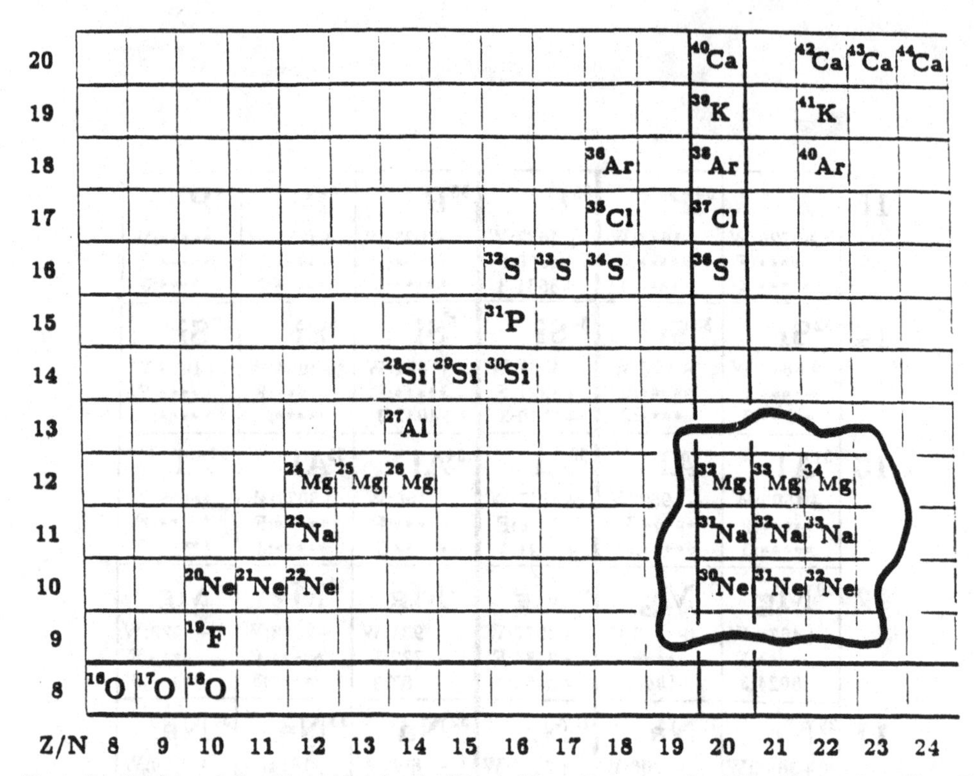

Figure 9. Partial periodic table highlighting the "island of inversion" centered at ^{32}Na. The magic numbers $Z = 8$ and $N = 20$ are emphasized with darker lines.

keV in ^{40}Ca to 5115 keV in ^{28}O. This is in contrast to the results of Storm et al.,[28] which predict the gap actually becomes negative for ^{28}O. We see from Fig. 5 that our $0\hbar\omega$ mass predictions (and hence our ESPE) are in good agreement with experimental $N = 19$, 20 and 21 binding energies from $Z = 20$ to $Z = 13$. Thus, we have some confidence in the correctness of our extrapolation from $Z = 12$ to $Z = 8$. We conclude that the decrease in the *sd-pf* gap contributes to, but is not the primary cause of, the inversion.

For proton excitations the gap between $d_{5/2}$ and $f_{7/2}$ is very large for ^{28}O (14.7 MeV), and the *sd-pf* excitations for protons can be ignored. Of course, the *sd-pf* gap is the same for protons and neutrons in ^{16}O and ^{40}Ca, and both excitations must therefore be considered in these regions. However, we find that the low-lying $2\hbar\omega$ states in ^{38}Ar are already dominated by the neutron excitations.

The excitation energies of the $1\hbar\omega$ neutron $1p$-$1h$ states in the $N = 20$ isotones (see Fig. 8) are close to the gap energy E_{gap} in ^{28}O, ^{34}Si, and ^{35}P. The lowering of the $1\hbar\omega$ excitation to 2404 keV in ^{31}Na and its increase back up to 4747 keV in ^{34}Si is an aspect of the correlation energy (but on a smaller scale), which will now be discussed in more detail for the $2\hbar\omega$ excitations in $N = 20$ and the $1\hbar\omega$ excitations in $N = 19$ and 21.

First we consider the $2\hbar\omega$ excitations for $N = 20$. It can be seen from Fig. 7 that the excitation energy of the $2\hbar\omega$ neutron excitations is always much lower than two times the ESPE gap. Both the neutron-neutron interaction energy E_{nn} and the proton-neutron interaction energy E_{pn} contribute to this effect. Since the mononpole interaction is taken into account by the changes in the ESPE, E_{nn} is primarily due to the residual "pairing" interaction. E_{nn} can be estimated from the calculated excitation energy of the 2*p*-2*h* state in ^{28}O, with

$$E_{nn} = Ex - 2x(E_{gap}) = 3038 - 2x(5115) \text{ keV} = -7192 \text{ keV}.$$

(with the weak coupling estimate for *Ex*). Since E_{nn} depends only on the neutron configurations, we expect E_{nn} to be approximately independent of *Z*. Thus, the proton-neutron contribution E_{pn} as a function of *Z* can be estimated from

$$E_{pn}(Z) = Ex(Z) - 2x(E_{gap}) - E_{nn}, \text{ with } E_{gap} = 5115 \text{ keV and } E_{nn} = -7192 \text{ keV}.$$

For $Z = 9$, 10, 11, 12, 13, and 14 we thus obtain $E_{pn} = -1752$, 3736, −3540, −3954, −2184, and −1222 keV, respectively. These turn out to be qualitatively similar to the "correlation" energy discussed by Poves and Retamosa.[27] However, a quantitative comparison cannot be made, since the correlation energy is not itself quantitatively defined.

In summary, we find that there are three mechanisms which combine to give the inversion of $2\hbar\omega$ relative to $0\hbar\omega$: the (small) reduction in the ESPE gap, the pairing energy E_{nn}, and the proton-neutron interaction energy E_{pn}. The only one of which has a strong *Z* dependence is E_{pn}.

For all $N = Z$ nuclei in the *sd* shell, there are low-lying (1.4-2.2 MeV) collective 2^+ states. When both protons and neutrons fill the beginning of a major shell, there are strong ^{4}He-type correlations which lead to well-defined prolate deformations, as in the case of ^{20}Ne and ^{24}Mg in the *sd* shell. The prolateness of ^{20}Ne and ^{24}Mg is reinforced by the small energy gap between the $d_{5/2}$ and $s_{1/2}$ orbits in the lower part of the *sd* shell. In $N = Z$ nuclei in the middle to end of a major shell, there is a competition between prolate and oblate deformations, as in the case of ^{28}Si, ^{32}S, and ^{36}Ar in the *sd* shell. For these nuclei the type of collectivity is particularly sensitive to the spacing of the single-particle orbits in the major shell.

For the $2\hbar\omega$ configurations in $N = 20$, we note that the $f_{7/2}$ and $p_{3/2}$ orbits are close for all *Z* values (see Fig. 4). However, the situation for the two-proton configuration is quite different. At $N = 20$ and $Z = 8$ the $d_{5/2}$ and $s_{1/2}$ proton ESPE are close in energy (see Fig. 10), and this reinforces the collectivity of the (neutron-pf)^2 (proton-sd)^2 configuration near ^{28}O. Hence, E_{pn} is large for ^{30}Ne. However, the gap between the $d_{5/2}$ and $s_{1/2}$ proton ESPE increases to about 6 MeV for ^{34}Si. In this case the $d_{5/2}$ protons tend to form a closed shell, and the energy of the $0\hbar\omega$ 2^+ state in ^{34}Si is very high (4.9 MeV). Thus, in the weak-coupling model the contribution to E_{pn} is greatly reduced and would be zero in the limit of a $(d_{5/2})^6$ proton closed-shell configuration (since only the monopole term can contribute in this case). This is the reason for the strong *Z* dependence in E_{pn} and for the end of the island of inversion in ^{34}Si. As a consequence of the large $d_{5/2}$-$s_{1/2}$ proton splitting in ^{34}Si, the $d_{5/2}$ proton truncation assumed by Poves and Retamosa may be partly useful for nuclei just below ^{34}Si (with some effective interaction changes), but closer to ^{28}O the explicit contribution from the proton $s_{1/2}$ orbit must be taken into account, and for the Gamow-Teller decay properties the $d_{3/2}$ orbit will always be important.

The E_{nn} contribution to the $1\hbar\omega$ excitation energy in the $N = 19$ nuclei can be estimated from ^{27}O, as

$$E_{nn} = Ex - E_{gap} = 2110 - 5115 \text{ keV} = -2077 \text{ keV}.$$

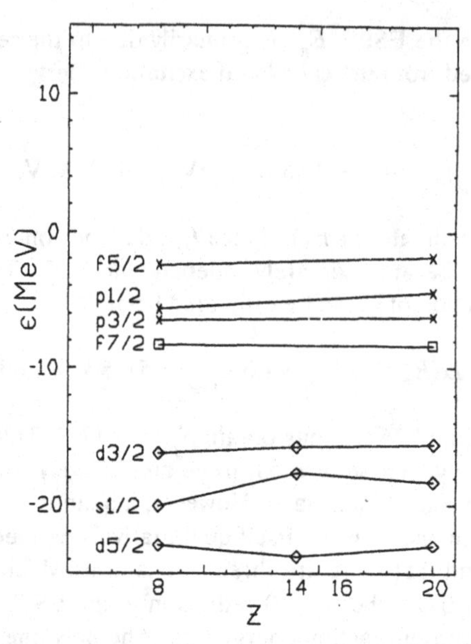

Figure 10. Proton single-particle energies for the $N = 20$ isotones as a function of proton number. The values at $Z = 14$ and 16 were obtained in the extreme j-j coupling limit.

E_{pn} as a function of Z is then about −1448, −1899, −1804, −2102, −1151, and −1175 keV for ^{28}F, ^{29}Ne, ^{30}Na, ^{31}Mg, ^{32}Al, and ^{33}Si, respectively. The results for the $1\hbar\omega$ states of the $N=21$ nuclei are E_{nn} = −4187 and E_{pn}(Z) = −304, −1837, −1736, −1862, −1033, and −47 for ^{30}F, ^{31}Ne, ^{32}Na, ^{33}Mg, ^{34}Al and ^{35}Si, respectively. The lowering of these states is due to the ^{3}He(^{3}H)-type correlation energy, and the relative magnitudes of these $E_{pn}(1\hbar\omega)$ relative to the $E_{pn}(2\hbar\omega)$ due to ^{4}He-type correlations discussed at the beginning of this section are reasonable. The inversion of the $2\hbar\omega$ excitations relative to $0\hbar\omega$ appears when the $0\hbar\omega$ is itself not very collective. The reason why there is not an inversion of the $3\hbar\omega$ relative to $1\hbar\omega$ in $N = 19$ and 21 is probably because the $1\hbar\omega$ are themselves already fairly collective. For ^{31}Na we do find an inversion of $3\hbar\omega$ relative to $1\hbar\omega$.

It is well known that $n\hbar\omega$ spectra often lie very low in excitation energy for the reasons discussed above. The low-lying $4\hbar\omega$ $4p$-$4h$ state in ^{16}O and its explanation in terms of a weak-coupling model[32] is perhaps the most famous example. In fact, if the p-sd single-particle gap were only about 1 MeV smaller than what it actually is, this would probably be the ground state of ^{16}O, and the island of inversion near ^{31}Na would not be so unique. The same can be said of ^{40}Ca. But, by the time we reach $A = 80$, the nominal shell closure of the pf shell is lost by the lowering of the $g_{9/2}$ orbit, and beyond this point it is well known that magic numbers are no longer those of the major harmonic-oscillator shells. The mechanisms behind our island of inverion are also responsible for the low-lying intruder states in heavy nuclei.[33] This suggests obvious applications of the weak-coupling approximation to these heavy nuclei. Relationships to the deformed shell-model approach have also been pointed out.[34] In addition, we note that there is a similar intruder state problem in the region around ^{12}Be, where the ground state of ^{11}Be has a well-known $1\hbar\omega$ $1/2^{+}$ configuration.

20.3.C Comments on the Comparisons to Experiment

In Ref. 5 we discuss many examples where the qualitative aspects of our weak-coupling calculations agree with experiment. Here we give a few examples and some comments on places where more experimental data are needed and/or where more calculations are needed:

a) The low-excitation energy of both the $1\hbar\omega$ (Fig. 8) and $2\hbar\omega$ (Fig.7) configurations in ^{31}Mg are consistent with the recent data which shows a large level density at low excitation energy.[35] However, we are not able at present to calculate the precise order of these configurations or the order of the states within a given configuration. We note that ^{31}Al has experimental β-decay properties, as well as low-lying levels, which agree with the *sd* calculation, but that ^{31}Mg has experimental β-decay properties which are in complete disagreement with the *sd* calculations.[36] This indicates that the ground state of ^{31}Mg is not the expected $3/2^+$ *sd* configuration, but is a $1\hbar\omega$ or $2\hbar\omega$ configuration. More experiments and calculations are needed to sort out the possibilities.

b) In ^{34}Si, the 2^+ state seen experimentally at 3.33 MeV is close to where we would expect our 2^+ $2\hbar\omega$ state. The negative-parity states near 4.3 MeV are close to where we predict the $1\hbar\omega$ excitations to start (see Fig. 8) and are also consistent with the β decay of ^{34}Al.[30] The Poves-Retamosa calculations are also in reasonable agreement for these known states.[37] However, we predict a 0^+ state from the $2\hbar\omega$ configuration around 1.5 MeV (see Fig.7), whereas the Poves-Retamosa calculation puts this excited 0^+ state near 4 MeV. This state has not yet been observed experimentally.

c) In ^{32}Mg the known 2^+ state at 0.88 MeV is close to where we expect the first excited state of the $2\hbar\omega$ ground-state band. Because the $1\hbar\omega$ and $2\hbar\omega$ configurations in ^{32}Na are close in energy (and both below the $0\hbar\omega$ configuration), the states around 3.0 MeV in ^{32}Mg seen in ^{32}Na β decay have several interpretations in our model. If ^{32}Na has a $2\hbar\omega$ ground state, then it would have a negative parity and would decay to the negative-parity states expected around 3 MeV in ^{32}Mg. If ^{32}Na has a $1\hbar\omega$ ground state, we predict that this would be 0^+ and then it would decay to 1^+ states in ^{32}Mg, which could also lie around 3 MeV (we have not yet made calculations for these 1^+ states). The excitation energy in ^{32}Mg of our predicted 0^+ state from the $0\hbar\omega$ configuration is around 1 MeV (see Fig.7). In contrast, the Poves-Retamosa calculations predict a 4^+ state at about this excitation energy. More experimental information is needed to test these predictions.

d) The most serious disagreement with our present calculations is the underbinding of ^{32}Na and ^{33}Na. Even though our intruder states comes below the $0\hbar\omega$ configurations (see Figs. 7 and 8) for these nuclei, it is not enough to explain the magnitude of the discrepancy indicated in Fig. 5. It would be very useful to have new experimental data for the masses of the neutron-rich Ne and Na isotopes. Continued disagreement with experiment may indicate, for example, that our extrapolated *pf* single-particle energies are wrong.

e) For the secondary-beam experiments which may eventually be possible, we would like to suggest that one-nucleon transfer experiments will provide the most valuable spectroscopic information for improving the calculations. Theoretical extrapolations of the single-particle energies in nuclei far from stability are difficult and uncertain. They are also the most important single ingredient for the understanding of more detailed aspects of the structure.

f) Calculations are needed for the β decay properties of the intruder states. In particular, the unusual half-life and decay properties[36] of ^{31}Mg must be understood.

20.4 β-DECAY PROPERTIES

Beta-decay and electromagnetic properties provide exacting tests of the wavefunctions. For the nuclei close to stability, the $0\hbar\omega$ shell-model calculations have been well tested. In the *sd* shell essentially all data have been compared for Gamow-Teller β decay,[38] $M1$ γ decay and μ moments,[39,40] and $E2$ γ decay and Q moments.[2,41] With simple effective operators, the comparison between experiment and theory is excellent. For the half-life and Gamow-Teller strength functions discussed below, the most important aspect of the effective operator from the known *sd*-shell data is the overall quenching factor of about 0.60 in the B(GT) values.[40] (That is, the experimental B(GT) values are 60% of that expected theoretically.) There are some orbit and mass dependences found and expected in the effective Gamow-Teller operator.[40] In particular, one does not expect so much quenching for the *p*-shell nuclei. However, for uniformity we include this overall quenching factor of 0.60 in all the comparisons for exotic nuclei discussed below.

20.4.A Neutron-Rich *sd* Model Space

In conjunction with our study of individual Gamow-Teller matrix elements near the $N = Z$ line,[38] we calculated the half-lives and final-state branching ratios for all neutron rich *sd*-shell nuclei.[36] The half-life comparison is shown in Fig. 11 for an older set of data.[36,42] The new half-lives measured since we published these predictions[43-46] are in reasonable agreement with the calculations. The major disagreement in the half-life and decay of ^{31}Mg noted in Ref. 38 still stands and was discussed in terms of the intruder states in the last section.

Several new data have appeared on the details of the final-state γ-decay spectra seen in these β decay. The decay of ^{22}O has been studied and compared with the *sd*-shell

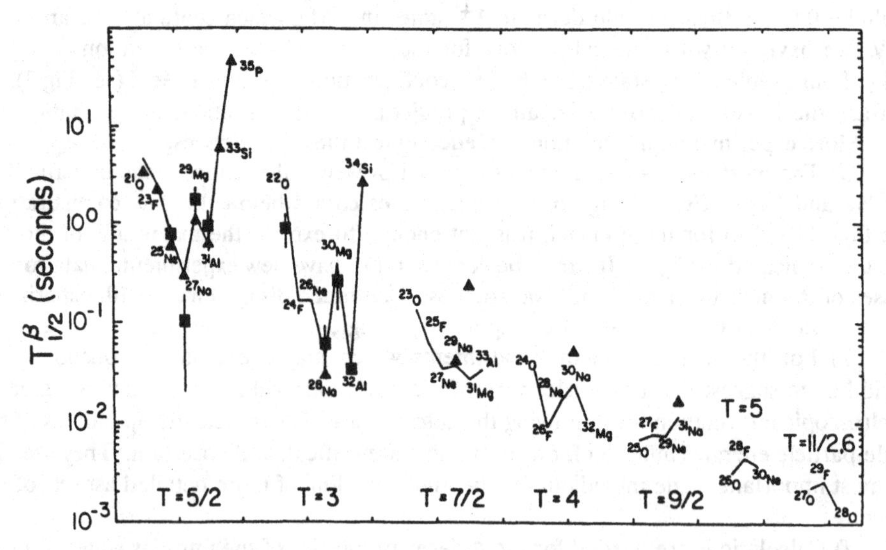

Figure 11. Comparison of half-lives calculated for the Gamow-Teller β decay of *sd*-shell nuclei with five or more excess neutrons with experimental data.[36] Experimental data from Ref. 42 are shown as squares and other experimental data summarized in Ref. 36 are shown as triangles. The calculated values are connected by the solid lines. Newer data from Ref. 43-46 not shown are in reasonable agreement with the calculations.

calculations.[43] These data when combined with the transfer data[47] provide a spectrum for ^{22}F which is in excellent agreement with the calculations. [The apparent disagreement between excitations energies obtained from transfer reactions[48] and β decay has been attributed to the doublet structure of the ^{22}F ground state[43,47]]. For the β branching, it is particularly interesting to note that the decay to the lowest 1^+ state is predicted to be very weak. It is also weak in the experiment but perhaps not as much so as predicted. In the extreme *j-j* coupling picture we would expect the lowest 1^+ states to have a $d_{5/2}$ proton-particle neutron-hole configuration and to have a reasonably large strength. Its counterpart in the *pf* shell is well known as the lowest 1^+ state in ^{48}Sc and is strong in the ^{48}Ca(*p, n*) reaction.[49] In contrast, the simple particle-hole state in ^{22}F is the second state at 2.37 MeV in the calculation, and the lowest 1^+ state has a completely different nature. This state turns out to have a strong overlap with the ^{20}O + dueteron configuration.

The recent detailed study of the ^{29}Na (Ref. 50) and ^{30}Na (Ref. 51) β decays provides much new information on the Gamow-Teller strength functions and the final state spectra, which are in reasonable agreement with the *sd*-shell calculations. The quenching observed for ^{29}Na is consistent with the 0.6 factor discussed above, but the quenching observed for ^{30}Na is about a factor of two larger (that is, the theory has to be multiplied by a factor of about 0.3 rather than 0.6). Again, this may relate to the intruder state problem.[51] Beta-decay calculations including the intruder states for these nuclei have not yet been carried out. Alternatively, it may indicate that the Gamow-Teller strength function is systematically moved to a higher energy compared to the calculation.

Qualitative aspects of the calculated β-delayed-neutron probabilities can be inferred from the tables in Ref. 36. (In some cases there is the possibility that not enough excited states were calculated for the tabulations of Ref. 36. However, in the following examples this is not the case.) For example, a *Pn* value of 1.9% for ^{23}O can be obtained by assuming that all β decays to levels above the neutron-decay threshold in ^{23}F of 7.55 MeV will neutron decay. (The two-neutron decay threshold of 12.74 MeV is above the β decay Q value of 11.63 MeV.) This is in disagreement with a recent experimental result of $Pn = 31 \pm 7\%$.[46] Similarly, for ^{24}O the calculated *Pn* of 15.3%, obtained with a neutron decay threshold of (3.9±0.2) MeV for ^{24}F,[22] is in disgreement with the experimental value of 58±12%.[46] (Since a calculated level at 4.1 MeV in ^{24}F accounts for 10% of this *Pn* value, the *Pn* would be reduced to 5.3% if this level turned out to be below the neutron-decay threshold.) It is not clear to us why these disagreements are so large.

20.4.B Neutron-Rich *p-sd* and *sd-pf* Model Spaces

Calculations and comparison to experiment for the half-lives and branching ratios have also been made for neutron-rich *p-sd* nuclei[45,52–55] and *sd-pf* nuclei.[29] The parent nuclei are assumed to have $0\hbar\omega$ ground-state configurations and then the GT leads to $1\hbar\omega$ excited-state configuration in the daughter nucleus. The decay to the lower-lying $0\hbar\omega$ states in the daughter nuclei is first-forbidden. Except in unusual circumstances, the first-forbidden branchings are negligibly small in these two mass regions[55] and have been ignored in some calculations.[52,53]

For both the *p-sd* and *sd-pf* regions, the calculated half-lives are within about a factor of two of those measured. In particular, as we mentioned in Ref. 53, the experimental half-lives for ^{11}Li and ^{12}Be are about a factor of two larger than those calculated, giving some hint of intruder-state mixing. As was mentioned above for the binding energies in the region, the calculations could probably be improved with a better effective interaction. It would also be interesting to investigate the intruder state problem in the region around ^{12}Be. In the future we plan to extract the delayed neutron probabilities from the Gamow-Teller strength functions calculated in the *p-sd* region in order to compare with experiment.[45,46,54] In the *sd-pf* region, in addition to the theoretical work mentioned above which needs to be done, it is imporant for the element synthesis aspects of astrophysics to extend the β-decay calculations to the region of $16 \leqslant Z \leqslant 20$ and $22 \leqslant N \leqslant 30$.

20.4.C Proton-Rich *sd* Model Space

Data on the proton-rich side of the sd shell provides complimentary information to that obtained on the neutron-rich side. Because of the coulomb interaction, the stable nuclei do not extend out so far in *N–Z* as on the neutron rich side, but β-decay *Q* values are larger allowing for study of the Gamow-Teller strength function up to and just above the isobaric analogue state. The proton-rich stable nuclei do not extend far enough in *N–Z* to run into the intruder state problem observed on the neutron-rich side. Recent experiments have probably reached the limit of stability on the proton-rich side[56,57] and delayed proton spectra have been or are currently being analyzed for most of these.[56,58]

The half-lives of the proton-rich nuclei are in excellent agreement with theory.[59,60] Experimental determination of the complete Gamow-Teller strength functions requires a complex analysis of all delayed gammas and protons. Several of these cases for $T_z = -1$ and $-3/2$ were included in the analysis of Ref. 38. In the few more recent cases,[61] those of ^{32}Ar and ^{33}Ar where such a complete analysis has been carried out, the agreement with the *sd*-shell theory is good.

It turns out that the *Q* value for these β decays allows for measurable transitions to final states only up to about the middle of the Gamow-Teller strength function in the Cl nuclei. Thus, the comparison between experiment and theory is sensitive to the energy of the Gamow-Teller resonance. It is difficult to distinguish between energy shift effects and quenching when only part of the Gamow-Teller strength function can be measured. This comment applies generally to all β^+ and β^- decays discussed here (see the comment above concerning the ^{30}Na decay.)

In many cases there is experimental information only on the delayed protons. One thing one learns is the energy of the isobaric analogue state. This can then be combined with the masses of the other members of the isobaric multiplet to test and study the isobaric-mass-multiplet equation, (IMME). Recently isospin nonconservation (INC) has been put into the *sd*-shell calculations, with the result that one can calculate the coefficients of the IMME and compare to experiment. The comparison is good expecialy for the "c" coefficients.[60,62]

The delayed-proton spectra are complicated. The protons often go to intermediate states which themselves can proton decay, so that one has to sum over a large number of intermediate states. Furthermore, the delayed protons from the isobaric analogue state are all isospin forbidden so that one must use an INC interaction to carry out the calculation. All of these calculations are now possible but time-consuming. Such calculations for the decay of ^{28}S were in reasonable agreement with experiment.[60] For the proton rich *p* and *p-sd* regions there can also be β delayed ^{3}He and ^{4}He emission, which are complicated because cluster overlaps have to be considered. However, we have made such calculations for the decay[63] of ^{9}C, which are not inconsistent with the experimenal data.

An interesting aspect in the combination of neutron and proton rich data is observation of mirror asymmetry in Gamow-Teller transitions. These provide limits on "second-class currents" in the weak interaction, after corrections for the standard INC nuclear structure effects are taken into account.[64]

20.5 LARGE RADII AND COULOMB BREAK-UP CROSS-SECTIONS

One of the first reaction studies with exotic beams was to measure the reaction cross-sections on light stable targets.[65,66] The relatively large reaction cross-section found for the most neutron-rich nuclei has been interpreted in terms of an extended matter (neutron) density for these nuclei,[65] which is not easy to understand in terms of conventional Hartree-Fock calculations.[67] This has led to a number of less conventional models, such as the dineutron model,[68] being proposed. We were thus led to examine more carefully the assumptions which go into the more standard calculations.[69]

We carried out a new set of calculations for the total cross-sections with the following assumptions:

a) The radial wavefunctions were obtained from standard Hartree-Fock calculations using the SGII Skyrme-type interaction.[3]

b) The orbit occupancies were constrained by the type of $0\hbar\omega$ shell-model calculations described above (except for the ground state of ^{11}Be for which a $1\hbar\omega$ shell-model space was used).

c) The binding energy of the last neutron orbit was constrained to reproduce either: *i*) the one-neutron separation energy in the case of a closed neutron shell (e.g., ^{11}Li and ^{12}Be) or a single valence neutron (e.g., ^{11}Be), or *ii*) one-half of the two-neutron separation energy in the case where two-neutrons are in the valence shell (e.g., ^{14}Be.) These binding energies were constrained by multiplying the central part of the Hartree-Fock potential by a constant.

d) A Glauber-type model using a zero-range and a finite range nucleon-nucleon interaction with cross-section parameters of σ_{nn} = 40 m*b* was then used to calculate the cross-sections. The results are compared to experiment in Fig. 12.

The general trend of the experiment is well reproduced by the calculation. The key ingredient in getting the cross-section trend correct is in fixing the binding energy of the valence neutron as described in *c*), above. The potential normalizations of 0.82 to 0.86 which were required are not unrealistic. They reflect the uncertainties introduced from reducing the many-body problem to a mean-field problem and by using a Skyrme-type interaction adjusted to properties of nuclei near the valley of stability. Of course, it would be better to have a more fundamental approach. But we believe that our calculations reflect the essential physics of the situation.

More recently, the reaction cross-section experiments were extended to heavy stable targets.[70] The *Z* and *A* dependence of the total and partial cross-sections were then used to infer the amount due to coulomb excitation and subsequent breakup. The inferred cross-sections for this coulomb-excitation process turned out to be surprisingly large and has led one to speculate on the existence of a low-lying "soft-giant dipole state."[70] It is important to study what is expected from conventional nuclei structure models, and we will make a few comments in this regard on ^{11}Be and ^{11}Li.

The structure of ^{11}Be is thought to be relatively well understood. The $1/2^+$ ground state is, in fact, a $1\hbar\omega$ intruder state of the type discussed above for the *sd-pf* region. The only other bound state is the $1/2^-$ state at 0.32 MeV, which is associated with the $0\hbar\omega$ *p*-shell configuration. The low-lying nature of the $1/2^+$ state can be qualitatively understood in terms of the weak-coupling model discussed above or in terms of the Nilsson model.

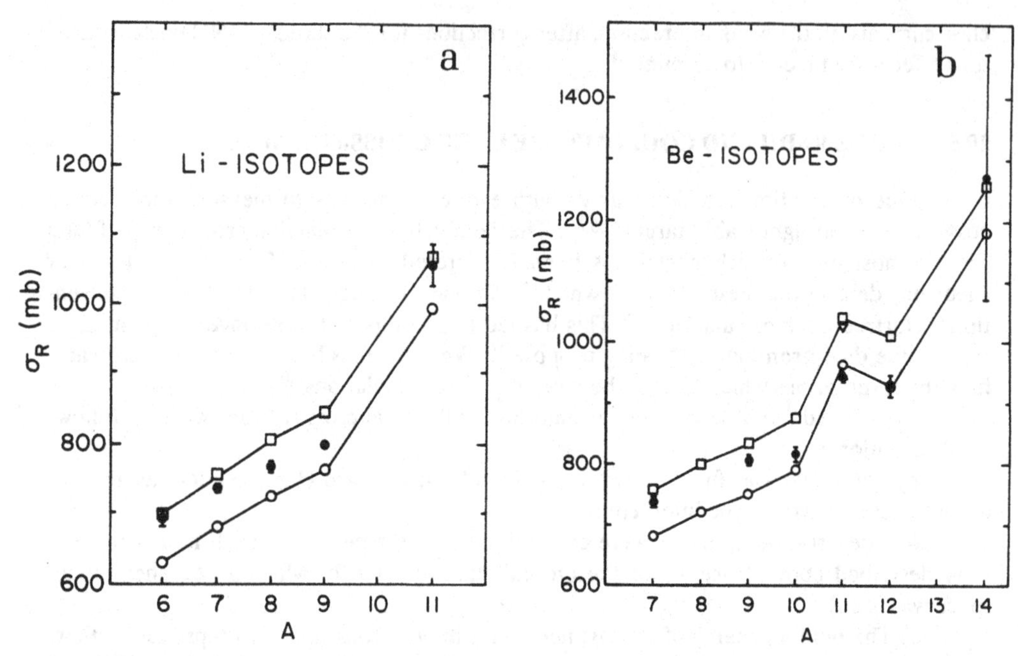

Figure 12. *a*) The reaction cross-sections for the Li isotopes on a ^{12}C target.[69] The solid line connecting the boxes (circles) shows the theoretical cross-sections obtained with the finite-range interaction (zero-range interaction) version of the Glauber-model calculations. The nucleon-nucleon cross-section parameter is taken as 40 mb. The data are taken from Ref. 65. *b*) Same as in *a*) but for the Be isotopes.

The low-lying spectrum of ^{11}Be is also easy to reproduce in a $0\hbar\omega$ plus $1\hbar\omega$ shell-model calculation.[11,71] The $1/2^+$ to $1/2^-$ $E1$ transition, with a $B(E1) = 0.12\ e^2\text{fm}^2 = 0.36$ wu, is one of the strongest known between bound states.[71] The theoretical value is very sensitive to the radial wave-functions used. But the experimental $B(E1)$ value can be reproduced if the radial wavefunctions are generated from potentials that reproduce the experimental separation energies.[71] In the calculation, the only other E1 transition of the type $1\hbar\omega$ to $0\hbar\omega$ is to an unbound $3/2^-$ state at 3 MeV in excitation. The calculated $1/2^+ \rightarrow 3/2^-$ $B(E1)$ is again about 0.13 $e^2\text{fm}^2$. However, this is over an order of magnitude too small to account for the inferred experimental coulomb excitation cross-section on Pb of 0.6 to 1.0 *b*n.[72] The remaining $E1$ strength lies in the $1\hbar\omega$ (ground state) $\rightarrow$ $2\hbar\omega$ transitions, and preliminary shell-model calculations in this space put most of the dipole strength above 10 MeV in excitation, which is much too high in energy to account for the experimental cross-section.

Bertsch and Foxwell have recently examined the case of ^{11}Li in the framework of the RPA model.[73] The method used takes into account the continuum aspects of the problem[74] which is important in describing the $E1$ strength down to the neutron decay threshold. Unlike ^{11}Be, the experimental magnetic moment and $3/2^-$ spin value confirm the expected valence $p_{3/2}$ proton configuration[75] for this nucleus. However, the neutron structure of the ^{11}Li ground state may not be well understood. The neutrons may have a $0\hbar\omega$ closed-shell configuration with the valence nucleons in the p shell (as was assumed above for the total-reaction cross-section calculations), or they may have some $(sd)^2$ configuration from the intruder state, or something in between. (A measurement of the quadrupole moment would be sensitive to the deformation and hence more sensitive to the neutron configuration.) With the same assumptions that went into the total cross-section calculations described above (a $p_{1/2}$ valence configuration with a 1.0 MeV binding energy), the cross-section for calculated for ^{11}Li on Pb was 0.19 b — much smaller than the value of 0.9±0.1 b inferred from experiment.[70] In the extreme case where the valence neutrons were assumed to be in the $s_{1/2}$ orbit with a small separation energy of 0.19 MeV, a cross-section of 0.54 b was obtained — still smaller than experiment.

Thus, in summary, it seems that the large coulomb-excitation cross-sections inferred from experiment are difficult to understand in a conventional approach. There is, however, some model dependence in subtracting the nuclear part of the interaction to obtain the coulomb excitation cross-section and this should be examined more carefully. A consistent calculation should be able to reproduce the total and partial cross-sections as well as the momentum distributions of the fragments,[76] and there is more work to be done.

ACKNOWLEDGMENT

This work was supported in part by the U. S. National Science Foundation grant number PHY 87-14432.

REFERENCES

1. A. deShalit and I. Talmi, Nuclear Shell Theory, (Academic Press, 1963).
2. B. A. Brown and B. H. Wildenthal, Ann. *Rev. of Nucl. and Part. Sci.* **38**, 29 (1988).
3. Nguyen van Giai and H. Sagawa, *Phys. Lett.* **106B**, 379 (1981).
4. B. A. Brown, *Workshop on Microscopic Models in Nuclear Structure Physics*, (World Scientific, 1989), pp. 337-355.
5. E. K. Warburton, J. A. Becker, and B. A. Brown, *Phys. Rev. C* **41**, 1147 (1990).
6. S. Cohen and D. Kurath, *Nucl. Phys.* **73**, 1 (1965), Nucl. Phys. **A101**, 1 (1967).
7. D. J. Millener, private communication.
8. B. H. Wildenthal, *Progress in Particle and Nuclear Physics 11*, Ed. by D. H. Wilkinson (Pergamon, Oxford, 1984), p.5
9. B. A. Brown, W. A. Richter, R. E. Julies, and B. H. Wildenthal, *Ann. Phys.* **182**, 191 (1988).
10. J. B. McGrory, *Phys. Rev. C* **8**, 693 (1973).
11. N. Poppelier, L. D. Wood, and P. W. M. Glaudemans, *Phys. Lett.* **B157**, 120 (1985); N. Poppelier, Thesis, Univ. of Utrect, 1989.
12. D. J. Millener and D. Kurath, *Nucl. Phys.* **A255**, 315 (1975).
13. E. K. Warburton, J. A. Becker, D. J. Millener, and B. A. Brown, BNL Report 40890 (1987).
14. J. Stevenson et al., *Phys. Rev. C* **37**, 2220 (1988).
15. K. K. Seth et al., *Phys. Rev. Lett.* **58**, 1930 (1987).
16. D. Guillemaud-Mueller et al., *Phys. Rev. C* **41**, 937 (1990).
17. M. Langevin et al., *Phys. Lett.* **150B**, 71 (1985); F. Poughoen et al., *Europhys. Lett. 2*, 505 (1986).
18. C. Detraz and D. J. Vieira, *Ann. Rev. of Nucl. Part. Phys.* **39**, 407 (1989).
19. D. J. Vieira and J. M. Wouters, *AIP Conference Proceedings* **164**, 1 (1988).
20. P. Moeller and J. J. Nix, *At. Data Nucl. Data Tables* **39**, 231 (1988).
21. A. H. Wapstra, G. Audi, and R. Hoekstra, *At. Data Nucl. Data Tables* **39**, 281 (1988).
22. D. J. Vieira et al., *Phys. Rev. Lett.* **57**, 3253 (1986); A. Gillibert et al., *Phys. Lett.* **192B**, 39 (1987).
23. M. Lickert et al., *Z. Phys.* A331, 409 (1988).
24. C. Thibault et al., *Phys. Rev. C* **12**, 644 (1975); C. Detraz et al., *Nucl. Phys.* **A394**, 378 (1983); C. Detraz et al., *Phys. Rev. C* **19**, 164 (1979).

25. B. H. Wildenthal and W. Chung, *Phys. Rev. C* **19**, 164 (1979).
26. X. Campi et al., *Nucl. Phys.* **A251**, 193 (1975).
27. A. Poves and J. Retamosa, *Phys. Lett.* **B184** 311 (1987).
28. M. H. Storm, A. Watt, and R. R. Whitehead, *J. Phys.* **G9**, L165 (1983); A. Watt, R. P. Singhal, M. H. Storm, and R. R. Whitehead, *J. Phys.* **G7**, L145 (1981).
29. E. K. Warburton et al., *Phys. Rev. C* **34**, 1031 (1986); J. W. Olness et al., *Phys. Rev. C* **34**, 2049 (1986); E. K. Warburton, *Phys. Rev. C* **35**, 2278 (1987); E. K. Warburton, J. A. Becker, B. A. Brown, and D. J. Millener, *Ann. Phys.* **187**, 471 (1988); E. K. Warburton and J. A. Becker, *Phys. Rev. C* **35**, 1851 (1987); *Phys. Rev. C* **39**, 1535 (1989); *Phys. Rev. C* **40**, 2823 (1989).
30. E. K. Warburton and J. A. Becker, *Phys. Rev. C* **37**, 754 (1988).
31. R. D. Lawson, *Theory of the Nuclear Shell Model*, (Clarendon Press, Oxford, 1980).
32. L. Zamick, *Phys. Lett.* **19**, 580 (1965).
33. K. Heyde et al., *Physics Reports* **102**, 291 (1983); *Phys. Rev. C* **38**, 984 (1984); *Nucl. Phys.* **A466**, 189 (1987); *Nucl. Phys.* **A484**, 275 (1988)
34. K. Heyde et al., *Phys. Lett.* **218B**, 298 (1989).
35. P. Baumann et al., unpublished.
36. B. H. Wildentha, M. S. Curtin, and B. A. Brown, *Phys. Rev. C* **28**, 1343 (1983).
37. P. Baumann et al., *Phys. Lett.* **228B**, 458 (1989).
38. B. A. Brown and B. H. Wildenthal, *At. Data Nucl. Data Tables* **33**, 347 (1986).
39. M. C. Etchegoyen, A. Etchegoyen, B. H. Wildenthal, B.A. Brown, and J. Keinonen, *Phys. Rev. C* **38** (1988) 1382 (1988).
40. B. A. Brown and B. H. Wildenthal, *Nucl. Phys.* **A474**, 290 (1987).
41. M. Carchidi, B. H. Wildenthal, and B. A. Brown, *Phys. Rev. C* **34** (1986) 2280.
42. M. J. Murphy et al., *Phys. Rev. Lett.* **49**, 455 (1982).
43. F. Hubert et al., *Z. Phys.* **A333**, 237 (1989).
44. J. P. Dufour et al., *Z. Phys.* **A324**, 487 (1988)
45. D. J. Vieira, private communication.
46. A. C. Mueller et al., *Nucl. Phys.* **A513**, 1 (1990).
47. N. A. Orr et al., *Nucl. Phys.* **A491**, 457 (1989).
48. N. M. Clarke et al., *J. Phys.* **G14**, 1399 (1988).
49. B. D. Anderson et al., *Phys. Rev. C* **31**, 1161 (1985).
50. P. Baumann et al., *Phys. Rev. C* **36**, 765 (1987).
51. P. Baumann et al., *Phys. Rev. C* **39**, 626 (1989).
52. M. Samuel et al., *Phys. Rev. C* **37**, 1314 (1988).
53. M. S. Curtin et al., *Phys. Rev. Lett.* **56**, 34 (1986).
54. M. Lewitowicz et al., *Nucl. Phys.* **A496** (1989).
55. E. K. Warburton, *Phys. Rev. C* **38**, 935 (1988) and E. K. Warburton and D. J. Millener, *Phys. Rev. C* **39**, 1120 (1989).
56. J. Aystro and J. Cerny, in *Treatise on Heavy-Ion Science*, Vol. 8, Ed. by D. A. Bromley, Plenum Press, New York, (1985).
57. M. Langevin, et al., *Nucl. Phys.* **A455**, 149 (1986); M. G. Saint-Laurant et al., *Phys. Rev. Lett.* **59** (1987).
58. V. Borrel et al., preprint (IPNO-DRE-87-18) and private communication.
59. B. A. Brown and B. H. Wildenthal, unpublished.
60. F. Pougheon et al., *Nucl. Phys.* **A500**, 287 (1989).
61. Bjornstad et al., *Nucl. Phys.* **A443**, 283 (1985); M. J. G. Borge et al., *Phys. Scr.* **36**, 218 (1987).
62. W. E. Ormand and B. A. Brown, *Nucl. Phys.* **A440**, 274 (1985); *Nucl. Phys.* **A491**, 1 (1989).
63. D. Mikolas et al., *Phys. Rev. C* **37**, 766 (1988).
64. I. S. Towner, *Nucl. Phys.* **A216**, 589 (1973).

65. I. Tanihata et al., *Phys. Rev. Lett.* **55**, 2676 (1985); *Phys. Lett.* **160B**, 380 (1985); and unpublished.
66. W. Mittig et al., *Phys. Rev. Lett.* **59**, 1889 (1987).
67. H. Sata and Y. Okuhara, *Phys. Lett.* **162B**, 217 (1985).
68. P. G. Hansen and B. Jonson, *Europhys. Lett.* **4**, 409 (1987).
69. G. F. Bertsch, B. A. Brown, and H. Sagawa, *Phys. Rev. C* **39**, 1154 (1989).
70. T. Kobayashi et al., *Phys. Lett.* **B232**, 51 (1989).
71. D. J. Millener et al., *Phys. Rev. C* **28**, 497 (1983).
72. T. Kobayashi et al., *Proceedings of the International Nuclear Physics Conference*, Sao Paolo, Brazil, 1989.
73. G. F. Bertsch and J. Foxwell, *Phys. Rev. C* **41**, (1990) 1300 and errata, to be published.
74. S. Shlomo and G. Bertsch, *Nucl. Phys.* **A243**, 507 (1975); and G. Bertsch, *Computational Nuclear Physics*, to be published by Springer, (1989).
75. E. Arnold et al., *Phys. Lett.* **B197**, 311 (1987).
76. T. Kobayashi et al., *Phys. Rev. Lett.* **60**, 2599 (1988).

21. Recent Results on Superdeformation

Marie-Agnes D. Deleplanque
Nuclear Science Division
Lawrence Berkeley Laboratory
1 Cyclotron Road
Berkeley, CA 94720, USA

ABSTRACT

I report on recent advances in the study of superdeformed nuclei: possible new mass regions of superdeformation, determination of the deformation by lifetime measurements (doppler-shift-attenuation method), and identification of the configuration of superdeformed bands using the measured dynamic moments of inertia of these bands. Finally, it will be emphasized that the future of these studies resides in the opportunity to use the next generation of γ-ray detector systems such as GAMMA-SPHERE.

21.1 INTRODUCTION

The nucleus can have a variety of shapes as shown in Fig. 1, where the "landscape" of potential energy as a function of the deformation parameters β and γ is plotted, as calculated[1] for the nucleus ^{152}Dy at spin 80. There are several minima at lower deformation, some oblate, some prolate or triaxial. At larger deformation ($\beta \approx 0.6, \gamma \approx 0°$), there is a deep minimum, which corresponds to a rather elongated shape; namely, that of an ellipsoid with a ratio of axes equal to 2:1. Nuclei with such a shape are called "superdeformed nuclei," as were the shapes found some twenty years ago in the actinide nuclei. At still larger deformation ($\beta \approx 0.6, \gamma \approx 0°$), corresponding to a 3:1 ratio of axes, a shallower minimum is predicted. Nuclei with such a large deformation, called "hyperdeformed nuclei" have not been found yet. This calculation has been done for a very high spin (80). High angular momentum acts through the centrifugal force to stabilize these very elongated shapes against lower deformations favored by the surface energy. Typically, superdeformed shapes are populated around spin 40–60 in a wide range of nuclei, whereas the predicted hyperdeformed shapes would be populated only around spin 70–80—very close to the fission limit. In contrast, superdeformation occurs at spin 0 in the actinides, since the coulomb force stabilizes large deformations in these nuclei. There are several possible definitions of "superdeformation": one would be a shape corresponding to a 2:1 ratio of axes, as above; another is any shape corresponding to a second minimum in the potential-energy curve as a function of deformation. I will refer here to superdeformed nuclei as those with "large" deformation, whose ratio of axes is greater that 1.3:1.

I will first make some general comments on the origin of superdeformation, then I will discuss recent results obtained in two mass regions, $A \approx 190$ and 130. I will conclude

Exotic Nuclear Spectroscopy, Edited by W. C. McHarris
Plenum Press, New York, 1990

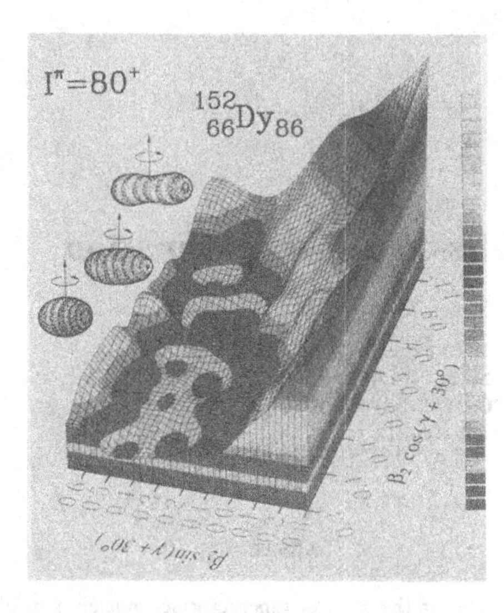

Figure 1. A contour plot of the potential energy for the nucleus ^{152}Dy calculated at spin 80. The inserts show the shapes of the nucleus in the deepest minima. The left axis is parallel to the $\gamma = 60°$ line and the right axis to the $\gamma = -30°$ line. The deepest minima fall close to the $\gamma = 0°$ line.

with some comments on the progress made in understanding some features of superdeformed nuclei.

21.2 ORIGIN OF SUPERDEFORMATION

Superdeformation results from a combination of shell effects reinforced by "flattened" liquid drop curves (as a function of deformation). In the simple harmonic oscillator, clear "gaps" (see Fig. 2) appear when the ratio of axes is an integer, giving favorable shell energy for these deformations. But the spin-orbit force and anharmonicities in the potential energy tend to wash out these effects. Nevertheless, in realistic calculations,[1] using a Woods-Saxon potential and the Strutinsky renormalization, there still remain "families" of gaps for particular deformations and nucleon numbers as shown in Fig. 3. These gaps can be related to the existence of pseudo-$SU(3)$ symmetries obeyed by the normal-parity states.

The study of superdeformed nuclei is interesting, for we observe the nucleus in conditions very different from usual. One expects the coriolis force to be less important in this case, since the coupling to deformation is larger and also the frequencies are small at a given spin because of the larger moment of inertia. The configurations at these large deformations involve very high-j orbitals (two shells above the valence shell), and by observing them we should learn about the position of these high-lying shells. Also, these long bands are smoother than usual, and, so far, no sharp band crossing has been observed within them. This is a different behavior, which needs to be understood. Finally, Mottelson has predicted that excited superdeformed bands might exist corresponding to a "bending" mode produced by octupole effects. This would be very interesting, as such a mode has never been observed.

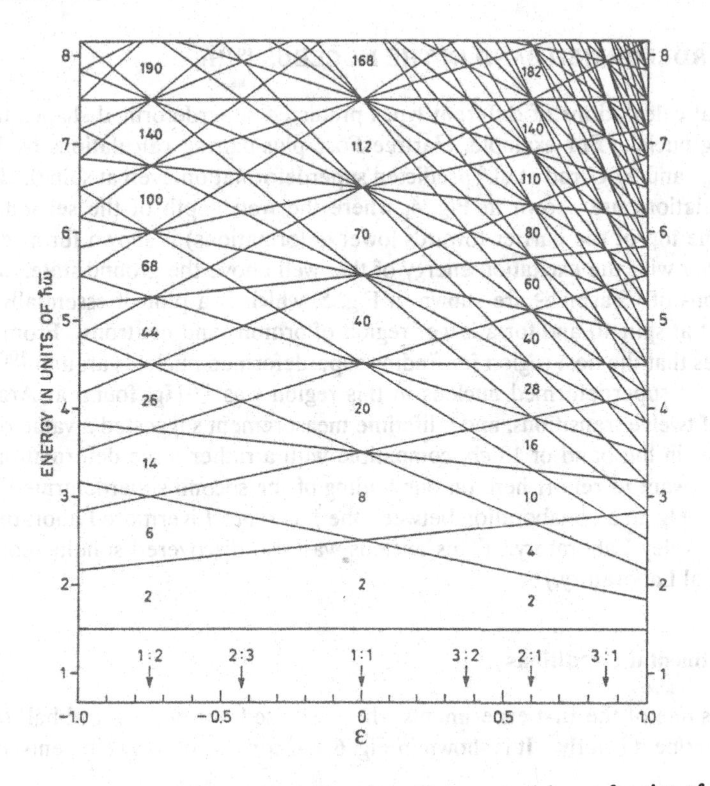

Figure 2. Single-particle levels of the axially-symmetric oscillator potential as a function of the elongation coordinate ϵ.

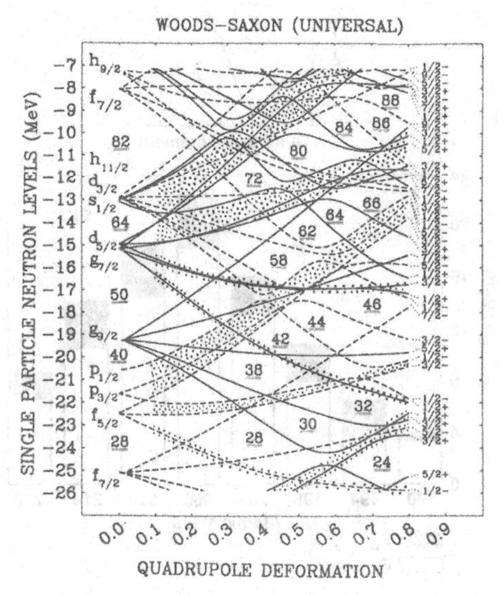

Figure 3. The single-particle spectrum in a realistic potential (the Woods-Saxon potential in this case). Higher level-density regions are shaded.

21.3 SUPERDEFORMED BAND IN THE NUCLEUS ^{192}Hg

Several calculations of different types predicted superdeformed shapes to occur in the heavy Hg nuclei. For example, Hartree-Fock-plus-pairing calculations by Weiss,[2] by Bonche et al.,[3] and by Girod et al.[4] predicted superdeformation even at spin 0. The results of such calculations are shown in Fig. 4, where the well depth of the second minimum (relative to the top of the barrier towards lower deformations) is shown for a series of Hg nuclei, together with the excitation energy of that well above the ground state. The cranking calculations of Chasman[5] are shown in Fig. 5, which is a plot of essentially the same quantities but at spin 40 and for a larger region of protons and neutrons. From these one also concludes that the best region for finding superdeformed nuclei is around $^{190\text{-}194}$Hg.

The first superdeformed nucleus in this region was ^{191}Hg, found at Argonne.[6] It had a band of twelve transitions, and a lifetime measurement suggested a value of the transition moment in the band of 18 *e*b, compatible with a rather large deformation (ratio of axes, 1.6:1). I want to report here on our finding of the second superdeformed nucleus in that region, ^{192}Hg, in a collaboration between the Lawrence Livermore Laboratory and the Lawrence Berkeley Laboratory. (This nucleus was also discovered simultaneously at Argonne National Laboratory.)

21.3.A Experimental Conditions

This is one of the first experiments where we used our new central ball for HERA, and I will describe it briefly. It is shown in Fig. 6 and consists of forty elements arranged in

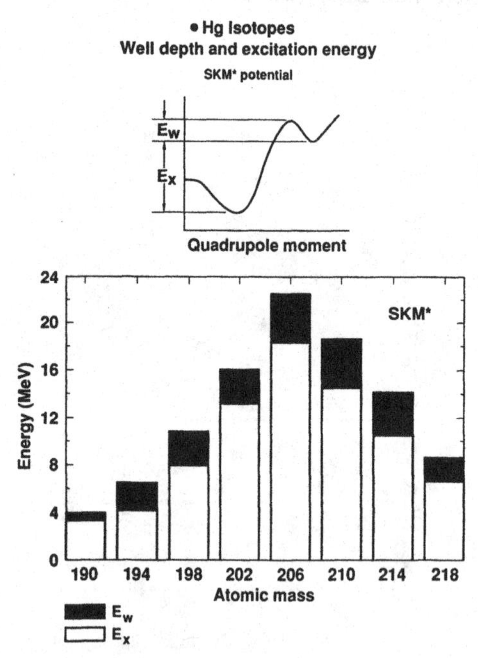

Figure 4. Hartree-Fock plus pairing calculations of the excitation energy (light areas) and well depth (dark areas) for superdeformed shapes for various Hg nuclei at spin 0. These two quantities are defined in the top part of the figure (from Ref. 2).

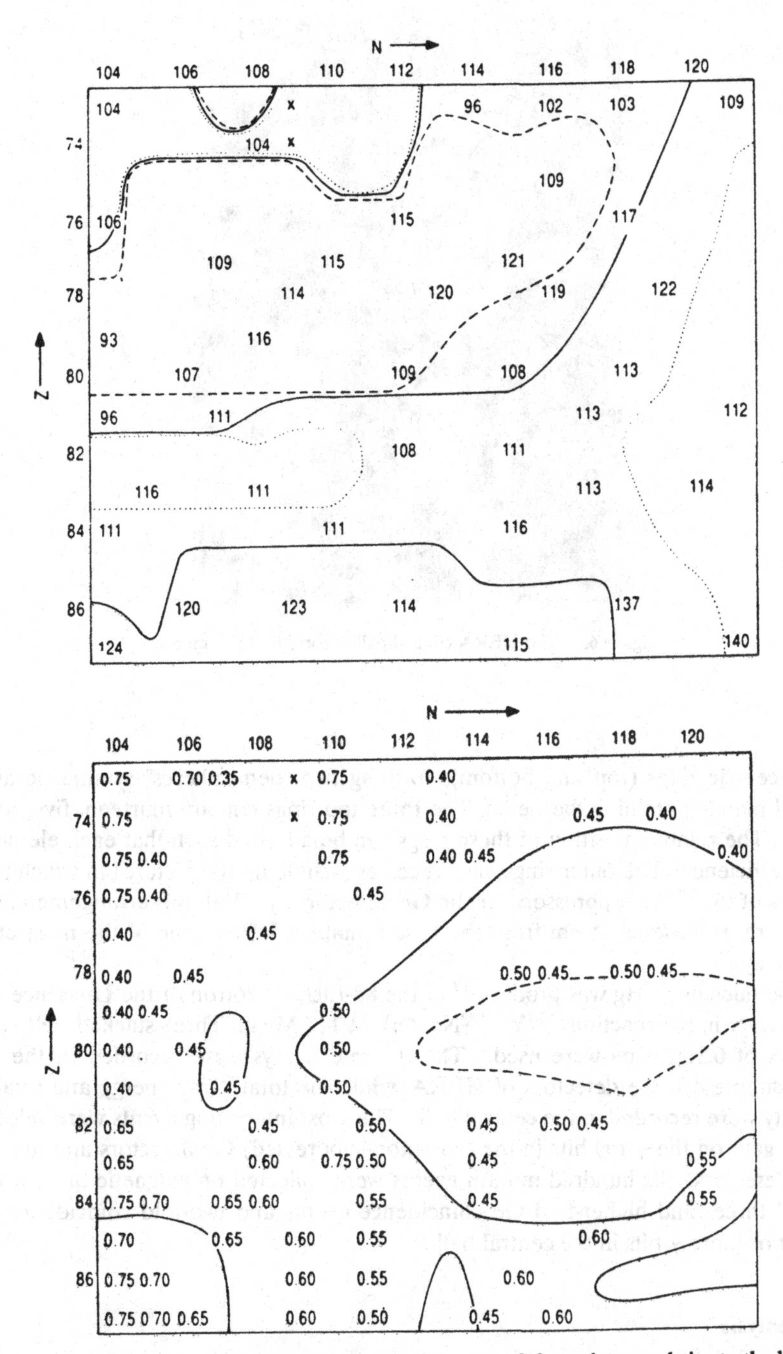

Figure 5. Top: Contour plot of the excitation energy of the superdeformed state relative to the lowest state from cranking calculations at spin 40 as a function of N and Z in the Hg nuclei region (from Ref. 5). The contour levels are 0.5 MeV (dashed), 2 MeV (solid), and 4.5 MeV (dotted). The numbers are moments of inertia for the corresponding (N, Z) nuclei. Bottom: Same contour plot for the well depth. The contour levels are 2 MeV (solid) and 3.5 MeV (dashed). The numbers are the quadrupole deformation parameter for the *SD* minimum.

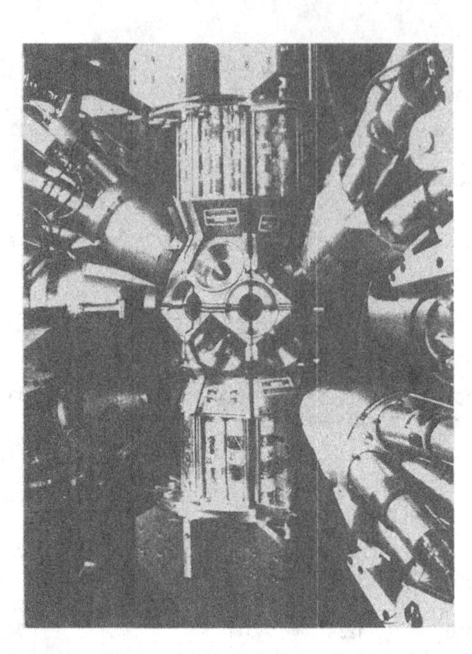

Figure 6. The HERA central ball of forty BGO elements.

three concentric rings (top and bottom), forming two "hemispheres" symmetric about a horizontal plane containing the beam. The three top rings contain fourteen, five, and one elements. The relative position of these rings can be adjusted such that each element has the same efficiency. The outer rings have recesses (visible on the picture) in which the NaI nose cones of the BGO suppressors for the Ge detectors fit. With this arrangement, the Ge detectors are as close as 15 cm from the target, making HERA one of the most efficient existing arrays.

The nucleus ^{192}Hg was produced[7] at the 88-Inch Cyclotron of the Lawrence Berkeley Laboratory in the reaction, ^{176}Yb (^{22}Ne, $6n$), at 122 MeV. Three stacked, self-supporting targets of 0.5mg/cm^2 were used. The discrete γ rays were recorded in the twenty compton-suppressed Ge detectors of HERA, while the total γ-ray energy and total γ-ray multiplicity were recorded in the central ball. The most interesting events were selected by hardware gate on the γ-ray hits in the (compton-suppressed) Ge detectors and in the central ball detectors. Six hundred million events were collected on magnetic tape, which included all three- and higher-fold Ge coincidence events and two-fold coincidence events with eight or more γ hits in the central ball.

21.3.B Analysis

The analysis was done using various double(Ge)-coincidence sortings, requiring a multiplicity of at least 11 and a sum energy of at least 6 MeV in the central ball, plus triple (Ge)-coincidence sortings. Fig. 7 shows the superdeformed band found in ^{192}Hg in a triple-coincidence spectrum, where one gate was always the (cleanest) transition at 496 keV (see caption) and the other was (almost) any other transition in the band. It consists of seven-

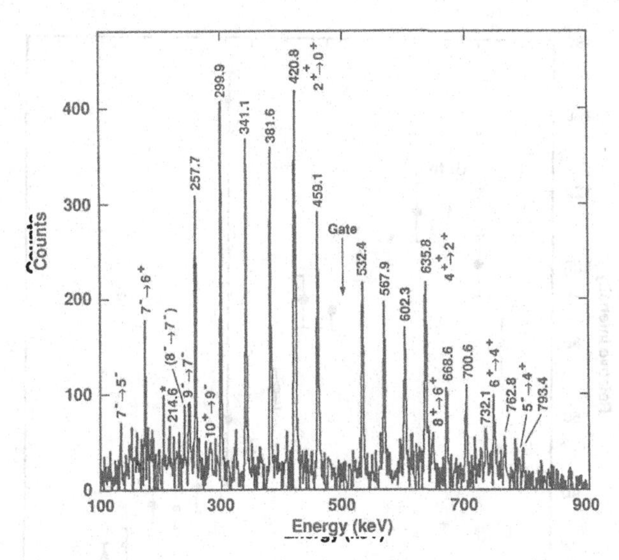

Figure 7. Spectrum in coincidence with the transition E_γ = 496 keV and the sum of all the band members (except for the 215- and 732-keV members). Known ^{192}Hg low-lying transitions are identified by $I_i \rightarrow I_f$.

teen transitions whose separation in energy varies smoothly from 43 to 31 keV, in a manner fairly similar to that of ^{191}Hg. Transitions up to spin 10 in the yrast sequence are also seen in that spectrum. To find the linking transitions, a triple-coincidence sort was made by constructing matrices gated by the cleanest superdeformed (*SD*) band members. No linking transition was found up to 4 MeV. The relative intensities (for a gate at 341 keV, see Fig. 8), corrected for the detector efficiency and internal conversion, show that the feeding of the *SD* band occurs over the upper half of the band. This is rather similar to the ^{191}Hg case. The angular-correlation data are consistent with the assignment of stretched quadrupole transitions for the *SD* band members.

The feeding of the *SD* band into the yrast sequence is shown in Fig. 9. From the intensities of the yrast transitions populated by the decay of the *SD* band, an average entry spin of $6\hbar$ into the yrast sequence is deduced. The decay out of the *SD* band occurs mostly out of the 257-keV transition, and, assuming, as is usually done, that two units of spin are carried out between the decay out of the *SD* band and the entry into the yrast sequence, a spin of $8\hbar$ is deduced at the bottom of the 257-keV line. However, we thought of another method to determine the spin in these *SD* bands, a method which should work particularly well for low-spin bands, and I would like to explain and use that method now.

21.3.C Spin Determination

The plot of the dynamic moment of inertia $\Im^{(2)}$ as a function of ω^2 (see Fig. 10*a*) is almost a straight line, so that we can expand $\Im^{(2)}$ in powers of ω^2,

$$\Im^{(2)} = \alpha + \beta\omega^2 + \gamma\omega^4, \tag{1}$$

with $\omega = E_\gamma/2$ and $\Im^{(2)} = dI/d\omega$. A simple integration of Eq. (1) with respect to ω then gives the spin,

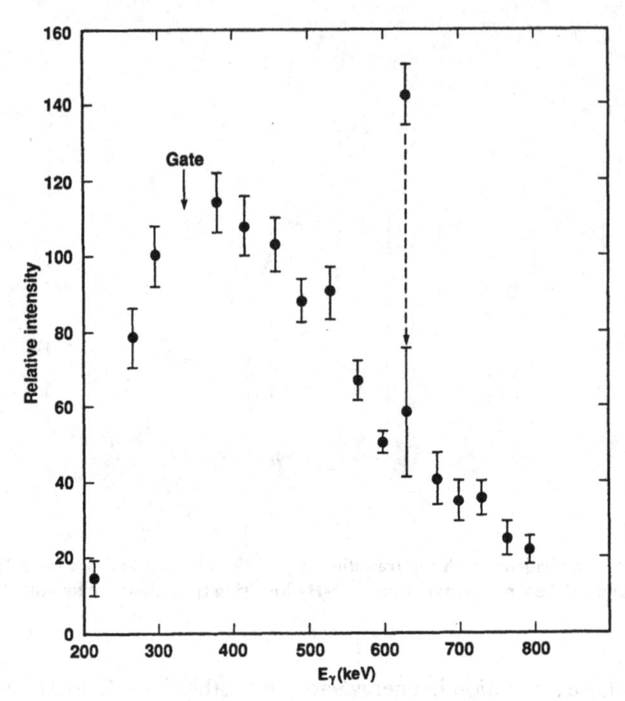

Figure 8. Relative population of the *SD* band in ^{192}Hg normalized to the yield of the 299.9-keV transition. An estimated contribution from the 4 → 2 transition at 635 keV is indicated.

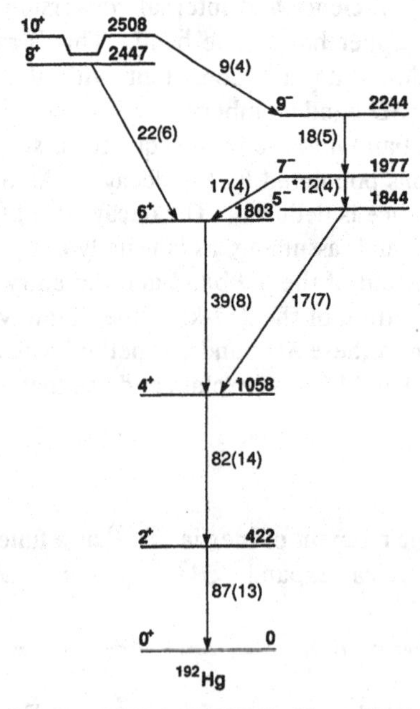

Figure 9. The intensity of low-lying transitions in ^{192}Hg coincident with the decay of the *SD* band (normalized to 100). Errors are statistical.

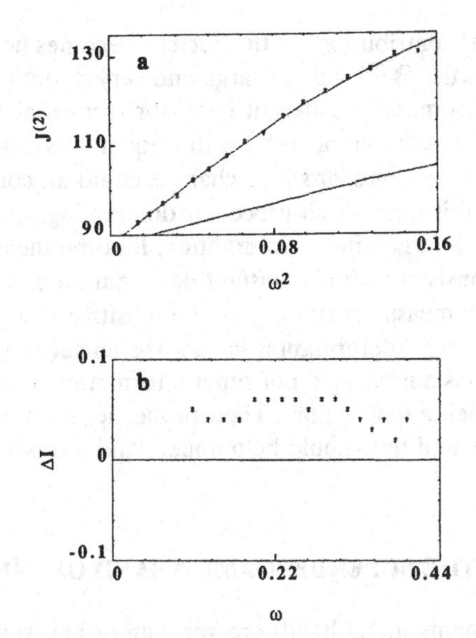

Figure 10. *a*) **Dynamic moment of inertia as a function of ω^2 for the *SD* band in ^{192}Hg and its best fit with the expansion given in Eq (1).** *b*) **Deviation from integers for spins obtained from Eq (2).**

$$I + 1/2 = \alpha\omega + \beta\omega^3/3 + \gamma\omega^5/5 \ , \tag{2}$$

where I is the intermediate spin for the transition of frequency ω: $(I + 1 \rightarrow I - 1)$. This assumes that, near spin 0, the frequency can be written as $\omega = \hbar^2/2\Im(2I + 1)$, so that for $\omega = 0$, $I = -1/2$. This formula takes into account gradual alignments occurring as a function of frequency, but does not take into account constant alignments i that would be completed at a frequency below the lowest one where the *SD* band was observed.

In the case of ^{192}Hg, this method gives a spin 8 at the bottom of the 215-keV transition, which differs by two units from the previous determination. It is unlikely that a constant alignment occurs below that spin in the unseen part of the *SD* band, since this would add (from what is known about first backbends in this mass region) about six units of spin and it would be difficult to understand why the *SD* band feeds the yrast sequence at spin 10. We can then assume that $i = 0$. The spins are determined without ambiguity with that method, as shown in Fig. 10*b*. That figure shows the deviation from integer values for the spins obtained from Eq. (2). This deviation is very small, only a few hundredths of a unit, showing that a rotational-type sequence fits the data well. If one tries to use a spin sequence starting at a different spin, the rms deviations from the best fit become large, showing that the spins are determined uniquely.

21.3.D Some Remarks

The spins found in the *SD* band of ^{192}Hg(≈10–40) are intermediate between those in the mass-150 region (≈25–65) and in the actinides (≈0). This regular evolution could reflect the effect of the coulomb force, which stabilizes the elongated shapes more and more easily (with less centrifugal contribution) as the nucleus becomes heavier.

The moment of inertia $\Im^{(2)}$ shows a large and very smooth increase with frequency (see Fig. 10*a*). This is a somewhat different behavior from that of the *SD* bands in the mass-150 region, where the variation of $\Im^{(2)}$ with frequency was smaller and less regular. Pairing effects, slow alignments, and/or shape changes could all contribute to such a variation, but it is not clear at this time which process is dominating. Alignments alone are not expected to produce such a large effect. In addition, lifetime measurements[8] reported by the Argonne group are consistent with a constant deformation corresponding to a 1.6:1 ratio of axes, although such a measurement may not be sensitive to a gradual stretching.

The observation of superdeformation in two Hg nuclei suggests that the region of nuclei of mass around 190 is a new region of superdeformation, and there will no doubt be more searches for *SD* nuclei in that region. Their properties seem to differ somewhat from those of the lighter nuclei, and this should help understand the behavior of superdeformed nuclei.

21.4 LINESHAPES IN THE SUPERDEFORMED BAND OF ^{135}Nd

Lifetime measurements in *SD* bands are very important, since they are the only direct measure of the deformation of the nucleus in those bands. In these bands, the transitions are fast and, therefore, the Doppler-Shift-Attenuation Method (DSAM) is used. However, if the side-feeding into these bands is not negligible in the region of sensitivity, that is, where the lines are partially doppler-shifted, a simple centroid measurement will not be sufficient to determine the lifetimes within the band. In this case, the decay time of a state might be complicated, reflecting a composite feeding pattern, and only a lineshape analysis can separate the components.

21.4.A Experimental Measurement

The *SD* band in ^{135}Nd, found[9] a few years ago, is shown in Fig. 11. The relative intensity of the *SD* lines shows that there is side-feeding into most of the states of that band. The DSAM was used in a reaction where an ^{40}Ar beam of 175 MeV bombarded a 1-mg/cm^2 Au-backed target. The data were sorted into various matrices. In particular, the two most useful ones were obtained from the coincidences of all detectors against the forward ones and all detectors against the backward ones. In this way the best forward and backward spectra were obtained by using all detectors to gate on the cleanest stopped lines of the *SD* band; namely, the 546- and 676- keV lines. Such gated spectra are shown in Fig. 12.

21.4.B Lineshape Analysis

To determine the transition quadrupole moment of the *SD* band, the lineshapes of the transitions of the band (in the region of sensitivity) are calculated[10] and fitted to the data of Fig. 12. A constant quadrupole moment is assumed within the *SD* band, and this is generally satisfactory. The slowing down of the recoils in the target and backing materials is calculated, using the latest estimates for the electronic stopping power and the nuclear stopping power. In addition, for the nuclear stopping, a Monte-Carlo method is used to determine the direction and velocity of the recoiling nucleus after the nuclear collision.

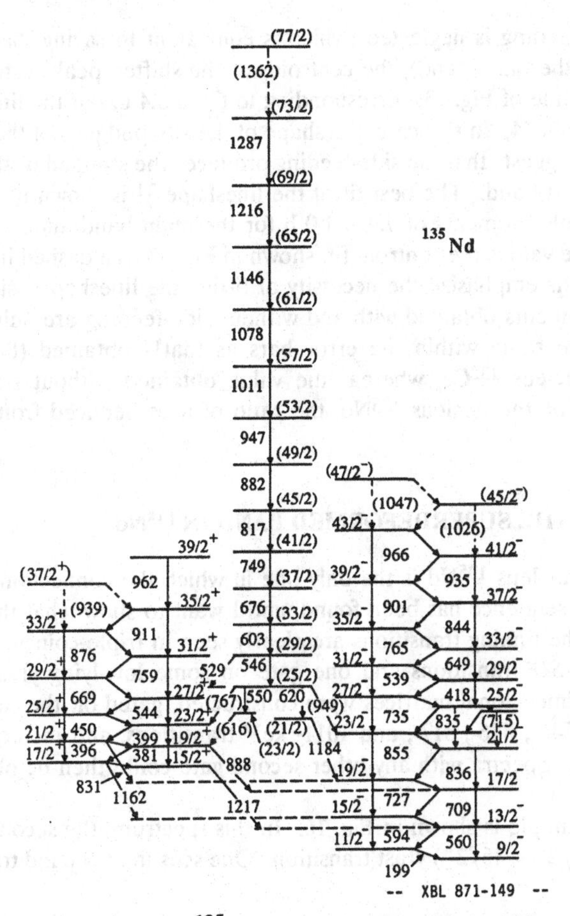

Figure 11. Level scheme of ^{135}Nd, showing the *SD* band (spins 25/2 → 77/2).

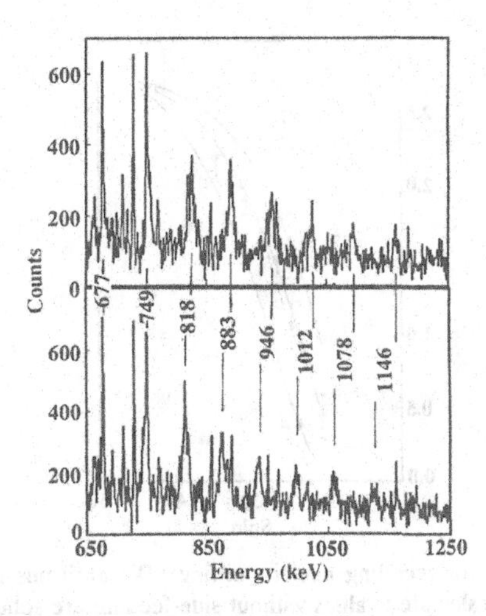

Figure 12. Sum of coincidence spectra gated on the clean *SD* lines at 546 and 676 keV on four forward (top) and four backward (bottom) detectors.

If the side-feeding is neglected (which is equivalent to saying that it has the same lifetime as that of the main band), the centroids of the shifted peaks can still be fitted, as shown by the solid line of Fig. 13, corresponding to Q_t of 5.4 b, but the lineshape fit is then poor, as shown in Fig. 14. In this case, the shape of the stopped part of the peaks cannot be reproduced. This suggests that the side-feeding produces the stopped peak and is therefore slower than the main band. The best fit of the lineshapes[11] is shown in Fig. 15. It corresponds to quadrupole moments of 7.4 ± 1.0 b for the main band and ≈3.5 b for the side-feeding. With those values the centroid fit, shown in Fig. 13 as a dashed line, is also slightly better. These results emphasize the necessity of fitting the lineshapes, since the values of the quadrupole moments obtained with and without side-feeding are quite different. The correct value is the same within the error bars as that[12] obtained (8.8±1.7 b) for the neighboring *SD* nucleus ^{132}Ce, whereas the value obtained without side-feeding is significantly lower. For the nucleus ^{135}Nd, the ratio of axes deduced from these measurements is 1.42:1.

21.5 DECAY OF THE SUPERDEFORMED BAND IN ^{135}Nd

So far, the nucleus ^{135}Nd is the only one in which the connection between the *SD* band and the yrast sequence has been found, and I want to show what the evidence is for this connection. The linking transitions are clearly seen in triple-coincidence spectra with one gate on some *SD* transitions and one gate on some low-lying yrast transition. To achieve this, two-dimensional matrices were constructed, gated on the cleanest *SD* transitions; namely, the 546-, 676-, 817-, and 1011- keV transitions, and then added after background subtraction. Spectra with any other second gate could then be obtained from this matrix.

A typical example is shown in Fig. 16. In this spectrum, the second gate was taken on the 727-keV ($19/2^- \rightarrow 15/2^-$) yrast transition. One sees the *SD* band transitions, (with a

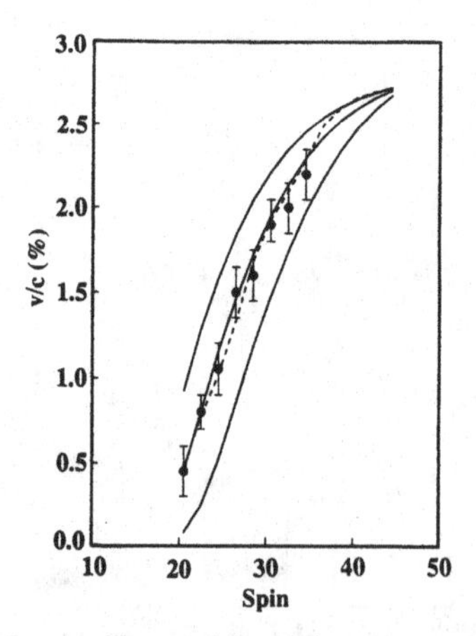

Figure 13. Average value of v/c for recoiling nuclei emitting *SD* transitions from the state of spin *I*. The experimental values are dots, the calculated values without side-feeding are solid lines (Q_t = 4.0, 5.4, and 7 b, respectively), and the best value (Q_t = 7.4 b) is obtained with a side-feeding about four times as slow as than the main band (dashed line).

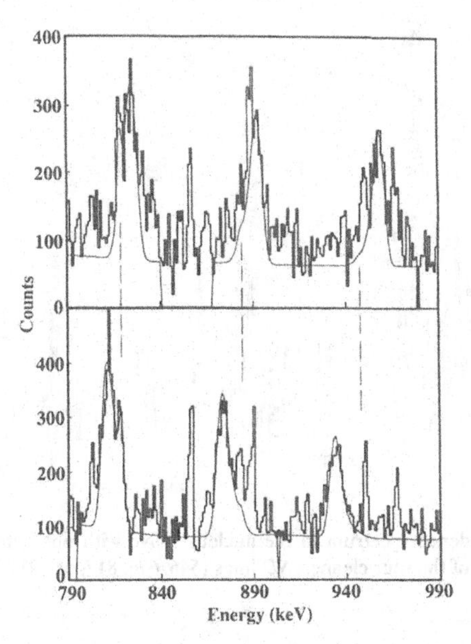

Figure 14. Lineshapes of the 818-, 883-, and 949-keV transitions for forward (top) and backward (bottom) detectors: experimental (histogram) and calculated (line) for Q_t = 5.4 b and no side-feeding. The dashed lines indicated the position of the stopped peaks.

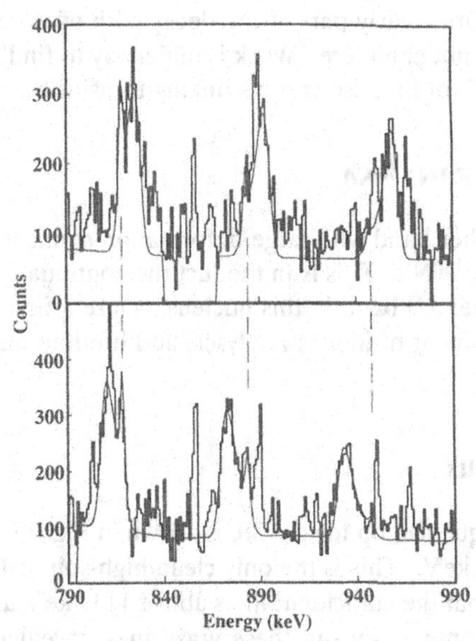

Figure 15. Same as Fig. 15 but for calculations with Q_t = 7.4 b for the main *SD* band and Q_t decreasing with spin from 4.0 to 3.3 b for the side-feeding.

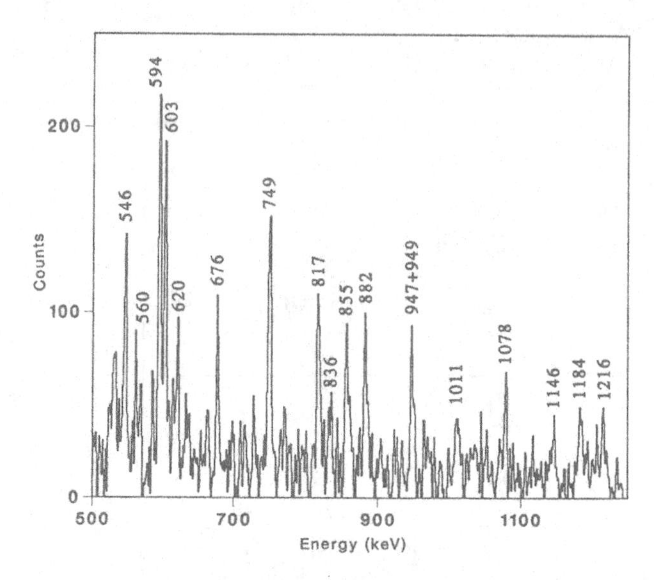

Figure 16. Partial triple-coincidence spectrum in the nucleus ^{135}Nd with one gate on the low-lying 727- keV transition, and one gate on one of the four cleanest *SD* lines (546, 676, 817, or 1011 keV).

lower intensity for the transitions used as gates) the low-lying yrast transitions at 594 and 855 keV, and the linking 620-, 949- (double with the 945-keV *SD* transition), and 1184-keV transitions. This spectrum is very clean, and spectra gated on other low-lying transitions are of the same quality. All the identified linking transitions can be seen in Fig. 11. The transitions at 529 and 550 keV are clearly part of the decay path of the *SD* band, but the link to the low-lying transitions is not complete. Work is underway to find it. About half of the *SD* band intensity is accounted for in these various linking transitions.

21.6 YRAST SEQUENCE IN ^{136}Nd

There may be another band with large deformation connected to the yrast sequence in the neighboring nucleus ^{136}Nd. This is in the fact the continuation of the yrast sequence, which may turn into another *SD* band in this nucleus, where a first one is already known.[13] This result is based on moment of inertia analyses, and lifetime measurements are needed to establish the deformation.

21.6.A Experimental Results

The entire yrast sequence, up to spin 30, is shown in Fig. 17, where the gate is set on the (double) peak at 1139 keV. This is the only clean high-spin gate. The spectrum is very clean, but it is not clear what the coincident lines above 1139 keV are. A triple-coincidence sorting, using clean gate combinations of these yrast lines, revealed three peaks which are tentatively assigned to the continuation of this yrast sequence; namely, transitions at 1161, 1223, and 1283 keV. The moments of inertia, plotted on Fig. 18, show that after three backbends, both the $\Im^{(1)}$ and $\Im^{(2)}$ moments of inertia in the yrast sequence have values

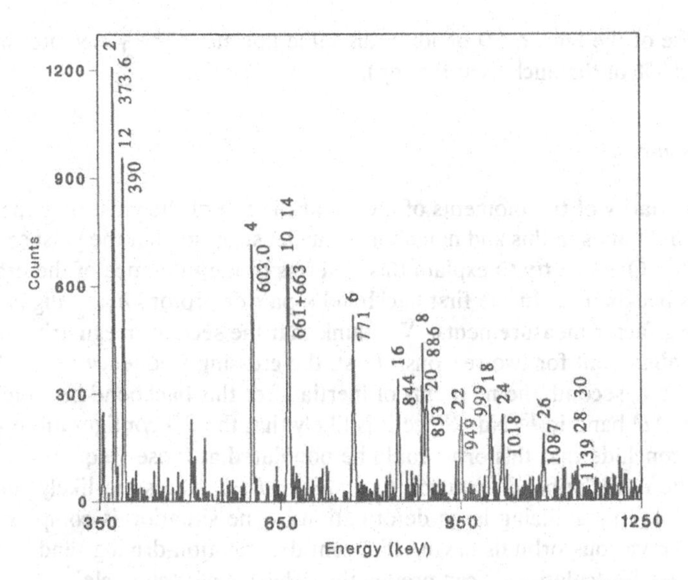

Figure 17. Double-coincidence spectrum in the nucleus ^{136}Nd, gated by the 1139-keV line, showing the entire yrast sequence up to spin 30. The 1139-keV line is double (initial spins 28 and 30).

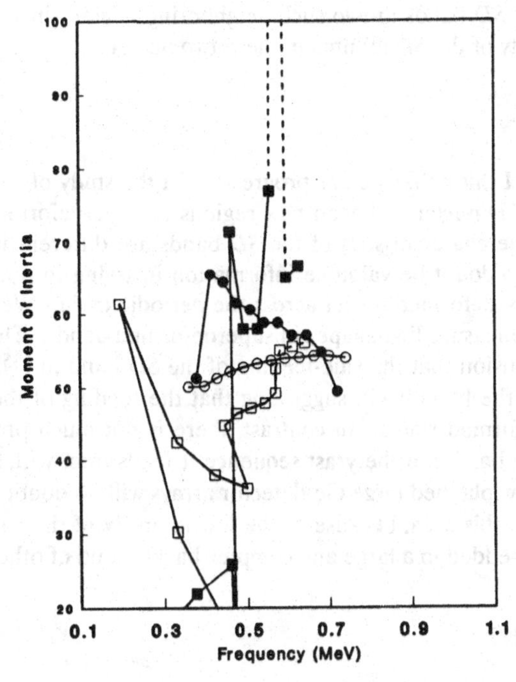

Figure 18. Moments of inertia in the nucleus ^{136}Nd for the known *SD* band (circles) and for the yrast sequence (squares). Dynamic moments of inertia (2) are indicated as full symbols, and kinematic moments of inertia(1) are shown as open symbols. The dashed lines are in the region of the tentative new transitions above spin 30.

similar to those of the known *SD* band in this same nucleus. The γ-ray intensities are also very similar (0.5% of the nucleus at the top).

21.6.B Discussion

The similarity of the moments of inertia at the top of the yrast sequence with those of the known *SD* bands in this and neighboring nuclei suggests that the yrast sequence turns into a *SD* band. One can try to explain this in terms of the influence of the orbits involved in the various backbends. In the first backbend a pair of protons $h_{11/2}$ aligns, as has been shown[14] from *g*-factor measurements. We think that the second irregularity may involve a neutron $i_{13/2}$ alignment for two reasons. First, the crossing frequency ($\hbar\omega \approx 0.45$ MeV) is roughly right, and, second, the moments of inertia after this backbend become very similar to those of the *SD* band in ^{137}Nd. Since it is likely that the *SD* configurations contain $i_{13/2}$ neutrons, we conclude that this orbit could be populated at those frequencies in these two nuclei. For the third "upbend", neutron $h_{9/2}$ and/or $h_{11/2}$ states are likely candidates and could play a role in stabilizing large deformations. The situation is complex in this mass region, with the various orbitals having different deformation-driving tendencies, and it is not clear how well calculations[15] can predict the behavior of such nuclei.

Lifetime measurements are necessary to determine the deformation. We do have DSAM data (from the same experiment as for ^{135}Nd; see above), but the data are very poor, since no clean gate can be found for the stopped peaks at reasonably high spins. We do see doppler shift for the lines at 893 keV and above, but it will be very difficult to get a value for the quadrupole moments. It would be very interesting if one could compare the excitation energies of *SD* bands in two such neighboring nuclei, since this could give clues on the relative stability of the *SD* minima in these two nuclei.

21.7 CONCLUSION

In conclusion, I think that we are progressing in the study of superdeformed nuclei. We are finding new *SD* nuclei and even new regions of superdeformation, as exemplified by the Hg nuclei. The characteristics of the *SD* bands are different in these different regions, and this will no doubt be valuable information in trying to understand the general behavior of these very deformed nuclei across the periodic table. We can now, in reasonably favorable cases, measure line-shapes in superdeformed bands. This actually led to the very interesting conclusion that the side-feeding of the *SD* band in ^{135}Nd nucleus was four times slower than in the band itself, suggesting that the feeding of the *SD* band proceeds through normally-deformed states. In contrast, there is not much progress in finding the links between the *SD* band and the yrast sequence it feeds into, with still only one or two cases known. The new planned large Ge detector arrays will no doubt be essential to make significant progress in this area, because of the low intensity of the studied γ rays and the fact that they are embedded in a large and complex background of other lines.

ACKNOWLEDGMENT

Part of this work (on ^{192}Hg) was done in collaboration with a Livermore group of J. Becker, E. Henry, and N. Roy. Various part of the work I reported here were done in our Berkeley group of F. S. Stephens and R. M. Diamond, together with postdoctoral fellows, students, and visitors. I would like to mention particularly C. W. Beausang and J. E. Draper. This work was supported by the Director, Office of Energy Research, Office of High Energy Research, Div. of Nuclear Physics of the Office of High Energy and Nuclear Physics Division of the U.S. Department of Energy under Contract DE-AC03-76SF00098.

REFERENCES

1. J. Dudek, W. Nazarewicz, Z. Szymanski, and G. A. Leander, *Phys. Rev. Lett.* **59**, 1405 (1987).
2. M. S. Weiss, UCRL-96773 (1987); *Proc. O. E. LASE88 Conference*, SPIE 875, Ed. by C. Randol Johns.
3. P. Bonche, S. J. Krieger, P. Quentin, M. S. Weiss, J. Meyer, M. Meyer, N. Redon, H. Flocard, and P. H. Heenen, *Nucl. Phys.* **A500**, 308 (1989).
4. M. Girod, J. P. Delaroche, D. Gogny, and J. F. Berger, *Phys. Rev. Lett.* **62**, 2452 (1989).
5. R. R. Chasman, *Phys. Lett.* **B219**, 227 (1989).
6. E. F. Moore, R. V. F. Janssens, R. R. Chasman, I. Ahmad, T. L. Khoo, F. L. J. Wolfs, D. Ye, K. B. Beard, U. Garg, M. W. Drigert, Ph. Benet, Z. W. Gabrowski, and J. A. Cizewski, *Phys. Rev. Lett.* **63**, 360 (1989).
7. J. A. Becker, N. Roy, E. A. Henry, M. A. Deleplanque, C. W. Beausang, R. M. Diamond, J. E. Draper, F. S. Stephens, J. A. Cizewski, and M. J. Brinkman, *Phys. Rev. C*, in press; D. Ye et al., *Phys. Rev. C*, in press.
8. T. L. Khoo, This Book, Chapter 19.
9. E. M. Beck, F. S. Stephens, J. C. Bacelar, M. A. Deleplanque, R. M. Diamond, J. E. Draper, C. Duyar, and R. J. McDonald, *Phys. Rev. Lett.* **58**, 2182 (1987).
10. J. C. Bacelar, private communication, 1986.
11. R. M. Diamond, C. W. Beausang, A. O. Macchiavelli, J. C. Bacelar, J. Burde, M. A. Deleplanque, J. E. Draper, C. Duyar, R. J. McDonald, and F. S. Stephens, to be published.
12. A. J. Kirwan, G. C. Ball, P. J. Bishop, M. J. Godfrey, P. J. Nolan, D.. Thornley, D. J. G. Love, and J. L. Kinsey, *Phys. Rev. Lett.* **58**, 467 (1987).
13. E. M. Beck, R. J. McDonald, A. O. Macchiavelli, J. C. Bacelar, M. A. Deleplanque, R. M. Diamond, J. E. Draper, and F. S. Stephens, *Phys. Lett.* **B195**, 531 (1987).
14. E. S. Paul, C. W. Beausang, D. B. Fossan, R. Ma, W. F. Piel, Jr., P. K. Weng, and N. Xu, *Phys. Rev. C* **36**, 153 (1987).
15. R. Wyss, J. Nyberg, A. Johnson, R. Bengtsson, and W. Nazarewicz, *Phys. Lett.* **B215**, 211 (1988).

22. Nuclear Collectivity and Complex Alignment Mechanisms in Light W and Os Nuclei

Noah R. Johnson
Oak Ridge National Laboratory
Oak Ridge, TN 37831

ABSTRACT

Potential-energy-surface calculations of the light-mass W and Os nuclei indicate that this is a region where both shape-coexistence effects and shape-evolution tendencies should be observed. To explore the predicted richness in these nuclear structure effects at a microscopic level, we have carried out lifetime measurements on some of these nuclei utilizing the doppler-shift recoil-distance technique and have done γ-γ coincidence measurements with our Compton Suppression Spectrometer System. Determining the collective behavior as a function of the rotational frequency in ^{170}W, ^{172}W, and ^{172}Os and comparing the results with theory have enabled us to establish that the early rotation-alignment processes occurring in each of these nuclei can probably be attributed to $i_{13/2}$ quasi-neutrons. The data on ^{172}Os present an especially interesting case, in that there appear to be three nearly degenerate band crossings by $i_{13/2}$ neutrons at low rotational frequencies. A discussion of both the experimental results and a comparison with theory will be presented.

22.1 INTRODUCTION

During the past few years there have been significant advances in our understanding of nuclei excited into states of high angular momentum. The development of large multi-detector arrays for γ-γ coincidence spectroscopy studies has propelled the amount of new experimental information available, and this has been complemented by notable advances in the theoretical treatments of high-spin phenomena. To provide yet a more detailed understanding of the structure of these high-spin states and to provide a stringent test of these models, we have resorted to measurements of their dynamic electromagnetic multipole moments, which are a direct reflection of the collective aspects of the nuclear wave-functions. For the most part, these multipole moments are obtained by lifetime measurements utilizing doppler-shift techniques. Let me stress that the great value of lifetime measurements is that they provide the transition matrix elements without the necessity to rely on nuclear models.

One important result from the measurements of these electromagnetic matrix elements is that they have provided us new insight into the effects of quasiparticle alignment on the nuclear collectivity. We are finding that many nuclei which have prolate shapes in their ground states exhibit a tendency to move toward triaxial and oblate shapes at higher spins. This is in part due to the alignment of angular momentum along the rotation axis, an effect induced by the coriolis forces as the nucleus rotates at high frequencies. At these higher spins such a nucleus may exhibit yrast structures which are more single-particle than collective in nature.

Exotic Nuclear Spectroscopy, Edited by W. C. McHarris
Plenum Press, New York, 1990

Numerous lifetime measurements have been performed[1-6] on nuclei in the $N = 90$ region, where the fermi surface lies near the bottom of the $i_{13/2}$ neutron shell, and the results are understood—at least qualitatively—in terms of Cranked-Shell-Model (CSM) and Cranked-Hartree-Fock-Bogoliubov (CHFB) theories. After the backbend, these nuclei undergo a shape change (driven by the aligned $i_{13/2}$ quasi-neutrons) to a triaxial shape that is oriented so as to reduce the collectivity of the rotation, i.e., to positive γ values. It was in ^{160}Yb that we first observed[2-4] this loss of collectivity at high rotational frequencies. From the calculated[7] systematics of triaxial shape-driving orbitals, it is expected that if one moves to where the fermi surface is near the middle of the $i_{13/2}$ neutron shell, the minimum in the quasiparticle energy becomes shallower and moves to negative γ, reaching $\gamma = -120°$ (noncollective rotations of a prolate ellipsoid) when the whole shell is below the fermi surface. Thus, in this picture, the equilibrium γ value adopted by a nucleus with excited quasiparticles depends on an interplay between the driving forces of the quasiparticles, the hydrodynamic behavior of the core, and the prolate-oblate energy difference. As a consequence, the nuclei near $N = 96$–98 would not be expected necessarily to show a reduction in collectivity in their s bands. A few years ago, we turned to studies of some of the light W and Os nuclei to investigate this idea. To this point we have carried out lifetime measurements on ^{170}W (Ref. 8), ^{172}W (Refs. 9, 10), and ^{172}Os (Ref. 11) and have performed detailed γ-γ coincidence spectroscopy measurements on ^{172}Os (Ref. 12). These studies are the primary topics of this paper.

22.2 EXPERIMENTAL DETAILS

22.2.A Recoil-Distance Experiments

The lifetime measurements were carried out by the Recoil-Distance (RD) technique, and the recoil-distance apparatus ("plunger") is described in detail in Ref. 13. Basically, the apparatus involves a thin target foil stretched tightly and held in a fixed position, together with a parallel and movable catcher foil (usually lead) to stop the recoiling nuclei following a heavy-ion-induced nuclear reaction in the target. Distances are measured both by a digital micrometer and by capacitance. At a series of target-stopper separations, γ-ray spectra are recorded with a BGO compton-suppressed Ge detector (25% efficiency) at 0° to the beam direction under the coincidence requirement that at least one of an array of six large-volume Ge detectors (positioned at 90° to the beam) be triggered. The target-detector separation is 9.7 cm and 5.9 cm for the 0° and 90° detectors, respectively. A picture of the experimental apparatus is shown in Fig. 1. With this setup, the experimental data are taken in the γ-γ coincidence mode in order to resolve any problems introduced by side feeding to the levels of interest. In addition, γ-γ coincidence-gated spectra can, in many cases, eliminate interfering lines and background complexities in the spectra.

High-spin states in the nuclei discussed here were produced in fusion-evaporation reactions with beams of heavy ions produced by the tandem accelerator at the ORNL Holifield Heavy Ion Research Facility. The reactions employed were ^{124}Sn(^{52}Cr,4*n*)^{172}W, ^{122}Sn(^{52}Cr, 4*n*)^{170}W, and ^{144}Nd(^{32}S, 4*n*)^{172}Os. The energy of the ^{52}Cr beam was 225 MeV in each case, while that of the ^{32}S beam was 162 MeV. Typically in these experiments, the number of target-stopper separations for which spectra are measured ranges from eighteen to twenty-four, with the separations being chosen so as to best define the lifetimes of all states of interest. The coincidence information along with the capacitance were stored in

Figure 1. Experimental arrangement for the doppler-shift recoil-distance measurements. The target is located in the annular end cap, and the entire detector array is on a movable support so that it can be positioned appropriately about this end cap. The 0° Ge detector (only partially shown in this picture) is housed within a large BGO compton-suppression shield.

event-by-event mode on magnetic tape. The spectra at zero separation were obtained with lead-backed targets.

To illustrate the shifted and unshifted peaks in the total-projected spectra, we show in Fig. 2 the 400-600 keV region in the ^{172}W data for four of the nineteen target-stopper distances measured.

22.2.B Gamma-Ray Spectroscopy Experiments

To carry out a proper analysis of lifetime data, it is important to have a detailed knowledge of the level properties of the nucleus under investigation. Since little was known about the high-spin level properties of ^{172}Os, we have carried out γ-γ coincidence measurements on this nucleus using our Compton Suppression Spectrometer System (twenty BGO-suppressed Ge detectors) in conjunction with the Spin Spectrometer, a 4π array of NaI detectors. The experimental apparatus is shown in Fig. 3. In the present usage, we replaced nineteen of the NaI units with compton-suppressed Ge units and used the remaining fifty NaI units for reaction-channel enhancement by utilizing the γ-ray multiplicity and total decay energy. However, in this case it turned out that the reaction-channel selection was not greatly enhanced and, therefore, we chose to use all of the 5.6×10^7 coincidence events collected to improve the statistical quality of the spectra.

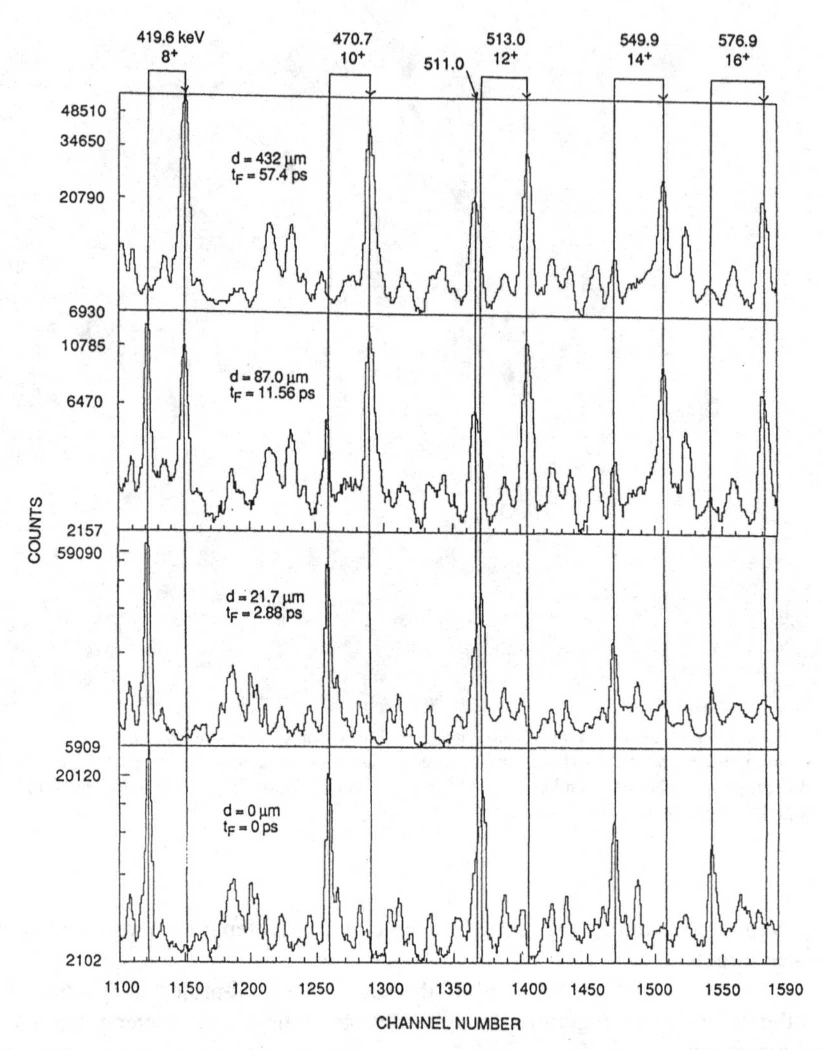

Figure 2. Illustrative "total-projected" coincidence spectra of ^{172}W covering the 400-600 keV region for four of the nineteen distances measured. These spectra in the 0° compton-suppressed detector were generated by coincidence events between any of the six 90° detectors and the 0° detector.

22.3 DATA ANALYSES AND RESULTS

22.3.A Recoil-Distance Experiments

In the lifetime experiments on each nucleus the tapes were scanned to build a 4096 x 4096 channel coincidence matrix (on one axis the 0° detector and on the other the sum of the 90° detectors) for each target-stopper separation. Later these matrices could be analyzed in any desired manner. Lifetimes were determined from spectra generated in three ways: 1) by gating on the γ-ray transition above a group of several transitions of interest [Gated-Above (GA) type data]; 2) from the sum of spectra obtained by gating on in-

Figure 3. The Oak Ridge Compton-Suppression Spectrometer System integrated with the Spin Spectrometer. In this arrangement, the nineteen compton-suppressed Ge detectors replace a corresponding number of NaI elements in the Spin Spectrometer.

dividual γ rays in cascade below the transitions of interest [Sum-Gated-Below (SGB) type data]; and 3) with the entire 90° spectrum below 1.5 MeV as the gate [Total-Projected (TP) type data]. The data from the GA-type spectra are of relatively poor statistical quality, but they provide decay curves free from contributions of direct side feeding to the levels under study. For the analysis of the spectra from the SGB and TP data, wherein the problems of the unknown side-feeding are not eliminated, two-step-cascade side-feeding was modeled to each of a sequence of levels under study. In all cases, a rotational band was used to model the feeding into the top level of the sequence.

The first step in analyzing the data is to get the shifted and unshifted γ-ray intensities. To first order, the intensities of the shifted $I(s)$ and unshifted $I(u)$ peaks are related exponentially to the mean life τ of a state by

$$R = e^{-\lambda T} = e^{-T/\tau} = e^{-D/v\tau}, \tag{1}$$

where

$$R = \frac{I(u)}{I(u) + I(s)} = \frac{I_{\propto}(s) - I_D(s)}{I_{\propto}(s)} = \frac{I_D(u)}{I_0(u)}. \tag{2}$$

In the above expressions T = flight time between target and stopper, v = average recoil velocity, and D = target-stopper separation. Actually, once the shifted and unshifted areas are obtained, there are numerous corrections that must be applied in extracting the lifetime of a state. This we do with the computer program LIFETIME.[14] Knowing the general features of the level scheme, one supplies the program with a feeding scheme for a given level. The program solves the Bateman equations while adjusting lifetimes and initial populations of levels in search of best fits. Fits to both the shifted and unshifted intensities are made where the peaks are clean enough to permit it. To illustrate, the experimental decay curves for the 4^+ through 20^+ states in the yrast sequence in ^{172}Os and the best fits are shown in Fig. 4.

The $B(E2)$ values—the reduced electric quadrupole transition probabilities—are extracted in a straightforward manner from the measured lifetimes. Next, the transition quadrupole moments (Q_t) are computed by the following equation

$$B(E2; I \rightarrow I-2) = \frac{5}{16\pi} \langle I\, 2\, K\, 0 | I-2\, K \rangle^2 Q_t^2, \tag{3}$$

where the term in brackets represents a clebsch-gordan coefficient.

To this point only the total-projected data and the SGB data for ^{170}W have been analyzed, and the preliminary Q_t values extracted from these are shown in Fig. 5. Analyses

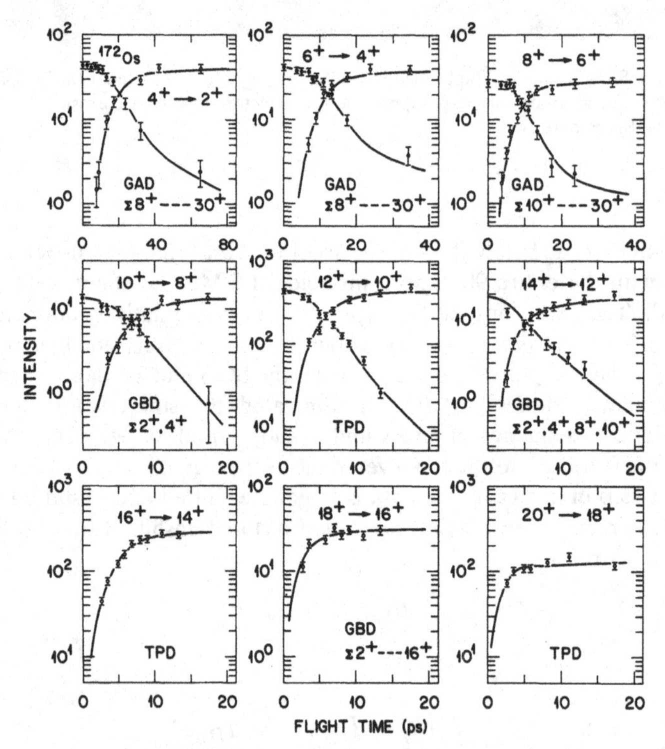

Figure 4. Experimental data (points) and fitted decay curves (solid lines) for the 4^+ through 20^+ members of the yrast sequence in ^{172}Os. The abscissa shows the recoil flight time in ps between the target and the lead stopping medium where the recoil velocity is $4.767 \pm 0.060\ \mu$m/ps.

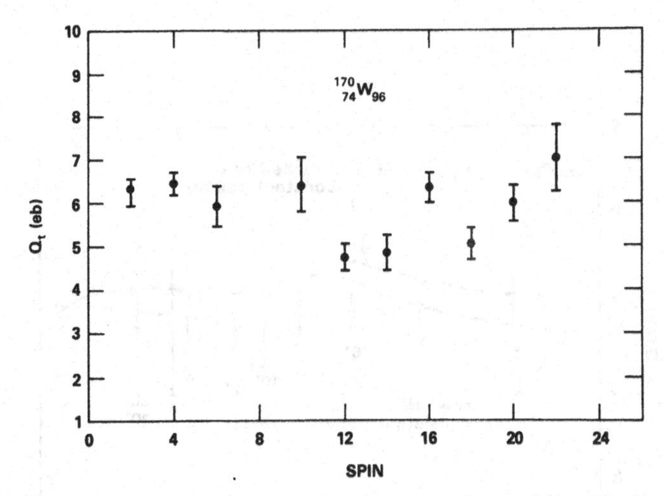

Figure 5. Preliminary Q_t values for the yrast sequence in ^{170}W, based on analysis of the "total-projected data" and the "sum-gated-below data."

of all the coincidence data for ^{172}W and ^{172}Os have been completed and plots of these Q_t values are shown in Figs. 6 and 7, respectively. Note that the Q_t values in Fig. 6 have been plotted as a function of the rotational frequency, $\hbar\omega \approx E_\gamma/2$.

22.3.B Gamma-Ray Spectroscopy Analysis

From the large matrix of γ-γ coincidence information on ^{172}Os it was possible to extract any desired spectrum in coincidence with either single gating γ-ray transitions or with sums of gating transitions. To illustrate the data, we show in Fig. 8 the γ-ray spectrum in coincidence with 228-, 378-, 448-, 471-, and 499-keV γ rays in ^{172}Os.

The coincidence relationships and γ-ray intensities from all of the coincidence data enabled us to formulate the complex level scheme consisting of five well-developed bands as shown in Fig. 9. In addition, the angular locations of the Ge detectors enabled us to extract angular correlation information from the γ-γ coincidence data and in turn to establish the spins for many of the states. Detailed information on this work can be found in Ref. 12. Prior to the current studies, only the 2^+ through 24^+ levels of the yrast sequence were well established.[15,16]

22.4 DISCUSSION

As already pointed out, when the fermi surface is near the middle of the $i_{13/2}$ neutron shell (i.e., the outermost orbital occupied is near mid shell), the triaxial shape driving influence of the $i_{13/2}$ aligned quasi-neutrons is much reduced. Further, the triaxial driving is to negative γ values which is a more collective effect. Thus, from this general Cranked-Shell-Model (CSM) picture we might expect that, contrary to the drop in Q_t values observed[2-4] beyond the backbend in ^{160}Yb, the 96-neutron nuclei ^{170}W and ^{172}Os and 98-neutron nucleus ^{172}W will show fairly constant Q_t values through the spin range of $I \approx 2^+ - 20^+$.

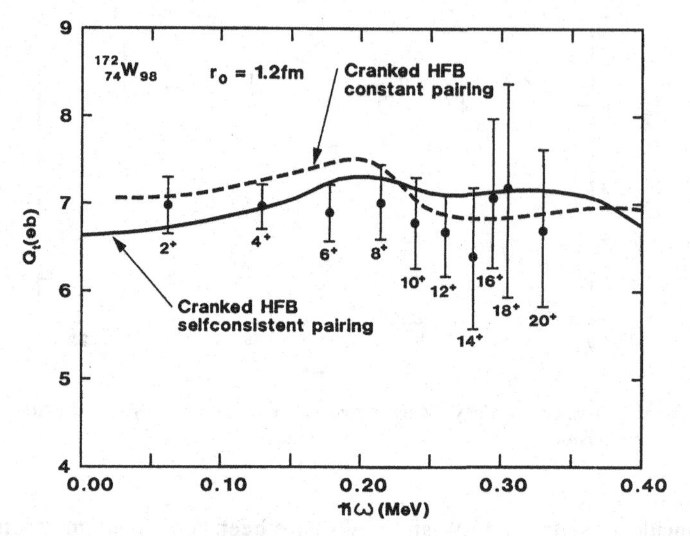

Figure 6. Q_t values for the yrast sequence in ^{172}W determined from an average of the three types of experimental data discussed in the text. The results from two different CHFB calculations[9,19] are shown (see text for details).

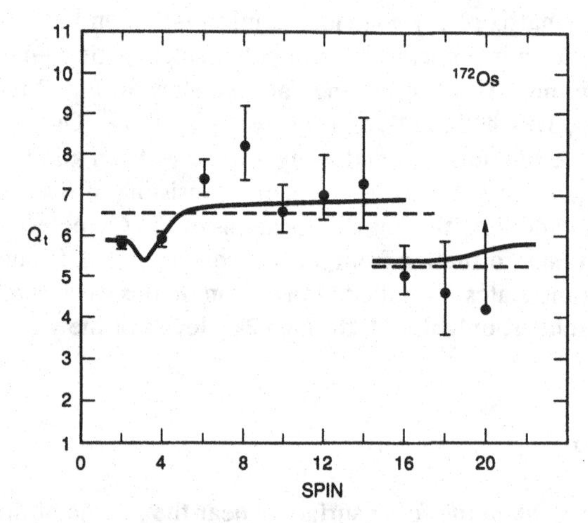

Figure 7. Q_t values for the yrast sequence in ^{172}Os determined from an average of the three types of experimental data discussed in the text. The dashed curve shows the trend of Q_t values based on CHFB calculations[12] and the solid curve shows values based on overlap matrix calculations,[24] both of which are discussed in the text.

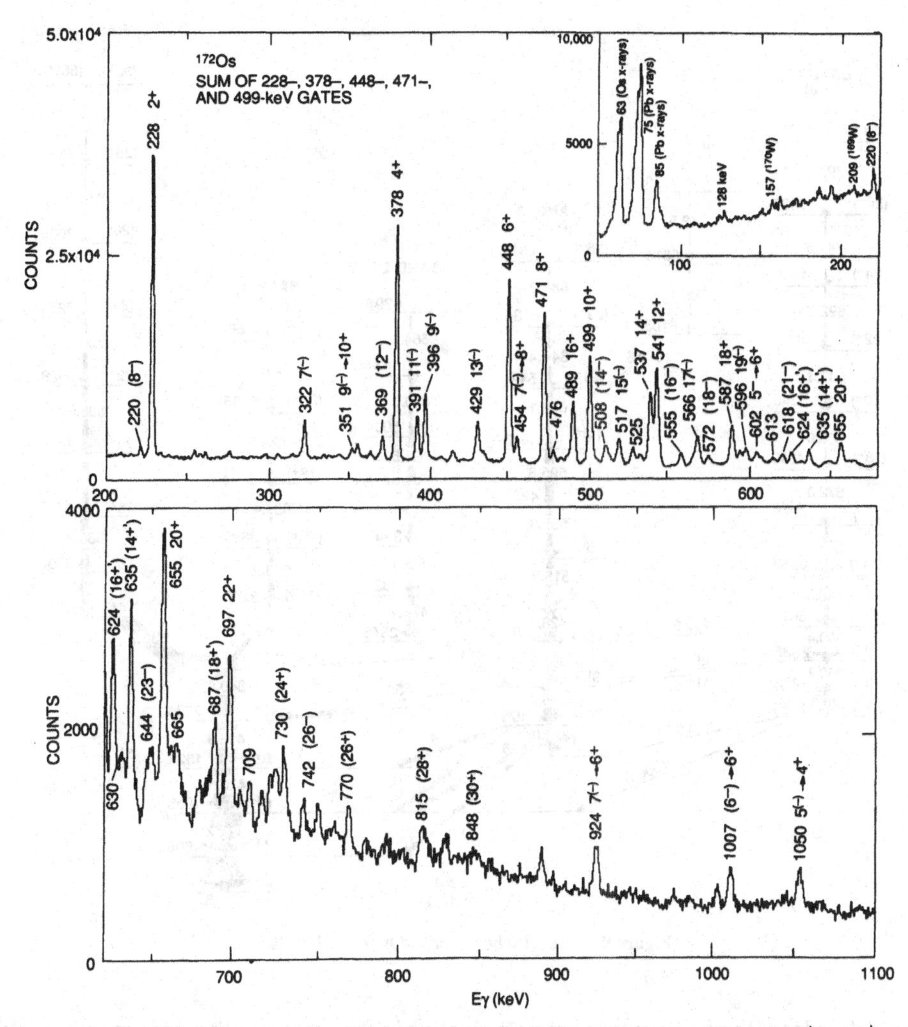

Figure 8. ^{172}Os γ-ray spectrum in coincidence with the sum of gates on transitions from the $2^+ - 10^+$ states.

However, for some time there has been much discussion about the possibility that the $h_{9/2}$ proton intruder orbital, the $1/2^-[541]$ Nilsson orbital, plays an active role in rotation alignment at rather low rotational frequencies in the light mass Os, W, Ir, and Pt nuclei. The arguments are built primarily on observed large alignment gains (more than that expected from the rotation alignment of $i_{13/2}$ neutrons alone) and on $B(M1)/B(E2)$ ratios before and after the backbend in odd-mass nuclei. Further, there exists considerable evidence that this proton orbital is either a ground state or a low-lying excited state for many nuclei in this region. It should be pointed out, however, that neither with the Nilsson potential nor with the Woods-Saxon potential can this orbital be brought down in energy to where it crosses the fermi level at rotational frequencies $\hbar\omega \leqslant 0.3$ MeV in this region if a realistic pairing gap parameter $\Delta\pi$ is used.

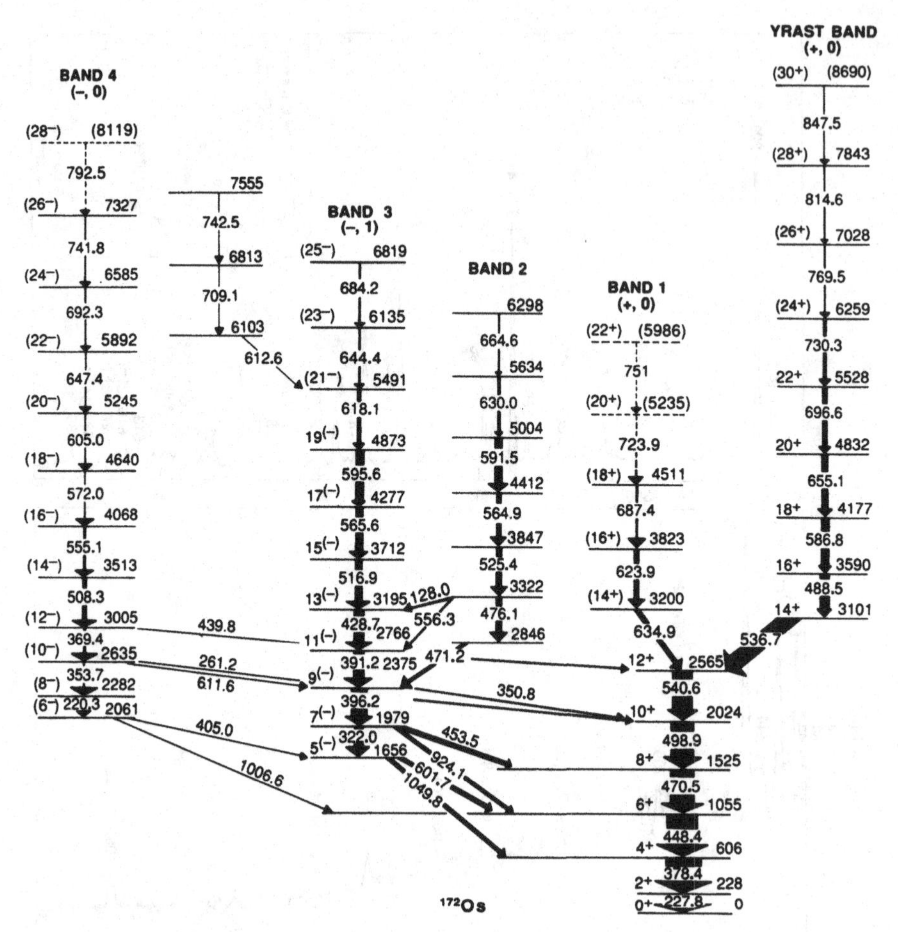

Figure 9. Level scheme of ^{172}Os from Ref. 12.

If, indeed, the $h_{9/2}$ protons are participating in the rotation alignment in such nuclei as ^{170}W and ^{172}W and ^{172}Os, then one expects a loss in collectivity reflected through reduced Q_t values at high spins. This is because the $h_{9/2}$ proton orbital drives the asymmetry parameter γ to positive values which give rise to lowered collective behavior. An examination of the Q_t values of ^{170}W in Fig. 5 reveals a generally flat trend, which is fully consistent with the general pattern predicted by CSM theory for $i_{13/2}$ neutrons. Further, Recht et al.[17] have carried out refined CHFB self-consistent calculations which indicate that at frequencies $\hbar\omega \leqslant 0.35$ MeV only $i_{13/2}$ neutrons play a role in the description of the rotation alignment properties of this nucleus.

In the case of ^{172}W with $N = 98$, the fermi surface is yet nearer the middle of the $i_{13/2}$ neutron shell. An examination of the Q_t values for this nucleus in Fig. 6 reveals only modest fluctuations from a constant value for states of $I = 2^+ - 20^+$.

Using the self-consistent CHFB model, the authors in Ref. 18 proceeded to test these results by theoretical calculations of the deformation parameters for the yrast band of ^{172}W. Their calculations were based on the Woods-Saxon potential, but without particle number projection, and the results indicate only $i_{13/2}$ neutrons are active in rotation alignment in the frequency range covered here. Briefly, in this process, the total routhian was minimized with respect to the deformation parameters, β_2, β_4, and γ, for a number of rotational frequencies up to 400 keV. From the calculated deformations, Q_t values were determined by use of the formula,

$$Q_t = \frac{6}{\sqrt{15\pi}} Zer_0^2 A^{2/3} \beta_2 (1 + 0.360\beta_2) \cos(30° + \gamma), \tag{4}$$

where r_0 = 1.20 fm. The theoretical Q_t values determined in this manner are shown in Fig. 6 as a solid line. The dashed curve in Fig. 6 is based on the CHFB calculations of Wyss et al.[19] They used a reduction to 75 percent for the neutron pairing strength over the entire frequency range. This choice is appropriate for the frequencies above the backbend. The proton pairing was held at full strength over the entire range; however, the proton role in this frequency range is minimal. We note that the dynamic moment of inertia $\Im^{(2)}$ suggests multiple band crossings at low $\hbar\omega$, and we still must consider this somewhat remote possibility—it is just that there seems to be no strong reason to argue that the $h_{9/2}$ protons are playing a role.

The nucleus ^{172}Os with N = 96 appeared to be another good case on which to perform lifetime measurements as a further test of these CSM predictions for nuclei near the middle of the $i_{13/2}$ neutron shell. As already pointed out, however, the level features of ^{172}Os were not well determined. Previous spectroscopy studies[15,16] of ^{172}Os had defined the yrast sequence up to I = 24$^+$, but they provided only very limited information on side bands. Both Durell et al.[15] and Wells et al.[16] reported two anomalies in the yrast sequence moment of inertia below a rotational frequency of $\hbar\omega$ = 0.27 MeV. The first of these anomalies occurs at unusually low spin (I = 8$^+$) and at a rotational frequency $\hbar\omega$ = 0.24 MeV. The second is at I = 16$^+$ and $\hbar\omega$ = 0.27 MeV. One of the suggestions [16,20] for the anomaly at I = 8$^+$ is that there may be early alignment of $h_{9/2}$[541]1/2 protons.

The rather rich experimental data we obtained recently on ^{172}Os has been complemented by CHFB calculations done by J. Dudek and reported in Ref. 12. In these, a Woods-Saxon potential was used and both the related particle-number projection technique and the effects of pairing were incorporated. The results of these calculations for the routhians of the $i_{13/2}$ quasi-neutrons with K quantum numbers of 5/2, 3/2, and 1/2 are shown in Fig. 10*a*. Note that the order of these orbitals has been inverted once pairing is included self-consistently. A most striking feature of these routhians is the strongly-repulsive interaction between these low-Ω orbitals and the narrow frequency range over which this takes place. As seen, the AB crossing occurs first at $\hbar\omega_c \approx$ 0.25 MeV, followed very closely by the BC and AD crossings at $\hbar\omega_c \approx$ 0.26 MeV. This figure also shows the effects of rotation on the pairing gap (a drop) in the vicinity of the band crossings. A similar calculation for the proton system gives rise to the quasiparticle spectrum shown in Fig. 10*b*. Based on this evidence that $h_{9/2}$ quasiprotons still appear inactive at $\hbar\omega$ < 0.4 MeV, we must conclude that it is unlikely that they play a significant role in the properties of the yrast sequence below $I \approx$ 20.

Because of the strong mixing expected between the low-lying quasineutron configurations in Fig. 10*a*, it may be somewhat difficult to give a unique interpretation to each of the nearly degenerate crossing frequencies. However, we feel that it is possible to understand this complex situation for the (+,0) configurations in ^{172}Os by using the simple standard band-crossing picture that follows.

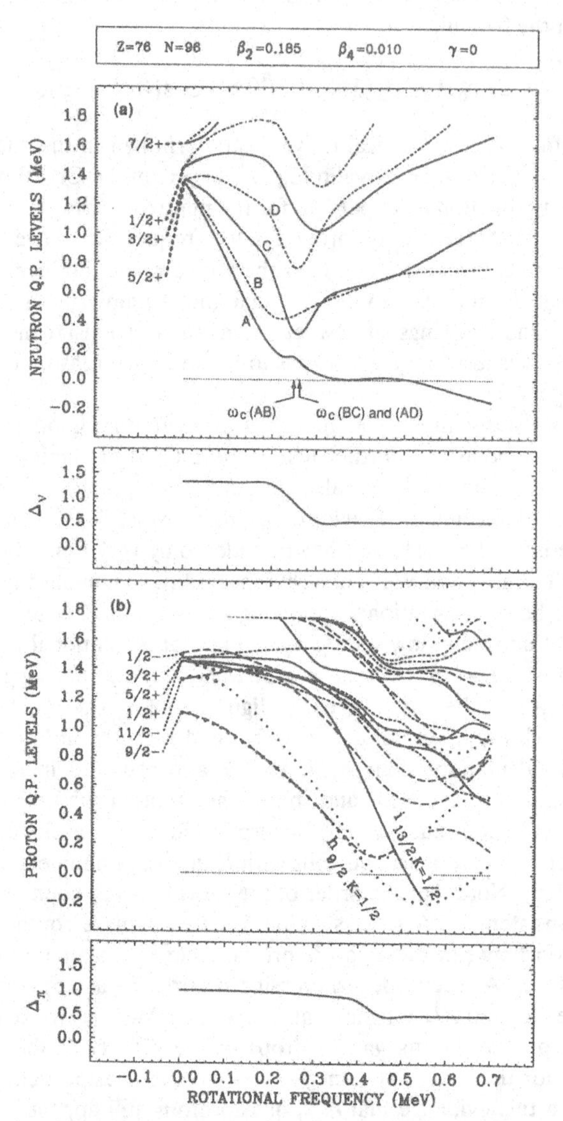

Figure 10. *a*) The quasiparticle orbitals in the rotating coordinate frame (routhians), where the effects of pairing have been included for the neutrons at N = 96. For simplification, only the lowest positive parity levels are shown here. The bottom frame shows the properties of the pairing gap Δ calculated using the CHFB theory, but without particle number projection. *b*) similar to that in *a*), but for the protons. In contrast to *a*), all the quasiparticle levels are shown here.

In this mass region, the first crossing of the ground-state rotational band is by the most alignable pair of $i_{13/2}$ neutrons, usually denoted as the *s* or *AB* band. In most cases it has not been possible to observe the continuation of the ground band through this first crossing. In a few cases, however, this continuation has been followed (e.g., see Refs. 21-23). In these it was found that another crossing occurs soon after the first *AB* crossing and this was attributed to the *BC* and *AD* alignments. In ^{172}Os, we have also observed the continuation of the ground band through the first crossing (*AB*) region (labeled Band 1 in Fig. 9). The high alignment gain seen in these states indicates that the *BC* (or *BC* and *AD*) alignments probably have already taken place and at a very similar frequency to the band crossings in the yrast sequence. In the negative-parity sidebands the *BC* crossing is seen at $\hbar\omega \approx 0.28$ MeV. It is interesting that the ground-*BC* (or *BCAD*) configuration changes occur at slightly lower frequency than the corresponding BC alignments in the negative parity bands (e.g., *AE* → *AEBC*). This same situation was seen for ^{158}Er (Refs. 21, 23), where the frequency difference is $\hbar\omega = 0.03$ MeV.

In summary, this coming together of excited (+,0) quasiparticle bands at low spin and the observation of a strong interaction strength of the *BC* alignment leads us to the following proposal for the explanation of the two low-spin anomalies in ^{172}Os. The first (*AB*), second (*BC*), and third (*AD*) band crossings occur at very similar frequencies ($\hbar\omega$ = 0.26–0.28 MeV). However, because the interaction strength of the *BC* crossings is strong, the frequency range for the ground-*BC*(*AD*) crossing is smeared out sufficiently so as to perturb the ground band slightly below the *AB*-crossing region. This would explain the first anomaly in the yrast sequence at about spin 8 and $\hbar\omega = 0.24$ MeV. The second anomaly at $\hbar\omega = 0.26$ MeV is proposed to be the standard crossing by the *s* (*AB*) band.

In light of the nearly degenerate crossings of three bands in ^{172}Os, it is interesting to speculate on what effect this will have on the lifetimes of the states in this vicinity of rotational frequency. This effect is revealed in the ^{172}Os Q_t values, plotted as a function of spin in Fig. 7. As seen, there is an enhancement in the collectivity in the vicinity of the 6^+ and 8^+ states where the first anomaly in $\mathfrak{J}$ occurs, while for the states above 14^+ there is a decrease in collectivity. The fact that we find an enhanced collectivity for these states goes counter to earlier suggestions that the $h_{9/2}$ protons in the $1/2^-[541]$ orbital may be responsible for the behavior of the moment of inertia at these spins. The influence of this orbital is to drive the shape toward positive γ deformation (noncollective triaxiality), which should produce a drop in Q_t values. The dashed lines, corresponding to Q_t values computed from the deformations obtained in the CHFB calculations and Eq. 3 mark the respective trends predicted[12] for the shape evolution.

In further theoretical considerations of the ^{172}Os Q_t values, Donau and Zhang (to be reported in Ref. 24) have calculated the overlap matrix of the wave-functions of these states in the yrast sequence of ^{172}Os, where alignment of $i_{13/2}$ neutrons only is assumed. For this, the generator-coordinate method was used to obtain the information on the dynamic aspects of the shape changes in a band, followed by solving the Hill-Wheeler equation numerically in two-dimensional space, i.e., in ω and β_2. With the new deformation parameters emerging from these calculations, we have determined Q_t values and these give rise to the solid curve shown in Fig. 7. It is interesting that the trend of stretching in the 6^+, 8^+, 10^+, and 12^+ states is reasonably well reproduced from these calculations, which also indicate a drop in collectivity beyond the backbend, the same behavior as shown by the experimental data.

In summary, the major conclusion to be drawn from this work is that the experimental data are reasonably consistent with the cranking theories, which predict that only the $i_{13/2}$ neutrons play a major role in the rotation-alignment processes in the frequency ranges covered in the ^{170}W, ^{172}W, and ^{172}Os measurements. Further, the complex behavior of ^{172}Os in the frequency range of $\hbar\omega \approx 0.25$ MeV is accounted for in the refined CHFB calculations; viz., there are three nearly degenerate band crossings, all by $i_{13/2}$ quasi-neutrons. Finally, the lifetime data for ^{172}Os are semiquantitatively accounted for by the calculations of overlap matrix elements with the Hill-Wheeler equation, where again only $i_{13/2}$ quasi-neutrons are invoked.

ACKNOWLEDGMENTS

I would like to express thanks to my numerous colleagues who have made major contributions to this work. These include F. K. McGowan, J. C. Wells, A. Virtanen, M. N. Rao, I.-Y. Lee, C. Baktash, M. Oshima, J. W. McConnell, A. Larabee, L. L. Riedinger, R. Bengtsson, Z. Xing, Y. S. Chen, P. B. Semmes, G. A. Leander, M. Kortelahti, V. P. Janzen, M. A. Riley, J. Dudek, F. Dönau, and J. Y. Zhang. Oak Ridge National Laboratory is operated by Martin Marietta Energy Systems, Inc., for the U. S. Department of Energy under contract No. DE-AC05-840R21400.

REFERENCES

1. H. Emling, E. Grosse, R. Kulessa, D. Schwalm, and H. J. Wollersheim, *Nucl. Phys.* **A419**, 187 (1984).
2. N. R. Johnson, in *Proceedings of the INS International Symposium on the Dynamics of Nuclear Collective Motion, Mt. Fuji, 1982*, Ed. by K. Ogawa and K. Tanabe (University of Tokyo, Tokyo, 1982), p 144.
3. M. P. Fewell, N. R. Johnson, F. K. McGowan, J. S. Hattula, I.-Y. Lee, C. Baktash, Y. Schutz, J. C. Wells, L. L. Riedinger, M. W. Guidry, and S. C. Pancholi, *Phys. Rev. C* **31**, 1057 (1985).
4. M. P. Fewell, N. R. Johnson, F. K. McGowan, J. S. Hattula, I.-Y. Lee, C. Baktash, Y. Schutz, J. C. Wells, L. L. Riedinger, M. W. Guidry, and S. C. Pancholi, *Phys. Rev. C* **37**, 101 (1988).
5. M. Oshima, N. R. Johnson, F. K. McGowan, C. Baktash, I.-Y. Lee, Y. Schutz, R. V. Ribas, and J. C. Wells, *Phys. Rev. C* **33**, 1988 (1986).
6. F. K. McGowan, N. R. Johnson, Y. Schutz, C. Baktash, A. J. Larabee, J. C. Wells, and I.-Y. Lee, *ORNL Physics Division Progress Report for Period Ending September 30, 1988, ORNL-6508*, p 77; full paper to be submitted to *Physical Review C.*
7. G. A. Leander, S. Frauendorf, and F. May, in proceedings of conference on *High Angular Momentum Properties of Nuclei, 1983*, Ed. by N. R. Johnson (Harwood Academic Publishers), p 281.
8. F. K. McGowan, N. R. Johnson, M. N. Rao, I.-Y. Lee, C. Baktash, J. C. Wells, M. Kortelahti, and V. P. Janzen, *ORNL Physics Division Progress Report for Period Ending September 30, 1987, ORNL-6508*, p 87.
9. M. N. Rao, N. R. Johnson, F. K. McGowan, I.-Y. Lee, C. Baktash, M. Oshima, J. W. McConnell, J. C. Wells, A. Larabee, L. L. Riedinger, R. Bengtsson, Z. Xing, Y. S. Chen, P. B. Semmes, and G. A. Leander, *Phys. Rev. Lett.* **57**, 667 (1986).
10. F. K. McGowan, N. R. Johnson, I.-Y. Lee, and C. Baktash, *ORNL Physics Division Annual Report for Peiod Ending September 30, 1988, ORNL-6508*, p 67.
11. A. Virtanen, N. R. Johnson, F. K. McGowan, M. A. Riley, I.-Y. Lee, C. Baktash, J. Dudek, and J. C. Wells, *ORNL Physics Division Progress Report for Period Ending September 30, 1988, ORNL-6508*, p 65.
12. J. C. Wells, N. R. Johnson, C. Baktash, I.-Y. Lee, F. K. McGowan, M. A. Riley, A. Virtanen, and J. Dudek, *Phys. Rev. C* **40**, 725 (1989).
13. N. R. Johnson, J. W. Johnson, I.-Y. Lee, J. E. Weidley, D. R. Haenni, and J. R. Tarrant, *ORNL Physics Division Progress Report for Period Ending November 1981, ORNL-5787*, p 147.

14. J. C. Wells, M. P. Fewell, and N. R. Johnson, *ORNL Technical Memo ORNL-TM-9105*, 1985 (unpublished).
15. J. L. Durell, G. D. Dracoulis, C. Fahlander, and A. P. Byrne, *Phys. Lett.* **115B**, 367 (1982); *Australian University Annual Report* ANU-p/284, 25 (1981).
16. J. C. Wells, N. R. Johnson, C. Baktash, I.-Y. Lee, F. K. McGowan, M. N. Rao, L. L. Riedinger, V. Janzen, W. C. Ma, Shuxian Wen, Ze-Min Chen, P. B. Semmes, G. A. Leander, and Y. S. Chen, *Phys. Rev. C* **36**, 431 (1987).
17. J. Recht, Y. K. Agarwall, K. P. Blume, M. Guttormsen, H. Hubel, H. Kluge, K. H. Maier, A. Maj, N. Roy, D. J. Decman, J. Dudek, and W. Nazarewicz, *Nuclear Physics* **A440**, 336 (1985).
18. M. N. Rao, N. R. Johnson, F. K. McGowan, I.-Y. Lee, C. Baktash, M. Oshima, J. W. McConnell, J. C. Wells, A. Larabee, L. L. Riedinger, R. Bengtsson, Z. Xing, Y. S. Chen, P. B. Semmes, and G. A. Leander, *Phys. Rev. Lett.* **57**, 667 (1986).
19. R. Wyss, R. Bengtsson, and A. Johnson, private communication.
20. G. D. Dracoulis, R. A. Bark, A. E. Stuchbery, A. P. Byrne, A. M. Baxter, and F. Riess, *Nucl. Phys.* **A486**, 414 (1988).
21. M. A. Riley, J. Simpson, J. F. Sharpey-Schafer, J. R. Cresswell, H. W. Cranmer-Gordon, P. D. Forsyth, D. Howe, A. H. Nelson, P.J. Nolan, P.J. Smith, N.J. Ward, J.C. Lisle, E. Paul, and P. M. Walker, *Nucl. Phys.* **A486**, 456 (1988).
22. J. Simpson, P. A. Butler, P. D. Forsyth, J. F. Sharpey-Schafer, J. D. Garrett, G. B. Hagemann, B. Herskind, and L. P. Ekstrom, *J. Phys.* **G10**, 383 (1984).
23. J. Simpson, M. A. Riley, J. R. Cresswell, D. V. Elenkov, P. D. Forsyth, G. B. Hageman, D. Howe, B. M. Nyako, S. Ogaza, J. C. Lisle, and J. F. Sharpey-Schafer, *J. Phys.* G13, 847 (1987).
24. A. Virtanen, N. R. Johnson, F. K. McGowan, M. A. Riley, I.-Y. Lee, C. Baktash, J. Dudek, J. C. Wells, F. Dönau, J. Y. Zhang, and L. L. Riedinger, to be published.

23. Discrete-Line Spectroscopy in Superdeformed Nuclei

V. P. Janzen, J. K. Johansson, J. A. Kuehner,
D. Radford, J. C. Waddington, D. Ward,
and the 8π Collaboration
AECL Research, Chalk River Laboratories
Chalk River, Ontario, Canada K0J1J0
and
Tandem Accelerator Laboratory,
McMaster University
Hamilton, Ontario, Canada L8S 4K1

ABSTRACT

The discrete-line spectroscopy of collective structures in nuclei having superdeformed shapes has progressed rapidly over the past three years, and we are now in a position to test the present nuclear models in a systematic manner. This talk will focus on developments from the 8π Collaboration, including the discovery of multiple superdeformed bands within the same nucleus and the use of high-resolution two- and three-dimensional γ-ray correlation techniques to extract band structures having very low intensity.

23.1 INTRODUCTION

Relative to the length of time that nuclear physics has been a distinct discipline, the study of collective structures in the atomic nucleus has had a fairly long history. For instance, nuclear systems having a quadrupole deformation with an exotic 2:1 axis ratio have been known since 1967.[1] However, the discovery[2-3] of an extensive rotational band built on an intrinsic "superdeformed" state in the nucleus ^{152}Dy, signalled a very important advance in the field. It not only provided a striking example of the unique possibilities open to a large-scale quantal system, but also served to emphasize that we now have the technical ability to probe the nucleus to limits once thought beyond reach.

In the four and a half years since that first discovery, a great deal of progress has been made, much of it using the techniques of discrete-line γ-ray spectroscopy. This Chapter will concentrate on recent developments in the spectroscopy of superdeformed nuclei, more specifically those related to the efforts of the 8π Collaboration. The member institutions are Atomic Energy of Canada's Chalk River Nuclear Laboratories, McMaster University, Universite de Montreal, and the University of Toronto. Experiments described here have been performed using the 8π spectrometer,[4] located at the Chalk River Tandem Accelerator Superconducting Cyclotron (TASCC) facility. Data analysis has been carried out at Chalk River and at the universities involved.

The 8π instrument is an example of the second generation of γ-ray spectrometer arrays, in which nearly 4π sr of solid angle are covered by a combination of high- and low-resolution detectors, grouped around a central target. As shown in Fig. 1 the 8π array con-

Figure 1. The 8π spectrometer, opened to show one-half of the inner BGO array. The support structure and some of the outer Ge detectors, which view the target through the small holes in the BGO ball, are also visible.

sists of an inner spherical shell of seventy-two bismuth germanate (BGO) crystals, together with twenty outer germanium (Ge) detectors which view the target through holes in the BGO ball. Each cylindrical Ge crystal is itself virtually surrounded by a separate multi-crystal configuration of BGO detectors, which is used to detect (and subsequently veto the acquisition of) Compton-scattered γ rays.* Typically, an overall γ-ray multiplicity (fold or K), summed energy (H), and a pattern of the detector "hits" are taken from the inner ball for each event having a respectable fold (normally $K \geqslant 10$). These data, in addition to high-resolution γ-ray energies and times from the Ge detectors, are recorded on tape for events in which two or more Ge units have fired.

There are presently eight spectrometers in the world with capabilities comparable to those of the Chalk River system. On the one hand, the 8π Ge array has a very good resolving power, because of *a*) good compton suppresion and *b*) a geometry which leads to a

*The term 8π refers to the goal of 4π coverage for γ rays emitted from the target, plus 4π detection of γ rays compton-scattered from the Ge crystals.

small doppler spread in γ-ray energies. On the other hand, the absolute efficiency is only fair. This last feature limits the statistical accuracy of our triple Ge coincidence data, especially when compared to the HERA array (at LBL). However, as shown in Section 3 we have been able to acquire triples data with useful statistics by running for a two-week period. Partly because of its large number of elements and high efficiency, the inner BGO shell gives very good information on reaction channel characteristics. It surrounds a relatively large volume (14-cm radius) in which additional detection systems may be installed.

One of the more important features of the 8π spectrometer is the associated instrumentation for setting up and extracting information from the device. There are likely few other systems operating at this level of complexity which are easier to use, a characteristic that will hopefully be present in the next generation of multi-detector γ-ray spectrometers. The existing (recently approved) proposal for a 110-Ge plus 360-BGO element GAMMASPHERE array is outlined by I.-Y. Lee in Chapter 21. A second design, from a European collaboration, calls for approximately seventy Ge units in a somewhat different configuration, with alternative proposals having as many as 500 Ge detectors in a more radical geometry.

At the time of this writing there are nine known cases of superdeformation in the $A \approx 150$ mass region, as shown in Table I.† The discovery of a discrete-line superdeformed band[7] in ^{149}Gd pointed out the importance of using a relatively "cold" reaction. By comparing results from $4n$ and $5n$ reactions leading to the same final nucleus, Haas et al. were able to show that the population trapped in the superdeformed well is greatest when the reaction entry region in excitation energy vs angular-momentum space overlaps the yrast line. (The term *yrast* refers to the minimum excitation energy at a given nuclear spin.) ^{149}Gd was the first case to be discovered after ^{152}Dy, and subsequent examples in this mass region have been studied with the same population mechanism.

Another region of superdeformation has been discovered in the $A \sim 190$ mass region by the Argonne/Notre Dame group,[12-14] and confirmed by a Berkeley/Livermore[15] collaboration. Section 2 of the present paper draws some preliminary comparisons be-

Table I. Nuclei Exhibiting Superdeformed Bands in the $A \sim 150$ Region

Nucleus	Reaction	E_{beam} (MeV)	Spectrometer	$Q_0(eb)$	Ref.
^{146}Gd	(^{40}Ar,$4n$)	175-180	OSIRIS	12 ± 2	5
^{148}Gd	(^{29}Si,$5n$)	150	HERA		6
^{149}Gd	(^{30}Si,$5n$)	150	8π	17 ± 2	7
^{150}Gd	(^{26}Mg,$6n$)	145	TESSA3	17 ± 2	8
^{150}Tb	(^{31}P,$5n$)	160	HERA		9
^{151}Tb	(^{27}Al,$6n$)	150	TESSA3		8
^{151}Dy	(^{34}S,$5n$)	175	ANL-ND		10
^{152}Dy	(^{48}Ca,$4n$)	205	TESSA3	19 ± 3	2,3
^{153}Dy	(^{34}S,$5n$)	165	8π		11

† Note that, in this chapter, we will use the term "superdeformed" to refer to structures which have been inferred to have a prolate shape with a deformation parameter $\beta_2 \sim 0.6$, corresponding to an ellipsoidal axis ratio of $\sim$ 2:1. This somewhat-arbitrary distinction excludes many examples of larger-than-normal deformations in the $\beta_2 \sim 0.4$ range but is probably an accurate reflection of the current terminology.

tween observed characteristics of the structures in the mass-150 and -190 regions.

Not surprisingly, there have been failed attempts to identify superdeformed band structures, for example, in ^{153}Ho,[16] ^{147}Eu,[17] and ^{144}Gd.[18-19] However, it is possible those structures do exist but have not been seen, due to very low γ-ray intensity and/or inappropriately chosen reaction characteristics.

The standard method used in searching for band structures involves the construction of two-dimensional Ge $E\gamma_1$-$E\gamma_2$ matrices, and subsequent gating or slicing to reveal coincidence relationships. Figure 2 shows a ^{149}Gd γ-ray spectrum obtained by slicing just such a matrix, which had been extracted from ^{124}Sn(^{30}Si,*xn*) data with BGO summed energy and fold (*H,K*) windows chosen to enhance the 5*n* reaction channel. Here a gate on a single peak, which is shown in the inset, provided a very clean spectrum of the regularly-spaced sequence of γ rays corresponding to a superdeformed rotational band. In the $A \approx 150$ region, typically fifteen to eighteen such $E2$ transitions in a cascade remove $\approx$20 MeV of excitation energy and 30–36$\hbar$ of angular momentum from the rotating nucleus. The intensities are typically $\approx$1% of the total reaction channel intensity and close to the limits of conventional discrete-line spectroscopy.

The spacing between consecutive $E2$ transitions in a band can be related directly to the rotational parameter $\Im^{(2)}$, or dynamic moment of inertia, via $\Im^{(2)} = 4/\Delta E\gamma \ \hbar^2/\mathrm{MeV}$. Assuming the nucleus has a prolate ellipsoidal shape, we may use $\Im^{(2)}$ to infer a deformation parameter β_2. In this case, the average energy spacing of 52 keV corresponds to a dynamic moment of inertia $\Im^{(2)} = 77\hbar^2/\mathrm{MeV}$, indicating that $\beta_2 \geqslant 0.5$. With good data from a solid target the forward-backward doppler-shift difference due to the recoil of the residual nuclei can be used to measure an average quadrupole moment, Q_0. This value represents a model-independent measure of the collectivity and is related directly to β_2. So far this measurement has been possible in six nuclei; namely, ^{152}Dy, ^{149}Gd, ^{150}Gd, ^{146}Gd, ^{191}Hg, and ^{192}Hg. For ^{191}Hg some of the γ-ray lineshapes have also been analysed. In all

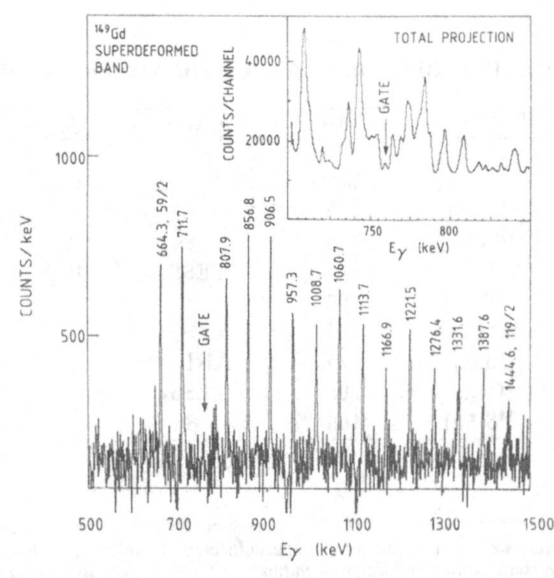

Figure 2. Coincidence γ-ray spectrum gated on the 759.5 keV member of the ^{149}Gd superdeformed band cascade.[7] The inset contains a portion of the total γ-ray projection spectrum, showing the gating transition.

six cases, the average quadrupole moments are extremely large (cf. Table I) leading to transition matrix elements $B(E2) \approx 2 \times 10^3$ single-particle units, which are indicative of an extremely high degree of collectivity and deformations in the range $\beta_2 = 0.5$–0.6. The results are consistent with those extracted from the dynamic moments of inertia, giving us some confidence that we can use the $\Im^{(2)}$ values to extract nuclear structure information.

23.2 PRELIMINARY RESULTS FROM THE FIRST 8π TRIPLE-COINCIDENCE EXPERIMENT

The ^{149}Gd spectrum shown in Fig. 2 is of unusually high quality and is not representative of most spectra obtained for superdeformed bands. There are cases where indications of superdeformation have only been found in γ-ray continuum ridge structure (e.g., Ref. 20). There are also data that probably contain useful but weak discrete-line information; however, with standard gating techniques it is impossible to extract a clear set of mutually-coincident transitions. For example, the weak superdeformed peaks are often obscured by unrelated transitions having much higher intensity. In this section, we discuss some novel methods for dealing with such problems.

A most striking feature of the superdeformed rotational bands is the long cascades of coincident transitions with very regular energy spacing (corresponding to a relatively constant moment of inertia), and a number of methods for examining the associated correlations have been proposed (e.g., Refs. 21,22). In this article we concentrate on the procedures developed within our collaboration, principally by J. Kuehner at McMaster University, and later extended by D. Radford at Chalk River. In an $E\gamma$-$E\gamma$ matrix the ^{149}Gd superdeformed band appears as an array of sharp "mountains" of intensity, placed on a regular grid as shown in Fig. 3. With the "Band-Aid" technique of Kuehner et al.,[24] it is possible to examine the matrix for such periodic coincidence patterns in two dimensions. The contribution of uncorrelated background events is first removed. Then, the net intensity arising from a two-dimensional lattice of gates is extracted from the data, as the lattice "slides" along the $E\gamma_1 = E\gamma_2$ diagonal axis of the matrix. The result is a *Diagonal Figure of Merit* (DFM) spectrum (an example from the ^{153}Dy three-dimensional analysis is shown in Fig. 4). Since only weak coincidence peaks are expected, an upper limit is placed on the contribution to the *DFM* from any single gate. This prevents a small number of very intense peaks from dominating the correlation. A cascade of m γ rays results in $m(m-1)$ peaks in the two-dimensional data, and for the correct energy spacing a strong signal is possible. Clearly, the longer the cascade of transitions, the stronger the two-dimensional correlation signal is relative to the noise.

Once an unambiguous correlation signal has been detected, γ-ray spectra can be extracted for one-dimensional gates chosen appropriately. In the last analysis, convincing proof of a rotational band requires that every transition be shown in coincidence with every other transition. For weak bands that may not be possible; however, strong evidence of such a relationship may be found by examining the relative contributions to the DFM from each of the m rows of gates which make up the test lattice.

The procedure is designed such that *a*) it is highly interactive and incorporates a high degree of visual feedback, and *b*) it will run on a microcomputer with reasonable speed. A comprehensive explanation of the method and its more sophsticated features will be given in Ref. 24.

The extension of this method to three dimensions is a natural step in the evolution of discrete-line correlation techniques. It has been shown that an increase in the dimen-

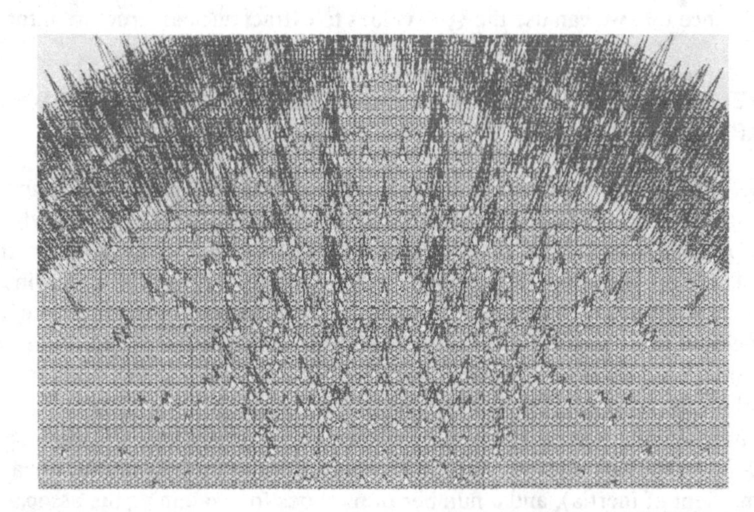

Figure 3. Isometric projection of the ^{149}Gd γ-γ matrix, showing a region above 1 MeV where some of the individual superdeformed coincidence peaks are visible. Uncorrelated background events have been subtracted using the method described in Ref. 2-3. A "valley" with very few events can be seen along the $E_{\gamma 1} = E_{\gamma 2}$ axis, which, in this view, runs from the top center (lower energy) to the bottom center (higher energy) of the diagram. Intense coincidence features due to non-superdeformed structures appear in the top half of the figure. (Original by M. Thompson.)

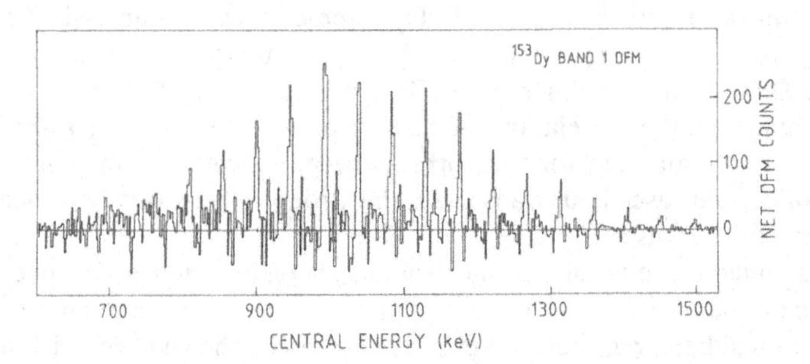

Figure 4. *Diagonal Figure of Merit* vs central energy, for a trial correlation lattice of 15 consecutive transitions having a 46-keV spacing. The upper threshold for discriminating out individual gate contributions, due to intense impurities, has been set to two counts.

sionality of γ-ray coincidence data should result in a corresponding increase in effective resolving power by a factor of 5–10 (see, for example, I.-Y. Lee, Chapter 15 in this book). A cube of γ-γ-γ coincidence data should therefore offer more posibilities than a square of γ-γ concidence data, provided its features are statistically significant.

The triples data may be analysed either through a selective reduction to one or two dimensions[21,22] or by truly three-dimensional methods.[21,25] There is clearly a great deal one may accomplish with a high-quality set of γ-γ-γ data, but we will restrict the present discussion to the potential for detecting and measuring the characteristics of superdeformed bands with the correlation approach outlined above.

In the γ-γ-γ case a cubic trial lattice is tested on a three-dimensional coincidence matrix. In order to store the data on our computer, we truncate the energy range and dispersion to 600-1500 keV and 2 keV/channel, respectively. A DFM spectrum is determined by the extraction of net counts from narrow cubic gates at the lattice points for each point along the $E\gamma_1 = E\gamma_2 = E\gamma_3$ axis of the cube. As in the two-dimensional case, an upper limit is placed on the contribution to the DFM from any one gate. By experience, we have set this limit to two counts. Although this seems surprisingly small, we can estimate that if a matrix contains $\approx 10^8$ events, then a single gate in a band of fifteen transitions carrying 0.5% of the total γ-ray flux has on the average a net count of the order of *one*. That a clear signal can be extracted from the data is an indication of the highly correlated nature of these bands.

Ideally, a peak in the DFM spectrum corresponds to the identification of a coincidence feature in the data, with a characteristic spacing equal to that of the trial lattice. In reality, it is possible to have spurious peaks that result from intense unrelated transitions and/or imperfect background subtraction. However, other characteristics such as the existence of companion "anti-peaks" may be used to point out false signals. To carry out a complete search for band structures, trial lattices having different inter-gate spacings and numbers of gates may be used. One can envision improvements and variations on the procedure, such as allowing a degree of "stretching" in the inter-gate spacing corresponding to a variation in the band moment of inertia (already accomplished in "Band-Aid"). The three-dimensional procedure is more difficult to implement on a microcomputer than was "Band-Aid" since the computational power needed is quite a bit greater. The current triples DMERIT code runs non-interactively on a mainframe computer, although a microcomputer-based three-dimensional version of "Band-Aid" is under development.[26]

The DFM spectrum shown in Fig. 4 has been extracted from our ^{153}Dy triple-coincidence cube. The experiment and results of the conventional γ-γ analysis are described in the following section. One can see that the most intense superdeformed band in that nucleus ($\approx$0.25% of the total flux into ^{153}Dy) produces a very clear correlation signal. Judging by the lack of extraneous positive or negative peaks, an upper threshold of two counts for individual gate contributions, in conjunction with a simple background subtraction method seems to work fairly well.

The use of three-dimensional correlations to search out regular band structures is in its infancy. In ^{153}Dy the non-yrast superdeformed bands (see Section 3) were discovered independently by examining the two- and three-dimensional data, but the degree of correlation was stronger in three dimensions. Proper methods of selective double-gating and three-dimensional background subtraction have yet to be perfected, but the γ-γ-γ analysis of highly regular, correlated structures may in the future prove to be more sensitive than γ-γ techniques.

23.3 SPECTROSCOPY IN THE SUPERDEFORMED MINIMUM

A crucial step in the study of nuclei with superdeformation is to measure the characteristics of yrast *and* excited structures whilst the nucleus is trapped in the superdeformed well of the nuclear potential energy surface. The existence of a well-developed second potential minimum is well known in ^{152}Dy, and it is now accepted that its formation results from the occupation of "high-*N*" orbitals originating from $\pi i_{13/2}$ (principal oscillator quantum number $N = 6$) and $\nu j_{15/2}$ ($N = 7$) shell-model states. Thus, it has become common practice to label superdeformed bands in terms of the number of these intruder orbitals that are occupied.[27,28] In this nomenclature a $\pi 6^4 \nu 7^2$ configuration is calculated to be yrast in ^{152}Dy with $Z = 66$, $N = 86$, at $\beta_2 \approx 0.6$. It is also often useful to note the parity and signature quantum numbers, as is done in the standard cranked shell-model analysis.

Comparisons between theoretical and experimental moments of inertia, $\Im^{(2)}$, have made it possible to suggest individual band configurations in the $A \approx 150$ region.[27-30] According to those calculations there are a number of orbitals available beyond the ^{152}Dy energy gap for the eighty-seventh neutron and the sixty-seventh proton, so the coupling of either type of particle to the superdeformed "core" should lead to a number of superdeformed bands in ^{153}Dy and ^{153}Ho, respectively. The observation of multiple bands within the same nucleus, and measurements of their characteristic moments of inertia and intensities, should make it possible to identify the positions of individual orbitals more stringently than comparisons based on neighbouring nuclei. Such information may also be used to compare the different theoretical approaches.

High-spin states in ^{153}Dy were populated by the ^{124}Sn(^{34}S,5*n*) reaction at 165–167 MeV in two experiments. Additional details and a description of the results are given in Ref. 11. A total of 3.4×10^8 γ-γ events, gated on *H* and *K* to enhance the 5*n* channel, were accumulated. The second of the two experiments was specifically designed to acquire high-quality triple γ-ray coincidence data, and resulted in 7.0×10^7 *H*-, *K*-gated γ-γ-γ events. After initial searches with conventional gating methods failed to uncover any clear band structures, the resulting two- and three-dimensional matrices were analysed with the techniques discussed in the previous section. Three superdeformed rotational bands were found, as shown in Fig. 5. In fact, Bands 2 and 3 were discovered during independent analysis of the three- and two-dimensional data. Band 1, which is most strongly populated, is estimated to extend in spin from $61/2\,\hbar$ or $63/2\,\hbar$ to $109/2\,\hbar$ or $111/2\,\hbar$.

With intensities of only 0.25, 0.18, and 0.13% of the reaction channel, it is clearly not possible to measure the quadrupole moments of such structures. In fact only three gates, one in Band 1 and two in Band 2, are relatively uncontaminated and strong enough to show most of the other members of their respective cascades. The claim that they are superdeformed depends on extracted moments of inertia, $\Im^{(2)}$, as it does for the majority of instances in this mass region. In addition to being sensitive indicators of orbital configurations, the $\Im^{(2)}$ values have proven to be entirely consistent with Q_0 measurements when available.

The ^{153}Dy dynamic moments of inertia are shown in Fig. 6. From theory the three lowest negative-parity bands are predicted to have a nearly constant $\Im^{(2)}$ over the frequency range that is observed experimentally. The positive-parity configurations, involving $\pi 6^3$, are expected to have a strongly decreasing $\Im^{(2)}$ over this range. With this in mind, the assignments have been made as follows: Band 1, $\pi 6^4 \nu 7^3$ with negative parity and signature, $(-1/2)$; Bands 2 and 3 $(\pm 1/2)$ $\pi 6^4 \nu 7^2$. The $\Im^{(2)}$ predictions from Ref. 11 have been shown along with the experimental data in Fig. 6. One can see that the general trend has been

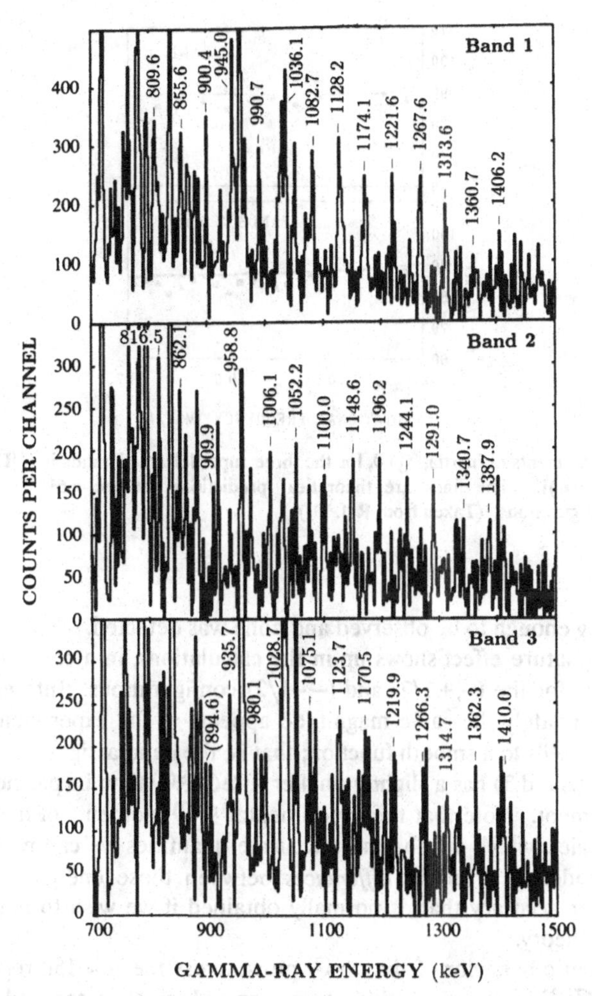

Figure 5. Efficiency-corrected coincidence summed spectra for the three superdeformed bands in ^{153}Dy. Transitions which are band members are labelled by their energies. (Taken from Ref. 11.)

calculated correctly, although the theory overestimates the values for $\Im^{(2)}$ slightly.

Since the $\pi 6^4$ configurations have been observed and not the $\pi 6^3$, the $Z = 66$ gap must be somewhat larger than that obtained in the calculations of Ref. 11. Otherwise, the $\pi 6^3$ configuration would have been yrast in the feeding region and would therefore have been populated. In fact the $Z = 66$ gap is larger in the Woods-Saxon model calculations of Ref. 29.

The similarity of the $\Im^{(2)}$'s for Bands 2 and 3 indicates that their assignment as signature partners is reasonable. Indeed, the combined series of transition energies forms a very regular sequence which has a spacing approximately half that of the individual band energy spacings. The predicted signature splitting at $\hbar\omega \approx 0.7$ MeV suggests that the $\pi 6^4 \nu 7^2$ (–, +1/2) band should be more strongly populated than its signature partner and therefore has been assigned to Band 2. No $M1$ transitions between these bands are ex-

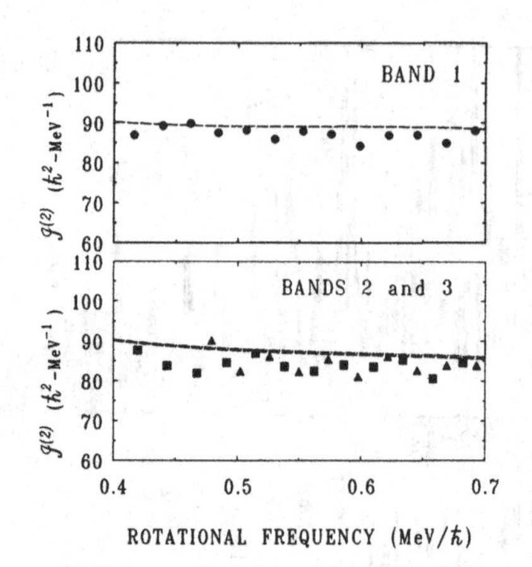

Figure 6. Dynamic moments of inertia, $\mathcal{J}^{(2)}$, for the three superdeformed bands in ^{153}Dy [Band 1(●), Band 2(■), and Band 3(▲)]. The lines are theoretical predictions for the $\pi 6^4 \nu 7^3$ (−, −1/2) and the $\pi 6^4 \nu 7^2$(−, ±1/2) configurations. (Taken from Ref. 11.)

pected to be strong enough to be observed and none was detected.

Another signature effect shows up in the calculations, in a comparison of the $\mathcal{J}^{(2)}$ moments of inertia for the (−,+1/2) and (−,−1/2) configurations. Interestingly enough, a splitting of approximately the same magnitude appears in the experimental values if one fits the extracted $\mathcal{J}^{(2)}$'s to a smooth function; that is, the band assigned $\alpha = -1/2$ from intensity arguments (Band 3) has a slightly smaller fitted $\mathcal{J}^{(2)}$ than its partner Band 2 in both theory and experiment. Note that the experimental $\mathcal{J}^{(2)}$ moments of inertia depend only on the γ-ray energies, which are the most accurate quantities we can measure. However, the specific dependence is on the *differences* between those energies; and, in fact, we require even better accuracy than is normally obtained if we wish to make such detailed comparisons with theory.

The systematic behavior of the $\mathcal{J}^{(2)}$ plots across the $A \approx 150$ region has been explained quite well[27-31] by the successive occupation of high-N proton and neutron orbitals, progressing from $Z = 64$ to 66 and $N = 82$ to 87. In ^{153}Dy all three bands include the four lowest orbitals of the N = 6 proton subshell; i.e., $\pi 6^4$. As discussed above, the neutron contributions have been assigned as $\nu 7^3$ (Band 1) and $\nu 7^2$ (Bands 2,3). Based on the theoretical systematics of Bengtsson et al. Bands 2 and 3 should therefore possess $\mathcal{J}^{(2)}$ moments of inertia that have a greater downward slope than does the Band 1 $\mathcal{J}^{(2)}$, as a function of rotational frequency. This trend agrees with the experimental results and also with the analogous comparison between Bands 2, 3, and the superdeformed band in ^{152}Dy, which has been assigned the same high-N configuration as Band 1 in ^{153}Dy ($\pi 6^4 \nu 7^2$). Recently, multiple superdeformed bands have been observed in the nucleus ^{149}Gd,[26] where it should be possible to make a similar comparison with regard to the successive occupation of neutron $N = 7$ orbitals.

The majority of known superdeformed bands in this mass region exhibit moments of inertia that are either effectively flat or decrease as a function of rotational frequency.*

*The ^{151}Dy band does depart from that trend but shows only a small rise in $\mathcal{J}^{(2)}$. There is also recent evidence in ^{146}Gd (Ref. 5) for a sudden increase in $\mathcal{J}^{(2)}$ at high frequency, which has been explained using arguments similar to those discussed here.

This tendency is more or less echoed by the Nilsson-Strutinsky cranking theory and its Woods-Saxon counterpart.[29] Indeed, within the framework of the former type of calculation, it is virtually impossible to obtain a rising $\Im^{(2)}$ without promoting particles into the top half of the intruder subshells. Even then a significant increase in the slope of $\Im^{(2)}$ may not result. As outlined in Ref. 28, to invoke higher-lying configurations is to drastically increase the dependence on individual parameters, since the high-N orbitals above the fourth member of the subshell become much more mixed with higher-lying configurations, precluding simple arguments based on the "purity" of the first four or so orbitals in the $\pi 6$ and $\nu 7$ subshells. Indications are that a substantial rise in $\Im^{(2)}$ as a function of $\hbar\omega$ is an unlikely consequence of the unpaired Nilsson-Strutinsky theory.

The behaviour of the dynamic moments of inertia for the recently-discovered Hg superdeformed bands is therefore unexpected, especially since the predicted configuration $\pi 6^4 \nu 7^4$ does not include any of the high-lying orbitals from either high-N subshell. Both ^{191}Hg (Ref. 6) and ^{192}Hg (Ref. 13) show a pronounced ($\approx$ 30%) rise in $\Im^{(2)}$. The only case where such a trend has been calculated in the $A \approx 150$ region is the yrast configuration of ^{148}Eu, for which the paired Woods-Saxon theory was used. There, a strikingly similar rise in the $\Im^{(2)}$ is a result not of the potential used nor of the position of individual orbitals but is caused by a large-interaction band crossing due to the alignment of a pair of $N = 6$ protons.[29] It would not be surprising if the Hg features were found to be caused by a similar type of band crossing, due to the alignment of a pair of either $N = 7$ protons or N = 8 neutrons, with the former probably favoured. Such an explanation is even more likely in the Hg region, since the superdeformed bands are observed in a lower spin regime (10 - 32$\hbar$) compared to the $A \approx 150$ region (24–60$\hbar$). As discussed elsewhere, (e.g., Refs. 14,15) pairing may thus play a role in the structure of these bands at low to medium spins. Clearly, additional experiments to show the variation of $\Im^{(2)}$ across the $A \approx 190$ region of superdeformation are required, so that blocking arguments, etc., can be used to draw more specific conclusions.

23.4 LOOKING AHEAD

The study of superdeformation in atomic nuclei with discrete-line γ-ray spectroscopy has progressed quite rapidly from its beginning just three and a half years ago. Our experimental capabilities have increased markedly, and we now have a good number of sophisticated detector arrays contributing high-quality data. New analysis techniques have pushed the limits of detectability to the 0.1% level for some superdeformed structures, once thought unreachable with the current generation of instruments.

But what lies ahead in the field? There are likely other examples of unusual structures having deformations of the same order as those discussed here which can test our current understanding of the physics involved. One example is the predicted large-deformation triaxial minima in the Hf isotopes.[31] Possibly the richest prize will be the discovery of rotational bands belonging to so-called hyperdeformed nuclei,[32] having prolate axis ratios of approximately 3:1. It may be that such features are only within the grasp of the next generation of spectrometers, such as the GAMMASPHERE facility described by I.-Y. Lee in Chapter 15. Certainly that instrument, with its one-hundred-ten large Ge and three-hundred-sixty BGO detectors, will enable the spectroscopy of extremely weak bands to be carried out much more easily. However, the possibility of detecting hyperdeformed bands before GAMMASPHERE or a European counterpart come on line is not that remote. After all, the literature contains statements, some quite recent, downplaying the possibility of detecting multiple superdeformed bands within the same nucleus. With matrices of very high statistical quality ($\sim 10^9$ events) it may be that hyperdeformed bands can be snatched from the jaws of fission, so to speak, with existing arrays and γ-γ or γ-γ-γ methods similar to those described here. But if that is not possible, then they may be revealed[32] by taking advantage of the enhancement of charged-particle emission which has been postulated. Projects using arrays of particle detectors coupled to existing γ-ray spectrometers are already underway at a number of laboratories (e.g., Ref. 33).

In addition, the analysis of multidimensional data can be carried much further. Efficient procedures for triple-coincidence γ-ray analysis are only now being developed;[25] the γ-γ-γ correlation methods discussed in Section 2 are just the beginning. Apart from the study of superdeformed bands, the term "complete spectroscopy" will undoubtedly take on a new meaning if we can push the detection limit of normal band structures below the 0.1% level. Certainly, as the next generation of spectrometers is under construction, the need for increasingly sophisticated multidimensional analysis is becoming more pressing.

In conclusion, there is already a large amount of intriguing information being gathered with the present instruments. The next two to three years, even before results from the new arrays start pouring in, should see a great deal of new and interesting physics emerging from the discrete-line spectroscopy of superdeformed nuclei.

ACKNOWLEDGMENTS

The results discussed in this talk represent the efforts of a great number of people, among them the unnamed members of the 8π Collaboration and the technical staff of the Chalk River TASCC facility. The 8π spectrometer has been funded by the Canadian Natural Sciences and Research Council and Atomic Energy of Canada Ltd. Research performed by the 8π Collaboration is also supported by those two organizations.

REFERENCES

1. V. M. Strutinsky, *Nucl. Phys.* **A95**, 420 (1967); **A122**, 1 (1968).
2. P. Twin et al., *Phys. Rev. Lett.* **57**, 811 (1986).
3. M. A. Bentley et al., *Phys. Rev. Lett.* **59**, 2141 (1987).
4. J. P. Martin et al., *Nucl. Instr. Meth.* **A257**, 301 (1987).
5. G. Hebbinghaus et al., KFA-Julich preprint (1989); G. Hebbinghaus et al., in *Proceedings of the International Workshop on High Spins, Bad Honnef*, March 1989.
6. M. A. Delcplanque et al., *Phys. Rev. Lett.* **60**, 1626 (1988).
7. B. Haas et al., *Phys. Rev. Lett.* **60**, 503 (1988).
8. P. Fallon et al., Phys. Lett. **B218**, 137 (1989); P. Fallon, *Proceedings of the International Workshop on High Spins, Bad Honnef*, March 1989.
9. M. A. Delaplanque et al., *Phys. Rev. C* **39**, 1651 (1989).
10. G. E. Rathke et al., *Phys. Lett.* **209B**, 177 (1988).
11. J. K. Johansson et al., *Phys. Rev. Lett.* **63**, 2200 (1989).
12. E. F. Moore et al., *Phys. Rev. Lett.* **63**, 360 (1989).
13. D. Ye et al., *Phys. Rev. C.* **41**, R13 (1990).
14. T. L. Khoo, Chapter 19 in this book.
15. M. A. Deleplanque, Chapter 21 in this book.
16. J. C. Waddington, private communication (1988).

17. N. Nadon, private communication (1989).
18. J. P. Vivien et al., *Phys. Rev. C* **33**, 2007 (1986).
19. Y. Schutz et al., *Phys. Rev. C* **35**, 384 (1987).
20. M. J. A. de Voigt et al., *Phys. Rev. Lett.* **59**, 270 (1987).
21. B. R. Mottelson, in *Proceedings of the Conference on High-Spin Nuclear Structure and Novel Nuclear Shapes*, Ed. by I. Ahmad et al. ANL-PHY-88-2 (Argonne, IL, 1988), p.1.
22. B. Herskind, in *Proceedings of the Conference on High-Spin Nuclear Structure and Novel Nuclear Shapes*, Ed. by I. Ahmad et al. ANL0-PHY-88-2 (Argonne, IL, 1988), p.179.
23. G. Palameta and J. C. Waddington, *Nucl. Instr. Meth.* **A234**, 476 (1985).
24. D. Radford, in Proceedings of the 1989 International Nuclear Physics Conference, (São Paulo, Brazil, Aug. 1989) in press.
25. J. A. Kuehner, to be published.
26. J. A. Kuehner, private communication (1989).
27. S. Aberg et al., in *Proceedings of the XXVI International Winter Meeting on Nuclear Physics, Bormio, Italy* (1988).
28. T. Bengtsson, I. Ragnarsson, and S. Aberg, *Phys. Lett* **B208** 39 (1988).
29. W. Nazarewicz, R. Wyss, and A. Johnson, *Nucl. Phys.* **A503** 285 (1989).
30. Y. R. Shimizu, E. Vigezzi, and R. A. Broglia, (Niels Bohr Institute, 1989) preprint.
31. P. J. Nolan, in *Proceedings of the Workshop on Microscopic Models in Nuclear Structure Physics*, Oak Ridge, TN (World Scientific, Singapore, 1989), p.139.
32. J. Kuehner et. al., to be published.
33. I. Ragnarsson, *Phys. Rev. Lett.* **62**, 2084 (1989).
34. J. Dudek, T. Werner, and L. L. Riedinger, *Phys. Lett.* **B211**, 252 (1988).
35. D. G. Sarantites et al., Chapter 30 in this book.

24. High-Spin Properties of Doubly-Odd Nuclei of Mass ≈130

E. S. Paul, C. W. Beausang, D. B. Fossan,
Y. Liang, R. Ma, W. F. Piel, Jr.,
S. Shi, and N. Xu
State University of New York at Stony Brook
Stony Brook, NY 11794

ABSTRACT

Light rare-earth nuclei of mass $A \approx 130$ are predicted to possess relatively flat potential-energy surfaces with respect to γ, the asymmetry parameter in the polar representation of rotating quadrupole shapes. Because of this γ softness, valence nucleons in high-j orbitals can greatly modify the nuclear shape. Whereas protons from the bottom of the $h_{11/2}$ subshell maintain the nuclear core near the collective prolate shape, $\gamma = 0°$, valence neutrons from the top of the $h_{11/2}$ subshell drive the nuclear core towards the collective oblate shape, $\gamma = 60°$. Such competition between the shape-driving forces of the valence nucleons makes the study of doubly-odd nuclei most interesting. Triaxial yrast bands built on the $\pi h_{11/2} \otimes \nu h_{11/2}$ configuration have systematically been observed in isotopes ranging from Cs to Eu ($55 \leq Z \leq 63$). The characteristic features of these bands will be discussed and their properties compared to theoretical expectations.

24.1 INTRODUCTION

The spectroscopy of the low-lying yrast states of nuclei in the mass $A \approx 130$ region of light rare-earth nuclei is dominated by $h_{11/2}$ intruder orbitals. The proton fermi surface lies near the bottom of the $h_{11/2}$ shell, while the neutron fermi surface lies in the upper $h_{11/2}$ midshell. Both odd-Z and odd-N nuclei exhibit yrast bands built on valence $h_{11/2}$ orbitals. The character of the bands is, however, somewhat different; the odd-Z nuclei exhibit decoupled $\Delta I = 2$ bands built on low-Ω $h_{11/2}$ orbitals ($\Omega = 1/2, 3/2$), while the odd-N nuclei exhibit strongly-coupled bands built on high-Ω orbitals ($\Omega = 7/2, 9/2$).

The intrinsic shape of a nucleus can adequately be described by the quadrupole asphericity β and the quadrupole shape asymmetry, or triaxiality parameter, γ. (The Lund convention[1] is used for the definition of γ throughout this report.) The nuclei of this mass region are predicted[2-4] to show soft energy potentials with respect to the γ deformation, i.e., they are γ soft. Furthermore, valence quasiparticles in high-j orbitals are predicted[5] to exert strong shape-driving (γ) forces on the nuclear core. The preferred γ deformation of these orbitals is strongly dependent on the position of the fermi surface within the high-j shell. Specifically for this mass region, the valence $h_{11/2}$ protons favour shapes close to the collective prolate shape ($\gamma = 0°$), while the valence $h_{11/2}$ neutrons favour the oblate but still collectively rotating shape ($\gamma = -60°$). The differing driving forces of the valence $h_{11/2}$ proton and neutron orbitals, and the dependence on the position of the fermi surface,

Exotic Nuclear Spectroscopy, Edited by W. C. McHarris
Plenum Press, New York, 1990

are illustrated in Fig. 1, which shows single-quasiparticle routhians e' as a function of the γ deformation.[6] The nuclear shape and sense of rotation are also shown for specific values of γ that correspond to axially-symmetric shapes. From this figure, the opposite driving forces of the intruder $h_{11/2}$ protons and neutrons can be seen. In contrast, the valence positive-parity proton orbitals are seen to be relatively flat with respect to the γ deformation and thus do not exert any significant shape-driving forces on the nuclear core.

For odd-Z nuclei, the valence $h_{11/2}$ protons stabilize prolate nuclear shapes, while, in the case of odd-N nuclei, the valence $h_{11/2}$ neutrons can drive the nuclear core to negative γ deformation. The signature splitting of the bands built on the $h_{11/2}$ orbitals gives an insight into the nuclear shape, being a sensitive function of the γ deformation. (Rotation or non-axial shape lifts the two-fold degeneracy of the Nilsson single-particle levels into two "signature" components. The energy difference between the components, denoted $\Delta e'$, is the signature splitting.) At low spins, the signature splitting of the low-Ω $h_{11/2}$ orbitals in odd-Z nuclei is typically[7-10] greater than 500 keV, in agreement with Cranked Shell-Model

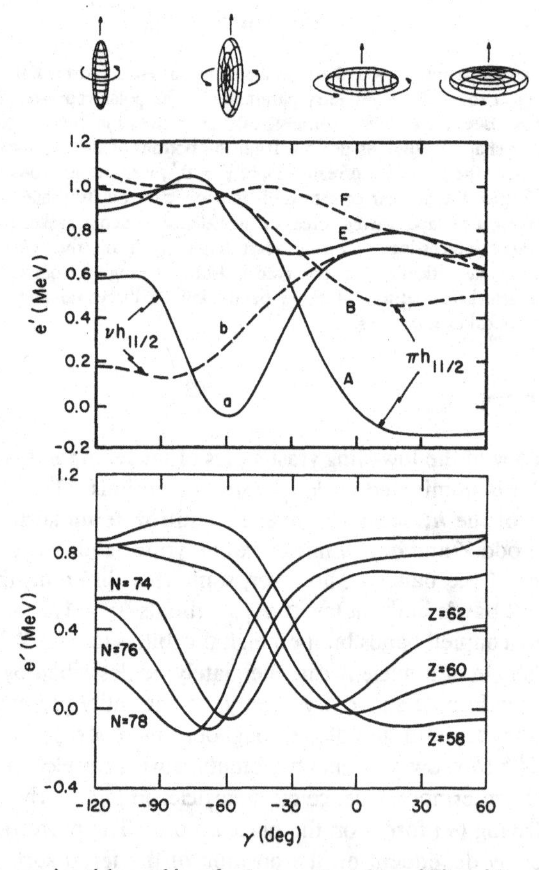

Figure 1. Top: Single-quasiparticle routhians for $h_{11/2}$ protons (A and B) and $h_{11/2}$ neutrons (a and b) as a function of the γ deformation. Positive-parity proton orbitals (E and F), based on $g_{7/2}$, $d_{5/2}$ admixtures are also shown for comparison. These cranked-shell-model calculations were performed at a rotational frequency $\hbar\omega$ = 250 keV and for particle numbers Z = 57 and N = 74. Bottom: $h_{11/2}$ routhians for different proton and neutron numbers.

(CSM)[6] calculations for prolate shapes ($\gamma \approx 0°$). Odd-N nuclei[11-15] also exhibit appreciable signature splittings of $\Delta e' \approx 100$ keV. However, CSM calculations show essentially no signature splitting for the high-Ω $h_{11/2}$neutron orbitals for prolate shapes and at moderate rotational frequencies. Indeed, the calculations require negative γ deformations of $\gamma \approx -20°$ to reproduce the observed signature splittings. Hence, whereas the valence $h_{11/2}$ protons stabilize near-prolate nuclear shapes ($\gamma \approx 0°$), the valence $h_{11/2}$ neutrons induce triaxial shapes with $\gamma \approx -20°$.

Since the yrast bands of odd-Z and odd-N nuclei are built on high-j $h_{11/2}$ orbitals, it is to be expected that the dominant structures in doubly-odd nuclei will be built on a $\pi h_{11/2} \otimes \nu h_{11/2}$ configuration, i.e., a low-Ω $\pi h_{11/2}$ orbital coupled to a high-Ω $\nu h_{11/2}$ orbital. Such bands built on this configuration have been predicted[16] in light La ($Z = 57$) and Pr ($Z = 59$) nuclei. The study of doubly-odd nuclei of this γ-soft mass region is thus most interesting because of the aforementioned competing shape-driving properties of the valence $h_{11/2}$ proton and neutron. This mass region is the heaviest where the proton and neutron fermi levels lie within the same high-j orbital. In the following sections, the results for ^{132}Pr will be presented as a typical example of a doubly-odd nucleus of this mass region.

24.2 HIGH-SPIN STATES IN DOUBLY-ODD ^{132}Pr

As an example for the band structures evident in a doubly-odd nucleus, the level scheme[17] of ^{132}Pr is presented in Fig. 2. This nucleus was populated via the ^{117}Sn(^{19}F, $4n\gamma$) fusion-evaporation reaction carried out at the SUNY Nuclear Structure Laboratory. Standard γ-ray coincidence techniques were employed to construct the level scheme. Bands 1 and 2, consisting of both signature components connected by strong $\Delta I = 1$ transitions, have been systematically observed in several doubly-odd nuclei in this mass region. Band 1 is associated with a $\pi h_{11/2} \otimes \nu g_{7/2}$ configuration, while Band 2 is associated with the $\pi h_{11/2} \otimes \nu h_{11/2}$ configuration.

The tentative spin-parity assignments are based on comparison with other doubly-odd nuclei and the assumed $\pi h_{11/2} \otimes \nu h_{11/2}$ configuration for Band 2. Perpendicular coupling of the low-Ω $h_{11/2}$ proton and high-Ω $h_{11/2}$ neutron yield a band-head spin $I_0 \approx 8$. In addition, the coupling of the Ω values of the proton and neutron, in accordance with the Gallagher-Moszkowski rule,[18] yields a K value 5^+ or 6^+ for the band-head, depending on which Nilsson orbitals are occupied. Signature splittings for the bands can be obtained by constructing experimental routhians using the standard techniques of Bengtsson and Frauendorf[6]; Band 1 exhibits no signature splitting, while Band 2 shows a signature splitting of $\Delta e' \approx 50$ keV, which is constant as a function of rotational frequency.

24.3 CHARACTERISTICS OF $\pi h_{11/2} \otimes \nu h_{11/2}$ BANDS

Whereas bands similar to Band 1 of ^{132}Pr have been observed in light La and Pr isotopes, bands similar to Band 2 have been established in nuclei ranging from Cs ($Z = 55$) up to Eu ($Z = 63$).[19-28] The systematics of these $\pi h_{11/2} \otimes \nu h_{11/2}$ bands are shown in Fig. 3 for the $N = 75$ isotones ^{132}La, ^{134}Pr, ^{136}Pm, and ^{138}Eu. The characteristic features of these $\pi h_{11/2} \otimes \nu h_{11/2}$ configurations are: 1) relatively strong $M1/E2$ $\Delta I = 1$ transitions compared to the $E2$ crossovers; and 2) a small signature splitting between the two $\Delta I = 2$ sequences. The observed signature splittings lie in the range $25 \leqslant \Delta e' \leqslant 50$ keV.

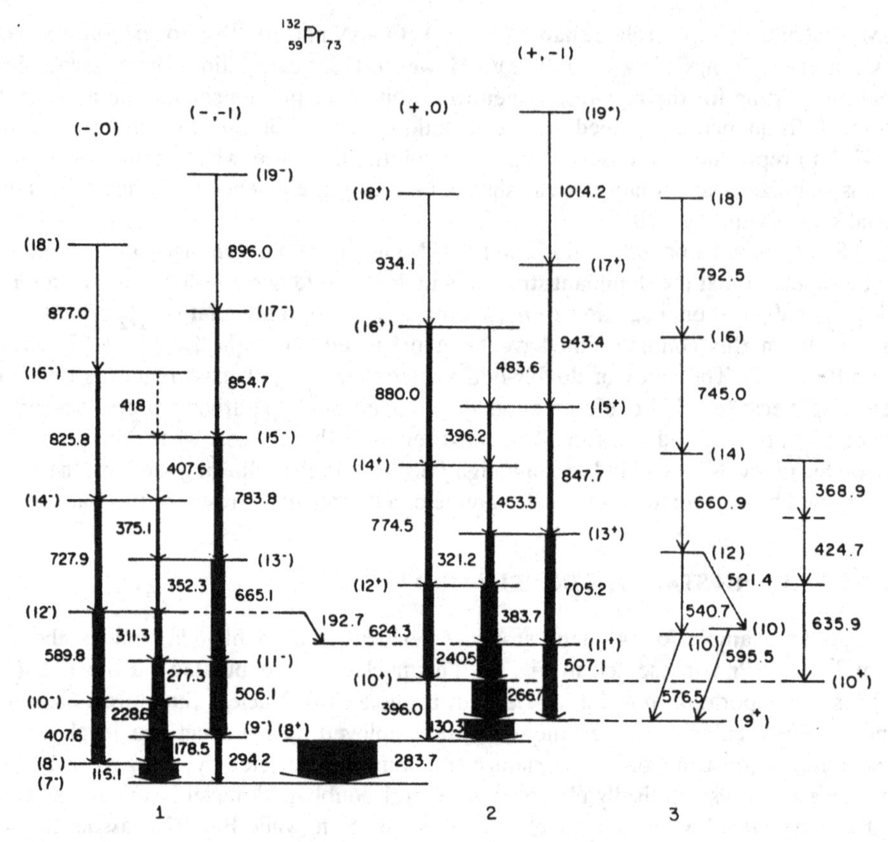

Figure 2. The decay scheme of doubly-odd ^{132}Pr deduced following the ^{117}Sn(^{19}F, $4n\gamma$) reaction. Band 1 is associated with the $\pi h_{11/2} \otimes \nu g_{7/2}$ configuration, Band 2 with the $\pi h_{11/2} \otimes \nu h_{11/2}$ configuration, and Band 3 with the $\pi h_{11/2} \otimes \nu i_{13/2}$ configuration.

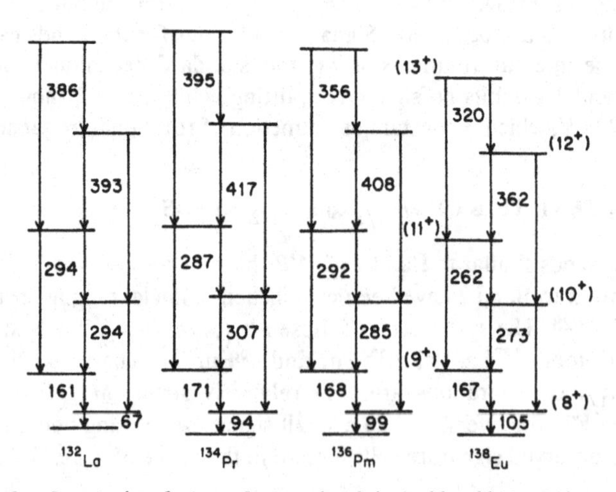

Figure 3. Systematics of $\pi h_{11/2} \otimes \nu h_{11/2}$ bands in doubly-odd N = 75 isotones.

24.4 CSM CALCULATIONS

Figure 1 shows only the single-quasiparticle routhians as a function of the γ deformation. In order to discuss the overall shape of the nucleus, it is possible to construct total routhians E' from the single-quasiparticle routhians e' following the methods proposed by Frauendorf and May.[29] Here the total routhian is given by

$$E'(\omega,\gamma) = \sum_{\mu} e'_{\mu}(\omega,\gamma) + \frac{1}{2}V_{po}\cos 3\gamma - \frac{2}{3}\omega^2\left(\Im_0 + \frac{1}{2}\omega^2\Im_1\right)\cos^2(\gamma + 30°). \quad (1)$$

The first term of Eq. 1 is simply a sum over the occupied quasiparticle orbitals, while the second term is a parameterization of the nuclear shape introducing V_{po}, the prolate-oblate energy difference. Values of V_{po} calculated for this mass region[1,3] are typically $|V_{po}| < 500$ keV. The third term of Eq. 1 represents the collective rotation of the nucleus and is obtained by multiplying the Harris[30] moment of inertia by a hydrodynamical γ dependence.

The total routhians corresponding to the $\pi h_{11/2} \otimes \nu h_{11/2}$ and $\pi h_{11/2} \otimes \nu g_{7/2}$ configurations in ^{132}Pr are shown in Fig. 4 as a function of the γ deformation. The routhians are labeled by parity and signature quantum numbers (π, α) with the (+, −1) and (+, 0) levels corresponding to the two signatures of the $\pi h_{11/2} \otimes \nu h_{11/2}$ band. (The signature for single-quasiparticle states may take the values $\alpha = \pm 1/2$, and signature is an additive quantum number. Thus the allowed values of α for two-quasiparticle configurations are $\alpha = 0, \pm 1$.) It can be seen that a minimum occurs at $\gamma \approx -10°$, where there is a small signature splitting of $\Delta e' \approx 50$ keV, which is in excellent agreement with the experimental value. The second pair of routhians, labelled (−, −1) and (−, 0), correspond to the $\pi h_{11/2} \otimes \nu g_{7/2}$ configuration. The calculations show an energy minimum at $\gamma \approx -5°$, where there is no signature splitting, again in agreement with experiment.

One failure of the CSM calculations is the prediction of the relative energies of the two configurations. The $\pi h_{11/2} \otimes \nu h_{11/2}$ configuration is predicted to lie several hundred keV below the $\pi h_{11/2} \otimes \nu g_{7/2}$ configuration; in contrast, the experimental results indicate that the $\pi h_{11/2} \otimes \nu g_{7/2}$ configuration lies lowest in energy. However, the CSM calculations to not include any residual interactions between the quasiparticles. A recent analysis of data for nuclei around ^{128}La[21,31] has indicated a more attractive p-n residual interaction

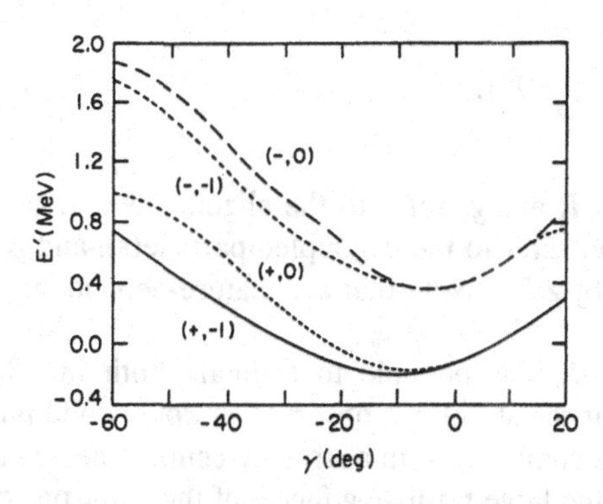

Figure 4. Total routhians calculated for ^{132}Pr at $\hbar\omega = 250$ keV.

between $h_{11/2}$ and $g_{7/2}$ particles than between similar $h_{11/2}$ particles. This somewhat surprising result could indeed lower the relative energy of the $\pi h_{11/2} \otimes \nu g_{7/2}$ configuration.

24.5 ELECTROMAGNETIC PROPERTIES

The ratio of reduced transition probabilities for competing $\Delta I = 1$ and $\Delta I = 2$ electromagnetic decay from a state of spin I is given by

$$\frac{B(M1; I \rightarrow I-1)}{B(E2; I \rightarrow I-2)} = \frac{|\langle I|M1|I-1\rangle|^2}{|\langle I|E2|I-2\rangle|^2}, \tag{2}$$

and the $E2/M1$ mixing ration for the $\Delta I = 1$ transitions by

$$\delta = \frac{\langle I|E2|I-2\rangle}{\langle I|M1|I-1\rangle}. \tag{3}$$

In a geometrical model[32,33] where one strongly-coupled particle (high-Ω) is combined with one or more decoupled particles (low-Ω), Dönau expresses the $M1$ matrix element as

$$\langle I|M1|I-1\rangle = \sqrt{\frac{3}{8\pi}}K\left[(g_1-g_R)\left\{\sqrt{1-\frac{K^2}{I^2}}-\frac{i_1}{I}\pm\frac{\Delta e'}{\omega}\right\}-(g_2-g_R)\frac{i_2}{I}\right], \tag{4}$$

while the $E2$ matrix elements are given by

$$\langle I|E2|I-1\rangle = q_0\frac{K}{I}\sqrt{\frac{3}{2}}\left[1-\frac{K^2}{I^2}\right]^{1/2} \tag{5}$$

and

$$\langle I|E2|I-2\rangle = q_0\sqrt{\frac{3}{8}}\left[1-\frac{K^2}{I^2}\right], \tag{6}$$

where the collective intrinsic quadrupole moment is given by

$$q_0 = \sqrt{\frac{5}{12\pi}}Q_0\cos(\gamma+30^\circ). \tag{7}$$

In the above expressions, i_1 and g_1 refer to the alignment[6] and gyromagnetic factor of the coupled particle, i_2 and g_2 refer to the decoupled particle(s), and g_R is the rotational g-factor given approximately by Z/A. Note that a signature-dependent term $\pm\Delta e'/\omega$ appears in Eq. 4.

Combining equations 2–7, it is possible to estimate both the $B(M1)/B(E2)$ ratios and $E2/M1$ mixing ratios for the structures observed in doubly-odd nuclei where a decoupled proton ($h_{11/2}$, $\Omega \approx 3/2$) is combined with a strongly-coupled neutron ($h_{11/2}$, $\Omega \approx 9/2$ or $g_{7/2}$, $\Omega \approx 7/2$). In particular, the large positive g factor of the $h_{11/2}$ proton ($g_2 \approx +1.2$) will reinforce the negative g factor of the $h_{11/2}$ neutron ($g_1 \approx -0.2$) such that the term within the square brackets of Eq. 4 will be large and negative. This will lead to relatively strong $\Delta I = 1$ $M1$ transitions linking the two signature components of the $\pi h_{11/2} \otimes \nu h_{11/2}$ configuration and will, in addition, give rise to negative $E2/M1$ mixing ratios (the sign convention of Krane[34] is used for δ) for the $\Delta I = 1$ transitions. In contrast, for the $\pi h_{11/2} \otimes \nu g_{7/2}$ configuration, the small positive g factor of the $g_{7/2}$ neutron ($g_1 \approx +0.25$) will tend to cancel with the positive proton g factor; hence, smaller $B(M1)/B(E2)$ ratios are expected for this configuration, although the $E2/M1$ mixing ratios are still predicted to be negative.

Experimentally, $B(M1)/B(E2)$ ratios can readily be obtained from the intensities of the observed $\Delta I = 1$ and $\Delta I = 2$ transitions within a rotational band using the expression,

$$\frac{B(M1; I \rightarrow I-1)}{B(E2; I \rightarrow I-2)} = 0.697 \times \frac{E_\gamma^5(I \rightarrow I-2)}{E_\gamma^3(I \rightarrow I-1)} \times \frac{I_\gamma(I \rightarrow I-1)}{I_\gamma(I \rightarrow I-2)} \times \frac{1}{1+\delta^2}. \quad (8)$$

Typically, the $E2/M1$ mixing ratio is such that $\delta^2 << 1$ and can thus be assumed to be zero when extracting empirical $B(M1)/B(E2)$ ratios. Results for ^{132}Pr are shown in Fig. 5 for the $\pi h_{11/2} \otimes \nu h_{11/2}$ and $\pi h_{11/2} \otimes \nu g_{7/2}$ configurations, where they are compared to theoretical predictions using the geometrical model outlined above. As expected, the larger ratios are obtained for the $\pi h_{11/2} \otimes \nu h_{11/2}$ band, being four times as large as those extracted for the $\pi h_{11/2} \otimes \nu g_{7/2}$ band. Ratios of a similar magnitude have been extracted for several doubly-odd nuclei in this mass region. The ratios obtained for the $\pi h_{11/2} \otimes \nu h_{11/2}$ configuration in ^{132}Pr are consistent with the signature dependence expected theoretically.

In the calculations of the $B(M1)/B(E2)$ ratios, accepted values for i_1, g_1, i_2, g_2, g_R, and Q_0 were used, while different K values were tried. The best fit to the data for the $\pi h_{11/2} \otimes \nu h_{11/2}$ configuration was obtained with $K = 4$, rather than the predicted $K = 5, 6$. However, K is not a completely good quantum number under rotation and for non-axial shapes, and hence an effective K value of $K = 4$ is not unreasonable. For the $\pi h_{11/2} \otimes \nu g_{7/2}$ configuration, the best fit was obtained with the expected K value of 2.

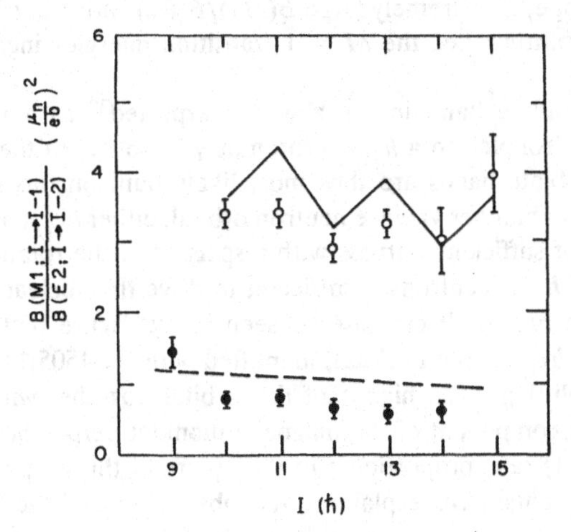

Figure 5. Experimental and theoretical ratios of reduced transition probabilities for the $\pi h_{11/2} \otimes \nu h_{11/2}$ band (open circles) and $\pi h_{11/2} \otimes \nu g_{7/2}$ band (filled circles) in ^{132}Pr.

Angular distributions have recently been measured for ^{134}Pr, enabling $E2/M1$ mixing ratios to be extracted for the $\pi h_{11/2} \otimes \nu h_{11/2}$ band. The results do indeed yield negative mixing ratios δ, as predicted by theory. In addition, negative mixing ratios have also been extracted[21] in ^{128}La for both the $\pi h_{11/2} \otimes \nu h_{11/2}$ and $\pi h_{11/2} \otimes \nu g_{7/2}$ configurations.

24.6 RECENT RESULTS: OTHER BAND STRUCTURES

24.6.A Doubly-Decoupled Bands

In addition to rotational bands showing both $\Delta I = 1$ and $\Delta I = 2$ transitions, such as Bands 1 and 2 in ^{132}Pr, there is increasing evidence for doubly-decoupled $\Delta I = 2$ bands in several doubly-odd nuclei of this mass region. Band 3 of ^{132}Pr (Fig. 2) is such an example, and a similar band has been observed in ^{134}Pm.[26] In order to produce a doubly-decoupled band, both the valence proton and neutron must occupy low-Ω orbitals such that the signature splitting of both orbitals is large. The proton fermi surface already lies at the bottom of the $h_{11/2}$ shell near the $\Omega = 1/2, 3/2$ components. For the neutrons, both $\nu i_{13/2}$, $\Omega = 1/2$, and $\nu h_{9/2}$, $\Omega = 1/2$, orbitals approach the fermi surface as the rotation increases or for enhanced quadrupole deformation. Indeed, the $\nu i_{13/2}$, $\Omega = 1/2$, orbital is thought[35] to be responsible for the enhanced-deformation structures observed in this mass region with $\beta \approx 0.4$, corresponding to a prolate rotor with a 3:2 axis ratio.

The $\pi h_{11/2} \otimes \nu i_{13/2}$ configuration would, therefore, drive the nuclear core to a larger quadrupole deformation. Furthermore, a highly-deformed band has recently been observed[36] in doubly-odd ^{130}La based on a possible $\nu i_{13/2} \otimes [\pi h_{11/2}]^n$ configuration.

24.6.B Collective Oblate Bands

Recently $\Delta I = 1$ bands have been established[37] in ^{128,130}La showing properties similar to a proposed collectively rotating oblate ($\gamma \approx -60°$) band[38] in odd-Z ^{131}La. These bands show: 1) small dynamic moments of inertia, $dI/d\omega < 20\hbar^2$ MeV^{-1}, indicative of a different collective shape; 2) extremely large $B(M1)/B(E2)$ ratios in excess of $10(\mu_N/eb)^2$; and 3) no signature splitting; i.e., the $\Delta I = 1$ transition energies increase smoothly with spin.

The collective-oblate band in ^{131}La was interpreted[38] as a pair of rotationally-aligned $h_{11/2}$ neutrons coupled to a $h_{11/2}$ proton at $\gamma \approx -60°$. In the nearby doubly-odd ^{128,130}La nuclei, the oblate bands are thus most likely built on this $\pi h_{11/2} \otimes [\nu h_{11/2}]^2$ configuration coupled to another valence neutron orbital, either $h_{11/2}$ or $g_{7/2}$.

For a nucleus of sufficient softness with respect to γ, the rotational alignment of a pair of upper-midshell $h_{11/2}$ neutrons is sufficient to drive the nuclear shape from prolate to oblate, as shown in Fig. 1. It can also be seen in this figure that for such an oblate ($\gamma \approx -60°$) shape, the $h_{11/2}$ proton orbital, identified with the [505]11/2$^-$ Nilsson orbital, shows no signature splitting. The high Ω of this orbital, together with the large g factor (+1.2) induces a large component of the magnetic moment perpendicular to the spin I of the nucleus. The $B(M1)$ rate, proportional to the square of this perpendicular component ($B(M1) \propto \mu^2$), is thus enhanced, explaining the observation of the large $B(M1)/B(E2)$ ratios.

24.7 CONCLUSIONS

The study of doubly-odd nuclei in the mass $A \approx 130$ region has blossomed over the past few years. With both the proton and neutron fermi surfaces lying within the $h_{11/2}$ shell, the dominant structure seen in nuclei throughout this mass region is built on the $\pi h_{11/2} \otimes \nu h_{11/2}$ configuration. The competing shape-driving effects of the $h_{11/2}$ proton and neutron induces a small degree of triaxiality $\gamma \approx -10°$, which is intermediate between the values for neighboring odd-Z ($\gamma \approx 0°$) and odd-N ($\gamma \approx -20°$) nuclei.

Recently several doubly-decoupled structures have been observed, which requires the neutron to occupy the β-driving $\nu i_{13/2}$ orbital, or possibly a $\nu h_{9/2}$ orbital. In ^{130}La, both highly-deformed[36] and collectively-rotating oblate[37] nuclear shapes have recently been found to coexist with the normally-deformed prolate shape.

ACKNOWLEDGMENT

This work is in part supported by the National Science Foundation.

REFERENCES

1. G. Andersson, S. E. Larsson, G. Leander, P. Moller, S. G. Nilsson, I. Ragnarsson, S. Aberg, R. Bengtsson, J. Dudek, B. Nerlo-Pomorska, K. Pomorski, and Z. Szymanski, *Nucl. Phys.* **A268**, 205 (1976).
2. I. Ragnarsson, A. Sobiczewski, R. K. Sheline, S. E. Larsson, and B. Nerlo-Pomorska, *Nucl. Phys.* **A233**, 329 (1974).
3. Y. S. Chen, S. Frauendorf, and G. A. Leander, *Phys. Rev. C* **28**, 2437 (1983).
4. B. D. Kern, R. L. Mlekodaj, G. A. Leander, M. O. Kortelahti, E. F. Zganjar, R. A. Braga, R. W. Fink, C. P. Perez, W. Nazarewicz, and P. B. Semmes, *Phys. Rev. C* **36**, 1514 (1987).
5. G. A. Leander, S. Frauendorf, and F. R. May, in *Proceedings of the Conference on High Angular Momentum Properties of Nuclei, Oak Ridge, 1982*, Ed. by N. R. Johnson (Harwood Academic, New York, 1983), p. 281.
6 R. Bengtsson and S. Frauendorf, *Nucl. Phys.* **A327**, 139 (1979).
7. L. Hildingsson, C. W. Beausang, D. B. Fossan, R. Ma, E. S. Paul, W. F. Piel, Jr., and N. Xu, *Phys. Rev. C* **39**, 471 (1989).
8. L. Hildingsson, C. W. Beausang, D. B. Fossan, and W. F. Piel, Jr., *Phys. Rev. C* **37**, 985 (1988).
9. T. M. Semkow, D. G. Sarantites, K. Honkanen, V. Abenante, L. A. Adler, C. Baktash, N. R. Johnson, I.-Y. Lee, M. Oshima, Y. Schutz, Y. S. Chen, J. X. Saladin, C. Y. Chen, O. Dietzsch, A. J. Larabee, L. L. Riedinger, and H. C. Griffin, *Phys. Rev. C* **34**, 523 (1986).
10. N. Xu, C. W. Beausang, R. Ma, E. S. Paul, W. F. Piel, Jr., D. B. Fossan, and L. Hildingsson, *Phys. Rev. C* **39**, 1799 (1989).
11. R. Aryaeinejad, D. J. G. Love, A. H. Nelson, P. J. Nolan, P. J. Smith, D. M. Todd, and P. J. Twin, *J. Phys.* **G10**, 955 (1984).
12. R. Ma, E. S. Paul, C. W. Beausang, S. Shi, N. Xu, and D. B. Fossan, *Phys. Rev. C* **36**, 2322 (1987).
13. W. F. Piel, Jr., C. W. Beausang, D. B. Fossan, L. Hildingsson, and E. S. Paul, *Phys. Rev. C* **35**, 959 (1987).
14. R. Ma, C. W. Beausang, E. S. Paul, W. F. Piel, Jr., S. Shi, N. Xu, D. B. Fossan, J. Burde, M. A. Deleplanque, R. M. Diamond, A. O. Macchiavelli, and F. S. Stephens, *Phys. Rev. C* **40**, 156 (1989).
15. R. Ma, K. Ahn, D. B. Fossan, Y. Liang, E. S. Paul, and N. Xu, *Phys. Rev. C* **39**, 530 (1989).
16. A. J. Kreiner and M. A. J. Mariscotti, J. Phys. G6, L13 (1980).

17. S. Shi, C. W. Beausang, D. B. Fossan, R. Ma, E. S. Paul, N. Xu, and A. J. Kreiner, *Phys. Rev. C* **37**, 1478 (1988).
18. C. J. Gallagher, Jr. and S. A. Moszkowski, *Phys. Rev.* **111**, 1282 (1958).
19. E. S. Paul, D. B. Fossan, Y. Liang, R. Ma, and N. Xu, *Phys. Rev. C* **40**, 619 (1989).
20. P. J. Nolan, R. Aryaeinejad, P. J. Bishop, M. J. Godfrey, A. Kirwan, D. J. G. Love, A. H. Nelson, D. J. Thornley, and D. J. Unwin, *J. Phys.* **G13**, 1555 (1987).
21. M. J. Godfrey, Y. He, I. Jenkins, A. Kirwan, P. J. Nolan, D. J. Thornley, S. M. Mullins, and R. Wadsworth, *J. Phys.* **G15**, 487 (1989).
22. E. S. Paul, C. W. Beausang, D. B. Fossan, R. Ma, W. F. Piel, Jr., N. Xu, and L. Hildingsson, *Phys. Rev. C* **36**, 1853 (1987).
23. J. R. B. Oliveira, L. G. R. Emediato, M. A. Rizzutto, R. V. Ribas, W. A. Seale, M. N. Rao, N. H. Medina, S. Botelho, and E. W. Cybulska, *Phys. Rev. C* **39**, 2250 (1989).
24. R. Ma, E. S. Paul, S. Shi, C. W. Beausang, W. F. Piel, Jr., N. Xu, D. B. Fossan, T. Chapuran, D. P. Balamuth, and J. W. Arrison, *Phys. Rev. C* **37**, 1926 (1988).
25. C. W. Beausang, L. Hildingsson, E. S. Paul, W. F. Piel, Jr., P. K. Weng, N. Xu, and D. B. Fossan, *Phys. Rev. C* **36**, 1810 (1987).
26. R. Wadsworth, S. M. Mullins, P. J. Nolan, A. Kirwan, P. J. Bishop, and M. J. Godfrey, *University of Liverpool Nuclear Structure Group Annual Report* (1988), p 36.
27. C. W. Beausang, P. K. Weng, E. S. Paul, W. F. Piel, Jr., N. Xu, and D. B. Fossan, *Bull. Am. Phys. Soc.* **31**, 1212 (1986).
28. Y. Liang, K. Ahn, R. Ma, E. S. Paul, N. Xu, and D. B. Fossan, *Phys. Rev. C* **38**, 2432 (1988).
29. S. Frauendorf and F. R. May, *Phys. Lett.* **125B**, 245 (1983).
30. S. M. Harris, *Phys. Rev.* **138**, B509 (1965).
31. J. D. Garrett, J. Nyberg, C. H. Yu, J. M. Expino, and M. J. Godfrey, in *Proceedings of the Conference on Contemporary Topics in Nuclear Structure Physics, Cocoyoc, Mexico, 1988*, (World Scientific, Singapore, 1988).
32. F. Dönau and S. Frauendorf, in *Proceedings of the Conference on High Angular Momentum Properties of Nuclei, Oak Ridge, 1982*, Ed. by N. R. Johnson (Harwood Academic, New York, 1983), p 143.
33. F. Dönau, *Nucl. Phys.* **A471**, 469 (1987).
34. K. S. Krane, *At. Data Nucl. Data Tables* **16**, 384 (1975).
35. R. A. Wyss, J. Nyberg, A. Johnson, R. Bengtsson, and W. Nazarewicz, *Phys. Lett.* **B215**, 211 (1988).
36. M. J. Godfrey, Y. He, I. Jenkins, A. Kirwan, P. J. Nolan, R. Wadsworth, and S. M. Mullins, *J. Phys.* **G15**, L163 (1989).
37. M. J. Godfrey, Y. He, I. Jenkins, A. Kirwan, P. J. Nolan, D. J. Thornley, S. M. Mullins, R. Wadsworth, and R. A. Wyss, *J. Phys.* **G15**, 671 (1989).
38. E. S. Paul, C. W. Beausang, D. B. Fossan, R. Ma, W. F. Piel, Jr., N. Xu, L. Hildingsson, and G. A. Leander, *Phys. Rev. Lett.* **58**, 984 (1987).

25. Rotational-Band Structure in Odd-Odd Re Nuclei

Wade A. Olivier, Wen-Tsae Chou, Aracelys Rios, and Wm. C. McHarris
National Superconducting Cyclotron Laboratory
and
Departments of Chemistry and Physics/Astronomy
Michigan State University
East Lansing, MI 48824

Rahmat Aryaeinejad
Idaho National Engineering Laboratory/EG&G
Idaho Falls, ID 83415

ABSTRACT

Odd-odd ^{174,176}Re nuclei were studied via in-beam γ-ray spectroscopy with a variety of *HI*-induced reactions at both Michigan State University NSCL and SUNY Stony Brook. One of the most prominent features of each is the existence and very strong population of a doubly-decoupled band. These $K = 1$ bands result from the coupling of the $h_{9/2}$ $1/2^-[541]$ proton and an $i_{13/2}$ $1/2^-$ neutron state. Such doubly-decoupled bands exist in many nuclei in this region. The minimal variations in these structures from nucleus to nucleus implies a relatively weak coupling of the odd particles to the core. This offers hope of a more detailed understanding of *p-n* residual interactions in relatively high-mass systems. Other information about the character of the doubly-decoupled structures, backbending, and reaction mechanisms can also be learned from these nuclei. Our studies also yielded information on two rotational bands in ^{175}Re, which had been entirely uncharacterized prior to our investigations. The bands are built on the decoupled $h_{9/2}$ $1/2^-[541]$ proton and a $\Delta I = 1$ band based on the $h_{9/2}$ $9/2^-[514]$ proton.

25.1 INTRODUCTION

Until recently, odd-odd nuclei were not investigated rigorously because of apprehension about the complexity of the spectra. However, in the past ten years, as more were studied, it has been found that it is possible to decipher and understand the relationships among their γ rays and to interpret structural details from these spectra. Studies of odd-odd nuclei have now come into their own, encouraged by a variety of phenomena that have been observed. Some of the motivating factors include: a desire for completeness of the data base, filling the odd-odd holes; an opportunity to observe two odd-particle interactions in a nuclear system; the odd-particle effects on backbending and shape transitions at higher spins; and perhaps ultimately a view into the character of *p-n* residual interactions in relatively high-mass systems.

Our group has specialized in studying Re nuclei, since it has been demonstrated that they are very feasible for in-beam γ-ray spectroscopy studies.[1] *HI*-induced reactions populate a relatively small number of the available states, which then tend to deexcite via a few

Exotic Nuclear Spectroscopy, Edited by W. C. McHarris
Plenum Press, New York, 1990

"klokast" rotational bands. Other motivations for our studies include our success at interpreting the data from our experiments in this region, the observation by our and other groups of the doubly-decoupled rotational bands,[2,3] and our observation of backbending[4] in these nuclei. We are also encouraged by the fairly complete data base which exists for neighboring nuclei. This allows better comparisons of band structures when trying to interpret the deviations of the doubly-decoupled bands from their associated even-even core rotational bands. These deviations, at least to zeroth order, can be ascribed to the *p-n* interaction of the odd particles.

25.2 THEORETICAL CONSIDERATIONS

Re lies in a region of well-deformed nuclei. Their ground-state structure exists as a prolate spheroid, with $\beta \approx 0.25$. Therefore, the physical model we have chosen to depict these odd-odd systems includes a prolate spheroid core with two odd particles coupled to it. The odd proton and odd neutron can occupy a variety of orbits, which include $d_{5/2}$ $5/2^+[402]$, $h_{11/2}$ $9/2^-[514]$, and $h_{9/2}$ $1/2^-[541]$ for the proton and $i_{13/2}$ $7/2^+[633]$, $p_{3/2}$ $1/2^-[521]$, $h_{9/2}$ $5/2^-[512]$, and $f_{7/2}$ $7/2^-[514]$ for the neutron, depending on the specific deformation and thus the shape of the fermi surface at this point. Figure 1 is a pictorial representation of this model. This specific figure illustrates the so-called "doubly-decoupled" case. In this structure, the odd proton and odd neturon both occupy $K = 1/2$ orbitals. The most prominently-populated orbitals in our studies were those which were both highly aligned and highly deformed.

For this type of coupling scheme, we can write the hamiltonian for the system as[5]:

$$H = H_{int} + H_{rot} + H_{cor} + H_{pn} + H_{irrot} \tag{1}$$

Then the expression for energy of rotational bands in the nucleus can be written:

$$E_I = E_0(K) + AI(I + 1) + BI^2(I + 1)^2 + \ldots \tag{2}$$

Here the A term is usually negative, resulting in a compression of the band, and the B term is non-negligible and positive. This is characteristic of coriolis-type distortions in the band structure.

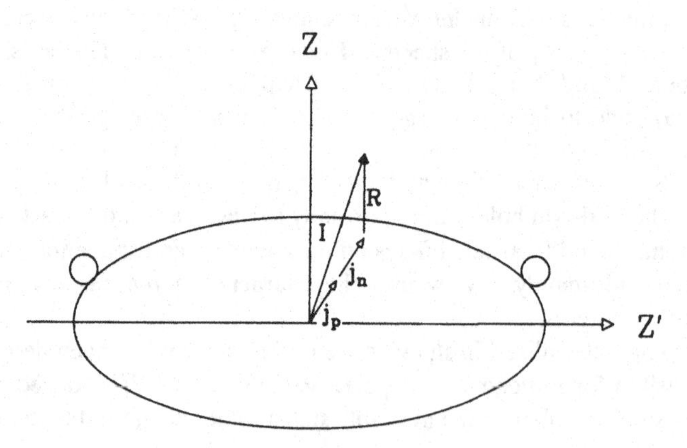

Figure 1. "Physical" representation of the doubly-decoupled coupling in odd-odd Re nuclei.

A special case is that of the Doubly-Decoupled rotational Bands (DDB). In the analogous cases for decoupled odd-mass nuclei, $K = 1/2$ rotational bands, it is necessary to include a term that incorporates the decoupling characteristics, including a "decoupling" parameter and a signature-splitting factor. Doing this for the odd-odd case, the resulting energy expression is empirically of the form:

$$E_I = E_0(K) + AI(I+1) + (-1)^I a_{pn} (I + C) + BI^2(I+1)^2 + \ldots \quad (3)$$

The third term on the right side of this equation results from effects of the decoupling of odd-particle orbitals. This term drives the even members of this coupling to higher energies and the odd members to lower energies, or vice versa, depending on the sign of the so-called decoupling parameter, a_{pn}. [For $K = 1/2$ odd-mass bands, the term $(I + C)$ is usually written as $(I + 1/2)$, but for odd-odd bands, we prefer, at this point, to leave C as an empirical constant.]

The odd protons in odd-mass "decoupled" systems in this region are "highly decoupled," with decoupling constants on the order of 5. This introduces a very large splitting between the signatures of the DDB couplings. To date, only the odd members of any DDB have been observed.

25.3 EXPERIMENTAL DETAILS

The studies of odd-odd ^{176}Re and ^{174}Re were performed at Michigan State University NSCL and the State University of New York at Stony Brook. They consisted of in-beam γ-ray spectroscopy, using the ^{159}Tb(^{22}Ne,5$n\gamma$) reaction at 108 and 112 MeV at NSCL to produce ^{176}Re and the ^{159}Tb (^{20}Ne,5$n\gamma$) and ^{139}La (^{40}Ar,5$n\gamma$) reactions at 105 and 191 MeV, respectively, to produce ^{174}Re; also, the ^{165}Ho (^{16}O,5$n\gamma$) and ^{165}Ho (^{16}O,7$n\gamma$) at 99.6 MeV and 137.5 MeV, respectively, to produce ^{176}Re and ^{174}Re at SUNY Stony Brook. Targets were self-supporting, rolled foils ranging from 0.5 mg/cm^2 to 1.2 mg/cm^2 for the NSCL experiments. A thick target, 52 mg/cm^2, was used in the Stony Brook experiments.

Gamma rays were detected with two coaxial 15%-efficient HPGe detectors positioned $\pm 90°$ to the beam in the NSCL experiments. The detectors were shielded with 20.3x20.3-cm and 30.5x30.5-cm NaI(Tl) annuli to suppress compton events. In the later experiments a planar Ge detector (LEPS) was used for low-energy transition identification and located at either 45° or 125°. Figure 2 illustrates the detector arrangement used in the MSU-NSCL experiments.

The detection of γ rays at Stony Brook was performed with their γ-ray detector array[6] using four HPGe detectors. The detectors were positioned at $\pm 35°$, 98°, and $-158°$ with respect to the beam.

Figure 3 shows a singles spectrum from the ^{165}Ho(^{16}O,7$n\gamma$) reaction at 138 MeV. The principal products of this bombardment on a thick target are $^{174\text{-}177}$Re and ^{174,176}W. One can observe the relatively high compton background even with suppression as a result of the high-n-out channel. Also, note the very strong population of the decoupled band in ^{175}Re, even in a "singles" spectrum.

Figure 4 is an integral coincidence spectrum from the ^{159}Tb(^{22}Ne,5$n\gamma$) reaction to produce ^{176}Re. Again, note the relatively strong population of a "decoupled-type" band, here the DDB.

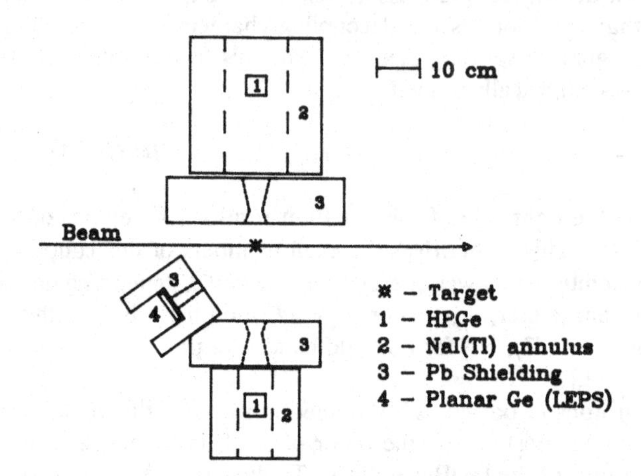

Figure 2. Detector arrangement used in MSU-NSCL experiments.

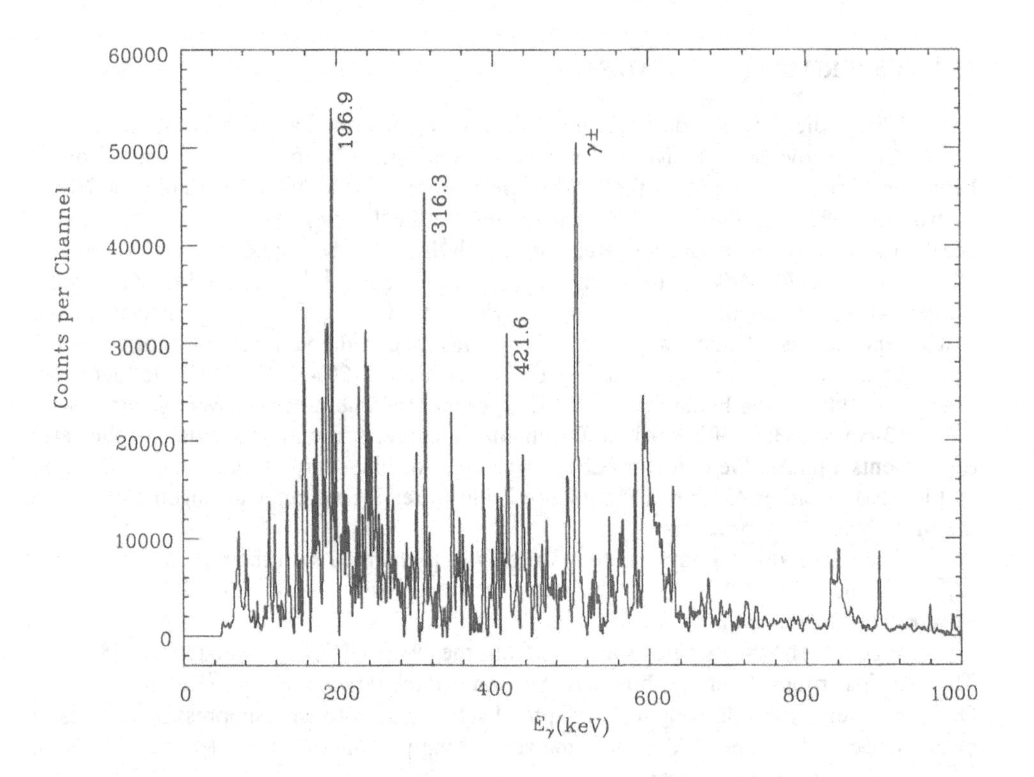

Figure 3. ^{174}Re "singles" spectrum from the ^{165}Ho(^{16}O,$7n\gamma$) reaction at 138 MeV on a thick target.

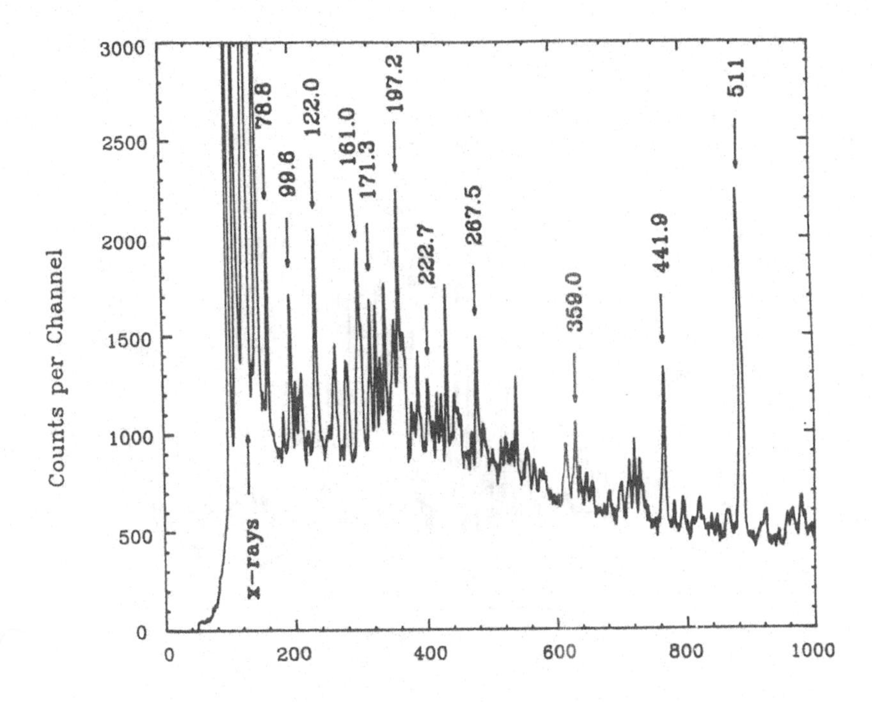

Figure 4. ^{176}Re "integral-coincidence" spectrum.

Figures 5 and 6 show gated coincidence spectra of ^{174}Re and ^{176}Re. Both nuclei display a decoupled band, in addition to a more normal coupling with $\Delta I = 1$ spacings in the band. It can be observed, even with the relatively simple detection system that was used at MSU-NSCL, very good quality spectra were obtained, though somewhat lacking in statistics.

As a side product of our reactions to form ^{174}Re, we also produced ^{175}Re. This nucleus had been uncharacterized prior to our studies. Although we did not optimize beam energies for its production and study, the novel character of the data warrants its inclusion in this contribution. The odd-mass, neutron-deficient Re nuclei have been well characterized by several groups[7-9] from ^{171}Re up to stability, with the exception of ^{175}Re. Figure 7 shows gated coincidence spectra for the two rotational bands we identified in ^{175}Re.

While further studies are necessary to elicit all the low-energy details of the structures and connecting transitions, Figs. 8–10 show the partial level schemes of these neutron-deficient Re systems.

25.4 ANALYSIS

HI-induced reactions introduce very large amounts of angular momenta into the compound systems. The most effective manner for the prolate compound nucleus to carry

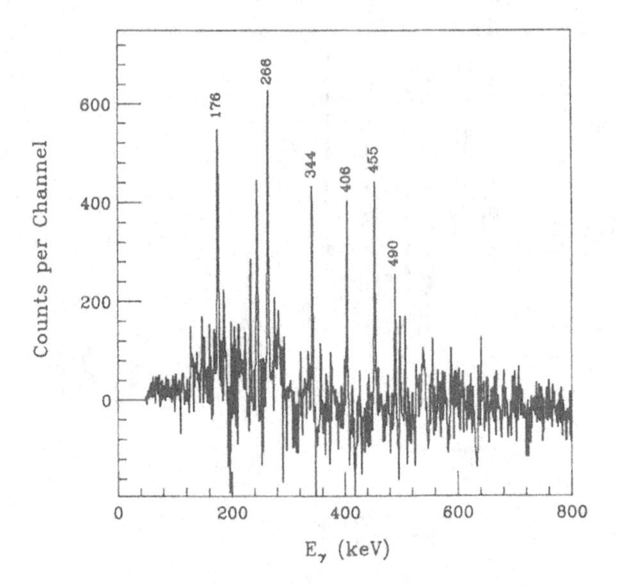

Figure 5a. Spectrum gated on DDB transitions in ^{174}Re.

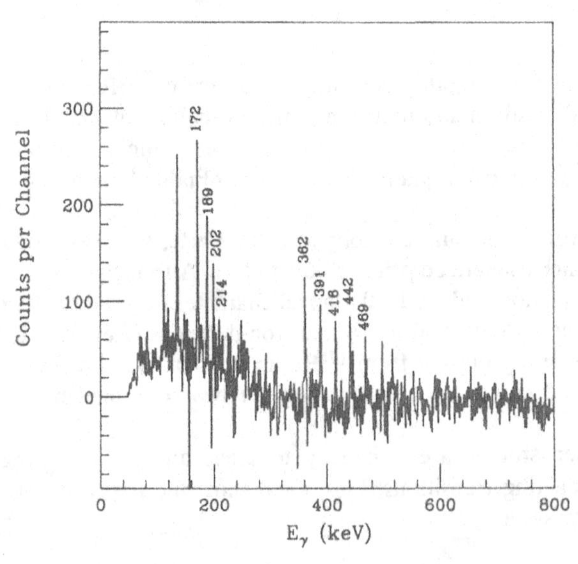

Figure 5b. Spectrum gated on 158-keV transition in ^{174}Re.

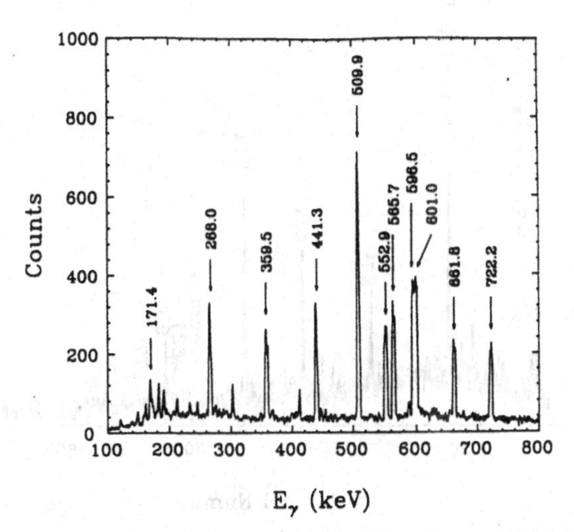

Figure 6a. Summed spectrum of gates in the DDB in the ^{176}Re.

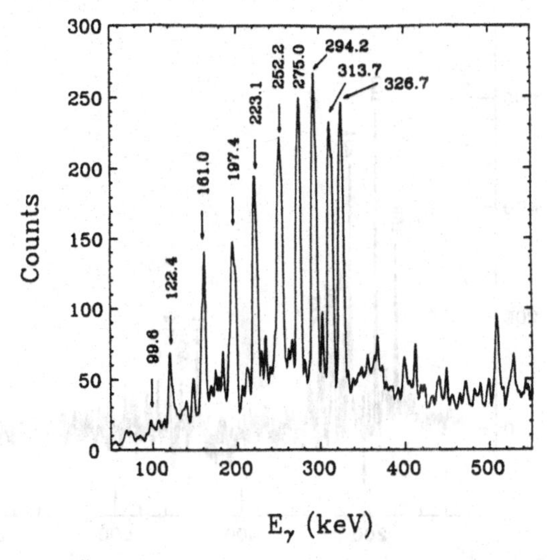

Figure 6b. ^{176}Re spectrum gated on transitions in the ΔI = 1 rotational band.

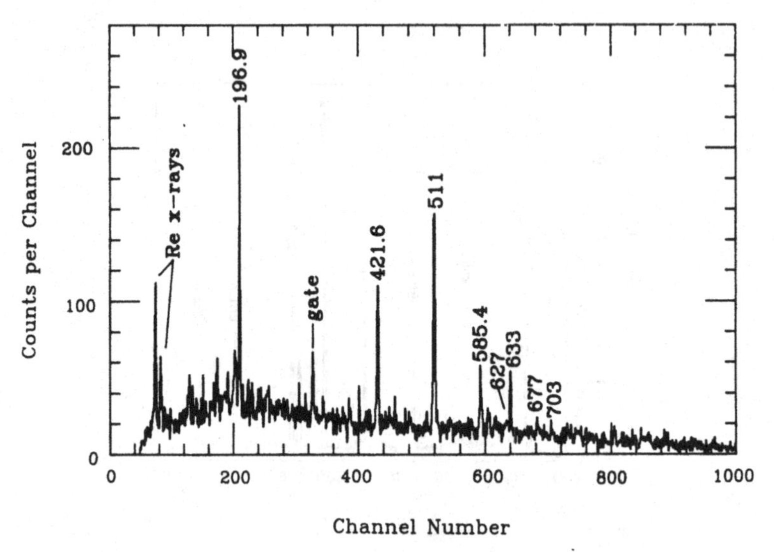

Figure 7a. Spectrum gated on 316.3-keV transition in ^{175}Re.

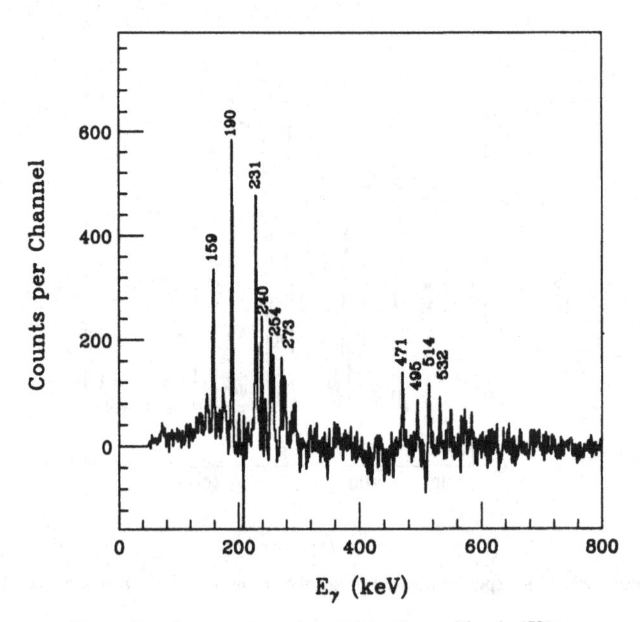

Figure 7b. Spectrum gated on 279-keV transition in ^{175}Re.

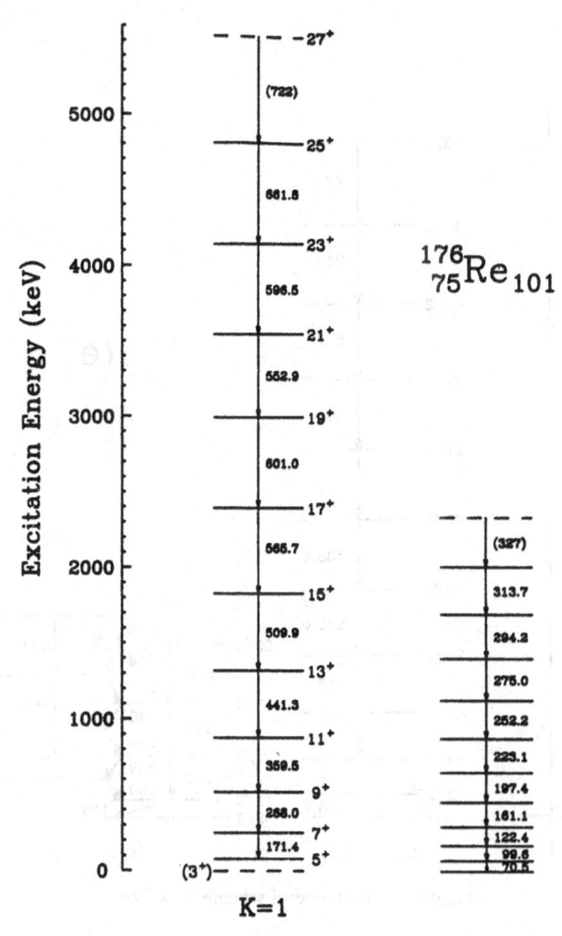

Figure 8. Partial level scheme for ^{176}Re.

this is to rotate perpendicular to its axis of symmetry. The additional alignment of any quasi-particles in this rotational frame can also assist in carrying the angular momentum. We can thus argue that the potential energy surface for reaction is depressed for these configurations, and the reactions are driven to (and seemingly do) populate preferentially nuclei in highly-aligned and -deformed configurations.

Another interesting aspect of these nuclei is the fact that they exhibit backbending. The observation of backbending in ^{176}Re is only the second time this phenomena has been observed in odd-odd systems. ^{176}Re exhibits a band crossing at a frequency of 0.29 MeV, while ^{174}Re does not show such a distinct backbend, but it does have a very strong upbend (possibly a slight backbend) at a frequency of 0.25 MeV. Figure 11 shows the alignment of the angular momenta in the DDB in ^{174,176}Re and the DB in ^{175}Re. The crossing frequency of the bands in the odd-odd nuclei is a "compromise" of the effects of the two odd-particle states which form the band.[2] Both the ^{176}Re and ^{174}Re DDB are built on couplings of the $h_{9/2}$ $1/2^{-}[541]$ proton and $i_{13/2}$ neutron orbitals, with the neutron orbitals

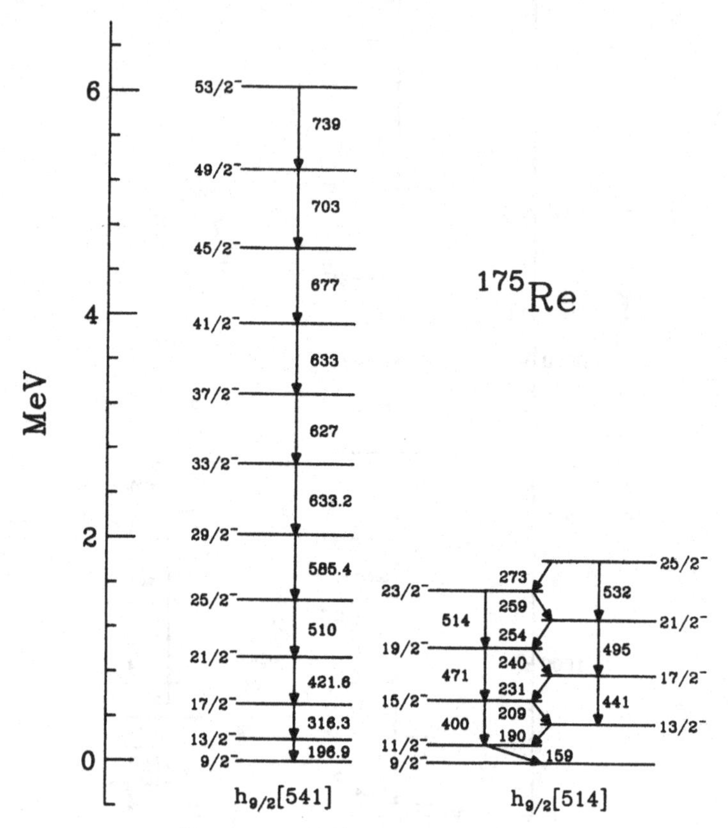

Figure 9. Partial level scheme for ^{175}Re.

weakening the pairing force, thus lowering crossing frequencies, while the proton orbital drives toward greater deformations and increasing crossing frequencies.

One of the most interesting observations of the population of the DDB is that they seem to populate independent of the angular momentum induced in the system. In the case of ^{174}Re, which we studied with a wide variety of projectile/target combinations—including ^{16}O on ^{165}Ho, ^{20}Ne on ^{159}Tb, and ^{40}Ar on ^{139}La—we observed essentially the same population of the DDB in each system. Each was populated within statistical errors up to approximately 27-29 $\hbar\omega$. One plausible explanation for this behavior is illustrated in Fig. 12. It is likely that the DDB lies at some position not far above the yrast band in the phase space of the *HI*-induced reaction. In a geometry such as this, the DDB acts as a trap for the deexcitation of the compound nuclei.

Perhaps the most interesting aspect of all the studies of the odd-odd Re isotopes is the hope that they will provide that "window" by which to view the *p-n* interaction in a fairly massive system. The systematic deviations from the "core" rotational bands in the even-even analogies is encouraging. Figure 13 demonstrates the reasonably high degree of correlation amongst these bands in several cases. Though extensive analysis of these correlations is not complete, the consistent parabolic shape of the deviations inspires one

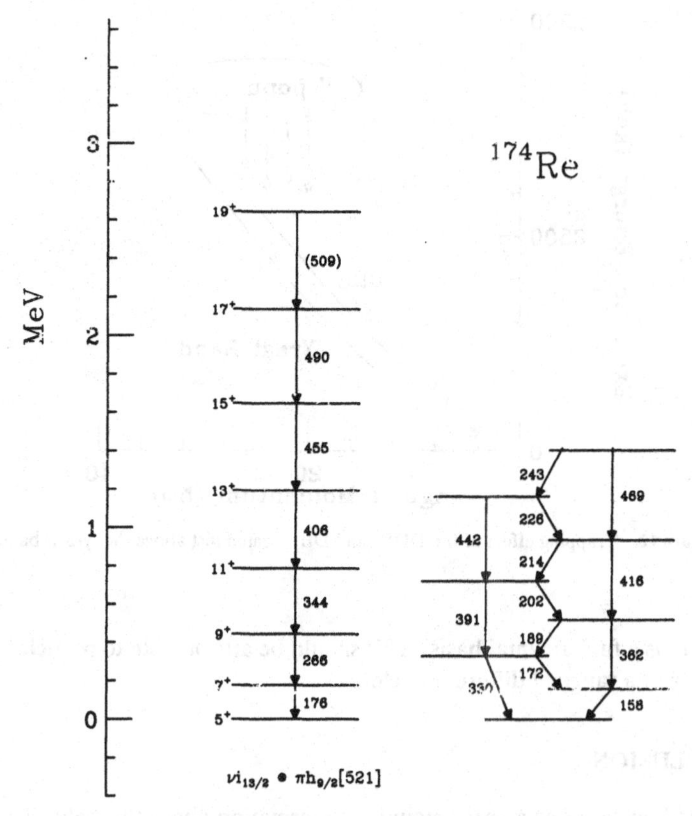

Figure 10. Partial level scheme for ^{174}Re.

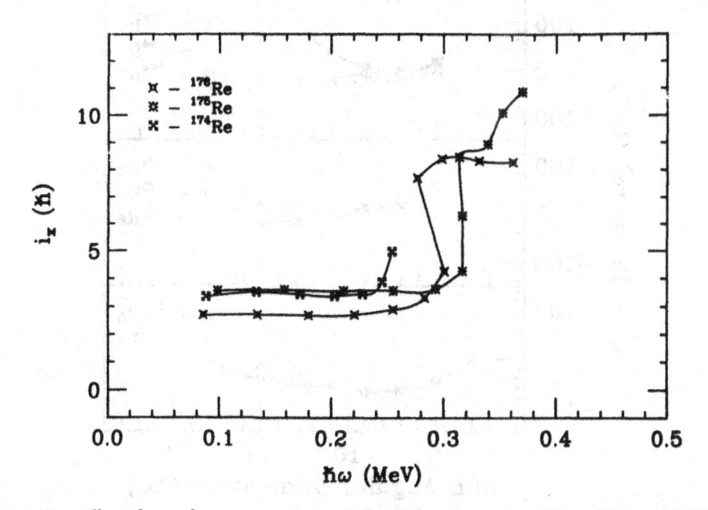

Figure 11. Aligned angular momenta vs rotational frequency for ^{176}Re, ^{175}Re, and ^{174}Re.

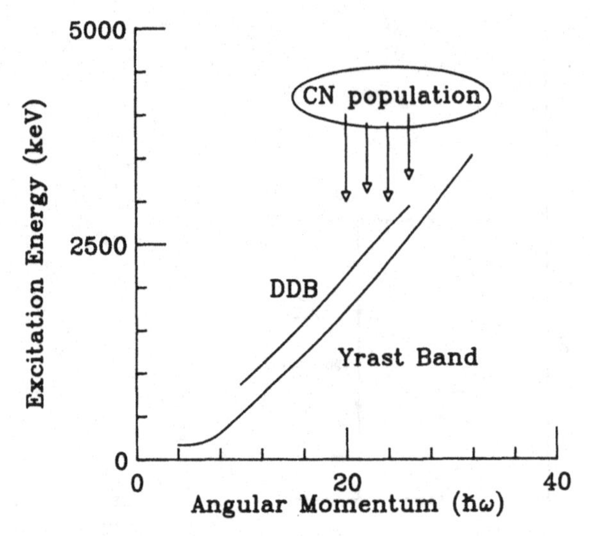

Figure 12. Trapping effect of the DDB (and DB) located just above the "yrast" band.

to believe there is a fundamental basis. This should be attributable to particle effects, since it is observed over a range of different nuclei.

25.5 CONCLUSION

Odd-odd nuclei offer a vast amount of information about the behavior of odd- and quasi-particles in nuclear systems. They encourage us in our desire to more completely understand the *p-n* interactions in relatively high-mass nuclei at or near ground-state energies.

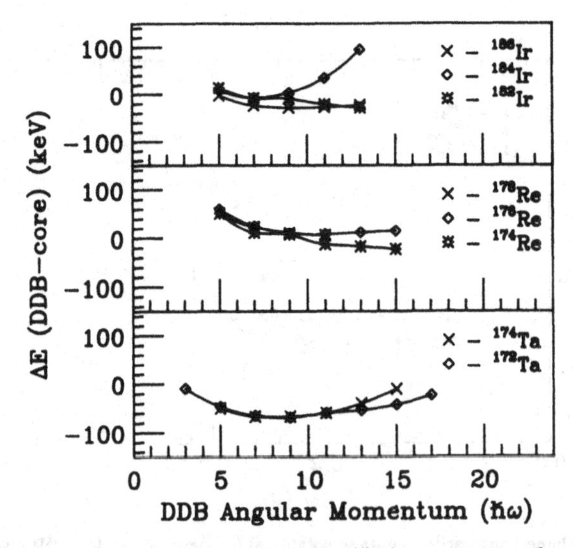

Figure 13. Plot of ΔE (DDB-analogous core transition) vs average angular momentum.

The consistencies observed in the DDB and core rotational bands offer convincing evidence that such particle states can be treated to a fair degree as independent of particle-core interactions; thus, the deviations can be attributed to odd-particle effects. An adjunct to this work has been IBAFFA calculation for states in these deformed nuclei. The reader is referred to Chapter 34 in this book.

ACKNOWLEDGMENTS

We acknowledge the help of Yves Dardenne and Christine Hampton in performing the experiments at MSU. We also thank Drs. E. S. Paul and D. B. Fossan for providing us the opportunity to perform experiments at SUNY Stony Brook and to R. Ma, N. Xu, and Y. Liang for their help with the experiments. This work was supported in part by the U. S. National Science Foundation, under Grant PHY-85-19653.

REFERENCES

1. M. F. Slaughter, R. A. Warner, T. L. Khoo, W. H. Kelley, and Wm. C. McHarris, *Phys. Rev. C* **29**, 114 (1984).
2. W.-T. Chou, Ph.D. Thesis, Michigan State University, 1989.
3. D. Santos, A. J. Kreiner, J. Davidson, M. Davidson, M. Debray, D. Hojman, and G. Falcone, *Phys. Rev. C* **39**, 902 (1989).
4. W.-T. Chou, W. A. Olivier, Aracelys Rios, Wm. C. McHarris, R. Aryaeinejad, E. S. Paul, R. Ma, N. Xu, and Y. Liang, MSU-NSCL Report MSUCL-717, submitted to *Phys. Rev. C.* (1990).
5. Wm. C. McHarris, Ph.D. Thesis, University of California, Berkley, 1965.
6. L. Hildingsson, C. W. Beausang, D. B. Fossan, W. F. Piel, Jr., A. P. Byrne, and G. D. Dracoulis, *Nucl. Instrum. Methods* **A252**, 91 (1986).
7. J. R. Leigh, J. O. Newton, L. A. Ellis, M. C. Evans, and M. J. Emmott, *Nucl. Phys.* **A183**, 177 (1972).
8. W. Walus, L. Carlen, S. Jonsson, J. Lyttkens, H. Ryde, J. Konacki, W. Nazarewicz, J. C. Bacelar, J. Dudek, J. D. Garrett, G. B. Hagemann, B. Herskind, and C. S. Yang, *Phys. Scr.* **34**, 710 (1986).
9. R. A. Bark, G. D. Dracoulis, A. E. Stuchbery, A. P. Byrne, A. M. Baxter, F. Riess, and P. K. Weng, *Nucl. Phys.* **A501**, 157 (1989).

26. Double Decoupling in Deformed Doubly-Odd Nuclei

Andrés J. Kreiner
Departmento de Física
Comisión Nacional de Energía Atómica
Buenos Aires, Argentina

ABSTRACT

When decoupled proton and neutron orbitals combine in doubly-odd nuclei, one obtains a special type of collective structure called a "doubly-decoupled" band. A particle-plus-rotor and cranking model analysis provide a qualitative understanding of all the special features of these bands, showing that strong coriolis effects dictate their main behavior, while the *p-n* residual force seems to be much less important. They display additivity of striking accuracy for alignments and inertia parameters. Several of these bands have been followed beyond the first band crossing, showing again additivity for the shift in crossing frequency. These properties may raise questions concerning the deformation-driving tendency of the $\pi h_{9/2}$ orbital and on the exclusive $\nu i_{13/2}$ origin of the first backbend.

26.1 INTRODUCTION

The study of near-yrast states in doubly-odd deformed nuclei is increasingly attracting the attention of high-spin spectroscopists. The results from this effort are starting to reveal a wealth of new and interesting structures, most of which deviate drastically from a normal rotational behavior because of strong coriolis effects. Actually, a general classification of the different coupling modes of two distinguishable particles between themselves and to a rotor is emerging.[1-3]

In this Chapter I shall deal exclusively with one of the best understood cases; namely, the double-decoupling scheme,[4-6] reviewing all the available information and all its known properties. This structure actually turns out to be the equivalent of a decoupled band[7] (originally described in an odd nucleus) but in a doubly-odd system, and it involves both the valence proton and neutron occupying predominantly $\Omega = 1/2$ orbitals with large decoupling parameters. The interest in decoupled rotational bands derives, in particular, from the fact that the features of the collective and intrinsic motion can be clearly identified and separated (as, at the other extreme, in rigid, strongly-coupled bands). The constancy of the intrinsic structure in the decoupled case manifests itself in the constancy (in certain regions of angular momentum I or angular velocity $\hbar\omega$) of quantities such as the alignment[8] i and also allows one to extract reliably the inertia parameters characterizing the motion of the core in the presence of quasi-particles.

In Section 2 we shall try to reach an understanding of this structure in terms of the two-quasiparticle-plus-rotor model[9,10] (TQPRM), particularly of its deformation dependence. Section 3 presents an analysis in terms of the cranking model[8] (CM) in order to ex-

Exotic Nuclear Spectroscopy, Edited by W. C. McHarris
Plenum Press, New York, 1990

tract alignments and inertia parameters revealing striking additivity properties.[11] Finally, Section 4 studies higher frequency phenomena such as shifts in crossing frequencies of doubly-decoupled bands (DDB's) with respect to the same quantity in related structures in neighboring odd-mass nuclei.[12]

26.2 PARTICLE-PLUS-ROTOR ANALYSIS

One of the most interesting features revealed by in-beam spectroscopy work[4,5,13-15] on $^{182-186}$Ir is a band of $\Delta I = 2$ character. This "decoupled"[7] cascade is shown in Fig. 1 for ^{186}Ir along with the ground-state band[16] (GSB) of ^{184}Os. This figure beautifully illustrates the very clear signature that decoupled bands have. They follow (qualitatively) the spacing of the GSB in neighboring even-even nuclei. They cannot be understood with the $I(I+1)$ law, but, rather, $R(R+1)$ must be used, R being the collective angular momentum. This observation can be cast in a quantitative way following the method proposed in Ref. 4. One assumes that the 5^+ ground state has a remnant nonvanishing R_0 for the expectation value of the collective angular momentum (only in the spherical limit one can expect $R_0 = 0$ for the "band-head state"), which increases in steps of two units along the band, and that the level energies follow the expression, $AR(R+1)$. ($A = \hbar^2/2$ is the rotational constant.) R_0 and A can be deduced from the two first transition energies and, in turn, used to predict the upper ones. One obtains, $R_0 = 0.40\hbar$ and $A = 15.46$ keV. The predicted ratios, $R_I = (E_I - E_5)/(E_7 - E_5)$ are compared to the experimental values in Table I, and the agreement is excellent. (Similar ratios are also given for ^{184}Os, showing a larger deviation from the asymptotic limit than the values for ^{186}Ir, in turn reflecting a larger rigidity for the 5^+ band structure as compared to the ^{184}Os GSB.) In terms of transition energies this represents, for instance, 365 keV for the $11^+ \rightarrow 9^+$ line, as compared to 362 keV measured. One can easily see the inadequacy of the $I(I+1)$ expression. For instance, if A is extracted from the $7^+ \rightarrow 5^+$ spacing, using this expression one predicts 153.8 vs 241.3 keV for the $9^+ \rightarrow 7^+$ transition.

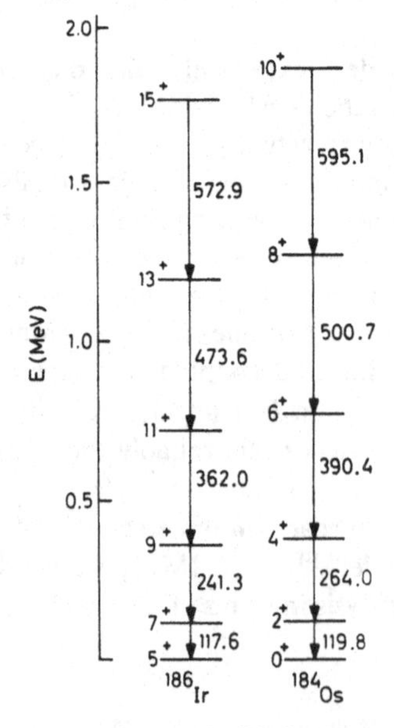

Figure 1. The DDB band based on the 5^+ state in ^{186}Ir is compared to the ^{184}Os ground-state band.

Table I. Experimental and Calculated Ratios $R_I = (E_I - E_5)/(E_7 - E_5)$ for ^{186}Ir, and $R_I = E_I/E_2$ for ^{184}Os

^{186}Ir			^{184}Os		
I	R_I^{exp}	R_I^{calc}	I	R_I^{exp}	R_I^{calc}
11	6.13	6.16	6	6.46	7.00
13	10.16	10.31	8	10.64	12.00
15	15.03	15.52	10	15.61	18.33

A less phenomenological analysis will now be carried out within the TQPRM.

In order to understand the decoupling process in a deformed situation, one has to recall some facts more related to the properties of the individual Nilsson orbits rather than of the whole high-j multiplets, as in transitional cases. (The term decoupling is used here for a situation in which a normal band splits into two $\Delta I = 2$ sequences, displaced in energy against each other, so that the normal spin sequence ($I, I+1, \ldots$,) is altered; namely, that the $I_0 + 2, I_0 + 4, \ldots$, favored members lie below the $I_0 + 1, I_0 + 3, \ldots$, unfavored members. This is specifically a quantal feature related to the reflection symmetry of the deformation.[23] Classically (or quantum mechanically but without reflection symmetry), a particle can decouple from the binding to the deformed mean field because of a strong coriolis force but will not give rise to the situation just described.

As it is well known, in odd and axially symmetric nuclei the decoupling is related to a single diagonal signature dependent matrix element; namely,

$$-A\delta_{K,1/2}a(-1)^{I+1/2}(I+1/2)\ , \tag{1}$$

a being the decoupling parameter, defined as

$$a = \sum_j c_j^2(-1)^{j+1/2}(j+1/2)\ . \tag{2}$$

It is easy to see that decoupling occurs from a critical value for a on; namely, for $|a| \gtrsim 1$. If I is such that $(-1)^{I+1/2} = -1$ (i. e., I = even integer $+1/2$), the difference in consecutive transition energies in units of the rotational constant is

$$[\Delta E(I \rightarrow (I-1))]/A = 2I(1+a)\ . \tag{3}$$

Hence, if $a = -1$, the pairs of states with $I = 3/2, 5/2; 7/2, 9/2; \ldots$, etc. are degenerate. One may define[23] a new quantum number, called signature, as $\sigma = (-1)^{I-1/2}$. Hence, the original $\Delta I = 1$ band splits into two sequences of opposite signature: $I = 1/2, 5/2, 9/2, \ldots$, ($\sigma = +1$) and $I = 3/2, 7/2, \ldots$, ($\sigma = -1$). To be sure, decoupling occurs because of the presence of $K = \Omega = 1/2$ states. These states play a special role, since they have nonvanishing diagonal matrix elements of the coriolis operator. This operator is related to $\mathbf{I}^{\perp} \cdot \mathbf{J}^{\perp}$ (where $\perp$ denotes the component of the total angular momentum $\mathbf{I}$ and of the intrinsic angular momentum $\mathbf{J}$ perpendicular to the symmetry axis).

One has

$$\langle K = 1/2|\mathbf{I}^{\perp} \cdot \mathbf{J}^{\perp}|\overline{K = 1/2}\rangle = (a/2)(-1)^{I+1/2}(I + 1/2) \ . \tag{4}$$

If we divide out the length of $\mathbf{I}^{\perp}$, which is $\approx[(I + 1/2)^2 - (1/2)^2] \approx I + 1/2$, we obtain for the projection of $\mathbf{J}^{\perp}$ (or $\mathbf{J}$) on $\mathbf{I}^{\perp}$ approximately $(a/2)(-1)^{I+1/2} = -(a/2)\sigma$. This quantity may be called alignment. For $a = -1$ and $\sigma = +1(-1)$, the alignment is $1/2(-1/2)$, so that an isolated $\Omega = 1/2$ state acts as having an effect $J_{eff} = 1/2$.
In doubly-odd nuclei, there is no such *diagonal* signature-dependent matrix element. However, in this case, an investigation[4] of the rotational hamiltonian shows that the equivalent role is played by a nondiagonal matrix element of the coriolis operator connecting $K = 0$ and 1 states of a special kind (degenerate if the proton-neutron residual interaction is neglected); namely, $K = \Omega_n + \Omega_p = 1/2 + 1/2 = 1$ and $K = \Omega_n - \Omega_p = 1/2 - 1/2 = 0$. This matrix element becomes

$$-A(a_p+(-1)^{I+1}a_n)[I(I+1)]^{1/2} \approx -A(a_p+(-1)^{I+1}a_n)(I+1/2) \ . \tag{5}$$

In order to have decoupling here, it is necessary that both decoupling parameters satisfy $|a_n|, |a_p| \geqslant 1$. This is illustrated in Fig. 2, where the excitation energies of several states are plotted against the ratio $x = a_n/a_p$ (for the particular value $a_p = 4.4$, which, in absolute value, is appropriate for the $h_{9/2}$ orbital, the relevant proton excitation, as we shall see below) for the simplest two-band $K = 0$ and 1 system. Decoupling occurs only beyond the $x_0 = 1/a_p$ point (i.e., $a_n = 1$), and we speak of double decoupling. It is worth noting that in order to have the odd-spin states as favored states (for $I > 1$, see Ref. 4), both decoupling parameters have to be of the same sign.

In transitional odd-odd nuclei involving high-j excitations for both protons and neu-

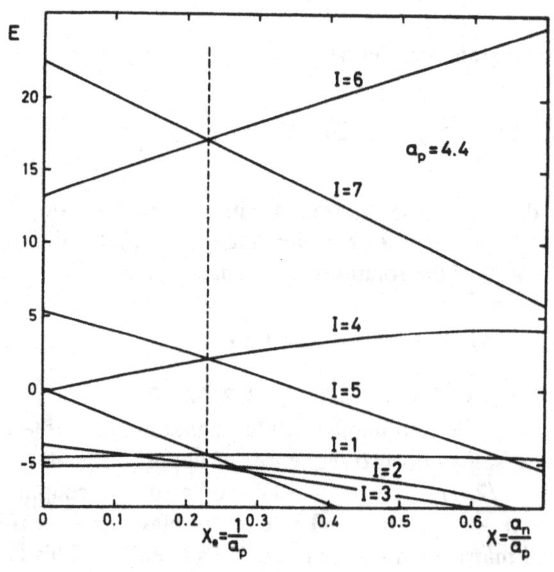

Figure 2. Energy of yrast states (in units of $A = \hbar^2/2\mathfrak{J}$) as functions of the ratio of neutron and proton decoupling parameters, $x = a_n/a_p$, for a two-band ($K = 1 = 1/2 + 1/2$, $K = 0 = 1/2 - 1/2$) doubly-odd system.

trons, double decoupling also exists[9,10] only if both particles satisfy the conditions for decoupling; namely, that both fermi levels lie nearest to the corresponding $\Omega = 1/2$ states. Hence, also in the deformed case, in order to have double decoupling it is necessary that both participating particles decouple individually, the mechanism being in fact the same as in the transitional case.

The conclusion from the previous discussion is that one would have to look for proton and neutron $\Omega = 1/2$ orbitals near the respective fermi levels with $|a| \geqslant 1$ in order to find cases where the double-decoupling scheme is actually realized. There are in fact not many possibilities in this region. The two relevant building blocks turn out to be the $1/2^-[541]$ proton orbit, which gives rise to decoupled bands in many odd-Z isotopes, and the $1/2^-[521]$ neutron orbit, known to originate $\Delta I = 2$ sequences in several odd-N nuclei. Figs. 3, 4, and 5 show all DDB's known to date, along with the $\pi h_{9/2}$ decoupled bands in neighboring odd-Z nuclei.

We shall now discuss[17] why the 5^+ state lies lowest in ^{186}Ir while it is crossed by the 3^+ state in the lighter Ir isotopes (in ^{182}Ir the 3^+ is most likely below the 5^+ state) and why the $3^+ \rightarrow 5^+$ spacing increases in going to lower Z values (i.e., Re and Ta nuclei).

Figure 6 shows the excitation energies in units of the rotational constant A for some favored (odd-spin) states of the $\pi h_{9/2} \otimes \nu 1/2^-[521]$ system referred to the 3^+ state as func-

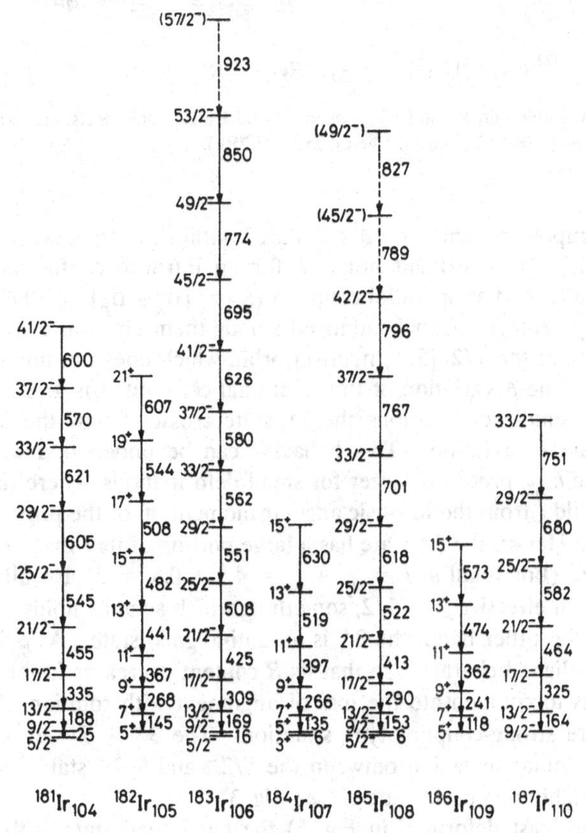

Figure 3. DDB's in doubly-odd Ir and $\pi h_{9/2}$ bands in odd Ir isotopes, Refs. are 22, 24 (^{181}Ir); 5 (^{182}Ir); 25, 26 (^{183}Ir); 5, 14 (^{184}Ir); 25, 27, 28 (^{185}Ir); 4, 13 (^{186}Ir); 29 (^{187}Ir).

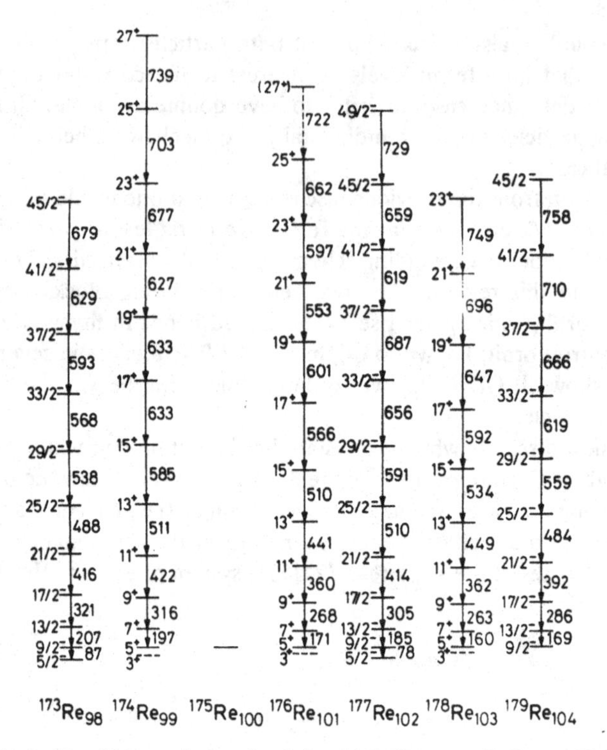

Figure 4. DDB 's in double-odd Re and $\pi h_{9/2}$ bands in odd Re isotopes. Refs. are: 30 (^{173}Re); 31 (^{174}Re); 17, 31, 32 (^{176}Re); 33, 34 (^{177}Re); 17, 32, 35 (^{178}Re); 33, 35 (^{179}Re).

tions of the quadrupole deformation β calculated within the framework of the TQPRM. $\pi h_{9/2} \otimes \nu 1/2^-[521]$ is a shorthand notation for an intrinsic configuration space which comprises the parallel and antiparallel coupling ($K = |\Omega_n \pm \Omega_p|$) of the five Nilsson proton orbits of $h_{9/2}$ parentage (strongly admixed among themselves, as evidenced by the $h_{9/2}$ decoupled bands) and the $1/2^-[521]$ neutron orbit, which does not mix with other states, thus acting alone. The β variation of the rotational constant A is given by $204/(\beta^2 A^{7/3})$. One sees that for small deformations the 5^+ state crosses below the 3^+, in qualitative agreement with the Ir situation. This behavior can be understood by noting that the "alignability" of the $h_{9/2}$ proton is larger for small deformations, where the system prefers energetically to build I from the intrinsic angular momentum of the proton ($\approx 4.5\hbar$) instead of core rotation R. Hence, the 5^+ state has a large portion of its I made out of proton and neutron alignment. (For small β, $J = j_p + j_n = 4.5 + 0.5 = 5$. Recall that the neutron behaves as having an effective $j = 1/2$, something which actually holds over a large range of β values.) On the other hand, the 3^+ is an anti-aligned state. As β increases, the 3^+ state loses its anti-aligned character so that its R content decreases faster than that for the 5^+ state which has to compensate the loss of alignment with rotation. This causes it to evolve into a more strong-coupling-type situation. The $3^+ \rightarrow 5^+$ crossing is intimately connected with a similar inversion between the $5/2^-$ and $9/2^-$ states of decoupled $h_{9/2}$ proton bands in neighboring odd Ir nuclei (see Fig. 3).

In ^{187}Ir (the least deformed in Fig. 3) the band-head state is $9/2^-$ ($R \rightarrow 0$ for $\beta \rightarrow 0$), while the $5/2^-$ state (which lies higher) is antialigned ($R \rightarrow 2$ for $\beta \rightarrow 0$; $I = 5/2 =$

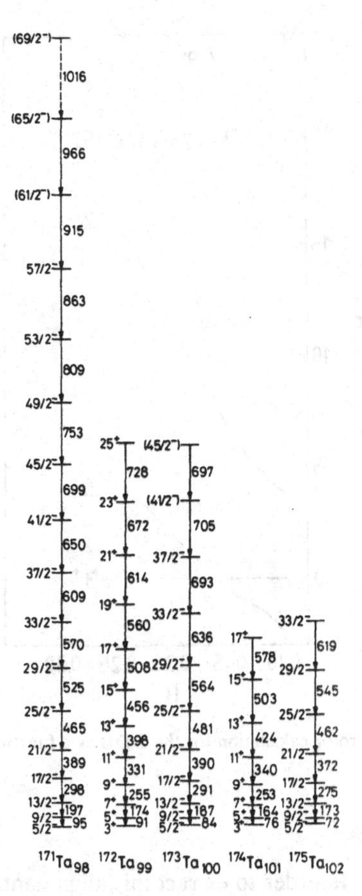

Figure 5. DDB's in doubly-odd Ta and $\pi h_{9/2}$ bands in odd Ta isotopes. Refs. are 34, 36 (^{171}Ta); 37 (^{172}Ta); 36 (^{173}Ta); 11 (^{174}Ta); 38 (^{175}Ta).

$-R + j = -2 + 9/2$); and, because it involves rotation, it is shifted to higher energies. In the odd-Re case, the 5/2 is the ground state while the $9/2 \rightarrow 5/2$ transition has only been seen in ^{173}Re, probably because of a very high internal conversion for low energy $E2$'s. We believe this to be the reason which also prevents the observation of the $5^+ \rightarrow 3^+$ transition in the Re case.

26.3 CRANKING-MODEL ANALYSIS

Still another method of analysis can be used[8] in order to gain more information on the structure of the DDB's. It consists in obtaining an alignment (related to the quantity defined in Section 2) which is supposed to characterize the intrinsic structure in a specific way. This alignment is the angular momentum associated with the intrinsic motion in the

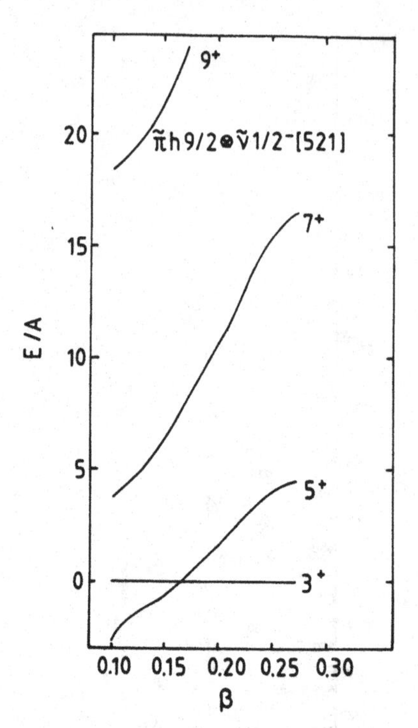

Figure 6. Two quasi-particle plus rotor calculation for the DDB as a function of the quadrupole deformation β.

direction of the rotation axis. In order to extract this alignment, the rotation frequency has to be obtained for each state involved, which is done using the following expression:

$$\frac{\partial E}{\partial I_x}(I) \approx [E(I+1) - E(I-1)/[I_x(I+1) - I_x(I-1)] \approx \hbar\omega(I) \qquad (6)$$

where

$$I_x(I) = [(I+1/2)^{1/2} - \langle K^2 \rangle]^{1/2} \qquad (7)$$

is the total transverse angular momentum and $< K^2 >$ denotes the expectation value of the square of the angular momentum component on the nuclear symmetry axis.

We used $K = 1/2$ for bands in odd-neutron and odd-proton nuclei. Also $K = 1/2$ was used for the $3^+(5^+)$ band in accord with the fact that this band has predominantly $K = 0$ and 1 components.[4-6]

The total transverse angular momentum I_x is composed of two parts:

$$I_x = (R + 1/2)(\omega) + i(\omega) , \qquad (8)$$

where R denotes the collective part of the angular momentum (of the associated even-even core) and i the alignment.

From here on, two different procedures can be followed. One of them consists in associating with a particular odd nucleus a certain neighboring even-even core. One would then represent the rotational-core angular momentum through a second-order cranking expansion (making a two-parameter fit),

$$(R+1/2)(\omega) = (\Im_0/\hbar^2 + \Im_1/\hbar^4(\hbar\omega)^2)\hbar\omega \ , \tag{9}$$

and obtain a frequency-dependent alignment for the odd nucleus through

$$i(\omega) = I_x(\omega) - (R+1/2)(\omega) \tag{10}$$

for the experimental frequencies determined with Eq. (6) in this odd nucleus.

The alternative procedure consists in allowing for a modification of the core parameters in the odd (or doubly-odd) nucleus, i.e., in presence of valence nucleons, by fitting $\Im_0$ and $\Im_1$ directly to the experimental bands. We adhere here to this philosophy. Again, the parameters are fixed through the minimization of the following quantity:

$$\chi^2 = \sum_{j=1}^{n}(I_x^{exp}(\omega_j) - [(\Im_0/\hbar^2)(\hbar\omega_j) + (\Im_1/\hbar^4)(\hbar\omega_j)^3 + \langle i\rangle]^2 \ , \tag{11}$$

where j runs over the experimental points and $\langle i \rangle$ represents an average value for the alignment in a given band. ($\langle i \rangle = i(\omega) = 0$ for the ground bands of even-even nuclei, because only points below the backbending are included.) Instead of applying a fitting procedure, one may also take the lowest three frequencies (in the odd cases) and solve for the parameters. It should be stressed that these parameters are obained independently for each nucleus.

Table II lists the ground-state cranking-model moments of inertia and the alignments for most of the DDB's and for the corresponding even-even (*ee*) and odd-mass nuclei displayed in Figs. 3–5.

A striking additivity of the deviations of the inertia parameters with respect to those of an appropriate *ee* core (here the nucleus with one proton and one neutron fewer) is documented in Table II to be a systematic feature of DDB's (compare Columns 4 and 5). In other words, $\Im_{onp}^{exp}$ determined experimentally is equal, with a high degree of precision, to

$$\Im_{onp}^{exp} = \Im_{oee}^{exp} + \delta\Im_{on}^{exp} + \delta\Im_{op}^{exp} \tag{12}$$

where

$$\delta\Im_{oi}^{exp} \equiv \Im_{oi}^{exp} - \Im_{oee}^{exp} \ , \tag{13}$$

with $i = n$ or p. $\Im_0$ tends to follow an analogous rule[11] but with less precision. A similar additivity property holds for the alignments; namely,

$$i_{np}^{exp} = i_n^{exp} + i_p^{exp} \tag{14}$$

(as in Columns 6 and 7).

Let us illustrate the procedure for the case of ^{178}Re. The core is always the *ee* nucleus with a proton and a neutron fewer, in this case ^{176}W, with parameters $\Im_0 = 27.20$

Table II: Inertia Parameters and Alignments Extracted from the First Three Transitions in GSB's (even-even nuclei), $\nu 1/2^-[521]$ (Odd-N), $\pi h_{9/2}$ (odd-Z), and DDB's (odd-odd nuclei).

Nucleus	I^π	α	$\Im_0^{exp}$ ($\hbar^2$MeV^{-1})	$\Im_{0np}^{calc}$ ($\hbar^2$MeV^{-1})	i^{exp} ($\hbar$)	$i_n + i_p$ ($\hbar$)	$\hbar\omega_c$ [a] (keV)
^{170}Hf	0+	0	29.26				265
^{172}Hf	0+	0	31.32				
^{171}Hf	1/2−	1/2	39.14		0.41		220
^{173}Hf	1/2−	1/2	38.90		0.42		
^{171}Ta	5/2−	1/2	38.94		2.14		305
^{173}Ta	5/2−	1/2	39.21		2.35		
^{172}Ta	3+	1	48.63	48.82	2.27	2.55	250
^{174}Ta	0+	1	46.42	46.79	2.72	2.77	
^{174}W	0+	0	26.05				290
^{176}W	0+	0	27.20				
^{175}W	1/2−	1/2	34.89		0.44		255
^{177}W [b]	1/2−	1/2	33.49		0.445		
^{179}W	1/2−	1/2	32.09		0.45		
^{177}Re	9/2−	1/2	30.64		3.11		320
^{176}Re	5+	1	39.45		3.07		
^{178}Re	5+	1	36.95	36.93	3.55	3.56	295
^{180}Os	0+	0	21.98				265
^{182}Os	0+	0	23.28				
^{181}Os	1/2−	1/2	28.35		0.53		210
^{183}Os	1/2−	1/2	29.49		0.49		
^{181}Ir	9/2−	1/2	22.40		3.81		295
^{183}Ir	9/2−	1/2	24.29		3.88		
^{182}Ir	5+	1	26.72	28.77	4.48	4.34	265
^{184}Ir	5+	1	30.73	30.50	4.41	4.37	

a Ground-state → S-band crossing frequency; its uncertainty is estimated at ≈5-10 keV.
b Values for ^{177}W are interpolated from ^{175}W and ^{179}W.

and $\Im_1 = 134.47$. Since the $1/2^-[521]$ band is unknown in ^{177}W, we interpolate its parameters from ^{176}W and ^{179}W, obtaining $\Im_0 = 33.49$, $\Im_1 = 128.87$, and $i_n^{exp} = 0.445$. For ^{177}Re, we have $\Im_0 = 30.64$, $\Im_1 = 61.98$, and $i_p^{exp} = 3.11$. The deviations with respect to ^{176}W of the parameters of ^{177}W and ^{177}Re are:

$$\delta\Im_{on}^{exp} = 6.29,\ \delta\Im_{1n}^{exp} = -5.60, \delta\Im_{op}^{exp} = 3.44,\ \delta\Im_{1p}^{exp} = -72.49\ , \qquad (15)$$

respectively. Hence, we obtain

$$\Im_{onp}^{calc} = 27.20 + 6.29 + 3.44 = 36.93\ (\text{vs}\ \Im_{onp}^{exp} = 36.96)\ , \qquad (16)$$

$$\Im_{1np}^{calc} = 134.47 - 5.60 - 72.49 = 56.38\ (\text{vs}\ \Im_{1np}^{exp} = 54.28)\ , \qquad (17)$$

and $$i_n^{exp} + i_p^{exp} = 0.445 + 3.11 = 3.56 = i_{np}^{calc} \text{ (vs } i_{np}^{exp} = 3.55) . \tag{18}$$

The agreement is certainly impressive in this case. It suggests that neutron and proton "fluids" behave largely in an independent way.

It is worth noting that $\Im_{onp}$ for ^{178}Re is 36% larger than $\Im_{oee}$ (for ^{176}W). This can mainly be traced to the rather large blocking of pairing correlations by the odd proton and neutron quasiparticles. The large decrease of $\Im_{1np}$ (meaning a significant gain in rigidity as already discussed in Section 2 for ^{186}Ir) is also consistent with this interpretation.

There is still another quantity which shows the same additivity property; namely, the relative routhian e' (from here on, we drop the superindex *exp* and use $\hbar = 1$); but, as we shall see, this is not independent of the previous two.[12] The routhian of the nucleus[8] is

$$R_i(\omega) = E_i(\omega) - \omega I_{xi} \tag{19}$$

(i = c, n, p, or np), and the relative routhian with respect to a given core (actually only its rotational part) is

$$e' = R_i(\omega) - R_c = E_i - E_c - \omega(I_{xi} - I_c) . \tag{20}$$

For instance, the core routhian has the following expression:

$$\begin{aligned} R_c = E_c - \omega I_c &= \frac{1}{2}\omega^2\left(\Im_0 + \frac{3}{2}\Im_1\omega^2\right) - \omega\left(\Im_0\omega + \Im_1\omega^3\right) \\ &= -\frac{1}{2}\omega^2\Im_0 - \frac{1}{4}\omega^4\Im_1 \end{aligned} \tag{21}$$

If the core parameters are determined in each nucleus separately (as in Table II), one has $E_i \approx E_c$ and $e' = \omega i_i$. So, in this version the additivity of the routhian is merely a consequence of the additivity of the alignments. If, alternatively, one refers the odd nuclei to the *ee* core, one gets, for instance, for the routhian of the doubly-odd nucleus:

$$\begin{aligned} R_{np} - R_c &= -\frac{1}{2}\omega^2\Im_{onp} - \frac{1}{4}\omega^4\Im_{1np} - \omega i_{np} - (np \to c) \\ &= -\frac{1}{2}\omega^2\delta\Im_{onp} - \frac{1}{4}\omega^4\delta\Im_{1np} - \omega i_{np} = e'_{np} \end{aligned} \tag{22}$$

Here the additivity property,

$$e'_{np} = e'_p + e'_n , \tag{23}$$

follows from both the additivity of deviations of inertia and alignments together.

It should be stressed that these relative routhians are only related to the rotational energies. To compare total routhians, one has to resort to mass data as is done below in connection with the proton-neutron (*p-n*) residual interaction in DDB's.

Figure 7 shows an example of this additivity rule for the ^{174}Ta case.[11] The relative rotational Routhian e_{np}'(^{174}Ta) is almost indistinguishable from e_p'(^{173}Ta) + e_n'(^{173}Hf).

A puzzling aspect of these DDB's is that in none of the cases the unfavored part of this structure is known. Since the neutron decoupling parameter is near unity, there should

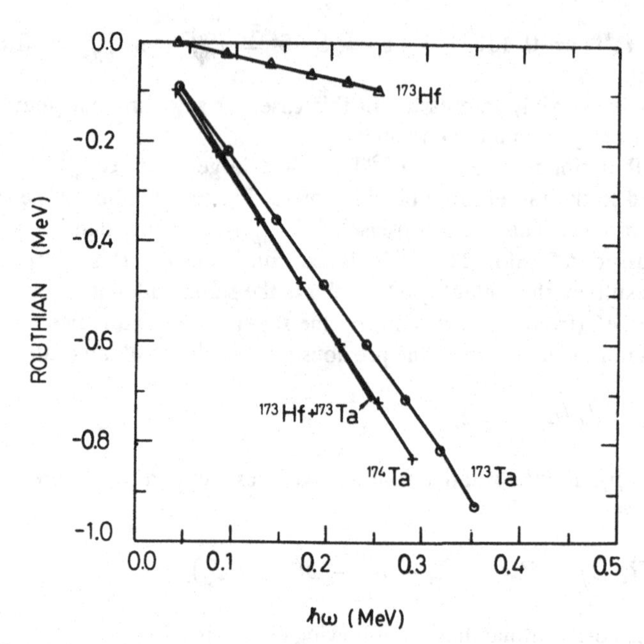

Figure 7. Experimental and calculated (line without symbols) relative rotational routhians as functions of $\hbar\omega$.

be a near-degeneracy of the favored and unfavored members (see Section 2), as is the case for the $\nu 1/2^-[521]$ bands in neighboring odd-N nuclei. There is, however, a cause which may shift the unfavored part of this structure to higher energies, thus making it non-yrast and difficult to observe in a heavy-ion induced fusion evaporation reaction. This cause is the residual p-n force. The expectation value V_{np} of this interaction in the lowest state of this structure (3^+ or 5^+) can be obtained from:

$$\begin{aligned} V_{np} &= S_n(N+1,Z) + S_p(N,Z+1) - S_{np}(N+1,Z+1) \\ &= B(N+1,Z) + B(N,Z+1) - B(N+1,Z+1) - B(Z,N) \end{aligned} \quad (24)$$

where $S(B)$ denote separation (binding) energies for the excitations involved.

V_{np} turns out systematically to be of the order of ≈ -0.5MeV. If V_{np} is significantly different for the unfavored part of the band (which has a different intrinsic structure), we would have found a reason for its non-observation. This may occur if V_{np} is mainly of the Newby type.[13,18] This is a special matrix element of the p-n force connecting a $K = 0$ state with its time reversed and has the form $(-1)^J C\delta_{K,0}$ (C would hence be ≈ 0.5MeV to have the right sign to lower odd-spin states).

Let us briefly discuss another potential implication of the additivity properties which bears on the question of core shape polarization.

It is the current opinion[19] that the prolate $h_{9/2}$ quasi-proton drives the nuclear shape to larger deformations as compared to those of the ee core. Now, if the odd-proton system would indeed have a significantly larger deformation, should one expect such precise additivity properties? The moment of inertia is a complex quantity, which depends on pairing and deformation (and possibly quite sensitively on both variables). The odd-neutron system has approximately the same deformation as the *ee* one, while the odd-proton and, thus, also the doubly-odd system would be more deformed if there is shape polarization due to the $h_{9/2}$ proton. The quantities related to the neutron in the odd-*N* system do not contain the increased deformation information, while they should be affected in the doubly-odd one, if the deformed field is something felt equally by all nucleons. On the other hand, if only blocking is active (i.e., there is no deformation increase), the effects should be additive because proton and neutron pairing correlations are largely decoupled in heavy nuclei. A similar remark holds for the alignments.

So far the CM analysis has been phenomenological. We can also explore theoretically[6] the conditions for double decoupling within this frame. One starts from a single-quasi-particle hamiltonian (for instance, cranked-Nilsson-BCS):

$$h' = h - \omega\hbar j_1 \tag{25}$$

If h is reflection symmetric, it implies that h' commutes with

$$R_1 = e^{-i\pi j_1} \tag{26}$$

(rotation of 180° around the intrinsic 1-axis). Hence the eigenstates of h and h' can be classified according to this operator:

$$R_1(\pi)|\alpha\rangle = e^{-i\pi\alpha}|\alpha\rangle \tag{27}$$

where α is called the signature[8] (different from the one defined in Section 2 but similar in meaning). For single quasi-particle states, α can take only the two values $\pm 1/2$, and it is an additive quantity. Let us discuss the relevant case of $\Omega = 1/2$ states. For $\omega = 0$, the two states $\Omega = \pm 1/2$ are degenerate. One can transform to a basis of good α, having again two degenerate states; namely, those with $\alpha = \pm 1/2$. The point is that, once the system is set into rotation, the axial symmetry is lost and Ω is no longer a good quantum number, although α still is. However, the degeneracy present at $\omega = 0$ is now removed. Actually, the splitting between $|\alpha = \pm 1/2\rangle$ is linear in ω, as can be easily verified:

$$\langle\alpha = \pm\frac{1}{2}|h'|\alpha = \pm\frac{1}{2}\rangle = \varepsilon_0 \pm \frac{1}{2}\hbar\omega a \tag{28}$$

If $a < 0$, $|\alpha = 1/2\rangle$ is the favored state (our sign convention for the decoupling parameter a is opposite to the one in Ref. 23).

Due to the reflection symmetry, which imposes the restriction[23]

$$e^{-i\pi I} = e^{-i\pi J}\ , \tag{29}$$

$\alpha = 1/2$ orbitals intervene in the description of $I = 1/2, 5/2, 9/2, \ldots$, while $\alpha = -1/2$ orbitals only participate in $I = 3/2, 7/2, \ldots$ states.

In the doubly-odd case and the most schematic situation (both proton and neutron in $\Omega = 1/2$ orbitals at $\omega = 0$), we have in the signature representation four different intrinsic states:

$$|np\rangle = |\pm\frac{1}{2}\rangle_n|\pm\frac{1}{2}\rangle_p \tag{30}$$

The states

$$|A\rangle = |+\frac{1}{2}\rangle_n|+\frac{1}{2}\rangle_p ,$$

$$|D\rangle = |-\frac{1}{2}\rangle_n|-\frac{1}{2}\rangle_p \tag{31}$$

belong to the signature $\alpha = 1$ (or -1 since $e^{\pm i\pi} = -1$), while

$$|B\rangle = |+\frac{1}{2}\rangle_n|-\frac{1}{2}\rangle_p ,$$

$$|C\rangle = |-\frac{1}{2}\rangle_n|+\frac{1}{2}\rangle_p \tag{32}$$

belong to $\alpha = 0$.

All these states are orthogonal among themselves and the $\alpha = 1$ subspace describes states of the total system with I = 1, 3, 5 . . ., (odd spins), while $\alpha = 0$ applies to $I = 0, 2, 4$, . . ., (In the TQPRM picture one also has two states for each I, namely $K = 0$ and 1, and the reflection symmetry is incorporated into the total wave-function).

The CM hamiltonian is, in this case:

$$H' = h'_n + h'_p = h_n + h_p - \omega\hbar(j_{1_n} + j_{1_p}) \tag{33}$$

The DDB's we have been discussing so far (a_n, $a_p<0$) are described by state $|A>$, and the expectation value of H' becomes

$$\langle A|H'|A\rangle = \varepsilon_{0_n} + \varepsilon_{0_p} + \frac{1}{2}\hbar\omega(a_n + a_p) \tag{34}$$

The unfavored part would correspond to state $|C>$).

The contact to the laboratory system is established through the constraint,

$$[I(I+1) - \langle J_3^2\rangle]^{1/2} \approx [(I+\frac{1}{2})^2 - \langle J_3^2\rangle]^{1/2} = \Im(\omega)\omega + \langle J_1\rangle \tag{35}$$

$$J_1 = j_{1_n} + j_{1_p} , \quad J_3 = j_{3_n} + j_{3_p} . \tag{36}$$

It can be shown that $<J_3^2> = 1/2$, and ω can then be fixed from Eqs. (34) and (35); namely,

$$\Im\omega = [(I+\frac{1}{2})^2 - \frac{1}{2}]^{1/2} + \frac{1}{2}(a_n + a_p) \tag{37}$$

Actually, $\Im\,\omega$ is the collective angular momentum (ω is obtained with some assumption on $\Im$). There is no self-consistency question here, since the intrinsic state (and hence expectation values derived from it) does not vary with ω (This model is the simplest version of the one employed in Refs. 20 and 21). The phenomenological procedure used previously went the other way round; ω and $\Im(\omega)$ were determined from the data, and Eq. (35), hence, gives the alignment,

$$\langle J_1 \rangle \equiv i_{n+p}\,. \tag{38}$$

Once ω is obtained from Eq. (37), one can calculate the collective energy $1/2\ \Im\ \omega^2$ (in this context, $\Im$ is a parameter). It is easy to verify that the same conditions for double decoupling as in the TQPRM are obtained here.

From $a_n = -1$ onwards (i. e., increasing in absolute value), the odd-spin states (I = 1, 3, 5, . . .,) cross below the even-spin states (I = 0, 2, 4, . . .,). At $a_n = -1$, the pairs of states I = (0, 1); (2, 3); (4, 5); . . ., are degenerate, as can be seen from the following expression:

$$\Im\omega \approx I + \frac{1}{2} - \frac{1}{2}(\pm|a_n| + |a_p|) \tag{39}$$

(+ for odd spins, – for even spins.)

Using R+ 1/2 = $\Im\omega$, Eq. (9), and, for instance, $a_n = -1$ and $a_p = -5$, one gets $R = I-3$ for odd-spin states and x $R = I-2$ for even spins. States with I = 0 and 1 correspond to $R = -2$ (anti-aligned), I = 2, 3 to R = 0 (band heads), I = 4, 5 to R = 2, etc. (compare to Fig. 2). This schematic model would hold for very large deformations (provided the $\pi 1/2^-[541]$ remains pure $h_{9/2}$). For small deformations one would rather have $\langle j_{1p} \rangle = 9/2$, which gives $R = I-5$ (odd I) and $R = I-4$ (even I). Here I = 4, 5 would be band-head states, in accord with the situation in Ir nuclei.

26.4 FIRST BANDCROSSING (BACKBENDING)

Some of the DDB's have been measured to high enough angular velocities to reach states well beyond the first backbend[12] (FBB), allowing the extraction of crossing frequencies.[19] Figure 8 shows the relative routhians for ^{182}Ir, ^{172}Ta and ^{178}Re, along with the information for related bands. These relative routhians, e', have been constructed referring each nucleus to its own "local" reference extracted from the first few transitions along each band (see Table II). It is well known that the ground-to-S-band crossing can be obtained from the intersection of the two slopes in the e' vs $\hbar\omega$ plots before and after the FBB. These crossing frequencies, $\hbar\omega_c$, are also given in Table II.

The current view on the structure of the S band in this region of the periodic table is that of a pair of aligned $i_{13/2}$ neutrons.[19] Its crossing frequency would correspond to the value given in Table II for the *ee* nuclei. The rather low $\hbar\omega_c$ value for odd-neutron nuclei is interpreted as a particular blocking effect. The neutron pairing correlations are decreased

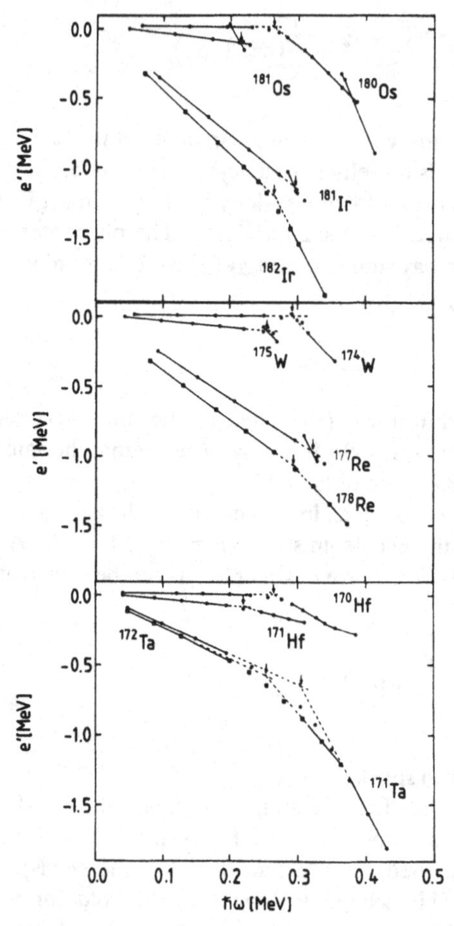

Figure 8. Routhians, e', as functions of angular velocity $\hbar\omega$ for respective GSB's in ^{170}Hf, ^{176}W, and ^{180}Os, the $1/2^-[521]$ band in ^{171}Hf, ^{175}W, and ^{181}Os, (Refs. 39, 40, and 41, respectively), the $\pi h_{9/2}$ band in ^{171}Ta, ^{177}Re, and ^{181}Ir, and the DDB's in ^{172}Ta, ^{178}Re, and ^{182}Ir. The reference parameters are indicated in Table II. For the odd nuclei, the quantities shown are actually $e' - \Delta_n$, $e' - \Delta_p$ and $e' - \Delta_n - \Delta_p + V_{np}$ [where $\Delta_{n(p)}$ is the neutron (proton) pairing gap]. References for odd-odd and odd-Z nculei, see Figs. 3 – 5.

in the odd-N system with respect to the even-even one, since the pair of time reversed orbits associated with $1/2^-[521]$ is blocked. This means that the neutron pairing gap, Δ_n, is smaller, and, hence, the energy to break a pair of $i_{13/2}$ neutrons is smaller, leading to a smaller $\hbar\omega_c$. The effect is thought to be particularly large because both the $1/2^-[521]$ and the highly-alignable $1/2^+[660]i_{13/2}$-parentage orbital responsible for the FBB are both prolate, this being a manifestation of quadrupole pairing.[19] On the other hand the rather high value of $\hbar\omega_c$ for the $h_{9/2}$ proton is interpreted as a deformation effect.[19] This highly prolate quasi-proton configuration is believed to drive the nucleus to larger deformation, increasing the spacing among the highly-alignable low-Ω quasi-neutron orbitals and also to the fermi surface, thus hindering the action of the coriolis force on the pair of $i_{13/2}$ neutrons and resulting in larger crossing frequencies. This conjecture is, however, not supported by lifetime measurements[22] in the decoupled $\pi h_{9/2}$ band in ^{181}Ir. In addition, cranked-shell-model calculations performed here, with a realistic position for the

$\pi 1/2^-[541]$ orbit (namely, right at the fermi surface as indicated by the odd-proton spectra), give equal crossing frequencies for both a pair of $h_{9/2}$ protons and a pair of $i_{13/2}$ neutrons, of about 0.3 MeV. (A standard Nilsson potential with $\beta = 0.25$, $\kappa_n = 0.063$, $\mu_n = 0.411$, $\kappa_p = 0.063$, $\mu_p = 0.605$, and $\Delta_n = \Delta_p = 0.8$ MeV has been used.) Another argument is that the quasiparticle character of the $1/2^-[541]$ excitation tends to quench its quadrupole moment (since the single-quasiparticle contribution is is $u^2 - v^2$ times the single-particle one, u and v being the usual BCS occupation amplitudes,[23] which are about equal in this case), thus reducing a possible polarization tendency.

An alternative interpretation[12,35] for the behavior of $h_{9/2}$ bands could be at least partial participation of a pair of $h_{9/2}$ protons (together with a pair of $i_{13/2}$ neutrons) in the structure of the S band. Here the FBB would be delayed, because the highly-alignable $1/2^-[541]$ orbital would be blocked in the odd-proton system. This interpretation would require some kind of coupling or linkage between the two S-band configurations (namely, $\pi h_{9/2}$ and $\nu i_{13/2}$); otherwise, if they are independent, one should observe two distinct backbendings.

The crossing frequency of the DDB is intermediate (actually, almost the average) between the FBB frequency of neighboring odd-neutron and odd-proton nuclei. In an approximately equivalent way, one may say that the shift in crossing frequency of the DDB (with respect to that of the associated even-even core) is almost the sum of the shifts in crossing frequencies for the neighboring odd-N and odd-N nuclei.

As an example, one has:

$$\hbar\omega_{cee}(^{174}\mathrm{W})+\delta\hbar\omega_{c,n}(^{175}\mathrm{W})+\delta\hbar\omega_{c,p}(^{177}\mathrm{Re}) = 285 \quad (40)$$

(vs $\hbar\omega_{c,np}(^{178}\mathrm{Re}) = 295$) and

$$\frac{1}{2}(\hbar\omega_{c,p} + \hbar\omega_{c,n}) = 288\,. \quad (41)$$

This would be consistent with the increased-deformation picture, since the presence of the $1/2^-[521]$ neutron would bring the crossing frequency down, while the increased deformation would bring it up.

However, the $\hbar\omega_c$ value for the DDB is also consistent with the alternative interpretation. The presence of the $1/2^-[541]$ proton would block the $\pi h_{9/2}$ component while the occupation of the $1/2^-[521]$ orbital would facilitate the decoupling of a pair of $i_{13/2}$ neutrons as indicated above.

26.5 SUMMARY AND CONCLUSIONS

This paper describes a special type of collective structure identified in doubly-odd deformed nuclei in the heavy rare-earth region.

This structure, called doubly decoupled, which is the equivalent to the decoupled band in odd nuclei, involves both proton and neutron in a decoupling situation.

In spite of the expected complexity of the excitation spectrum of heavy-mass doubly-odd nuclei, relatively simple and pure structures are shown to exist, and a transparent model like the particle-plus-rotor (or the cranking) is shown to be able to provide a qualitative understanding of all the special features revealed by these bands. Strong coriolis effects are found to dictate their main behaviors, while the p-n force does not seem to play an important role in determining their structure.

The DDB has been followed in several cases beyond the first backbend. Its crossing frequency is almost the average between the values found for the participating quasi-particle states in neighboring odd-N and -Z nuclei. This linear superposition of odd-N and odd-Z effects appears again in the form of a striking additivity of alignments and deviations of inertia parameters. These results may set a question mark on earlier assertions on the deformation driving tendency of the $\pi h_{9/2}$ orbital and on the exclusive $\nu i_{13/2}$ origin of the first backbending in this region of the chart of nuclides.

REFERENCES

1. A. J. Kreiner, J. Davidson, M. Davidson, D. Abriola, C. Pomar, and P. Thieberger, *Phys. Rev. C* **36**, 2309 (1987); **37**, 1338 (1988).
2. A. J. Kreiner, in *Contemporary Topics in Nuclear Structure Physics*, Ed. by R. Casten et al. (World Scientific, Singapore, 1985), p. 521, and references therein.
3. A. J. Kreiner, in *Proc. XII Workshop on Nuclear Physics*, Ed. by C. Cambiaggio et al. (World Scientific, Singapore, 1990).
4. A. J. Kreiner, D. E. DiGregorio, A. J. Fendrik, J. Davidson, and M. Davidson, *Phys. Rev. C* **29**, R1572 (1984).
5. A. J. Kreiner, P. Thieberger, and E. K. Warburton, *Phys. Rev. C* **34**, R1150 (1986).
6. A. J. Kreiner, *Proc. IX Workshop on Nuclear Physics. Buenos Aires* (World Scientific, Singapore, 1986), p. 337.
7. F. S. Stephens, *Rev. Mod. Phys.* **47**, 43 (1975).
8. B. Bengtsson, and S. Frauendorf, *Nucl. Phys.* **A314**, 27 (1979); **A327**, 139 (1979).
9. A. J. Kreiner, M. Fenzl, S. Lunardi, and M. A. J. Mariscotti, *Nucl. Phys.* **A282**, 243 (1977).
10. A. J. Kreiner, *Z. Phys.* **A288**, 373 (1978).
11. A. J. Kreiner and D. Hojman, *Phys. Rev. C* **36**, R2173 (1987), and references therein.
12. A. J. Kreiner, *Proc. Conf. on High-Spin Nuclear Structure and Novel Nuclear Shapes*, Argonne National Laboratory — PHY-88-2, 297 (1988).
13. A. J. Kreiner, D. E. DiGregorio, A. J. Fendrik, J. Davidson, and M. Davidson, *Nucl. Phys.* **A432**, 451 (1985).
14. A. J. Kreiner, J. Davidson, M. Davidson, P. Thieberger, E. K. Warburton, J. Genevey, and S. Andre. *Nucl. Phys.* **A489**, 525 (1988).
15. A. BenBraham, C. Bourgeois, P. Kilcher, R. Roussiere, J. Sauvage, M. H. Porquet, and A. J. Kreiner, *Nucl. Phys.* **A482**, 553 (1988).
16. C. M. Lederer, and V. S. Shirley, Eds., *Table of Isotopes* (New York, Wiley, 1978).
17. J. Davidson et al., *Z. Phys.* **A324**, 363 (1986).
18. N. D. Newby, *Phys. Rev.* **125**, 2063 (1962).
19. J. D. Garrett, in *Proc. Conference on High Angular Momentum Properties of Nuclei, Oak Ridge, Tennessee*, p. 17 and references therein.
20. P. Ring and H. J. Mang, *Phys. Rev. Lett.* **33**, 1174 (1974).
21. A. J. Kreiner, *Phys. Rev. Lett.* **42**, 829 (1979).
22. R. Kaczarowski et al., *Proc. Intern. Conf. on Nuclear Physics. Florence, Italy*, Vol. 1, (1983) p. 181.
23. A. Bohr, and B. R. Mottelson, *Nuclear Structure*, Vol. 2 (Benjamin, Reading, Mass., 1975).
24. U. Garg et al., *Phys. Lett.* **B151**, 335 (1985).
25. S. André, J. Genevey-Rivier, J. Treherne, J. Jastrzebski, R. Kaczarowski, and J. Lukasiak, *Phys. Rev. Lett.* **38**, 327 (1977).
26. V. P. Janzen et al., *Phys. Rev. Lett.* **61**, 2073 (1988).

27. S. André et al., *Nucl. Phys.* **A328**, 445 (1979).
28. D. L. Balabanski et al., *Z. Phys.* **A332**, 111 (1989).
29. S. André et al., *Nucl. Phys.* **A243**, 229 (1975).
30. R. A. Bark et al., *Nucl. Phys.* **A501**, 157 (1989).
31. Wm. C. McHarris, W.-T. Chou, and W. A. Olivier, in *Proc. XII Workshop on Nuclear Physics*, Ed. by C. Cambiaggio et al., (World Scientific, Singapore, 1990).
32. D. Santos et al., *Phys. Rev. C* **39**, 902 (1989).
33. J. R. Leigh, J. O. Newton, L. A. Ellis, M. C. Evans, and M. J. Emmott, *Nucl. Phys.* **A183**, 177 (1972).
34. C. X. Yang et al., *Phys. Lett.* **B133**, 39 (1985).
35. A. J. Kreiner et al., *Phys. Rev. C* **40**, R487 (1989).
36. J. C. Bacelar, et al., *Nucl. Phys.* **A442**, 547 (1985).
37. A. J. Kreiner, D. Hojman, J. Davidson, M. Davidson, M. Debray, G. Falcone, D. Santos, C. W. Beausang, D. B. Fossan, E. S. Paul, R. Ma, S. Shi, and N. Xu, *Phys. Lett.* **B215**, 629 (1988).
38. C. Foin, Th. Lindblad, B. Skanberg, and H. Ryde, *Nucl. Phys.* **A195**, 465 (1972).
39. G. D. Dracoulis and P. M. Walker, *Nucl. Phys.* **A330**, 186 (1979).
40. P. M. Walker, G. D. Dracoulis, A. Johnston, J. R. Leigh, M. G. Slocombe, and I. F. Wright, J. Phys. G: *Nucl. Phys.* **4**, 1655 (1978).
41. Neskakis, R. M. Lieder, M. Muller-Veggian, H. Beuscher, W. F. Davidson, and C. Mayer-Böricke, *Nucl. Phys.* **A261**, 189 (1976).

27. Deformed Odd-Odd Nuclei: Matrix Elements for the Residual *p-n* Interaction and Patterns of Alternating Perturbations in Level Spacings

R. W. Hoff
Lawrence Livermore National Laboratory
Livermore, CA 94550

A. K. Jain
Department of Physics
University of Roorkee
Roorkee 247667, India

J. Kvasil
Department of Nuclear Physics
Charles University
CS 180 00 Prague, Czechoslovakia

P. C. Sood and R. K. Sheline
Florida State University
Tallahassee, FL 32306

ABSTRACT

The application of a simple semi-empirical model is discussed in terms of interpreting experimental nuclear structure data for twelve of the best characterized odd-odd deformed nuclei. An essential part of this modeling is to calculate values of the Gallagher-Moszkowski splittings and Newby shifts, the observables that arise from the *n-p* residual interaction in odd-odd nuclei. Assumptions regarding the form for this *n-p* force are traced historically. The predictive power of a favored form of the *n-p* force, one that includes a central force with short- and long-range components, a tensor force, and some effects of core polarization, is examined in light of experimental data obtained since its formulation. A data set of forty-two experimentally-determined Newby shifts has been reviewed as to the reliability of each entry. Exceptions to a recently proposed rule for the a priori determination of the sign of a Newby shift are discussed. Evidence is presented for the existence of an odd-even staggering or signature effect in the rotational spacings of many K^- bands (with $K > 0$). By use of coriolis-coupling calculations, it has been possible to reproduce the staggering observed in some of the K^- rotational bands of ^{156}Tb, ^{168}Tm, ^{176}Lu, ^{182}Ta, and ^{182}Re.

The complexity found in the decay schemes of deformed odd-odd nuclei arises naturally from several modes of excitation, both single-particle and collective, that occur in these nuclei and from the many possible interactions between these modes. Even though the experimental spectroscopic probes of today are providing measurements of nuclear emissions with greater precision and sensitivity than ever before, it is still essential that the data collected for these nuclei be interpreted with the use of a theoretical model of nuclear

Exotic Nuclear Spectroscopy, Edited by W. C. McHarris
Plenum Press, New York, 1990

structure. So-called model-independent approaches to understanding these experimental data permit one to sketch only the broadest outline of decay schemes that have proven otherwise to be exceedingly rich in detail. Thus, we are fortunate that a very useful model exists for use in predicting the level structure and, in some cases, the important transitions of odd-odd deformed nuclei.

The model has already played a major role in our understanding of these nuclei, having been proposed almost twenty-five years ago in papers by Struble, Motz, et al.[1] The essential assumption they made was that if the *n-p* interaction energy in an odd-odd deformed nucleus was small compared with the energy with which the odd nucleons are bound to the core, the excitation can be calculated by a simple extension of the odd-A model, and the interaction energy can be treated separately as a perturbation. This assumption implies that the two unpaired nucleons move independently of one another to a first approximation, and thus their contributions to the total single-particle energy or to the total nuclear moment of inertia are additive.

From these assumptions, it follows that the excitation energies of levels in an odd-odd deformed nucleus can be calculated by use of the following expression,

$$\begin{aligned} E_I &= E^p_{qp} + E^n_{qp} + \hbar^2/2\theta_{o-o}[I(I+1) - K^2] \\ &\quad -(1/2 - \delta_{\Sigma,0}) + \delta_{K,0}(-1)^{I+1}[E_N + \hbar^2/2\theta_{o-o}a_p \times a_n], \end{aligned} \tag{1}$$

where the E_{qp} terms denote quasiparticle energies for the proton and neutron orbitals, θ represents the nuclear moment of inertia, E_{GM} and E_N denote matrix elements arising from the residual *n-p* interaction, arising from the Gallagher-Moszkowski (G-M) splitting and the Newby shift, respectively, and the a's represent decoupling parameters for $\Omega = 1/2$ orbitals.

It is clear from the above expression that we have included only single-particle excitations, energy from rotational motion, and the effects of a residual *n-p* interaction in describing the total excitation. Neglected in this treatment are several other forms of excitation, e.g., collective motion in the form of vibrations and quasiparticle excitations for states that are produced by breaking one or more pairs of nucleons. Also, we have neglected, so far, any interactions between the various forms of motion. We will treat in an explicit manner the band mixing due to the coriolis force, but this has been done in separate calculations that will be discussed later.

An important detail in the application of Eq. 1 is that we obtain the quasiparticle excitation energies from empirical data on these excitations in neighboring odd-mass nuclei, whenever possible. An appropriate correction for contributions from rotational motion is applied before inserting these energies into Eq. 1. In a similar way, contributions to the total nuclear moment of inertia from the presence of the two unpaired particles are deduced from empirical data on their contributions to the moments of inertia of odd-mass nuclei, relative to that of an even-even core.[2] Thus, empirical data play an important role in this modeling technique. The value of the approach has been demonstrated previously.[3,4]

An essential part of this modeling is to calculate values for the G-M splitting and Newby shifts. Although we will discuss the calculations that have been made for these matrix elements and, in the case of the Newby shifts, compare them with experimental measurements later, for now we need only to refer to comprehensive compilations of calculated matrix elements for this modeling application. Such compilations are available for odd-odd nuclei in both the rare-earth [5] and actinide [3] regions of deformation.

A comparison of experimental and calculated band-head energies, rotational parameters, and G-M splittings is presented in Table I for twelve well-characterized odd-odd nuclei. Recent experimental data for certain of these species can be found in Refs. 6-12. From the average deviations listed in columns 4 and 5 of Table I, it can be seen that our simple model can produce good correlations with experimental data, and, for nuclei such as ^{238}Np, ^{244}Am, and ^{250}Bk, the correlations are outstanding. At the other extreme, our model provides a considerably poorer correlation with experiment for ^{154}Eu. This is a case where the band density is very high, even at relatively modest excitation energies. As a consequence, mixing effects are quite important, especially for the neutron states, and the poor correlation can be understood because the model does not treat band mixing. Nevertheless, it still holds true that the results of the modeling listed in Table I, where empirical data for the quasiparticle energies and for the rotational parameters were employed, are far superior to those obtained from purely theoretical estimates for these quantities. For example, the mean deviation for the calculated band-head energies of ^{238}Np is 32 keV in Table I, whereas the mean deviation is 147 keV if the quasiparticle energies are calculated by assuming a modified harmonic-oscillator (Nilsson) potential. To the extent that the coriolis mixing found in odd-mass nuclei also occurs in related odd-odd nuclei, the use of empirical quasiparticle energies and rotational parameters as input to the model provides at least a partial accounting for band-mixing effects. As an example, the compression and

Table I. Odd-Odd Nuclei in Actinide and Rare-Earth Regions: Comparison of Experimental and Calculated Bandhead Energies, Rotational Parameters, and G-M Splittings

Nucleus	No. of Bands	Energy Range (keV)	$\langle E_{exp} - E_{calc} \rangle$ (keV)	$\langle A_{exp} - A_{calc} \rangle$ (keV)	E_{GM} exp/calc
^{154}Eu	32	0 - 550	91		43% mean dev.
^{160}Tb	8	0 - 380	41	0.61 (8.1%)	1.03,1.07,1.13
^{166}Ho	10	0 - 560	47	0.74 (8.7%)	0.80,1.08,1.31
^{170}Tm	5	0 - 450	63	0.46 (5.2%)	2.04,0.98
^{174}Lu	10	0 - 640	43	0.33 (3.1%)	0.89,0.94,1.14
^{176}Lu	12	0 - 840	58	1.0 (9.2%)	1.14,0.48,1.01, 0.91,0.39
^{182}Ta	17	0 - 600	47	0.47 (3.9%)	0.94,0.97,1.14
^{186}Re	11	0 - 610	55	2.77(16.7%)	0.84,1.14,1.17, 0.97
^{238}Np	13	0 - 570	32	0.14 (3.2%)	1.17,0.87,0.93, 1.08,0.04
^{242}Am	13	0 -1020	56	0.21 (4.4%)	0.08,0.99,0.46, 1.05
^{244}Am	16	0 - 680	19	0.28 (7.4%)	1.15,0.14,0.96
^{250}Bk	14	0 - 570	17	0.20 (4.7%)	1.11,0.96,2.87, 1.39,1.32

expansion of level spacings caused by the coriolis effect is sometimes well reproduced in the odd-odd nucleus; one often finds good agreement between predicted and experimental moments of inertia.

Regarding the limitations of this simple model, we note that Balodis et al.[13] have recently reported the identification of a vibrational band at 544 keV in ^{166}Ho with the configuration $K = [K_0 - 2] = 2^-$ that is based upon the ground state band $0^-\{7/2[523]p - 7/2[633]n\}$. An obvious extension of our present model would be to include a provision for predicting vibrational states in odd-odd nuclei, if such predictions can be made with acceptable accuracy. Afanasjev et al.[14] have described an appropriate theoretical treatment for these vibrational-rotational excitations.

In calculating values for the G-M splittings (and Newby shifts) in odd-odd nuclei, a crucial question is the assumption made regarding the form of the force used to describe the n-p interaction. These matrix elements can be expressed, as follows:

$$E_{GM} = \langle \chi_p^k \chi_n^{-k} | V_{np} | \chi_p^k \chi_n^{-k} \rangle - \langle \chi_p^k \chi_n^k | V_{np} | \chi_p^k \chi_n^k \rangle \tag{2}$$

and

$$E_N = \langle \chi_p^k \chi_n^{-k} | V_{np} | \chi_p^{-k} \chi_n^k \rangle \tag{3}$$

Here χ^k are the intrinsic single-particle wave functions, and V_{np} denotes the effective n-p interaction. In the earliest theoretical treatments of this residual interaction,[15,16] a central force of zero range (CFZR) or δ force was assumed. The calculated G-M matrix elements showed a reasonable fit to experimental data, particularly as the body of experimental data was enlarged. From a data set of twenty-seven G-M matrix elements in the rare-earth region,[5] a best fit (rms dev = 40 – 45 keV) was obtained with αW = 0.79 MeV; from a data set of nineteen G-M matrix elements in the actinide region,[17] a best fit (rms dev = 46 keV) was obtained with αW = 1.03 MeV.

Although a CFZR produces a reasonable fit to the body of G-M matrix-element data, its application to the data for Newby shifts does not produce an acceptable fit because many of the calculated values are too large. Jones et al.[18] provided a solution to this problem by introducing a central force of finite range (CFFR) and a tensor force and assuming a gaussian radial dependence. With this approach (four adjustable parameters), they produced a reasonable fit (rms dev = 57 keV) to a data set of twenty-seven G-M matrix elements and seven Newby shifts. The introduction of a tensor force was necessary because the CFFR calculations of the Newby shifts for triplet (spin) configurations were too small in magnitude and of the opposite sign to the experimental data. As Newby[19] has stated, in the asymptotic limit of large nuclear deformation, no central force, regardless of its exchange character, can produce the scattering required for a Newby shift in triplet configurations. Only a tensor force is capable of transferring angular momentum between spin and orbital spaces, as required for this shift.

In 1976, Elmore and Alford[20] chose to employ, in addition to the forces already described, a central force with a gaussian radial dependence that had a combination of short (r = 1.4 fm) and long (r = 4 fm) range parts. They obtained a best fit (rms dev = 41 keV) with this short-range/long-range force (four adjustable parameters) to a data set of thirty G-M matrix elements.

Bennour et al.[21] modeled the levels of odd-odd nuclei using the Hartree-Fock approximation. They determined the ingredients of a Bohr-Mottelson unified-model description from the Skyrme III interaction through the adiabatic limit of a time-dependent Hartree-Fock-Bogolyubov approximation. In their description of nuclear states in a deformed nucleus, the single-particle wave-functions and energies, the static-equilibrium core properties, and the residual *n-p* interaction are all determined from the same effective interaction. This effective nucleon-nucleon force has been derived from the saturation properties (energy, radius) of closed-shell nuclei. In a detailed comparison of the experimental level schemes of ^{160}Tb, ^{174}Lu, and ^{238}Np, Bennour et al. found good correspondence with their calculations. Their calculated G-M matrix elements are compared with a data set of nineteen experimental values for actinide nuclei in Ref. 17 (rms dev = 71 keV). The agreement with experiment is really quite remarkable, considering the absence of any ad hoc renormalization.

The most comprehensive treatment of the experimental and theoretical aspects of the residual *n-p* interaction is that of Boisson et al.[5] In addition to testing several assumptions regarding the nature of the *n-p* force, these authors have also tested the use of both modified harmonic-oscillator wave-functions and those derived from a Woods-Saxon potential. Although use of the latter proved slightly superior, they found little difference in the results between these wave-functions. We have cited several papers where the authors have assumed a finite-range central force for the residual interaction. The usual result was only a modest improvement in the correspondence with experiment, as compared with the zero-range calculations (e.g., see Table II, where the rms deviations are all about 40 keV). Boisson et al. produced a very significant improvement in the rms deviation for their data "Set A" by introducing some effects of the polarization of the core by the odd particles (rms dev = 17 keV). These effects were included by polarizing the intrinsic-spin operators and by adding a long-range force and a tensor force (seven adjustable parameters). We have compared the predictions obtained from this force (designated CPTL) with experimental data for twenty new G-M matrix elements that have been measured in the rare-earth region. The results are given in Table II, along with predictions for a central force with zero range and for central forces with finite ranges. It can be seen in Table II that the predictive powers of the calculations assuming various forms of the *n-p* force for this newer set of data are all in the range of rms deviations of 50 to 100 keV. The magnitude of these deviations is somewhat surprising, in view of the low deviation obtained with the CPTL force by fitting the earlier data.

Since the publication of the last two comprehensive papers[5,20] on experimental and theoretical aspects of the *n-p* interaction, the body of E_{GM} data has grown appreciably. In Table II, there are indicated twenty experimental matrix elements that were not included in the Boisson et al.[5] data "Set A" (twenty-seven entries). In addition, there are nineteen matrix elements[17] for nuclei in the actinide region that have not been included in any attempt to define the parameters of a complex *n-p* force. A fit to all of the existing data with our knowledge of *n-p* force today yields rms deviations that are 35 to 60% of any grand mean value for G-M matrix elements. There appears to be no consensus among authors on the choice of an optimal parameter set to describe the best effective *n-p* force. Further research to define the nature of this force seems warranted.

Calculation of the Newby shift seen in $K = 0$ bands has yielded results that were often at odds with experiment. Calculated values from some of these efforts are listed in Columns 6-10 of Table III. Frisk[22] has recently examined this problem. He has fit the parameters of a phenomenological type of *n-p* force, one that includes a tensor component and is described by six parameters, to experimental Newby shifts for twenty "most reliable" bands in Table III (rms dev = 16 keV). A distinctive feature of this force is its strongly-at-

Table II. G-M Matrix Elements, Rare-Earth Region Calculated vs Experimental Values for Newly Measured Cases

Experimental Data before 1977

	Calculated Values[a] (keV)			
	CFZR	**CFFR**	**CPTL**	**EA**
Data Set A (BPO),[b] rms deviation:	43	40	17	
# Entries:	27	27	27	
Data Set, Table 1a (EA),[c] rms deviation:				41
# Entries:				30

Experimental Data after 1976[d]

Proton	Neutron	Nucleus	Expt.	**CFZR**	**CFFR**	**CPTL**	**EA**
				Calculated Values (keV)			
3/2[411]	7/2[633]	^{166}Ho	+191	+87	+94	+146	+132
1/2[411]	1/2[521]	^{168}Tm	+192	+130	+130	+94	+125
5/2[402]	5/2[512]	^{174}Lu	+129	+150	+145	+169	+206
1/2[411]	7/2[514]	^{176}Lu	+123	+180	+171	+321	
9/2[514]	7/2[514]	^{176}Lu	−68	−260	−239	−141	
Data listed above, rms deviation:				105	95	108	69
# Entries:				5	5	5	3

Proton	Neutron	Neutron	Nucleus	Calculated Values (keV)		
5/2[413]	1/2[400]	3/2[402]	^{154}Eu	15 E_{GM}'s reported		
5/2[532]	3/2[651]	5/2[642]				
	1/2[530]	3/2[521]			**CPTL**	**EA**
	3/2[532]	5/2[523]		rms deviation:	51	50
	11/2[505]			# entries:	11	15

[a] First three calculated values from Ref. 5; CFZR = central force, zero range, CFFR = central force, finite range (1.4 fm), CPTL = central and tensor forces with spin polarization and a long-range component. Fourth calculated value (EA) from Ref. 20, central force with both short (1.4 fm) and long (4 fm) range components.
[b] Data "Set A," as listed in Ref. 5.
[c] Data set from Table *1a* in Ref. 20.
[d] Data from Refs. 6-12.

tractive space-exchange term. As with previous studies, this force is not useful for calculating G-M splittings.

Frisk has proposed an empirical rule for the sign of Newby shifts when the angular momenta of the proton and neutron orbitals are good quantum numbers. His rule is the following: The favored spins in a $K = 0$ band, I_F, where the angular momentum of the proton and neutron orbitals (j_p, j_n) are good quantum numbers, are given by the expression $I_F = (j_p + j_n)_{mod2}$. Experimental determinations of E_N that follow this rule are indicated in Column 6 of Table III following a listing of Frisk's calculated E_N. The rule seems to work well, with fourteen of twenty-two configurations tested in Table III conforming to it. Some caveats apply to the use of Frisk's rule: If the configuration is of mixed singlet-triplet

Table III. Newby Shifts: Experimental and Calculated Values, Summary of Experimental Evidence for the $K = 0$ Bands

No.	Configuration	Nucleus	R[a]	Energy, Newby Shift[c] (keV)						Rot. Band[d]		Summary[e]	
				Exp	Fsk.[b]	BPO	EA	BQ	JOHS	$E(I)$	# Lvls.	Expt. Data	
	2	3	4	5	6	7	8	9	10	11	12	13	14
2)	5/2[413]p-5/2[642]n T	^{154}Eu	B	13	40 Y					287(0)	4	GM	ng dp
3)		^{156}Eu	C	3	40 Y	13				0(0)	2+	??	bd
4)		^{160}Tb	B	−3	40 N	13	24			222(0)	5	G?	ng
5)	5/2[413]p-5/2[523]n S	^{154}Eu	C	−8	13 N					415(0)	4	GM	ng arc dp
6)	3/2[411]p-3/2[651]n S	^{154}Eu	C	36	−14 N					342(0)	3	GM	ng dp
7)	3/2[411]p-3/2[521]n S	^{154}Eu	A	8	29 Y					279(0)	5	GM	ng arc dp
8)		^{156}Tb	C	−10	29 N		12			100(1)	4		hd at
9)		^{158}Tb	C	−8	29 N	50	11		74	110(0)	5		dt
10)		^{160}Tb	A	17	29 Y	50	10			79(0)	5	GM	ng dt
11)	3/2[411]p-3/2[402]n T	^{158}Tb	C	32	31 Y				34	420(0)	3		dt
12)	7/2[523]p-7/2[633]n S	^{166}Ho	A	32	31 Y	0	10		13	0(0)	8	GM	ng arc dp ta
13)	7/2[402]p-7/2[633]n T	^{166}Ho	C	75	36 Y					803(0)	7		ta
14)		^{170}Lu	A	42	36 Y	29	16			0(0)	3	GM	ec
15)		^{172}Lu	B	56	36 Y	29			24	65(0)	2	GM	ec at hd
16)		^{174}Lu	A	44	36 Y	29	16			281(0)	10	G	png lng dt at
17)		^{176}Ta	B	42	36 Y	29				100(0)	3	GM	ec
18)	1/2[411]p-1/2[521]n S	^{168}Tm	C	28	30 Y	57	18		75	167(0)	4		dt
19)		^{170}Tm	A	37	31 Y	57	18		76	150(0)	6	G?	ng dp ta
20)		^{172}Tm	B	+	31 Y					475(0)	2	GM	bd
21)		^{170}Lu	C	+	31 Y	57	19			407(0)	2	GM	ec
22)	1/2[411]p-1/2[400]n T	^{168}Tm	C	25	16 Y		47			1127(1)	2		??
23)	1/2[541]p-1/2[521]n S	^{172}Lu	C	−92	−34 Y					232(0)	4		ec at hd
24)	5/2[402]p-5/2[512]n S	^{174}Lu	C	−28	12 N					522(1)	2		??
25)	7/2[404]p-7/2[514]n S	^{176}Lu	A	−69	−3 N	−35	−70		−19	127(1)	8	GM	ng dp ta

Table III. (continued)

No.	Configuration	Nucleus	R[a]	Energy, Newby Shift[c] (keV)						Rot. Band[d]		Summary	
				Exp	Fsk.[b]	BPO	EA	BQ	JOHS	$E(I)$	# Lvls.	Expt. Data	
	2	3	4	5	6	7	8	9	10	11	12	13	14
26)	7/2[523]p-7/2[514]n T	^{176}Lu	C	154	68 Y					1057(0)	6		ta
27)	7/2[404]p-7/2[503]n T	^{182}Ta	A	−26	−36 Y	−25	−1			558(1)	6	G?	ng rc dp
28)		^{184}Ta	C	–	−35 Y		−1			228(1)	2	G	bd
29)	9/2[514]p-9/2[505]n T	^{188}Re	C	54	70 Y	65	47			500(3)	2	G	ng dp
30)		^{190}Re	C	+	70 Y					162(0)	2	G	bd it
31)	1/2[530]p-1/2[631]n T	^{234}Pa	A	40	−27 N			13		74+(0)	3	GM	bd
32)		^{236}Pa	C	47	−26 N			16		0(0)	3	G	bd
33)		^{238}Np	A	44	−25 N			17		218(0)	5	GM	ng arc
34)	5/2[523]p-5/2[622]n T	^{238}Np	B	−23	−27 Y			−13		300(1)	4	G	ng arc
35)		^{240}Am	B	−28	−27 Y			−13		346(1)	7		dt
36)		^{242}Am	A	−27	−27 Y			−13		0(1)	7	GM	ng dp dt
37)		^{244}Am	A	−26	−27 Y			−13		286(0)	5	GM	ng
38)	1/2[400]p-1/2[631]n T	^{238}Np	B	−3	−17 N					244(1)	4	GM	ng
39)	5/2[642]p-5/2[622]n S	^{238}Np	C	−49	−29 Y			−73		431(4)	3		dp
40)		^{242}Am	A	−59	−29 Y			−73		230(1)	4+	GM	ng
41)	7/2[633]p-7/2[613]n S	^{250}Bk	B	−25	−19 Y			−63		175(1)	2	GM	ng
42)	7/2[633]p-7/2[624]n T	^{244}Am	A	33	63 Y			15		374(0)	5	GM	ng
	RMS (All Entries):				36	26	20	25	42				
	RMS (Set A):				33	28	19	19	38				

Notes to Table III:

[a] Ranking as to reliability of the supporting evidence for the value of the Newby shift where A indicates most reliable.
[b] A Y (for "yes") in this column indicates that the sign of the experimental value agrees with that predicted by Frisk's rule (Ref. 22).
[c] Experimental data are listed in Col. 5; calculated data listed in Cols. 6-10 were taken from Refs. 22, 5, 20, 21, and 18, respectively. For the signs of the Newby terms in this paper, we have adopted the convention of Frisk[21]; namely, a positive value of E_N implies that even spins are favored. Note that this convention sometimes differs from that of other authors; e.g., Boisson et al.[5] reverse the sign of E_N, but only for configurations with negative parity. Other authors (Refs. 20, 18) completely reverse the sign of E_N relative to the convention adopted here.
[d] Listed for each rotational band in Cols. 11 and 12 are band-head energy in keV, angular momentum of the lowest energy level (in parenthesis), and number of levels identified.
[e] G denotes observation of depopulating γ rays; M denotes multipolarities from conversion coefficient measurements. Mode of populating band levels: *ng* = thermal neutron capture; *dp, hd, at, dt, ta* = single nucleon transfer reactions where p = proton, d = deuteron, t = triton, h = ^{3}He, and a = ^{4}He; *bd* = β decay; *arc* = average resonance neutron capture; *ec* = electron-capture decay, *png* = primary neutron-capture γ ray; *lng* = (^{7}Li, $n\,\gamma$) reaction; *rc* = resonance neutron capture; *it* = isomeric transition.

character, or if the dominating part of the configuration involves an s-shell, the rule is not generally useful.

We have examined the experimental evidence supporting the configuration assignments in Table III. The goal was to identify those cases where the existence of the band and the spins and parities of its levels could be determined without influence from systematic behavior in other nuclides or from theoretical calculations. In other words, we wished to find band assignments that were completely independent of any but the simplest nuclear models. The experimental data were examined for 1) the number and type of reactions used to populate band levels, 2) the presence of data on γ-ray transitions that populate or depopulate band levels, including experimentally determined multipolarities, 3) γ-γ coincidence data, 4) the total number of levels identified as belonging to the band, and 5) the band-head energy, because rotational bands are more easily characterized at low excitation energies. The entries in column 4 of Table III are ranked A, B, or C, according to the reliability of their supporting experimental data, which is summarized in Columns 11-14.

Fourteen of the entries in Table III were given the most reliable (A) ranking; ten different configuration are represented. The identification of this subset produces some possible clarifications where multiple determinations of a matrix element exist. For the data sets of the {3/2[411]p-3/2[521]n} and {7/2[404]p-7/2[633]n} bands, the A data show less spread than the others; in the former case, it only the A values that exhibit the same sign as the calculated values. On the other hand, restricting the data set to A-ranked values does not produce any dramatic improvements in the data fit, as indicated by the rms deviations listed at the bottom of Table III.

Frisk[22] noted that certain experimental E_N values violate this rule and also have a sign opposite to that of his calculated value. Of these few, he had the most difficulty explaining the data for the 0^-{1/2[530]p-1/2[631]n} band in ^{234}Pa, ^{236}Pa, and ^{238}Np. He suggested a re-examination of the data. We find that this band has been best characterized in ^{238}Np where five levels have been identified.[23] Interpretation of the data for ^{234}Pa and for ^{236}Pa, especially, has been unclear regarding the levels in this band. In our modeling we find the $0 \rightarrow 2$ spacing predicted to be 32–33 keV. Our spin assignments (Fig. 1) conform to this prediction, although the ground state of ^{236}Pa has previously been considered an I = 1 level,[24] thereby suggesting a negative value for E_N if the ground state

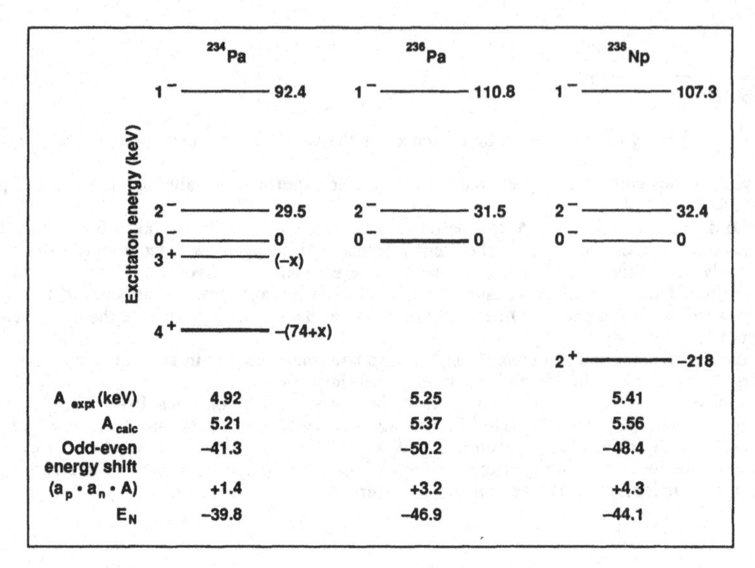

Figure 1. Experimental levels of the 0^-{1/2[530]-1/2[631]} band in ^{234}Pa, ^{236}Pa, and ^{238}Np. The spin assignments for the negative-parity levels are those of the authors. Please note the use of the opposite sign convention on E_N in this figure.

is part of a $K = 0$ band. However, because we believe it is very likely that the γ transitions associated with the β decay of ^{236}Th will populate each of the $I = 0$, 1, and 2 levels in a low-lying band, the most logical placement of the $I = 0$ and 2 levels of the $0^-\{1/2[530]p\text{-}1/2[631]n\}$ configuration is as the lowest two levels in ^{236}Pa. A similar band exists in ^{234}Pa where the evidence[25] is judged to be more conclusive that for ^{236}Pa. Thus, we find logical reasons for considering the E_N values for this configuration to be positive.

It is now well recognized that the coriolis force plays an important role in influencing the structure of deformed nuclei at both low and high spin. We have examined the question of the degree and the extent to which coriolis distortion is present in rotational bands of odd-odd nuclei.[26] As a criterion, we have chosen the presence of substantial deviations from regular behavior of the level spacings in a rotational band. For this purpose, we defined regular behavior as level energies that follow the expression,

$$E(I) = AI(I+1) + BI^2(I+1)^2. \tag{4}$$

An examination of the empirical data for odd-odd nuclei revealed that practically all $K^- = (k_p - k_n)$ bands, where $k = |\Omega|$, exhibit some odd-even staggering, i.e., forty-three out of forty-five K^- bands in our sample. Of these, twenty-five were $K = 1$ bands and the remainder were $K = 2$, 3, or 4 bands.

In our calculations we used a two-quasiparticle-plus-rotor model[5] for the description of the low-lying energy spectra of deformed odd-odd nuclei. Since the model is well known, we will present only our results here. We have performed calculations that fit known level energies for rotational bands in ^{168}Tm, ^{176}Lu, ^{182}Ta, and ^{182}Re. The hamiltonian used in this model included the elements discussed at the beginning of this paper plus terms for rotation-particle coupling and particle-particle coupling. The total hamiltonian matrix was diagonalized for each value of angular momentum. Generally, we have succeeded in making precise fits to the levels of rotational bands in the nuclei listed previously, even when the bands exhibit appreciable staggering.

Among the bands shown in Fig. 2, the $4^-\{1/2[541]p\text{-}9/2[624]n\}$ band in ^{182}Re is of particular interest. The nine experimentally determined levels in this bands were fit in a calculation that included twelve rotational bands, six each in the $K^+\{1/2[541]p + [i_{13/2}]n\}$ and the $K^-\{1/2[541]p - [i_{13/2}]n\}$ series. The odd-even staggering observed in the 4^- band is the product of the Newby shift in a $0^-\{1/2[541]p\text{-}1/2[660]n\}$ band that we estimate to exist at 2000 keV. Since this band has $k_p = k_n = 1/2$, its Newby shift will include contributions both from the usual matrix element and from a term of the form a_p times a_n times $\hbar^2/2\mathfrak{J}$. The latter quantity is estimated at 200–250 keV, much larger than the usual E_N matrix element. Although we cannot provide a full discussion of these results here, it should be noted that the odd-even shift of this $K = 0$ band is transmitted to the levels of the 4^- band via coriolis mixing in both the K^- *and* K^+ series of bands mentioned above. Surprisingly, the more significant source of the odd-even staggering in the 4^- band appears for arise from its interaction with bands in the K^+ series. These bands display an interesting property not found in the K^- series, namely that the lower-lying bands in the series show odd-even staggering of one phase and the higher lying members of the series show odd-even staggering of the *opposite phase*. For ^{182}Re, the interaction between the experimentally known $K = 4$ band of the K^- series with that of a low-lying $K = 5$ band in the K^+ series is dominant and is responsible for the observed odd-even staggering.

Another $K = 4$ band that exhibits staggering has been determined experimentally in ^{156}Tb by Bengtsson et al.[27] This rotational band, whose $I = 4$ occurs at 49.7 keV, has been assigned the configuration $4^+(3/2[411]p + 5/2[642]n)$. The energies of levels in this band were determined up to $I = 23$ by use of in-beam spectroscopy with the ^{150}Nd(^{11}B, 5n) reaction. An examination of the level spacings produces the staggering plot shown in Fig. 3. It can be seen that there is some mild odd-even variation at low angular momentum, that the

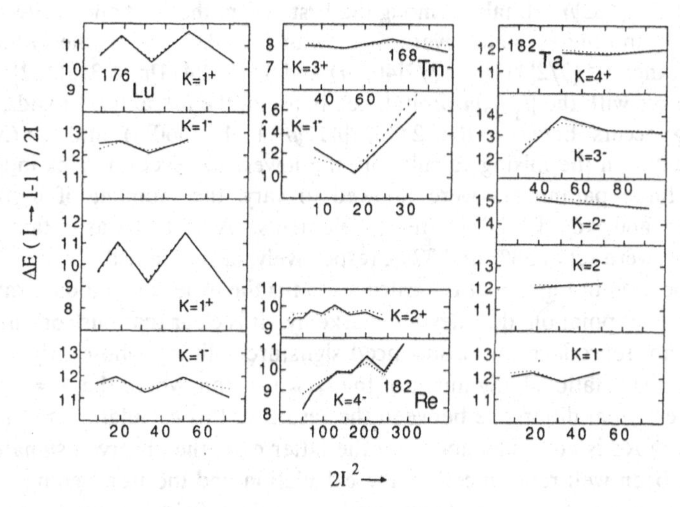

Figure 2. Experimental data (solid lines) that show odd-even staggering for level spacings of rotational bands in four odd-odd deformed nuclei. The calculated level spacings (dotted lines) were obtained by diagonalizing a hamiltonian that includes coriolis mixing terms.

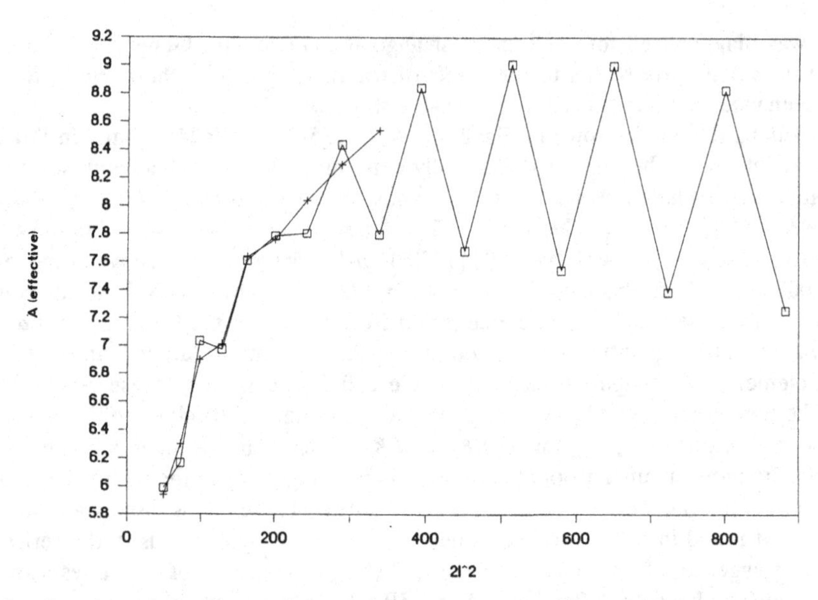

Figure 3. Experimental data (boxes) from the level spacings of the 4^+(3/2[411]*p* + 5/2[642]*n*) rotational band in ^{156}Tb. The calculated level spacings (crosses) were obtained by diagonalizing a hamiltonian that includes coriolis mixing terms.

staggering changes phase around $I = 9$, and that a very pronounced signature splitting is seen at higher angular momentum.

In order to fit the experimental data for this band, a coriolis mixing calculation was made that included a sequence of bands with the configurations (3/2[411]*p* ± $[i_{13/2}]$*n*) plus two bands with 3/2[402]*n* orbitals. Among the first group, the K^+ bands were those with $K = 2$, 3, 4, and 5 and the K^- bands were those with $K = 0$, 1, and 1; the latter two bands were the following: 0^+(3/2[411]*p* – 3/2[402]*n*) and 3^+(3/2[411]*p* + 3/2[402]*n*). Between the two sequences with the $[i_{13/2}]$*n* orbital, i.e., between the K^+ and K^- bands, the only $K = 1$ coupling occurs between the 2^+(3/2[411]*p* + 1/2[660]*n*) and 1+(3/2[411]*p* – 1/2[660]*n*) bands. In the mixing calculation, the lowest six level energies in the 4^+ band were fit and three parameters were allowed to vary, the moment of inertia and the $\langle 651|j^+|660\rangle$ and $\langle 642|j^+|651\rangle$ matrix elements. A best fit was obtained when the matrix elements were set at 58% and 37%, respectively, of theoretical.

It can be seen in Fig. 1 that the calculation fits the experimental data rather well up to $I = 10$; from that point on, they deviate markedly. It is clear that our coriolis mixing calculation does not reproduce the pronounced signature splitting whose origin is generally accepted to be the rotational alignment of the $i_{13/2}$ neutron, where the $\alpha = 1/2$ levels are favored. The essential difference between the results of this calculation and those for the $K = 4$ band in ^{182}Re is not understood. In the latter case, the observed signature splitting seems to have been well reproduced by the calculation and the trend going to higher angular momentum seems to be one of increased staggering. This is certainly not the case for this example in ^{156}Tb. Additional cases will be studied.

ACKNOWLEDGMENTS

Work performed under the auspices of the U. S. Department of Energy by the Lawrence Livermore National Laboratory under contract W-7405-ENG-48. It is a pleasure to thank J. Kern and G. L. Struble for helpful discussions regarding the research described in this paper.

REFERENCES

1. G. L. Struble, J. Kern, and R. K. Sheline, *Phys. Rev.* **137B**, 772 (1965); H. T. Motz et al., *Phys. Rev.* **155**, 1265 (1967).
2. G. Scharff-Goldhaber and K. Takahashi, *Bull. Acad. Sci. USSR, Phys. Ser.* **31**, 42 (1967).
3. R. W. Hoff et al., *Proceedings of International Symposium, Knoxville, TN on Capture Gamma-Ray Spectroscopy and Related Topics - 1984*, Ed., S. Raman, (AIP, New York, 1985, Conf. Proc. No. 125), p. 274.
4. R. W. Hoff, *Bull. Acad. Sci. USSR, Phys. Ser.* **50**, 41 (1986).
5. J. P. Boisson, R. Piepenbring, and W. Ogle, *Phys. Rpts.* **26**, 99 (1976).
6. M. K. Balodis et al., *Nucl. Phys.* **A472**, 445 (1987).
7. R. A. Dewberry, R. K. Sheline, R. G. Lanier, and R. Lasijo, *Z. Phys.* **A307**, 351 (1982); P. C. Sood, R. K. Sheline, and R. S. Ray, *Phys. Rev. C* **35**, 1922 (1987).
8. A. Bruder et al., *Nucl. Phys.* **A474**, 518 (1987); *ibid.*, **A467**, 1 (1987).
9. R. A. Dewberry et al., *Phys. Rev. C* **24**, 1628 (1981); M. K. Balodis et al., *Nucl. Phys.* **A194**, 305 (1972).
10. R. G. Lanier et al., *Phys. Rev.* **178**, 1919 (1969); D. Glas, *Z. Phys.* **255**, 175 (1972).
11. J. L. Salicio et al., *Phys. Rev. C* **37**, 2371 (1988).
12. T. von Egidy et. al., *Phys. Rev. C* **29**, 1243 (1984).
13. M. K. Balodis, A. V. Afanasjev, P. T. Prokofjev, and Yu. Ya. Tamberg, *Izv. AN SSSR, Ser. Fiz.* **52**, 2117 (1988).
14. A. V. Afanasjev, T. V. Guseva, and J. J. Tambergs, *Bull. Acad. Sci. USSR, Phys. Ser.* **52**, 121 (1988); *Izv. AN SSSR, Ser. Fiz.* **53**, 54 (1989).
15. N. I. Pyatov, *Bull. Acad. Sci. USSR, Phys. Ser.* **27**, 1409 (1963); N. I. Pyatov, and A. S. Chernyshev, *Izv. AN SSSR, Ser. Fiz.* **28**, 1173 (1964).
16. L. A. Neiburg, P. T. Prokofjev, and J. J. Tambergs, *Bull. Acad. Sci. USSR, Phys. Ser.* **36**, 2220 (1972).
17. R. W. Hoff, A. K. Jain, P. C. Sood, and R. K. Sheline, Lawrence Livermore National Laboratory Report UCAR 10062-88, (1988) p 151; R. W. Hoff, A. K. Jain, P. C. Sood, and R. K. Sheline, Lawrence Livermore National Laboratory Report UCRL-97683, June 1988.
18. H. D. Jones, N. Onishi, T. Hess, and R. K. Sheline, *Phys. Rev.* **C3**, 529 (1971).
19. N. D. Newby, *Phys. Rev.* **125**, 2063 (1962).
20. D. Elmore and W. P. Alford, *Nucl. Phys.* **A273**, 1 (1976).
21. L. Bennour, J. Libert, M. Meyer, and P. Quentin, *Nucl. Phys.* **A465**, 35 (1987); L. Bennour, Ph.D. thesis, Univ. de Bordeaux, Gradignan, France (1987).
22. H. Frisk, *Z. Phys.* **A330**, 241 (1988).
23. R. W. Hoff, *J. Phys. G: Nucl. Phys.* **14 Suppl.**, 343 (1988).
24. S. Mirzadeh et al., *Phys. Rev. C* **29**, 985 (1984).
25. Y. A. Ellis-Akovali, *Nucl. Data Sheets* **40**, 523 (1983).
26. A. K. Jain, J. Kvasil, R. K. Sheline, and R. W. Hoff, *Phys. Lett.* **B209**, 19 (1988); A. K. Jain, J. Kvasil, R. K. Sheline, and R. W. Hoff, *Phys. Rev. C* **40**, 432 (1989).
27. R. Bengtsson et al., *Nucl. Phys.* **A389**, 158 (1982).

28. Proton-Neutron Interactions in the $A = 100$ Nuclides

Richard A. Meyer*
Montgomery College
Takoma Park, MD 20012

Iain Morrison
University of Melbourne
Parkville, Victoria 3052, Australia

William B. Walters
University of Maryland
College Park, MD 20742

ABSTRACT

Evidence is presented and reviewed for the role of the proton-neutron interaction in dominating the low-energy nuclear structure of the medium-mass nuclei near $A=100$. We show that the interaction of the $1g_{9/2}$ protons and $1g_{7/2}$ neutrons provide the major driving force for the onset of deformation, while the $1h_{11/2}$ orbital only serves a secondary role to stabilize the resulting deformation.

One of the most important topics of discussion in the area of nuclear structure concerns the source of deformation in nuclides. The development of the Interacting-Boson Model (IBM) in which proton and neutron pairs are treated as bosons has highlighted the fact that a number of nuclear properties and phenomena change as a simple function of neutron or proton number. The role of proton-neutron interactions has been quantitatively explored in the work of Casten[1] and others,[2,3] who have identified numerous properties of nuclides that scale with the product of the number of valence protons and valence neutrons. Among the important features of these revelations has been the ability to describe the effects of the $Z = 64$ subshell closure in the neutron-rich nuclides with $50 \leqslant Z \leqslant 64$ and $N > 82$ and to be able to account for the effects of the changing number of valence protons as the subshell closure breaks down.[4,5]

The question arises as to whether there is sensitivity to the orbitals that the neutrons and protons are occupying. That is, is the *p-n* interaction spatially sensitive to the direct overlap between the radial wave-functions of the interacting protons and neutrons, or is the interaction of such long range that it simply reflects the total number of interacting protons and neutrons? That there is a sensitivity has been demonstrated simply by noting that the energy of the first 2^+ level in $^{132}Te_{80}$ lies at 973 keV, while the energy of the first 2^+ level of $^{136}Te_{84}$ possibly lies at 608 keV.[6] Both of these nuclides have $2g_{7/2}$ protons beyond the

*Permanent address: ER-23/GTN Division of Nuclear Physics, U. S. Department of Energy, Washington, DC 20545

Exotic Nuclear Spectroscopy, Edited by W. C. McHarris
Plenum Press, New York, 1990

closed Z = 50 proton shell. ^{136}Te has two neutrons in the $f_{7/2}$ orbital beyond the N = 82 closed neutron subshell, while ^{132}Te has two neutron holes distributed between the $d_{3/2}$ and $h_{11/2}$ orbitals. The drastic change in energies suggests that there is a strong sensitivity of the *p-n* interaction to the occupancy of the underlying orbitals. The systematics of the N = 84 and N = 86 2^+ levels shown in Fig. 1 may also be contrasted, as the energies decline with increasing Z for N = 80 shown in Fig. 2, while increasing with increasing Z for N = 84.

Given that there is a sensitivity, two viewpoints on the notion that the orbitals that are occupied do have a difference have emerged, one in support of the idea that it is the interaction between the highest-possible-j orbitals that is driving deformation,[7] and the other that it is the interaction between spin-orbit-partner orbitals that is driving deformation.

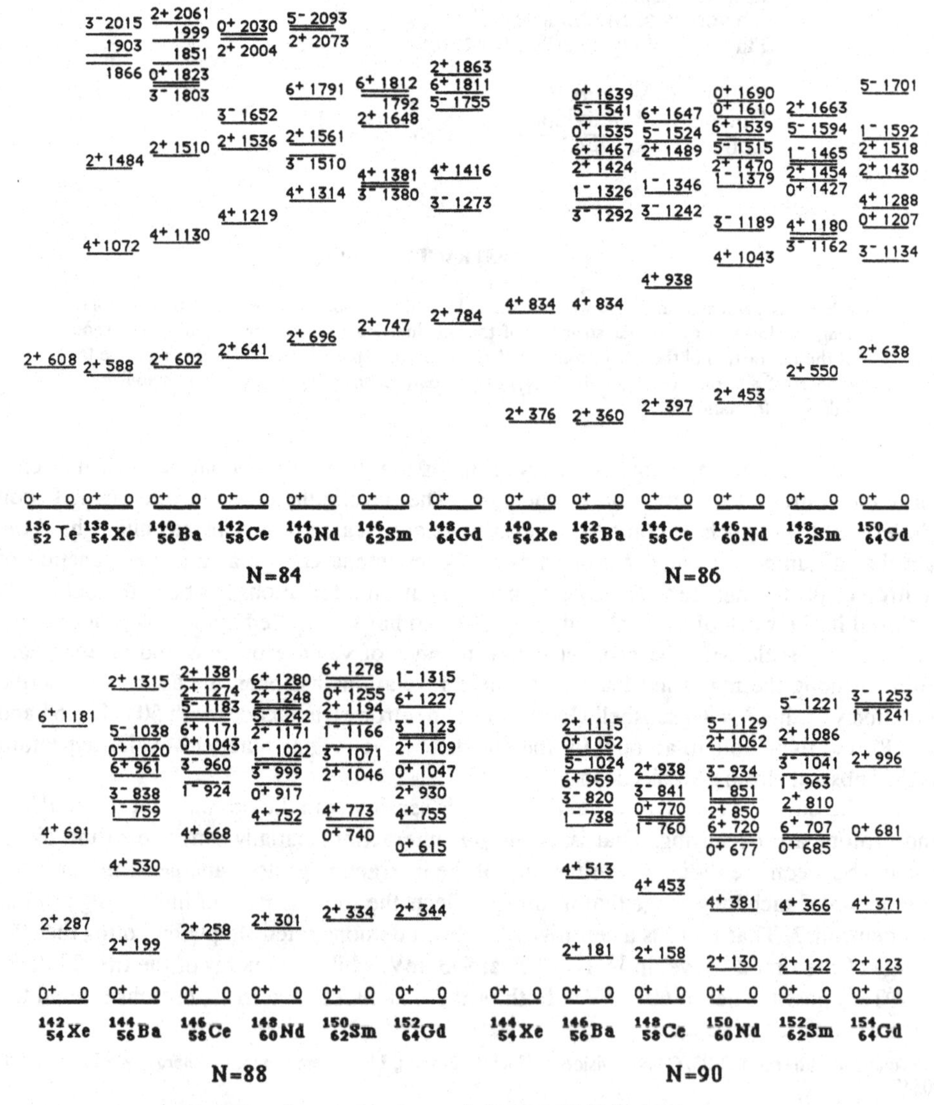

Figure 1. Systematics of states in N = 84, 86, 88, and 90 isotones.

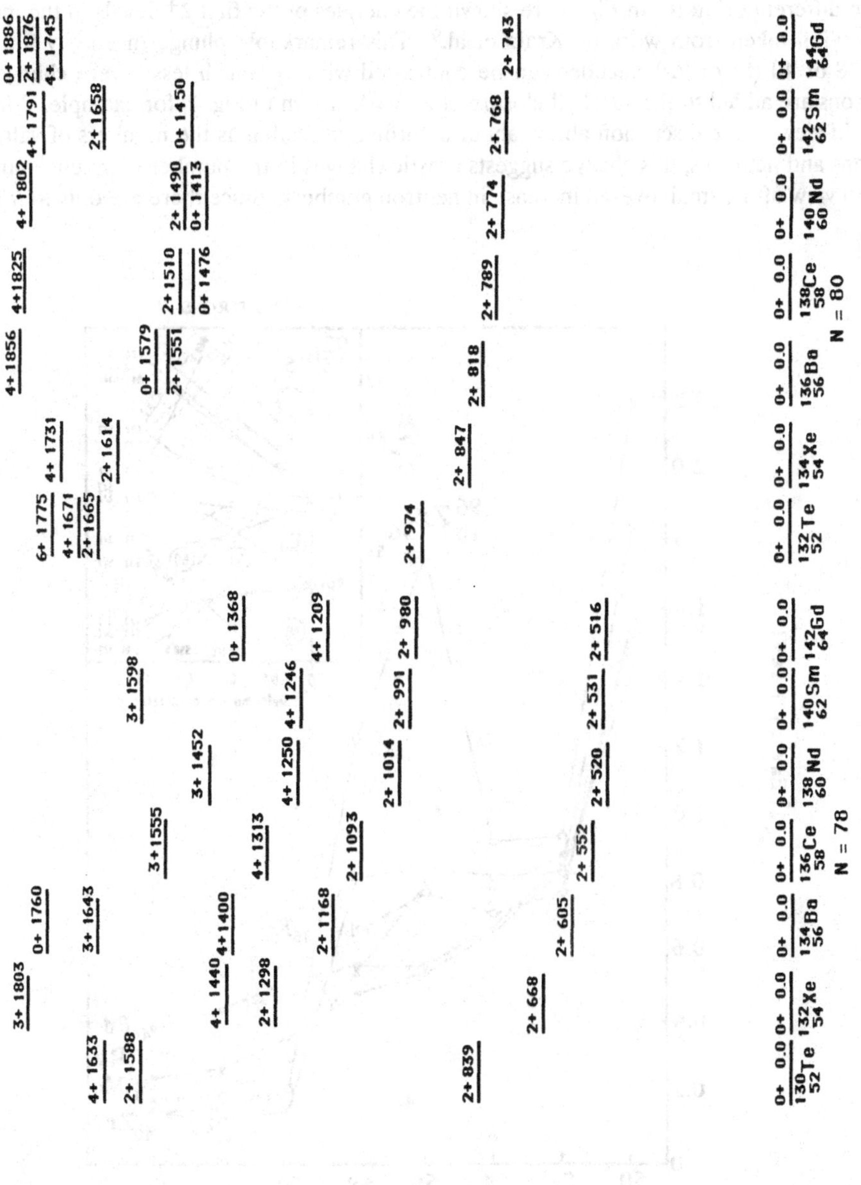

N = 78

$^{130}_{52}$Te: 0+ 0.0; 2+ 839; 2+ 1588; 4+ 1633

$^{132}_{54}$Xe: 0+ 0.0; 2+ 668; 2+ 1298; 4+ 1440; 3+ 1803

$^{134}_{56}$Ba: 0+ 0.0; 2+ 605; 2+ 1168; 4+1400; 3+ 1643; 0+ 1760

$^{136}_{58}$Ce: 0+ 0.0; 2+ 552; 2+ 1093; 4+ 1313; 3+1555

$^{138}_{60}$Nd: 0+ 0.0; 2+ 520; 2+ 1014; 4+ 1250; 3+ 1452

$^{140}_{62}$Sm: 0+ 0.0; 2+ 531; 2+ 991; 4+ 1246; 3+ 1598

$^{142}_{64}$Gd: 0+ 0.0; 2+ 516; 2+ 980; 4+ 1209; 0+ 1368

N = 80

$^{132}_{52}$Te: 0+ 0.0; 2+ 974; 2+1665; 4+ 1671; 6+ 1775

$^{134}_{54}$Xe: 0+ 0.0; 2+ 847; 2+1614; 4+ 1731

$^{136}_{56}$Ba: 0+ 0.0; 2+ 818; 2+ 1551; 0+ 1579; 4+1856

$^{138}_{58}$Ce: 0+ 0.0; 2+ 789; 0+ 1476; 2+1510; 4+1825

$^{140}_{60}$Nd: 0+ 0.0; 2+ 774; 0+ 1413; 2+ 1490; 4+ 1802

$^{142}_{62}$Sm: 0+ 0.0; 2+ 768; 0+ 1450; 2+1658; 4+ 1791

$^{144}_{64}$Gd: 0+ 0.0; 2+ 743; 4+ 1745; 2+ 1876; 0+ 1886

Figure 2. Systematics of states in $N = 78$ and 80 isotones.

The latter idea was described quantitatively by Federman and Pittel, who presented calculations showing that, in the mass-100 region, it is the interaction between the neutrons in the $g_{7/2}$ orbitals and the protons in the $g_{9/2}$ orbitals that are responsible for the rapid change in nuclear structure in the Zr nuclides between the double subshell closure at ^{96}Zr and the highly-deformed nuclide ^{100}Zr.[8] One approach to this question is to examine structures where either the protons or neutrons are in the same orbitals and the other nucleons are in different orbitals. In Fig. 3 are shown the energies of the first 2^+ levels in the mass-100 region taken from work by Kratz et al.[9] This remarkable plunge in energy for the $Z = 38$ or 40 (Sr or Zr) nuclides can be contrasted with the much less severe change as neutrons are added to the subshell closure at $Z = 64$, shown in Fig. 1, for example. Moreover, in view of the discussion above about deformation scaling as the numbers of valence protons and neutrons, this change suggests drastic changes in the number of valence nucleons, in view of the small overall increase in neutron numbers. Since there are only four ad-

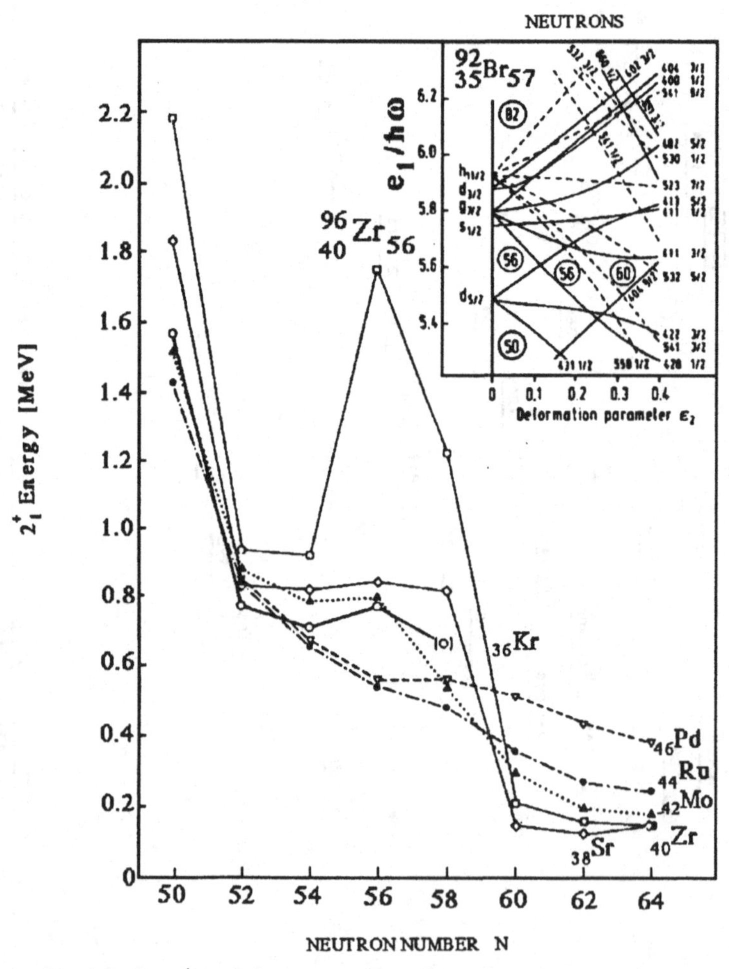

Figure 3. Energies of the first 2^+ levels in the mass-100 region. (Taken from Ref. 9.) The inset shows the Nilsson proton orbitals for this region.

ditional neutrons in $^{100}Zr_{60}$, as contrasted with $^{96}Zr_{56}$, all of which likely occupy the $g_{7/2}$ orbital, the additional valence neutrons must have come from the shifting of existing nucleons from full orbitals to empty orbitals; namely, for the neutrons from the $d_{5/2}$ orbital, which is filled at N = 56, shifted into the $h_{11/2}$ orbital. For the protons whose Nilsson orbitals are shown in Fig. 3, this would mean shifting from the negative parity $f_{5/2}$, $p_{3/2}$, and $p_{1/2}$ orbitals into the $g_{9/2}$ orbitals. Thus, the importance of the $h_{11/2}$ neutron orbital lies not in its high j value, but in its large occupancy possibility.

It must be noted that, once the $g_{7/2}$ neutron orbital has four neutrons in it and the $g_{9/2}$ proton orbital has five protons in it, the interaction will be at a maximum. Additional neutrons or protons that occupy these orbitals will curtail the space into which proton and neutron pairs can scatter. So it is that ^{98}Sr and ^{100}Zr are nearly at maximum deformation. Additional neutrons would appear to begin to fill the $g_{7/2}$ neutron orbitals beyond half full and reduce the interaction. But, with the large space available for additional neutrons in the $h_{11/2}$ orbital, it is possible for five or six additional neutrons to occupy the downsloping Nilsson orbitals from the $h_{11/2}$ single-particle orbital without having much effect on the occupancy of the important deformation-driving positive-parity orbitals from the $g_{7/2}$ single-particle neutron orbital. In like manner, as additional protons are added, they first refill the vacant positions in the $f_{5/2}$, $p_{3/2}$, and $p_{1/2}$ orbitals before beginning to reduce the vacancies in the $g_{9/2}$ orbitals and hence reduce deformation. In Fig. 4 are shown the yrast

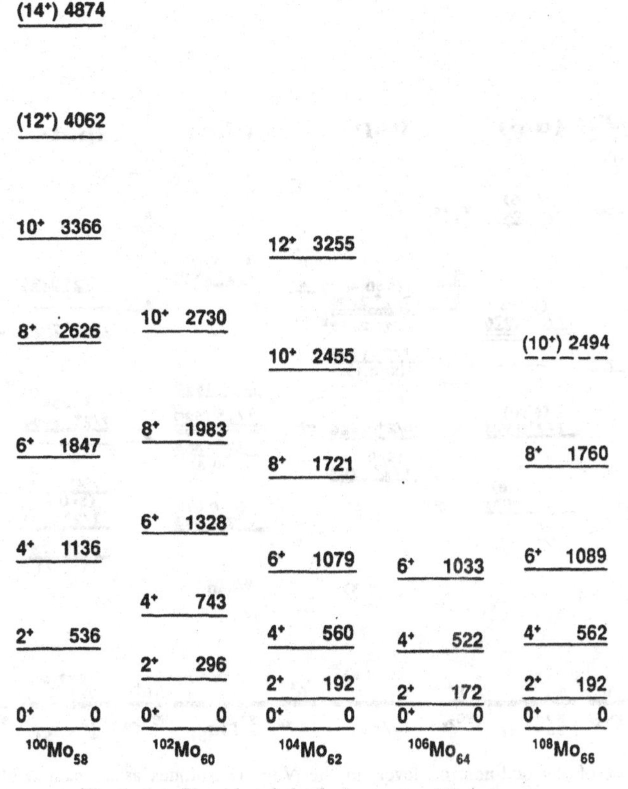

Figure 4. Yrast bands in the even-even Mo isotopes.

bands for the even-even Mo isotopes. The minimum energy occurs for $N = 64$ rather than at higher N, where the $h_{11/2}$ orbital would be half full.

Now we would like to show the scope of the changes in the single-particle energies that occur with changing orbital occupancies, which also bring about changes in the size of the nucleus. Consequently, the changes shown are a result of both the filling of orbitals and the size changes that the filling brings about. First, in Fig. 5 we show the movement of the odd-neutron single particle levels in the $N = 51$ isotones as proton numbers change and, in particular, as $g_{9/2}$ protons are added to the nucleus. The largest shift is that for the $g_{7/2}$ neutron level, which drops from the 3-MeV region toward near equality with the $d_{5/2}$ orbital by the time ^{101}Sn is reached. In Fig. 6 we show the changes in single-proton energies in the Sb nuclides that accompany the filling of the neutron shell up to $N = 82$. Since there is no particular effect from the p-n interaction, these changes must arise largely from monopole interactions and changes in nuclear size. In Fig. 7 the changes are summarized by showing the single-neutron orbitals in ^{91}Zr, as contrasted with those in ^{131}Sn.

The mechanism by which the spin-orbit partner interaction acts is the change of position of the affected orbital. That is, the way that occupancy changes are accomplished is by moving one partner orbital down in energy in the presence of the occupied other partner orbital. This effect is shown in Fig. 8. The top part of Section A shows the situation that would occur if there were no strong spin-orbit-partner orbital interaction, while the bottom half shows a smeared-out occupancy pattern, and Section D shows the spin-orbit partner orbitals actually moved down in energy. The drop in energy places these single-particle

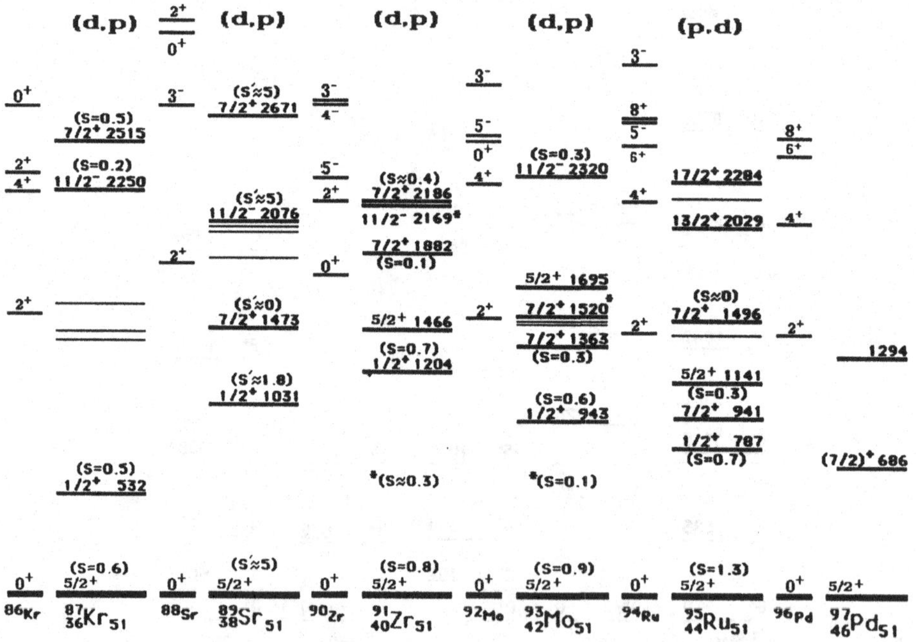

Figure 5. Movement of the odd-neutron levels in the $N = 51$ isotones as a function of changing proton number. The even-even cores are also shown for comparison.

levels much closer to each other, the usual condition for deformation. This change also produces somewhat of a gap above these levels, and considerable stability for these deformed shapes. It is important to note that even in $^{96}Sr_{58}$ the deformed state is present, but because of the small number of protons, it lies at an elevated energy.

The idea that the unique-parity orbitals do not drive deformation in nuclei is an important conclusion of the Fermion Dynamical Symmetry Model (FDSM) described elsewhere in this volume.[10] In that model, it is the occupancy and interactions among the major-parity orbitals—for the mass-100 region, the $d_{5/2}$, $g_{7/2}$, $d_{3/2}$, and $s_{1/2}$ neutrons and $g_{9/2}$ protons that determine the shape of the nucleus. We extend that idea to note that it is the capacity of the $h_{11/2}$ neutron orbitals to absorb neutrons that stabilizes the deformation. Without the capacity to absorb neutrons, once the major-parity orbitals are half filled, the deformation would change very quickly with neutron number. Because there are no comparable unique-parity orbitals for the protons, deformation changes much more rapidly with changing proton number than with neutron number, as shown in Fig. 9 for the N = 59 isotones. Once protons have refilled the negative-parity orbitals and must start filling the active $g_{9/2}$-proton orbital, deformation decrases very rapidly as the Z = 50 closed shell is approached.

In summary, we have presented data to ilustrate that deformation in the mass-100

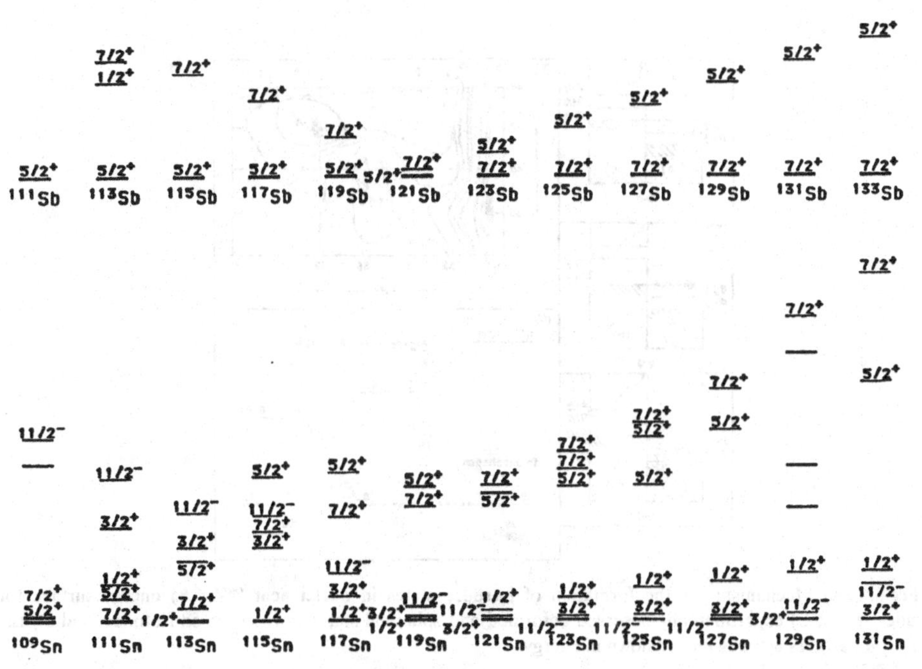

Figure 6. Changes in single-proton energies in Sb isotopes as a function of filling the neutron shell up to N = 82.

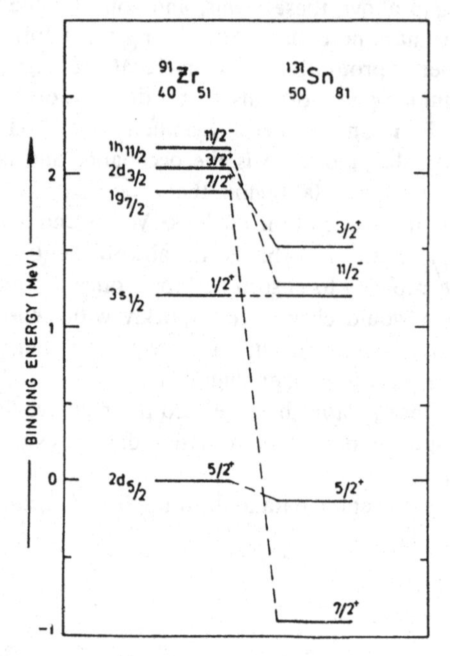

Figure 7. Positions of single-neutron orbitals in ^{91}Zr contrasted with those in ^{131}Sn. These demonstrate primarily the affects of monopole interactions and nuclear size on the *p-n* interaction.

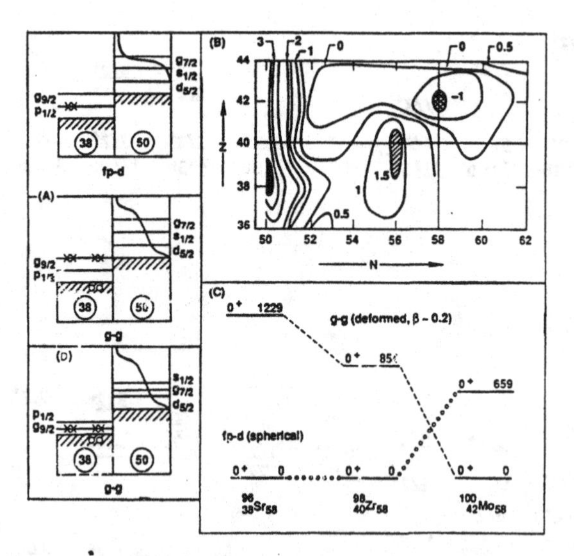

Figure 8. *A*) Mechanism for the formation of intruder states in nuclei near ^{98}Zr; *B*) energy surface for intruder states; *C*) experimentally-observed deformed and spherical states in the N = 58 isotones; and *D*) the spin-orbit partners actually moved down in energy.

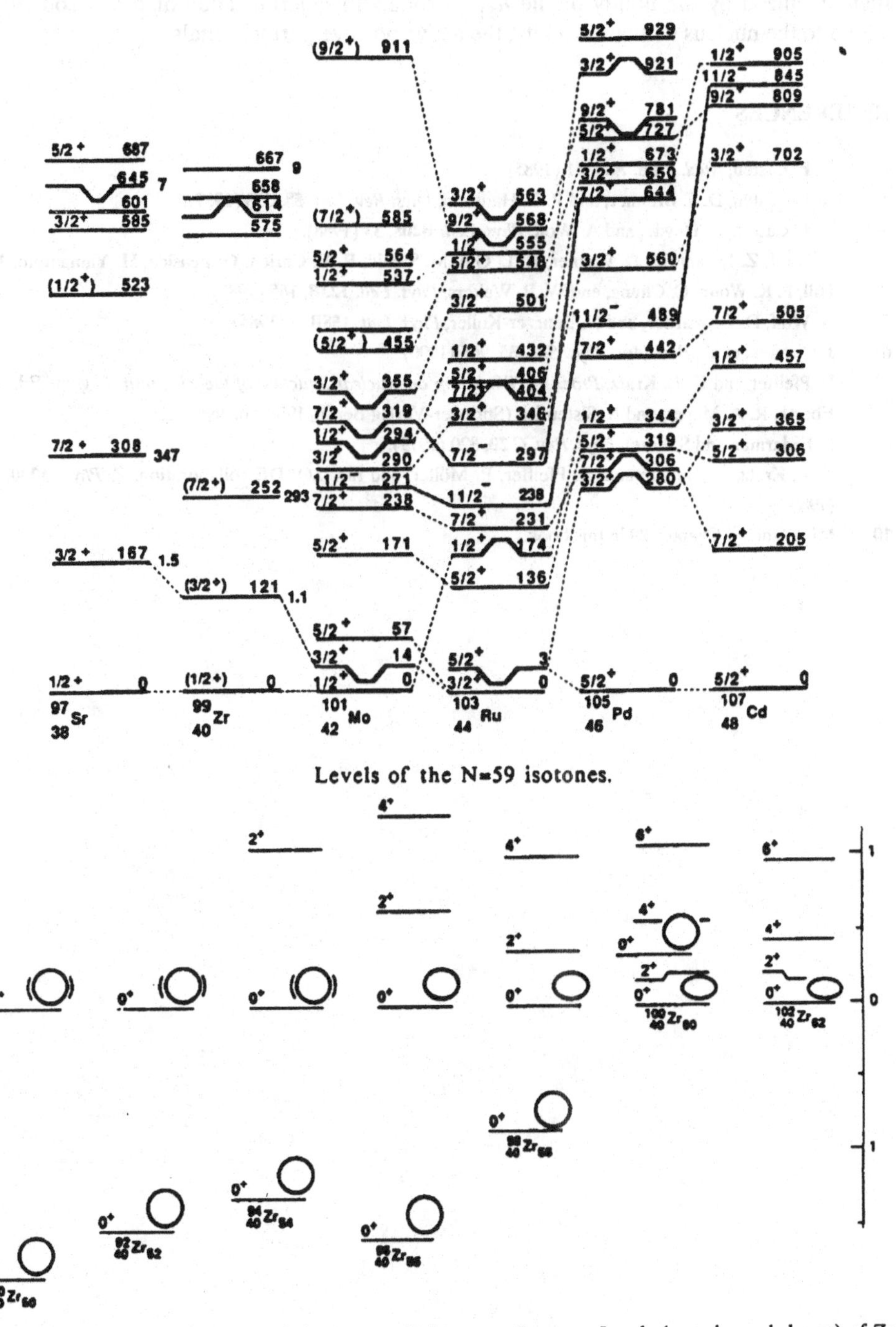

Figure 9. Top: Systematics of levels in the N = 59 isotones. Bottom: Levels (energies and shapes) of Zr with respect to the O^+ ($1g_{9/2}$ proton) states.

region is driven by the *p-n* interaction between the proton $g_{9/2}$ and neutron $g_{7/2}$ orbitals, then stabilized by the ability of the $h_{11/2}$ orbitals that permit additional neutrons to be added to the nucleus without blocking the active positive-parity orbitals.

REFERENCES

1. R. F. Casten, *Nucl. Phys.* **A443**, 1 (1985).
2. R. F. Casten, D. S. Brenner, and P. E. Haustein, *Phys. Rev. Lett.* **58**, 658 (1987).
3. R. F. Casten, K. Heyde, and A. Wolf, *Phys. Lett.* **B208**, 33 (1988).
4. A. Wolf, Z. Berant, D. D. Warner, R. L. Gill, M. Shmid, R. E. Chrien, G. Peaslee, H. Yamamoto, J. C. Hill, F. K. Wohn, C. Chung, and W. B. Walters, *Phys. Lett.* **123B**, 165 (1983).
5. A. Wolf, D. D. Warner, and N. Benczer-Koller, *Phys. Lett.* **158B**, 7 (1985).
6. J. Cizewski et al., *Bull. Am. Phys. Soc.* **35**, 966 (1990).
7. B. Pfeiffer and K.-L. Kratz, *Proc. Int. Workshop on Nuclear Structure of the Ziconium Region,* Ed. by J. Eberth, R. A. Meyer, and K. Sistemich (Springer-Verlag Berlin, 1988), p. 368.
8. P. Federman and S. Pittel, *Phys. Rev. C* **20**, 820 (1979).
9. K. L. Kratz, H. Gabelmann, B. Pfeiffer, P. Möller, and the ISOLDE collaboration, *Z. Phys.* **A330**, 229 (1988).
10. Mike Guidry, Chapter 29 in this book.

29. Dynamical Symmetries for Fermions

Mike Guidry
Department of Physics
University of Tennessee
Knoxville, TN 37996-1200

ABSTRACT

An introduction is given to the Fermion-Dynamical-Symmetry Model (FDSM). The analytical symmetry limits of the model are then applied to the calculation of physical quantities such as ground-state masses and $B(E2)$ values in heavy nuclei. These comparisons with data provide strong support for a new principle of collective motion, the Dynamical Pauli Effect, and suggest that dynamical symmetries which properly account for the pauli principle are much more persistent in nuclear structure than the corresponding boson symmetries. Finally, we present an assessment of criticisms which have been voiced concerning the FDSM and a discussion of new phenomena and "exotic spectroscopy" which may be suggested by the model.

29.1 INTRODUCTION

I would like to introduce to you, in this chapter, a new approach to nuclear structure physics. The method I wish to discuss

- Uses principles of dynamical symmetry but has no bosons;
- Has no explicit deformations, or moments of inertia, or other macroscopic "spontaneous breakers of symmetry"; yet it reproduces the observable matrix elements of collective-model physics;
- Represents a solution to a problem generally thought intractable (the spherical shell model for heavy collective nuclei); yet in many cases the solutions are not only feasible, they are *analytical*;
- Is a model for which many physically-interesting problems are so ridiculously simple that you can obtain approximate solutions with a hand calculator; nevertheless, in some cases these results are in better or comparable accord with data than large-scale computer calculations with standard approaches;
- Represents a novel perspective; therefore, it suggests important new ways of understanding nuclear structure which may not easily be seen in standard models.

The model to which I refer is the Fermion Dynamical Symmetry Model (FDSM), of course.[1,2] In the space I have available, I would like at least to make plausible the above assertions by providing a simple introduction to the basic features of the model, by applying

Exotic Nuclear Spectroscopy, Edited by W. C. McHarris
Plenum Press, New York, 1990

the model to the calculation of physically interesting quantities in a variety of systems, by discussing (objectively, I hope) criticisms that have occasionally been leveled at the model, and finally—for the sake of our organizer—by addressing the question of "exotic spectroscopy."

29.2 A SUMMARY OF RECENT FDSM RESULTS

Before introducing the FDSM, I want to motivate the discussion by providing a compressed summary of what I consider to be the most important accomplishments of the FDSM within the past five years. After all, if you are going to learn something new, you have some right to demand that the new thing deliver some "bang for the buck," so to speak.

1. The IBM is the large-degeneracy limit of the FDSM—*Phys. Rev. C* **34**, 2269 (1986); **37**, 2789 (1988).
2. The Particle-Rotor Model is the large particle number limit of the FDSM—*Phys. Lett.* **193B**, 163 (1987); **198B**, 119 (1987).
3. HFB theory is a coherent-state approximation to the FDSM in the truncated model space—*Phys. Rev. C* **37**, 1281 (1988).
4. The FDSM is identically the spherical shell model if all heritages ("broken pairs") are included—*Phys. Rev. C* **36**, 1157 (1987).
5. The Nilsson model is recovered for deformed nuclei—*Phys. Rev. C* **37**, 1739 (1988); and to be published.
6. The FDSM prescribes a connection, Valence Shell → Dynamical Symmetry → Dominant Collective Modes, which is supported by nuclear systematics—*Phys. Lett.* **192B**, 253 (1987).
7. Three kinds of vibrations, differing in pauli effects—*Phys. Rev. C* **36**, 1157 (1987).
8. A new transitional symmetry, $SO(7)$, predicted and observed—*Phys. Rev. Lett.* **56**, 2578 (1986).
9. Two kinds of rotation:
   - $Sp(6) \supset SU(3)$, axially symmetric
   - $SO(8) \supset SO(6)$, γ-soft rotor—*Phys. Lett.* **168B**, 313 (1986).
10. $B(E2)$ values in deformed nuclei comparable in accuracy with large-scale numerical calculations:
    - Analytical $B(E2;\ 2^+ \to 0^+)$ of all heavy deformed nuclei with three parameters—*Phys. Lett.* **B205**, 156 (1988).
    - Analytical description of $E2$ collectivity loss in heavy nuclei; algebraic description of band termination—*Phys. Lett* **187B**, 210 (1987); and to be published.
11. Analytical description of actinide masses comparable in accuracy with large-scale Strutinsky calculations—*Phys. Lett.* **194B**, 447 (1987); and to be published.
12. Shell-model derivation of the VMI effect: stretching and coriolis antipairing—*Phys. Rev. C* **40**, 2844 (1989), and to be published.
13. Shell-model description of basic high-spin phenomena: band crossings and backbending—*Phys. Lett.* **176B**, 1 (1986).
14. Identification of a "new" principle in collective motion: The Dynamical Pauli Effect. Widespread evidence in transition matrix elements and masses—*Phys. Lett.* **205B**, 156 (1988).
15. Possible origin of coriolis attenuation problem of Particle-Rotor Model in finite par-

ticle-number effects—*Phys. Rev. C* **37**, 1739 (1988).

16. Simple algebraic explanation for the dominance of prolate over oblate geometry in heavy, deformed nuclei—Submitted to *Phys. Lett. B*.
17. Relation of γ-instability in high-spin states to pauli representation restrictions on the dynamical symmetry—*Phys. Rev. C* **38**, 1475 (1988).
18. A preliminary FDSM effective interaction for actinide nuclei which gives a good description of a variety of observables:
    - ground-state masses
    - 2^+ energies
    - odd-even mass differences ("pair gap")
    - β, γ band-head energies
    - energies of yrast states
    - $B(E2)$ values on yrast line—to be published
19. Suggested connection between dynamical symmetry breaking and occurrence of quantum chaos in nuclear structure—*Phys. Rev. Lett.* **61**, 2167 (1988).
20. A suggestion that ground-band fluctuations in $B(E2)$ values by 5–10% from rotational are CAP effects – *Phys. Lett.* **176B**, 1 (1986).
21. Possible explanation of recently observed large suppression of two-neutron transfer to $I \approx 5$–$10\hbar$ states in heavy-ion collisions—to be published.
22. Demonstration of equivalence of algebraic and geometrical descriptions of vibrational bands in deformed nuclei:
    - FDSM $\underset{n\to\infty}{\to}$ Alaga values for $B\,(E2)$ ratios
    - 1:1 correspondence in state labeling—Submitted for Publication to *Fizika*.
23. First stage of FDSM code (valid for no broken pairs) now available. Numerical calculation of observables [spectrum, Q, $B(E2)$, $B(M1)$, . . .] away from symmetry limits. Stage 2 (broken pairs) in progress—*Phys. Rev. C* **39**, 1066 (1989).
24. Evidence for important influence of the Dynamical Pauli Effect on the behavior of nuclear g factors with angular momentum, and on the behavior of the VMI modification of the nuclear moment of inertia as a function of particle number—to be published.
25. A new algebraic model of superdeformation which seems already to have the correct quantitative, and even quantitative, features in the analytical symmetry limit—to be published.

This is blatant propaganda, of course, but it is of the sort that we generally consider acceptable in physics because it is (for the most part) documented by literature references. Indeed, I hope that you will not take the preceding assertions at face value, but rather that you will go to the published papers and preprints and critically assess them for yourself.

The weight of the evidence cited above indicates that the FDSM provides a microscopic link between the common approaches to the structure of strongly collective nuclei and the spherical shell model: the FDSM is a truncation of the spherical shell model problem, which yields the Particle-Rotor Model and the Cranked-Shell Model in the limit of large particle number, the Interacting Boson Model in the limit of large shell degeneracy (neglect of pauli principle), and the Hartree-Fock Bogoliubov theory in the coherent state approximation. The situation is summarized in Fig. 1.

Let's now turn to a more concrete examination of all of this. We begin by introducing the basic ideas of the FDSM in a form which will facilitate the later discussion. A more comprehensive description can be found in Refs. 2 and 3.

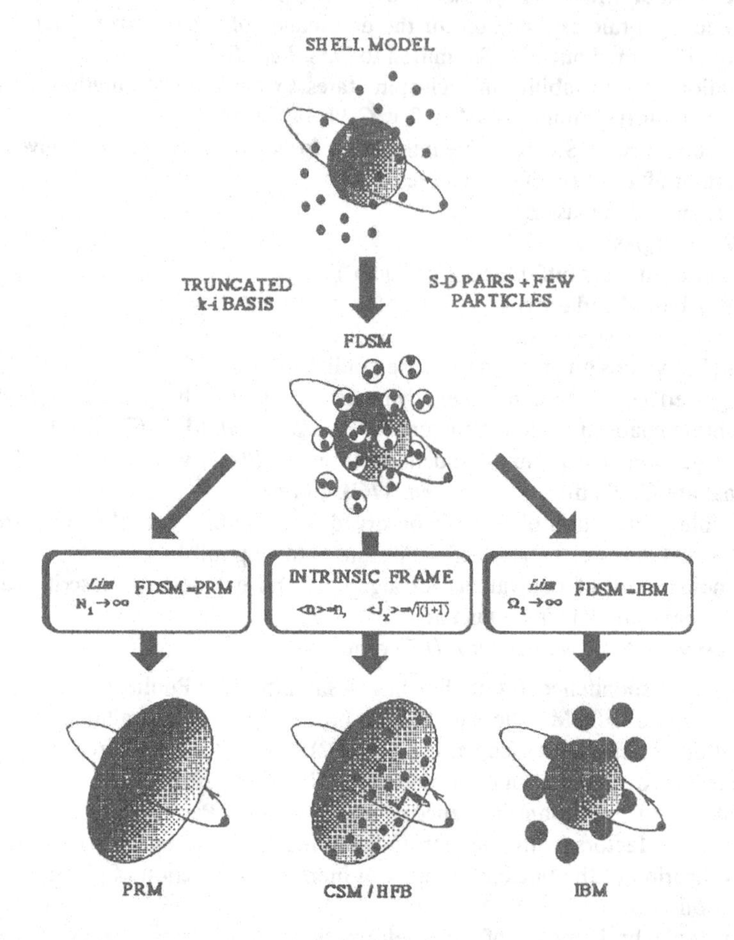

Figure 1. A schematic illustration of how the FDSM is related to standard models. Generally, the FDSM is derived from the spherical shell model through symmetry-dictated truncation. Then, geometry-based models represent the infinite particle-number limit of the FDSM, algebraic-boson models represent the infinite-shell-degeneracy limit of the FDSM, and Hartree-Fock models correspond to mean-field approximations to the FDSM in the truncated space. The abbreviation CSM stands for Cranked-Shell Model, PRM stands for Particle-Rotor Model, and IBM stands for Interacting Boson Model.

29.3 THE FERMION-DYNAMICAL-SYMMETRY MODEL

The most useful way to view the FDSM is as a particular truncation of the spherical shell model. The FDSM truncation is unusual because it is a truncation according to principles of dynamical symmetry, a *symmetry-dictated truncation*, if you will. That is, in more normal truncations of the shell-model space—and such trunctions are necessary at some level in any shell model calculation—the truncation is done according to energy considerations; the typical restriction of the shell model to a single major shell is one simple example of such an energy-dictated truncation. The truncation of the shell model space implied by the FDSM is qualitatively different. We adopt a point of view made familiar by the IBM approach to nuclear structure: Collective modes correspond to dynamical symmetries in the underlying algebraic structure of many-body theories. However, in the FDSM we are going to look for these symmetries directly in the fermion space, without recourse to boson degrees of freedom or mapping procedures; this is illustrated in Fig. 2. It is these fermion symmetries which will guide the FDSM truncation of the shell model space.

The starting point for the FDSM is the Ginocchio Model,[4] where the angular momentum of a single-particle orbital is separated into an integer pseudo-orbital part k and a half-integer pseudo-spin part i. The coupling scheme for the Ginocchio model is shown in Table I. In the Ginocchio coupling scheme the $k = 1$ coupling may be termed *k-active coupling*, while the $i = 3/2$ coupling may be termed *i-active coupling*. These couplings, and the normal coupling of the pairing model (which we will term $k = 0$ coupling), are illustrated in Fig. 3.

These coupling schemes give rise to a nice group structure which can lead to dynamical symmetries. [A dynamical symmetry results if the hamiltonian for a system can be written in terms of the invariants (Casimir operators) of a group and a chain of subgroups. The dynamical symmetry chains listed here all end in the group $SO(3)$ corresponding to conservation of total angular momentum.] In terms of monopole pair operators S, quadrupole pair operators D, and multiple operators P^r of rank r, we may define the closed algebras shown in Table II. The angular momentum labeling in these expressions corresponds to

$$\mathbf{i}_1 + \mathbf{i}_2 = \mathbf{I} \quad \mathbf{k}_1 + \mathbf{k}_2 = \mathbf{K} \quad \left[b^\dagger_{k_1 i_1} b^\dagger_{k_2 i_2} \right]^{KI}_{\mu_K \mu_I} . \tag{1}$$

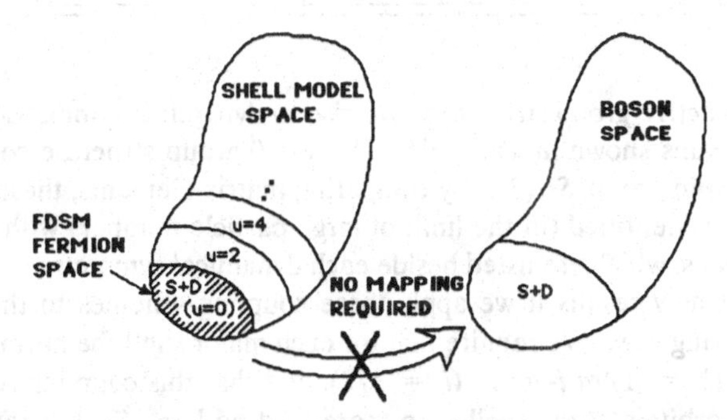

Figure 2. The FDSM is a symmetry-dictated truncation of the *fermion* shell-model space. The inspiration for associating dynamical symmetries with collective modes is derived from algebraic-boson models; however, the FDSM works at all times in the fermion space, so there is no need of a mapping procedure between the fermion and boson spaces. The FDSM truncates the shell-model space with respect to the heritage quantum number u, which is the number of particles which do not form (collective) S and D fermion pairs. If all heritages are included, the FDSM model space becomes the original shell-model space. For light nuclei this is computationally feasible, and the FDSM is the spherical shell model at whatever level of truncation one wishes. For heavy nuclei in the middle of valence shells it is only possible to include low heritages in realistic calculations; however, there is considerable evidence that low-energy nuclear structure is dominated by such configurations. This is the algebraic form of the conventional idea that nuclear structure is well approximated by the geometrical collective modes plus a few broken pairs.

Table I. The Ginocchio Model[4]

GINOCCHIO MODEL

Generalized pseudo-orbit, pseudo-spin representation

$$\mathbf{j} = \mathbf{k} + \mathbf{i}$$

$$a^{\dagger}_{jm} = \sum_{m_k m_i} \langle k m_k i m_i \mid jm \rangle \, b^{\dagger}_{k m_k i m_i}$$

MOTIVATION: Generalize quasispin to include D-pairs

ASSUMPTION: S and D pairs play a crucial role in nuclear structure.

Ginocchio noted that two choices of k and i ensured that S and D (and *only* S and D) pairs were formed:

$$i_1 = i_2 \equiv i = 3/2$$

$$k_1 = k_2 \equiv k = \text{ Anything consistent with } \mathbf{j} = \mathbf{k} + \mathbf{i}$$

$$\mathbf{K} = \mathbf{k}_1 + \mathbf{k}_2 = 0$$

$$k_1 = k_2 \equiv k = 1$$

$$i_1 = i_2 \equiv i = \text{ Anything consistent with } \mathbf{j} = \mathbf{k} + \mathbf{i}$$

$$\mathbf{I} = \mathbf{i}_1 + \mathbf{i}_2 = 0$$

The i-active and k-active group structures give rise to dynamical symmetries corresponding to the subgroup chains shown in Table III. The $k = 0$ group structure corresponds to the well-known quasi-spin group, $SU(2)$. By comparing matrix elements, these dynamical symmetry chains can be identified (in the limit of large particle number) with various geometrical collective modes, which are listed beside each dynamical symmetry.

The FDSM now results if we apply these coupling schemes to the spherical shell model in the following way. We require that for each major shell the normal-parity orbitals either be k-active ($k = 1$) or i-active ($i = 3/2$), and that this coupling reproduce exactly the normal-parity orbitals of the shell—no more, and no less. Such a reclassification appears to be unique, as illustrated in Fig. 4. The requirement that there be only one abnormal-parity level in each shell means that each abnormal-parity level must correspond to $k = 0$ coupling, as also illustrated in Fig. 4.

This unique assignment of coupling schemes to each shell, and the group structure already discussed for these coupling schemes, implies that the FDSM associates particular dynamical symmetries, and therefore particular allowed collective modes, with each valence shell. The allowed symmetries are shown in the row labeled "sym" in Fig. 4; the corresponding collective modes have been given above.

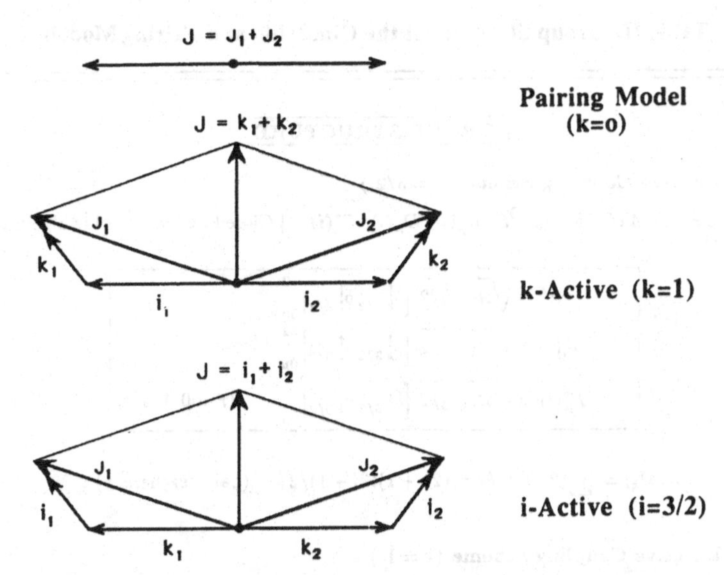

Figure 3. Illustration of the basic pair coupling schemes used in the Fermion Dynamical Symmetry Model for heritage-zero configurations. The k- and i-active schemes are used for the normal-parity orbitals in the collective S–D subspace. The $k = 0$ coupling scheme is used for the heritage-zero sector of the abnormal-parity orbital space. In each diagram, the part of the angular momentum coupled to zero contribution to the pair angular momentum is termed the *inert* part of the single-particle angular momentum; the part coupled to give finite angular momentum for the pair is termed the *active* part of the single-particle angular momentum. For configurations where the heritage is non-zero, the inert part of the angular momentum may be recoupled to a non-zero resultant angular momentum, which corresponds physically to the angular momentum of broken pairs. In configurations with non-zero heritage, this angular momentum of the broken pairs then couples to the active angular momentum of the S and D pairs to give the total angular momentum.

This shell-model reclassification leads to a unified, algebraic fermion description of collective motion, as illustrated in Fig. 5. I won't bother to describe the details of this somewhat formidable diagram, as they can be found in Ref. 1. The essential point for our discussion is that in the FDSM the valence shell dictates the coupling scheme, which in turn defines the allowed dynamical symmetries. This leads to a rich variety of collective modes (corresponding to geometrical rotations and vibrations) with broken pairs of partides, or single particles, coupled to them.

29.4 SOME PHYSICAL QUANTITIES CALCULATED WITH THE FDSM

We now turn to some examples of applying the FDSM to the calculation of physical quantities. The FDSM program can be implemented at two general levels. In the *symmetry limits*, one obtains analytical formulas and a systematic overview and classification of nuclear structure. However, real nuclei do not have perfect dynamical symmetries, so the second stage corresponds to using the symmetry limits as a starting point for numerical calculations with symmetry-breaking terms. Although both levels of investigation are being pursued, I will concentrate here on analytical calculations performed in the symmetry limits (or in perturbation theory around the symmetry limits). I do this for two reasons. First,

Table II. Group Structure of the Ginocchio and Pairing Models

GROUP STRUCTURE

(1) i-Active Coupling Scheme ($i = 3/2$)

GENERATORS: $S^\dagger$, S, $D^\dagger_\mu(i)$, $D_\mu(i)$, $P^r_\mu(i)$ [Closed under $so(8)$ Lie algebra]

$$S^\dagger = \sqrt{\Omega_{k\,3/2}/2}\,\left[b^\dagger_{k\,3/2}\,b^\dagger_{k\,3/2}\right]^{00}_{00}$$
$$D^\dagger_\mu(i) = \sqrt{\Omega_{k\,3/2}/2}\,\left[b^\dagger_{k\,3/2}\,b^\dagger_{k\,3/2}\right]^{02}_{0\mu}$$
$$P^r_\mu(i) = \sqrt{\Omega_{k\,3/2}/2}\,\left[b^\dagger_{k\,3/2}\,\tilde{b}_{k\,3/2}\right]^{0r}_{0\mu} \qquad (r = 0,1,2,3)$$

$$\Omega_{ki} = \sum_j (2j+1)/2 = (2k+1)(2i+1)/2 \qquad \text{(pair degeneracy)}$$

(2) k-Active Coupling Scheme ($k = 1$)

GENERATORS: $S^\dagger$, S, $D^\dagger_\mu(k)$, $D_\mu(k)$, $P^r_\mu(k)$ [Closed under $sp(6)$ Lie algebra]

$$S^\dagger = \sqrt{\Omega_{1i}/2}\,\left[b^\dagger_{1i}\,b^\dagger_{1i}\right]^{00}_{00}$$
$$D^\dagger_\mu(k) = \sqrt{\Omega_{1i}/2}\,\left[b^\dagger_{1i}\,b^\dagger_{1i}\right]^{20}_{\mu 0}$$
$$P^r_\mu(k) = \sqrt{\Omega_{1i}/2}\,\left[b^\dagger_{1i}\,\tilde{b}_{1i}\right]^{r0}_{\mu 0} \qquad (r = 0,1,2)$$

(3) Monopole Pair Coupling ($k = 0$).

GENERATORS: $\mathcal{S}^\dagger$, $\mathcal{S}$, $\mathcal{P}^0$ [Closed under $su(2)$ Lie algebra]

$$\mathcal{S}^\dagger = \sqrt{\Omega_{j_o}/2}\,\left[b^\dagger_{0j_o}\,b^\dagger_{0j_o}\right]^{00}_{00}$$
$$\mathcal{P}^0 = \sqrt{\Omega_{j_o}/2}\,\left[a^\dagger_{j_o}\,\tilde{a}_{j_o}\right]^{0}_{0}$$

this illustrates the basic points that I wish to leave you with in the simplest fashion; second, we will find that already in the symmetry limit the results are quantitatively in agreement with data for a variety of physical observables.

Before discussing some specific calculations, there is one conceptual and practical point which may have been bothering you and which requires some clarification. The discussion in the previous section has assumed a single kind of nucleon, and it presumably has not escaped your attention that nuclei contain both valence protons and neutrons. Furthermore, there is widespread agreement that quadrupole collective modes are a consequence primarily of the quadrupole-quadrupole interaction between neutrons and protons. How then do n–p interactions enter the FDSM?

The complete answer to this question would require more space than is available

Table III. FDSM Dynamical Symmetries

DYNAMICAL SYMMETRIES

(1) i-Active Coupling ($i = 3/2$)

$SO(8) \supset SO(5) \times SU(2) \supset SO(5) \supset SO(3)$	(Vibrator)
$SO(8) \supset SO(6) \supset SO(5) \supset SO(3)$	(γ-Unstable Rotor)
$SO(8) \supset SO(7) \supset SO(5) \supset SO(3)$	(Vibrator)

(2) k-Active Coupling ($k = 1$)

$Sp(6) \supset SU(3) \supset SO(3)$	(Axially-Symmetric Rotor)
$Sp(6) \supset SU(2) \times SO(3) \supset SO(3)$	(Vibrator)

(3) Monopole pair coupling ($k = 0$)

$SU(2)$	(Quasispin Symmetry)

here; it has already been discussed to some degree in Refs. 1 and 2, and will be discussed fully in a review article which is in preparation.[5] Here, I will make only the following qualitative remarks. Generally, we assume a wave-function that is a product of neutron and proton wave-functions, and a hamiltonian that contains an interaction term H_{np}, which couples the neutrons and protons strongly through an n–p quadrupole interaction. In the simplest calculations we assume that the product wave-function is symmetric for the lowest-lying states. In some situations, such as in the actinides, where the neutrons and protons have the same highest symmetry, it is possible to obtain a dynamical symmetry corresponding to a coupled neutron-proton system. For example, well-deformed actinides can be described by the symmetry $SU(3)^{\pi} \times SU(3)^{\nu} \supset SU(3)^{\pi+\nu}$, where π and ν denote protons and neutrons, respectively. In other cases, the neutrons and protons cannot be coupled like this, and there is no dynamical symmetry. This is the case for the rare-earth nuclei—where neutrons are $Sp(6)$ and protons are $SO(8)$. However, numerical calculations indicate that even when formally there is no dynamical symmetry for the neutron-proton system, there often is a practical dynamical symmetry. For example, numerical studies of well-deformed rare-earth nuclei indicate that they behave like the FDSM $SU(3)$ symmetry limit for many properties.[6]

Let's begin our survey of some FDSM results by looking at the FDSM calculation of ground-state masses in the actinide region. The details may be found in Ref. 7, and we present the results in Fig. 6. Even though the FDSM calculation is analytical, it fits the 332 masses with thirteen adjustable parameters and an rms error of 0.23 MeV.

In Fig. 7 we show an FDSM calculation of the $B(E2; 2^+ \rightarrow 0^+)$ for even-even nuclei

No.	1	2	3	4	5	6	7	8
n	0	1	2	3	3 4	4 5	5 5 6	6 6 7
k	0	1	1	0	1 0	2 0	1 1 0	1 1 0
i	$\frac{1}{2}$	$\frac{1}{2}$	$\frac{3}{2}$	$\frac{7}{2}$	$\frac{3}{2}$ $\frac{9}{2}$	$\frac{3}{2}$ $\frac{11}{2}$	$\frac{1}{2}$ $\frac{7}{2}$ $\frac{13}{2}$	$\frac{3}{2}$ $\frac{9}{2}$ $\frac{15}{2}$
CONFIGURATION	$s_{1/2}$	$p_{1/2}$ $p_{3/2}$	$s_{1/2}$ $d_{3/2}$ $d_{5/2}$	$f_{7/2}$	$p_{1/2}$ $g_{9/2}$ $p_{3/2}$ $f_{5/2}$	$s_{1/2}$ $h_{11/2}$ $d_{3/2}$ $d_{5/2}$ $g_{7/2}$	$p_{1/2}$ $f_{5/2}$ $i_{13/2}$ $p_{3/2}$ $f_{7/2}$ $h_{9/2}$	$s_{1/2}$ $g_{7/2}$ $j_{15/2}$ $d_{3/2}$ $g_{9/2}$ $d_{5/2}$ $i_{11/2}$
SYM			$G_6G_8G_3$		$G_6G_8G_3$	G_8	G_6	G_6
Ω_0	0	0	0	0	5	6	7	8
Ω_1	1	3	6	4	6	10	15	21
n	2	8	20	28	50	82	126	184

G_6= Sp6×$\mathbb{SU2}$ (k-active)

G_8= SO8× $\mathbb{SU2}$ (i -active)

G_3= SU3 ×SO6 × $\mathbb{SU2}$ (k/i-active)

Figure 4. FDSM reclassification of the shell model. The first two rows give the shell number and the oscillator quantum number. The row labeled "configuration" gives the normal shell-model classification of orbitals. Rows 3 and 4 reclassify according to the FDSM quantum numbers k and i. The quantities Ω_0 and Ω_1 are pair degeneracies of the normal- and abnormal-parity orbitals of the shell respectively. Normal-parity orbitals of a shell are either k active ($k=1$) or i active ($i=3/2$), or (in light nuclei) k/i active ($k=1$ and $i=3/2$); abnormal-parity orbitals are required to be $k=0$ (and, therefore, $j=i$) by the constraint that there be only one abnormal-parity orbital within a shell. The group structure implied for each valence shell is shown in the row labeled "sym," with the symbols G_6, G_8, and G_3 defined below the table. The symbol "*SU2*" in slant font denotes the symmetry of the abnormal-parity orbital.

in the rare-earth region. This result I consider to be rather remarkable, since it is again obtained analytically, and only a single overall scale parameter is adjusted—*the shape of the curve is completely fixed by the theory*! A similar-quality description has been obtained for actinide nuclei and for nuclei in the 50–82 shells: We find that all havy even-even nuclei can be described with an accuracy comparable to that in Fig. 7 using only three adjustable parameters, which just set the scale of the $B(E2)$ values. The details may be found in Ref. 9.

In Fig. 8 we show an example of another anlytical FDSM calculation – in this case the behavior of $B(E2)$ values at high angular momentum. Here we see the ability of the FDSM to reproduce the measured loss of $E2$ collectively at high angular momentum, even when the moment of inertia stays relatively constant as in the case of ^{166}Yb. This is because the essential loss of collective strength in the FDSM is a dynamical effect, not a shape change (the shape can change, too, but that is a separate effect); thus, the $B(E2)$ can be attenuated while the moment of inertia remains almost constant. The details of this calculation may be found in Ref. 10.

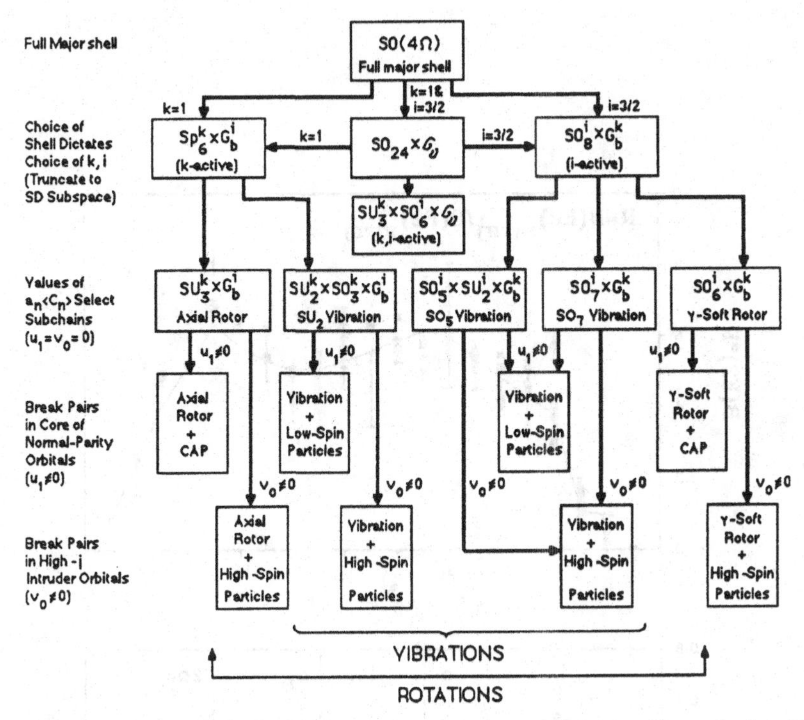

Figure 5. Overall group structure of the FDSM within a single major shell. Details may be found in Refs. 1 and 2. CAP stands for coriolis antipairing, and $G_b{}^{k,i}$ stands for the group structure associated with the abnormal-parity orbitals.

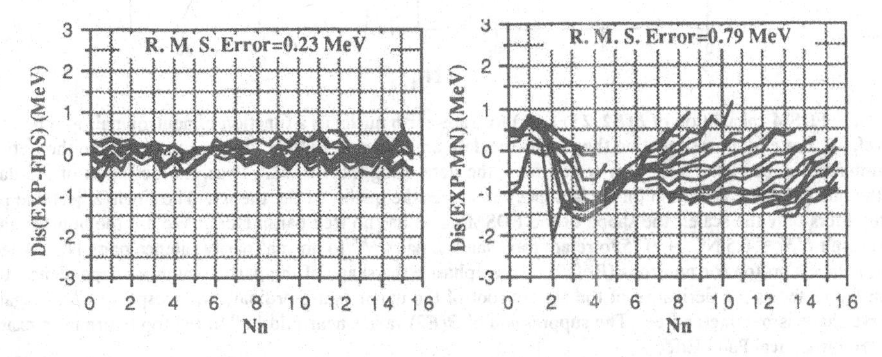

Figure 6. Calculated ground-state masses of actinide nuclei. The difference between theoretical and experimental masses is plotted as a function of valence neutron pair number N_n. The FDSM calculation on the left is analytical, and uses thirteen adjustable parameters to fit the 332 known masses in this region with an rms deviation of 0.23 MeV. The details of the FDSM calculation may be found in Ref. 7. A corresponding calculation using the Strutinsky method is shown on the right (Møller and Nix, Ref. 8).

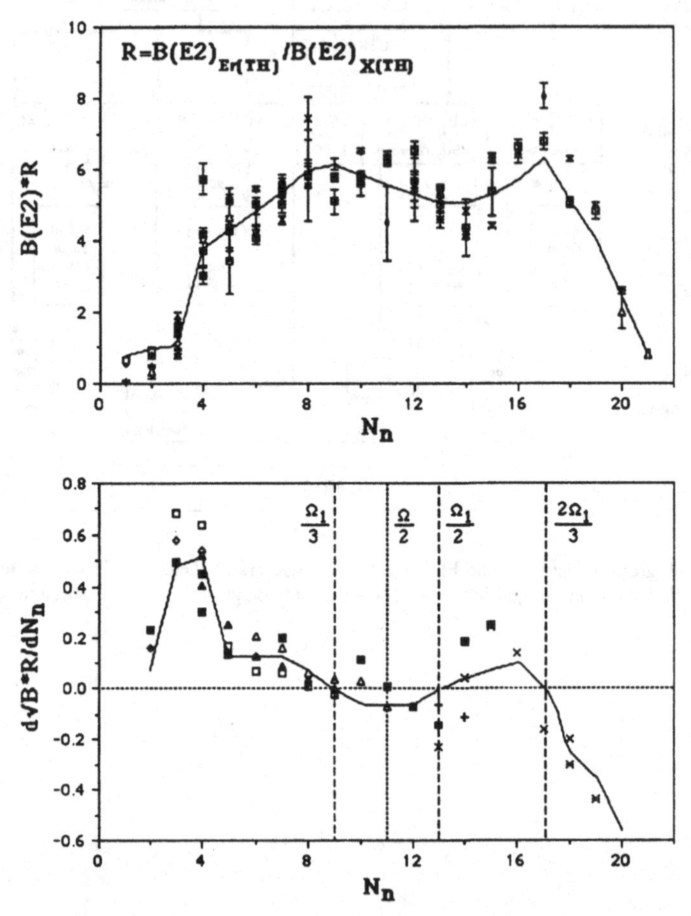

Figure 7. FDSM calculation of $B(E2, 2^+ \to 0^+)$ for rare-earth nuclei as a function of neutron pair number, N_n (see Ref. 9). For ease in assimilating the results the data are shown as follows: In the top figure the theoretical proton-number dependence has been scaled from the data using the FDSM. Thus, the deviation of the data from the line (FDSM calculation for Er isotopes) measures the quality of the theory. The single adjustable parameter alters only the scale: the shape of the FDSM curve has no free parameters. We use the approximate FDSM result $N_1^\sigma = 0.5\, N^\sigma + 0.75$ to relate total valence pairs N^σ to normal-parity valence pairs N_1^σ, where $\sigma = \pi, \nu$ labels protons or neutrons (Ref. 9). To emphasize the shape of the particle-number dependence, the bottom figure shows the derivative of the square root of the upper figure ordinate with respect to N_n. Again, the curve shape is parameter-free. The suppression of $B(E2)$ values near midshell in the top figure is an example of the Dynamical Pauli Effect.

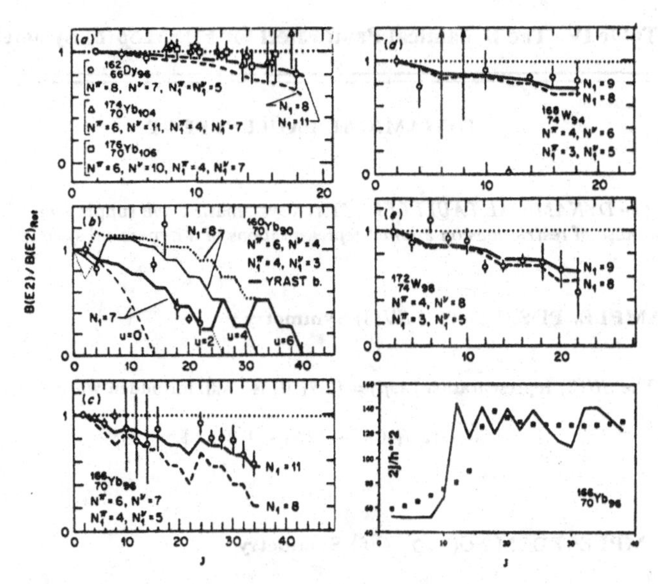

Figure 8. The ratio $B(E2)/B(E2)_{Rot}$ as a function of angular momentum for several rare-earth nuclei is shown in Parts *a*) — *e*). The curves are two FDSM calculations with slightly different assumptions about distribution of particles in the valence shells. An $SU(3)$ symmetry was assumed, and the FDSM results are analytical; details may be found in Ref. 10. Part *b*) exhibits a complete yrast band termination in the calculation near spin 40; in Part *b*) only, the heavy line represents the $B(E2)$ values on the yrast sequence, and the lighter lines are non-yrast $B(E2)$'s. The bottom right figure shows a corresponding FDSM calculation of the moment of inertia for ^{166}Yb as a function of angular momentum. This figure illustrates that in the microscopic FDSM theory the moment of inertia may remain nearly constant, even when there is a significant change in the $B(E2)$ value as a function of angular momentum. The local fluctuations of the moment of inertia seen in this figure are a property of the dynamical symmetry limit. Realistically, small symmetry-breaking terms in the hamiltonian would tend to smooth the fluctuations which appear above angular momentum 10.

29.5 THE DYNAMICAL PAULI EFFECT

The results presented in Figs. 6–8 all have contained within them the action of an important new principle of collective motion which we term the *Dynamical Pauli Effect:*

DYNAMICAL PAULI EFFECT. *A constraint on allowed representations of a dynamical symmetry, brought about by the competition between building collectivity and the pauli principle, which is more restrictive than that imposed by the particle-hole symmetry.*

An example is shown in Table IV: in the FDSM $SU(3)$ symmetry there are various representations near midshell which are forbidden by the pauli principle; thus, $SU(3)$ collectivity is restricted near midshell. This is a property of a group-theoretical solution of a fermion many-body system, but it has a simple physical interpretation. Building quadrupole collectivity requires maximal overlap of valence particles with an attractive effective interaction. As the shell is initially filled, there are many free orbitals, and this collectivity can be accomplished without undue complaint from the pauli principle. However, as the shell begins to be more filled, the desire to build collectivity comes into conflict with the

Table IV. The Dynamical Pauli Effect for FDSM $Sp(6)$ Symmetry

DYNAMICAL PAULI EFFECT

The DYNAMICAL PAULI EFFECT is a systematic Pauli suppression of collectivity because of restrictions on allowed representations of a dynamical symmetry

EXAMPLE: FDSM $Sp(6) \supset SU(3)$ Symmetry

The $SU(3)$ representation $(\lambda, \mu) = (2N_1, 0)$ is Pauli forbidden when $N_1 > \Omega_1/3$.

$$(2N_1, 0) \longrightarrow (2N_1 - 4, 2) \ldots]$$

EXAMPLE: FDSM $SO(8) \supset SO(6)$ Symmetry

NO DYNAMICAL PAULI EFFECT; the usual lowest energy representation is allowed until $N_1 = \Omega_1/2$ (the normal particle-hole symmetry).

pauli principle. The difficulty is exacerbated for collective states relative to single-particle states because the required correlations for collective states can't be generated by putting particles into arbitrary orbitals.

Thus, collectivity will be supressed by the pauli principle at some percentage of shell filling. Where this occurs is a property of the details of the collective mode; that is, it is a property of the particular dynamical symmetry. For example, in the FDSM $SO(6)$ symmetry this happens at midshell (for the normal-parity orbitals). Since this is just the ordinary particle-hole symmetry, it causes no effects distinguishable in the low-lying states from that expected for the pauli principle for independent-particle motion. In this case we say there is no dynamical pauli effect. However, for the $SU(3)$ symmetry the situation is different. There the collectivity begins to be suppressed when the normal-parity shell is one-third full, not at midshell, and there is a dynamical pauli effect between $N_1 = \Omega_1/3$ and $2\Omega_1/3$.

The suppression of collectivity near midshell displayed in Fig. 7 is a consequence of such a dynamical pauli effect. The calculations indicate that the derivative curve shown in the lower part of Fig. 7 should pass through zero at $N_n = 9$, 14, and 17, as a consequence of this dynamical pauli effect (corresponding to the predicted peaks and valleys in the upper figure). The data fluctuate, but the average behavior is clearly in agreement with this prediction (which, I remind you, has no adjustable parameters available to shift zeros). On the other hand, if one makes the same sort of plot, but with respect to the *proton* number, the collectivity peaks near normal-parity *midshell*, as expected since there is no dynamical pauli effect for the $SO(6)$ protons.

These dynamical pauli effects are found in many properties of nuclei, two notable ones being $B(E2)$ values and ground-state masses (effects in the spectrum will be discussed in Section 6). In general, one finds that the explicit fermion nature of the generators of the dynamical symmetry improves the range of validity of the dynamical symmetries relative to those for pure boson models. [By pure boson models, I mean the boson dynamical symmetries without pauli factors generated by additional procedures such as boson-fermion mapping. If one does mapping, or boson expansions, it is, of course, possible to bring the pauli effects into a boson theory at some level of approximation.] For example, the pure $SU(3)$ boson theory fails to describe properly $B(E2)$ values above angular momentum $\approx 10\hbar$, but the $SU(3)$ limit of the FDSM, because it incorporates fermion degrees of freedom, can describe electromagnetic transitions to much higher angular momentum (Fig. 8). Therefore, an important qualitative FDSM result is that dynamical symmetries appear to be more prevalent in nuclei than would be thought from simple boson models, provided the generators of the symmetry incorporate the richer physics implied by the fermion nature of the problem.

29.6 CRITICAL REVIEW OF OBJECTIONS TO THE FDSM

I have presented a few examples of the power of the FDSM to describe physical observables in an economical manner in the previous section. These are only representative examples; there are many more that we could discuss if space allowed (see the summary in Section 2). However, as in any theory which employs a new approach, there are also criticisms, some a result of simple misunderstanding, but others of a more substantial nature.

In this section I would like to briefly discuss some of these. In most cases I will only have space to sketch the objection, and my reply. A more complete discussion of all of this is in preparation.[5]

• 1) **Isn't the separation of the normal and abnormal-parity orbitals into a direct-product structure artificial?**

This separation presumably is artificial at some level. It is implemented in the FDSM as a mathematical convenience so that simple dynamical symmetries emerge. *However, there are strong physical reasons to believe that the normal and abnormal-parity orbitals enter low-lying collective nuclear structure on very different footings.* For example, consider the Nilsson model. The high-j abnormal-parity orbitals remain largely unmixed, even at normal nuclear deformations; this is easily seen by consulting the Nilsson wave-functions, or by noting that the abnormal-parity orbitals in a Nilsson diagram exhibit little curvature with deformation. [Incidentally, it is important to distinguish the *splitting* of the abnormal-parity orbitals, which is quite large, from the *mixing* of the abnormal-parity orbitals with other orbitals, which is quite small.] On the other hand, the normal-parity orbitals are strongly mixed at larger deformation, as seen from their wave-functions or their curvature in the Nilsson diagram.

This difference in behavior is associated with the removal of the abnormal-parity orbital from its fellows by the spin-orbit interaction: It has no nearby orbitals of the same parity with which to mix. As a consequence, the abnormal-parity orbital tends to behave more like a spherical shell-model orbital at large deformation than do the normal-parity orbitals. But this is precisely what is suggested by the FDSM quasispin symmetry for the abnormal-parity orbital (implying a dominance of pairing), and $Sp(6)$ or $SO(8)$ symmetry for the normal-parity orbitals (implying a potentially large role for multipole interactions).

Finally, we note that a large amount of data is correlated by the FDSM with the assumption that in the $SU(3)$ limit the normal parity valence particle number n_1 and the abnormal-parity valence particle number n_0 play different roles in nuclear structure, and that deformation systematics depend primarily on the effective value of n_1. For example, the $B(E2)$ systematics in Figs. 7 and 8, and the mass systematics in Fig. 6, which reproduce the data at a level comparable with much more complicated tranditional calculations, all employ this assumption.

Of course, the coupling between the normal- and abnormal-parity orbitals does matter in realistic cases. For example, the odd-mass spectrum for the unpaired particle in an abnormal-parity orbital requires such a coupling, which breaks the dynamical symmetry and must be handled numerically. This we have described in Ref. 11. However, the main requirement is not that dynamical symmetries describe nuclei perfectly, but that they be sufficiently good caricatures of nuclear structure that they are useful starting points for more sophisticated calculations. The FDSM separation of normal and abnormal-parity orbitals seems to fulfill this condition rather well.

• 2) **Don't the FDSM wave-functions sometimes have poor overlap with "realistic wave-functions"?**

Yes; and a good thing, too! The FDSM must be understood as a microscopic theory of effective interactions within a truncated model space, and effective interactions always depend on the model space. The severely symmetry-truncated model space of the FDSM is certainly quite different from the model space of standard theories. Therefore, we may generally expect that the comparison of wave-functions between the FDSM and other model spaces is *meaningless*. The relevant comparison is of matrix elements, for that is the best that we can measure in quantum mechanics. But it is quite clear that we can have $\langle\Psi_1|O|\Psi_2\rangle = \langle\Psi_1'|O'|\Psi_2'\rangle$, even if $\Psi_1 \neq \Psi_1'$ and $\Psi_2 \neq \Psi_2'$ and $O \neq O'$.

As we have seen in the preceding section, the FDSM gives a very good account of a variety of matrix elements (actually, squares of matrix elements in these examples). Since the FDSM operators are chosen for their symmetry properties, they generally differ from corresponding operators in other theories. Thus, it is *imperative* that the FDSM wave-functions differ from the wave functions in other theories, if the matrix elements are to be equivalent, and the comparison of wave-functions in different model spaces (see for example, Refs. 12–13) is a red herring. The only relevant test of the FDSM is to compare with its *matrix elements* calculated using an effective interaction matched to the FDSM model space. The details of such a comparison constitute a long-term project, but the preliminary results are quite encouraging.

• 3) **Don't the FDSM symmetries work poorly in the *sd* shell?**

Actually, as we have emphasized in the caption to Fig. 2, the FDSM is *identically* the shell model in the *sd* shell if all heritages are included. Since this is quite feasible technically for the light nuclei, this is more than an academic statement. Thus, the question is not whether the FDSM is "correct" in the *sd* shell (it is); the question rather is how large a heritage is required to approximate the correct phsyics in the *sd* shell. Just as in the discussion above, this comparison must not be sidetracked by comparisons of wave-functions; it is the matrix elements which are relevant.

However, as can be seen in Fig. 4 and as we have emphasized from the beginning,[1,2] in the *sd* shell the FDSM allows the nucleons to be both *k* and *i* active. In that case there is a no a priori reason to prefer *k*- or *i*-active coupling, and the most reasonable assumption as a starting point is that both operate at the same time. Then the lowest symmetry is the full shell symmetry, which is $SO(24)$, and we would expect the collective motion in this shell to be complicated. It is only in the heavy nuclei that the symmetries simplify and the *k*- or *i*-active coupling become unique.

• 4) **Doesn't the FDSM ignore the *n–p* interaction, which we all know to be the microscopic basis of deformation?**

This objection is based on a misconception. As we have already explained in Section 4, the *n–p* interaction is *essential* to all the realistic FDSM calculations presented there. Indeed, the preliminary FDSM effective interaction indicates that in the truncated *SD* subspace the effective *n–p* quadrupole interaction is large and attractive, but the like-particle quadrupole-quadrupole interaction is weak and may even be repulsive. *The FDSM absolutely requires a strong n–p quadrupole effective interaction to make sense of the properties of low-lying nuclear states.* Otherwise, the collective spectrum would have far too many states at low energy.

• 5) **Doesn't the FDSM suggest that the abnormal-parity orbitals play no role in deformation systematics, and doesn't this contradict the conventional wisdom from the deformed shell model?**

This statement also is based on a misconception and is the basis for a large amount of confusion. It is important to understand that the abnormal-parity orbital does play an enormous role in deformation systematics in the FDSM: Agreement between FDSM calculations and data would be destroyed if the abnormal-parity orbitals were removed. However, in the symmetry limits, the abnormal parity orbital plays a pivotal but indirect role; it doesn't influence the deformation so much by deforming as by serving as a kind of "sink" for particles driven from the normal-parity orbitals by the competing desires to maximize collectively and at the same time satisfy the pauli principle (the Dynamical Pauli Effect at work again). This allows the quadrupole collectivity to persist over a much broader range of nucleon number than would otherwise be possible, as required by the data, and has been likened to the action of chemical buffer, or the role played by *f*-electron orbitals in the filling of the lanthanide series of elements. [These marvelous analogies were suggested by Bill Walters; I won't explain it further because this is a chemistry conference, and chemists are supposed to understand these things!] Of course, the abnormal-parity orbital will also deform in the presence of the normal-parity deformed field (see Ref. 11). Our results indicate that this is a secondary effect, however. The *origin* of normal deformation lies primarily in the effective quadruple-quadrupole interaction operating in the normal-parity orbitals. (This is no longer true for superdeformation; however, this is another story which I have no time to tell).

Our detailed investigations cannot be reproduced here (see Ref. 14, forthcoming), but there appears to be no basic contradiction between the FDSM and Nilsson pictures of deformation systematics. Indeed, we have shown how to *derive* the Nilsson model from the FDSM in the limit of large particle number, so it would be surprising if the result were otherwise.

• 6) **In the FDSM the rare-earth protons have $SO(8)$ symmetry, but the neutrons have $Sp(6)$ symmetry; isn't that a bit strange?**

As we discussed in Section 4, this implies that there is no dynamical symmetry in the rare-earth region when the *n–p* interaction is turned on. However, numerical investigations indicate that pairing-plus-FDSM-quadrupole interactions diagonalized in an $SO(8)^{\pi}$ x $Sp(6)^{\nu}$ basis give $SU(2)$-like observables near closed shells and $SU(3)$-like observables away from closed shells[6]; thus an *effective dynamical symmetry* is present in the rare-earth region. Furthermore, FDSM calculations of physical observables like spectra and $B(E2)$ values agree well with the data for rare-earth nuclei. In particular, the $B(E2)$ systematics displayed in Fig. 7 provide strong evidence for an approximate $SO(8)^{\pi}$ x $Sp(6)^{\nu}$ structure for low-lying rare-earth states when analyzed as a function of proton number and neutron number (see Ref. 9).

• 7) **You have emphasized the Dynamical Pauli Effect and its influence on properties like nuclear masses and electromagnetic transitions strengths; what about the effect on the spectrum?**

In the $SU(3)$ symmetry limit, the disappearance of representations in midshell which was discussed in Section 5 will also have a profound influence on the low-lying spectrum. Probably the most dramatic consequence is that the "γ band" should drop to very low energy relative to the ground state but not the "β band."

To my knowledge, this does not happen in the rare-earth region. However, there are several things that one should keep in mind. First, these are pure symmetry-limit predictions which will be modified by symmetry-breaking in realistic cases. In the rare-earth region this may be very important, because we have already argued that the different symmetries for the protons and neutrons in the rare earths mean there is no dynamical symmetry at all there. We have suggested that there is a *practical* dynamical symmetry with respect to quantities like $B(E2; 2^+ \rightarrow 0^+)$ values; however, the numerical studies which show this have not yet been applied to a calculation of the spectrum in the pauli-forbidden region. In particular, we note that the assumption of completely symmetric states for the *np* system, which we employed so successfully for the ground and first (ground-band) 2^+ state in the rare earths, is likely to break down for excited states. Therefore, the symmetry-limit prediction of the excited-state spectrum in the rare-earths may fail, even if successful predictions for the low-lying ground-state band survive. There seems to be insufficient evidence to tell one way or the other so far.

The optimal place to see such effects should be the heavy actinides, since the FDSM predicts that deformed actinides will have $SU(3)$ symmetries for both protons and neutrons, and therefore a coupled $SU(3)$ symmetry will be favored by the n–p quadrupole interaction. Unfortunately, one begins to run out of stability at the $\Omega_1/3$ boundary for both protons and neutrons in the actinides. However, for the little data available there is a substantial drop in the position of the γ band-head beyond $\Omega_1/3$ for neutrons and/or protons (for example, in the heaviest known Fm isotopes, ^{254}Fm and ^{256}Fm). This drop is about 50% relative to a straight-line extrapolation of data well before the $\Omega_1/3$ boundary. This isn't as much as predicted by the simplest symmetry limit calculation, but it is certainly a dramatic effect in the right direction, and the inclusion of symmetry-breaking terms like pairing will likely modify the theory in the direction of the data.

Therefore, I believe that the comparison with data is inconclusive in the rare-earths, until the effect of the expected symmetry-breaking is analyzed, and that there is a tantalizing hint of this effect in the actinides, but we need data on several more nuclei to confirm it. I emphasize that these arguments don't prove that the FDSM is correct on this point; they just show that the jury isn't yet in. From my own perspective, I find it difficult to believe that the Dynamical Pauli Effect could show up so clearly in quantities like transition rates and masses, yet not show up in the spectrum for actinide nuclei; therefore, I consider the resolution of this matter to be one of the most important tasks facing the theory.

It appears that part of this resolution could come from better data in a region which requires special techniques and persistence on the part of the experimentalists. Thus we come to the topic of *exotic spectroscopy*.

29.7 EXOTIC SPECTROSCOPY?

I hope that I have at least made it plausible that the FDSM can, with only modest effort, describe a broad range of conventional nuclear properties. It then follows that the theory may be particularly well suited to the description of nuclei far from the beaten path, for in those cases it may be necessary to go beyond the conventional assumptions. The theory which is able to describe the conventional data with the simplest starting point is likely to fare best in this endeavor. Therefore, I will conclude this chapter with a brief discussion of the FDSM and "exotic spectroscopy." I will define this loosely, with *exotic* meaning "out of the ordinary" and spectroscopy meaning "the measurement of any properties of nuclear states." [The dictionary offers "outlandish" as a definition for "exotic"; that has almost, but not quite, the desired connotation.] Space limitations preclude more than a cursory listing of some of the ideas that have emerged from the FDSM; many of these are unpublished, and most have not yet been observed experimentally.

- 1) The FDSM mass formulas predict the occurrence of stable superheavy elements which may differ somewhat in particle number from standard predictions and which are bound by about 30% more than in standard predictions.
- 2) The accuracy of the FDSM mass formula bodes well for predictions of masses far from stability. We are investigating possible astrophysical implications for this in *r*-process production of heavy elements.
- 3) The FDSM predicts unusual effects in the collective motion associated with the Dynamical Pauli Effect for nuclei in the $Z \gg 100$ nculei. These would appear in the excited, vibrational-band spectrum and in quantities like electromagnetic transition rate, g factors, and so on [see Point (7) in Section 6].
- 4) The FDSM suggests that there may be interesting properties associated with the Coriolis Antipairing Effect (CAP) in a variety of nuclei: low-spin fluctuations in $B(E2)$ values, disappearance of the CAP at certain nucleon numbers, anomalous quenching of two-particle transfer strength at very low angular momentum, and previously unexpected angular momentum dependence of g factors in the spin region below $10\hbar$.
- 5) The FDSM suggests the possible existence of a new kind of nuclear state, the heritage isomer, which is an unusual, many-body configuration for which "breaking" pairs lowers the energy.
- 6) The FDSM suggests that an exotic kind of collective motion might develop in nuclei which have many like-particles outside a closed shell, but very weak n–p interaction. If the like-particle quadrupole-quadrupole effective interaction is weakly repulsive, as it may be, the $SU(3)$ representations could be inverted relative to their more normal order. The resulting collective motion would be quite different from normal rotation for the low-lying states.
- 7) The FDSM has been used to construct a model of superdeformation which seems to give a good account of the data and which may allow a unified algebraic theory of superdeformations, fission isomers, and giant resonances.

29.8 SUMMARY

In conclusion, I hope that this introduction to the Fermion Dynamical Symmetry Model has convinced you of at least one thing: Dynamical symmetries which emphasize fermion degrees of freedom are quite common in complicated many-body systems. This allows a fresh conceptual perspective on the nuclear structure problem; at the same time, the presence of the approximate fermion symmetries suggests new ways to make practical microscopic calculations in these systems. Although the emphasis here has been on normal spectroscopy, I have presented arguments that such an approach may be even more useful for exotic spectroscopy. This is a fitting place to end, for such was our original motivation for gathering in this Symposium.

ACKNOWLEDGMENTS

The research presented here has been done in collaboration with several people: I especially want to acknowledge the contribution of Cheng-Li Wu, at the University of Tennessee and Drexel University; Xiao-Ling Han and Zhen-Ping Li, at the University of Tennessee; and Da Hsuan Feng and Jin-Quan Chen, at Drexel University.

REFERENCES

1. C. L. Wu, D. H. Feng, X. G. Chen, J. Q. Chen, and M. W. Guidry, *Phys. Lett.* **168B**, 313 (1986).
2. C. L. Wu, D. H. Feng, X. G. Chen, J. Q. Chen, and M. W. Guidry, *Phys. Rev. C* **36**, 1157 (1987).
3. J. Q. Chen, X. G. Chen, D. H. Feng, C. L. Wu, J. N. Ginocchio, and M. W. Guidry, *Phys. Rev C* **40**, 2844 (1989).
4. J. Ginocchio, *Ann. Phys.* **126**, 234 (1980).
5. C. L. Wu, D. H. Feng, M. W. Guidry, and J. Q. Chen, "The Fermion Dynamical Symmetry Model," in preparation for *Advances in Nuclear Physics*.
6. H. Wu and M. Vallieres, *Phys. Rev. C* **39**, 1066 (1989).
7. C. L. Wu, X. L. Han, Z. P. Li, M. W. Guidry, and D. H. Feng, *Phys. Lett.* **194B**, 447 (1987).
8. P. Møller and J. R. Nix, *Atom. Nucl. Data Tables* **39**, 213 (1988).
9. D. H. Feng, C. L. Wu, M. W. Guidry, and Z. P. Li, *Phys. Lett.* **205B**, 156 (1988).
10. M. W. Guidry, C. L. Wu, Z. P. Li, D. H. Feng, and J. N. Ginocchio, *Phys. Lett.* **187B**, 210 (1987); and to be published.
11. H. Wu, D. H. Feng, C. L. Wu, Z. P. Li, and M. W. Guidry, *Phys. Lett.* **193B**, 163 (1987).
12. P. Halse, *Phys. Lett* **189B**, 119 (1987).
13. P. Halse, *Phys. Rev. C* **39**, 1104 (1989.
14. Z. P. Li, M. W. Guidry, C. L. Wu, and D. H. Feng, "The Origin of Deformation in the Nilsson Model and the Fermion Dynamical Symmetry Model," to be published.

30. α Particles as Probes of Nuclear Shape in the Rare Earths and Structure Effects on Proton Emission in the Mass-80 Region

D. G. Sarantites, N. G. Nicolis, V. Abenante,
Z. Majka, and T. M. Semkow
Washington University
St. Louis, MO 63130, USA

C. Baktash, J. R. Beene, G. Garcia-Bermudez,
M. L. Halbert, D. C. Hensley, N. R. Johnson, I.-Y. Lee,
F. K. McGowan, M. A. Riley, and A. Virtanen
Oak Ridge National Laboratory
Oak Ridge, TN 37830, USA

H. C. Griffin
University of Michigan
Ann Arbor, MI 48109, USA

ABSTRACT

Low-emission barriers and large subbarrier anisotropies in the α-particle decay, with respect to the spin direction, of Sn and rare-earth compound nuclei are examined in the light of recent calculations incorporating deformation. For the rare-earth systems deformation which increases with spin is necessary to explain the data. Energy spectra and angular correlations of evaporated protons from the $^{52}Cr(^{34}S, 2p2n)^{82}Sr$ reaction were measured in coincidence with discrete transitions. Large changes in the shape of the proton spectra were observed when high-spin states in different rotational bands are populated. These effects cannot be explained by phase space arguments that do not include explicitly structure in the deexcitation process. They are interpreted as resulting from near-yrast to near-yrast stretched proton emission, which preferentially populates the yrast band by subbarrier protons.

30.1 INTRODUCTION

The study of nuclear shapes at high angular momentum and excitation energy is a topic of current extensive theoretical and experimental interest in heavy-ion physics. It is well known that collective nuclei near the yrast line are deformed and their structure is well described by liquid-drop-Strutinsky cranked-shell-model calculations. A question of interest is the evolution of these shapes as the spin and the excitation energy (temperature) are increased. There is already considerable experimental evidence for the existence of superdeformed nuclei (β = 0.6) at high spins.[1] Theoretical calculations that explain these highly deformed shapes predict even higher deformations ("hyperdeformed" structures with $\beta \approx$ 0.9) for nuclei close to the fission stability limit.[2,3] Temperature-induced noncollective rotation in nuclei, as well as shape changes, have also been discussed in connection with

Exotic Nuclear Spectroscopy, Edited by W. C. McHarris
Plenum Press, New York, 1990

predictions of mean field theories.[4] A number of experimental studies have tried to explore the effect of high excitation and/or angular momentum degrees of freedom on the nuclear shapes. This is made, on one hand, by exploiting the γ-decay properties (for example, study of giant resonances built on excited states) of the deexciting compound nuclei.[5] On the other hand, extensive searches are being made to find signatures of shape effects in the charged-particle-decay properties of such systems.[6-12]

The motivation for light-charged-particle studies lies in the well-established fact that (fission-stable) compound nuclei, with the highest possible angular momentum, often decay by emitting α particles and protons. If the deexciting nucleus is deformed, it exhibits a lower evaporation barrier along the longer axis for charged-particle emission, compared to the spherical case. This results in strong enhancements of α and proton decay along the long axis, especially in the energy region below the evaporation coulomb barrier.[6,7,12] Simulation studies along these lines have motivated a number of experiments consisting of the observation of α-particle spectra in heavy-ion fusion-evaporation reactions in a singles mode or in coincidence with evaporation residues.[7,11,12] The inability to reproduce the subbarrier part of the observed α spectra with statistical-model calculations assuming spherical emission shapes has been used as an indication of a deformation effect.

The desire for enhancing the observed anisotropies for α-particle emission in order to study these effects led to development of the spin-alignment method with the Spin Spectrometer,[13] a highly-segmented 4π γ-ray detector system. In this method, the magnitude and orientation of the spin of the residual nuclei is deduced on an event-by-event basis. This makes possible detailed studies, such as the measurement of α-particle angular distributions with respect to the estimated spin direction.[14] Furthermore, the γ-multiplicity selection with the spin spectrometer allows us to study these decay characteristics as a function of the evaporation residue spin, which is closely correlated to the compound-nucleus spin. Therefore, the α-decay properties of different compound nuclear systems can be studied in detail.[8-10]

In the first part of this paper, we report on the results of an earlier survey study[10] concerning α-particle energy spectra and angular distributions with respect to the spin direction for a number of compound nuclear systems. The two typical cases of the closed-shell $^{114}Sn^*$ and the rare-earth $^{170}Yb^*$ deexciting compound nuclei are compared. The α-emission properties of these systems are described by the anisotropy coefficients of the α-particle emission with respect to the spin direction as a function of the α-particle energy and γ-ray multiplicity. Differences in the emission patterns in the energy region below the evaporation coulomb barrier suggest nearly spherical and deformed emission shapes in the cases of $^{114}Sn^*$ and $^{170}Yb^*$, respectively.

Detailed statistical-model calculations have been performed to clarify the deviations of the "deformed" versus the "spherical" behavior. In the sub-barrier region, the predictions of the statistical model for charged-particle emission are sensitive to barrier-penetration effects. These effects are expressed in terms of transmission coefficients resulting from an optical-model calculation. Although our statistical-model calculations described closely the decay of $^{114}Sn^*$, we observe discrepancies in the other system. These discrepancies, appear as *a*) an underestimate of the subbarrier 90° CM spectra, *b*) underestimates of the multiplicity-gated 90° CM spectra, which increase with spin, and *c*) deviations in the trend of the anisotropy coefficients from the one predicted in the $^{114}Sn^*$ case, which also increase with spin. We show that a simulation of deformation effects due to transmission accounts for these discrepancies. A detailed comparison with the $^{170}Yb^*$ data is made, which shows the angular-momentum dependence of the deformation effect in the α spectra. However, more elaborate calculations are required to describe the corresponding effect in the anisotropy coefficients of these systems. The above findings are corroborated by both the known ground-state properties and data of giant resonances built on excited states of similar compound-nuclear systems.

In the second part of this paper, we report on a correlation between the shape of the proton-emission spectrum and nuclear structure. Proton spectra from the ^{52}Cr(^{34}S, $2p2n$)^{82}Sr reaction at 130 MeV have been measured in coincidence with discrete γ transitions of selected exit channels. The proton spectra were observed with the Dwarf-Ball array,[15] a seventy-two unit CsI(Tl) 4π detector, in coincidence with eighteen compton-suppressed Ge detectors. This system was operated in conjunction with the Spin Spectrometer, which recorded the associated γ-ray multiplicity. We found significant changes and shifts in the proton-energy spectra as we selected gating transitions from different bands or transitions from states of different spin in the same band. Substantial differences were also seen as a function of the γ-ray multiplicity. The above results cannot be explained by phase-space arguments, as shown by statistical-model calculations that do not explicitly include nuclear structure. They can be explained by stretched proton emission from near-yrast to near-yrast states, which preferentially populate the yrast band by subbarrier proton emission.

30.2 α-EMISSION PROPERTIES AND DEFORMATION EFFECTS

The experiments in this work were performed at the Oak Ridge *H*olifield *H*eavy-*I*on *R*esearch *F*acility (HHIRF). The compound systems studied were ^{110}Sn*(E^* = 93.9 MeV), ^{114}Sn* (79.5 MeV), ^{138}Nd*, ^{164}Yb* (67.2 MeV), and ^{170}Yb* (134.8 MeV). The details of the experimental method and some of the general features of the data can be found in the current literature.[8-10,14] The complete study has been submitted for publication. Below, we present only the distinct features shown in the deexcitation data of ^{114}Sn* and ^{170}Yb*.

The compound nuclei ^{114}Sn* and ^{170}Yb* were produced in the reactions, ^{64}Ni (250 MeV) + ^{50}Ti $\rightarrow$ ^{114}Sn* and ^{20}Ne (176.6 MeV) + ^{150}Nd $\rightarrow$ ^{170}Yb*, respectively. In this study, self-supporting targets of high isotopic enrichment in each of the isotopes were used. The α particles emitted in the deexcitation of the above compound nuclei were recorded by Si surface barrier telescopes positioned at the laboratory angles corresponding to $\approx$90° in the center-of-mass system.

The ΔE detectors had thickness of 65 μm and an acceptance cone of $\approx$6° half angle. The E detectors were 1500-μm thick and served as the triggers of the spin spectrometer. The spin spectrometer served as the γ-ray detector and measured simultaneously the γ-ray multiplicity, M_γ, the total γ-ray deexcitation energy, and the γ-ray angular correlations. In these experiments the spin spectrometer provided a coverage of 95.8% of 4π sr.

For each compound nuclear system (A_{CN}, Z_{CN}), the α-particle events were transformed in the center-of-mass system for α + (A_{CN-4}, Z_{CN-2}), using two-body kinematics. The method used for determining the estimated spin direction is based on the emission of γ radiation with a particular angular relationship to the spin direction.[8,14] The γ cascades from rotational nuclei formed in heavy-ion fusion-evaporation reactions have a preponderence of stretched $E2$ transitions which exhibit a doughnut-like pattern about the spin axis. The spin direction is identified with the short symmetry axis of this pattern. This is close to the compound nucleus spin, i.e., perpendicular to the beam, provided that the misalignment caused by particle emission is small. The γ pattern for each event was projected on a plane perpendicular to the beam direction, and centroid-searching methods were used to determine the angle between the short symmetry axis and the direction of the emitted α particle.

In the following analysis, α-particle events corresponding to emission angles near θ_{CM} = 90° were sorted, imposing different γ-coincidence fold (k_γ gates) or angle with respect to spin (β gates). Some useful remarks can be made from an inspection of multiplicity-decomposed spectra. Figure 1*a*) shows for ^{114}Sn* the experimental θ_{CM} = 90° spectra corresponding to k_γ = 11–14 and 27–33. The closed squares on the bottom show the total 90° center-of-mass spectrum integrated over k_γ (for $k_\gamma \geq 11$) and β. The corresponding α-particle spectra for ^{170}Yb*, under the same gating conditions, are shown in Fig.1*b*). In both cases the selected k_γ bins correspond to α-particle emission from nuclei with an average spin of ≈34ℏ and ≈64ℏ, respectively, deduced from statistical model calculations.[8,10]

A common trend in both cases is that the k_γ = 11–14 spectra are slightly harder than the k_γ = 27–33 ones. This can be understood in terms of the higher excitation energy selection made by the low-k_γ gate. However, the subbarrier trends of the spectra are quite different. For ^{114}Sn*, the low-k_γ compared to the high-k_γ spectrum has an excess of subbarrier α particles, whereas these two regions in the ^{170}Yb* spectra are very similar. The solid lines in Fig. 1*a*) are the result of a statistical model calculation with standard parameters.[8,16] We see that there is a good agreement with the data in the whole energy range,

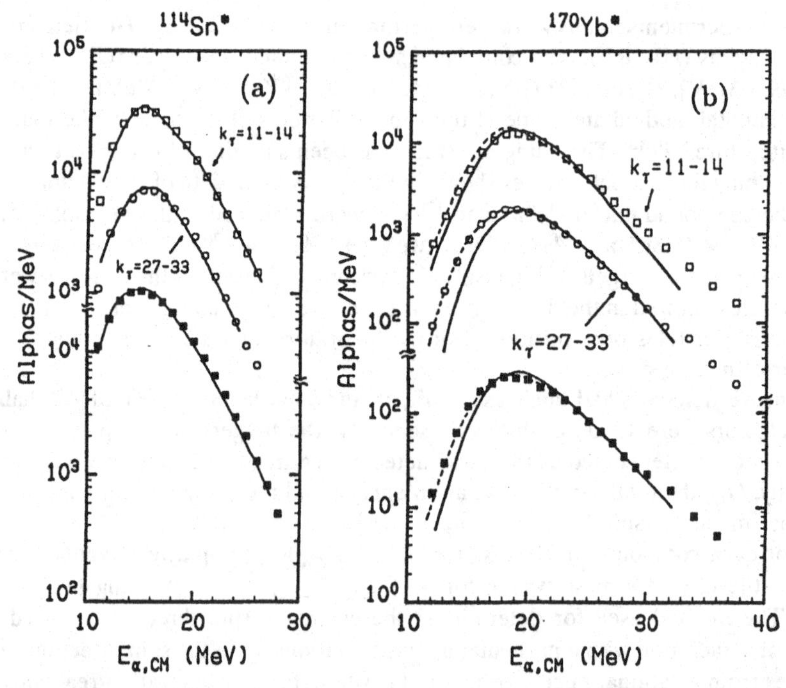

Figure 1. *a*) γ-fold gated 90° center-of-mass α-particle spectra from the deexcitation of ^{114}Sn*. The open squares correspond to k_γ = 11–14 and the open circles to k_γ = 27–33. The closed squares show the 90° center-of-mass spectrum integrated over k_γ and β. *b*) γ-fold gated 90° center-of-mass α-particle spectra from ^{170}Yb*. The squares correspond to k_γ = 11–14 and the circles to k_γ = 27–33. The closed squares show the 90° center-of-mass spectrum integrated over k_γ and β. The solid and dashed lines are the results of calculations described in the text.

for all of the gating conditions. The corresponding calculation for $^{170}Yb^*$ is shown on Fig. 1*b*) by the solid lines. The k_γ = 11–14 spectrum is underpredicted in the subbarrier region as well as at high energies. In the k_γ = 27–33 bin, the discrepancy is only in the subbarrier region and has increased. Similarly, the total-α spectrum is underpredicted at subbarrier energies. In summary, although the behavior of the α-particle spectra from $^{114}Sn^*$ is well described by statistical model calculations, we observe a systematic underprediction of the sub-barrier parts of the $^{170}Yb^*$ spectra in a manner which increases with spin.

A striking difference is observed in the trend of the anisotropy coefficients of α emission with respect to the spin direction. The anisotropy coefficients A_2 (of a Legendre-polynomial expansion) are plotted for the two systems as a function of $E_{\alpha,CM}$ for the indicated k_γ bins in Fig. 2. For $^{114}Sn^*$ we have monotonically decreasing A_2 coefficients (increasing anisotropies) with increasing E_α in each of the k_γ bins. In contrast with these findings, the A_2 coefficients for $^{170}Yb^*$ have maximum value (minimum anisotropy) at the evaporation coulomb barrier ($\approx$20 MeV for $^{170}Yb^*$ assumed spherical) and become more negative (larger anisotropies, with stronger emission perpendicular to the spin direction) at lower and higher E_α values. Compared with the almost linear decrease with E_α observed for $^{114}Sn^*$, we see a deviation in the trend of the experimental correlations below the cou-

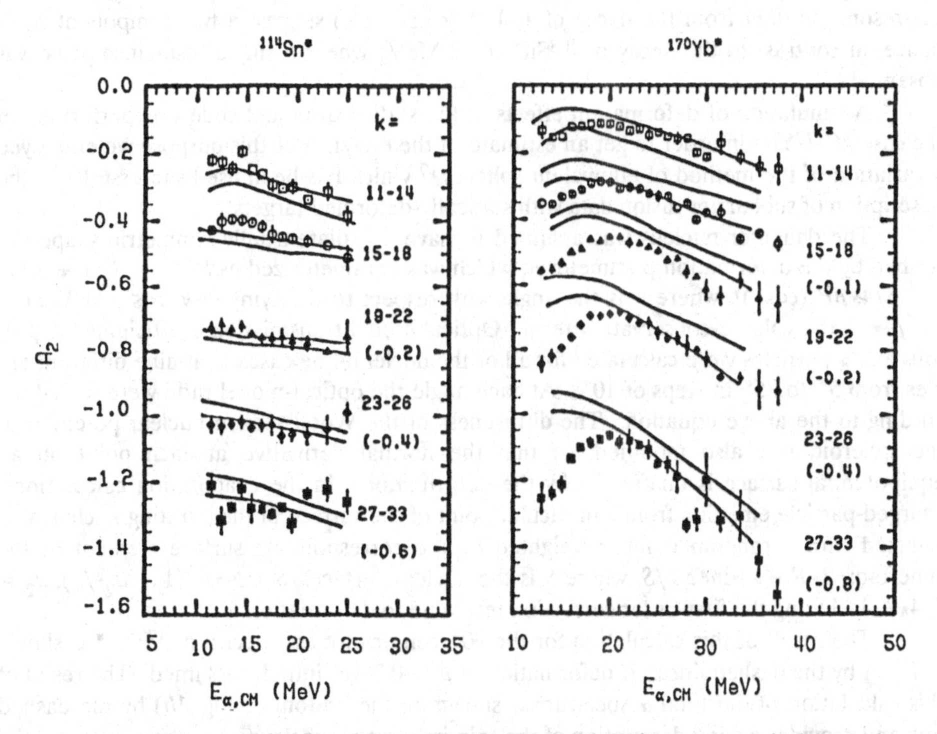

Figure 2. A_2 coefficients as a function of $E_{\alpha,CM}$ from the $^{114}Sn^*$ and $^{170}Yb^*$ systems. In both cases, the open squares, circles, closed triangles, diamonds and closed squares correspond to the k_γ bins of 11–14, 15–18, 19–22, 23–26 and 27–33, corresponding to $< I >_\alpha$ for α emission of 34, 43, 51, 59, and 64 $\hbar$, respectively. In some cases the data have been shifted along the A_2 axis by the indicated amount. The pairs of curves are FWHM boundaries of the A_2 coefficients from a statistical model calculation using transmission coefficients from a spherical-optical-model potential.

lomb barrier. The $^{170}Yb^*$ data suggest enhanced anisotropies, which become larger with decreasing E_α. This enhancement increases with increasing spin.

Both of the data sets are compared with the results of a statistical-model calculation with standard parameters in Fig. 2. For $^{114}Sn^*$ we see that the trend of the A_2 coefficients and their absolute magnitude are reproduced by the calculation.

The calculated A_2 coefficients for $^{170}Yb^*$ agree well with the monotonic decrease of the experimental A_2 values above the coulomb barrier but do not reproduce the decrease of A_2 at low E_α.

The fact that the observed deviation occurs at emission energies sensitive to barrier penetration effects has suggested the nuclear deformation as a possible factor for the decrease of the measured A_2 coefficients at low E_α. If the emitting system is deformed with its longest axis perpendicular to the spin direction, the subbarrier α particles will be emitted preferentially along this direction (because of the lower coulomb barrier). This leads to decreasing A_2 coefficients with decreasing E_α. On the other hand, α particles above the barrier would not be affected much by the deformation, since their emission is mainly determined by the level densities. Therefore, the observed deviation can be interpreted as a deformation effect which increases with spin. The same interpretation accounts also for the discrepancies observed in the multiplicity-gated spectra of Fig. 1.

An interesting comparison of the above systems has been made with data from the decay of giant resonances built on excited states of similar compound-nuclear systems. Giant-resonance data from the decay of $^{166}Er^*$ (61.5 MeV) suggest a two-component resonance, in contrast to the decay of $^{108}Sn^*$ (61.2 MeV) where a single resonance peak was observed.[5]

A simulation of deformation effects in the statistical-model code was performed in the case of $^{170}Yb^*$ in order to get an estimate of the effect. For this purpose we employed a variation of the method of equivalent spheres,[17] which has been used successfully in the description of subbarrier fusion data with statically-deformed targets.

The daughter nucleus was assumed to have a prolate-axially-symmetric shape described by the deformation parameter β, which was parameterized as $R(\theta) = R_0[1 + X(\beta) + \sqrt{5/4\pi}\beta P_2(\cos\theta)]$, where θ is the angle with respect to the symmetry axis and $X(\beta) = -\beta^2/4\pi$ is the volume-conservation term. Optical-model transmission coefficients for protons and α particles were calculated for all of the nuclei in the cascade at nine different angles from 5° to 85° in steps of 10°. At each angle the optical-model radii were scaled according to the above equation. The diffuseness of the Woods-Saxon nuclear potential of the spheroid was also modified, so that the normal derivative at each point on an equipotential surface is unaffected by the deformation. In the evaporation calculations, charged-particle emission from a particular point of the surface of the emitting nucleus was selected with a random number weighted by the corresponding surface element of the spheroid: $2\pi R^2(\theta)\sin\theta\Delta\theta/S$, where S is the nuclear surface: $S = 4\pi R^2(1 + a_2^2/5)$, $a_2 = \sqrt{5/4\pi}\beta$, including the first-order correction term due to deformation.

The result of this calculation for the 90° center-of-mass spectra of $^{170}Yb^*$ is shown in Fig. 1 by the dashed lines. A deformation of $\beta = 0.2$ was initially assumed. The result of this calculation of the total α spectrum is shown on the bottom of Fig. 1*b*) by the dashed line and provides a good description of the spin integrated spectrum.

On the same plot the dashed line for the $k_\gamma = 11$–14 bin ($\langle I \rangle_\alpha = 34\hbar$), shows the calculated spectrum with $\beta = 0.2$. The subbarrier data points lie between the curves, $\beta = 0$ and 0.2. In this case, a deformation somewhat smaller than $\beta = 0.2$ is required to fit the spectrum. On the other hand, for the high spin case $k_\gamma = 27$–33 ($\langle I \rangle_\alpha = 64\hbar$), $\beta = 0.2$

was insufficient to account for the excess of subbarrier α particles. The dashed curve in the figure corresponds to $\beta = 0.35$ and fits the spectrum closely. These calculations demonstrate the extent and the angular-momentum dependence of the effect. The originally-deduced $\beta = 0.2$ was based on the total 90° center-of-mass spectrum and represents an average over different deformed shapes.

The calculated A_2 coefficients, using the above logic, show a trend approaching the experimental data at subbarrier energies. However, no quantitative statement was made from such comparisons, because our present formalism for the calculation of the angular correlations is limited to spherically-symmetric emitters. More refined calculations are required for this purpose.[20]

Summarizing, the distinct differences in the α-decay properties of the compound nuclei ^{114}Sn and ^{170}Yb have been interpreted in terms of deformation effects. For ^{170}Yb* our calculations have demonstrated the angular momentum dependence of the effect in the 90° center-of-mass spectra, besides the one observed in the trend of the anisotropy coefficients of α emission with respect to the spin direction. One should keep in mind that the α-particle-emission probe for studying nuclear shapes at high (E^*, I) involves a broad range of initial excitation energies which contribute in low-energy particle emission.[10]

30.3 STRUCTURE EFFECTS AND PROTON-EVAPORATION SPECTRA

Compound nuclei that decay to residual nuclei with large deformations, such as superdeformed nuclei with discrete-level structures, may be expected themselves to have significant deformations, which persist to sufficiently high excitations of the order of at least one nucleon binding energy above the yrast line. In this case, particle emission can be significantly influenced by nuclear-shape effects. Furthermore, it is quite possible that structure effects can be observed on the shape of the charged-particle-evaporation spectra particularly near and below the emission barrier. The experimental observation of the existence of such effects is important both on its own merit and also for providing important information about methods of populating the so called "hyperdeformed" structures in nuclei. In particular, proton emission may be more important than α emission, since protons do not remove as much angular momentum and can probe deformation with minimal perturbation of the emitting system.

In this work we address the issue of structural influences on the shapes of the proton spectra. We have selected for study ^{82}Sr as the final nucleus, because it has been predicted to be a good candidate for superdeformation.[18] We have observed a strong dependence of the probability for subbarrier proton emission on the nature of the rotational bands being populated in the final nucleus.

The experiment was performed at HHIRF. The ^{82}Sr nuclei were produced by the ^{52}Cr(^{34}S, $2p2n$)^{82}Sr reaction by bombarding a stack of self-supporting ^{52}Cr target foils with a 130-MeV ^{34}S beam. The experimental setup consisted of the ORNL Compton-Suppression Spectrometer, having eighteen Ge detectors which recorded the discrete γ-ray spectra from the reaction. The associated total γ-ray multiplicity and total energy were recorded with the Spin Spectrometer. The protons and α particles were detected with the 4π CsI(Tl) Dwarf Ball.[15] This system provided both high-resolution γ-ray spectroscopic information and definite exit-channel selection. The seventy-two-element Dwarf Ball also provided light-charged-particle spectra and angular-correlation information. The apparatus was triggered by two and more Ge detectors firing in coincidence with any element of the Dwarf Ball. A total of 1.6 x 10^8 such events were collected and processed.

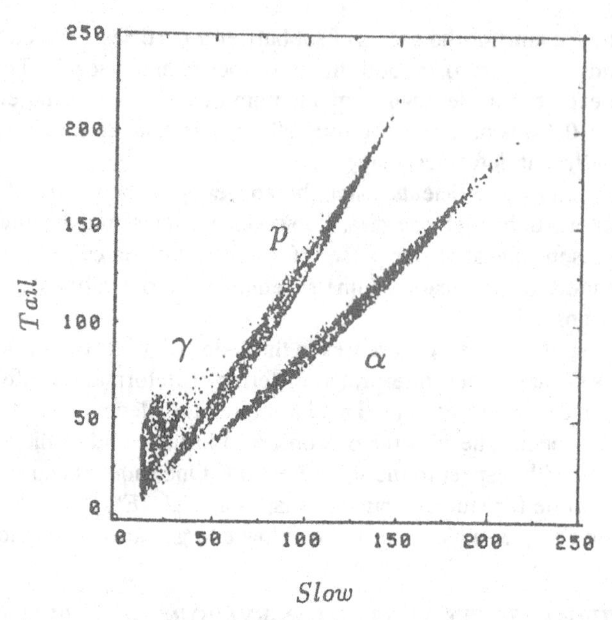

Figure 3. Scatter plot of the slow-vs-tail light output of a CsI(Tl) detector in the Dwarf Ball showing the α, proton, and γ-ray identification. This was achieved by placing two time gates, 400 ns (slow) and 1500 ns (tail) wide, starting at times 0 and 1500 ns from the front of each pulse, respectively, and integrating the corresponding charge.

Figure 3 shows a scatter plot of the slow-vs-tail map from a 42° CsI(Tl) detector. The γ rays, protons, and α particles are well separated.

Excellent separation for all energies was achieved between proton and α pulses from each other and from γ rays or neutrons for the detectors forward of 102° in the laboratory (≈120° center-of-mass). Protons with sub-barrier energies and yields ≈1/20 of that at the most probable value could be clearly identified. Proton spectra sorted in this way contained ≈65% of the total proton yield. At larger angles, because of kinematic forward focussing, some of the subbarrier protons cannot be distinguished from α particles, and therefore the detectors at these angles were used for channel selection but not for particle spectra. The measured overall detection efficiency for protons was 85% of 4π (sixty-seven detectors out of seventy-two were used). This resulted in ≈28% of the events with two protons to be identified as involving only one proton. The Ge energy spectra coincident with two protons involved primarily ^{82}Sr γ rays with no contamination from the 1*pxn* or the αxn emitting channels. In the most forward detectors the counting rate was ≈7000 c/s. A close examination of proton and α-particle spectra recorded on different tapes revealed gain shifts up to 8% in the tail component, while the slow component was considerably more stable. Gain shift corrections were applied to the tail component for every 2 x 10^6 events. This ensured good particle identification for the complete data set. The more-stable slow component was energy calibrated using the ^{12}C(p, p') reaction at 9.0 and 20.0 MeV. The energy-calibration uncertainties were estimated to be less than 3% for all but the 24° detectors, for which the uncertainty may be as large as 6%.

The particle energies were converted event by event to the center-of-mass system. The centroid angles for each detector in the laboratory system were used in deriving the center-of-mass energy and angle, assuming two-body kinematics. Proton energy spectra were sorted using the detectors forward of 102°, when two protons were identified in the complete Dwarf Ball and for three k_γ gates of 3–9, 10–14, and 15–25. Further selection was made by placing gates on discrete γ rays associated with various rotational bands. The background to the proton spectra associated with the underlying compton contribution in the Ge peaks were subtracted by placing equal width gates near each γ peak. Care was exercised to avoid peaks that are known to be doublets. Thus, for the four main bands, gates placed on single γ-ray peaks for transitions up to spin 10 gave spectra of good statistical quality. For higher spins, the proton spectra for two or three transitions were added to provide spectra of improved statistical quality, but in each case the spectra for each gate were examined for consistency.

Angular correlations of the coincident protons were recorded at 24°, 42°, 50°, 63°, 68°, and 78° in the laboratory, corresponding to angles ranging from ≈30° to 95° in the center of mass.

The level scheme for ^{82}Sr was constructed from a γ-γ matrix obtained by requiring that at least one proton was detected and that the γ-ray multiplicity exceeded 10. The matrix thus constructed was dominated by γ rays from ^{82}Sr. Figure 4 shows a partial level scheme for ^{82}Sr constructed from these data. Two new bands were established and four previously-known bands were extended from 20 to 27 $\hbar$. The even-parity Band 4 is yrast for spins between 10 and 22 $\hbar$. The odd-spin Band 2 becomes yrast for $I \geqslant 23$.

Proton spectra coincident with the $2^+ \rightarrow 0^+$ ground transition are shown in Fig. 5 for the three k_γ gates. For purposes of comparison, the spectra are shown normalized to the same total counts. For the higher k_γ gates, the spectra shift to lower energies and become narrower, but the higher-energy slopes (above ≈15 MeV) for all k_γ gates are similar. This is understood in terms of the decreasing available thermal energy as k_γ or, equivalently, the spins of the entry region are increased (the three k_γ gates correspond to $I \approx$ 4–19, 17–29, and 26–45 $\hbar$ and to average yrast energies of ≈7, 14, and 23 MeV, respectively). A statistical-model calculation with the evaporation code PACE[16] for this system reproduces these features at least qualitatively. In contrast to this the proton spectra associated with the 2^+ level of Band 4 (non-yrast at this spin, but yrast at spins 12 to 22 $\hbar$) show similar peak positions for the three k_γ gates (Fig. 5*b*). Again, as k_γ is increased the spectra also become narrower, but at higher energies the slopes are similar. These differences between the spectra coincident with two different 2^+ states suggest that some structural effects and/or feeding patterns may be responsible.

In order to explore any dependence of the proton-emission probability on the structure of the final nucleus, we have compared in Fig. 6*a*) the proton spectra coincident with the γ rays from the 2^+, 4^+, and 6^+ states in the ground Band 3 and for the high-k_γ gate. Clearly, the spectra are very similar. This is understood from the fact that the 2^+, 4^+, and, for the most part, 6^+ levels receive practically all their feeding from all the bands in ^{82}Sr.

In contrast, the 8^+ level in the same Band 3 is fed from higher-spin members of the ground band and to a significant fraction by the yrast Band 4. Levels above 8^+, however, are only populated by the decay of the levels of the continuation of the ground band. The proton spectra coincident with the 8^+ and the sum of 10^+ and 14^+ levels are compared in Fig. 6*b* with the 2^+ spectrum. It is seen that the spectra for the 8^+ and (10^+ + 14^+) levels have most probable values that shift toward lower and higher energies, respectively. These unexpected shifts are consistent with the behavior of the spectra associated with the high spin states in the four major bands in ^{82}Sr.

If we further examine the proton spectra from individual levels within each band

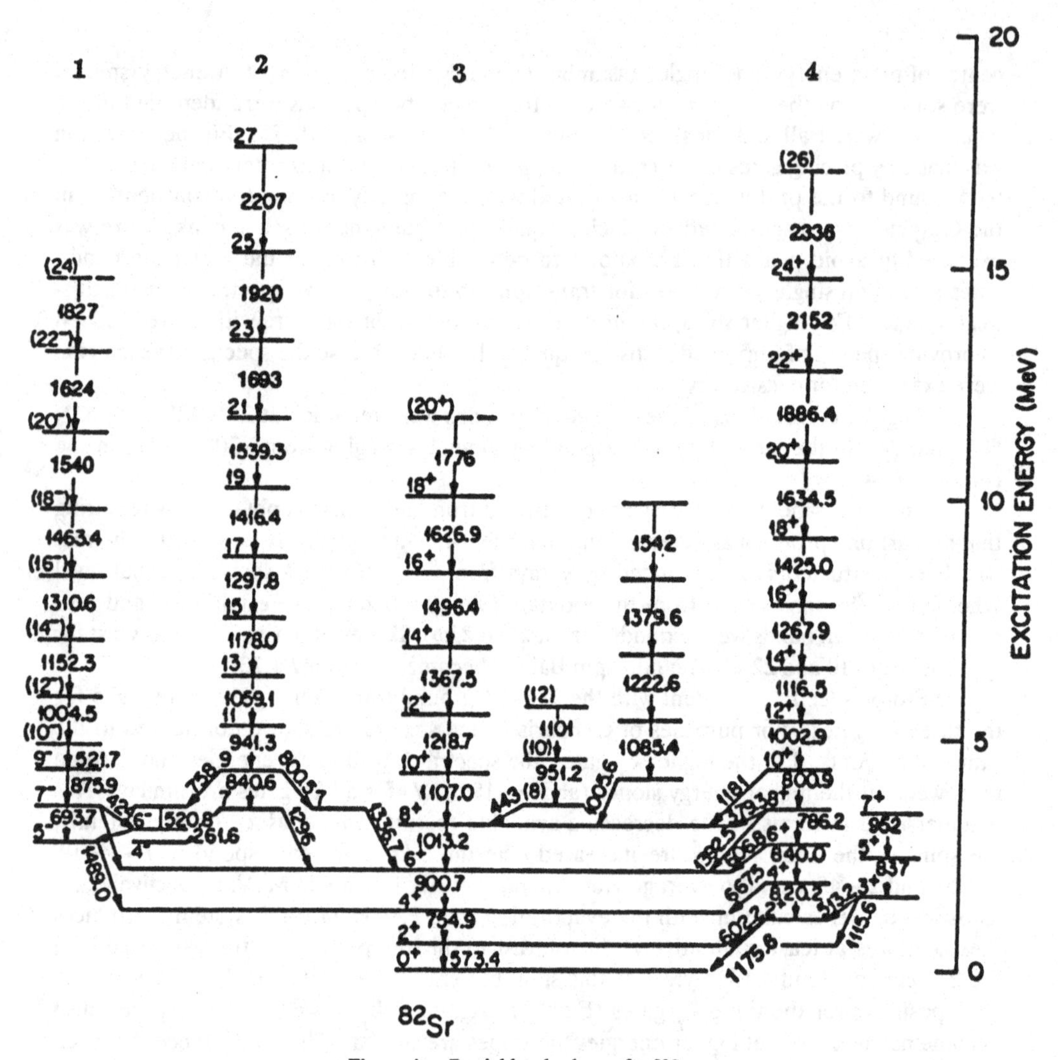

Figure 4. Partial level scheme for ^{82}Sr.

having spins higher than 10ħ, we find them to be similar to each other, but they differ considerably if they are associated with different bands. Thus, the proton spectra associated with transitions from the levels with spin 11, 13, and (19 + 23) in the odd-spin Band 2 of ^{82}Sr were found to be essentially identical in shape.

In contrast, when we compare proton spectra associated with high-spin levels ($\geqslant 10\hbar$) at comparable excitation energies from different Bands, large differences are found for all three k_γ gates. this is clearly seen in Fig. 7, where proton spectra associated with the sum of the 10^+ and 14^+ levels in the ground Band 3 (dashed line), the I = 13 level of Band 2 (thin line), and the sum of the 14^+ and 16^+ levels of the yrast Band 4 (thick line) are shown for the high-k_γ gate. The striking feature in these spectra is that for the same k_γ gate, the peak in the proton spectrum shifts down by 1 MeV in going from Band 3 to Band 2 and

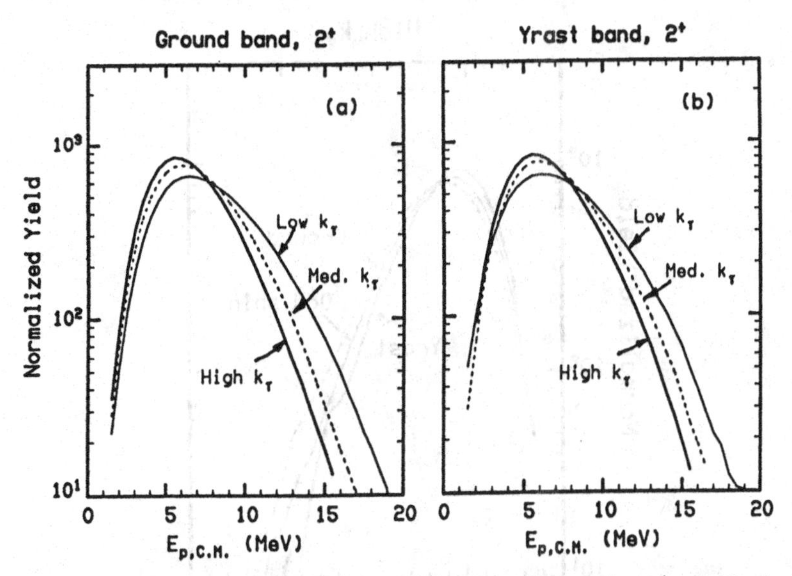

Figure 5. Panel *a*) shows the center-of-mass proton spectra coincident with $2^+ \rightarrow 0^+$ transition in the ground band (Band 3) for the three k_γ gates. The spectra are normalized to the same total counts for comparison. It is seen that the proton-emission barrier decreases as k_γ increases. Panel *b*) shows the center-of-mass proton spectra coincident with the 2^+ level in Band 4 (see Fig. 4) for the three k_γ gates. Here the proton-emission barrier is essentially independent of the k_γ gating, although the high-energy yield decreases as k_γ increases.

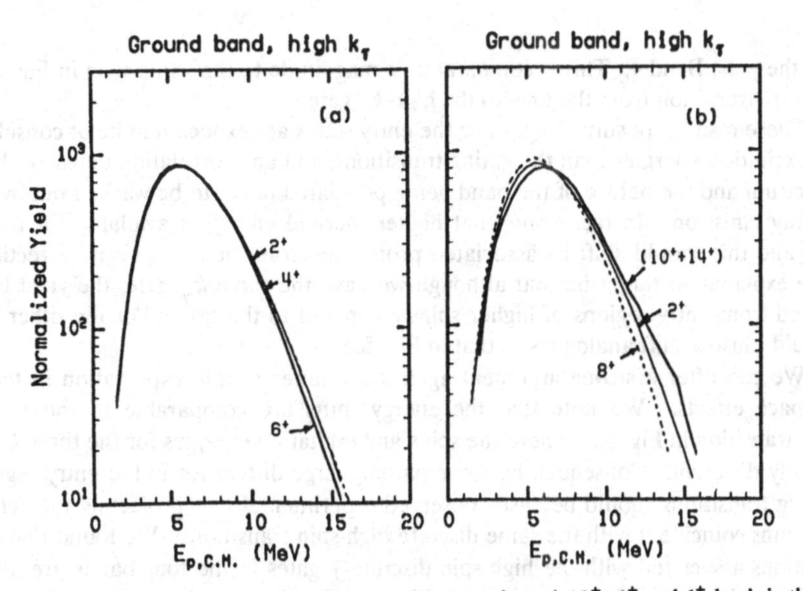

Figure 6. Panel *a*) shows proton spectra coincident with γ-rays from the 2^+, 4^+ and 6^+ levels in the ground Band 3 for the high-k_γ gate. The spectra are normalized to the same total counts for ease of comparison. The spectra are remarkably similar. Panel *b*) shows proton spectra coincident with the 8^+ and the sum of $(10^+ + 14^+)$ levels in the same ground Band 3 and for the high-k_γ gate. The 2^+ spectrum is also shown for comparison. Significant differences are seen (see text).

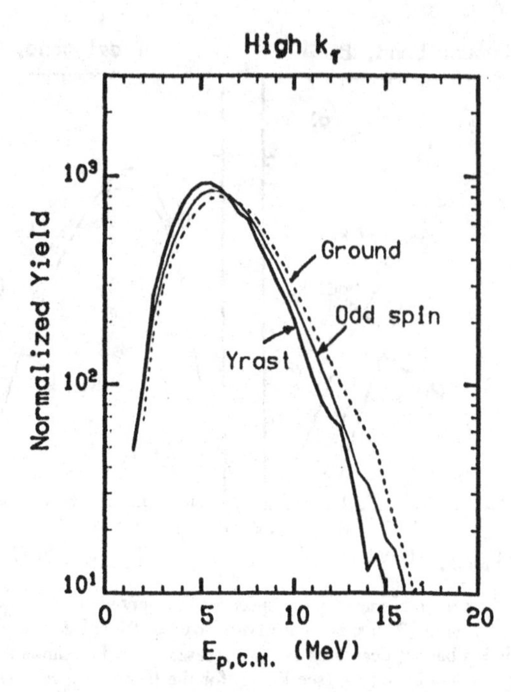

Figure 7. Proton spectra from the high-k_γ gate coincident with transitions from the $(10^+ + 14^+)$ levels of the ground Band 3 (dashed line), from the $I = 13$ level of the odd-spin Band 2 (thin line), and from the $(14^+ + 16^+)$ levels in the yrast Band 4 (thick line). Shifts as large as 1 MeV are seen in going from the ground to the yrast band gates.

then to the yrast Band 4. This is comparable in magnitude to the shifts seen in Fig. 5*a*) for the $2^+ \rightarrow 0^+$ transition from the low- to the high-k_γ gate.

These results are surprising, since the entry states are expected to lie at considerably higher excitation energies than the gating transitions, and any correlation between the proton spectrum and the nature of the band being populated ought to be washed out by statistical γ-ray emission. In fact, somewhat higher thermal energy is available for the yrast Band 4 and this should shift its associated proton spectrum in the opposite direction. A possible explanation might be that although we have the same k_γ gate, the yrast band is populated from entry regions of higher spins compared to the ground or the other bands. This could cause a shift analogous to that in Fig. 5*a*).

We can offer a strong argument against the latter simple explanation in terms of phase-space effects. We note that the energy shifts are comparable to those for the $2^+ \rightarrow 0^+$ transition in Fig. 5*a*), where the spins and excitation energies for the three k_γ gates are greatly different. Consequently, a comparably large difference in the entry regions of the gating transitions should be easily observed experimentally by projecting the actual k_γ distributions coincident with the same discrete high-spin transitions. We found that the k_γ distributions associated with the high-spin discrete-γ gates in the four bands are identical (the average k_γ values for the ground, odd spin, and the yrast bands were found to be 15.5 ± 0.2, 15.5 ± 0.2, and 15.4 ± 0.2, respectively). In addition, a statistical-model calculation in which the high-multiplicity gate was moved up by one unit produced a considerably smaller shift in the proton spectrum than observed in Fig. 7.

Another unlikely possibility would be that through fractionation the ground band samples the high total-γ-energy part of the entry region, while the yrast band samples the lower one. This would give the yrast and ground bands less and more thermal energy, respectively. According to a statistical-model calculation, this does not affect the peak and subbarrier portions of the proton spectra but shifts the high-energy part in the opposite direction than observed.

A reasonable explanation for the observed large shifts in the proton spectra is suggested by the steep experimental yrast line and consequently the entry line in ^{82}Sr and the energy balance in this reaction, which places the entry line only a few MeV above the yrast line. This is also confirmed by statistical-model calculations. These observations and the data suggest that a significant fraction of the population of the rotational bands in ^{82}Sr and, in particular the yrast band, occurs by direct proton emission from near-yrast to near-yrast states with the second proton playing that role. This mechanism could explain many features of the data and would in addition require that the last proton emission be stretched in character, particularly at and below the emission barrier, where it should exhibit a strong anisotropy, $R(0°/90°) \equiv W(0°)/W(90°)$, with respect to the beam direction. This is indeed what we observe in Fig. 8, where the anisotropy ratios for the high-spin gates of Fig. 7 of the ground and the yrast bands are plotted relative to the anisotropy of the $2^+ \rightarrow 0^+$ ground transition as a function of the proton center-of-mass energy for the high-k_γ gate. This further supports this explanation, since the direct population of the ground band has a higher coulomb barrier compared to that for the population of the yrast band.

Proton emission from the near-yrast states may be associated with a predicted mechanism based on instability toward nucleon emission at large spins, connected with the

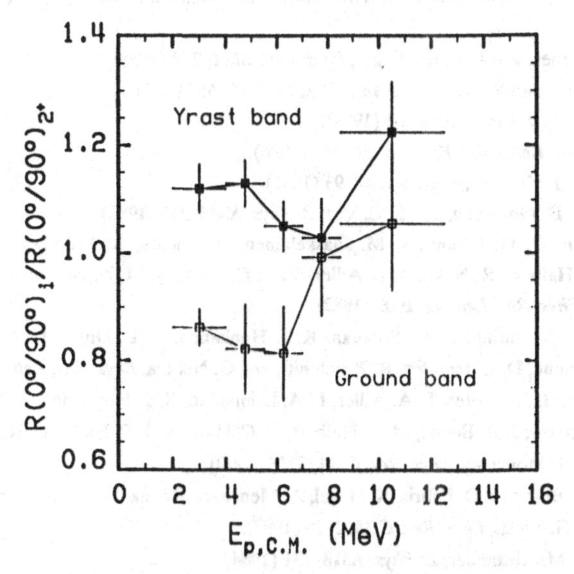

Figure 8. Anisotropy ratios for the protons from high-k_γ gate coincident with the high spin transitions (Fig. 7) in the yrast Band 4 (full squares) and the ground Band 3 (open squares), relative to that for the $2^+ \rightarrow 0^+$ transition to the ground state. The horizontal error bars give the widths of the proton-energy gates used. These were chosen to give comparable statistics in the correlations. Note that a ratio smaller than unity means that the anisotropy is smaller than that for the ground $2^+ \rightarrow 0^+$ transition.

population of $h_{11/2}$ resonance states in nuclei in this region.[19] These predicted yrast proton transitions were expected to have enhanced anisotropies as observed in our work.

In summary, we have observed a strong dependence of the proton-energy spectra on the nature of the final high-spin states belonging to different rotational bands in ^{82}Sr, which cannot be accounted for in terms of phase-space effects. The large shifts toward lower energies and the stronger anisotropies when subbarrier protons lead to high spin states in the yrast band compared to the ground band are interpreted as resulting from near-yrast emission of high-ℓ protons. These results suggest that spectra and angular distributions of subbarrier protons may provide a sensitive probe of the structure of excited, highly spinning nuclei.

ACKNOWLEDGMENTS

The authors acknowledge fruitful discussions with T. Dossing. The assistance of X. T. Liu during the data acquisition for the last experiment is appreciated. M. A. R. and A. V. acknowledge support from the Joint Institute for Heavy Ion Research. This work was supported in Part by the U. S. Department of Energy under grant No. DE-FG02-88ER-40406. Oak Ridge National Laboratory is operated by Martin Marietta Energy Systems, Inc., under Contract No. DE-AC05-84OR21400 with the U. S. Department of Energy.

REFERENCES

1. P. J. Twin, B. M. Nyako, A. H. Nelson, J. Simpson, M. A. Bentley, H. W. Cranmer-Gordon, P. D. Forsyth, D. Howe, A. R. Mokhtar, J. D. Morrison, J. F. Sharpey-Schafer, and G. Sletten, *Phys. Rev. Lett.* **57**, 811 (1986); P. J. Nolan and P. J. Twin, *Ann. Rev. Nucl. Part. Sci.* **38**, 533 (1988), and references therein.
2. J. Dudek, T. Werner, and L. L. Riedinger, *Phys. Lett.* **B211**, 252 (1988).
3. S. Cohen, F. Plasil, and W. Swiatecki, *Ann. Phy., N. Y.* **82**, 557 (1974).
4. L. A. Goodman, *Phys. Rev. C* **37**, 2162 (1988).
5. K. A. Snover, *Ann. Rev. Nucl. Part. Sci.* **36**, 545 (1986).
6. M. Blann and T. T. Komoto, *Phys. Scr.* **24**, 93 (1981).
7. J. M. Alexander, D. Guerreau, and L. C. Vaz, *Z. Phys.* **A305**, 313 (1982).
8. F. A. Dilmanian, D. G. Sarantites, M. Jaaskelainen, H. Puchta, R. Woodward, J. R. Beene, D. C. Hensley, M. L. Halbert, R. Novotny, L. Adler, R. K. Choudhury, M. N. Namboodiri, R. P. Schmitt, and J. B. Natowitz, *Phys. Rev. Lett.* **49**, 1909 (1982).
9. Z. Majka, D. G. Sarantites, L. G. Sobotka, K. J. Honkanen, E. L. Dines, L. A. Adler, Ze Li, M. L. Halbert, J. R. Beene, D. C. Hensley, R. P. Schmitt, and G. Nebbia, *Phys. Rev. Lett.* **59**, 322 (1987).
10. N. G. Nicolis, D. G. Sarantites, L. A. Adler, F. A. Dilmanian, K. J. Honkanen, Z. Majka, L. G. Sobotka, Z. Li, T. M. Semkow, J. R. Beene, M. L. Halbert, D. C. Hensley, J. B. Natowitz, R. P. Schmitt, D. Fabris, G. Nebbia, and G. Mouchaty, *Phys. Rev. C* **41**, 2118 (1990).
11. Z. Majka, M. E. Brandan, D. Fabris, K. Hagel, A. Menchaca-Rocha, J. B. Natowitz, G. Nebbia, G. Prete, B. Sterling, and G. Viesti, *Phys. Rev. C* **35**, 2125 (1987).
12. L. C. Vaz and J. M. Alexander, *Z. Phys.* **A318**, 231 (1984).
13. M. Jaaskelainen, D. G. Sarantites, R. Woodward, F. A. Dilmanian, J. T. Hood, R. Jaaskelainen, D. C. Hensley, M. L. Halbert, and J. H. Barker, *Nucl. Instr. Meth.* **204**, 385 (1983)
14. K. J. Honkanen, F. A. Dilmanian, D. G. Sarantites, and S. P. Sorensen, *Nucl. Instr. Meth.* **257**, 233 (1987).
15. A CsI(T1) version of the Dwarf Ball system described in: D. G. Sarantites, L. G. Sobotka, T. M. Semkow, V. Abenante, J. Elson, J. T. Hood, Z. Li, N. G. Nicolis, D. W. Stracener, J. Valdes, and D. C.

Hensley., *Nucl. Instr. Meth.* **A264**, 319 (1987).

16. A. Gavron, *Phys. Rev. C* **21**, 230 (1980); modification PACE2S by J. R. Beene.
17. R. G. Stokstad and E. E. Gross, *Phys. Rev. C* **23**, 281 (1981).
18. W. Nazarewicz et al., *Nucl. Phys.* **A345**, 397 (1985), and private communication.
19. T. Dossing, S. Frauendorf, and H. Schulz, *Nucl. Phys.* **A287**, 137 (1977).
20. V. P. Aleshin, *J. Phys. G: Nucl. Part. Phys.* **16**, 853 (1990).

31. Fast β Transitions and Intrinsic Structures in Deformed Nuclei

P. C. Sood
Physics Department
Banaras Hindu University
Varanasi 221005, India

R. K. Sheline
Florida State University
Tallahassee, FL 32306, USA

ABSTRACT

Selection rules for allowed β transitions in deformed nuclei are re-examined. All presently known one hundred twenty-two allowed transitions with log*ft* values $\leqslant$ 5.2 are ascribed to a single particle spin-flip transformation between an $h_{11/2}$ proton and an $h_{9/2}$ neutron. In the asymptotic-quantum-number $[Nn_3\Lambda\Sigma]$ representation the underlying transition in all cases is $p[5(5-\Lambda)\Lambda\uparrow] \longleftrightarrow n[(5-\Lambda)\ \Lambda\downarrow]$ with $\Lambda = 2 - 5$. Multi-particle states are populated in decay modes wherein the transforming nucleon comes from breaking a $[5(5-\Lambda)\Lambda]^2$ pair; in such cases the unpaired nucleon(s) configuration is unaltered. This consideration provides a "strong" rule for configuration assignments to isomers in odd-odd nuclei, one- and three-quasiparticle states in odd-mass nuclei, and the high-lying two- and four-quasiparticle states in even-even nuclei.

31.1 INTRODUCTION

While checking the validity of the Alaga selection rules[1] for β decays in odd-mass deformed nuclei, Mottelson and Nilsson[2] observed that the allowed spin-flip transitions, not bringing in any change in the asymptotic quantum numbers $[Nn_3\Lambda]$, form a distinct category of fast β decays with log*ft* < 5. Such transitions are usually referred to as allowed unhindered (*au*). The role of these transitions in the identification of single- as well as multi-particle configurations has been discussed in a number of reviews.[3-8] In particular, Bunker and Reich[4] pointed out that only two orbital pairs; namely, [523] and [514], give rise to *au* transitions in odd-mass rare-earth nuclei and that these orbitals occur as low-lying states in entirely different mass regions; they used this observation to emphasize the "strong rule" for unambiguous configuration assignments to states connected by *au* transitions. Soloviev[8] extended the applicability of the "strong rule" to include three-quasiparticle (3*qp*) states populated in odd-mass decays and two- and four-quasiparticle (2*qp* and 4*qp*) states populated in even-mass decays. Our recent survey[6] of the *au* decays involving 2*qp* states lists configuration assignments for forty-two 2*qp* states in the odd-odd and thirty-two connected 2*qp* states in the even-even nuclei of the rare earth region. Examination of even-mass decays has further revealed[6,9] the appearance of two more orbital pairs; namely, [532] and [505], connected by *au* transitions, plus certain other features not observed in the odd-mass decays. No *au* decays have been identified in the nuclei of the actinide region.[10]

Fujita et al.[11] had pointed out another category of allowed decays which are only slightly retarded in comparison with the *au* decays, but faster than other allowed-hindered (*ah*) decays. In the present paper we review the current situation in respect of these non-*au* fast decays.

Another domain of related fast decays in proton-rich nuclei beyond the $^{146}Gd_{82}$ doubly-magic structure has been extensively investigated in recent years.[7] These decays of odd-mass, even-even, and odd-odd spherical nuclei with $N \leqslant 86$ have log*ft* $\leqslant$ 5.0 and proceed through the underlying single-particle spin-flip transition of an $h_{11/2}$ proton to an $h_{9/2}$ neutron. Further, the $(2j + 1)$ degeneracy of the shell-model orbits in these spherical nuclei results in possibly the only known instances of β decays from the aligned 3*qp* parent state.[12] We discuss these and other specific features of *au* decays of these spherical nuclei.

In Section 2 we briefly present an outline of the basic processes. A comparison of odd-mass and even-mass *au* decays is presented in Section 3, followed, in Section 4, by a discussion of the new features appearing in even-mass decays. Decays to 3*qp* and 4*qp* states in deformed nuclei are discussed in Section 5. The non-*au* fast decays suggested by Fujita et al.[11] are examined in Section 6, followed by a discussion of *au* decays of spherical nuclei with $65 \leqslant Z \leqslant 70$ and $82 \leqslant N \leqslant 86$ in the final section. The source references for the experimental data, if not given herein, may be found in our recent review paper,[7] and/or in the latest *Nuclear Data Sheets*.

31.2 BASIC PROCESSES

The selection rule for allowed unhindered (*au*) β transitions in deformed nuclei is stated in terms of the single-particle Nilsson-model asymptotic quantum numbers $\Omega\ [Nn_3\Lambda]$ as follows:

$$\Delta\Omega = 1; \Delta N = \Delta n_3 = \Delta\Omega = 0 \tag{1}$$

The transition is basically a spin-flip exchange,

$$\Omega_p\,(= \Lambda + 1/2) \longleftrightarrow \Omega_n\,(= \Lambda - 1/2), \tag{2}$$

with all other quantum numbers $[Nn_3\Lambda]$ unchanged. In the rare-earth region, these orbitals satisfy the conditions,

$$N = 5:\ (n_3 + \Lambda) = 5; \tag{3}$$

$$[Nn_3\Lambda\,] = [5(5 - \Lambda)\Lambda]. \tag{4}$$

The orbital pairs with $\Lambda = 2$ and $\Lambda = 3$; namely, [523] and 514], were identified with the *au* transitions quite early.[2-5] As explicitly stated by Bunker and Reich,[4] an observation of an allowed transition with log*ft* $\leqslant$ 5 unambiguously points to the [523] or [514] component in the β connected states, thus providing the basis for the strong rule for the configuration assignments for *au* decays. Our recent surveys[6,7,9] of even-mass decays found evidence for *au* transitions involving the [532] orbital pair in the lighter rare earths and the [505] orbital

Table I. Allowed Unhindered (Spin-Flip) β Decays Involving the [532] Transition in the Lighter Rare Earths and the [505] Transition in A ~190 Nuclei

INITIAL STATE			Transforming Orbitals	$\beta^{\pm}$	logft	FINAL STATE			Spectator Orbital
^{A}X	E(keV)	I^{π}				^{A}X	E(keV)	I^{π}	
(a) $p5/2[532] \leftrightarrow n3/2[532]$									
^{152}Tb	502	8+	[532]	β^+	4.8	^{152}Gd	2394	7+	$n11/2[505]$
^{154}Ho	0+x	8+	[532]	β^+	4.87	^{154}Dy	2473	7+	$n11/2[505]$
^{154}Nd	0.0	0+	[532]	β^-	4.9	^{154}Pm	152	1+	$n3/2[532]$
					4.8		850	1+	
(b) $p11/2[505] \leftrightarrow n9/2[505]$									
^{190}W	0.0	0+	[505]	β^-	5.12	^{190}Re	320	1+	$n9/2[505]$
^{190}Re	0.0	2−	[505]	β^-	5.18	^{190}Os	1387	3+	$p5/2[402]$
^{190}Ir	175	11−	[505]	β^+	4.7	^{190}Os	1705	10−	$n11/2[615]$
^{191}Os	0.0	9/2−	[505]	β^-	5.30	^{191}Ir	171	11/2−	

pair in nuclei with $A \approx 190$. This evidence is summarized in Table I, which also includes the only known instance[13] of odd-mass decay involving a [505] nucleon transformation. An instance of odd-mass *au* decay involving the [532] orbital pair leading to the population of a 3*qp* state is presented later in this paper.

One may expect to see similar *au* transitions involving $N = 6$ positive-parity orbital pairs in the actinide region. However, the $\Omega_p = (\Lambda + 1/2)$ and $\Omega_n = (\Lambda - 1/2)$ orbitals with the same $[Nn_3\Lambda]$ configuration, which are connected through *au* transitions, are found to be too far apart on either side of the fermi surface to be reached within the available Q_β values in this region. Another possibility of the occurrence of an *au* transition connecting positive-parity orbitals in the rare-earth region may be visualized as involving the [402]-orbital pair in transitional nuclei with $A \approx 150$. However, the $\Delta N = 2$ mixing of the $3/2^+[402]$ and $3/2^+[651]$ neutron orbitals complicates the situation, and presently no clear evidence of any $3/2[402]_n \leftrightarrow 5/2[402]_p$ transition can be extracted. It is an interesting challenge to look for the *au* transition between positive-parity orbitals possibly in even-mass nuclei of the actinide region and/or in multi-particle products from precise, high-Q_β decay studies.

In our recent survey[7] of 122 β decays with $\log ft \leq 5.2$ in $147 \leq A \leq 190$ nuclei, we found that essentially the same single-particle spin-flip transition discussed above populated the 0, 1, 2, 3, and 4 *qp* states in all these decays. The schematic diagrams for these *au* decays are presented in Figs. 1 and 2. In Fig. 1 we include the cases wherein the number of unpaired particles (*qp*) remains unchanged in the decay process. The transition connects either an unpaired nucleon in the parent to an unpaired nucleon in the daughter (in cases

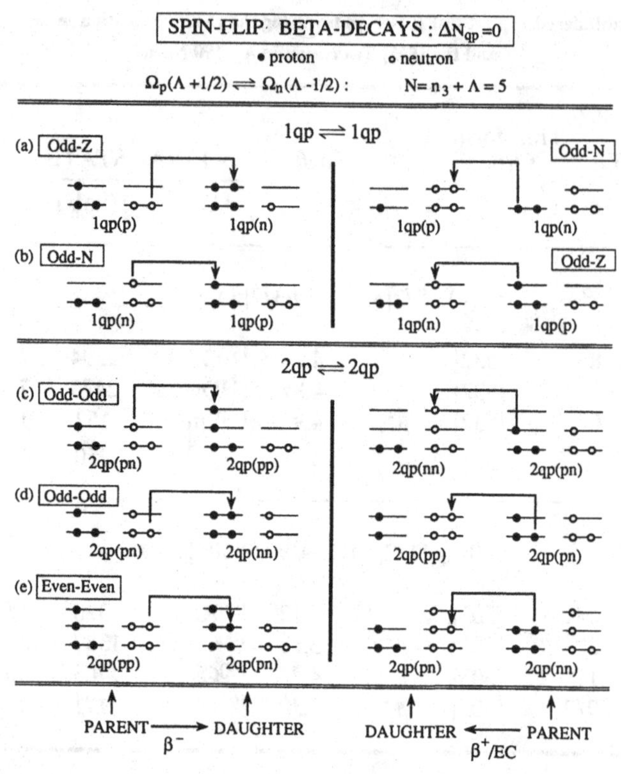

Figure 1. Schematic drawings of β-decay processes in which the number of paired nucleons remains unchanged. The nucleons involved in the spin-flip allowed-unhindered *au* transitions, shown connected by an arrow, have asymptotic quantum numbers $[Nn_3\Lambda] = [5(5-\Lambda)\Lambda]$, with the proton (denoted by filled circles) having $\Omega_p = (\Lambda + 1/2)$ and the neutron (open circles) $\Omega_n = (\Lambda - 1/2)$. The β^- decays are shown on the left and the β^+/EC decays on the right. The number N of quasi-particles and the corresponding nucleonic composition is indicated for each.

wherein the valence nucleon has the $[5(5-\Lambda)\Lambda]$ configuration) as in Fig. 1*b* and *c*, or it results in breaking the coupled $[5(5-\Lambda)\Lambda]^2$ pair in the parent to form a coupled $[5(5-\Lambda)\Lambda]^2$ pair in the daughter, as in Fig. 1*a*, *d*, and *e*. The 2*qp* states in even-even nuclei occur above the pairing gap, usually at excitation energies E_x 1 MeV. Presently only one instance is known of the decay from the 2*qp* state of an even-even nucleus, as depicted in Fig. 1*e*; it is the β^- decay of the 61.5-m 1173- keV 8^- isomer of ^{182}Hf.

The processes depicted in Fig. 2, populating 2, 3, and 4 *qp* states, are observed in cases wherein the $[5(5-\Lambda)\Lambda]$ orbital is occupied by a coupled pair of nucleons in the parent nucleus and its spin-flip counterpart orbital lies above the fermi surface in the daughter nucleus. In these processes the number of unpaired particles changes by 2 in the transition. The decays of the $1^+\{\Omega_p[5(5-\Lambda)\Lambda\uparrow] - \Omega_n[5(5-\Lambda)\Lambda\downarrow]\}$ states of the odd-odd nuclei to the 0^+ ground states of the even-even nuclei, depicted in Fig. 2*b*, also fall in this category. The processes leading to multiparticle states ($E_x \to$ 1–3 MeV) can only occur in high-Q_β decays.

In addition to the *au* decays discussed above, Fujita et al.[11] examined the selection rules for allowed β transitions of deformed nuclei, based on the projection-operator

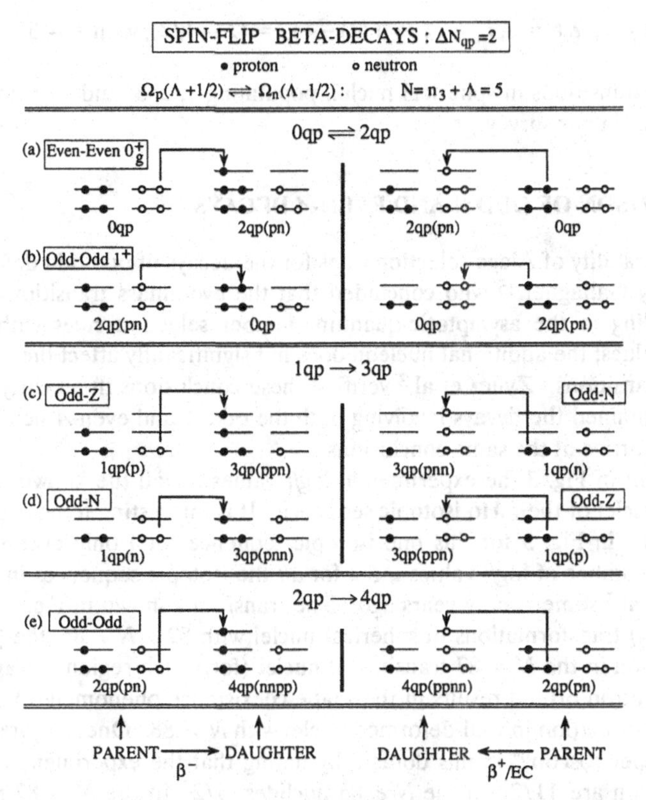

Figure 2. Same as Fig. 1 but for decay processes in which the number of unpaired nucleons (N_{qp}) changes by two.

method to introduce an effective operator including an $i(\ell \times \sigma)$ term which obeys the selection rule,

$$\Delta N = 0, \ |\Delta n_3| = |\Delta\Omega| = \pm 1 \text{ or } 0. \tag{5}$$

They[11] surveyed the then-available data to conclude that these non-*au* fast transitions have log*ft* values intermediate between those for the *au* and the *ah* transitions. Meijer[5] had also included this category in his survey. We include in our survey the presently available information on the log*ft* values for the following transitions from this category: *p*5/2[532] ↔ *n*3/2[521], *p*5/2[532] ↔ *n*5/2[523], and *p*7/2[532] ↔ *n*5/2[512], in comparison with those for the *p*[523] → *n*[523] *au* transitions, wherever available, between the same pairs of nuclei.

The Gamow-Teller (GT) β transitions of proton-rich nuclei beyond the doubly-magic structure, $^{146}Gd_{82}$, have been the subject of intense study over the recent years. As described in another paper[14] in these proceedings, the valence protons in these nuclei occupy the $h_{11/2}$ shell-model orbit. The β^+/EC decays of these nuclei proceed through the transformation of an $h_{11/2}$ proton into an $h_{9/2}$ neutron. In parallel with the selection rules (1–4) for the deformed nuclei, these single-particle spin-flip decays of spherical nuclei obey the shell-model quantum-number $|n\ell j\rangle$ selection rules,

$$\Delta n = 0,\ \Delta \ell = 0, j_p = (\ell + 1/2) \longrightarrow j_n = (\ell - 1/2), \text{ with } \ell = 5. \qquad (6)$$

These *au* transitions in spherical nuclei, populating 1, 2, 3, and 4 *qp* configurations, are discussed later in our survey.

31.3 COMPARISON OF ODD-*A* AND EVEN-*A* DECAYS

The applicability of Alaga selection rules for the decays of odd-odd deformed nuclei was examined by Gallagher,[15] who concluded that the even-mass transitions can also be classified according to the asymptotic-quantum-number selection rules within the same ranges of log*ft* values; the additional nucleon does not significantly affect the corresponding single-particle transitions. Zylicz et al.[3] verified these conclusions in an early survey. Ekstrom et al.[16] examined the decays involving both the odd-*A* and even-*A* neutron-deficient $_{67}$Ho isotopes to arrive at the same conclusions.

We present in Fig. 3 the experimental log*ft* values for all the known *au* transitions connecting the nuclei of the $_{67}$Ho isotopic sequence. It is interesting to note that the number of data points in Fig. 3 for this one isotopic sequence total only twenty-nine, which equals the total number of log*ft* values $\leqslant$ 5.2 for *all* the isotopic sequences in the compilation of Zylicz et al.[3] some twenty years ago. The transitions shown in Fig. 3 relate to the $p(h_{11/2}) \to n(h_{9/2})$ transformations in spherical nuclei with $82 \leqslant N \leqslant 86$, the [532] orbital-pair transformation in the $N = 87$ transitional nuclei (for which region the appearance of the 11/2[505] neutron orbital results in the shape-coexistence phenomenon) and the [523] orbital-pair transformation in well-deformed nuclei with $N > 88$. One may view the play of "exotic nuclear spectroscopy" in this domain by noting that the experimental spin-parities for the 67th proton are $11/2^-$ in the $N \leqslant 86$ nuclides, $5/2^-$ in the $N = 87$ nuclides, and $7/2^-$ in the $N \geqslant 89$ nuclides. The high-spin isomers, both in ^{152}Ho and in ^{158}Ho, have $I^\pi = 9^+$; however, in ^{152}Ho it results from the coupling of an $11/2^-$ proton with a $7/2^-$ neutron, while in ^{158}Ho it arises from the coupling of a $7/2^-$ proton with an $11/2^-$ neutron! Our twenty-nine log*ft* values shown in Fig. 3 include eleven odd-mass and eighteen even-mass cases. No systematic or significant difference is observed in the transition rates for *au* decays of odd-mass and even-mass nuclei. The same situation is observed in seventeen *au* transitions in nuclei with $173 \leqslant A \leqslant 182$, shown in Fig. 4. The *au* transitions in this region involve the [514] orbital-pair transformation. It is seen that while the 7/2[514] neutron orbital appears as the ground state in all the $N = 105$ isotones. The 9/2[514] proton orbital just misses being the ground state in the $Z = 71$ or 73 isotopes. As discussed later, only very recently evidence has been presented for the 9/2[514] ground state in Lu isotopes with $A \geqslant 181$.

31.4 EVEN-MASS DECAYS

The extension of the Alaga rules[1] to the even-mass decays was examined by Gallagher,[15] who concluded that the additional nucleon in the even-mass cases does not significantly affect the β transition rate. A survey[3] of both the odd-*A* and even-*A* *au* transitions throughout the rare-earth region and a more restricted survey[16] of β transitions involving neutron-deficient $_{67}$Ho and $_{68}$Er nuclei confirmed these conclusions. We have recently examined the even-mass decays[6,9] with a much wider data base to arrive at the same conclusions.

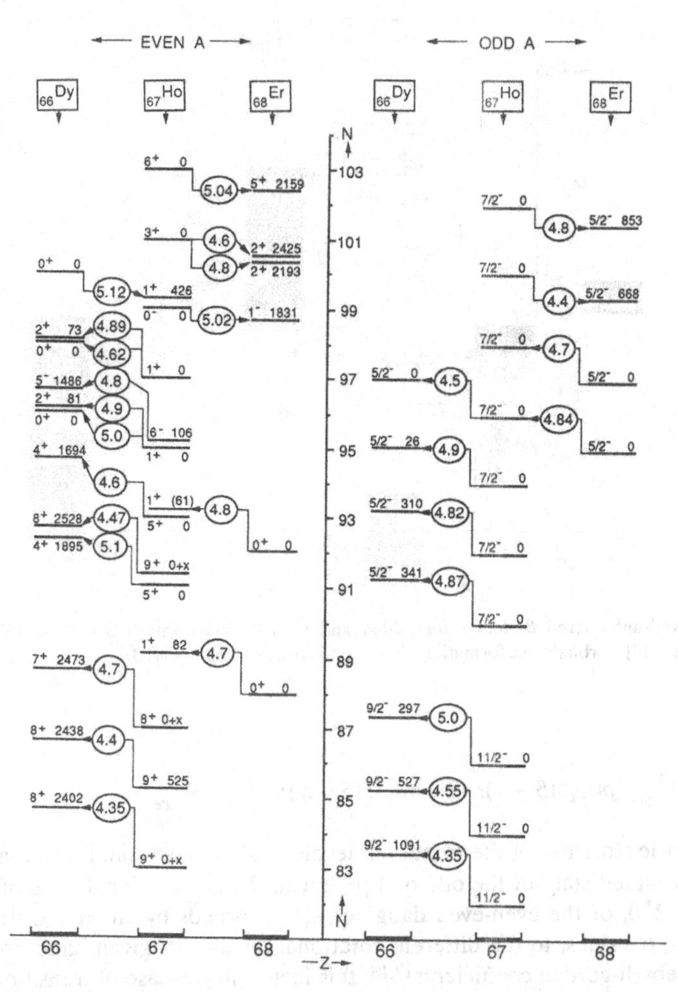

Figure 3. β decays with log*ft* ⩽ 5.2 involving $_{67}$Ho isotopes with 82 ⩽ N ⩽ 103. The numbers on the lines, representing energy levels, denote spin-parity $I\pi$ and level energy in keV. The level position approximately indicates the excitation energy. The circled numbers denote log*ft* values. The arrows toward the right represent β^- and those towards the left β^+/*EC* decays. The shaded areas represent β-stable nuclei.

Several features of the even-mass decays, not encountered in the odd-mass decays, have been brought out in various individual studies. In our recent exhaustive analysis[6] we have brought all these features into focus at one time. Here we briefly describe these aspects of even-mass *au* decays within the framework of the extended coverage attempted in this chapter.

The *au* decays of the even-mass deformed nuclei, connecting the odd-odd and even-even species, broadly fall in two divisions. Decays involving the ground state 0^+ (referred to as the zero-*qp* state) of the even-even nuclei connect only to the 1^+ coupled state of the $[5(5-\Lambda)\Lambda]$ orbital pair in the odd-odd nuclei, as shown in Fig. 2*a*, 2*b*. The transition for this group may be written as follows in terms of the asymptotic quantum numbers, Ω $[Nn_3\Lambda\Sigma]$:

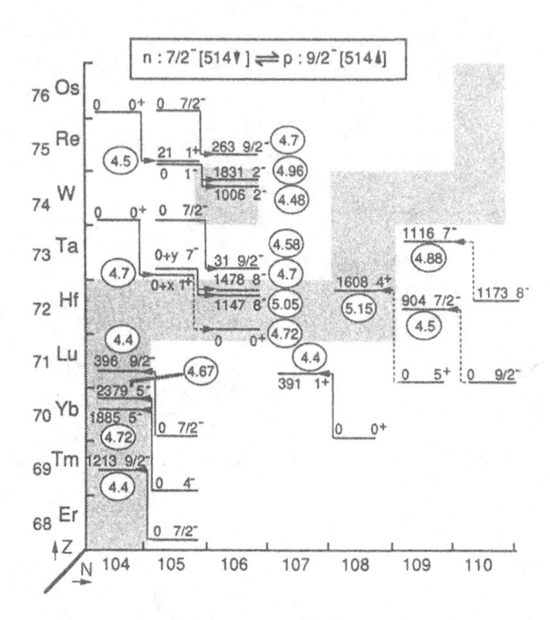

Figure 4. Allowed-unhindered β decays for odd-A and even-A nuclei with $173 \leqslant A \leqslant 182$, all of which correspond to the [514] - orbital transformation. Notation is the same as in Fig. 3.

$$1^{+}_{oo}\{p\Omega_p[5(5-\Lambda)\Lambda\uparrow]-n\Omega_n[5(5-\Lambda)\Lambda\downarrow]\}\longleftrightarrow 0^{+}_{ee} \qquad (7)$$

Since the intrinsic structure of the rotational levels in a given K^{π} band is the same, the decay of the 1^{+} coupled state of the odd-odd parent to the 2^{+} rotational state of the ground band, denoted $2^{+}0$, of the even-even daughter also proceeds by an *au* transition. The β transition rates, *ft* values, to the different rotational levels of a given band are related as ratios of the clebsch-gordan coefficients[17,18]; this factor, in the case of transitions to the 0^{+} and $2^{+}0$ states, is 2.0. The other category of even-mass decays involves the decays of the odd-odd nuclei, from the states not coupled to 1^{+}, feeding the 2*qp* excitation in the even-even daughters as shown in Fig. 1*c*, 1*d*. These 2*qp* excitations, lying above the pairing gap, occur usually at $E_x \geqslant 1$ MeV, and in many cases are found to have configuration admixtures which may arise either from vibrational components[8,19] or from $\Delta K = 0$ admixture of the *n*-*n* and the *p*-*p* two-particle configurations. Further, the odd-odd nuclei are known to have long-lived metastable states, which may decay by $\beta\pm/EC$ exclusively or in competition with electromagnetic decay. As already mentioned, presently there is only one known β^{-} decaying excited isomer of even-even nuclei, corresponding to the decay shown in Fig. 1*e*.

Even-mass decays provide a much wider field of study in comparison with odd-mass decays, as illustrated in Fig. 5 for *au* decays involving the $A = 161$ and $A = 162$ mass chains. In $A = 161$ we find only one instance each of *au* β^{-} and β^{+}/EC decay, whereas in $A = 162$ we have eight instances of such decays spread over the range of protons $_{64}$Gd to $_{72}$Hf and neutrons from 90 to 98. Mainly the pairing correlations in the 0^{+} ground state cause such a wide spread. Included in this plot are two instances of high-spin isomer decays.

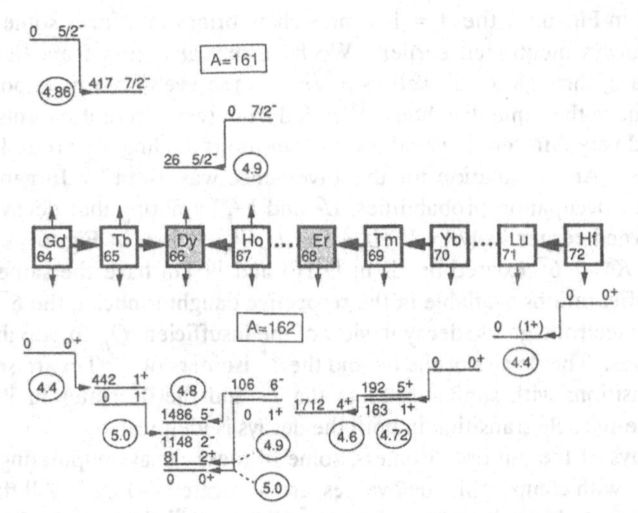

Figure 5. Allowed unhindered decays of A = 161 (top) and A = 162 (bottom) isobaric nuclei. Notation is the same as in Fig. 3.

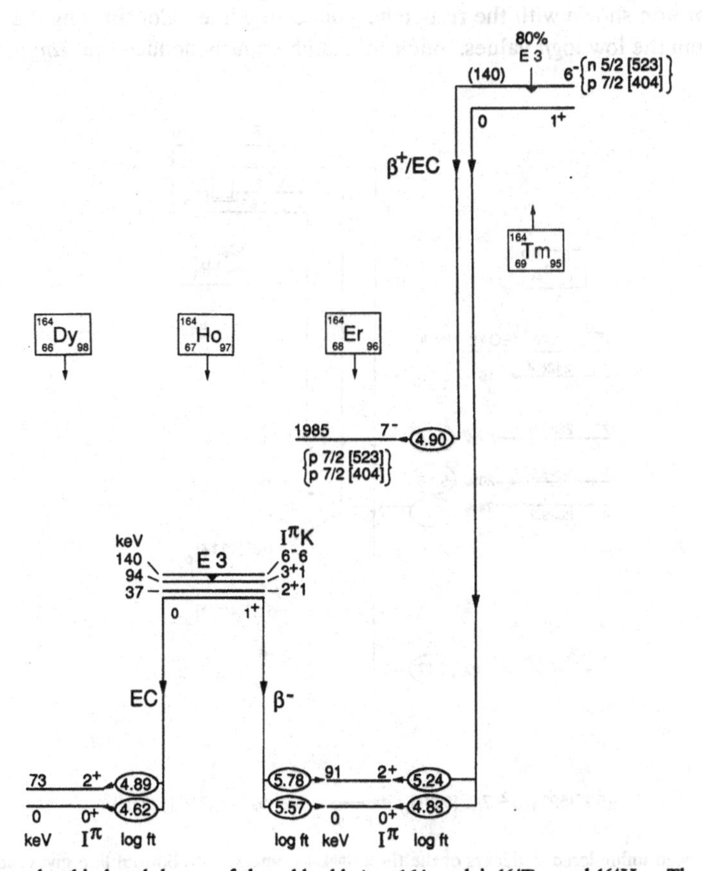

Figure 6. Allowed-unhindered decays of the odd-odd A = 164 nuclei, ^{164}Tm and ^{164}Ho. The decay modes, the 2qp configuration of the $K\pi$ = 6$^-$ isomers, and the decays of the 1$^+$ parent to both the 0$^+$ and 2$^+$0 rotational states are shown explicitly. Notation is the same as in Fig. 3.

The plot in Fig. 6 for the A = 164 mass chain brings into focus some other features of even-mass decays mentioned earlier. We have the *au* decays from the same parent, ^{164}Ho in this case, through β^- as well as β^+/EC to the even-even neighbors on opposite sides. Also we have the same daughter, ^{164}Er, fed from two different parents. In the decay of ^{164}Ho we find very different log*ft* values for transitions feeding the ground-band levels in ^{164}Dy and ^{164}Er. An explanation for this divergence was given by Jorgensen et al.[20] in terms of the pair occupation probabilities, U^2 and V^2, by noting that decays to ^{164}Er and ^{164}Dy are governed respectively by $V_p^2U_n^2$ and $U_p^2V_n^2$. Also in Fig. 6 we note that, although the two K^π = 6^- excited bands in ^{164}Ho and ^{164}Tm have the same 2*qp* structure with similar configurations available in the respective daughter nuclei, the 6^- ^{164}Ho isomer undergoes only electromagnetic decay in view of the insufficient Q_β to populate the higher-lying 7^- 2*qp* states. The decays of the 6^- and the 1^+ isomers of ^{164}Tm are seen to proceed through *au* transitions with similar rates to the 7^- and the 0^+ states in ^{164}Er, since the underlying single-particle transition in both the decays is identical.

The decays of the multiple isomers, some of these decays populating several states with the same I^π with comparable log*ft* values, are illustrated in Fig. 7. All three isomers of ^{166}Lu, having comparable half-lives, undergo β^+/EC decay[21] through the [523] orbital-pair transformation listed in the bottom line of the figure, with the configuration of the "spectator" proton shown with the respective connecting line. Considering the *au* nature of the decay from the low log*ft* values, one can unambiguously deduce the 2*qp* configurations

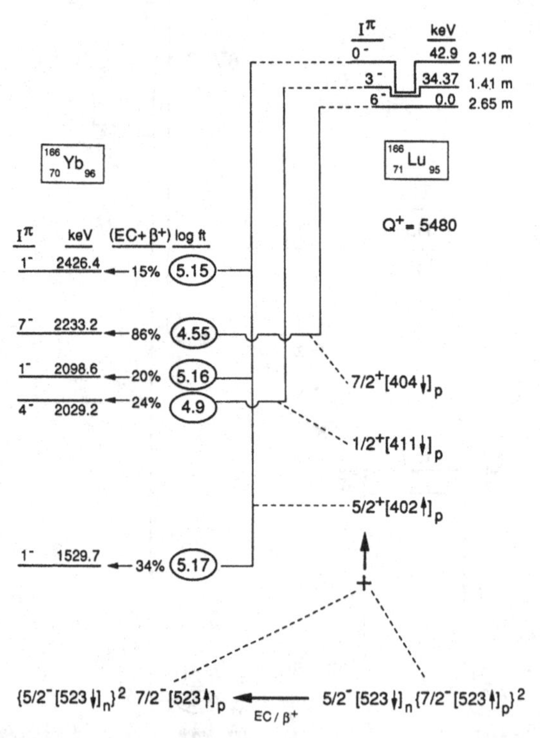

Figure 7. Allowed-unhindered β decays of the three ^{166}Lu isomers. The bottom line gives the configuration of the orbitals involved in the transition, and the configuration of the "spectator" proton for each isomer decay is indicated above it. Note that the 0^- decay populates three 1^- levels in ^{166}Yb with comparable intensity and similar log*ft* values.

of the 3^- and 6^- parent states and the 4^- and 7^- daughter states connected through these decays. The decays to the three different 1^- daughter states with comparable log*ft* values is understood in terms of the $\Delta K = 0$ mixing between various 2*qp* configurations for the $K^\pi = 1^-$ state. The *au* decay proceeds through the distributed strength of the $1^-\{p7/2[523] - p5/2[402]\}$ configuration in the three states. By combining a variety of decay and reaction studies, one can deduce the $\Delta K = 0$ percentage admixture in the given I^π states. An instance of this procedure is provided in the characterization[22-24] of the two 8^- levels in ^{178}Hf. The 1479- keV 8^- level in ^{178}Hf, fed from the ^{178}Ta 2.1-h 7^- isomer decays with log*ft* = 4.81, has been deduced as mainly a two-proton state with about 34% admixture of the two-neutron configuration, while the 1147- keV 8^- level, fed with log*ft* = 5.05 in the same decay, has the reverse admixture

The effectiveness of the "strong" rule in deducing the 2*qp* configurations can be illustrated through the example of ^{158}Ho isomers.[25] The 21-m ^{158}Ho high-spin isomer is seen to decay with log*ft* = 4.47 to the 2528-keV 8^+ level in ^{158}Dy; this observation admits of only $I^\pi = 9^+$ for the parent, with the unique $9^+\{p7/2[523] + n11/2[505]\}$ 2*qp* assignment. This assignment was confirmed through, the calculations[25] of the 2*qp* band-head energies in ^{158}Ho and ^{158}Dy and also by comparison with the analogous *au* decay log*ft* = 5.0 from the $5^+\{p7/2[523] + n3/2[521]\}$ ground state of ^{158}Ho to the 1895- keV 4^+ 2*qp* level in ^{158}Dy with log*ft* = 5.0. Recently the long-standing problem of the ^{180}Lu isomer, in the context of nucleosynthesis of the stable ^{180}Ta isomer,[26-29] has been investigated by us[30] on the basis of the β-transition rates, the recently identified single-particle energies in the A-1 isotope, and model calculations including hexadecapole deformations (recently found[29] to be decisive for describing the heavier Lu isotopes). We have concluded[30] that the ^{180}Lu ground state has the configuration, $5^+\{p9/2[514] + n1/2[510]\}$, which undergoes *au* decay to the 1608 keV $4^+\{n7/2[514] + n1/2[510]\}$ 2*qp* state in ^{180}Hf with log*ft* = 5.15 through the [514] orbital-pair transformation. A compilation of 2*qp* configurations deduced on the basis of *au* decays may be seen in our recent reviews.[6,7]

31.5 β DECAYS TO 3*qp* AND 4*qp* STATES

In cases wherein the *au* related $[5(5-\Lambda)\Lambda]$ orbital in the parent nucleus is not the valence orbital but is occupied by a coupled pair while the isospin and spin-flip orbital is unoccupied, the *au* transition may proceed by breaking the coupled pair in the parent to lead to two unpaired nucleons (*n* and *p*) in the daughter. The schematic drawing for these decays is shown in Fig. 2 *c-e*). In odd-mass decays, such decays populate 3*qp* states, while in even-mass decays, they populate 4*qp* states. Soloviev[8] had presented specific instances wherein such transitions may be observed.

Decays leading to the 3*qp* states were-reviewed by Bunker and Reich[4] and also by Meijer.[5] The present situation in respect to such decays is summarized in Table II; transitions involving the [523] orbital pair are listed in Table II*a* and those involving the [514] orbital pair are listed in Table II*b*. An examination of these tables reveals new features not revealed in the *au* transitions discussed in the previous sections. First, we observe that the excitation energies of the 3*qp* states range from 0.5 MeV in ^{165}Ho to over 2 MeV in several cases. The case of the low-lying ^{165}Ho 3*qp* state has been discussed in detail by Bunker and Reich[4]; they described it in terms of coupling of the 7/2[523] proton to the 2^+ γ vibration of the even-even core. (The microscopic 2*qp* decomposition of the vibrational states in even-even nuclei has been discussed in detail by Soloviev and co-workers.[8,19]) The core 2^+ vibration, in the case of the ^{165}Ho 3*qp* state, has a significant (≈26%) component of the

Table II. Single-Particle Spin-Flip β Decays Leading to Three-Quasiparticle States

${}^{A}_{Z}X_{N}$	E(keV)	$\Omega\pi\,[Nn_3\Lambda\Sigma]$		${}^{A}_{Z}X_{N}$	E(keV)	$I\pi$	log*ft*	$\Omega_1^{\pi} \neq I_{(12)}^{\pi}\,\{\Omega_2\Omega_3\}$
				II(a): $p7/2^-[523\uparrow] \leftrightarrow n5/2^-[523\downarrow]$				
${}^{163}_{65}\mathrm{Tb}_{98}$	0.0	p:3/2$^+$[422↑]	β^- →	${}^{163}_{66}\mathrm{Dy}_{97}$	884	1/2$^+$	5.01	
					935	3/2$^+$ 1/2	5.2	n5/2$^-$ -2$^-${p3/2p7/2}
					949	5/2$^+$ 1/2	5.7	
${}^{165}_{65}\mathrm{Tb}_{100}$	0.0	p:3/2$^+$[411↑]	β^- →	${}^{165}_{66}\mathrm{Dy}_{99}$	1773.1	(1/2)$^+$	5.1	n5/2$^-$ -2$^-${p3/2p7/2}
${}^{165}_{66}\mathrm{Dy}_{99}$	108	n:1/2$^-$[521↓]	β^- →	${}^{165}_{67}\mathrm{Ho}_{98}$	515.5	3/2$^-$	5.2	p7/2$^-$ -2+[n5/2n1/2]
${}^{163}_{70}\mathrm{Yb}_{93}$	0.0	n:3/2$^-$[521↑]	β^+ →	${}^{163}_{69}\mathrm{Tm}_{94}$	860	5/2$^-$	5.2	p7/2$^-$ -1+{n3/2n5/2}
${}^{163}_{69}\mathrm{Tm}_{94}$	0.0	p:1/2$^+$[411↓]	β^+ →	${}^{163}_{68}\mathrm{Er}_{95}$	1538.4	3/2$^+$	5.1	4$^-${p1/2p7/2} -n5/2$^-$
					1802.0	1/2$^+$	4.9	3$^-${p1/2p7/2} -n5/2$^-$
${}^{165}_{69}\mathrm{Tm}_{96}$	0.0	p:1/2$^+$[411↓]	β^+ →	${}^{165}_{68}\mathrm{Er}_{97}$	1427.4	3/2$^+$	5.4	4$^-${p1/2p7/2} -n5/2$^-$
${}^{165}_{71}\mathrm{Lu}_{94}$	0.0	p:7/2$^+$[404↓]	β^+ →	${}^{165}_{70}\mathrm{Yb}_{95}$	1734.1	5/2$^+$	4.4	
								n5/2$^-$ -0$^-${p7/2p7/2}
					2125.9	5/2$^+$	4.1	

(continued)

{n5/2[523]:n1/2[512]} 2*qp* configuration, which leads to the observed *au* transition to this level. Another interesting feature of these decays is revealed in the population of the 3*qp* configuration at 857 keV in ^{183}Ta with log*ft* = 5.88. This rather anomalously large log*ft* value for an indicated *au* transition has been explained by McIssac et al.[31] in terms of pair-occupation probabilities taken together with a two-neutron component in the vibrational phonon. The decay of ^{175}Tm presents a unique example of *au* decays to two different 3*qp* multiplets, one involving the [514] orbital pair (usual for $A \gtrsim 173$ decays) and the other involving the [523] orbital pair. Experimentally, one finds that in ^{173}Lu the 9/2[514] and 7/2[523] proton orbitals are separated by ≈285 keV, which is very close to the separation energy of the two 3/2$^+$ states in ^{175}Yb populated through the postulated *au* transitions.

Table II. (continued)

${}^{A}_{Z}X_N$	E(keV)	$\Omega\pi[Nn_3\Lambda\Sigma]$		${}^{A}_{Z}X_N$	E(keV)	$I\pi$	log*ft*	$\Omega_1{}^{\pi}\neq I_{(12)}{}^{\pi}\{\Omega_2\Omega_3\}$
				II(b): $p9/2^-[514\uparrow] \leftrightarrow n5/2^-[514\downarrow]$				
${}^{175}_{69}Tm_{106}$	0.0	$p{:}1/2^+[411\downarrow]$	$\beta^- \rightarrow$	${}^{175}_{70}Yb_{105}$	1497	$3/2^+$	5.0	$5^-\{p1/2p9/2\}$ $-n7/2^-$
					1891	$1/2^+$	4.8	$4^-\{p1/2p9/2\}$ $-n7/2^-$
					1792	$3/2^+$	5.5	$4^-\{p1/2p7/2\}$ $-n5/2^{-a}$
					2113	$1/2^+$	5.2	$3^-\{p1/2p7/2\}$ $-n5/2^{-a}$
${}^{177}_{70}Yb_{107}$	0.0	$n{:}9/2^+[624\uparrow]$	$\beta^- \rightarrow$	${}^{177}_{71}Lu_{106}$	1241	$7/2^+$	4.40	$8^-\{n9/2n7/2\}$ $-p9/2^-$
					1231	$11/2^+$	4.23	$p9/2^-$ $+1^-\{n9/2n7/2\}$
					1336	$7/2^+$	5.05	$p9/2^-$ $-1^-\{n9/2n7/2\}$
${}^{183}_{71}Lu_{112}$	0.0	$p{:}7/2^+[404\downarrow]$	$\beta^- \rightarrow$	${}^{183}_{72}Hf_{111}$	1125.3	$5/2^+$	5.3	$n7/2^-$ $-1^-\{p9/2p7/2\}$
${}^{183}_{71}Lu_{112}$	0.0	$p{:}7/2^+[404\downarrow]$	$\beta^- \rightarrow$	${}^{183}_{72}Hf_{111}$	1125.3	$5/2^+$	5.3	$n7/2^-$ $-1^-\{p9/2p7/2\}$
${}^{183}_{72}Hf_{111}$	0.0	$n{:}3/2^-[512\downarrow]$	$\beta^- \rightarrow$	${}^{183}_{73}Ta_{110}$	856.9	$5/2^-$	5.88	
								$p9/2^-$ $-2{+}[n3/2n7/2]$
					1543.4	$5/2^-$	5.84	
${}^{177}_{74}W_{103}$	0.0	$n{:}1/2^-[512\downarrow]$	$\beta^+ \rightarrow$	${}^{177}_{73}Ta_{104}$	1253.0	$3/2^-$	5.6	$p9/2^-$ $-3{+}[n7/2n1/2]$
					1512.3	$(1/2)^-$	5.4	$p9/2^-$ $-4{+}[n7/2n1/2]$
${}^{179}_{75}Re_{104}$	0.0	$p{:}5/2^+[402\uparrow]$	$\beta^+ \rightarrow$	${}^{179}_{74}W_{105}$	720.2	$3/2^+$	5.31	$n7/2^-$ $-2^-\{p9/2p5/2\}$
					1680.3	$7/2^+$	5.07	$7^-\{p5/2p9/2\}$ $-n7/2^-$

[a] This configuration arises from the breaking of a $7/2[523]_0+$ proton pair in the parent nucleus to yield a $5/2[523]$ neutron and a $7/2[523]$ proton coupled to the unpaired proton orbital from the parent state.

In Fig. 8 we present the experimental results[32] from ^{151}Nd decay. Unfortunately in this region $\Delta N = 2$ mixing complicates the situation. However, this region represents the area covered by the [532] orbital pairs which have been seen in Table I to be connected through *au* transitions. In Fig. 8, the transition to the 2304 keV state with very low log*ft* $\leq$ 4.91 is particularly striking and can be understood only by interpreting it as a [532] orbital-pair *au* transformation feeding the indicated 3*qp* configuration. Further careful and precise studies of the feeding of the excited states in the range of 1-3 MeV of this region is desired to reveal possible other instances of such 3*qp* decays.

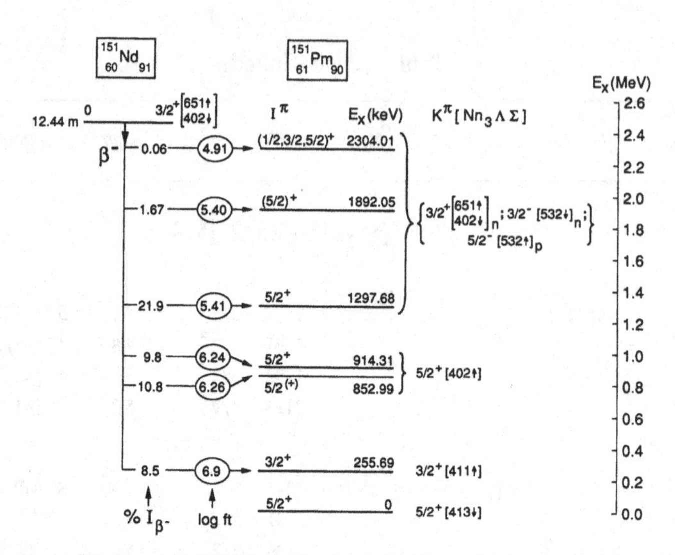

Figure 8. Experimental β decays of ^{151}Nd including levels in ^{151}Pm which are populated with > 5% intensity and/or have log*ft* ⩽ 5.5. The data are interpreted as evidence for *au* decays to 3*qp* states in ^{151}Pm involving the [532] orbital pair.

The population of 4*qp* states through *au* decay in ^{176}Hf[33,34] is detailed in Table III, wherein we also list the observed location, and, wherever applicable, β feeding of the constituent 2*qp* components of the 4*qp* configuration. We also list the high spin $K^\pi = 14^-$ member of the 4*qp* multiplet, which has been experimentally identified. These *au* decays with typical low-log*ft* values provide possibly the only means of assigning specific configurations to low-spin states at such high excitation energies. Another instance of the population of a 4*qp* state through *au* decay, not listed in Table III, is provided in the feeding of the 3013.7-keV (4^+) level in ^{164}Dy from ^{164}Tb decay with log*ft* = 5.0. This case involves the [523] orbital-pair transformation in the β-decay process. Again, more precise studies of β feeding of levels with $E_x \geqslant 3$ MeV can be reasonably expected to reveal other instances of such 4*qp* decays.

31.6 NON-*AU* FAST DECAYS

According to the modified Alaga rules for allowed β decay of deformed nuclei, discussed by Fujita et al.,[11] transitions connecting orbitals from the same oscillator shell ($\Delta N = 0$), but having $|\Delta n_3| = \pm 1$ and also $|\Delta\Lambda| = \pm 1$, may be expected to be retarded "by something like an order of magnitude" more than the *au* transitions discussed above. The three transitions corresponding to specific cases (one each) discussed by them, are listed by us in Section 2, above. It is seen that, in contrast with the only $\Delta I = \Delta\Omega = 1$ spin-flip transitions in *au* decays, here we have either spin-flip transitions with $\Delta I = \Delta\Omega = 0$ ($\Delta n_3 = \pm 1$ and $\Delta\Lambda = \mp 1$ with $n_3 + \Lambda = 5$) or no-spin-flip transitions with $\Delta I = \Delta\Omega = 1$ [$\Delta n_3 = \pm 1$ and $\Delta\Lambda = \pm 1$ with $\Delta(n_3 + \Lambda) = 2$].

In Table IV we list seventeen cases of such transitions observed in the lighter rare-earth nuclei as compared to three cases presented by Fujita et al.[11] and six cases tabulated by Meijer.[5] For the eight cases listed in Table IV*b* and *c*, we also observe an *au* transition

Table III. Four-Quasiparticle States Populated in Single-Particle Spin-Flip Allowed Unhindered β Decays

${}^{A}_{Z}X_{N}$	E(keV)	$\Omega\pi[Nn_3\Lambda\Sigma]$	${}^{A}_{Z}X_{N}$	E(keV)	$I\pi$	log*ft*	Configuration
			$p9/2^{-}[514\uparrow] \leftrightarrow n7/2^{-}[514\downarrow]$				
${}^{176}_{73}Ta_{103}$	0.0	$1+\{p_1{:}7/2^{+}[404\downarrow]^{a}$	${}^{176}_{72}Hf_{104}{}^{a}$				
		$-n_1{:}5/2^{-}[512\uparrow]\}$		1559	8^{-}		$p_1(7/2^{+}) + p_2(9/2^{-})$
				1722[b,c]	1^{-}		$p_2(9/2^{-}) - p_1(7/2^{+})$
				1333	6^{+}		
				1761	6^{+}		$n_1(5/2^{-}) + n_2(7/2^{-})$
				1672(c)	1^{+}		
				1863(c)	1^{+}		$n_2(7/2^{-}) - n_1(5/2^{-})$
				2866	14^{-}		$8^{-}[p_1+p_2]+6^{+}[n_1+n_2]$
				2944	2^{-}	4.3	$8^{-}[p_1+p_2]-6^{+}[n_1+n_2]$
				2912	$0^{-}0$	4.7	
				2920	$1^{-}0$	4.6	$1^{-}[p_2-p_1]-1^{+}[n_2-n_1]$
				2969	$2^{-}0$	4.8	

Notation:

$p_1{:}7/2^{+}[404\downarrow]$
$p_2{:}9/2^{-}[514\uparrow]$
$n_1{:}5/2^{-}[512\uparrow]$
$n_2{:}7/2^{-}[514\downarrow]$

[a] The levels between 1.5 and 2.0 MeV listed here are highly-admixed, two-quasiproton and two-quasineutron structures. The listed configuration is the component from which the 4*qp* states at ≈3 MeV are constructed.
[b] The 1722-keV state is likely an admixture of the $K^{\pi} = 1^{-}$ octupole vibrational state and the indicated two-quasiproton configuration.
[c] Log*ft* values for the 1722-keV 1^{-} and the 1672-keV and 1863-keV 1^{+} states are 7.74, 7.21, and 6.85, respectively, and do not refer to *au* decays.

in the same decay, thus providing a direct comparison of the respective transition rates.

It is observed in Table IV that, while all the *au* decays have log*ft* $\leqslant$ 4.9, the corresponding values for non-*au* decays are distributed over the range 5.6 ± 0.5. A comparison with experimental log*ft* values for the *ah* transitions shows only a marginal overlap of the lowest *ah* values with the values observed for the non-*au* fast decays discussed here, as predicted by Fujita et al.[11] So far we have limited our discussion to odd-mass decays and specifically those only for the lighter rare earths. Examinations of a wider data base, including both odd-*A* and even-*A* decays over the whole region, is in progress.

Table IV. Comparison of Non-*au* and *au* Fast Allowed β-Transition Rates (log*ft* Values) in Odd-*A* Decays

Parent	$\beta^{\pm}$	Daughter	non-*au* E_x (keV)	non-*au* log*ft*	*au* E_x (keV)	*au* log*ft*
	(a) $n3/2[521] \leftrightarrow p5/2[532]$					
^{153}Pm	β^-	^{153}Sm	36	5.4		
^{155}Sm	β^-	^{155}Eu	104	5.54		
^{155}Dy	β^+	^{155}Tb	227	6.25		
^{157}Dy	β^+	^{157}Tb	326	5.44		
^{157}Er	β^+	^{157}Ho	391	6.0		
^{159}Dy	β^+	^{159}Tb	364	6.1		
^{159}Er	β^+	^{159}Ho	624	5.61		
^{161}Er	β^+	^{161}Ho	827	5.40		
^{163}Yb	β^+	^{163}Tm	947	5.6		
	(b) $p7/2[523] \leftrightarrow n5/2[512]$				$n5/2[523]$	
^{157}Ho	β^+	^{157}Dy	897	5.27	341	4.87
^{159}Ho	β^+	^{159}Dy	1016	5.24	310	4.82
^{161}Ho	β^+	^{161}Dy	711	5.2	26	4.9
^{167}Ho	β^-	^{167}Er	347	5.8	668	4.4
	(c) $n5/2[523] \leftrightarrow p5/2[532]$				$p7/2[523]$	
^{161}Gd	β^-	^{161}Tb	480	6.1	417	4.86
^{163}Er	β^+	^{163}Ho	1114	5.4	0.0	4.84
^{165}Yb	β^+	^{165}Tm	1251	5.5	160	4.8
^{167}Yb	β^+	^{167}Tm	1527	$\approx$6.1	293	$\approx$4.6

31.7 β DECAYS OF SPHERICAL NUCLEI WITH $64 \leq Z \leq 70$ AND $82 \leq N \leq 86$

The region between $Z = 64$ and 82 is spanned by the $h_{11/2}$, $s_{1/2}$, and $d_{3/2}$ shell-model orbitals, with no clear-cut sequential order of filling in normal nuclei. Nuclei around the stability line in the beginning of this region have well-deformed shapes with the consequent breakdown of shell-model degeneracies into doubly-degenerate Nilsson orbitals. However, if we proceed away from the stability region toward more exotic nuclei, we encounter a doubly-closed shell structure in $^{146}Gd_{82}$. Nuclei just above this "doubly-magic" core constitute an island of spherical shape wherein, at least for nuclei with $65 \leq Z \leq 70$, the protons enter only the $h_{11/2}$ orbit until it is half-filled, while the neutrons between 83 and 86 occupy the $f_{7/2}$ orbit, with the $h_{9/2}$ neutron orbital (primarily unoccupied) located above the fermi surface . Accordingly, the following distinctive features appear in this "island of exotic nuclei":

1) Proton-rich odd-mass nuclei undergo *EC* with the spin-flip transformation of an $h_{11/2}$ proton into an $h_{9/2}$ neutron through fast GT transitions.
2) Even-even nuclei with an even number of $h_{11/2}$ protons decay through the same single-particle spin-flip transition by breaking a 0^+ $(h_{11/2})^2$ pair, leading to a 1^+ $(ph_{11/2} - nh_{9/2})$ state in the odd-odd daughter nuclei.
3) Odd-odd nuclei with the odd neutron in the $f_{7/2}$ orbit, have 9^+ high-spin and 2^+ low-spin isomers from the $(ph_{11/2} \pm nf_{7/2})$ configuration. The high-spin isomer again decays through the same single-particle spin-flip transformation to an 8^+ two-neutron state, with the $f_{7/2}$ odd neutron acting as a spectator.

The experimental situation with respect to such three-nuclei sets of the region is summarized in Fig. 9. [The two missing data, relating to the decays of the nuclei ^{154}Tm (9^+ isomer) and ^{155}Tm, should also show the same pattern and should be experimentally investigated to complete the set.]

$Z \geq 65 : p(h_{11/2}) \xrightarrow{\beta^+/EC} n(h_{9/2})$

	$p(h_{11/2})$ —SPECTATOR→ $n(f_{7/2})$	
$11/2^-$ –(log ft)→ $9/2^-$: keV	0^+ –(log ft)→ 1^+: keV	9^+ –(log ft)→ 8^+: keV
N : 82→83	82→83	83→84
^{147}Tb –(4.2)→ ^{147}Gd : 1398	^{148}Dy –(3.95)→ ^{148}Tb : 620	^{148}Tb –(4.5)→ ^{148}Gd : 2693
^{149}Ho –(4.35)→ ^{149}Dy : 1091	^{150}Er –(3.6)→ ^{150}Ho : 476	^{150}Ho –(4.35)→ ^{150}Dy : 2402
^{151}Tm –(4.3)→ ^{151}Er : 802	^{152}Yb –(3.5)→ ^{152}Tm : 482	^{152}Tm –(4.6)→ ^{152}Er : 2183
N : 84→85	84→85	85→86
^{149}Tb –(4.3)→ ^{149}Gd : 796	^{150}Dy –(4.07)→ ^{150}Tb : 397	^{150}Tb –(4.1)→ ^{150}Gd : 2554
^{151}Ho –(4.55)→ ^{151}Dy : 527	^{152}Er –(4.1)→ ^{152}Ho : 179	^{152}Ho –(4.4)→ ^{152}Dy : 2438
^{153}Tm –(~4.8)→ ^{153}Er : 299	^{154}Yb –(3.6)→ ^{154}Tm : 133	^{154}Tm – ?
N : 86→87	86→87	87→88
^{151}Tb –(4.83)→ ^{151}Gd : 380	^{152}Dy –(4.2)→ ^{152}Tb : 257	11/2 $[505]_n$
^{153}Ho –(~5.0)→ ^{153}Dy : 296	^{154}Er –(4.0)→ ^{154}Ho : 27	deformed orbitals
^{155}Tm – ? → ^{155}Er (227)	^{156}Yb –(3.78)→ ^{156}Tm : 115	COEXISTENCE

Figure 9. Experimental β decays with log*ft* < 5 for nuclei in the region $65 \leqslant Z \leqslant 70$ and $82 \leqslant N \leqslant 86$. The single-particle decays of the odd-mass nuclei, shown on the left, involve the transformation of an $h_{11/2}$ proton into an $h_{9/2}$ neutron. Decays from the 0^+ ground state of the even-even nuclei, shown in the middle, proceed to the 1^+ level in the odd-odd nuclei through the same transformation, with an $h_{11/2}$ proton acting as "spectator." Decays of the high-spin (9^+) isomer of the odd-odd nuclei, shown on the right, lead to the 8^+ $2qp$ state in the even-even daughter through the same transformation, with an $f_{7/2}$ neutron acting as "spectator."

The "deduced" log*ft* values for decays in this region are seen to vary considerably from one investigator to another mainly because of the large Q_{EC} values, which make missing decay information much more likely. For instance, the recent[35] "more precise measurements with substantially increased sensitivity allowing detection of gamma transitions with an intensity of as low as 10^{-4} per decay" revealed fourteen new ^{148}Tb energy levels with good I^π assignments in the ^{148}Dy decay, in contrast with only one γ transition reported earlier.[36] Similar new information on γ transitions missed earlier has been reported by Toth et al.[37] We may also point out the wide variation in log*ft* values, ranging from 3.6 to 4.6, quoted for ^{152}Tm (9^+) decay. In the case of the ^{154}Lu^m decay, earlier investigators[36] reported 100% feeding to only the 8^+ level in ^{154}Yb, with a deduced log*ft* = 3.3. A recent study[38] deduces < 55% feeding to this level, with a log*ft* > 4.3, the limit being set to account for the population of this level by "unobserved γ transitions from higher lying levels."

These conflicting reports[36,38] on the decay of ^{154}Lu^m are of particular interest in this context. All the high-spin isomers in the N = 83 isotonic chain were earlier reported[36] to populate only the 9^+ state in the daughter nuclei with progressively decreasing (with increasing Z value) log*ft* values, thus favoring an $I^\pi = 9^+$ assignment for the parent, corresponding to the $p(h_{11/2})n(f_{7/2})$ configuration. The recent report by Hoffman et al.[39] of a new isotope, ^{156}Ta, populating the 8^+ isomer in ^{156}Hf by β decay was seen as "a very natural extension of the known N = 83 series." However, Vierinen et al.[38] in a recent precision study conclude that the ^{154}Lu decay {earlier reported[36] as 100% (β^+/EC) decay to the 8^+ level in ^{154}Yb with log*ft* = 3.3}, is observed to have "strong direct feeding to the 8^+ and 6^+ levels," with log*ft* values > 4.3 and > 4.6, respectively, and implying a spin assignment of 7^+ for the ^{154}Lu parent. The changeover from the $9^+\{p(h_{11/2})n(f_{7/2})\}$ configuration observed for the lighter isotones to a different value for nuclei around Z = 70 was attributed to the "vanishing" of the strong two-particle coupling when the $h_{11/2}$ proton orbital is half-filled. Studies of McNeill et al.[40] on the relatively short-lived $27/2^-$ isomer in ^{153}Yb$_{83}$ are also understood in terms of this diminishing of the strong *n-p* interaction around $Z \approx 70$. In view of these observations, definitive studies of ^{154}Lu and ^{156}Ta decays[38,39] and also the decays of their odd-A N = 85 isotones are highly desirable to clarify the orbital filling order and the role of the *n-p* interaction.

Experiments have also revealed that the same single-particle spin-flip transition (in the case of high Q_{EC} decays) populates three-particle[12] and also four-particle[41] configurations in this region. However, individual branches to the multiplet members have not been investigated, and only an overall log*ft* value could be deduced for the multiplet as such. On the other hand, the degeneracy of the shell-model orbitals brings into play another mode of decay, which is not seen in the Nilsson scheme. A pair of protons in the $h_{11/2}$ orbital can align to yield a two-particle angular momentum of 10^+. Thus, in an even-Z, odd-A nucleus, a three-particle (*ppn*) aligned state with $I^\pi = 27/2^-$ is possible, which under favorable circumstances may competitively decay through β^+/EC to a three-particle (*pnn*) state in the daughter nucleus. The presently-known situation[12] for such decays is summarized in Table V. The recent study[41] on the decay of the 9^+ ^{152}Ho isomer to a four-particle multiplet has suggested the possible fast GT decay of the 2^- ^{152}Ho isomer to 4*qp* states with $I^\pi = 1^-$ to 3^- around E_x > 3.5 MeV in ^{152}Dy. It can be reasonably expected that experiments in the near future will furnish more precise and varied information on such exotic decays.

Table V. Spin-Flip β Decays to $3qp$ States in Spherical Nuclei with $Z > 65$ and $82 < N < 84$ (All are β^+/EC Decays Involving the $p(h_{11/2}) \rightarrow n(h_{9/2})$ Transition)

INITIAL STATE				logft	FINAL STATE			
${}^{A}_{Z}X_N$	E(keV)	I^π	Configuration		${}^{A}_{Z}X_N$	E(keV)	I^π	Configuration
				(a) $1qp \rightarrow 3qp$				
${}^{151}_{68}Er_{83}$	0.0	$7/2^-$	$(h_{11/2})^2_0 f_{7/2}$	4.5	${}^{151}_{67}Ho_{84}$	1833	$(5/2,7/2,9/2)^-$	$(h_{11/2}h_{9/2})_1 f_{7/2}$
${}^{151}_{69}Tm_{82}$	0.0	$11/2^-$	$(h_{11/2})^2_0 h_{11/2}$	4.2	${}^{151}_{68}Er_{83}$	2917) 3032)	$(9/2,11/2,13/2)^-$	$(h_{11/2}h_{9/2})_1 h_{11/2}$
				(b) $3qp \rightarrow 3qp$				
${}^{149}_{66}Dy_{83}$	2661	$27/2^-$	$(h_{11/2})^2_{10} f_{7/2}$	>4.3	${}^{149}_{65}Tb_{84}$	2303	$27/2^-$	$(h_{11/2}h_{9/2})_{10} f_{7/2}$
${}^{151}_{68}Er_{83}$	2585.5	$27/2^-$	$(h_{11/2})^2_{10} f_{7/2}$	4.6	${}^{151}_{67}Ho_{84}$	2098	$25/2^-$	$(h_{11/2}h_{9/2})_{10} f_{7/2}$
				5.0		2226	$27/2^-$	

31.8 CONCLUSIONS AND OUTLOOK

In conclusion, we find that single-particle spin-flip transitions, with all other particle quantum numbers and "spectator" nucleons configurations unaltered, proceed with log*ft* < 5.2 when populating 0, 1, 2, 3, or 4 *qp* states in the daughter nuclei. This observation is valid both for the deformed nuclei and spherical nuclei just beyond the ${}^{146}Gd_{82}$ doubly-magic structure. Thus, transitions with observed log*ft* < 5.2 can be confidently used to deduce configurations for the β-connected states. In the case of highly-excited daughter products, this configuration, providing a conduit for the fast β transitions, may just be a component of a complex state with appropriate spin-parity. As we progressively investigate more "exotic" nuclei with higher Q_β values, more transitions of this type, particularly those populating multi-particle states, are expected to be observed, yielding a better glimpse into the interactions valid for these regions, and they also may possibly bring into light more puzzling phenomena.

ACKNOWLEDGMENT

These investigations were supported by the National Science Foundation under contract No. PHY89-06613 with the Florida State University.

REFERENCES

1. G. Alaga, *Phys. Rev.* **100**, 432 (1955); *Nucl. Phys.* **4**, 625 (1957).
2. B. R. Mottelson and S. G. Nilsson, *Mat. Fys. Skr. Dan. Vid. Selsk.* **1**, no. 8 (1959).
3. J. Zylicz, P. G. Hansen, H. L. Nielsen, and K. Wilsky, *Ark. Fys.* **36**, 643 (1967).
4. M. E. Bunker and C. W. Reich, *Revs. Mod. Phys.* **43**, 348 (1971).
5. B. J. Meijer, *Z. Physik* **A278**, 365 (1976).
6. P. C. Sood and R. K. Sheline, *Phys. Scripta* (1989), in press.
7. P. C. Sood and R. K. Sheline, *Atomic Data & Nucl. Data Tables*, **43**, 259 (1989).
8. V. G. Soloviev, *Soviet Phys. JETP* **16**, 176 (1963); *Theory of Complex Nuclei* (Pergamon Press, Oxford, 1976).
9. P. C. Sood and R. K. Sheline, *Mod. Phys. Lett.* **4**, 1711 (1989).
10. Y. A. Ellis and M. R. Schmorak, *Nucl. Data Sheets* **B8**, 345 (1972).
11. J. Fujita, G. T. Emery, and Y. Futami, *Phys. Rev.* **C1**, 2060 (1970).
12. R. Barden, A. Plochocki, D. Schardt, B. Rubio, M. Ogawa, P. Kleinheinz, P. Kirchner, O. Klepper, and J. Blomqvist, *Z. Phys.* **A329**, 11 (1988).
13. E. Browne, *Nucl. Data Sheets* **56**, 709 (1989).
14. J. H. McNeill, J. Blomqvist, A. A. Chisti, M. A. C. Hotchkis, B. J. Varley, P. J. Daly, W. Gelletly, M. Piiparinen, and P. J. Woods, *Phys. Rev. Lett.* **63**, 860 (1989) and Chapter 4, This Book.
15. C. J. Gallagher, *Nucl. Phys.* **16**, 215 (1960).
16. C. Ekström, T. Noreland, M. Olsmats, and B. Wannberg, *Nucl. Phys.* **A135**, 289 (1969).
17. G. Alaga, K. Alder, A. Bohr, and B. R. Mottelson, *Mat.-Fys. Medd. Dan. Vid. Selsk.* **29**, no. 9 (1955).
18. V. M. Mikhailov and M. A. Mikhailova, *Bull. Acad. Sc. USSR Phys. Ser.* **43**, no. 11, 37 (1979).
19. E. P. Grigoriev and V. G. Soloviev, *Structure of Even Deformed Nuclei* (Nauka, Moscow, 1974).
20. M. H. Jorgensen, O. B. Nielsen, and O. Skilbreid, *Nucl. Phys.* **84**, 569 (1966).
21. F. W. N. deBoer, P. F. A. Goudsmit, P. Koldejwin, and B. J. Meyer, *Nucl. Phys.* **A225**, 317 (1974).
22. C. J. Gallagher and H. L. Nielsen, *Phys. Rev.* **126**, 1520 (1962).
23. R. G. Helmer and C. W. Reich, *Nucl. Phys.* **A114**, 649 (1968).
24. T. E. Ward and Y. Y. Chu, *Phys. Rev. C* **12**, 1632 (1975).
25. P. C. Sood, R. W. Hoff, and R. K. Sheline, *Phys. Rev. C* **33**, 2163 (1986).
26. W. Eschner, W. D. Schmidt-Ott, K. L. Gippert, E. Runte, H. Beer, G. Walter, R. Kirchner, O. Klepper, E. Roeckl, and D. Schardt, *Z. Phys.* **A317**, 281 (1984).
27. S. E. Kellogg and E. B. Norman, *Phys. Rev. C* **34**, 224 (1986).
28. K. T. Lesko, E. B. Norman, D. M. Moltz, R. M. Larimer, S. G. Crane, and S. E. Kellogg, *Phys. Rev. C* **34**, 2256 (1986).
29. K. Rykaczewski, K. L. Gippert, N. Kaffrell, R. Kirchner, O. Klepper, V. T. Koslowsky, W. Kurcewicz, W. Nazarewicz, E. Roeckl, E. Runte, D. Schardt, W. D. Schmidt-Ott, and P. Tidemand-Petersson, *Nucl. Phys. A* **499**, 529 (1989).
30. P. C. Sood, A. K. Jain, R. W. Hoff, R. K. Sheline, and D. G. Burke, *Proc. Conf. on Nuclear Structure in the Nineties*, Ed. by N. R. Johnson (Oak Ridge, 1990), Vol. I, p. 236.
31. L. D. McIssac, R. G. Helmer, and C. W. Reich, *Nucl. Phys.* **A132**, 28 (1969).
32. B. Singh, J. A. Szucs, and M. W. Johns, *Nucl. Data Sheets* **55**, 185 (1988).
33. F. M. Bernthal, J. O. Rasmussen, and J. M. Hollander, *Phys. Rev.* **C3**, 1294 (1971).
34. R. Broda, V. Valyus, J. Zwolski, J. Molnar, N. Nenov, E. Z. Fydina, U. M. Fainer, and P. Shoshev, *Bull. Acad. Sc. USSR Phys. Ser.* **35**, 648 (1971).
35. P. Kleinheinz, K. Zuber, C. Conci, C. Protop, J. Zuber, C. F. Liang, P. Paris, and J. Blomqvist, *Phys. Rev. Lett.* **55**, 2664 (1985).
36. W. Habenicht, L. Spanier, G. Korschinek, H. Ernst, and E. Nolte, *Proceedings of the Seventh International Conference on Atomic Masses and Fundamental Constants* (AMC07), Ed. by O. Klepper

(GSI, Darmstadt, 1984), p 244 and references therein.
37. K. S. Toth, D. C. Sousa, J. M. Nitschke, and P. A. Wilmart, *Phys. Rev. C* **35**, 310 and 620 (1987).
38. K. S. Vierinen, A. A. Shihab-Eldin, J. M. Nitschke, P. A. Wilmarth, R. M. Chasteler, R. B. Firestone, and K. S. Toth, *Phys. Rev. C* **38**, 1509 (1988).
39. S. Hofmann, P. Armbruster, G. Berthes, T. Faestermann, A. Gillitzer, F. P. Hessberger, W. Kurcewicz, G. Münzenberg, K. Poppensieker, H. J. Schott, and I. Zychor, *Z. Phys.* **A332**, 275 (1989).
40. J. H. McNeill, A. A. Chisti, P. J. Daly, M. A. C. Hotchkis, M. Piparinen, and B. J. Varley, *Z. Phys.* **A332**, 105 (1989).
41. A. Gadea, B. Rubio, J. L. Tain, P. Kleinheinz, C. F. Liang, and P. Paris, *Z. Phys.* **A333**, 407 (1989).

32. Intruder States, Coexistence, and Approaches to Deformation: The Study of ^{120}Xe and the = 66 Isotones

P. F. Mantica, Jr., B. E. Zimmerman,
C. E. Ford, and W. B. Walters
Department of Chemistry and Biochemistry
University of Maryland
College Park, MD 20742

D. Rupnik and E. F. Zganjar
Department of Physics and Astronomy
Louisiana State University
Baton Rouge, LA

H. K. Carter
Oak Ridge Associated Universities
Oak Ridge, TN 37831

J. Rikovska and N. J. Stone
Clarendon Laboratory
Oxford University
Oxford, OX13PU, UK

ABSTRACT

In this chapter, new results of γ-ray and conversion-electron spectroscopic studies of the structure of ^{120}Xe will be reported. These results will be described along with the results of previous studies of intruder structures in the Te and Xe mass region. The observed structures will be contrasted with other intruder structures identified in the Sn, Cd, and Pd isotopes. The role of intruder structures in the development of nuclear deformation will be discussed.

32.1 INTRODUCTION

We have recently completed a study of the low-energy levels of ^{120}Xe populated in the β/EC decay of the 60-s 2^- and 8 isomers of ^{120}Cs. The even-even nuclide ^{120}Xe has four protons above the Z = 50 closed shell and has a mid-shell neutron number, N = 66. In this chapter, we would like to discuss the structure observed in ^{120}Xe as it relates to the structures of the even-even Xe isotopes below the N = 82 closed shell and the structures of the N = 66 isotones above and below the Z = 50 closed shell.

The odd-mass nuclides in this mass region are distinguished by the presence of low-energy "intruder" $9/2^+$ states in the $_{51}$Sb and $_{53}$I nuclides that have been attributed to the promotion of a proton from below the Z = 50 shell gap. The promoted proton is paired with the odd-proton (either $g_{7/2}$ or $d_{5/2}$) of the "normal" configuration, leaving a $g_{9/2}^{-1}$

Exotic Nuclear Spectroscopy, Edited by W. C. McHarris
Plenum Press, New York, 1990

hole state.[1] The resulting two-particle–one-hole and four-particle–one-hole intruder states in Sb and I, respectively, lie at low excitation energies around the $N = 66$ mid-shell. Well-defined $\Delta J = 1$ bands built upon the $9/2^+$ hole state have also been identified.[2] For example, in $_{55}$Cs it has been shown that the six-particle–one-hole proton intruder state is the ground state.[3-5] Similar six-, four-, and two-hole–one-particle intruder states exist in the odd-Z nuclides below the $Z = 50$ closed shell in $_{45}$Rh, $_{47}$Ag, and $_{49}$In, respectively.[1,6,7]

For the even-even nuclides in this region, formation of intruder states involves the promotion of a pair of protons across the $Z = 50$ shell gap. Such excitations have been described in a shell-model framework by Heyde et al.[8] The energy needed to promote a pair of protons across the shell gap (approx. 10 MeV) is partly offset by a pairing energy gain from the formation of a proton particle pair across the shell and creation of a proton hole pair below the shell, also by a proton-neutron monopole correction that accounts for the change in the interaction of the promoted protons with the neutrons. The principal energy reduction has its origin with the quadrupole proton-neutron interaction. Because the quadrupole interaction energy is proportional to $N_\pi N_\nu$ the minimum in the energy of these intruder states is expected to lie at mid-neutron shell, close to $N = 66$.

The promotion of two protons across the $Z = 50$ proton closure can result in the formation of a 0^+ excited state at relatively low energy in even-even nuclides with mid-shell neutron numbers. The resulting intruder state, being a result of proton promotion into an orbital with a different angular momentum, should have a different radius, as compared to the normal proton configuration. As γ decay of the intruder 0^+ state to the ground state is forbidden, decay via the internal conversion process is allowed, with the electric-monopole strength proportional to the nuclear-radius difference between the unmixed initial and final states and the relative admixtures between the two configurations.[9] Without mixing between the normal and intruder proton configurations,[10] decay of the intruder 0^+ state via an $E0$ transition is inhibited as the one body operator cannot change the ℓ value of the proton in transition.

From experimental studies on the even-even Te isotopes about the $N = 66$ midshell, Walker et al.[11] identified several strong $E0$ transitions in ^{118,120,122}Te between states with $J^\pi = 0^+$ and also observed $E0$ strength above the expected $M1/E2$ conversion on transitions[12] of 496, 545, and 876 keV in ^{118}Te$_{66}$. The identification of the $E0$ transitions and the subsequent calculations by Rikovska et al.[13] have indeed suggested that two-proton-particle and four-proton-particle–two-proton-hole configurations mix to define the low-energy structure of the Te isotopes near mid-shell.

The structures of the Xe isotopes from mid-shell to the $N = 82$ closed shell have been the focus of numerous experiments and many theoretical calculations. The mean-square charge radii[14] of the Xe nuclides from the $N = 82$ closed shell through mid-shell at $N = 66$ suggest a gradual increase in deformation toward a value of $\beta = 0.3$ at ^{120}Xe, extracted assuming axially-symmetrical shapes for these nuclides. The low-energy positive-parity levels of the odd-mass Xe nuclides are shown in Fig. 1 and can be regarded as belonging to $s_{1/2}$ and $d_{3/2}$ particle-plus-core multiplets. The multiplets drop slowly in energy with the decrease in the $2_1{}^+$ core energies as neutrons are removed from the $N = 82$ closed shell. The single-neutron-hole $g_{7/2}$ state also decreases gradually with the removal of neutrons from ^{135}Xe, and retains its identity through to mid-shell. This is not the case for the single-neutron-hole $d_{5/2}$ state, which appears to be thoroughly mixed with other states and cannot be readily identified as a band head in any of the odd-mass Xe isotopes below the $N = 82$ shell. The levels of odd-mass ^{133}Xe have been described in the IBFM

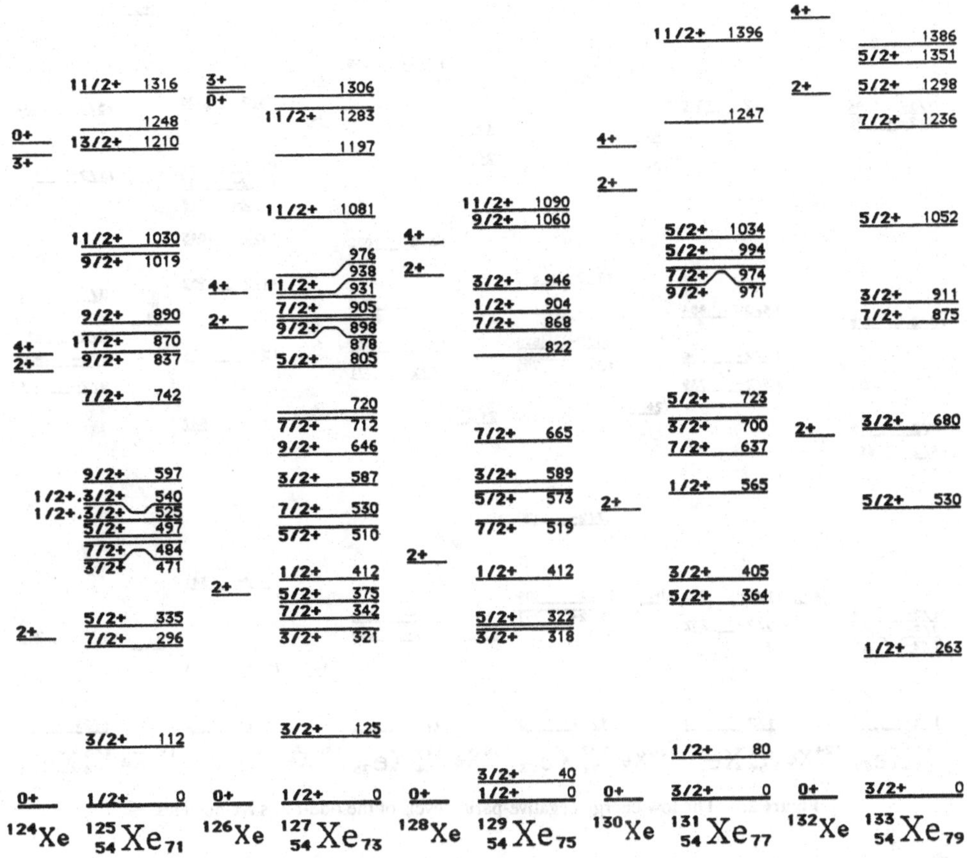

Figure 1. The low-energy positive-parity levels of the odd-mass Xe nuclides.

(Interacting Boson-Fermion Model) with some success,[15] and the use of spinor symmetry[16] for ^{131,133}Xe has also proven useful. Cunningham has also used the IBFM in a systematic approach to describe other odd-mass Xe nuclides,[17] which met with some success in describing the regular movement of the low-lying positive-parity levels as neutrons are removed from the $N = 82$ neutron-shell closure.

The low-energy negative-parity levels in the odd-mass Xe isotopes are shown in Fig. 2. The lowering of the $j-1$ and $j-2$ members of the $h_{11/2}$ plus phonon multiplet as neutrons are removed from ^{135}Xe distinguishes these nuclides. The extent of the lowering of the $9/2^-$ and $7/2^-$ members is such that the $9/2^-$ level becomes the metastable state in ^{127}Xe and the $7/2^-$ level becomes the metastable state in ^{123}Xe. Typical particle-rotor calculations can depress one or the other of these levels. Cunningham has been able to depress both in IBFM calculations but required significant admixtures of the $f_{7/2}$ and $h_{9/2}$ single-particle configurations from across the $N = 82$ shell.[18] Near mid-shell, the structure of the negative- and positive-parity levels in the odd-mass Xe isotopes becomes quite complex. Recent experimental data collected on these nuclides[19,20] may help resolve some of the intricacies of these nuclides.

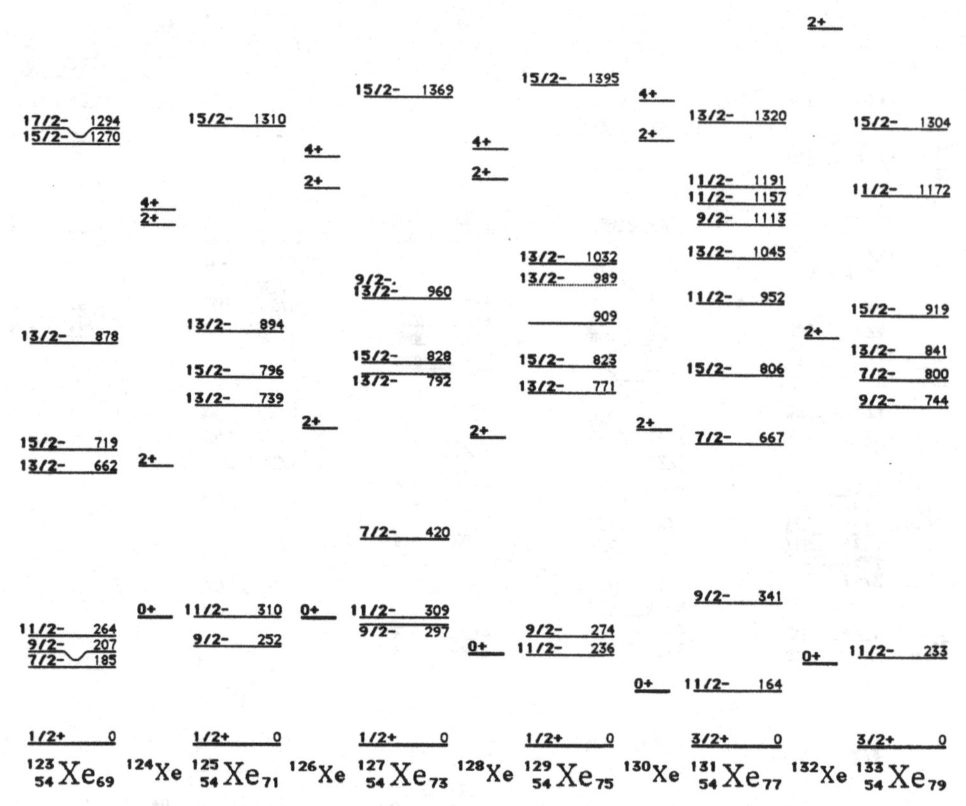

Figure 2. The low-energy negative-parity levels of the odd-mass Xe nuclides.

The even-even Xe nuclides whose structures are shown in Fig. 3 have been the subject of considerable theoretical work. An early interacting boson model (IBM) calculation by Puddu et al.[21] was used to describe level positions and decay characteristics of the Xe isotopes in the major shell from $N = 50$ to $N = 82$. Their calculations had the disadvantage of very little experimental work for comparison. Schneider et al.,[22] using IBM-2, calculated the level energies of the ^{128}Xe and ^{130}Xe nuclides with some success to reproduce successfully the elevation of the second 0^+ level in the heavier Xe isotopes. More recent IBM-1 calculations in the $O(6)$-limit have been reported by Casten and von Brentano,[23] who argued that the structure closely approaches that limit in ^{128}Xe and ^{130}Xe. Certain characteristics of the $O(6)$ limit were seen as carrying through to mid-shell at ^{120}Xe. Other models have also been used to describe the structures of the even-even Xe nuclides below $N = 82$; for example, a microscopic description using the MONSTER approach[24] and triaxial perturbations to the IBM $O(6)$ symmetry.[25]

Low-energy 0^+ states have been observed in the Xe nuclides in the $N = 82$ closed-shell nuclide ^{136}Xe and from ^{130}Xe through ^{118}Xe. No evidence for the presence of enhanced $E0$ strength has been found[26-28] in the nuclides, ^{124}Xe, ^{126}Xe, or ^{128}Xe. The absence of $E0$ strength in these nuclides reinforces the argument that they are indeed nearly representative of an IBM $O(6)$ symmetry. Enhanced monopole strength has been observed[29] for $2_3^+ \to 2_1^+$ and $6_2^+ \to 6_1^+$ transitions in ^{122}Xe and for $2_3^+ \to 2_1^+$ and $4_3^+ \to 4_1^+$ transitions in ^{118}Xe.

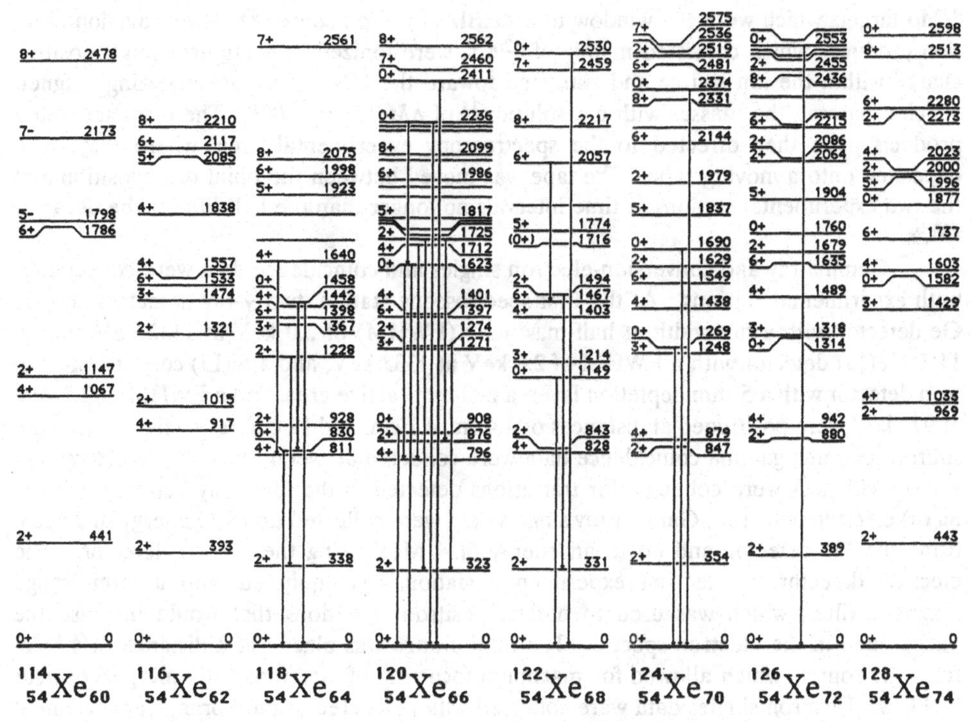

Figure 3. Systematics of the $Z = 54$ isotopes near mid-shell. The transitions mark decays which have been observed with enhanced $E0$ strength.

Previous studies on the structure of ^{120}Xe have used a range of techniques. Population of ^{120}Xe by decays of the ground and isomeric states of ^{120}Cs was reported by Genevey-Rivier et al.,[30] who identified transitions at 909 and 952 keV as having some $E0$ strength. They also identified a strong ground band up to spin 8 at 2099 keV. Two recent in-beam investigations have yielded similar results for high-spin states in ^{120}Xe. The study of Loewenich et al.[31] using the ^{110}Cd(^{13}C, $3n$)^{120}Xe and ^{111}Cd(^{12}C, $3n$)^{120}Xe reactions resulted in the identification of a well-defined ground band with positive- and negative-parity branches at $J > 8$ and a quasi γ-band starting with a spin parity 2^+ level at 876 keV. The quasi γ-band had several interconnecting γ rays to the ground band. The study by Rouabah et al.[32] using the ^{106}Cd(^{18}O, $4n$)^{120}Xe reaction also resulted in the identification of a ground band with one negative-parity and two positive-parity branches above $J = 8$. No quasi γ-band was identified in their work.

32.2 EXPERIMENTAL PROCEDURE

We carried out our experiment on the decay of $t_{1/2} \approx$ 60-s ^{120}Cs^{m,g} to levels in ^{120}Xe at the *UN*iversity *I*sotope *S*eparator at *O*ak *R*idge (UNISOR) at the *H*olifield *H*eavy *I*on *R*esearch *F*acility (HHIRF) located at Oak Ridge National Laboratory. Sources of ^{120}Cs^{m,g} were prepared by directing a 175-MeV ^{32}S tandem ion beam upon two 3-mg/cm^2

^{92}Mo targets which were the window to a FEBIAD-B2 ion source.[33] The heavy-ion reaction products, which consisted mainly of ^{120}Cs, were ionized to a uniform singly-positive charge within the ion source and extracted toward the UNISOR mass-analyzing magnet, which separated the masses with a resolution[34] of $\Delta M/M = 1000$. The mass separated products were then directed to the spectroscopy experimental area, where they were deposited onto a moving tape. The tape was moved between the point of deposition and the two experimental stations at time intervals no longer than the half-life for the decay of ^{120}Cs.

Gamma-ray and conversion-electron singles and coincidence data were collected at both experimental stations. At the first spectroscopy station, two γ-ray detectors, a 23% Ge detector with a full-width at half-maximum (FWHM) of 2.0 keV at 1332 keV and an 11% Ge(Li) detector with a FWHM of 2.8 keV at 1332 keV, and a Si(Li) conversion-electron detector with a 5-mm depletion layer, a 200-mm^2 active area, and a FWHM of 2.7 keV at 975 keV were positioned at distances of 4.4 cm, 4.7 cm, and 12 cm, respectively, from the source. Gamma-gamma coincidence data were collected between the γ-ray detectors, and e-γ coincidences were collected for transitions detected in the Ge γ-ray detector and the Si(Li) electron detector. Gamma-ray singles data were collected up to an energy of 2 MeV using the Ge detector and up to an energy of 4 MeV using the Ge(Li) detector. The electron detector in the first experimental station was equipped with a mini-orange magnetic filter, which was used to deflect positron radiations that would increase the background in the electron spectra. The mini-orange was placed at a distance of 4.5 cm from the source, which allowed for maximum focussing of electrons with energies of near 700 keV. Electron singles data were collected with this detector mini-orange arrangement to an electron energy of 3 MeV.

A γ-ray and a conversion-electron detector were positioned 4.6 cm and 7.0 cm, respectively, from the ^{120}Cs^{m,g} source at the second experimental station. The γ-ray detector was a 25% Ge detector with a FWHM of 2.1 keV at 1332 keV, while the Si(Li) electron detector had a 3-mm depletion layer and an active area of 200 mm^2. The electron detector had a FWHM of 3.0 keV and was also equipped with a mini-orange filter which was placed 2.5 cm from the source to allow for maximum electron detection efficiency at electron energies of approx. 300 keV. Electron-γ-ray coincidence data were collected between the Ge and Si(Li) detectors at this station, as were a γ-ray singles spectrum to an energy of 6 MeV and a conversion-electron singles spectrum to an energy of 2.5 MeV.

32.3 LEVEL SCHEME OF ^{120}Xe

The experimental conversion coefficients determined for some of the transitions in the decay of ^{120}Xe are listed in Table I and are shown graphically in Fig. 4. The conversion coefficients were normalized to a value of 27 x 10^{-3}. The theoretical lines for $M1$, $E1$, and $E2$ multipolarity were determined using values reported in Ref. 35. Relative-efficiency curves were plotted from data collected from a standard source of ^{207}Bi-^{133}Ba placed in a geometry identical to that of the ^{120}Cs^{m,g} sources. The relative electron intensities for transitions in ^{207}Pb and ^{133}Cs used in determining the relative efficiencies were taken from the *Table of Isotopes.*[36] A partial level scheme for the decay of ^{120}Xe is shown in Fig. 5, where transitions that are $E0$ or have significant $E0$ contributions are shown with heavy lines. The levels are arranged in the manner shown in Fig. 5 to draw attention to the possible presence of coexisting structures in ^{120}Xe that have different degrees of deformation. The observation of excess $E0$ strength for the 849-, 952-, and 1403-keV

Table I. Experimental Conversion Coefficients for Transitions Following the β/EC Decay of ^{120}Cs Which Have Excess Electron Intensity

Transition Energy (keV)		Conversion Coefficient[a] $\times 10^{-3}$	Theoretical *K*-Conversion[b] Coefficients $\times 10^{-3}$ *M*1	*E*2	*E*1
613.2	-*K*	*E*0			
714.6	-*K*	*E*0			
	-*L*	*E*0			
720.7	-*K*	*E*0			
849.4	-*K*	9.0(9)	2.76	2.02	0.82
908.7	-*K*	*E*0			
	-*L*	*E*0			
915.2	-*K*	21.3(25)	2.32	1.71	0.71
951.9	-*K*	11.7(12)	2.12	1.56	0.66
	-*L*	1.6(2)			
1402.8	-*K*	2.4(4)	0.89	0.69	0.33
1623.3	-*K*	*E*0			
2236	-*K*	*E*0			
2411	-*K*	*E*0			

a Conversion coefficients normalized to 322.5-keV transition *K*-conversion coefficient theoretical *E*2 value of 27×10^{-3} from Rosel et al.[35] The number(s) in parentheses is the error in the last digit(s) of the value for the experimental conversion coefficient.

b Calculated from Rosel et al.[35]

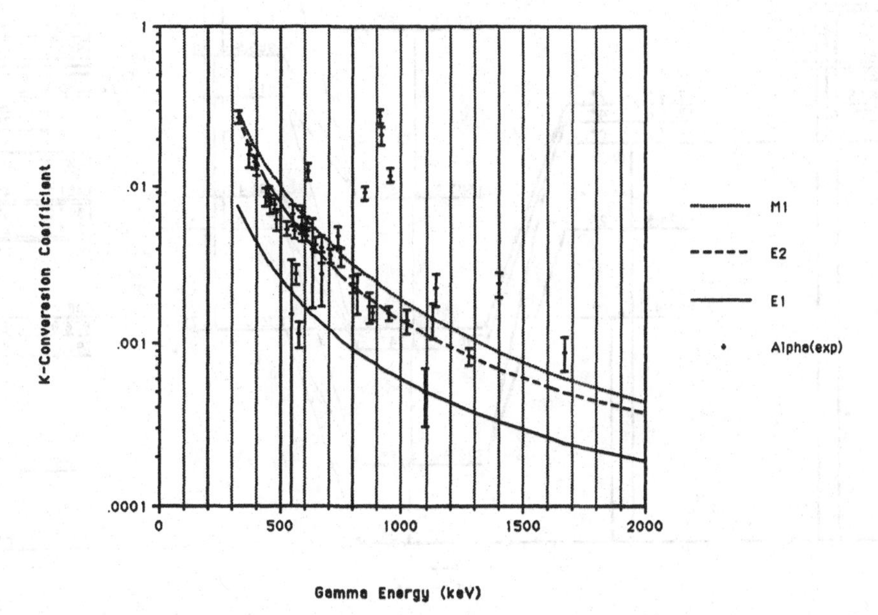

Figure 4. Experimental *K*-conversion coefficients plotted against transition energy for γ-rays observed in the decay of ^{120}Cs.

$2^+ \rightarrow 2^+$ transitions and on the 915-keV $4^+ \rightarrow 4^+$ transition serves as a basis for the separation of the structures based on the 909-keV 0^+ level from the ground band and quasi γ-band which, as noted earlier, had been observed in previous experiments.[26]

The levels lying above the 909-keV level in Fig. 5 have spin-parities and energy separations that can be interpreted as those of a vibrator. The fact that the levels of the vibrator decay into the ground band via transitions with $E0$ character also identifies this band, from discussions earlier in this text, as having a different radius from that of the ground band. The quasi γ-band starting with the 2^+ level at 876 keV does decay into the ground band, but not through transitions with enhanced $E0$ strength. The second 2^+ level of the "vibrator" band at 1725 keV, however, decays via a transition with a strong $E0$ component to the 876- keV 2^+ state, the band-head of the quasi γ-band. We have, therefore, interpreted the low-energy structure of ^{120}Xe as being composed of a more deformed rotation-like ground state band with $E_{4+}/E_{2+} \approx 2.5$ (typical of γ-soft nuclides and typical of the higher-mass Xe nuclides) and a quasi γ-band, both have the same radius, and a coexisting less deformed vibrational-like structure which lies 909 keV above the ground state. This structure has a radius different from the ground-state band and the quasi-γ band. The $E0$ transition strengths indicate mixing between these two systems despite the fact that their energies show little distortion.

We recognize that these same levels could be arranged as a simple phonon scheme as shown at the right in Fig. 5 with well defined first, second, and third phonon groupings.

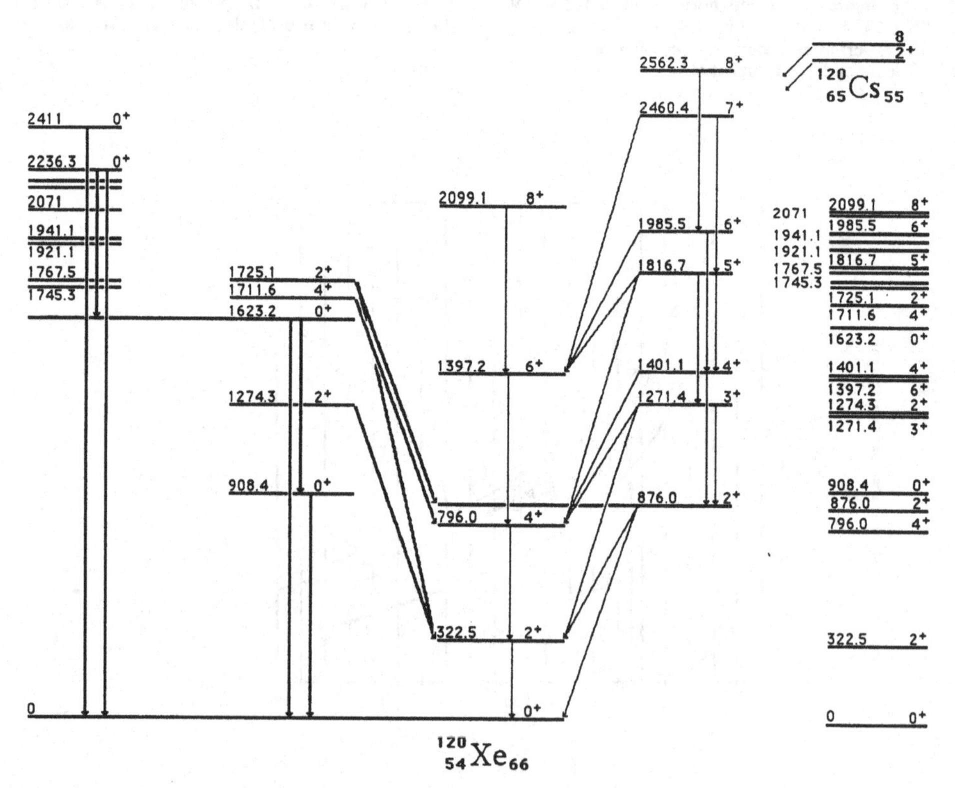

Figure 5. The level structure of ^{120}Xe. Only selected γ-ray transitions are shown. Heavy-lined transitions indicate the presence of $E0$ strength for that transition.

Even the fourth phonon excitation is relatively well defined, with two or three possible extra levels in that energy range. In this picture the transitions with excess $E0$ intensity are mainly $\Delta n = 2$ transitions as outlined by Colvin and Schreckenback.[37] Thus, the 915-keV $4_4^+ \to 4_2^+$ and 849-keV $2_4^+ \to 2_2^+$ transitions are both $\Delta n = 2$ transitions and both show strong $E0$ transitions. We comment, however, that the ratio $E(2_2^+)/E(2_1^+)$ (= 2.4) deviates strongly from the simple vibrator value of 2.0.

32.4 STRUCTURES OF THE N = 66 ISOTONES AROUND Z = 50

To begin the discussion of the N = 66 isotones (midshell for the N = 50 to 82 shell), we consider the structure of $^{116}Sn_{66}$ which is on the Z = 50 closed shell. The identification[38] of collective excitations built upon the 0_2^+ level at 1757 keV and the subsequent calculation[39] of absolute $E0$ and $E2$ transition rates for decays in ^{116}Sn allows interpretation of the levels in ^{116}Sn as shown in Fig. 6. The ground state structure is nearly vibrational, and the 0_2^+ state is a result of a two proton particle-two neutron hole intruder configuration upon which a regular ΔJ = 2 band is built. The collective intruder band has been reproduced using calculations which consider the coupling of the intruder configuration to the spherical quadrupole vibration in interaction with the low-lying quadrupole vibrational states.[40] The collective intruder band is very similar[41] to the ground bands observed in ^{120}Xe (Fig. 5) and in $^{112}Pd_{66}$. Strong support for the notion that

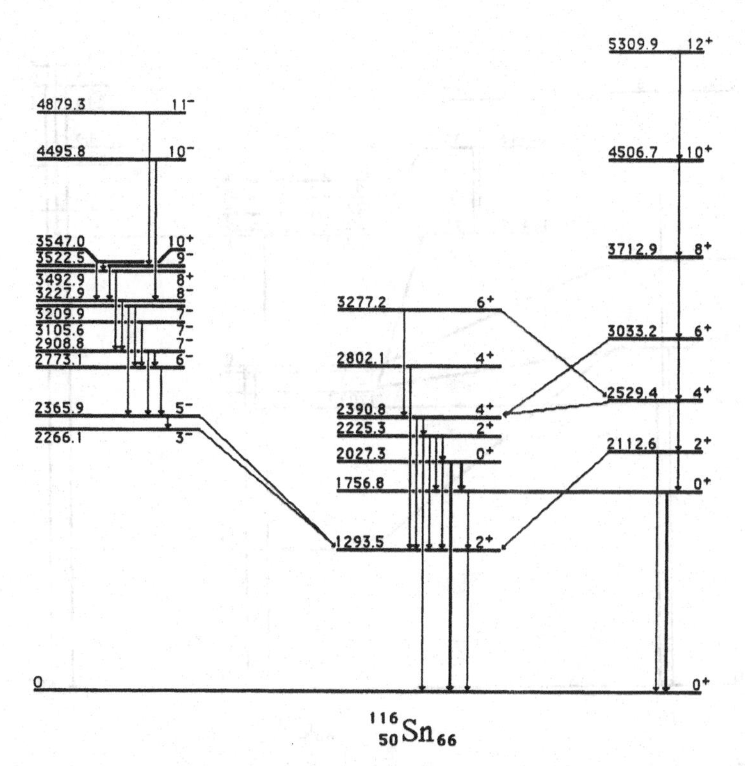

Figure 6. The level structure of ^{116}Sn. Heavy-lined transitions indicate the presence of $E0$ strength for that transition.

the band built on the 1756.8- keV level is a two-particle–two-hole band is found in the significant cross-section for the population of this level in the two proton transfer $^{114}Cd(^{3}He, n)$ reaction.[42]

The level structure of ^{114}Cd has been interpreted by Fahlander et al.[43] as arising strictly from a pure harmonic vibrator, where the extra 0^+ and 2^+ levels at 1307 and 1364 keV, respectively, are interpreted as members of the third quadrupole phonon. The wealth of conversion-electron data for transitions in $^{114}Cd_{66}$ that has been collected following the thermal-neutron capture of ^{113}Cd by Mheemeed et al.,[44] however, suggests that two-proton-particle–four-proton-hole intruder configurations play an important role in defining the low-energy structure. A partial level scheme has been arranged for ^{114}Cd similar to the level schemes for ^{116}Sn and ^{120}Xe and is shown in Fig. 7. Again, taking into account the notion that enhanced $E0$ strength arises in transitions between mixed states whose bases have differing radii, two coexisting systems with different deformation may be postulated in ^{114}Cd. The lowest structure is qualitatively of collective vibrational character, with nearly equal energy separations for the first, second, and third quadrupole phonons at 560, 1250, and 1930 keV, respectively. The coexisting system built upon the 0_2^+ level at 1134 keV is fairly well-defined $\Delta J = 2$ collective band and is accompanied by a $\Delta J = 1$ quasi γ-band beginning with the 2^+ level at 1842 keV. Both the collective band and quasi γ-band are connected to the vibrational structure by transitions showing enhanced $E0$ strength, but there are no transitions with enhanced $E0$ strength between the excited collective band and the quasi γ-band. The $E0$ strength observed for the $3_2^+ \rightarrow 3_1^+$ transition is particularly

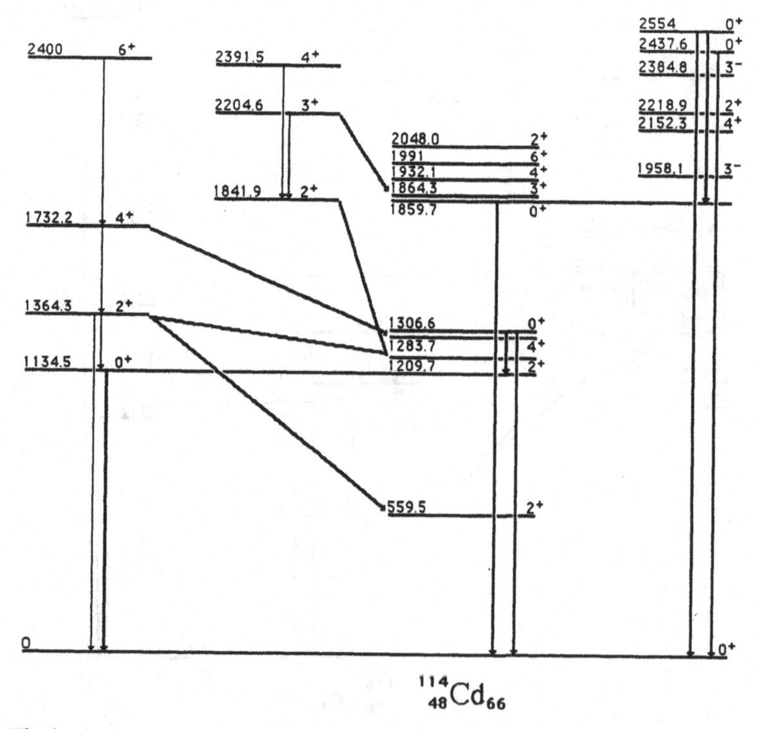

Figure 7. The level structure of ^{114}Cd. Only selcted γ-ray transitions are shown. Heavy-lined transitions indicate the presence of $E0$ strength for that transition.

valuable in making this analysis. IBM-2 calculations with configuration mixing were used to calculate energy levels and electromagnetic properties of ^{114}Cd with considerable success.[45,46] Clearly, this interpretation differs considerably from the purely vibrational alternative suggested by Fahlander.[43]

Above the Z = 50 closed shell, in ^{118}Te, the lowest 0^+ intruder four-particle–two-hole state is the major component of the 957- keV level.[12] It is possible to recognize the ΔJ = 2 band built on this level and the ΔJ = 1 quasi γ-band built on the 1482-keV 2^+ level, but all these structures are distorted by strong mixing. As before, the presence of $E0$ strength on the $2_2^+ \rightarrow 2_1^+$ and $4_2^+ \rightarrow 4_1^+$ transitions make possible this interpretation.

Following the systematics, it is tempting to interpret the level scheme of ^{120}Xe (Fig. 5) as a coexistence of a more deformed "intruder" six-particle–two-hole state that forms the ground state and the "normal" four-particle state, which lies at 909 keV and whose excitations are more vibrational-like. Apart from arguments based on the observed $E0$ transitions between the two systems, discussed in detail in Section 32.3, we stress that the ground state band has very similar character to those recognized as intruder bands elsewhere in the Z = 50 region, e.g., ^{116}Sn. We will discuss these interpretations further in Section 32.5. Here we only note that "intruder" ground states with "normal" coexisting configurations at higher energies are known in Mo and Pt isotopes.[47,48]

The systematics of the N = 66 isotones on either side of the Z = 50 closed shell are shown in Fig. 8.[49,50] the levels of the collective ΔJ = 2 band in each nuclide are labelled with a darkened circle, and levels attributed to the ΔJ = 1 quasi γ-band are labelled with a triangle. The collective ΔJ = 2 bands and the ΔJ = 1 quasi γ-bands move through the N = 66 isotones in similar fashion. The ΔJ = 2 band begins as the intruder band in ^{116}Sn at 1757 keV above the ground state decreases in energy on both sides of the Z = 50 closed shell and becomes the ground state band in ^{120}Xe and possibly in ^{112}Pd. Drawing any distinction about possible coexisting structures in ^{112}Pd is largely speculative, owing to the absence of measured $E0$ enhancements for the $2^+ \rightarrow 2^+$, $4^+ \rightarrow 4^+$, etc., transitions. The onset of larger deformation in ^{110}Ru, mirroring that in ^{122}Ba, means that any description in terms of coexisting "normal" and "intruder" structures of the type considered here loses relevance in these nuclides.

32.5 COEXISTING STRUCTURES IN ^{120}Xe

In Section 32.4, we interpreted the low-energy levels in ^{120}Xe using systematics of the N = 66 isotones. However, equally important is to consider the development of the coexisting structures as a function of neutron number. As can be easily seen in Fig. 3, energy levels in ^{118}Xe certainly and possibly in ^{122}Xe could be grouped in a similar way to those in ^{120}Xe. Comparable ground-state and quasi γ-bands can be identified, as well as somewhat distorted, but still recognizable, vibration-like structures at 830 keV (^{118}Xe) and 1149 keV (^{122}Xe). These distortions might be possibly attributed to configuration mixing, which also may account for weaker $E0$ strengths in both nuclei as compared to ^{120}Xe.[10] This analysis suggests that the vibrational-like structure is closest to the ground state in the mid-shell (800–900 keV) and increases in energy for N = 68. Unfortunately, this systematic approach cannot be extended further for lack of relevant experimental data. However, there is rather strong evidence that a coexisting structure does not affect the low-energy levels in ^{124}Xe and heavier Xe nuclides, as $E0$ strength practically disappears and these nuclei exhibit characteristic $O(6)$-like behavior.

We comment that there is no evidence so far in the Xe nuclides that the vibrational

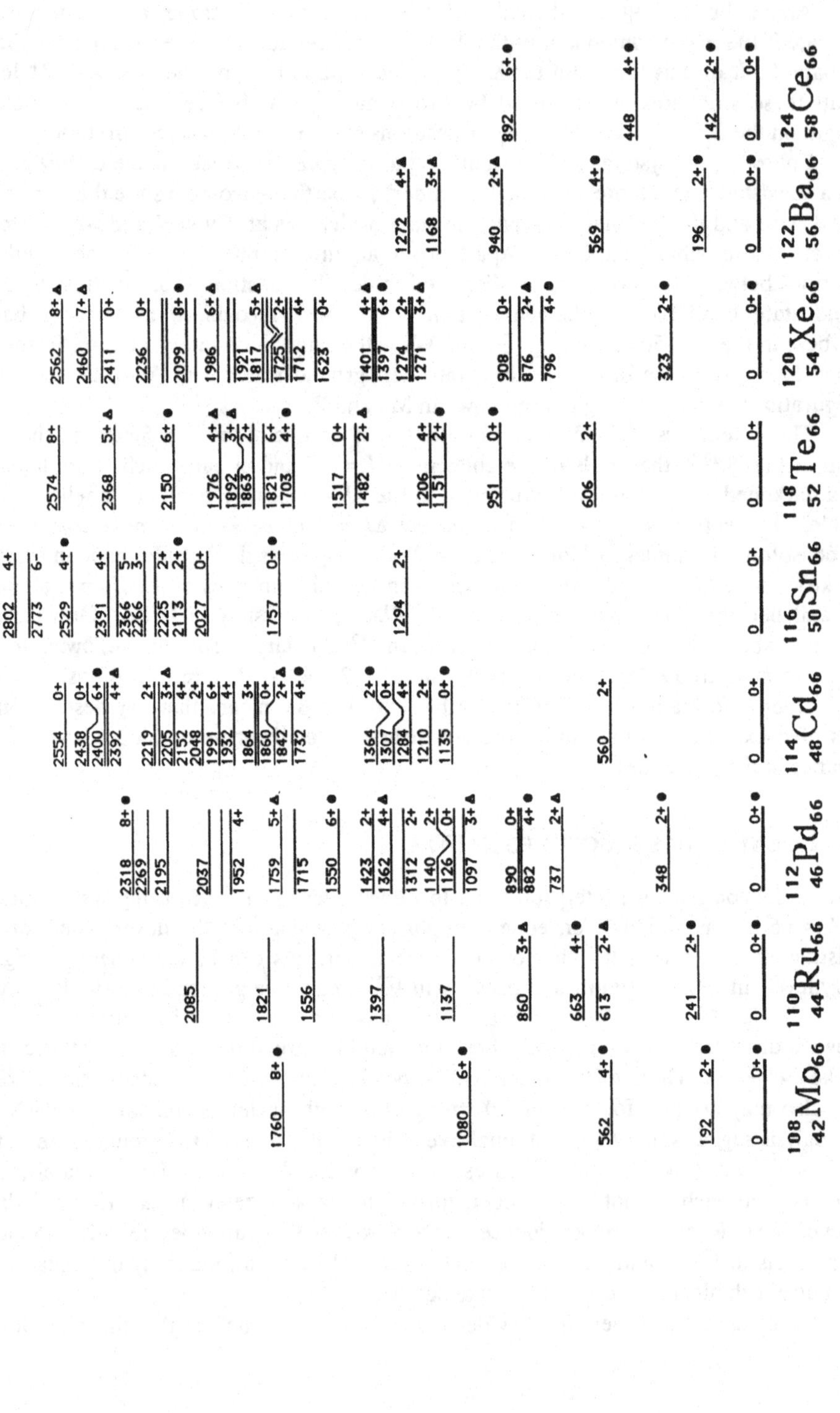

Figure 8. Systematics of the $N = 66$ isotones near the $Z = 50$ closed shell.

structures would become the ground state further from mid-shell as they do, e.g., in the Pt nuclei.[48] There the transition points are marked by a sudden increase in the energy of the 2_1^+ level energy. In the Xe nuclei there is little evidence for such a change for $N > 66$. It seems more likely, however, that the large jump in energy of the O_2^+ level between $^{120}Xe_{66}$ (908 keV) and $^{122}Xe_{68}$ (1149 keV) marks the transition from a configuration where the more deformed structure is the ground state to a configuration where the vibrational structure dominates the ground state. Moreover, for $N = 62$ and 60, the energy also rises more markedly. Thus, it is of great relevance to find more information on the 0^+ levels and $E0$ strength in the light Xe nuclides. If our concept of coexisting structures with different deformation in Xe isotopes is correct, the relative position and N dependence of these structures requires further elucidation. At this time it would appear that the more deformed structure dominates the ground state only for $^{118}Xe_{64}$ and $^{120}Xe_{66}$, with the deformed "intruder" structure rising rapidly as N drops to 62 or below or rises to 68 or above.

To place the discussion of deformed configurations in mid-shell even-A Xe nuclei in context, we note, briefly, evidence from neighboring odd-Z nuclides (and from the odd-A Xe nuclides). The role of strongly-sloping Nilsson orbitals in driving deformation has been widely discussed. For $Z > 50$ the relevant proton orbital is $9/2^+[404]$. The presence of the hole states in this orbital is direct evidence for strong deformation. In the odd-Z neighbors of ^{120}Xe we find the $(g_{9/2})^{-1}$ proton hole as the ground state in ^{121}Cs and the same deformed intruder has its minimum energy in I nuclides at ^{119}I (307 keV). These hole states are co-existent with the shell-model $(d_{5/2}{}^3)$ ground-state configuration in ^{119}I that is nearly spherical,[51] as evidenced by its magnetic moment, and the less-deformed $3/2^+[422]$ isomer in ^{121}Cs; neither involves the transfer of protons across the $Z = 50$ closed shell. For the higher-mass, odd-N Xe nuclides shown in Fig. 1, Mantica et al. have shown that their structure changes very slowly with increasing N and indicates only weak deformation.[52]

32.6 CONCLUSIONS

New measurements have revealed intense $E0$ activity in ^{120}Xe that can be contrasted with minimal $E0$ strength in ^{124}Xe and heavier Xe nuclides. The relatively stable structures found at and above mass 124 correlate well with the equally stable and not very deformed structures in Cs above mass 125 and I above mass 123. The more complex structures observed near mid-shell may be attributed to the coexistence of and mixing between normal four-proton structures and six-particle, two-hole configurations that involve the promotion of a pair of protons across the $Z = 50$ shell gap. More theoretical work and more data, in particular on the lighter isotopes ($A \leqslant 118$), are needed to understand this phenomenon fully.

ACKNOWLEDGMENTS

This work was supported by the U. S. Department of Energy under contract numbers DE-FG05-88ER 40418 (UM), DE-FG05-84ER40159 (LSU), DE-AC05-OR00033 (UNISOR), DE-AC05-84OR21400 (ORNL) and the Scientific and Engineering Research Council (S.E.R.C.) of the U. K. and by a NATO travel grant. WBW is pleased to acknowledge the support of the John Simon Guggenheim Foundation and the General

Research Board of the University of Maryland during the performance of these studies. The authors are pleased to acknowledge numerous fruitful discussions on the subject of intruder structures with Profs. K. Heyde and J. Wood and Dr. R. A. Meyer.

REFERENCES

1. K. Heyde, P. Van Isacker, M. Waroquier, J. L. Wood, and R. A. Meyer, *Phys. Rept.* **102**, 291 (1983).
2. M. Gai, D. M. Gordon R. E. Shroy, D. B. Fossan, and A. K. Gaigalas, *Phys. Rev. C* **26**, 1101 (1981).
3. T. L. Shaw, V. R. Green, C. J. Ashworth, J. Rikovska, N. J. Stone, P. M. Walker, and I. S. Grant, *Phys. Rev. C* **36**, 413 (1987).
4. C. Thibault, F. Touchard, S. Buttgenbach, R. Klapisch, M. DeSaintSimon, H. T. Duong, P. Jacquinot, P. Juncar, S. Liberman, P. Pillet, J. Pinard, J. Vialle, A. Pesnelle, G. Huber, and the ISOLDE Collaboration, *Nucl. Phys.* **A367**, 1 (1981).
5. L. G. Kostova, W. Andrejtscheff, K. K. Kostov, F. Dönau, L. Kaubler, H. Prade, and H. Rotter, *Nucl. Phys.* **A485**, 31 (1988).
6. E. W. Schneider, G. J. Mathews, S. V. Jackson, P. W. Gallagher, and W. B. Walters, *Phys. Rev.* C 13, 1624 (1976).
7. J. Rogowski, N. Kaffrell, D. DeFrenne, K. Heyde, E. Jacobs, M. N. Harakeh, J. M. Schippers, and S. Y. Van Der Werf, *Phys. Lett.* **B207**, 125 (1988).
8. K. Heyde, J. Jolie, J. Moreau, J. Ryckebusch, M. Waroquier, P. Van Duppen, M. Huyse, and J. L. Wood, *Nucl. Phys.* **A466**, 189 (1987).
9. E. L. Church and J. Weneser, *Phys. Rev.* **103**, 1035 (1956).
10. K. Heyde and R. A. Meyer, *Phys. Rev. C* **37**, 2170 (1988).
11. P. M. Walker, C. J. Ashworth, I. S. Grant, V. R. Green, J. Rikovska, T. L. Shaw, and N. J. Stone, *J. Phys.* **G13**, L195 (1987).
12. T. L. Shaw, Ph.D. Thesis, Oxford Univ. (1987).
13. J. Rikovska, N. J. Stone, P. M. Walker, and W. B. Walters, *Nucl. Phys.* **A505**, 145 (1989).
14. W. Borchers, E. Arnold, W. Neu, R. Neugart, K. Wendt, G. Ulm, and the ISOLDE Collaboration, *Phys. Lett.* **B216**, 7 (1989).
15. R. A. Meyer, V. Paar, S. Brant, E. A. Henry, S. Lane, R. Chrien, and B. K. S. Koene, preprint (1989).
16. M. M. Michailova, *J. Phys.* **G13**, L149 (1987).
17. M. A. Cunningham, *Nucl. Phys.* **A385**, 221 (1982).
18. M. A. Cunningham, *Nucl. Phys.* **A385**, 204 (1982).
19. P. F. Mantica, Jr., et al., private communication.
20. J. Blachot and C. E. N. Grenoble, private communication.
21. G. Puddu, O. Scholten, and T. Otsuka, *Nucl. Phys.* **A348**, 109 (1980).
22. E. W. Schneider, M. D. Glascock, W. B. Walters, and R. A. Meyer, *Phys. Rev. C* **19**, 1025 (1979).
23. R. F. Casten and P. von Brentano, *Phys. Lett.* **B152**, 22 (1985).
24. E. Hammaren, K. W. Schmid, F. Grummer, A. Faessler, and B. Fladt, *Nucl. Phys.* **A454**, 301 (1986).
25. A. Sevrin, K. Heyde, and J. Jolie, *Phys. Rev. C* **36**, 2631 (1987).
26. W. B. Walters, J. Rikovska, N. J. Stone, T. L. Shaw, P. M. Walker, and I. S. Grant, *Hyperfine Int.* **43**, 343 (1988).
27. P. F. Mantica, Jr., B. E. Zimmerman, W. B. Walters, D. Rupnik, E. F. Zganjar, W. L. Croft, Y.-S. Xu, and H. K. Carter, *Bull. Am. Phys. Soc.* **34**, 1236 (1989).
28. S. Ohya, N. Mutsuro, K. Katsumata, K. Miyazawa, T. Itoh, H. Yagi, Y. Fujita, and H. Kawakami, *INS Tokyo Prog. Rept.*, 40 (1980).
29. W. B. Walters, private communication.
30. J. Genevey-Rivier, A. Charvet, G. Marguier, C. Richard-Serre, J. D'Auria, A. Huck, G. Klotz, A. Knip-

per, G. Walter, and the ISOLDE Collaboration, *Nucl. Phys.* **A283**, 45 (1977).
31. K. Loewenich, K. O. Zell, A. Dewald, W. Gast, A. Gelberg, W. Lieberz, P. von Brentano, and P. Van Isacker, *Nucl. Phys.* **A460**, 361 (1986).
32. M. S. Rouabah, Th. Byrski, F. A. Beck, D. Curien, B. Haas, J. C. Merdinger, J. P. Vivien, J. Gizon, and B. Nayako, *Z. Phys.* **A328**, 493 (1987).
33. R. Kirchner, K. H. Burkard, W. Huller, and O. Klepper, *Nucl. Inst. Meth.* **186**, 295 (1981).
34. E. H. Spejewski, R. L. Mlekodaj, and H. K. Carter, *Nucl. Inst. Meth.* **186**, 71 (1981).
35. F. Rosel, H. M. Fries, K. Alder, and H. C. Pauli, *At. Data Nucl. Data Tables* **21**, 215 (1978).
36. C. M. Lederer and V. S. Shirley, Eds., *Table of Isotopes, 7th Ed.*, (John Wiley & Sons, Inc., New York 1978).
37. G. C. Colvin and K. Schreckenback, *Proc 5th Int. Sym. Capture Gamma-Ray Spectroscopy, AIP Conference Preceedings* **125**, 290 (1985).
38. J. Bron et al., *Nucl. Phys.* **A318**, 335 (1979).
39. J. Kantele, R. Julin, M. Luontama, A. Passoja, T. Poikolainen, A. Backlin, and N. G. Jonsson, *Z. Phys.* **A289**, 157 (1979).
40. G. Wenes, P. Van Isacker, M. Waroquier, K. Heyde, and J. Van Maldeghem, *Phys. Rev. C* **23**, 2291 (1981).
41. R. Estep, R. K. Sheline, D. J. Decman, E. A. Henry, L. G. Mann, R. A. Meyer, W. Stoeffl, L. E. Ussery, and J. Kantele, *Phys. Rev. C* **39**, 76 (1989).
42. H. W. Fielding, R. E. Anderson, C. D. Zafiratos, D. A. Lind, F. E. Cecil, H. H. Wieman, and W. P. Alford, *Nucl. Phys.* **A281**, 389 (1977).
43. C. Fahlander, A. Backlin, L. Hasselgren, A. Kavka, V. Mittal, L. E. Svensson, B. Varnestig, D. Cline, B. Kotlinski, H. Grein, E. Grosse, R. Kulessa, C. Michel, W. Spreng, H. J. Wollersheim, and J. Stachel, *Nucl. Phys.* **A485**, 327 (1988).
44. A. Mheemeed, K. Schreckenbach, G. Barreau, H. R. Faust, H. G. Borner, R. Brissot, P. Hungerford, H. H. Schmidt, H. J. Scheerer, T. von Edigy, K. Heyde, J. L. Wood, P. Van Isacker, M. Waroquier, G. Wenes, and M. L. Stelts, *Nucl. Phys.* **A412**, 113 (1984).
45. K. Schreckenbach, A. Mheemeed, G. Barreau, T. von Edigy, H. R. Faust, H. G. Borner, R. Brissot, M. L. Stelts, K. Heyde, P. Van Isacker, M. Waroquier, and G. Wenes, *Phys. Lett.* **B110**, 364 (1982).
46. K. Heyde, P. Van Isacker, M. Waroquier, G. Wenes, and M. Sambataro, *Phys. Rev. C* **25**, 3160 (1982).
47. M. Sambataro and G. Molnar, *Nucl. Phys.* **A376**, 201 (1982).
48. C. D. Papanicolopulos et at., *Z. Phys.* **A330**, 371 (1988).
49. A. C. Mueller, F. Buchinger, W. Klempt, E. W. Otten, R. Neugart, C. Ekstrom, J. Heinemeier, and the ISOLDE Collaboration, *Nucl. Phys.* **A403**, 234 (1983).
50. K. L. Ying, P. J. Bishop, A. N. James, A. J. Kirwan, D. J. G. Love, T. P. Morrison, P. J. Nolan, D. C. B. Watson, K. A. Connell, A. H. Nelson, and J. Simpson, *J. Phys. G: Nucl. Phys.* **12**, L211 (1986).
51. V. R. Green, N. J. Stone, T. L. Shaw, J. Rikovska, K. S. Krane, P. M. Walker, and I. S. Grant, *Phys. Lett.* **B173**, 115 (1986).
52. P. F. Mantica, Jr., B. E. Zimmerman, W. B. Walters, H. K. Carter, D. Rupnik, E. F. Zganjar, W. L. Croft, and Y.-S. Xu, *Physical Review C* **42**, 909 (1990).

33. Global Trends and Structural Consequences of the Proton-Neutron Interaction

D. S. Brenner
Clark University
Worcester, MA 01610, USA

R. F. Casten
Brookhaven National Laboratory
Upton, NY 11973, USA

C. Wesselborg
Brookhaven National Laboratory
Upton NY 11973, USA
and
University of Giessen
Giessen, West Germany

D. D. Warner
Daresbury Laboratory
Daresbury, Warrington WA4 4AD, England

J.-Y. Zhang
Clark University,
Worcester MA 01610, USA
and
Brookhaven National Laboratory,
Upton NY 11973, USA
and
Center of Theoretical Physics, (CCAST World Lab),
Institute of Modern Physics,
Lanzhou, PRC

ABSTRACT

The average *p-n* interaction energy of each of the last two protons with each of the last two neutrons can be computed from the double difference in binding energies,

$$\delta V_{pn}(Z+1,N+1) = \{ [B(Z+2,N+2) - B(Z+2,N)] - [B(Z,N+2) - B(Z,N)] \}/4.$$

Summation of δV_{pn} over all valence nucleons yields V_{pn}, the integrated valence *p-n* interaction, which is found to increase linearly with N_pN_n early in a shell and then saturates as mid-shell is approached. This saturation in V_{pn} is consistent with the saturation in collectivity near mid-shell as evidenced by experimental $B(E2{:}0^+ \rightarrow 2^+)$ values and thus provides empirical support for the N_pN_n scheme. Results for various shell and subshell regions are discussed.

Exotic Nuclear Spectroscopy, Edited by W. C. McHarris
Plenum Press, New York, 1990

33.1 INTRODUCTION

It has long been recognized that nuclei do not have deformed ground states or exhibit significant collective behavior unless they contain minimum numbers of *both* protons and neutrons beyond closed-shell configurations. In medium-mass and heavy nuclei, for example, it has been shown that each valence nucleon must have available four to ten valence nucleons of the opposite type before deformation sets in.[1,2] Empirically, it is found[3] that experimental observables such as $B(E2{:}2_1^+\rightarrow 0_1^+)$ values and $E_{4_1^+}/E_{2_1^+}$ ratios, which are sensitive indicators of the degree of collectivity and deformation, correlate to a remarkable extent with the valence-nucleon product, N_pN_n, which can be viewed as a rough first-order approximation to the integrated *p-n* interaction strength. Thus, it has become clear that the *p-n* interaction among valence nucleons is largely responsible for deformation, configuration mixing, and collectivity in nuclear matter. Because the N_pN_n scheme implicitly assumes that the *p-n* interaction is orbit independent, it is an oversimplification, especially in a deformed field, where the overlap of proton and neutron wave-functions depends on the spatial orientations of their orbits relative to each other. Nevertheless, the success of the N_pN_n scheme has refocussed attention on the critical role played by the attractive *p-n* force in nuclear structure, and we shall explore here several aspects of the interaction which have been highlighted by recent studies of empirical *p-n* interaction strengths.[4]

33.2 THE *p-n* INTERACTION

The *p-n* interaction differs from the *p-p* and *n-n* interactions in that it is comprised of both $T = 0$ and 1 components, in contrast to the latter two, where only $T = 1$ is allowed. Since, as we have already noted, nuclear deformation does not occur unless valence nucleons of both types are present, it is clear that the $T = 1$ interaction, which by charge independence is identical for *p-p*, *n-n*, and *p-n* cases, cannot be the cause of deformation. Otherwise, such a nucleus as ^{116}Sn, which has a closed proton shell but sixteen valence neutrons, would surely be deformed, which it is not. Therefore, the $T = 0$ component of the *p-n* interaction must be the critical factor which leads to deformation and collectivity, and, indeed, it has been recognized as such for many years.[5-6]

Efforts to extract *p-n*-interaction matrix elements empirically also have a long history. Both Schiffer and True[7] and Molinari et al.[8] have extracted $T = 0$ and 1 interaction energies from *p-n* multiplets near closed shells in order to develop a phenomenological parameterization of the residual interaction. Sakai[9] has undertaken a similar effort and Talmi[6] has discussed lucidly the global behavior of both components. Simply stated, the $T = 1$ component is found to be strongly attractive in the $J = 0$ (pairing) state. For higher J values, this component becomes repulsive because of the pauli principle. In contrast, the $T = 0$ interaction is both stronger and always attractive. Both interactions can be simulated by multipole expansions dominated by strong monopole and quadrupole components. To first order, a δ-function interaction provides a reasonable approximation, although repulsive components must be added, especially for the $T = 1$ channel.

33.3 THE N_pN_n SCHEME

Empirical results, such as those described above, are for cases where the interacting valence protons and neutrons reside in essentially pure shell-model configurations. Until recently, the systematic behavior of the *p-n* interaction for highly-collective nuclei, well removed from shell closures, has been largely unknown. Systematic nuclear-structure data, however, can provide a framework for ascertaining the importance of the *p-n* interaction in such regions, and it is from this approach that the phenomenological N_pN_n scheme has evolved. In this scheme, an observable is plotted as a function of the product of the number of valence protons or proton holes (N_p) and valence neutrons or neutron holes (N_n) as counted from the nearest closed shell or strong subshell. The success of this simple parameterization is clearly illustrated in Figs. 1 and 2, where observables which exhibit disparate behaviors, when plotted versus such traditional quantities as *Z*, *N*, or *A*, coalesce and become smoothly-varying functions of N_pN_n. While the empirical usefulness of the N_pN_n scheme, for example, in predicting the properties of unknown nuclei far from stability, should not be overlooked, it is important to keep in mind that it is presently a phenomenological formalism lacking a detailed microscopic foundation. In order to improve our understanding of the basis of the N_pN_n scheme, we have undertaken a global investigation of the *p-n* interaction strength using an approach, discussed recently in Refs. 10–12 and closely related to ones previously used in Refs. 13 and 14, which involves only nuclear binding energies and is, thus, usually free from assumptions concerning the structure of the nuclei involved or their wave functions.

33.4 VALENCE *p-n* INTERACTION ENERGIES

The method is as follows: Using experimental binding energies,[15] we isolate the *p-n* interaction energies of the *last* two protons with the *last* two neutrons by a double difference,

$$\delta V_{2p\text{-}2n}(Z+1,N+1) = [B(Z+2,N+2)-B(Z+2,N)]-[B(Z,N+2)-B(Z,N)]. \quad (1)$$

This is represented schematically in Fig. 3. Each binding energy (negative) in Eq. 1 is composed of *p-p*, *n-n*, and *p-n* contributions, as well as contributions to the average spherical field, where, by the latter usage, we mean that part of the interaction that is unrelated to the deformation-driving part of the *p-n* interaction. The first difference in Eq. 1 isolates the interaction of the last two neutrons with all $Z+2$ protons and with each other

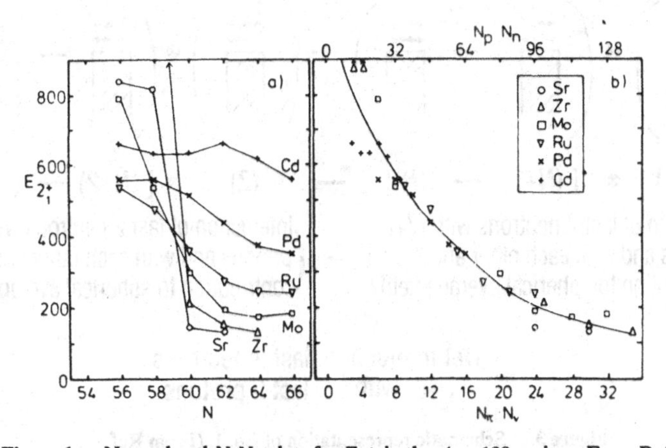

Figure 1. Normal and N_pN_n plots for $E_{2_1^+}$ in the A = 100 region. From Ref. 3.

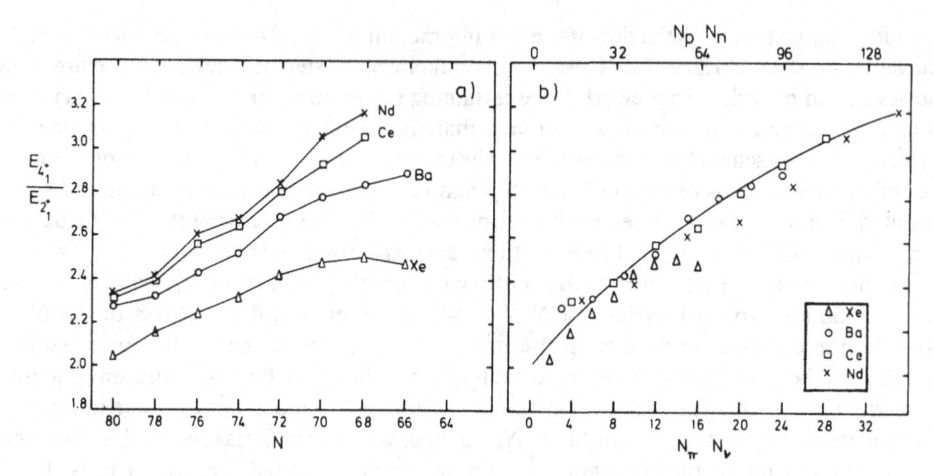

Figure 2. Normal and N_pN_n plots for $E_{4_1^+}/E_{2_1^+}$ in the $A = 130$ region. From Ref. 3.

and contains their contributions to the average spherical field. The second term does the same except that, in this instance, it is the interaction of the last two neutrons with Z protons. The difference of these two quantities cancels the p-p and n-n interactions as well as the mean field contributions, yielding the net interaction of the last two protons with the last two neutrons. In Eq. 1, we call this quantity $\delta V_{2p\text{-}2n}$ and denote it by the intermediate nucleus $(Z+1,N+1)$, recognizing that the p-n interaction isolated is actually that of protons $Z+1$, $Z+2$ with neutrons $N+1$, $N+2$. One can obtain the *average* interaction of *each* of the last two protons with *each* of the last two neutrons, δV_{pn}, by dividing by 4:

$$\delta V_{pn} = 1/4\ \delta V_{2p\text{-}2n} \tag{2}$$

For an attractive p-n interaction, both quantities are negative. Note that δV_{pn} is equivalent

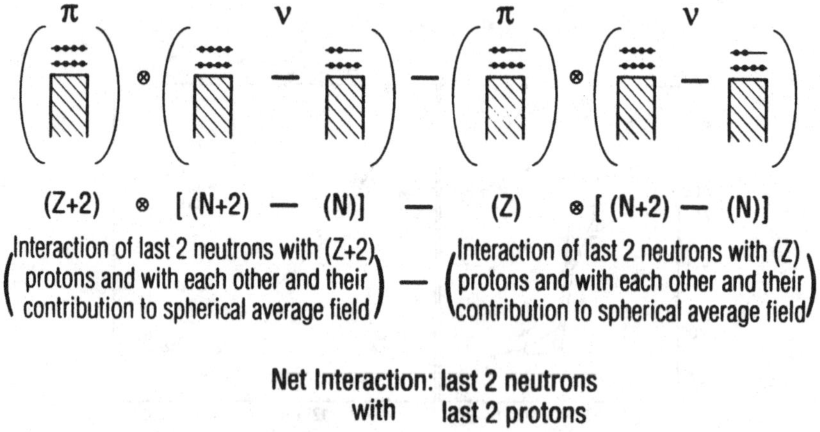

π ν π ν

(Z+2) ⊗ [(N+2) — (N)] — (Z) ⊗ [(N+2) — (N)]

(Interaction of last 2 neutrons with (Z+2) protons and with each other and their contribution to spherical average field) — (Interaction of last 2 neutrons with (Z) protons and with each other and their contribution to spherical average field)

Net Interaction: last 2 neutrons with last 2 protons

Figure 3. Schematic representation of Eq. 1. (From Ref. 4.)

to quantities defined in Refs. 11–12.

For comparison with the N_pN_n scheme, it is more convenient to obtain a *total, integrated* valence *p-n* interaction energy, V_{pn}, which is the sum of δV_{pn} values from the nearest doubly-magic nucleus. It is an algebraic exercise to show that

$$V_{pn}(Z+\delta_p, N+\delta_n) = 1/4\, \delta_p\delta_n[B(Z+2\delta_p, N+2\delta_n) - B(Z+2\delta_p, N_0)] \\ -[B(Z_0, N+2\delta_n) - B(Z_0, N_0)]\,, \qquad (3)$$

where (Z_0,N_0) specify the nearest magic numbers and $\delta_p(\delta_n)$ is +1 if the valence protons (neutrons) are particles and −1 if they are holes. Again, we have denoted V_{pn} by the intermediate odd-odd nucleus, but, in fact, its evaluation only involves the binding energies of even-even nuclei.

Unfortunately, this expression often involves binding energies of very unstable proton- and neutron-rich nuclei [i.e., $B(Z+2,N_0)$ or $B(Z_0,N+2)$], which are unavailable. However, in the proton-particle–neutron-hole region around A = 130, the data do exist to extract V_{pn} values for comparison with N_pN_n. For the A = 100 region (proton-hole–neutron-particle nuclei), the binding energies $B(50,50)$ and $B(48,50)$ are unknown but can be estimated by short extrapolation. Since they are constants for all nuclei in this region and all Cd nuclei, respectively, there is little or no effect on trends in V_{pn}. The results are shown in Fig. 4, where a remarkably-linear dependence of V_{pn} with N_pN_n is seen, extending to nuclei with nearly twenty valence nucleons. It should be noted that there are many overlapping data points corresponding to different combinations of valence neutrons and protons that have the same N_pN_n product and, in most instances, nearly identical V_{pn} values. [There are a few exceptions, of course, associated with extreme N_pN_n combinations. For example, at N_pN_n = 15, where the (N_p,N_n) = (1,15) combination yields a somewhat larger V_{pn} than the (3,5) and (5,3) cases.] The linearity of these results implies a constancy in δV_{pn}. As we shall see later, this is not always the case. Nevertheless, these results provide strong support to, and justification for, the rationale behind the N_pN_n scheme—that the *p-n* interaction strength is *linear* in N_pN_n, at least early in a shell.

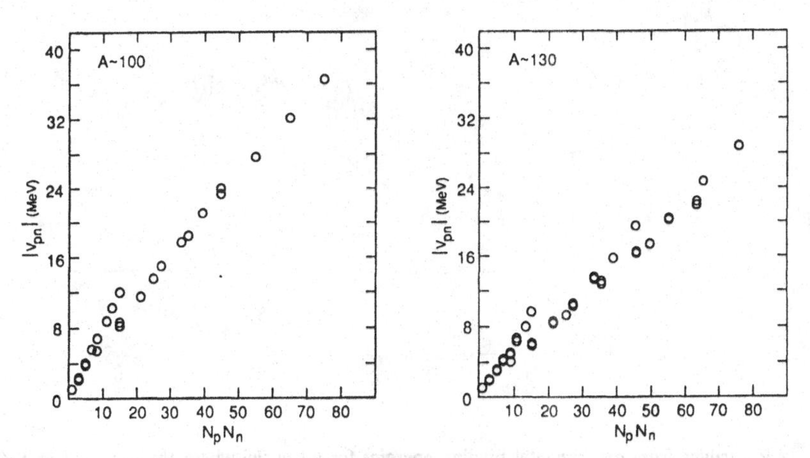

Figure 4. $|\delta V_{pn}|$ plotted against N_pN_n for two mass regions. Error bars are smaller than the points. (From Ref. 4.)

33.5 GLOBAL SYSTEMATICS OF δV_{pn}

Although, as we have seen, the N_pN_n scheme is very successful in systematizing observables and is linear with V_{pn} early in shells, a more extensive examination of δV_{pn} systematics reveals structural features (some subtle, others dramatic) that require insight into the orbital dependence of the *p-n* interaction strength. Figure 5 shows the global variation of δV_{pn} with N for all even-even nuclei where binding energies are known to a precision of 50 keV or better. We see that $|\delta V_{pn}|$ varies in a generally smooth manner, increasing from a minimum of $\approx$200 keV in the actinides to $\approx$700 keV near $N = 30$. Below $N = 30$, if one puts aside momentarily the obvious excursions at $N = Z$, this general trend continues and $|\delta V_{pn}|$ increases faster, although the scatter of individual δV_{pn} values is greater. The general trend, though not all the detailed structural features, has long been known[12-14] and, in particular, was discussed in Ref. 13 in the context of corrections to recursive partial difference mass formulas of generalized Garvey-Kelson[16] type.

Qualitatively, this behavior can be interpreted in terms of the radii of proton and neutron orbits. First, as N and Z increase and successive oscillator shells sequentially fill, these radii increase. Since the range of the residual interactions is roughly constant, as the average distance between the last proton and the last neutron increases, its strength decreases even if the two particles occupy equivalent orbits. As nuclei become heavier, N increases more rapidly than Z in order to mitigate the destabilizing coulombic repulsion among protons. This leads to a second important effect. In medium and heavy nuclei, the proton and neutron orbits are filled at different rates so that, as $N-Z$ increases, the radii

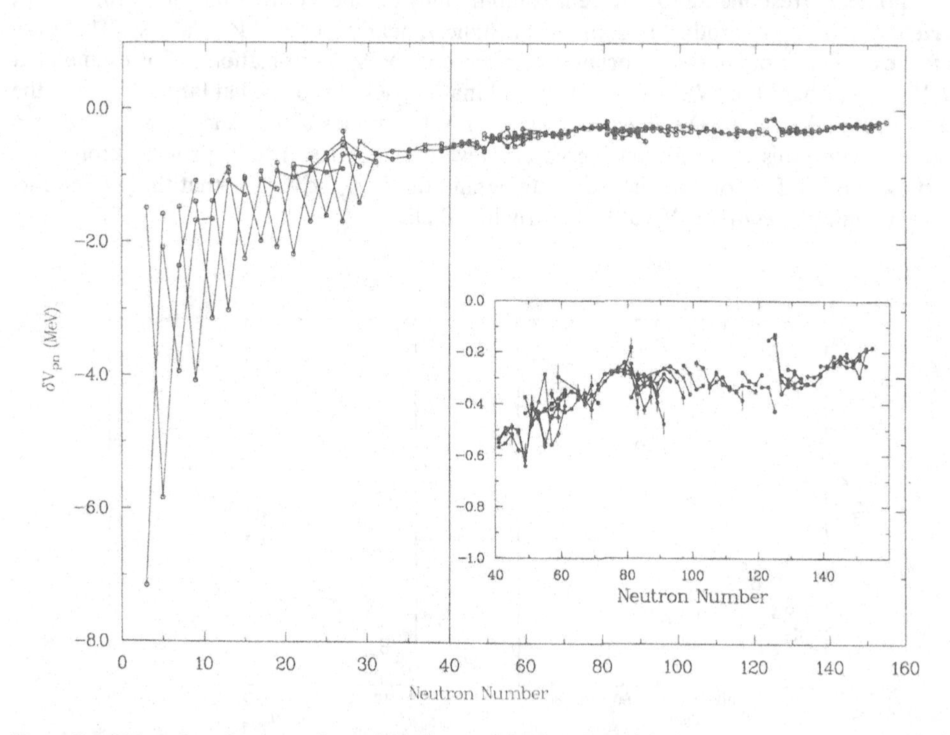

Figure 5. δV_{pn} values from experimental binding energies for all nuclei where the precision in δV_{pn} is 50 keV or better. Note the change in horizontal scale at $N = 40$. The inset gives an expanded view for $N > 40$. The sharp downward spikes for low N all occur at $N = Z$.

of the outermost neutrons grow faster than those for protons. This leads to smaller overlap and augments the first effect. Of course, one expects, superimposed upon this trend, structure due to effects arising from specific orbital occupancies which might be expected to modulate the *p-n* interaction strength. We shall see, later, that this is indeed so.

33.6 $N = Z$ NUCLEI

Undeniably, the most remarkable feature of Fig. 5 (shown in more detail in Fig. 6*a*) is the set of dramatic excursions occurring in all isotopic chains at $N = Z$ for $N < 30$. For most of these $N = Z$ nuclei, $|\delta V_{pn}|$ increases by a factor of 2–3 and ranges in value from ≈ 1.5 to nearly 6 MeV, pointing to special *p-n* interaction effects in such nuclei. In the clustering analysis of Ref. 17, a quantity related to the extra binding of *p-n* pairs also showed these $N = Z$ anomalies, although the effect was not discussed.

This phenomenon, in some sense, is well known and is traditionally dealt with in successful mass formulas by introduction of an asymmetry term which is usually of the form $(N-Z)^2/A$. This can be conceptualized in the following way.[18] There are $|N-Z|$ "excess" nucleons that occupy higher-lying orbits where the *p-n* overlap is diminished, resulting in lower nuclear binding. The fraction of the total nuclear volume affected is $|N-Z|/A$ so the total binding-energy effect is $(N-Z)^2/A$, as already noted. Such asymmetry terms are essentially of phenomenological origin and are parameterized to fit the data.

The dramatic variation of δV_{pn} for $N < 30$ and its regularity and persistence in all known $N = Z$ nuclei suggest a simple underlying microscopic origin. It is presumably closely related to evidence for α clustering previously adduced in quartet modelling[19-21] of nuclear mass systematics and thus to supermultiplet theory. In view of these points, it would seem reasonable that it might emerge as a characteristic feature of even simple, almost schematic, shell model calculations, where it should be possible to investigate the relative roles of the $T = 0$ and 1 interactions and the sensitivity to single-particle energies.

Our approach was to do a set of shell-model calculations for a single *j* shell—in this case the 2*s*-1*d* shell from *N* or $Z = 8$–20, using a surface-δ residual interaction.[22] Several sets of single-particle energies were tried, as well as various values of the interaction strengths $V_{T=0}$ and $V_{T=1}$. Examples of the results are shown in Fig. 6*b-e* using single-particle energies based on the low-lying states of ^{17}O and ^{17}N. All calculations produced larger $|\delta V_{pn}|$ values, specifically at $N = Z$. The effects are somewhat sensitive to the choice of single-particle energies, decreasing as $\epsilon(d_{5/2})$ increases. However, for reasonable V_T values (such as shown in Fig. 6*c*), the sensitivity to single-particle energies is minimal. The size of the effect grows as either $V_{T=0}$ or $V_{T=1}$ increases but is considerably more sensitive to $T=0$, as might be anticipated for an effect that is essentially singular at $N=Z$. A given value of $V_{T=0}$ produces a larger effect than the same value of $V_{T=1}$. In order to estimate a reasonable strength for $V_{T=1}$ in this region, the 0^+, 2^+, and 4^+ energies were fit for the two-neutron and two-proton nuclei, ^{18}O and ^{18}Ne. Good fits were obtained for $V_{T=1} = 0.8$ MeV. Since it is well known[19,23] that $V_{T=0} > V_{T=1}$ [and $V_{T=0} \sim (1.5\text{–}2) \times V_{T=1}$], we show results obtained with $V_{T=1} = 0.8$ MeV and $V_{T=0} = 1.6$ MeV in Fig. 6*c*, which indeed produces a large effect at $N = Z$.

It is clear that the large excursions at $N = Z$ result naturally from simple shell-model considerations, even though our calculations are highly schematic. We do not claim that these calculations are realistic; nevertheless, it is interesting to note that the spikes in Fig. 6*c* are within a factor of two of the magnitudes observed in Fig. 6*a*. Other consequences of the effect are revealed in the calculated ground-state wave functions. For

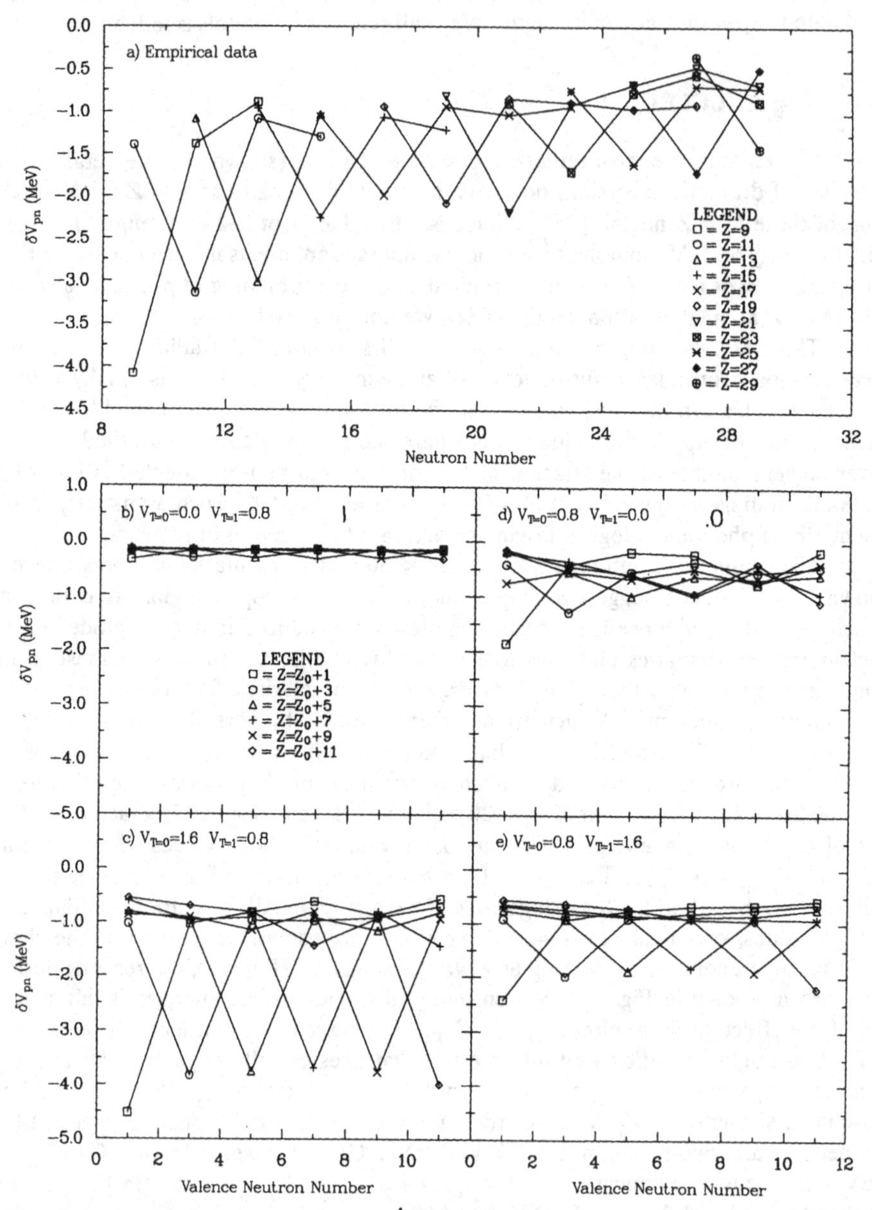

Figure 6. *a*) Expanded view of the data in Fig. 1 for light nuclei. *b-e*) Results of various shell-model calculations for the 2*s*-1*d* shell, corresponding to the N = 9–19 portion of Part *a*). The single-particle energies used were $\epsilon\,(d_{5/2}) = 0$, $\epsilon\,(s_{1/2}) = 0.5$ MeV, and $\epsilon\,(d_{3/2}) = 3.5$ MeV.

$N = Z$ nuclei, they are highly admixed, with strength distributed over many configurations. This is just another manifestation of the configuration mixing induced by the *p-n* interaction when its effects are maximized, because all proton and neutron pairs occupy orbits with a very high degree of spatial overlap.

The appearance of the effect in these highly-schematic calculations and its insensitivity to their details reflect its origin in very basic features (see, also Refs. 19–21) of the spatial symmetry character of two-nucleon wave functions. To see this, consider the simple examples of an $N = Z$ nucleus with two valence protons and two valence neutrons filling an $s_{1/2}$ orbit. Then, in the 0^+ ground state, the two-particle wave-functions have $T = 1, S = 0$. The requirement of total antisymmetry means that the spatial part of the wave function must be symmetric, giving a large interaction strength for a short-range (e.g., δ) force. For $N \neq Z$ in light nuclei, the last proton and neutron fill non-equivalent orbits. The *p-n* interaction is then a mixture of $T = 0$ and 1 parts, and both symmetric and antisymmetric spatial wave-functions occur. These arguments suggest that new data for heavier $N = Z$ nuclei would be highly interesting in order to see how long the $N = Z$ effect persists as the shell structure becomes more complex and coulomb effects alter the relative proton and neutron single-particle energies.

33.7 ORBIT SENSITIVITY OF THE *p-n* INTERACTION

The orbit sensitivity of the *p-n* interaction can be illustrated by examining the behavior of δV_{pn} around the $N = 126$ shell closure (Fig. 7*a*). In the region just prior to the neutron shell closure ($N \approx 121$–125), δV_{pn} is ≈ -300 keV for $Z = 81$ nuclei and considerably weaker, ≈ -150 keV for $Z = 83,85$. In contrast, for nuclei beyond the $Z = 82$ proton shell closure, namely $Z = 83,85$, $|\delta V_{pn}|$ increases sharply when N crosses to the new shell at $N > 126$. It has previously been noted[11] that the liquid-drop model does not account for these anomalies, so we must look to a shell-model, i.e., an orbit-dependent, explanation. This behavior can be understood qualitatively by considering the overlap of valence proton and neutron orbits. For $Z = 81, N < 126$, the proton hole occupies a low-*j* (e.g., $s_{1/2}$) orbit

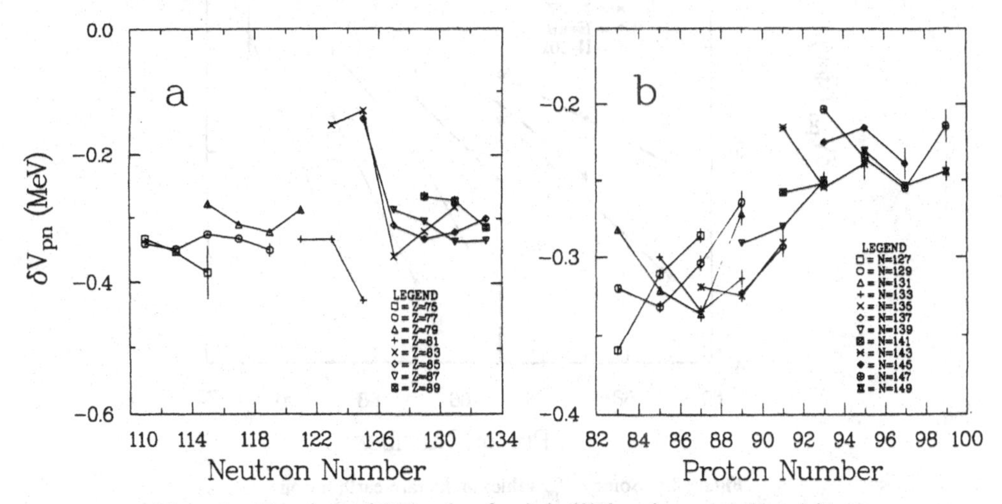

Figure 7. Experimental δV_{pn} values for *a*) the ^{208}Pb region and *b*) the actinides.

at the top of the 50–82 shell, while the neutron holes are also in low-j orbits (e.g., $p_{1/2}$) near the top of the 82–126 shell. These orbits have good spatial overlap, resulting in a strong p-n interaction. For Z = 83,85, in contrast, the valence protons are in high-j orbits (e.g., $h_{9/2}$, $f_{7/2}$), which overlap poorly with the $p_{1/2}$ neutrons, yielding a weak p-n interaction. Once the neutron shell is full (N > 126), the neutrons occupy $g_{9/2}$ orbits, and we expect a strong p-n interaction for Z = 83,85 with the $h_{9/2}$ protons. This is indeed what is seen.

Additional insight into the orbital dependency of the p-n interaction can be gained by examining the behavior of δV_{pn} in a deformed region such as the actinides. It is well known that many collective observables, e.g., $B(E2{:}2_1^+ \rightarrow 0_1^+)$ values, "saturate" near mid-shell and become asymptotic. To investigate whether this can be related to the p-n interaction, we can examine the systematics of δV_{pn} values shown in Fig. 7b, where isotonic data for Z > 82 are plotted versus Z. Here we see a clear trend toward smaller $|\delta V_{pn}|$ as Z increases, which again can be easily understood by considering orbital occupancy. All shells have the same characteristic structure: The first nucleons enter steeply-downsloping equatorial Nilsson orbits and then, as the shell fills, the orbital inclinations gradually increase, ending with the filling of polar orbits. In an approximate way, for a given fractional occupancy in each major shell, orbits of a similar slope are being filled. In the actinide region (Fig. 7b) the δV_{pn} values near Z = 84 result from neutrons near the beginning of the neutron shell such that both neutrons and protons occupy steeply-downsloping orbits with high overlap, yielding large $|\delta V_{pn}|$ values. Near Z = 94, N = 145, the fractional filling of the neutron shell is significantly greater than for protons, and consequently both the p-n valence orbital overlap and $|\delta V_{pn}|$ are smaller.

In the rare-earth region (Fig. 8), for all nuclei for which δV_{pn} values can be com-

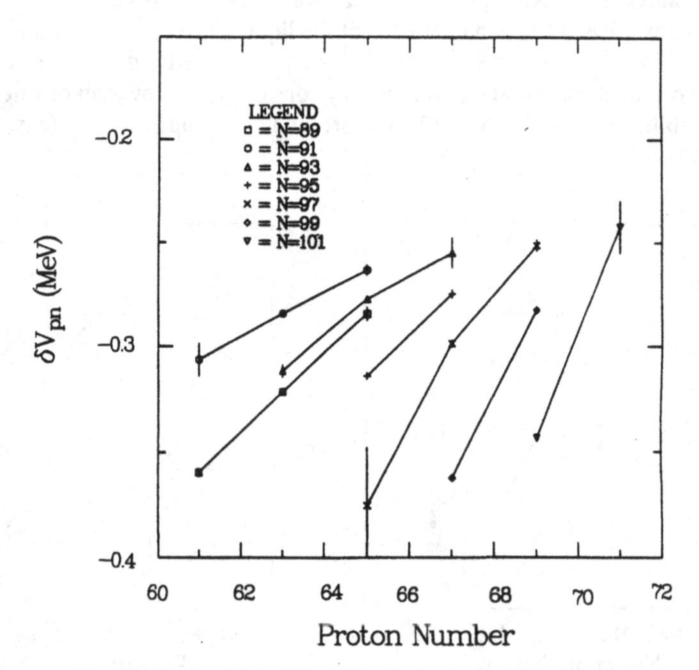

Figure 8. Some δV_{pn} values in the rare-earth region.

puted, the 50–82 proton shell is fractionally more filled than the 82–126 neutron shell. For these nuclei, the last proton occupies an orbit with relatively small overlap with that of the last neutron. This becomes even more so as Z increases and results in decreasing $|\delta V_{pn}|$ values. For a given Z, on the other hand, increasing N leads to the filling of neutron orbits with greater overlap with that of the last proton, and larger $|\delta V_{pn}|$ values result. Both of these trends are apparent in Fig. 8.

These decreasing trends in δV_{pn} towards mid-shell parallel the saturation in quadrupole collectivity observed experimentally. If we now interpret the near constancy of δV_{pn} values as mid-shell is approached (which is best seen in Fig. 7*b* for the actinide region) as reflecting the small values of the quadrupole component of the *p-n* interaction in these regions,[4] and if we assume that the only other significant component of the interaction is monopole, then the asymptotic value is an approximate measure of the latter. Therefore, we find that the critical monopole component of the *p-n* interaction is attractive with a strength ≈ -250 keV.

It seems clear from the examples cited above that orbital-occupancy patterns are responsible for much of the fine structure found for δV_{pn} values near shell closures and in deformed regions. In some cases, however, the explanation of the fine structure (see inset in Fig. 5) in δV_{pn} is not so obvious (e.g., near $N = 84$), and further study may be informative in regard to orbit sequences, subshell effects, etc. Such studies are in progress as is further work on δV_{pn} systematics for lighter nuclei.

33.8 SUMMARY

To summarize, a technique for extracting *p-n* interaction energies, δV_{pn}, has been utilized to investigate the behavior of the *p-n* interaction throughout the nuclidic chart. The smooth systematic decrease of the empirical *p-n* interaction strength (excepting $N = Z$ nuclei) with increasing A is understood as a consequence of the gradual decrease in valence proton and neutron orbital overlap due to the occupancy of shells with different average radii. The remarkable strength of δV_{pn} for $N = Z$ nuclei has been shown to result from enhanced *p-n* interactions in configurations with maximal spatial overlap, specifically at $N = Z$, as a consequence of the pauli principle for protons and neutrons in equivalent orbits. The effect results primarily from the $T = 0$ component of the effective *p-n* interaction. The behavior of δV_{pn} near the $N = 126$ shell and for deformed rare-earth and actinide nuclei has provided insight into the orbital occupancy dependence of the effective *p-n* interaction strength, an understanding of the variation of the collective quadrupole component as a shell fills, and a measure of the strength of the monopole component of the force. Finally, the results provide direct experimental support and a microscopic basis for the N_pN_n scheme, verifying the linearity of the *p-n* interaction strength against N_pN_n early in a shell and its subsequent saturation.

ACKNOWLEDGMENTS

This work was supported under contracts DE-FG02-88ER40417 and DE-AC02-76CH00016 with the U. S. Department of Energy and by the Humboldt Foundation.

REFERENCES

1. R. F. Casten, D. S. Brenner, and P. E. Haustein, *Phys. Rev. Lett.* **58**, 658 (1987).
2. Y. Y. Sharon, *Phys. Rev. C* **37**, 1768 (1988).
3. R. F. Casten, *Phys. Rev. Lett.* **54**, 1991 (1985).
4. J.-Y. Zhang, R. F. Casten, and D. S. Brenner, *Phys. Lett.* **227B**, 1 (1989).
5. A. de Shalit and M. Goldhaber, *Phys. Rev.* **92**, 1211 (1953).
6. I. Talmi, *Rev. Mod. Phys.* 34, 704 (1962).
7. J. P. Schiffer, *Ann. Phys.* **66**, 798 (1971); J. P. Schiffer and W. W. True, *Rev. Mod. Phys.* **48**, 191 (1976).
8. A. Molinari, et al *Nucl. Phys.* **A239**, 45 (1975).
9. M. Sakai, *Nucl. Phys.* **A345**, 232 (1980).
10. J.-Y. Zhang, C.-S. Wu, C.-H. Yu, and J. D. Garrett, *Contemp. Topics in Nucl. Structure Phys., Cocoyoc*, 1988, Abstract Vol., p. 109.
11. J. Janecke and E. Comay, *Nucl. Phys.* **A436**, 108 (1985).
12. N. Zeldes, M. Gronau, and A. Lev, *Nucl. Phys.* **63**, 1 (1965).
13. J. Janecke and H. Behrens, *Phys. Rev.C* **9**, 1276 (1974).
14. M. K. Basu and D. Banerjee, *Phys. Rev. C* **3**, 992 (1971).
15. A. H. Wapstra, G. Audi, and R. Hoekstra, *At. Data Nucl. Data Tables* **39**, 281 (1988).
16. G. T. Garvey and I. Kelson, *Phys. Rev. Lett.* **16**, 197 (1966).
17. G. G. Dussel, R. J. Liotta, and R. P. J. Perazzo, *Nucl. Phys.* **A388**, 606 (1982).
18. R. D. Evans, *The Atomic Nucleus* (McGraw-Hill, New York, 1955), p. 369.
19. A. Arima and V. Gillet, *Ann. Phys.* (N.Y.) **66**, 117 (1971).
20. M. Danos and V. Gillet, *Z. Phys.* **249**, 294 (1972).
21. Y. K. Gambhir, P. Ring, and P. Schuck, *Phys. Rev. Lett.* **51**, 1235 (1983).
22. Modified MSU shell-model code.
23. A. de Shalit and H. Feshbach, *Theoretical Nuclear Physics* (Wiley, New York, 1974).

34. IBFFA Calculations of Odd-Odd Nuclei

W.-T. Chou and Wm. C. McHarris
National Superconducting Cyclotron Laboratory
and Departments of Chemistry and Physics/Astronomy
East Lansing, MI 48824, USA
Olaf Scholten
Kernfysisch Versneller Instituut
9747 AA Groningen, The Netherlands

ABSTRACT

Odd-odd nuclei have been calculated before in the framework of the IBA model. However, in order to keep the calculations tractable, they were done by using supersymmetries or were limited only to unique-parity orbits. We used computer codes for the interacting-boson-fermion-fermion approximation (IBFFA) model, which adopted a very efficient model-space trunction scheme, so no restriction needed to be made and complete calculations of the excitation energy spectra were obtained. We calculated a series of odd-odd isotopes, $^{176-184}$Re. The calculations started from even-even Os cores, where a simplified hamiltonian for the *SU*(3) limit in the IBA-1 model was used. Both negative- and positive-parity calculations for nearby odd-mass Re and Os isotopes have also been carried out. The detailed results of excitation energy spectra of all three types of calculations are presented, also some selected quadrupole and magnetic moments, and *B*(*E*2) and *B*(*M*1) values. All calculations are compared with experimental data. One of the successes of the IBFFA model is the prediction of the doubly-decoupled bands, which have been verified from experiments.

34.1 INTRODUCTION

The properties of low-lying collective states in even-even nuclei can be described in terms of a system of interacting bosons. This is the interacting-boson approximation (IBA) model,[1] which has been proven to be able to give a rather accurate description of all different types of even-even nuclei. Likewise, its extension to odd-mass nuclei, the interacting-boson-fermion approximation (IBFA) model,[2] is capable of reproducing a large variety of properties in phenomenological calculations. We now introduce to you the further extension of the IBA model to odd-odd nuclei, where we call it the interacting-boson-fermion-fermion approximation (IBFFA) model. We show the results of deformed odd-odd Re isotopes as examples.

In order to calculate an odd-odd nucleus in the IBFFA model, we need to choose an even-even core plus neighboring odd-mass nuclei, then calculate them in the IBA and IBFA models first. For example, ^{182}Re. It has 107 neutrons (in the shell 82–126) and 75 protons (in the shell 50–82). Both particles lie in shells that are over half-filled; therefore, we should consider the number of holes. The appropriate even-even core is the one having one fewer neutron hole and one fewer proton hole, ^{184}Os. The corresponding odd-mass nuclei are those having either one fewer neutron hole or one fewer proton hole, ^{183}Re and ^{183}Os, respectively.

Exotic Nuclear Spectroscopy, Edited by W. C. McHarris
Plenum Press, New York, 1990

In order to make the calculations feasible, we truncate the model space.[3] It can be explained by the following example: First an IBFA calculation is done for ^{183}Os in which the odd-neutron hole is coupled to the even-even core. Then only the lowest fifteen states of each spin and parity of this calculation are used for coupling the odd-proton nucleus, ^{183}Re. The advantagve of this procedure is that one of the two odd particles is coupled in a complete basis. The truncation enters only in the coupling of the second particle. We can also calculate ^{182}Re by coupling in the reverse order: Calculate the odd-proton nucleus ^{183}Re in a complete basis first, then couple the odd-neutron ^{183}Os to only the lowest fifteen states of each spin and parity of ^{183}Re. The space truncation scheme has been tested[4] by comparing the results from these two different coupling methods. The energies for the first five states are almost identical in both calculations, and the differences for higher-energy states are within 20 keV of each other. We can further truncate the basis by coupling, say, to only the lowest twelve states of each spin and parity. By doing this, the results are the same within 10 keV.

The hamiltonian in the IBFFA model can be expressed as:

$$H_{\mathrm{IBFFA}} = H_{\mathrm{IBA}} + H_{n,\mathrm{F}} + H_{p,\mathrm{F}} + V_{n,\mathrm{BF}} + V_{p,\mathrm{BF}} + V_{n,p} \tag{1}$$

The first five terms are from pervious IBA and IBFA calculations. The last term is the neutron-proton residual interaction, which has two components and can be written as:

$$V_{np} = V_q Q\cdot Q + V_s\, \sigma\cdot\sigma \tag{2}$$

The first component is the quadrupole force. Since the major part of the neutron-proton interaction is already taken into account via the first five terms of the hamiltonian, we simply chose $V_q = 0$ for the odd-odd Re isotopes. The second component is the spin-spin interaction. Its effect is that it introduces a spin singlet-triplet splitting observed in the near-closed-shell nuclei. As can be seen here, no new parameter has been introduced in the IBFFA calculations. The method is simply to couple the odd-neutron and odd-proton nuclei together.

34.2 IBA AND IBFA CALCULATIONS

Before we show results of IBFFA calculations, we show examples of IBA and IBFA calculations since we need these to judge IBFFA calculations. Figure 1 shows an example of IBA calculations for even-even Os cores. Only states below 2 MeV in excitation energy are concerned, because only the structures of these lowest-lying states will enter into the later calculations for the odd-odd states. A description in terms of the IBA-1 model,[5] where no explicit distinction is made between proton and neutron excitations, is sufficient. A simplified hamiltonian for an axial rotor was used. (This makes our parameters a little different than other calculations done for the heavier Os isotopes, using the general hamiltonian). The parameters were adjusted in order to give a best overall agreement with experimental data. In addition, we ask that the parameters should vary smoothly from isotope to isotope.

Besides excitation energy spectra, we can also calculate electromagentic properties, such as quadrupole moments and reduced transition probabilities $B(E2)$, as shown in Fig. 2. We reach good agreement for the heavier isotopes, and for lighter isotopes the parameters

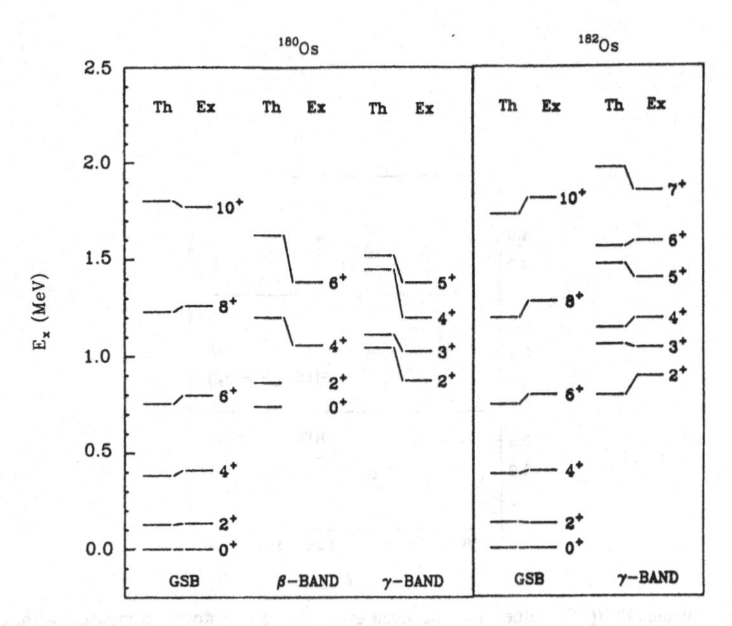

Figure 1. IBA-calculated excitation energy spectra for the even-even Os core isotopes compared with experimental data.

were simply chosen by systematics because of lack of experimental data.

In the IBFA method, positive- and negative-parity states are calculated separately, and we calculated unique-parity states first. All the available single-particle orbits in the shell were included in the calculations, so all the valence nucleons should be accounted for. Again, parameters varied smoothly from isotope to isotope. Figure 3 shows the IBFA calculations for negative-parity states of odd-mass Re isotopes. For these calculations, in addition to the $h_{11/2}$ orbit, which is the unique-parity orbit, we also included the $h_{9/2}$ orbit from the next major shell. Notice the band at the left side is built on the $h_{9/2}$ orbit and is a $K=1/2$ band. The coriolis force is usually very large in this kind of band and plays an important role on the order of the band members. This can be seen in Fig. 3, where some of the band members are missing and the level with $I=K=1/2$ is pushed higher than the $I=5/2$ and 9/2 levels.

Figure 4 shows the comparison of IBFA calculations with experimental results for positive-parity states of odd-mass Os isotopes. These positive-parity bands, originating from $i_{13/2}$, also suffer from large coriolis force. The IBFA calculations reproduce most of the features of these bands, but overestimate the staggering within the bands. Notice that there are no parameters in the IBFA calculations that control the magnitude of the coriolis force.

Figure 5 shows the results for negative-parity states of odd-mass Os isotopes. These calculations involve five single-particle orbits, and they are very time consuming. Therefore, we tried to keep the less important parameters unchanged. Also, we emphasized systematic variation of the parameters, which gave a reasonable overall view of all five isotopes, rather than try to obtain a good fit for individual isotopes.

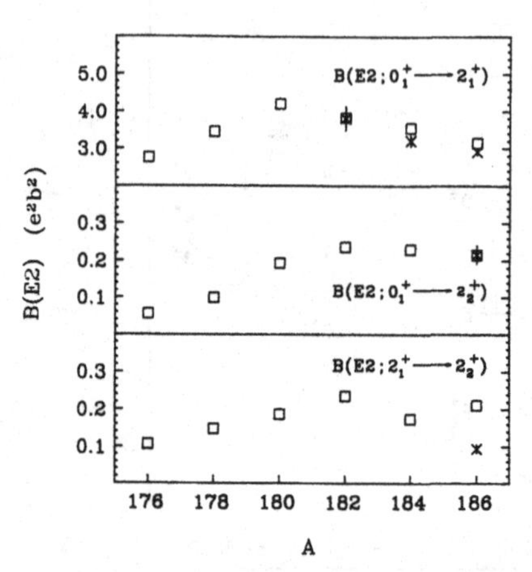

Figure 2. IBA-calculated $B(E2)$ values for the even-even Os core isotopes compared with experimental values. The squares are calculated values; the crosses (with error bars), experimental values.

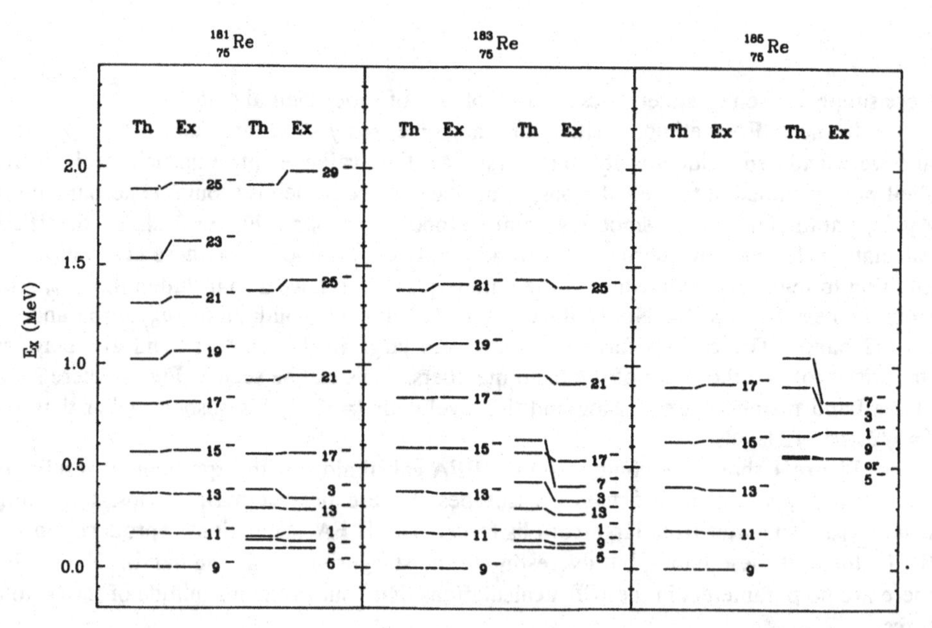

Figure 3. IBFA-calculated excitation energies for negative-parity states in the odd-mass Re isotopes compared with experimental data. The states are labeled with $2J$.

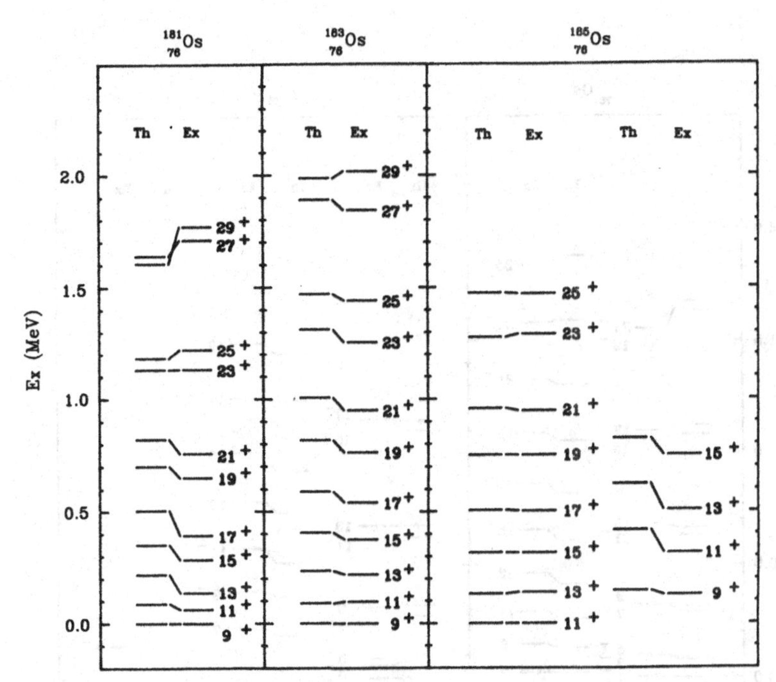

Figure 4. IBFA-calculated excitation energies for positive-parity states in the odd-mass Os isotopes compared with experimental data. The states are labeled with 2J.

Figure 6 shows the results for magnetic moments for odd-mass Re isotopes. The agreements with experiment are quite reasonable.

34.3 IBFFA CALCULATIONS

Among the five odd-odd Re isotopes we calculated, ^{182}Re is the one most studied[6]; therefore, we show it as a first example. The calculated results and experimental data are plotted in Fig. 7 in two parts because the relative position of the 7^+ triplet coupling and the 2^+ singlet coupling is not known experimentally. In the IBFFA calculation, the 2^+ singlet coupling was predicted to lie 38.6 keV higher than the 7^+ triplet coupling. We obtained good agreements for three out of the four bands shown in Fig. 7, but very large discrepancies for the 4^- band. The single particles for this 4^- band are the $1/2^-[541\downarrow]$ proton and the $9/2^+[624\uparrow]$ neutron, both having large coriolis matrix elements. In the IBFA calculations, we did not obtain particularly good spacings for the 1/2 band, and we overestimated the staggering for the 9/2 band. These effects apparently have been transmitted into the odd-odd states here. Also, in the calculations there is a close-lying 5^- singlet-coupling band. Because of the coriolis force, which is not attenuated in the present calculations, these two bands mix strongly, with the result that the lowest member of the $K^\pi = 4^-$ band is a 5^- state. We obtain quite reasonable values for the triplet-singlet splitting for the bands shown in Fig. 7, although this is not always true for some higher-lying bands.

Good tests of IBFFA calculations are the predictions for the doubly-decoupled bands[7] in ^{176}Re,[8,9] ^{178}Re,[10] and ^{180}Re.[11] As is well-known nowadays, a doubly-decoupled

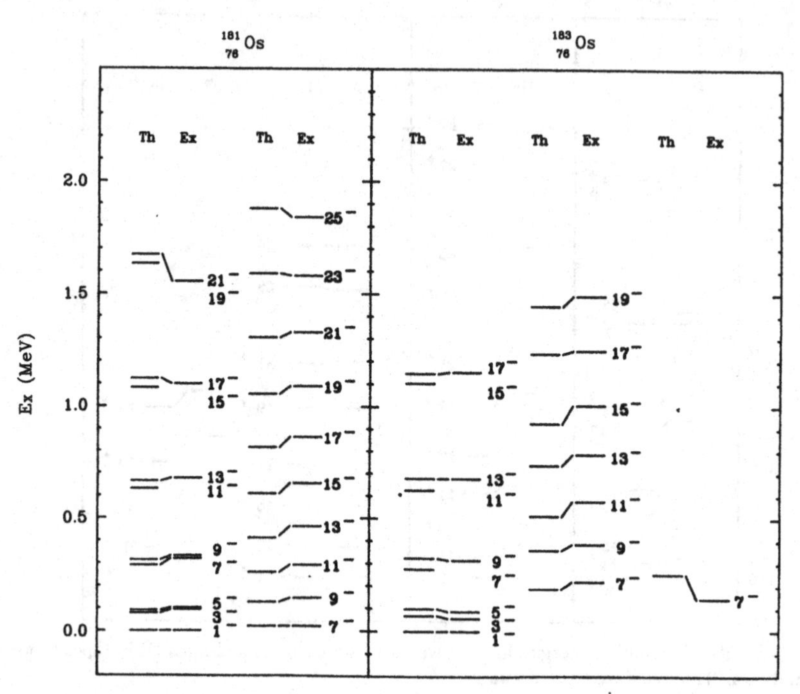

Figure 5. IBFA-calculated excitation energies for negative-parity states in the odd-mass Os isotopes compared with experimental data. The states are labeled with $2J$.

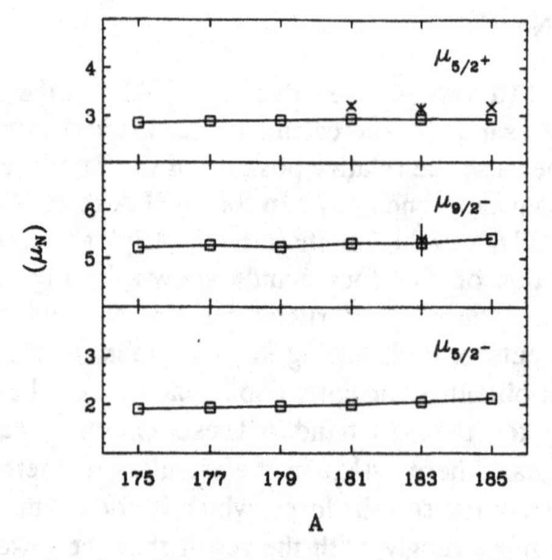

Figure 6. IBFA-calculated μ values for selected band-head states in the odd-mass Re isotopes compared with experimental data. The squares are calculated values; the crosses (with error bars), the experimental values.

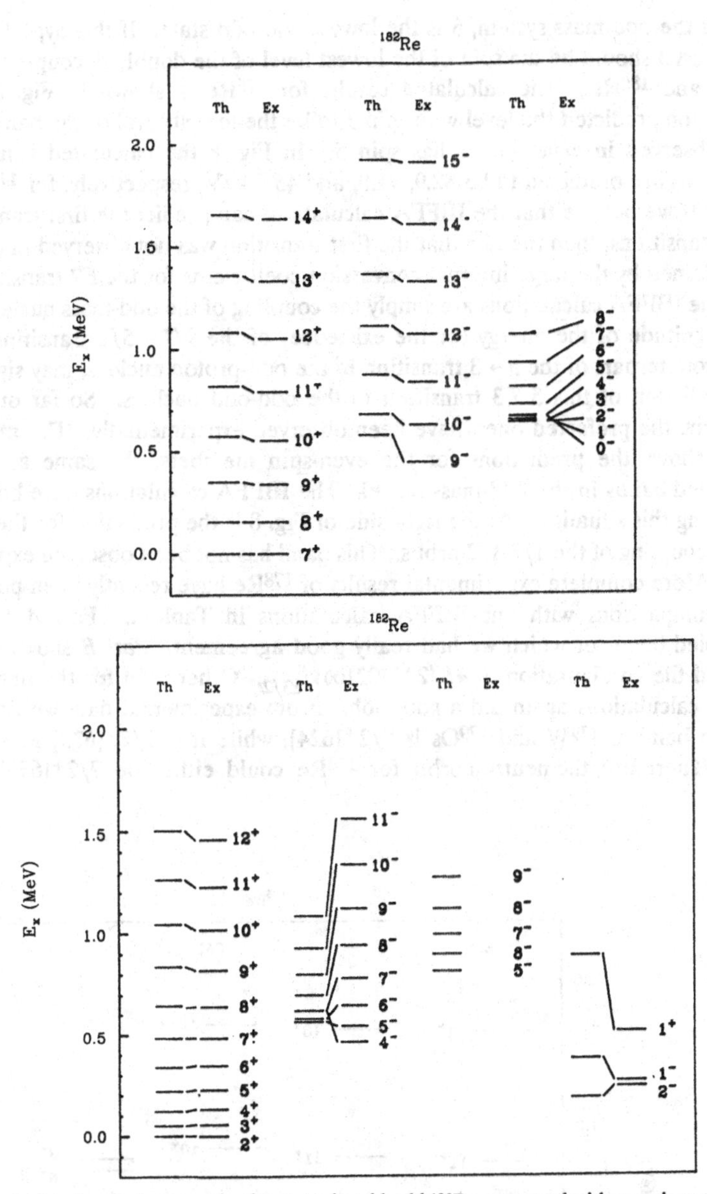

Figure 7. IBFA-calculated excitation energies for states in odd-odd ^{182}Re compared with experimental data.

band, with $K=1$, is built on an intrinsic state which is the triplet coupling of 1/2-1/2 *p-n* orbits. (Although K is not strictly a good quantum number for decoupled states, it remains a useful label.) Because of the large coriolis force in this kind of special coupling, the level having spin $I=K$ is not the lowest level of the bands. A survey of all the known doubly-decoupled band in this mass region shows that the spin of the lowest level of the doubly-decoupled bands seems to be correlated with that of the $K=1/2$ band of the corresponding odd-proton nucleus. It shows that when 5/2 is the I value of the lowest level of the $K=1/2$ band, 3 appears as the lowest state in the doubly-decoupled band; when 9/2 is the lowest state in the odd-mass system, 5 is the lowest odd-odd state. If this hypothesis is generally true, then 3 should be the spin of the lowest level of the doubly-decoupled bands of ^{176}Re, ^{178}Re, and ^{180}Re. The calculated results for ^{176}Re is shown in Fig. 8. The IBFFA calculation predicted the level with spin 3 to be the lowest level of the band, but the lowest level observed in experiments has spin 5. In Fig. 8 the calculated value for the $5 \rightarrow 3$ transitions are predicted to be 82.9, 71.0, and 45.7 keV, respectively, for ^{176}Re, ^{178}Re, and ^{180}Re. If we believe that the IBFFA calculations can predict the first transition as well as other transitions, then the fact that the first transition was not observed in experiments can be explained by the large internal-conversion coefficients for the $E2$ transitions. However, since the IBFFA calculations are simply the coupling of the odd-mass nuclei, ambiguities in the magnitude of the energy (or the existence) of the $9/2 \rightarrow 5/2$ transition in the $K=1/2$ band, counterpart of the $5 \rightarrow 3$ transition in the odd-proton nucleus, may significantly affect the prediction of this $5 \rightarrow 3$ transition in the odd-odd nucleus. So far only the odd-spin members, the preferred ones, have been observed experimentally. The middle portion of Fig. 8 shows the predictions for the even-spin members, the same as for the $K=1/2$ decoupled bands in the odd-mass nuclei. The IBFFA calculations have been successful in predicting this situation. At the right side of Fig. 8 is the prediction for the $K=0$ band, the singlet coupling of the 1/2-1/2 orbits. This band has not been observed experimentally.

More complete experimental results of ^{178}Re have recently been published, and we show comparisons with our IBFFA calculations in Table I. Part A lists the doubly-decoupled band for which we had really good agreement. Part B shows data for a band assigned the configuration of $\pi 5/2^+[402] \otimes \nu i_{13/2}$. Other than for the first transition, our IBFFA calculations again did a good job. From experimental data we find that the $i_{13/2}$ neutron band in ^{179}W and ^{179}Os is $9/2^+[624]$, while it is $7/2^+[633]$ in ^{175}W, ^{177}W, and ^{177}Os. Therefore, the neutron orbit for ^{178}Re could either be $7/2^+[633]$ or $9/2^+[624]$.

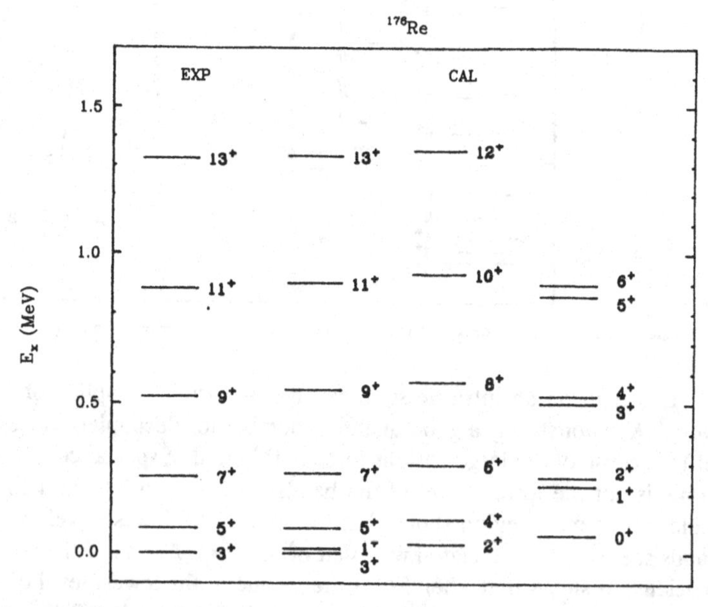

Figure 8. IBFA-calculated doubly-decoupled band of ^{176}Re compared with experimental data.

Table I. IBFFA Calculated Energy Spacings for ^{180}Re Compared with Experimental Data

A. Doubly-Decoupled Band

	Exp (keV)	IBFFA (keV)
7+ → 5+	159.9	170.1
9+ → 7+	263.2	270.2
11+ → 9+	361.5	364.6
13+ → 11+	449.1	445.2
15+ → 13+	533.5	531.7
17+ → 15+	591.8	600.6

B. $\pi 5/2^+[402] \otimes \nu i_{13/2}$ Band

	Exp (keV)	IBFFA (keV)
7+ → 6+	80.7	14.4
8+ → 7+	112.2	116.8
9+ → 8+	135.5	151.0
10+ → 9+	163.8	176.3
11+ → 10+	179.3	195.8
12+ → 11+	204.9	214.3
13+ → 12+	225.1	228.3
14+ → 13+	238.7	243.8
15+ → 14+	253.2	260.7

However, in the IBFFA calculations we only use ^{179}Os as the odd-neutron nucleus; therefore, the spin of the lowest member of the band would be 7^+ in our calculaton.

As mentioned earlier, the IBA-type calculations can also predict electromagnetic properties. Table II lists the magnetic moments for three odd-odd Re isotopes. Although not enough experimental data exist to test the calculations stringently, from Table II we can say that the IBFFA model has the potential to give reasonable predictions for electromagnetic properties.

The IBFFA model is able to give an accurate description of states in the odd-odd Re nuclei, as can be seen from the comparisons with experimental data. These odd-odd

Table II. Magentic Moments in Odd-Odd Re Isotopes

Isotope	J^π	μ_{IBFFA}(nm)	μ_{Exp}(nm)
^{180}Re	6^-	2.41	---
	1^-	2.35	---
^{182}Re	7+	2.33	2.76±0.17
	2+	3.28	3.16±0.22
^{184}Re	3^-	3.19	2.50±0.19
	8+	2.09	2.89±0.13

nuclei constitute a very stringent test of the model, because odd-odd nuclei do not provide the same sort of smoothly-varying systematics as do other types of nuclei. Also, errors tend to accumulate: any difficulties in describing the even-even cores and, particularly, the odd-mass states are compounded in the odd-odd systems. However, if good fits for the even cores and for the odd-mass states can be obtained, then good spacings in the odd-odd nuclei are likely. This can be seen in ^{182}Re calculations, where the $K^{\pi}=4^{-}$ band hadc the poorest fit. We had some problems with the positions of the band heads and with the triplet-singlet couplings. This points to difficulties in V_{np}, the neutron-proton residual interaction. To free V_q and V_s, allowing them to vary over a reasonable range, will certainly improve this situation. However, doing this would make the calculations much more complicated.

34.4 CONCLUSION

In conclusion, we would like to summarize why our IBFFA calculations are special. Odd-odd nuclei have been calculated before in the framework of the IBA model. The earlier calculations were done by assuming the hamiltonian of the system exhibits specifc symmetries. However, such symmetries are present in only relatively few nuclei. Or else the calculations had to be restricted to unique-parity orbits. In our IBFFA calculations, because of the particular, very-efficient model-space truncation scheme used, we do not assume any special symmetry in the hamiltonian, and there is no restriction to unique-parity orbits. (Also, we have performed calculations for the $N=80$ isotones.[12] This demonstrates the abilities of IBFFA calculations, not only for deformed nuclei, but also for near-spherical nuclei.)

ACKNOWLEDGMENT

We thank Dr. R. Aryaeinejad and Mr. Wade Olivier for help in carrying our the calculations. This work was supported by the U. S. National Science Foundation under Grant No. PHY-85-19653.

REFERENCES

1. A. Arima and F. Iachello, *Ann. Phys. (N.Y.)* **99**, 253 (1976);
A. Arima and F. Iachello, *Ann. Phys. (N.Y.)* **111** 201 (1979);
A. Arima and F. Iachello, *Ann. Phys. (N.Y.)* **123**, 468 (1979).
2. O. Scholten, in *Contemporary Research Topics in Nuclear Physics*, by D. H. Feng, M. W. Guidry, and L. L. Riedinger (Plenum, New York, 1985), p. 503.
3. O. Scholten, KVI Preprint, KVI-774 (1989).
4. W. T. Chou, Wm. C. McHarris, and O. Scholten, *Phys. Rev. C* **37**, 2834 (1988).
5. O. Scholten, Ph.D. Thesis, University of Groningen, 1980.
6. M. F. Slaughter, R. A. Warner, T. L. Khoo, W. H. Kelly, and Wm. C. McHarris, *Phys. Rev. C* **29**, 114 (1984).
7. A. J. Kreiner, D. E. Digregoria, A. J. Fendrik, J. Davidson, and M. Davidson, *Nucl. Phys.* **A432**, 451 (1985).
8. D. Santos, A. J. Kreiner, J. Davidson, M. Davidson, M. Debray, D. Hojman, and G. Falcone, *Phys. Rev. C* **39**, 902 (1989).

9. W. T. Chou, Ph.D. Thesis, Michigan State University, 1989.
10. A. J. Kreiner et al., *Phys. Rev. C* **40**, R487 (1989).
11. A. J. Kreiner et al., *Phys. Rev. C* **36**, 2309 (1987).
12. R. Aryaeinejad, W. T. Chou, and Wm. C. McHarris, *Phys. Rev. C* **40**, 1429 (1989).

35. Delayed-Neutron Emission Probabilities of Li–F Nuclides

P. L. Reeder, R. A. Warner, and W. K. Hensley
Pacific Northwest Laboratory
Richland, WA 99352

D. J. Vieira and J. M. Wouters
Los Alamos National Laboratory
Los Alamos, NM 87545

ABSTRACT

Delayed-neutron emission probabilities (P_n) have been measured for eighteen nuclides ranging from ^{9}Li to ^{25}F. Neutron-rich nuclides were produced by the reaction of 800-MeV p on a ^{232}Th target at the LAMPF accelerator at Los Alamos National Laboratory. Recoil nuclides were individually identified as to their mass, nuclear charge, and ionic charge by use of the Time-of-Flight Isochronous (TOFI) spectrometer. The distribution of time intervals between the arrival of a specific ion and the subsequent detection of a neutron was determined. The P_n was calculated from the total number of ions observed, the initial neutron count rate, the neutron counting efficiency, and the half-life. The technique is shown to be valid for half-lives ranging from 10 ms (^{15}B) to 4 s (^{17}N) and for P_n values as low as 0.3% (^{13}B). Delayed-neutron emission has been measured for ^{12}Be, ^{14}B, ^{17}C, ^{18}N, ^{21}N, and ^{25}F for the first time. A small branch for β-delayed two-neutron emission was observed in ^{15}B (P_{2n} = 0.4 ± 0.2 %).

35.1 INTRODUCTION

The decay properties of exotic nuclides in the light-mass region near the limits of particle-stable neutron-rich nuclides are currently being studied at the Time-of-Flight Isochronous (TOFI) spectrometer at the Los Alamos Meson Physics Facility (LAMPF). We are measuring the β-decay half-lives ($t_{1/2}$) and delayed-neutron-emission probabilities (P_n) of these nuclides and comparing them to the predictions of β-decay models. For many of the low-Z elements, masses have recently been measured for several nuclides beyond the region of known β half-lives.[1,2,3,4] Thus various β-decay models can be compared free from uncertainties concerning the decay energy.

Nuclides in this mass region have very-large decay energies (10–20 MeV), which allows β-delayed neutron emission to be a major decay process. By counting neutrons rather than β particles or γ rays, we can distinguish the low-yield very-neutron-rich nuclides from the abundant β/γ emitters produced close to the valley of stability. The large decay energy also allows decay via β-delayed two-neutron emission. The counting technique described below permitted the measurement of the probability for two-neutron emission (P_{2n}) as well as P_n.

Exotic Nuclear Spectroscopy, Edited by W. C. McHarris
Plenum Press, New York, 1990

The TOFI spectrometer does not physically separate the recoil nuclides at its focal point. It is necessary to identify each individual recoil as to its mass, nuclear charge, and ionic charge. Decay information is obtained from a delayed coincidence technique in which the arrival times of each identified ion and each neutron are recorded. Time-interval histograms constructed by software for each ion type contain an exponential component containing the β-decay half-life and delayed-neutron yield.

The following describes the experimental techniques and data analysis used to obtain the delayed neutron emission probabilities. Preliminary results on P_n values are presented. Final results are currently being obtained based on improved techniques for ion identification and will be published elsewhere.

35.2 ION PRODUCTION AND IDENTIFICATION

The TOFI spectrometer has been described previously.[5] The general layout of the TOFI beam transport system and spectrometer is shown in Fig. 1.

35.2.A. Ion-Beam Production and Transport

Neutron-rich nuclides were produced by bombardment of a ^{232}Th target with 800-MeV protons in the thin-target area at LAMPF. Fragmentation products recoiling from the target at about 90° were captured by a transport line consisting of four quadrupole triplets and a mass-to-charge filter and introduced into the TOFI spectrometer.

The transport line was tuned to accept ions with a momentum-to-charge of 195 MeV/c/Q. The mass-to-charge filter was set to A/Q = 3.1 to eliminate most of the high-yield uninteresting particles, such as neutrals, protons, deuterons, α particles, and other particles with $A/Q \leq 2.0$. Setting the mass-to-charge filter value to 3.1 enhanced high-yield delayed-neutron precursors, such as $^{9}Li^{3+}$ and $^{15}B^{5+}$, but was not optimum for exotic precursors, such as $^{11}Li^{3+}$ and $^{17}B^{5+}$.

A new technique was developed for this work to identify the atomic number based on the velocity change in a stack of thin degrader foils.[6] The degrader foils and new timing detector were placed midway between the first timing detector in the beam transport line and the timing detector at the entrance to the spectrometer. The velocity was then mea-

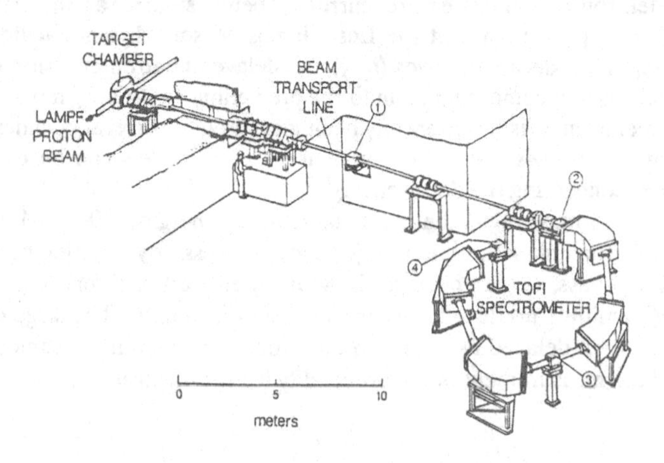

Figure 1. Layout of TOFI Spectrometer.

sured before the degrader and after the degrader from the known distances and the measured flight times. Further details on ion identification are given below.

35.2.B. TOFI Spectrometer

The spectrometer consists of four large dipole magnets arranged symmetrically so that ions are bent a total of 324°. Because of the energy loss in the degrader foil, the spectrometer was tuned for recoils with a momentum-to-charge of 157.5 MeV/c/Q to enhance neutron-rich species. Fast-timing detectors are located at the entrance focus and exit focus of the spectrometer. The spectrometer is isochronous such that ions of a given mass-to-charge ratio but different velocities all have the same flight time ($M/Q \propto t_{exit} - t_{ent}$). The mass resolution is thus determined by the high-precision measurement of the time-of-flight through the spectrometer.

In the present experiment a new beam tube was designed for the final ion-beam deposition point such that a Si total-energy (E) detector could be mounted inside a 7.62-cm-dia chamber inside the neutron detector. A thick Si detector (E_{rej}) was mounted behind the E detector to reject low-mass ions, such as protons or deuterons, which could interfere with the ion identification. The E detector provided both a fast-timing signal for the M/Q determination and a total-energy signal for the low-resolution mass determination.

35.2.C. Ion Identification

Each ion passing through the spectrometer was identified by its mass number (A), ionic charge (Q), and atomic number (Z). The preliminary analysis reported here determined the mass number from the low-resolution mass calculated from the measured total energy and velocity ($M_{low} = 2 \times E_{tot} / V^2$). The velocity was measured by two timing detectors in the transport line in front of the spectrometer. The ionic charge was determined from the low-resolution mass and the M/Q value measured by the flight time in the spectrometer [$Q = M_{low} / (M/Q)$].

The energy loss (ΔE) in the degrader foil was calculated from the low-resolution mass and the two velocities ($\Delta E = M \times (V_1^2 - V_2^2) / 2$). A two-parameter plot of the average velocity [$(V_1 + V_2) / 2$] versus ΔE showed the characteristic ridge lines as determined by the ion's dE/dX. It was empirically noted that a small correction to the calculated velocity before the degrader foil could lead to ΔE values approximately independent of the average velocity. This allowed the use of gates set directly on the ΔE value to determine the Z. As shown in Fig. 2, the ΔZ (FWHM) for the Li peak was 0.048, while that for the F peak was 0.178, corresponding to Z resolutions ($\Delta Z/Z$) of about 1.6% and 2.0%, respectively.

The Z resolution was quite adequate for nuclides with reasonable yields. However, a misidentification of Z did cause interference for the extremely-low-yield nuclides. This problem is illustrated in Fig. 3, which shows the low-resolution mass spectrum for B nuclides. Previous work has clearly shown that ^{16}B is particle unstable, so the few events observed at mass 16 in Fig. 3 must result from misidentification of the Z and presumably come from the much higher yield of ^{16}C. Other misidentified ions gave measurable peaks for ^{10}Li, ^{13}Be, ^{15}Be, ^{18}B, ^{21}C, ^{23}C, ^{24}N, and ^{25}O. These peaks were used to estimate the number of misidentified ions for correcting the P_n results discussed in Section 35.4. For the final analysis, the high-resolution mass based on the transit time within the spectrometer will be used, which should significantly improve the ion identification.

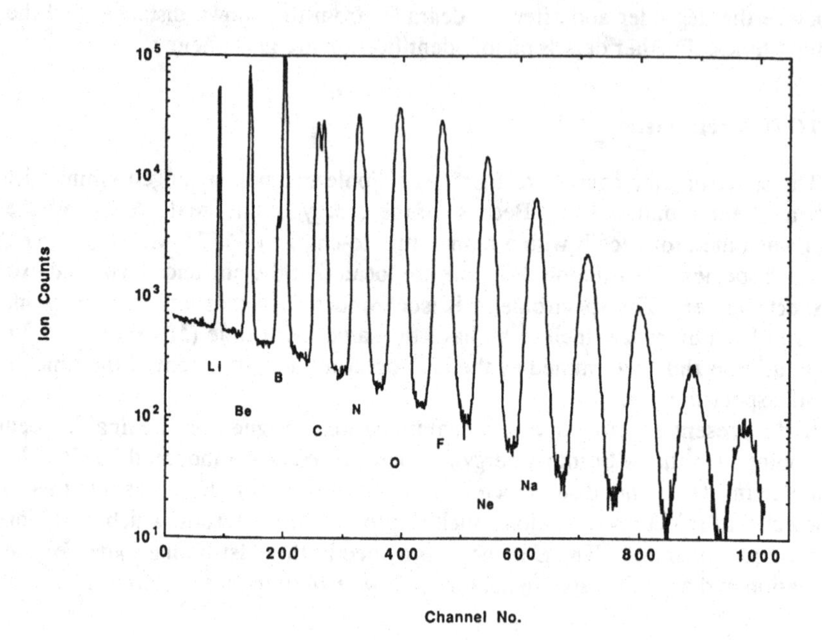

Figure 2. Distribution of ion yields for $Z = 3-15$.

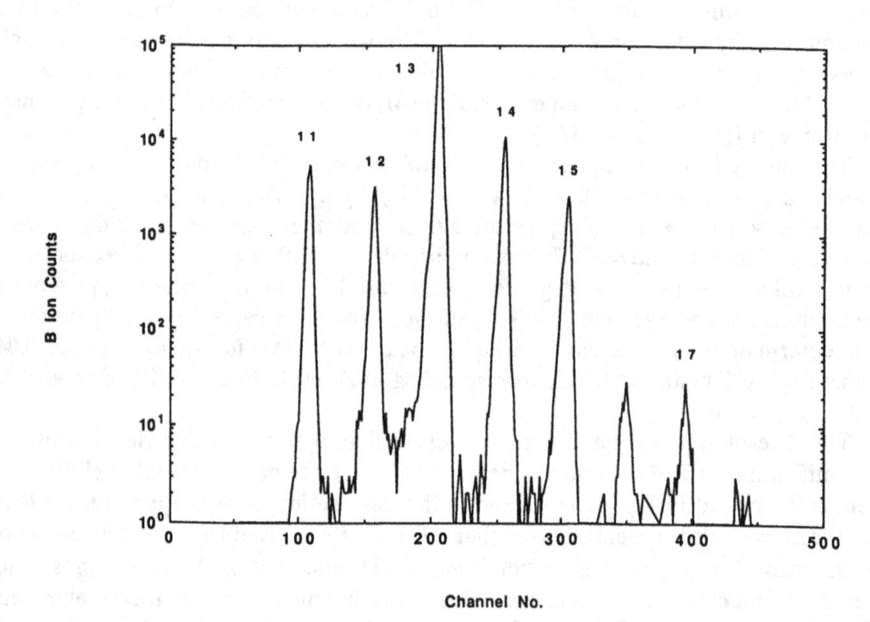

Figure 3. Distribution of B ($Z = 5$) ion yields for $A = 11-18$. Events at mass 16 and 18 are misidentified ions.

35.2.D. Data Acquisition

Event-mode data were obtained with a CAMAC-based data-acquisition system interfaced to a DEC MicroVAX-II computer. Histograms of mass, ionic charge, and Z distributions, plus histograms of ion-neutron time intervals for five identified nuclides were provided by the software analysis on-line. Time-interval distributions for other ions were obtained by replaying the event mode tapes off-line with different gates set for the ion identification.

The five identified ions were selected by setting gates on a single mass in the mass histogram, on a single Z in the atomic number histogram, and on a range of Q values, including the most abundant ionic charge states for a given Z.

35.3 HALF-LIFE AND EMISSION-PROBABILITY MEASUREMENTS

35.3.A. Neutron Counter

Neutrons were counted by a polyethylene-moderated counter containing forty ^3He proportional tubes. The tubes were placed in three concentric rings, with eight tubes in the inner ring and sixteen tubes in each of the middle and outer rings. A Monte-Carlo calculation with a simulation of the counter geometry indicated that the counting efficiency should decrease from 65% to 60% as the neutron energy increased from 20 keV to 1.0 MeV. At 3.0 MeV the calculated efficiency dropped to 45%. A PuBe source, which has a spectrum of energies averaging 4.6 MeV, gave a measured efficiency of 47%, which is consistent with the Monte-Carlo calculations. The neutron counter was located at the final focus of the TOFI spectrometer and was centered around the E and E_{rej} Si detectors.

The neutron background measured with LAMPF and all parts of the TOFI spectrometer operating, except for a beam valve which prevented ions from reaching the spectrometer, was 0.84 n/s between beam bursts. With the ion valve open, the neutron count rate was typically 2.5 n/s between beam bursts.

35.3.B. Data Analysis

The half-lives and emission probabilities were measured by a delayed-coincidence technique between the arrival times of identified ions and arrival times of delayed neutrons. Arrival times were measured with a real-time clock with 1-μs time resolution. An approach denoted Multiple Time Analysis (MTA) as described by Glatz and Lobner was used.[7] With this approach one histograms the time intervals of *all* stop events following each start event for a finite time range which is long compared to the lifetime of interest.

The probability distribution of time intervals of the MTA analysis is governed by

$$dP(t) = \epsilon_n P_n \lambda e^{-\lambda t} dt + B\, dt, \qquad (1)$$

where

λ = β-decay rate constant,
B = background rate,
ϵ_n = neutron counting efficiency,
and P_n = delayed-neutron emission probability.

The second term represents the background (chance) and is constant. The exponential term is independent of the background rate and thus can be resolved even when the background rate exceeds the β-decay rate constant. The half-life is determined from the slope of the exponential term and the P_n from the intercept.

35.3.C. Pulsing

The experimental data for an ion event consisted of all the parameters needed to identify an ion plus its time of arrival. The relevant data for a neutron event was just its arrival time. Ion events occurred only during the 0.8-ms time period of the LAMPF macropulses. There were 120 macropulses per second, i.e., 8.333 ms between the beginnings of macropulses. Because of the high-background rate during the macropulses, neutrons were counted only during a 7.1-ms period between macropulses. The dead time associated with gating the neutron counter off was accounted for by binning the data in integral units of a macropulse (8.333 ms) and appropriately correcting the neutron counting efficiency.

35.3.D. Background Correction

The time-interval distribution between neutrons and random ions was highly correlated with the LAMPF beam structure. By creating histograms for time intervals between neutrons and high-yield, non-delayed neutron precursors, the time-interval distributions for random coincidences could be precisely determined. The combined data for ^{11}Be, ^{14}C, ^{18}O, and ^{19}O were used for normalizing the background. The normalization factor was obtained by empirical adjustments until zero net counts were obtained when integrating over a selected time range, starting after at least seven times the expected half-life. The normalized background data were subtracted point by point from the time interval data for the nuclide of interest. This procedure eliminated the background data, leaving only a single exponential component due to the β half-life of the nuclide of interest.

35.3.E. Half-Life Results

We have observed sixteen nuclides for which there is at least one previous measurement in the literature. In addition, we have measured the half-lives for ^{21}N and ^{25}F for the first time. In Figures 4, 5, and 6 we show examples of the data for two well-known nuclides (^{9}Li and ^{15}B), two nuclides for which neutron emission was observed for the first time (^{14}B and ^{18}N), and two new nuclides (^{21}N and ^{25}F).

Table I gives the experimental half-lives measured here and the half-lives adopted for the P_n calculations. The uncertainties listed for our measurements include the statistical uncertainty of the chance data and the statistical uncertainty from the least-squares fitting. The uncertainty coming from the normalization factor was much smaller than these uncertainties and was ignored. No uncertainty was included for misidentified ions because, in most cases, a misidentified ion does not perturb the half-life measurement, as it probably is not a delayed-neutron precursor. Within the uncertainties, there is good agreement between our values and the most precise of the literature values.

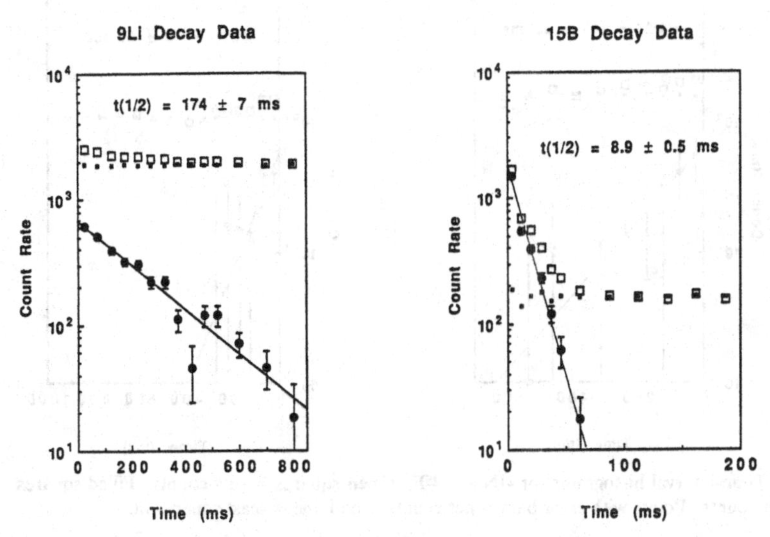

Figure 4. Time-interval histograms for ^{9}Li and ^{15}B. Open squares = raw counts. Filled squares = normalized random counts. Points with error bars = net counts. Solid line = least squares fit.

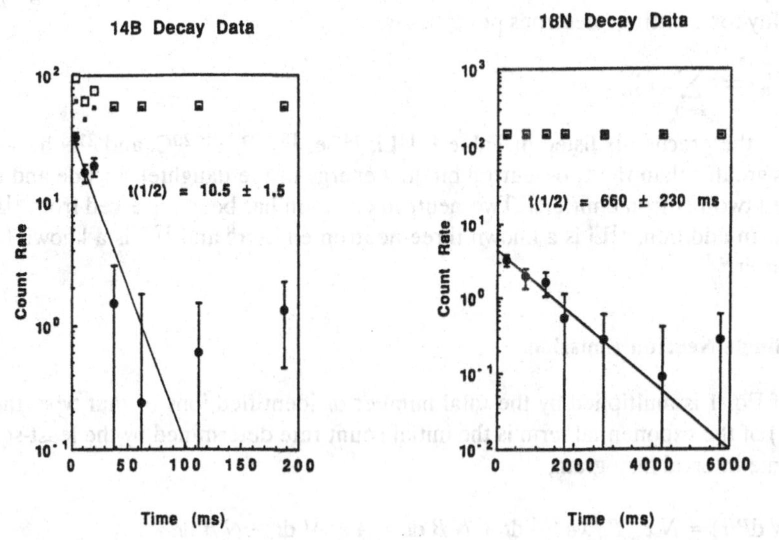

Figure 5. Time-interval histograms for ^{14}B and ^{18}N. Open squares = raw counts. Filled squares = normalized random counts. Points with error bars = net counts. Solid line = least squares fit.

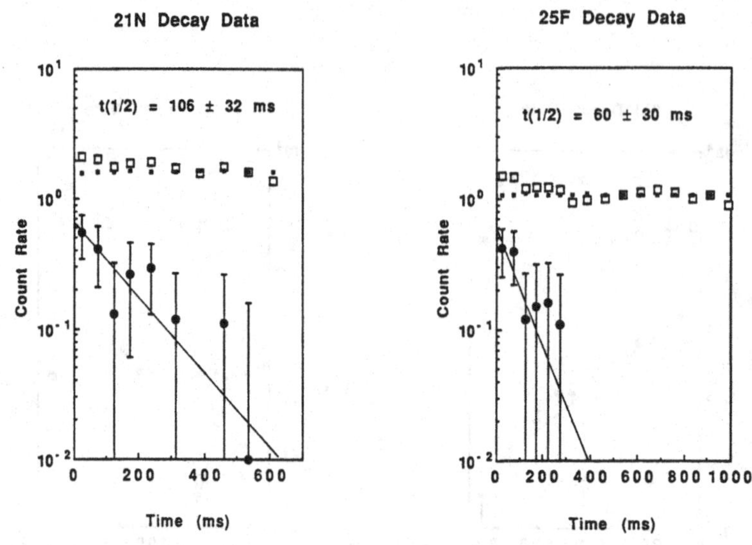

Figure 6. Time-interval histograms for ^{21}N and ^{25}F. Open squares = raw counts. Filled squares = normalized random counts. Points with error bars = net counts. Solid line = least squares fit.

35.4 EMISSION PROBABILITY

The delayed-neutron emission probability (P_n) is defined as the number of neutrons emitted per β decay. Multiple-neutron emission is accounted for by defining P_{in} as the probability for emitting i neutrons per β decay.

$$P_n = \sum_{i=0}^{i} iP_{in} \tag{2}$$

Of the precursors listed in Table I, ^{11}Li, ^{14}Be, ^{15,17}B, $^{18-20}$C, and ^{21}N have β-decay energies greater than the two-neutron binding energy of the daughter nuclide and could be β-delayed two-neutron emitters. Two-neutron emission has been observed from ^{11}Li, ^{14}Be, and ^{17}B. In addition, ^{11}Li is a known three-neutron emitter[8] and ^{17}B is a known four-neutron emitter.[9]

35.4.A. Single-Neutron Emission

If Eq. 1 is multiplied by the total number of identified ions of that type, the coefficient (A) of the exponential term is the initial count rate determined by the least-squares fit of the time interval histogram,

$$N\,dP(t) = N\,\epsilon_n\,P_n\,\lambda e^{-\lambda t}\,dt + N B\,dt = A\,e^{-\lambda t}\,dt + N B\,dt. \tag{3}$$

The expression for P_n is thus

$$P_n = A\,/\,(N\,\epsilon_n\,\lambda) = (A)\,(t_{1/2})\,/\,((N)\,(\epsilon_n)\,(\ln 2)). \tag{4}$$

The neutron-counting efficiency (ϵ_n) in Eq. 4 is an effective efficiency, which is the product of the intrinsic neutron-counter efficiency and the neutron-counter live-time. With

Table I. Half-Lives and Delayed-Neutron Emission Probabilities

Precursor	Half-life (ms)	$P_n^{a,c}$ (%)	Lit. P_n (%)	Ref.
^{9}Li	174 ± 0.7[a]	47.5 ± 1.0	50.0 ±4.0	d
	177.0[b]			
^{11}Li	6 ± 4[a]	>59 ±29	95 ±8	d
	8.7[b]	<92 ±45		
^{12}Be	40 ±30[a]	1.3±0.6	<1	e
	24[b]			
^{14}Be	2 ±3[a]	>28 ±12	86 ±3	f
	4.35[b]	<100 ±48		
^{13}B	17 ±11[a]	0.32 ±0.15	0.28 ± 0.04	g
	17.3[b]			
^{14}B	10.5 ±1.5[a,b]	6.2 ±0.5		
^{15}B	8.9 ±0.5[a,b]	76.9 ±2.7	>93	h
		P_{2n} = 0.4 ± 0.2		
^{17}B	3 ±8[a]	>7 ±9	70 ±30	i
	5.08[b]	<11 ±14	85 ±15	f
^{16}C	752 ±36	97.9 ±2.3	>98.8	j
	747[b]			
^{17}C	150 ±40[a,b]	24.3 ±2.9	<11	k
^{18}C	94 ±19[a,b]	37.7 ±3.7	25 ±4.5	k
			50 ±10	i
^{19}C	80 ±60[a]	>14 ±7	53 ±3	f
	49[b]	<21 ±11		
^{17}N	3700 ±400[a]	97.8 ±5.3	95 ±1	l
	4174[b]		95.1 ±0.7	j
^{18}N	660 ±230[a]	13.5 ±2.3		
	624[b]			
^{19}N	308 ±18[a,b]	60.6 ±1.9	33^{+34}_{-11}	k
^{20}N	121 ±17[a,b]	50.5 ± 4.1	53^{+11}_{-7}	k
^{21}N	106 ±32[a,b]	>35.3 ±6.2		
		<40.9 ±7.2		
^{25}F	60 ±30[a,b]	>21.0 ±8.9		
		<100 ±45		

a This work.
b Adopted value.
c In this work, the upper- and lower-limit signs indicate with or without correction for possible misidentified ions.
d T. Bjornstad, H. A. Gustafsson, P. G. Hansen, B. Jonson, V. Lindfors, S. Mattsson, A. M. Poskanzer, and H. L. Ravn, *Nucl. Phys.* **A359**, 1 (1981).
e D. E. Alburger, D. P. Balamuth, J. M. Lind, L. Mulligan, K. C. Young, Jr., R. W. Zurmuhle, and R. Middleton, *Phys. Rev. C* **17**, 1525 (1978).
f J. P. Dufour, R. Del Moral, F. Hubert, D. Jean, M. S. Pravikoff, A. Fleury, A. C. Mueller, K.-H. Schmidt, K. Summerer, E. Hanelt, J. Frehaut, M. Beau, and G. Giraudet, *Phys. Lett. B* **206**, 195 (1988).
g D. E. Alburger and D. R. Goosman, *Phys. Rev. C* **10**, 935 (1974).
h J. P. Dufour, S. Beraud-Sudreau, R. Del Moral, H. Emmermann, A. Fleury, F. Hubert, C. Poinot, M. Pravikoff, J. Frehaut, M. Beau, A. Bertin, G. Giraudet, A. Huck, G. Klotz, C. Miehe, C. Richard-Serre, and H. Delagrange, *Z. Phys. A* **319**, 237 (1984).
i M. Lewitowicz, Yu, E. Penionzhkevich, A. G. Artukh, A. M. Kalinin, V. V. Kamanin, S. M. Lukyanov, Nguyen Hoai Chau, A. C. Mueller, D. Guillemaud-Mueller, R. Anne, D. Bazin, C. Detraz, D. Guerreau, M. G. Saint-Laurent, V. Borrel, J. C. Jacmart, F. Pougheon, A. Richard, W. D. Schmidt-Ott, *Nucl. Phys.* **A496**, 477 (1989).
j D. E. Alburger and D. H. Wilkinson, *Phys. Rev. C* **13**, 835 (1976).
k A. C. Mueller, D. Bazin, W. D. Schmidt-Ott, R. Anne, D. Guerreau, D. Guillemaud-Mueller, M. G. Saint-Laurent, V. Borrel, J. C. Jacmart, F. Pougheon, and A. Richard, *Z. Phys. A* **330**, 63 (1988).
l M. G. Silbert and J. C. Hopkins, *Phys. Rev.* **134**, B16 (1964).

the beam-pulsing conditions discussed in Section 35.3.*C*, the neutron-counter live-time was 0.852. From the Monte-Carlo efficiency calculations (Section 35.3.*A*), the effective efficiency for 1-MeV neutrons was expected to be about 51%. If the P_n values for the well-known precursors ^{9}Li, ^{16}C, and ^{17}N are used in Eq. 4 to calculate effective efficiency, the value is only about 25%. It appears that a problem in the electronic setup and/or software caused only every other neutron to be recorded. This problem was not detected until after the end of the experiment. The evidence for this explanation is based on the apparent neutron count rate as recorded during a run being a factor of two less than the neutron count rate measured by separate scalers before and after the run.

The value of 25% for the effective neutron-counting efficiency was adopted for all nuclides. This assumes that the neutron energy spectrum for any specific precursor is not greatly different from those of ^{9}Li, ^{16}C, and ^{17}N. This assumption does not introduce much uncertainty in view of the slowly varying efficiency of the neutron counter up to 3 MeV.

35.4.B. Two-Neutron Emission

The identification of two-neutron emission was based on the observation of two neutrons within a time window governed by the residence time of neutrons within the polyethylene moderated neutron counter. Even though the two neutrons are emitted simultaneously following the β decay, they are thermalized independently in the neutron counter and are counted at different times. By histogramming the time-interval distribution between successive neutrons as measured by a time-to-amplitude converter, one can easily distinguish the coincident neutrons from random coincidences. The time-interval distribution has an exponential component with a short half-life due to the residence time of coincident neutrons in the counter and a second component with a very long half-life due to the random distribution from single events. This technique has been used previously in the discovery of β-delayed two-neutron emission in ^{11}Li and in measurements on other two-neutron precursors.[10,11,12]

The probability of two-neutron emission (P_{2n}) is the number of two-neutron events associated with a given precursor divided by the number of ions of that precursor. The number of two-neutron events was determined from the initial count rate found by least-squares fitting of the time-interval histogram recorded separately for two-neutron events.

In this experiment, the low yields of known two-neutron precursors prevented the observation of two-neutron emission. However, a small P_{2n} was observed for the first time in ^{15}B.

35.4.C. Emission-Probability Results

For those precursors with very-low ion yields, there is a contribution to the ion yield from misidentified ions. These misidentified ions are closer to stability and do not contribute to the initial delayed-coincidence count rate of the precursor of interest. The calculated P_n value will be a lower limit unless corrected for misidentified ions.

To correct for misidentified ions, it was assumed that the yield for a nuclide known to be particle unstable represented the amount of misidentification for that Z and A. This yield relative to the yield of its neighboring isobar was assumed to apply to nearby, low-yield, particle-stable nuclides as well. An alternative estimate based on the ratio of yield of a particle unstable nuclide to the total yield at that mass number was also determined. The two estimates generally agreed with each other.

The preliminary P_n values measured in this work and the P_n values from the literature are shown in Table I. In cases where the ion yield is low and misidentified ions may be a problem, the P_n uncorrected for misidentified ions is given as the lower limit and the P_n after correction is shown as the upper limit.

Among the nuclides measured here, only ^{15}B had a significant initial count rate in its two-neutron histogram. The initial count rate of 0.57 ± 0.34 had a large uncertainty, and the fit was only slightly better with two components than with just a background component. The P_{2n} calculated from these data is shown in Table I. Similar analysis of ^{13}B and ^{14}B did not give statistically-significant initial count rates, and the fit was not improved with two components. The ion yields of other known or potential two-neutron emitters were too weak to observe β-delayed two-neutron emission.

35.5 DISCUSSION

The P_n measured here for ^{9}Li, ^{13}B, ^{16}C, ^{17}N, ^{19}N, and ^{20}N have relatively-small uncertainties and are in agreement with previous measurements. The very-low ion yields for ^{11}Li, ^{14}Be, ^{17}B, and ^{19}C and the possibility of significant contributions from mis-identified ions means that the P_n values from this work should be given little weight in comparison with the more accurate values already published.

Previous work on ^{12}Be did not observe any neutron emission and set an upper limit of 1%. The present work has clear evidence of neutron emission with a P_n of 1.3 ± 0.6%, which is consistent with the published upper limit.

We have also observed delayed-neutron emission in two other nuclides for which P_n values have not been reported previously, ^{14}B and ^{18}N. Delayed-neutron emission from ^{14}B is interesting as a possible case for studying first-forbidden β decay. The first state above the neutron binding energy in ^{14}C is a 2^+ state at 8.32 MeV. This state would be populated by first-forbidden β decay from the 2^- ground state of ^{14}B. The lowest neutron-emitting state in ^{14}C populated by allowed β decay is a 3^- state at 9.80 MeV. Thus, a measurement of the neutron-energy spectrum from ^{14}B should clearly distinguish these states, and the intensity of the neutrons from the 2^+ state would be a direct measure of the first-forbidden β decay.

The P_n value of ^{18}N is of interest in regard to a recent paper on β delayed α emission in ^{18}N.[13] The recent work showed a $P_\alpha > 12.2\%$. That, coupled with the P_n of 13.5 ± 2.3% measured here, gives a total delayed-particle branching > 25.7%. This means that the assumption of 15% β-decay branching to *non*-γ emitting states used by Olness et al.[14] in their study of ^{18}N β-γ decay should be reconsidered. A puzzling feature of the delayed-α work was the observation of a broad α peak from states above the neutron binding energy. Apparently α emission and neutron emission compete with comparable intensity.

Our value for the P_n of ^{15}B is somewhat lower than the previously-published value. It is not likely that misidentified ions are significant for this nuclide. The discrepancy may be related to the fact that we measure a shorter half-life. In any case, one should be cautious in assuming a P_n of 100% for ^{15}B and normalizing other P_n values to it, as has been done in the literature.

Our P_n for ^{17}C has a relatively-small uncertainty, but is about twice the upper limit reported previously. This discrepancy should be checked.

Regarding the new P_n values for ^{21}N and ^{25}F, both the statistical uncertainty and the correction for misidentified ions are relatively small for ^{21}N thus giving confidence in the result. However, the large correction for misidentified ions for ^{25}F makes the P_n result suspect.

35.6 CONCLUSIONS

This first experiment has demonstrated the usefulness of the TOFI spectrometer for measuring both β decay half-lives and delayed-neutron emission probabilities. Although the spectrometer was tuned to emphasize the well-known precursors closer to stability, it was still possible to observe two previously-unmeasured half-lives and establish delayed-neutron emission in three other nuclides for the first time. The MTA analysis procedure worked extremely well and allowed the observation of emission probabilities of the order of 1% for nuclides with high ion yields.

The data on β decay half-lives and delayed-neutron emission probabilities provides valuable tests of the current models of β decay and point to particular nuclides where detailed spectroscopic measurements will provide more stringent information.

ACKNOWLEDGMENTS

We acknowledge K. E. G. Lobner for pointing out the advantages of the MTA data analysis procedure. Y. K. Kim and Z. -Y. Zhou provided assistance in setting up the velocity degrader ion identification technique. This work is supported by the U.S. Dept. of Energy (DOE) - Office of Medium Energy Physics and Office of High Energy and Nuclear Physics. Pacific Northwest Laboratory is operated for the DOE by Battelle Memorial Institute under Contract DE-AC06-76RLO 1830.

REFERENCES

1. J. M. Wouters, R. H. Kraus, Jr., D. J. Vieira, G. W. Butler, and K. E. G. Lobner, *Z. Phys.* **A331**, 229 (1988).
2. D. J. Vieira, J. M. Wouters, K. Vaziri, R. H. Kraus, Jr., H. Wollnik, G. W. Butler, F. K. Wohn, and A. H. Wapstra, *Phys. Rev. Lett.* **57**, 3253 (1986).
3. A. Gillibert, L. Bianchi, A. Cunsolo, B. Fernandez, A. Foti, J. Gastebois, Ch. Gregoire, W. Mittig, A. Peghaire, Y. Schutz, and C. Stephan, *Phys. Lett.* **B176**, 317 (1986).
4. A. Gillibert, W. Mittig, L. Bianchi, A. Cunsolo, B. Fernandez, A. Foti, J. Gastebois, Ch. Gregoire, Y. Schutz, and C. Stephan, *Phys. Lett. B* **192**, 39 (1987).
5. J. M. Wouters, D. J. Vieira, H. Wollnik, G. W. Butler, R. H. Kraus, Jr., and K. Vaziri, Nucl. Instr. Meth. **B26**, 286 (1987).
6. Y.-K. Kim, M. S. thesis, Utah State Univ. (1989).
7. J. Glatz and K. E. G. Lobner, Nucl. Instr. Meth. **94**, 237 (1971).
8. R. E. Azuma, T. Bjornstad, H. A. Gustafsson, P. G. Hansen, B. Jonson, S. Mattsson, G. Nyman, A. M. Poskanzer, and H. L. Ravn, *Phys. Lett.* **96B**, 31 (1980).
9. J. P. Dufour, R. Del Moral, F. Hubert, D. Jean, M. S. Pravikoff, A. Fleury, A. C. Mueller, K.-H. Schmidt, K. Summerer, E. Hanelt, J. Frehaut, M. Beau, and G. Giraudet, *Phys. Lett.* **B206**, 195 (1988).
10. R. E. Azuma, L. C. Carraz, P. G. Hansen, B. Jonson, K.-L. Kratz, S. Mattsson, G. Nyman, H. Ohm, H. L. Ravn, A. Schroder, and W. Ziegert, *Phys. Rev. Lett.* **43**, 1652 (1979).
11. P. L. Reeder, R. A. Warner, T. R. Yeh, R. E. Chrien, R. L. Gill, M. Shmid, H. I. Liou, and M. L. Stelts, *Phys. Rev. Lett.* **47**, 483 (1981).

12. P. L. Reeder, R. A. Warner, T. R. Yeh, R. E. Chrien, R. L. Gill, H. I. Liou, M. Shmid, and M. L. Stelts, *Proc. Int. Conf. Nuclei Far From Stability, Helsingor, June 7-13, 1981*, (CERN 81-09), p. 276.

13. Z. Zhao, M. Gai, B. J. Lund, S. L. Rugari, D. Mikolas, B. A. Brown, J. A. Nolen, Jr., and M. Samuel, Phys. Rev. **C39**, 1985 (1989).

14. J. W. Olness, E. K. Warburton, D. E. Alburger, C. J. Lister, and D. J. Millener, *Nucl. Phys.* **A373**, 13 (1982).

36. Exotic Decays at the Proton Drip Line

D. M. Moltz, T. F. Lang, and Joseph Cerny
Department of Chemistry
and
Lawrence Berkeley Laboratory
Berkeley, CA 94720

J. D. Robertson
Department of Chemistry
University of Kentucky
Lexington, KY 40506

J. E. Reiff
Sloan-Kettering Cancer Research Institute
New York, NY 10012

ABSTRACT

In this paper we review the general physics learned at the proton drip line. The extraordinary experimental problems which must be overcome are discussed in addition to several novel solutions. Finally, results of several recent experiments to look for ground state one- and two-proton emission are discussed.

36.1 INTRODUCTION

Studies of light proton-rich nuclei have attempted to answer many fundamental nuclear-physics questions. Many of these questions can be answered only because the proton drip line is accessible to experimental probes. This has permitted rigorous testing of nuclear models that derive quantities such as atomic masses, level structures, and half-lives. Experiments in these light proton-rich nuclei have progressed from observations of standard β and γ decay to the now commonplace β-delayed proton emission. This first observation of β-delayed particle decay has now been extended to β-delayed multi-particle decay. Major goals of these studies are to understand the decay mechanisms of these more exotic decays and to utilize their relative uniqueness to probe the underlying nuclear structure of nuclides which otherwise would be impossible to resolve because of the simultaneous and copious production of nuclei that only exhibit standard β-γ decay. Although these standard β-γ measurements are still extremely interesting, the frontier is now at the proton drip line where tests of the predictive powers of these nuclear models is most rigorous. Ground-state proton decay defines the drip line. Concerted searches for this rare decay mode have, however, yielded only four examples, ^{151}Lu,[1] ^{147}Tm,[2] ^{113}Cs, and ^{109}I[3]; these discoveries were made more than ten years after the original discovery of proton decay from a high-spin isomer in ^{53}Co.[4] Many experiments are currently searching for new examples of this decay mode, and some of these will be discussed herein.

Exotic Nuclear Spectroscopy, Edited by W. C. McHarris
Plenum Press, New York, 1990

The experimental difficulties associated with ground-state proton decay studies are magnified in searches for another extremely exotic and rare decay mode, ground-state two-proton decay; it has never been seen, only predicted.[5] To obtain adequate mass predictions to search for such a decay requires not only an extraordinary knowledge of the mass surface, but also some insight into the decay mechanism. However, this latter item is extremely difficult to predict because of the lack of any prior examples of a similar nature. In this paper we explore the methodology necessary to search for these exotic decays, and we review the status of experiments at the proton drip line.

36.2 EXOTIC DECAYS AT/NEAR THE DRIP LINE

Exotic decays of proton-rich nuclei can easily be divided into β-delayed-particle and direct-particle decays. Although recent reviews of β-delayed-proton decay[6,7] and general properties of proton-rich nuclei[8] cover these topics in much greater detail, we will give a general overview for completeness. Beta-delayed-particle emission can be viewed simply as a process where β decay proceeds to states in the daughter nucleus which are unbound to emission of that particle. In its simplest format, emission of this particle is governed solely by the coulomb and angular-momentum barriers, which must be traversed.

Figure 1 depicts the generic energetics of a β-delayed-proton emitter. Additionally, Fig. 1 shows the isobaric analog state unbound to proton emission. When this criterion is met, we say that that nuclide is a strong β-delayed-proton emitter; otherwise it would be considered a weak delayed-proton emitter. The first heavily-investigated sequence of strong β-delayed proton emitters was the $A = 4n + 1$, $T_z = -3/2$ series beginning with ^{9}C; the decays of this series have been reported extensively elsewhere.[6] More recently, this series has been extended to ^{61}Ge.[9] However, attempts to observe the next member of this series, ^{65}Se, failed.[10] Although examples of weak β-delayed proton emission abound in light nuclei, heavy delayed-proton emitters necessarily are weak because the isobaric ana-

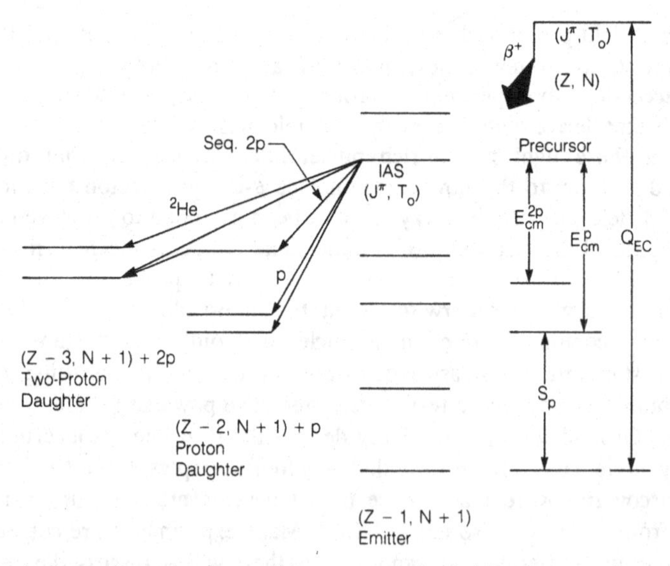

Figure 1. Generic decay scheme showing conditions necessary for β-delayed, one- and two-proton emission.

log state is energetically inaccessible via β decay. Although very useful information can be obtained from delayed-proton studies in heavy nuclei, the large density of states makes the spectroscopic information obtained less definitive, unlike the general information often obtained for specific nuclear states in light nuclei. Beta-delayed proton emission has also served to provide much insight into the structure of medium mass nuclei near the drip line, such as ^{65}Ge,[11] ^{73}Kr,[11] and ^{77}Sr.[12] Of course, β-delayed-proton emission is not the only possibility for β-delayed-particle decay in proton-rich nuclei.

Beta-delayed α decay has been known for some time in nuclei such as ^{8}B and ^{20}Na. However, there are few examples because of much larger α particle barriers; also, since α decay is generally well understood, we will not discuss this decay mode. One can also envisage such rare decays as delayed-^{3}He emission, but an even more exotic decay mode would be one where β decay is followed by the emission of more than one particle. The emission of two protons was first postulated by Gol'danskii[13]; Fig. 1 graphically depicts the energetics necessary for this decay mode. Close examination of the mass surface and the energies of the isobaric analog states showed that the $T_z = -2$ nucleus ^{22}Al would be an excellent candidate to exhibit this new decay mode. Since β-delayed single-proton emission had been used to discover ^{22}Al,[14] the mass of the isobaric analog state in ^{22}Mg was already known.

Following the development of suitable detectors, β-delayed two-proton decay was discovered[15] in ^{22}Al. Unlike the relatively well-understood weak decay followed by tunneling of a charged particle through a barrier problem, this new decay mode posed many new questions regarding the exact mechanism. Were the two protons emitted simultaneously as either a correlated or uncorrelated pair or sequentially via an intervening nuclear state? If the latter mechanism prevailed, then the additional kinematical complication of emission from a moving source would occur. Figure 2 shows two superimposed two-proton sum spectra taken with experimental set-ups designed to look at relative proton angles of

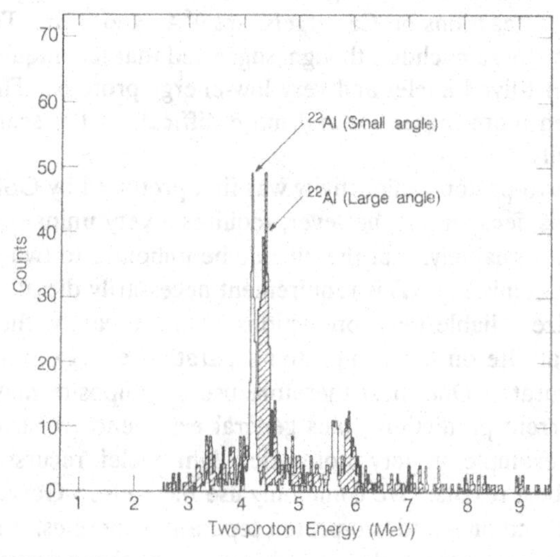

Figure 2. Superimposed ^{22}Al β-delayed two-proton sum spectra taken at average angles of 42° and 120°, dramatically showing the kinematic shift associated with sequential two-proton decay.

5–70° (small angle) and 70–170° (large angle). The observed kinematic shift in conjunction with the corresponding breakup spectra is consistent with a sequential decay process. More details and the exact kinematical formulae are given elsewhere.[16] The extremely interesting case of the emission of two correlated protons emitted in $^1S_0$ state (^{2}He) was also investigated in ^{22}Al decay[17]; results from this angular-correlation measurement were consistent with sequential emission, but a 10% ^{2}He branch could not, however, be excluded.

This new and unique decay mode provided a way to study spectroscopically and identify nuclei even further from β stability. Searching for β-delayed two-proton branches permitted the discovery of several new isotopes or the first decay studies of these species. These included ^{26}P,[18] ^{35}Ca,[19] the first stable $T_z = -5/2$ nuclide, and ^{31}Ar.[20] This last nuclide, along with several other $T_z = -5/2$ nuclides, were first identified[21] with the recoil-product separator LISE[22] at GANIL. Additionally, the βp decay of ^{31}Ar has also been studied[23] at GANIL. The ability to utilize a coincidence measurement to observe nuclei produced with very-low cross-sections has proven to be very effective. (A more complete review of β-delayed two-proton decay is given elsewhere.[24]) However, two problems make these observations even more difficult: the small detector solid angles necessary for adequate resolution and the fact that often only a few MeV of available energy must be shared between the two protons. These problems become even more severe as one attempts to study nuclei across the drip line, but both of these problems can, in principle, be solved simultaneously.

Nuclear decays at the proton drip line now become very short lived because of the lack of the slow weak-decay process. Only the coulomb and angular-momentum barriers served to impede the decay rate. (For a more complete description of the decay rates associated with proton decay, see Ref. 25.) A complete review of ground-state proton emission has recently been completed.[25] Since all ground-state proton emitters discovered to date are in the medium-to-heavy mass region where the coulomb barriers are much larger, searches for light-mass proton emitters are hampered by the much shorter lifetimes associated with the smaller coulomb barriers and the generally smaller angular-momentum barriers. Two prime candidates in the lighter masses, which can be produced in relatively large yield in heavy-ion reactions on Ca targets, are ^{65}As and ^{69}Br. The predicted proton separation energies for these nuclides, though, suggested that techniques were not available for observing very short-lived nuclei and very low-energy protons. This detection of low-energy protons is even more important and more difficult in the search for ground-state two-proton radioactivity.

Ground-state two-proton radioactivity was first proposed by Gol'danskii[5] more than twenty years ago. This decay mode, however, requires a very unique situation to occur on the atomic mass surface; namely, that the nuclide be unbound to two-proton emission but *bound* to single-proton emission. This requirement necessarily dictates that any search for this decay mode utilize reliable mass predictions. Unfortunately, the severe exponential dependence of the half-life on the two-proton separation energy makes all existing mass predictions too inaccurate. One must therefore use a composite mass-estimation system based upon many current predictions, plus general experience regarding certain types of mass formulae. For example, in very proton-rich light nuclei, recursive formulae seem to give consistently the best results. We commonly use the Kelson-Garvey mass relation[26] to obtain mass estimates and one- and two-proton separation energies. Figure 3 (from Ref. 8) graphically depicts the use of the Kelson-Garvey mass relation to predict these separation energies for *a*) $T_z = -2$, *b*) $-5/2$, and *c*) -3 nuclei.

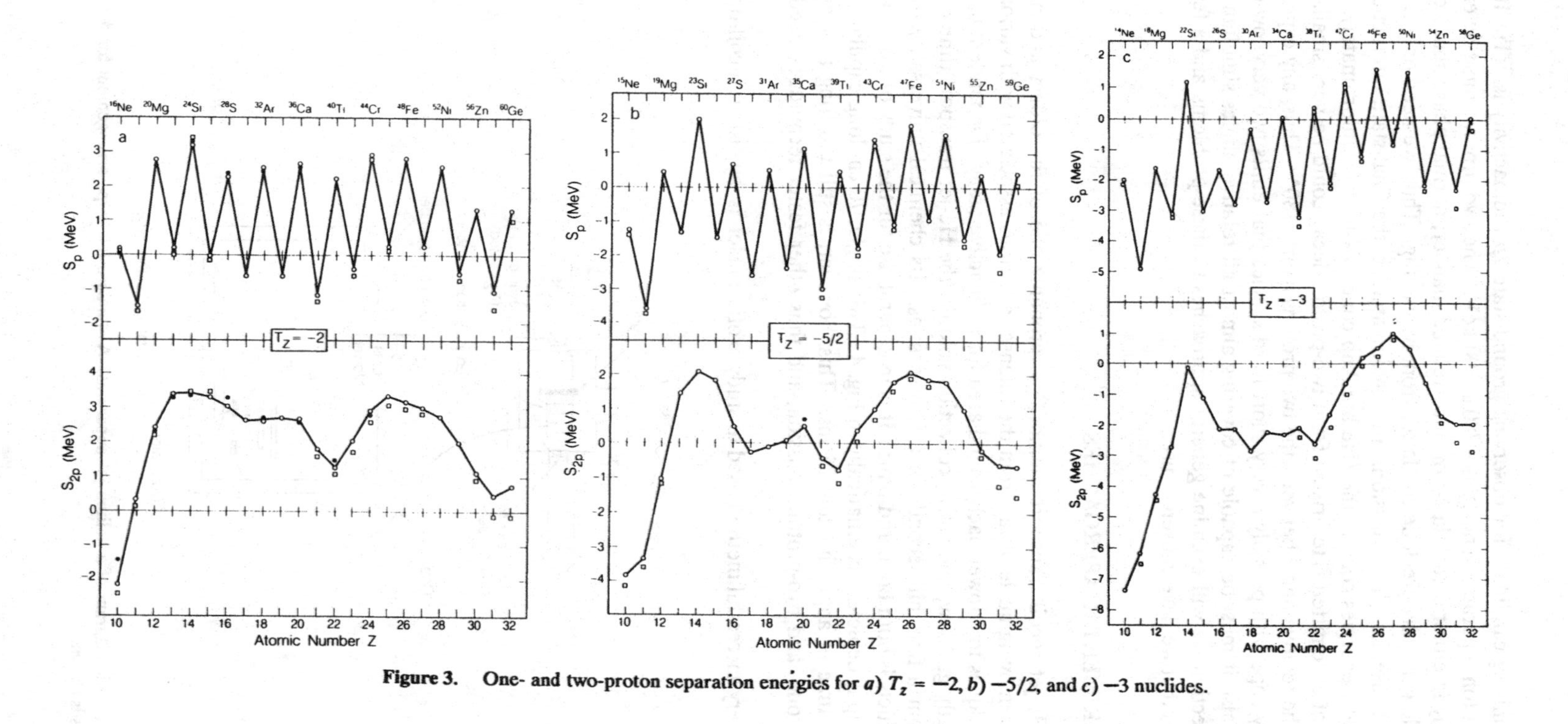

Figure 3. One- and two-proton separation energies for *a*) $T_z = -2$, *b*) $-5/2$, and *c*) -3 nuclides.

A promising candidate for observing ground-state $2p$ radioactivity is ^{39}Ti. Its predicted two-proton separation energy is –780 ± 300 keV. The two protons must be emitted simultaneously in either a correlated or uncorrelated manner; normal phase space considerations would generally preclude the latter from happening. Thus, we need only consider the correlated case, i.e., ^{2}He emission. In reaction studies, the final-state interaction and the large kinetic energies confine the ^{2}He breakup cone to ≈40°.[27] Unfortunately, the kinetic energy of an emitted ^{2}He arising from two-proton decay could be very small, which could make the relative angle between the two protons nearly 180°. Thus, any experiment must not only cope with possibly a very short-lived species that emits two very-low-energy protons, but also it must be capable of covering almost all relative angles simultaneously. In the next section we will examine general approaches to these problems and a few specific examples that we have chosen to use.

36.3 EXPERIMENTAL APPROACHES

The He-jet recoil transport method[28] is probably one of the oldest and simplest methods for removing nuclei away from the intense radiations associated with various production techniques to a lower-background area for radioactive assay. Dependent upon the capillary length, gas flow, and active collection volume, the He-jet transport time can vary anywhere from a few milliseconds to a few seconds. Its chemical universality for non-gaseous products is both boon and bane. It is in general use for the study of exotic nuclear decays; a setup is depicted schematically in Fig. 4. This single, 70-cm-long capillary system has a transit time of approximately 25 ms. This, however, is still too long to search for ground-state one- and two-proton emission, and thus other faster techniques need to be employed.

One very successful method used to study exotic nuclei is by using recoil-product

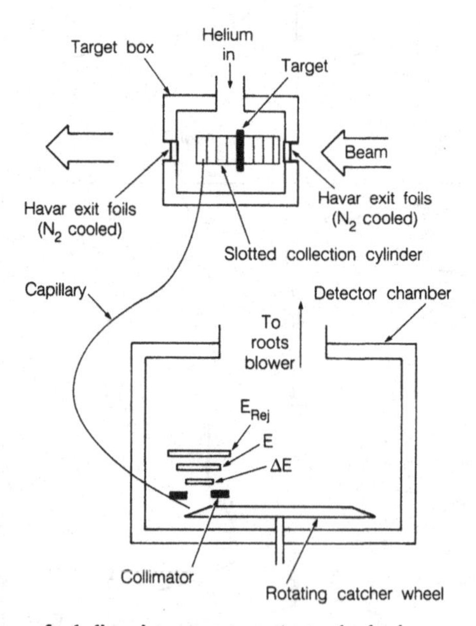

Figure 4. Schematic diagram of a helium-jet apparatus. A standard telescope arrangement for detecting β-delayed protons is shown.

separators. These devices can operate anywhere from a few MeV/nucleon all the way up to several hundred MeV/nucleon. Examples of these types of devices are SHIP[29] at GSI and LISE[23] at GANIL. The primary advantage of this type of device is the rapid (typically a few hundred nanoseconds) physical separation of the products of interest. A disadvantage of such systems is that for decay studies the primary beam must be turned off to await the decay of an identified nucleus. If the half-life is very short, this poses little problem, but if the half-life exceeds a few tens of milliseconds, then the overall yield can be significantly reduced. Moreover, implantation of the products (even in a detector) makes the detection and identification of quite low-energy protons difficult.

Another method involves catching recoils in foils for subsequent observations of their radioactive decay, typically under low-duty-factor accelerator conditions. Because there is no physical separation of the products, large backgrounds due to similar decays from competing reaction products could easily mask any signal. This technique is best suited, therefore, for proton searches in regions of the nuclidic surface where few β-delayed-proton emitters could be formed that have low-energy proton groups. Proton decays from ^{113}Cs and ^{109}I (Ref. 25) were discovered using this general technique. It is important to note that β-delayed-proton emitters in this mass region generally exhibit no protons with energies below $\approx$1 MeV. Thus any single, low-energy peaks are more easily identified.

On-line isotope separators (ISOL) are widely used to separate nuclei of interest from products of competing nuclear reactions. Many ion-source techniques have been utilized with ISOL systems, but most of these techniques involve significant sublimation or diffusion times. Although rapid-release techniques[30] have been developed for some elements, in general, fast ISOL systems are He-jet based. These include He-jet-coupled ion-source systems such as the Berkeley-88 RAMA system[31] and the ion-guide system[32] originally developed at Jyvaskylaa. The primary hold-up time for the RAMA system[33] is due to the capillary transit time ($\approx$ 200 ms); the ion-source hold-up time is short because of the rapid but low efficiency use of charge exchange with He^{1+} ions. The ion guide relies on the high first ionization potential of He; in principle, product nuclei remain in the +1 charge state once the recoils have thermalized. The He is skimmed off, and any charged atoms are accelerated to a final energy suitable for mass separation (typically 40–60 kV). Both of these techniques have little or no chemical selectivity. The generally lower efficiencies, though, make studies at the drip line nearly impossible, given the typically infinitesimally-small production cross-sections.

Although the above techniques are very useful for many experiments, the beam structure of the LBL 88-Inch Cyclotron and the desire for rapid removal of recoil products on the sub-millisecond time scale led to the development of the fast-rotating-wheel system depicted schematically in Fig. 5. The general idea is that some recoils from the target are caught in the Al catcher foils (the percentage is dependent upon the Al foil thickness). These catcher foils are rotated (continuously) between pairs of detector telescopes suitable for the appropriate decay measurements. The arrival of the radioactivity at the detector location coincides with the time when the beam is turned off during a 50% on/50% off cycle and is independent of the wheel speed; the wheel speed can be varied from 20 to 5000 rpm, corresponding to 250 to 1 ms cycle times. The entire system is rotated 70° from normal to permit a threefold increase in stopping material for a unit traverse by the emitted decay particles. (Details are given in Ref. 20.) This fast-rotating wheel partially solved the short-lifetime problem by permitting studies of nuclides with half-lives down to 100 μs. The difficult experimental problem of detecting low-energy protons remained,

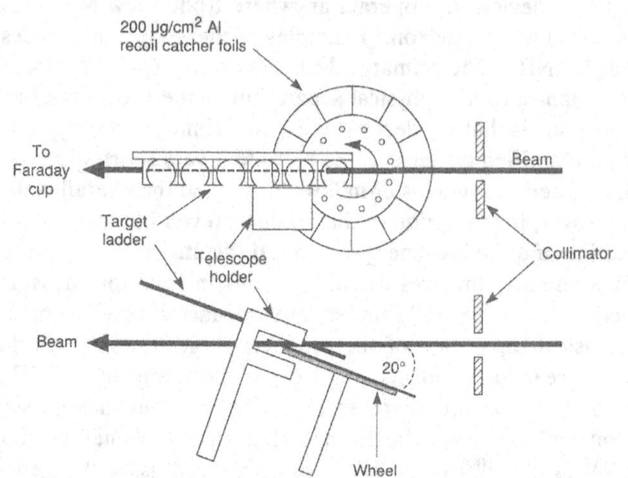

Figure 5. **Schematic diagram of the fast-rotating wheel system.**

however. A review of detectors which could be used for this purpose is the subject of the next section.

Most prior studies of low-energy proton emission (< 1 MeV) have been performed with single Si counters. Particle identification was generally accomplished on a peak-by-peak basis by comparing the measured energies with and without a thin degrader foil. This technique does not work, however, for very-low-yield experiments. One must identify on an even-by-event basis all emitted particles; this requires a telescope.

Three general types of telescopes can be envisioned: all-Si, gas-Si hybrid, and all-gas telescopes. The first type could only be realized because of the recent development of epitaxially grown Si crystals 1-3 μm thick. Unfortunately, the large-area (for large, solid angles) wafers needed for low-count-rate experiments have such large capacitances that the resolution is sufficiently poor to preclude their use. The last type of detector telescope can encompass both Bragg-curve spectrometers (see Ref. 3, for example) and proportional counters. These types of systems generally have higher thresholds ($\approx$0.5 MeV) due to the thick window needed to withstand the high gas pressures necessary to stop the low-energy protons.

Gas-Si hybrid detectors thus became an attractive option for experimental development. All-gas counters generally suffer from very slow charge-collection times. This problem was overcome by designing the gas-silicon detector depicted in the upper part of Fig. 6. The small, active gas volume has charge collected from the center of the detector. When combined with the use of CF_4 gas, the majority of the charge can be collected in 1 μs; this timescale is also typical for Si counters and is thus ideal for a hybrid system. Figure 7 shows a two-dimensional spectrum arising from products of the 40-MeV ^{3}He + Mg reaction: it was obtained with a gas-Se detector telescope. The proton peaks clearly evident in this spectrum are all attributable to the β-delayed-proton decay of ^{25}Si. This detector has essentially unit efficiency for protons with energies of 250–6000 keV. The lower part of Fig. 6 shows the six-telescope systems constructed for use with the fast-rotating wheel. A more complete description of these detectors is in preparation.[34]

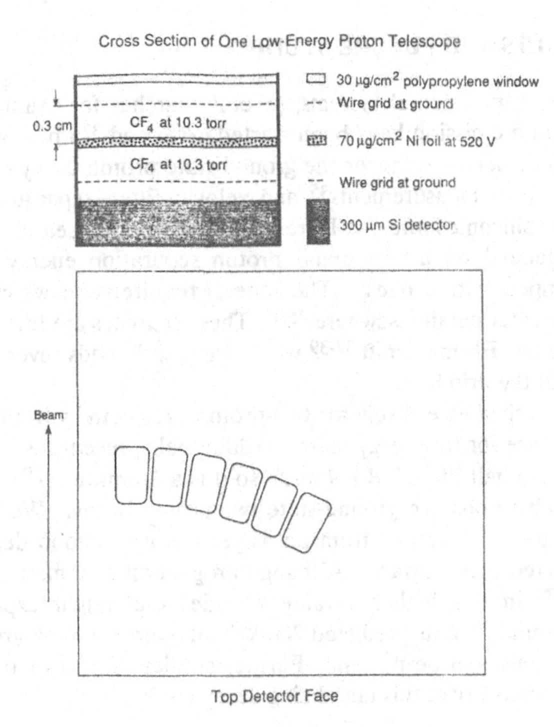

Figure 6. Top) cross-section of a single gas-silicon telescope developed to detect protons down to 250 keV. Bottom) external view of one of the six-telescope arrays constructed for use with the fast-rotating wheel.

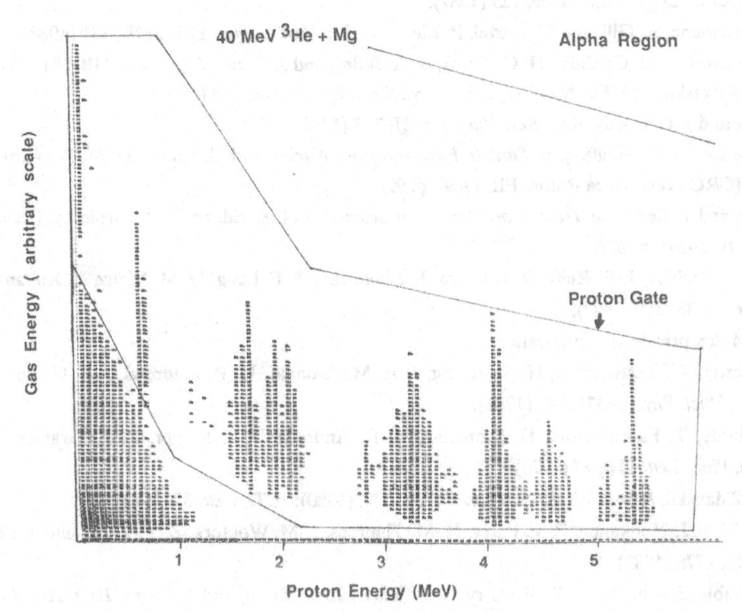

Figure 7. Two-dimensional plot of the silicon energy vs the differential energy loss in the gas counter. The proton band is clearly visible. (A small, zero suppression has been used in this figure—the β tail can interfere with the observed low-energy protons.)

36.4 RECENT RESULTS AND FUTURE STUDIES

Using these new experimental systems, several searches for examples of ground-state, one- and two-proton emission have been started. ^{28}Si and ^{32}S bombardments of Ca targets have yielded no evidence so far for the ground-state proton decay of either ^{65}As or ^{69}Br in both our fast-wheel measurements[35] and velocity-filter, separated-product measurements (with a single silicon counter) at Daresbury.[36] Successful searches for light-mass proton emitters will depend on a very small proton separation energy must be large enough, though, to compete with β decay. The general requirements which are necessary have been covered in greater detail elsewhere.[25,35] These searches are further complicated by the well known Thomas-Ehrman shift,[37-39] which frequently adds several hundred keV of stability to nuclides at the drip line.

We have also searched extensively for two-proton radioactivity from ^{39}Ti. To date, we have found no evidence for this decay mode. Additionally, recent results from GANIL have shown ^{39}Ti to have a half-life of 28 ± 9 ms,[40] so it is a β-emitter. ^{39}Ti was one of the best candidates in which to observe ground-state two-proton decay. We believe that its non-observation more probably results from an as yet not understood decay mechanism rather than a mispredicted mass surface. Although on general systematics (or on a weak Thomas-Ehrman shift[39] in this higher ℓ-value nuclide) one might expect ^{39}Ti to be ≈150–200 keV better bound than its predicted 780-keV unbound, a weak ground-state two-proton-decay branch should still be present. Further studies of ^{39}Ti or other candidates will hopefully yield the discovery of this tantalizing decay mode.

REFERENCES

1. S. Hofmann et al., *Z. Phys.* **A305**, 111 (1982).
2. O. Klepper et. al., *Z. Phys.* **A305**, 125 (1982).
3. T. Faestermann, A. Gillitzer, K. Hartel, P. Kienle, and E. Nolte, *Phys. Lett.* **137B**, 231 (1984).
4. K. P. Jackson, C. U. Cardinal, H. C. Evans, N. A. Jelley, and J. Cerny, *Phys. Lett.* **33B**, 281 (1970).
5. V. I. Gol'danskii, *Usp. Fiz. Nauk* **87**, 255 (1965), *Sov. Phys. Usp.* 8, 770 (1966).
6. J. Cerny and J. C. Hardy, *Ann. Rev. Nucl. Sci.* **27**, 333 (1977).
7. J. C. Hardy and E. Hagberg in *Particle Emission from Nuclei*, Vol. 3, Ed. by D. M. Poenaru and M. S. Ivascu, (CRC Press, Boca Raton, FL, 1989), p. 99.
8. J. Aysto and J. Cerny, in *Treatise on Heavy Ion Science*, Vol. 8, Ed. by D. A. Bromley, (Plenum Press, New York, 1989), p. 207.
9. M. A. C. Hotchkis, J. E. Reiff, D. J. Vieira, F. Blonnigen, T. F. Lang, D. M. Moltz, X. Xu, and J. Cerny, *Phys. Rev. C* **35**, 315 (1987).
10. D. M. Moltz, private communication.
11. J. C. Hardy, T. Faestermann, H. Schmeing, J. A. MacDonald, H. R. Andrews, J. S. Geiger, and R. L. Graham, *Nucl. Phys.* **A371**, 349 (1981).
12. J. C. Hardy, T. Faestermann, H. Schmeing, H. R. Andrews, J. S. Reiger, R. L. Graham, and K. P. Jackson, *Phys. Lett.* **63B**, 27 (1976).
13. V. I. Gol'danskii, *Pis'ma Zh. Eksp. Teor. Fiz.* **32**, 572, (1980); *JETP Lett.* 32, 554 (1980).
14. M. D. Cable, J. Honkanen, R. F. Parry, H. M. Thierens, J. M. Wouters, Z. Y. Zhou, and J. Cerny, *Phys. Rev. C* **26**, 1778 (1982).
15. M. D. Cable, J. Honkanen, R. F. Parry, S. H. Zhou, Z. Y. Zhou, and J. Cerny, *Phys. Rev. Lett.* **50**, 404 (1983).
16. M. D. Cable, J. Honkanen, E. C. Schloemer, M. Ahmed, J. E. Reiff, Z. Y. Zhou, and J. Cerny, *Phys. Rev. C* **30**, 1276 (1984).

17. R. Jahn, R. L. McGrath, D. M. Moltz, J. E. Reiff, X. Xu, J. Aysto, and J. Cerny, *Phys. Rev. C* **31**, 1576 (1985).
18. M. D. Cable, J. Honkanen, R. F. Parry, S. H. Zhou, Z. Y. Zhou, and J. Cerny, *Phys. Lett.* **133B**, 146 (1983).
19. J. Aysto, D. M. Moltz, X. Xu, J. E. Reiff, and J. Cerny, *Phys. Rev. Lett.* **55**, 1384 (1985).
20. J. E. Reiff, M. A. C. Hotchkis, T. F. Lang, D. M. Moltz, J. D. Robertson, and J. Cerny, *Nucl. Instr. Meth.* **A276**, 228 (1989).
21. M. Langevin et al., *Nucl. Phys.* **A455**, 149 (1986).
22. V. Borrel et al., *Nucl. Phys.* **A473**, 331 (1987).
23. R. Anne, D. Basin, A. C. Mueller, J. C. Jacmart, and M. Langevin, *Nucl. Instr. Meth.* **A248**, 267 (1986).
24. D. M. Moltz and J. Cerny, in *Particle Emission from Nuclei*, Vol. 3, Ed. by M. Poenaru and M. S. Ivascu (CRC Press, Boca Raton, FL, 1989), p. 133.
25. S. Hofmann, in *Particle Emission from Nuclei*, Vol. 2, Ed. by D. M. Poenaru and M. S. Ivascu (CRC Press, Boca Raton, FL, 1989), p. 25.
26. I. Kelson and G. T. Garvey, *Phys. Lett.* **23**, 689 (1966).
27. R. Jahn, D. P. Stahel, G. J. Wozniak, R. J. deMeijer, and J. Cerny, *Phys. Rev. C* **18**, 9 (1978).
28. R. D. MacFarlane, R. A. Gough, N. S. Oakey, and D. F. Torgerson, *Nucl. Instr. Meth.* **73**, 285 (1969).
29. S. Hofmann, G. Munzenberg, F. P. Hessberger, and H. J. Schott, *Nucl. Instr. Meth.* **223**, 312 (1984).
30. R. Kirchner, *Nucl. Instr. Meth.* **B26**, 204 (1987).
31. D. M. Moltz, R. A. Gough, M. S. Zisman, D. J. Viera, H. C. Evans, and J. Cerny, *Nucl. Instr. Meth.* **172**, 507 (1980).
32. J. Arje, J. Aysto, H. Hyvonen, P. Taskinen, V. Koponen, J. Honkanen, and K. Valli, *Nucl. Instr. Meth.* **247**, 431 (1986).
33. D. M. Moltz, J. M. Wouters, J. Aysto, M. D. Cable, R. F. Parry, R. D. von Dincklage, and J. Cerny, *Nucl. Instr. Meth.* **172**, 519 (1980).
34. D. M. Moltz, J. D. Robertson, J. E. Reiff, T. F. Lang, and J. Cerny, to be submitted to *Nucl. Instr. Meth.*
35. J. D. Robertson, J. E. Reiff, T. F. Lang, D. M. Moltz, and J. Cerny, to be submitted to *Phys. Rev. C.*
36. M. A. C. Hotchkis, private communication.
37. R. G. Thomas, *Phys. Rev.* **80**, 136 (1950); **88**, 1109 (1952).
38. J. B. Ehrman, *Phys. Rev.* **81**, 412 (1951).
39. E. Comay, I. Kelson, and A. Zidon, *Phys. Lett. B* **210**, 31 (1988).
40. C. Detraz, invited talk, *International Nuclear Physics Conference, Sao Paulo, Brazil, 1989*, GANIL Preprint 89.16.

37. Delayed Fission of Light Am Isotopes

H. L. Hall and D. C. Hoffman
Department of Chemistry
University of California
and
Nuclear Science Division
Lawrence Berkeley Laboratory
Berkeley, CA 94720

ABSTRACT

Fission characteristics of the delayed-fission decay mode in light Am nuclei have been investigated. Total kinetic energy and mass-yield distributions were measured for ^{232}Am and for ^{234}Am, and delayed-fission probabilities of 6.9 x 10^{-4} and 6.6 x 10^{-5}, respectively, were determined. The total kinetic energy and the asymmetric mass-yield distributions are typical of fission of mid-range actinides. No discernible influence of the anomalous triple-peaked mass division characteristic of the Th-Ra region was detected. Measurements of the time correlation between the *EC* x rays and the subsequent fission confirm that the observed fissions arise from the *EC* delayed-fission mechanism.

37.1 INTRODUCTION

Delayed fission (DF) is a nuclear decay process in which a decaying nucleus populates excited states in its daughter nucleus, which then fission. These states can be above the fission barrier(s) of the daughter (yielding prompt fission), within the second well of the potential energy surface (a fission shape isomer), or within the first well of the potential energy surface (an electromagnetic isomer). This is illustrated schematically in Fig. 1. This decay mode is believed to influence the production yields of heavy elements in multiple neutron capture-processes[1-5] followed by β decay, such as in the astrophysical *r* process and in thermonuclear weapons tests. Delayed fission processes may also provide a sensitive probe of fission barriers in the heavy-element region.[6]

The probability of this decay mode, P_{DF}, can be expressed in terms of experimentally measurable quantities as

$$P_{DF} = \frac{N_{if}}{N_i}\,, \tag{1}$$

where N_i is the number of the type of decays of interest (e.g., β^- or *EC*) and N_{if} is the number of those decays leading to delayed fission. P_{DF} can be derived from statistical considerations as

Exotic Nuclear Spectroscopy, Edited by W. C. McHarris
Plenum Press, New York, 1990

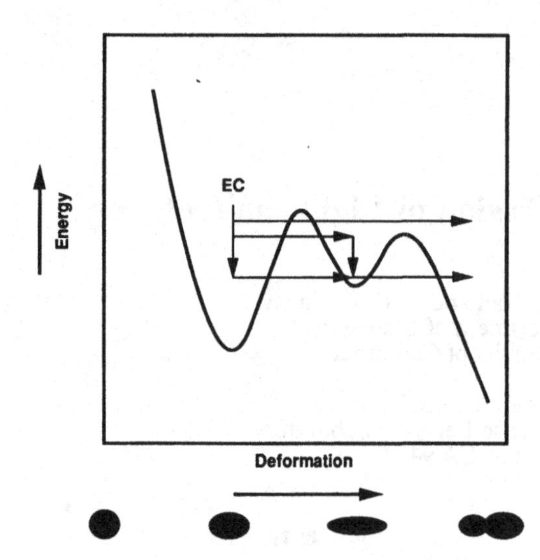

Figure 1. Schematic illustration of the delayed-fission process.

$$P_{DF} = \frac{\int_0^{Q_i} W_i(Q_i - E)\frac{\Gamma_f}{\Gamma_f+\Gamma_\gamma}(E)\mathrm{d}E}{\int_0^{Q_i} W_i(Q_i - E)\mathrm{d}E}, \tag{2}$$

where $W_i(E)$ is the transition probability function for the decay of interest, $[\Gamma_f/(\Gamma_f+\Gamma_\gamma)](E)$ is the ratio of the fission width of excited levels within the daughter nucleus to the total depopulation width of these states, E is the excitation energy of the daughter nucleus, and Q_i is the Q value for the decay mode of interest. It is assumed that no decay channels are open to the excited states other than fission and γ decay, so that $(\Gamma_\gamma + \Gamma_f)$ is the total decay width of the excited states.

P_{DF} is strongly dependent on the energy available for the decay and the structure of the fission barrier, primarily due to the fission-width term, $[\Gamma_f/(\Gamma_f+\Gamma_\gamma)](E)$. The γ-decay term can be estimated from semi-empirical relationships[7] to be

$$\Gamma_\gamma = \frac{C_\gamma \Theta^4 e^{E/\Theta}}{2\pi\rho}, \tag{3}$$

where ρ is the nuclear level density, C_γ is a constant, and Θ is the nuclear temperature. The fission width, derived from the penetrability of the fission barrier with several simplifying assumptions,[8] is qualitatively described by

$$\Gamma_f = \frac{R_B}{2\pi\rho}\left(1 + e^{\frac{2\pi(B_f - E)}{\hbar\omega_f}}\right)^{-1}, \tag{4}$$

where R_B is the penetrability of the outer fission barrier from the lowest-lying state in the second well, ρ is the level density in the inner well, B_f is the height of the inner fission barrier, and $\hbar\omega_f$ is the energy associated with the inner barrier curvature.

As a result, Γ_f is expected to be exponentially dependent on the difference between the fission barrier and the Q value (which enters as the upper limit of the integrals in Eq. (2). Hence, for the study of *EC*-delayed fission in the actinide region, it is necessary to choose nuclei for which Q_ε is comparable to the fission barrier (about 4–6 MeV). This requires study of nuclei far from the valley of β stability, which introduces a number of experimental difficulties in the production and characterization of these nuclei.

It should be noted that the above equations neglect the contribution of discrete nuclear structure in the daughter nucleus to the delayed-fission probability. The simplifications discussed above and in more detail in Ref. 8 are based on the assumption that decay to the daughter nucleus proceeds to sufficiently energetic states that the system can be treated statistically. However, the low-lying structure[9] of the daughter can either promote or hinder the delayed-fission mechanism. For example, β-delayed fission in $^{256}Es^m$ has been observed[10] to proceed from a single level at 1425 keV in the daughter ^{256}Fm. One would not normally expect fission from a level so close to the daughter's ground state, but γ decay from this level was heavily hindered (level half-life~70 ns) so that fission was able to complete successfully with γ decay. As a result, the P_{DF} for $^{256}Es^m$ was observed to be 2×10^{-5}. On the other hand, a nucleus with a high Q_ε might be expected to have a large P_{DF}, but if *EC* to the ground state is super-allowed ($\Delta I^{\Delta\pi} = 0^{no}$), essentially no high-lying states might be populated. Recent theoretical models[5,9] incorporate structural information in the β-strength function [a term in $W_i(E)$].

37.2 HISTORY

Anomalous fission activities were first observed[11,12] in the light americium and Np regions as early as 1966. In 1969, Berlovich and Novikov[13] noted that this region met the conditions required for delayed fission, although the observed fissions were not specifically attributed[14] to delayed-fission processes until 1972. A 55-s fission activity, attributed to εDF in ^{232}Am, was reported by Habs et al.[15] in 1978, and the P_{DF} for this isotope was reported to be on the order of 1%. An εDF branch has been tentatively assigned[16] to ^{242}Es, again with a P_{DF} on the order of 1%. Recently, εDF has been reported[17] outside the actinide elements, in the region of ^{180}Hg.

Most studies to date have reported only half-life and fission cross-section (σ_f) data measured without any separation of the delayed-fissile species from other reaction products. The *EC* cross-section (σ_ε), when reported, has generally been extracted from theoretical calculations or systematics, not measured experimentally. Gangrskii et al.[7] report delayed-fission probabilities for several transcurium nuclei using the measured α decay of the *EC* daughter to estimate σ_ε, by assuming the observed fission activity does arise from the same parent. All reports of εDF are summarized in Table I.

β-delayed fission (βDF) has been postulated to play a role in multiple neutron-capture processes. Burbidge, Burbidge, Fowler, and Hoyle[1] discuss the yield of heavy elements produced in supernovae, and βDF is one possible explanation as to why superheavy elements are not found in nature.[2,3] βDF had once been predicted to significantly influence heavy-element yields in thermonuclear weapons tests[2,3]; however, analyses of experimental data[20,21] show that the predicted delayed-fission effects are seriously overestimated.

The first report of an observed fission activity attributed to βDF appeared in 1978

Table I. Summary of Reported Observations of *EC*-Delayed Fission (Does not include measurements presented in this work)

Nuclide[a]	$t_{1/2}$[b]	P_{DF}[c]	Reference
^{250}Md	52 sec.	2×10^{-4}	Gangrskiĭ 1980 [7]
^{248}Es	28 min.	3×10^{-7}	Gangrskiĭ 1980 [7]
^{246}Es	8 min.	3×10^{-5}	Gangrskiĭ 1980 [7]
^{244}Es	37 sec.	10^{-4}	Gangrskiĭ 1980 [7]
^{242}Es?	5 - 25 sec.	$(1.4 \pm 0.8) \times 10^{-2}$	Hingmann 1985 [16]
^{240}Bk	4 min.	10^{-5}	Gangrskiĭ 1980 [7]
^{234}Am	2.6 ± 0.2 min.	NR[d,e]	Skobelev 1972 [14]
^{234}Am	2.6 ± 0.2 min.	NR	Somerville 1977 [18]
^{232}Am	1.4 ± 0.25 min.	NR[e]	Skobelev 1972 [14]
^{232}Am	0.92 ± 0.12 min.	$1.3^{+4}_{-0.8} \times 10^{-2}$	Habs 1978 [15]
^{228}Np	60 ± 5 sec.	NR	Skobelev 1972 [14]
^{180}Tl?	$0.70^{+0.12}_{-0.09}$ sec.	$\approx 10^{-6}$	Lazarev 1987 [17]

[a] The parent nuclide undergoing *EC* decay to excited states in the daughter which then fission is given.
[b] Half-life is given as reported, or converted to a common unit when multiple references exist.
[c] Errors limits are given, if reported.
[d] Not reported.
[e] Kuznetsov [19] subsequently used the reported fission cross-sections and calculated P_{DF} for ^{232,234}Am using an evaporation code to estimate the *EC* cross-section. The values obtained were 6.96×10^{-2} and 6.95×10^{-5}, respectively.

when ^{236}Pa and ^{238}Pa were reported by Gangrskii et al.[22] to exhibit βDF with probabilities of about 10^{-10} and $10^{-6.2}$, respectively. Gangrskii et al. performed no chemical separation of the two Pa isotopes produced in irradiations of U foils. Subsequently, Bass-May et al.[23] studied ^{238}Pa using automated chemical separation procedures and observed no βDF from this isotope. They set an upper limit on the delayed-fission probability for ^{238}Pa of $P_{DF} \leq 2.6 \times 10^{-8}$. This failure to confirm βDF in ^{238}Pa cast considerable doubt on the earlier report[22] of a βDF branch in ^{236}Pa, since both ^{236}Pa and ^{238}Pa were measured in a similar fashion. ^{256}Es^m is the most recently identified[10] β-delayed fissile species, and is also the first case in which the fissioning isomeric level in the daughter nucleus has been assigned. A summary of experimental reports of βDF is presented in Table II.

Table II. Summary of Reported Observations of β-Delayed Fission

Nuclide[a]	$t_{1/2}$[b]	P_{DF}	Reference
^{256m}Es	7.6 hour	2×10^{-5}	Hall 1989 [10]
^{238}Pa[c]	2.3 min.	6×10^{-7}	Gangrskiĭ 1978 [22]
^{238}Pa[d]	2.3 min.	$\approx 10^{-8}$	Gangrskiĭ 1978 [22]
^{238}Pa[c]	2.3 min.	$\leq 2.6 \times 10^{-8}$	Baas-May 1985 [23]
^{236}Pa[e]	9.1 min.	$\approx 10^{-9}$	Gangrskiĭ 1978 [22]
^{236}Pa[f]	9.1 min.	3×10^{-10}	Gangrskiĭ 1978 [22]

[a] The parent nuclide undergoing β decay to excited states in the daughter which then fission is given.
[b] Half-life is given as reported, or converted to a common unit when multiple references exist.
[c] Produced via ^{238}U(14.7-MeV *n,p*).
[d] Produced via ^{238}U(8–20-MeV *n,p*).
[e] Produced via ^{238}U(27-MeV γ, *np*).
[f] Produced via ^{238}U(18-MeV *d*,α).

37.3 SELECTION OF THE Am REGION

The neutron-deficient Am region (illustrated in Fig. 2) was selected for the present study for several reasons. First, there are already two isotopes in this region with reported delayed-fission branches (See Table I), ^{232}Am and ^{234}Am. The εDF branch in ^{232}Am was reported[15] to be approximately 1%, while no measurement of P_{DF} was reported[14] for ^{234}Am although Kuznetsov[19] had estimated it to be 7x10^{-5} from the data in Ref. 14. Also, using the systematic approach of Habs et al.[15] and a Q_ε of 3.96 MeV for ^{234}Am (calculated using the masses of Moller, Myers, Swiatecki, and Treiner,[24]) P_{DF} for ^{234}Am was estimated[8] to be on the order of 10^{-4} to 10^{-5}.

Second, these two isotopes are reported to have half-lives long enough (reported as 55 s[15] and 2.6 min,[14] respectively) that rapid radiochemical separations can be performed on them. With rapid radiochemical separations, an Am fraction can be purified sufficiently to allow observation of the *K*-capture x rays from the decay of Am to Pu without excessive γ interference. This would allow determination of the *EC* cross-section experimentally, yielding half of the data required to determine P_{DF} by Eq. (1).

Third, the recent development[25] of the Light Ion Multiple Target System (LIM target system) allows the use of multiple targets with high yield of the reaction products, so up to twelve ^{237}Np targets can be irradiated at a time. This target system is illustrated in Fig. 3. Since the fission production rate increases linearly with the number of targets irradiated, this would allow detection of a sufficient number of fissions to measure both the total kinetic energy (TKE) and mass-yield distributions of the εDF mode.

Measurement of the TKE and mass-yield distributions is important because εDF has the potential to vastly expand the number of nuclei in which low-energy fission[26–29] can be studied. This very low excitation-energy fission mode is essentially inaccessible for neutron-deficient species this far from stability with common techniques such as (*n,f*) and charged-particle reactions, unless the nuclei in question spontaneously fission. This is not the case for the light actinides. Low-excitation-energy fission data may assist in understanding the dynamics of the fission process, as the excitation energy of the fissioning nucleus approaches zero, leading to ground-state fission.

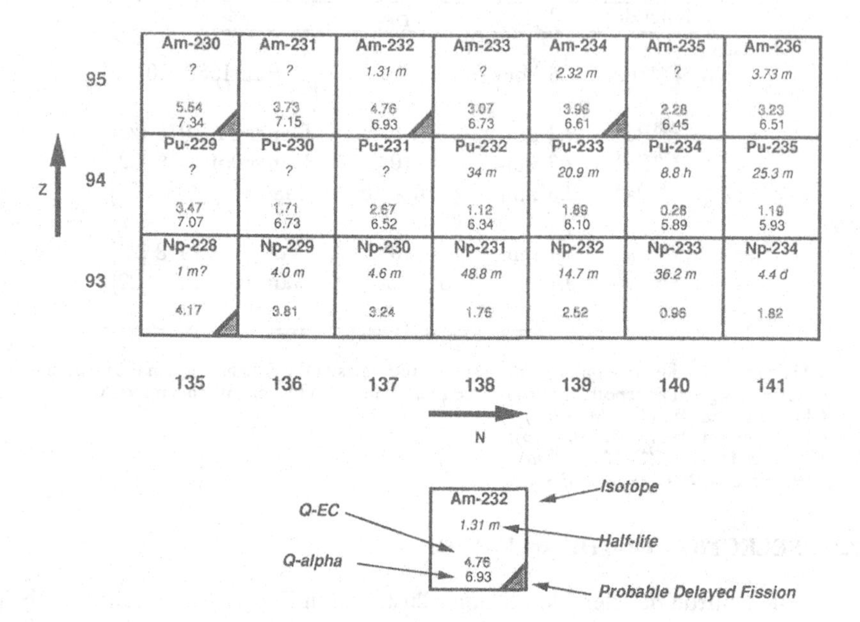

Figure 2. The neutron-deficient Am region of the chart of the nuclides.

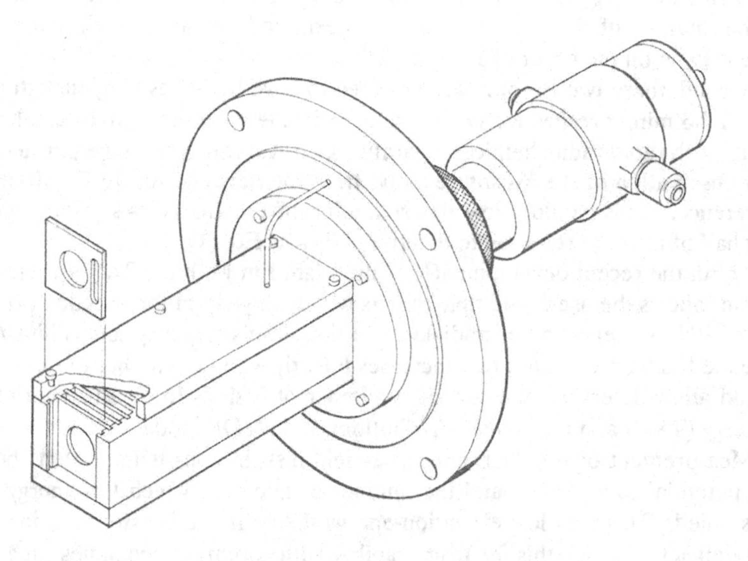

Figure 3. Illustration of the multiple-target system developed[25] for these studies.

Finally, the light Am region is nearly isotonic with nuclei displaying the "Ra anomaly"; i.e., triple-humped mass-yield distributions[30–33] from neutron-induced fission. Since εDF in the light Am region may be in the transition region between the Ra anomaly and the "normal" double-humped mass-yield distributions of the mid-range actinides, its fission properties may provide clues to understanding the Ra anomaly. Since εDF cannot bring more excitation energy into the nucleus than the Q_ε value, the influences of excitation energy on the fission properties are minimized.

37.4 EXPERIMENTAL

37.4.A Targets and Irradiations

^{237}Np targets ranging in thicknesses from 125 μg/cm^2 to 200 μg/cm^2 were mounted in the LIM[25] Target System, with a spacing of approximately one centimeter between the targets. A 25-μm Be foil served as the volume limiting foil, and another 25-μm Be foil served as the vacuum window for the system. For the initial studies of the fission properties of ^{232}Am the target backings were 2.5-μm Mo. However, because of the highly $\beta\gamma$-active by-products from the reactions of the beam with the Mo target backings, a set of targets on 25-μm Be foil was used for all subsequent measurements.

The α-particle beams used for this work were provided by the Lawrence Berkeley Laboratory 88-Inch Cyclotron. For the production of ^{232}Am by the ^{237}Np(α,9n) reaction, the α-particle enegy was 99±1.5 MeV on target (all energies are given in the laboratory frame of reference). ^{234}Am was produced via the ^{237}Np(α,7n) reaction, and the α-particle energy was 72±2MeV on target. The beam intensity was 2-7 pμA for all irradiations. The recoiling reaction products were collected on KCl aerosols in helium, which swept out the volume behind each target continuously. The activity-laden aerosols were transported via a polyvinyl chloride capillary tube to either the rotating-wheel system (Section 37.4.B, below) or to a collection site in a chemistry laboratory (Section 37.4.C, below).

37.4.B On-line Measurements

For on-line measurements of the fission properties of ^{234}Am, the KCl aerosols were transported about 5 m via a capillary tube and collected on thin (~40± 15 μg/cm^2) polypropylene foils placed on the periphery of a wheel. At preset intervals the wheel rotated 4.5°, passing the polypropylene foils through a series of six detector stations (each consisting of a pair of ion-implanted passivated Si detectors on either side of the wheel). This arrangement allowed detection of coincident fission fragments with an efficiency of approximately 60%. Each detector station could also detect α particles, again with a total efficiency of about 60%. Under the conditions of these experiments, the α particle energy resolution was about 40 keV. The detectors were calibrated for the fission measurements with a ^{252}Cf source on a thin polypropylene foil.

As the data were digitized, each event was tagged with a time and a detector marker, and then written to magnetic tape in list (event-by-event) mode. Subsequent sorting and histogramming were performed on the data to extract α spectra, fission-fragment spectra, coincidence data, and decay information. The rotating wheel is known as the "Merry Go-around" (MG), and the controlling computer system and its affiliated electronics are known as the Realtime Acquisition Graphics System (RAGS), hence the acronym MG-RAGS.

37.4.C Chemical Procedures

Two different chemical separations were performed on the reaction products of these irradiations. One separation was designed to assign the *Z* of the fissioning species to Am (or fission of Pu following *EC* in Am), and the other was used to produce an Am sample suitable for measurement of the Pu *K* x rays from the *EC* decay of the Am parent. Measurement of the *EC* decay in conjunction with the ϵDF branch would allow determination of P_{DF} experimentally.

37.4.C(1) Chemical Procedure for Elemental Assignment

In the separation designed to assign the *Z* of the fissioning activity produced by these reactions, the activity-laden aerosols were transported about 5 m via a capillary tube and collected on a Ta foil. The activity and KCl were then dissolved in 20 μL of 8*M* HNO_3. The resulting solution was passed through a 1-mm x 10-mm anion-exchange column (Bio-Rad AG 1-X8, 200-400 mesh). Under these conditions all trivalent actinides will pass through the column, while the higher valence actinides are adsorbed by the resin. The column was washed with ≈100 μL of 8*M* HNO_3, and the eluant was collected on a Ta foil, dried, flamed, and counted with a Si surface-barrier (SSB) detector for α particles and fissions. The column was then washed with ≈100 μL of 3*M* HCl - 0.1*M* HF to elute Np and Pu. This eluant was also collected on a Ta foil, dried, flamed, and counted. A flow chart of this separation procedure is given in Fig. 4. Data from the SSB detectors were stored using RAGS. The total time required for this separation was about 90 s.

The first fraction contains only the trivalent actinides produced in this reaction, while the second contains any Np, Pu, U, Pa, or Th produced. Fr, Ra, and Ac would follow the Am in this procedure, as would the lanthanides. However, the amount of Fr, Ra, and

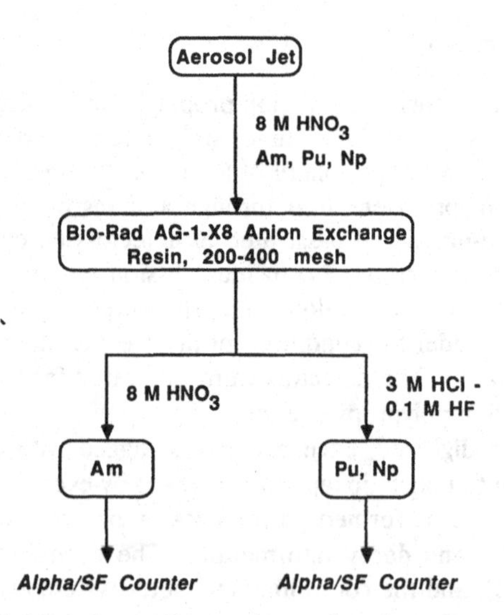

Figure 4. Flow chart of the chemical separation used to confirm the assignment of the fission activity produced in these reactions to Am.

Ac produced in this reaction was observed (from the on-line α spectra) to be very small, and the lanthanides are unlikely to fission. Hence, Am is the only reasonable elemental assignment for any fission activity observed in the first fraction.

37.4.C(2) Chemical Procedure for P_{DF} and σ_ε Measurements

This separation procedure had to be more specific for Am, since it was necessary to separate Am from highly γ-active fission products formed with production cross-sections on the order of barns. High purity was achieved by using a stacked-column technique. In this technique, a single column is made with two types of resin packed sequentially in a glass column. For this experiment, the column consisted of a 3-mm x 50-mm column of cation-exchange resin (Bio-Rad AG-MP-50, 200–400 mesh) on top of a 3-mm x 10-mm column of anion-exchange resin (Bio-Rad AG 1-X8, 200–400 mesh).

For this procedure, the activity was transported via capillary about 80 m to a collection site in the chemistry laboratory at the LBL 88-Inch Cyclotron. The activity and KCl were dissolved in 20 μL of 0.5*M* HCl to which a known quantity of ^{241}Am ($t_{1/2}$ = 432 y) had been added as a yield tracer. The resulting solution was passed through the stacked column. Elution with concentrated HCl allowed Am to be separated from monovalent fission products, divalent fission products, and the lanthanides in the top portion (cation exchange) of the column, and then Pu and Np were absorbed by the bottom portion (anion exchange) of the column.

The fraction containing Am was collected, and Am was coprecipitated with CeF_3. The precipitate was filtered, washed and then counted with an intrinsic-Ge γ-spectroscopy system. In the case of ^{232}Am, the final coprecipitation step was omitted to minimize the delay between the end of the irradiation and the start of the counting. The ^{232}Am was then directly γ counted as a liquid sample. A flow chart of this separation procedure is shown in Fig. 5. The total time required for this procedure was approximately 4 min if the coprecipitation was performed, or 90 s without it.

Fission from the respective εDF branches was measured on an alternating basis with the γ samples from the chemical separation. Samples for the fission measurements were produced by collecting the aerosols on a Ta foil in the same collection apparatus as used in the chemical separations. The Ta foil was flamed to red heat and counted in a windowless 2π gas flow proportional counter. The efficiency of this detector for fissions was determined to be 98.6% with a calibrated ^{252}Cf source.

By measuring the fission production rate and the *EC* decay on an alternating basis, any unknown values cancel out in the calculation of P_{DF} provided these values oscillate more slowly than the rate of the experiments (6–12 per hour). Therefore, values which would normally have to be estimated, such as gas-yet yield or effective target thickness, are time-averaged out of the calculation of P_{DF}. This increases the reliability of the measurement.

37.4.D X-Ray–Fission Correlation Procedure

The time correlation between the *K*-capture x ray and the subsequent fission was measured using aerosols collected directly without any chemical separation. The aerosols were collected on a thin substrate for a suitable interval and the collector was placed before a light-tight transmission-mounted 300-mm^2 Si surface-barrier detector operated in air. In most experiments, the SSB detector and foil were sandwiched between two Ge γ

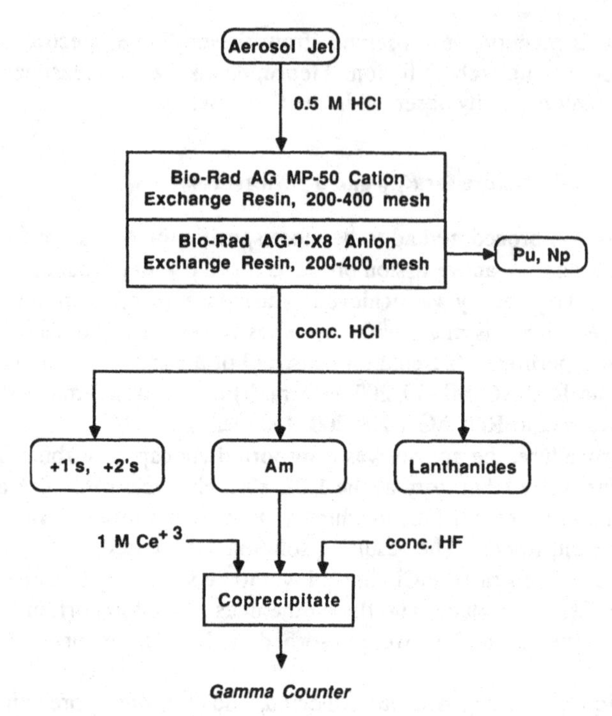

Figure 5. Flow chart of the procedure used to isolate Am from the ^{237}Np irradiations in a form suitable for γ counting.

detectors. In one measurement on ^{234}Am, a NaI(Tl) detector was added to provide better timing resolution.

Since fission produces about 10 prompt γ rays from the fission fragments,[26,34] a high overall γ-detection efficiency would reject many of the true x-ray events by summing the x-ray pulse with a pulse arising from prompt γ rays. On the other hand, if the overall γ efficiency is too low, the observed x-ray detection efficiency would be reduced and, hence, the observed correlation rate would be reduced. By measuring the prompt γ rays from spontaneous fission in a source of ^{252}Cf, the spacing between the γ detectors and the sample was adjusted to bring the summing rejection level to 50%. As long as the γ multiplicity of the Am εDF decay mode is not grossly different than that of ^{252}Cf, this would maximize the number of detected correlations. In the final configuration, each detector subtended a solid angle of about 6.7% of 4π. A 50% summing rejection level gives an overall correlation detection efficiency, using both Ge detectors, of 6.7% for each detected fission.

The signal from the SSB detector provided a common start for up to three electronic time-to-amplitude converters (TAC's). The stop signals for the first and second TAC's were provided by the first and second Ge γ detectors, respectively, and the stop signal for the third TAC (in the last measurement) was provided by the NaI(Tl) detector. The time window on the TACs was $\pm$500 ns. Calibrations were obtained using the prompt γ rays from the fission of ^{252}Cf and the γ rays in coincidence with the α particles from the decay of ^{249}Cf. The timing resolution of the Ge detectors was $\approx$12 ns full-width at half-maximum (FWHM), and the energy resolution of these detectors was $\approx$1.2 keV FWHM in the Pu K-x-ray region. The timing resolution of the NaI(Tl) detector was ≈3 ns FWHM, and its energy resolution was ≈30 keV FWHM in the 100-keV region.

37.5 RESULTS AND DISCUSSION

37.5.A Elemental Assignment

37.5.A(1) ^{232}Am

Using the chemical procedure described in 37.4.C(1), twenty-six samples were processed and counted over about 3 h. In each case, the aerosols were collected for 3 min. and then subjected to the chemical separation. Each sample was counted continuously for approximately 18 min. Eleven fissions were observed in the Am fraction, and none was observed in the Np/Pu fraction.

Based on this distribution, the fission activity produced in the 99-MeV α irradiation of ^{237}Np was assigned to Am or delayed fission from an Am precursor.

37.5.A(2) ^{234}Am

Thirty-eight samples were processed through the short chemistry and counted over about 4 h. Again, the aerosols were collected for 3 min and the samples were counted for approximately 18 min. Twenty-seven fissions were observed in the Am fraction, and one was observed in the Np/Pu fraction. The one fission in the second fraction is consistent with the amount of Am expected to tail into this fraction. Prior tracer studies of this procedure had shown cross-contamination of each fraction to be about 2%. The 6.46-MeV α group attributed[8,35] to ^{234}Am was also observed in the Am fraction.

Based on these results, we have assigned the fission activity produced by 72-MeV α particles on ^{237}Np to Am (or the delayed fission from an Am precursor).

37.5.B Fission Properties

The εDF properties of ^{232}Am and ^{234}Am were measured using the MG-RAGS as described in Section 37.4.B. The MG wheel was stepped at preset intervals so that the samples would spend approximately six half-lives between the six detector pairs. Each detector registered α particles and fissions for the full interval, except the first detector station. In the first station, signals from the α particles were suppressed for the first 12 s following the wheel motion to allow the ^{8}B+^{8}Li ($t_{1/2}$ < 1 s) α activity produced from the Be in the target system to decay without causing excessive system deadtime. Fission signals from this detector were not seriously affected by these activities and were analyzed for the full interval. After one full revolution of the wheel (eighty positions), the wheel was replaced with a clean one so that any build-up of long-lived spontaneous fission activities was minimized.

The data were corrected for neutron emission using the method originated by Schmitt, Kiker, and Williams (SKW).[36] The ^{252}Cf calibration parameters were taken from Weissenberger et al.[37] The neutron emission function, $\bar{\nu}(A)$, was taken as similar to that of ^{252}Cf, normalized to $\bar{\nu}_T = 2.40$. This value was deduced from the systematics of $\bar{\nu}_T$ vs A.

37.5.B(1) ^{232}Am

A total of 2201 coincident fission-fragment pairs was observed in these measurements using a wheel-stepping interval of 1.0 min. From these events, the half-life was found to be 1.31±0.04 min, closer to the early half-life of 1.4±0.25 min reported by Skobelev[14] than the more recent value of 0.92±0.12 min reported by Habs.[15] The decay curve for this fission activity is shown in Fig. 6. Each point on the decay curve has been normalized to represent the same number of samples per detector station. This is necessary since, for each wheel, the first station sees eighty foils before the acquisition is stopped, while the second station sees seventy-nine, the third seventy-eight, and so on. The correction is fairly small (0% for the first station, rising to 12% for the last) but can significantly affect the measured half-life.

From the decay curve, an apparent fission cross-section was estimated for the ^{232}Am εDF mode from this reaction. The effective target thickness was estimated by extrapolating low-energy recoil ranges for the compound nucleus linearly to zero energy. Recoil ranges were taken from Northcliffe and Schilling[38], and extrapolated when necessary. This method gave an estimate of the effective target thickness of 100 μg/cm^2 per target. The efficiency of the aerosol transport system was taken at 100%, although it could be lower. These assumptions result in an apparent fission cross-section of about 2.5 nb.

Fission from ^{232}Am was observed to have a highly asymmetric mass distribution, with no trace of the Ra anomaly. The mass-yield distribution is clearly two-humped, with a well-defined valley (after correction for neutron emission using the SKW[36] method with the constants of Weissenberger,[37] with no evidence of a symmetric component. The total kinetic energy distribution is symmetric about 174±5 MeV, with no evidence of multiple components. The TKE and mass-yield distributions are presented graphically in Fig. 7.

The behavior of the TKE and $\overline{\text{TKE}}$ as a function of mass fraction is shown in the TKE contour[39] plot in Fig. 8. The data in this figure suggest that the average TKE of the

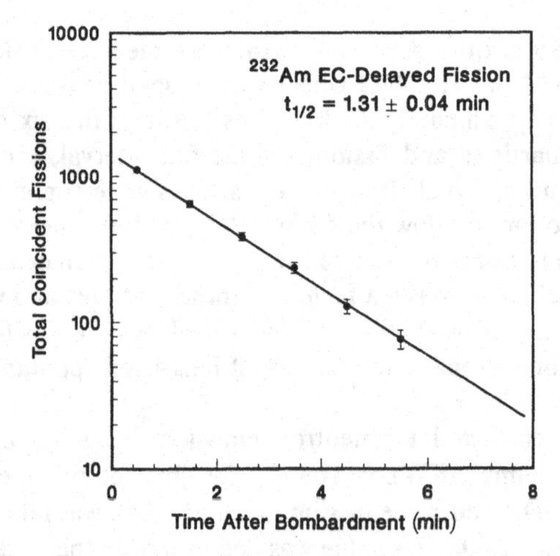

Figure 6. Decay curve of the ^{232}Am *EC*-delayed fission activity, as measured on MG-RAGS. The wheel-stepping time was 1.0 min per station.

^{232}Am εDF for symmetric mass division is about the same as it is for asymmetric division. This is unusual for low-energy fission of light actinides and hints that shell effects in the fission fragments of ^{232}Pu are influencing its fission. $\overline{TKE}$ for the near-symmetric division is based on only forty-six events, so the statistical significance of this behavior is small. The $\overline{TKE}$ value of 174 MeV for the ^{232}Am εDF is comparable to the predicted TKE[40,41] for ground state fission from ^{234}Pu, as shown in Fig. 9. The fission properties of the ^{232}Am εDF mode are summarized in Table III.

37.5.B(2) ^{234}Am

A total of 1188 coincident fission-fragment pairs was observed in these measurements using a wheel-stepping interval of 2.50 min. From these events, a considerably more accurate value of the half-life was obtained than previously[11,12,14,18] reported. The half-life was found to be 2.32±0.08 min, slightly shorter than found in the previous reports (See Table I). The decay curve for this fission activity is shown in Fig. 10. Each point on the decay curve has again been normalized to represent the same number of samples per detector station.

From the decay curve, we can estimate an apparent fission cross-section for the ^{234}Am εDF mode from this reaction. The effective target thickness was estimated (as before) to be 75μg/cm^2 per target. The efficiency of the aerosol-transport system was again taken as 100%. These assumptions result in an apparent fission cross-section of about 0.2 nb.

Fission from ^{234}Am was also observed to have a highly asymmetric mass distribution. Pre- and post-neutron emission values are given in Table III. Figure 11 shows the TKE and mass-yield distributions of the ^{234}Am εDF mode after corrections for neutron emission. The mass-yield distribution has a high peak-to-valley ratio, indicating highly asymmetric mass division. The TKE distribution is symmetric and shows only one compo-

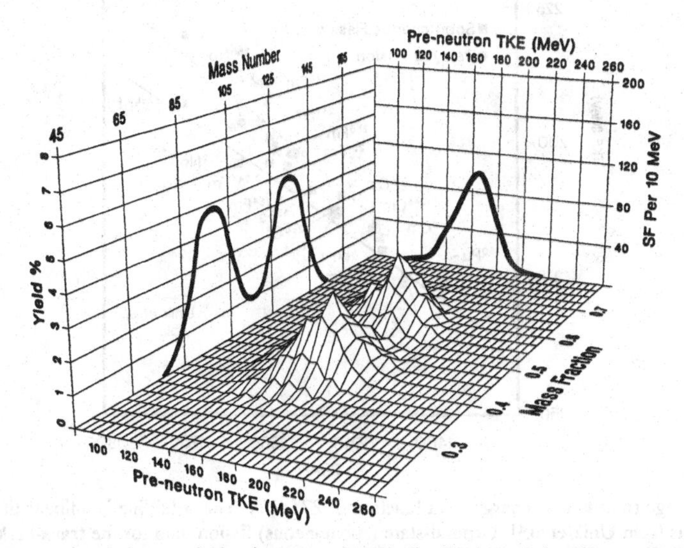

Figure 7. Pre-neutron emission total kinetic energy (TKE) distribution of the ^{232}Am εDF mode and pre-neutron emission mass-yield distribution.

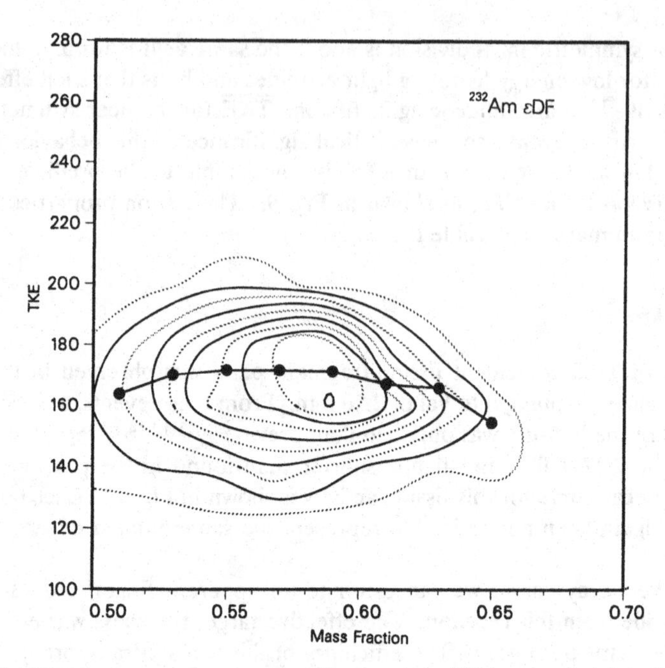

Figure 8. Total kinetic energy and average total kinetic energy of ^{232}Am, as a function of mass fraction.

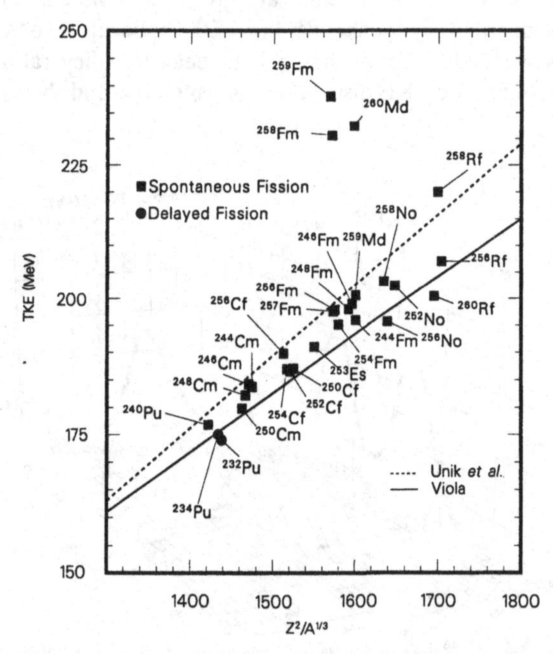

Figure 9. Average total kinetic energy, as a function of $Z^2/A^{1/3}$. The solid line is a linear fit of Viola,[40] and the dashed line is from Unik et al.[41] Ground-state (spontaneous) fission data for the trans-berkelium actinides are taken from Hoffman and Somerville,[34] and data for the lighter actinides are from Hoffman and Hoffman.[26] $Z^2/A^{1/3}$ for the Am-delayed fission is calculated for the Pu daughter, since that is the fissioning nucleus.

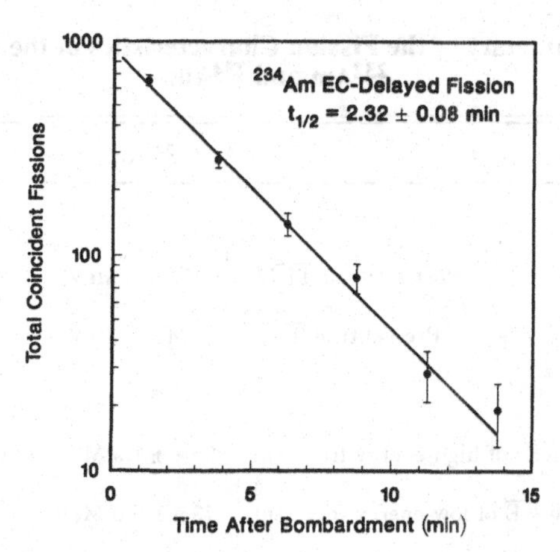

Figure 10. Decay curve of the ^{234}Am *EC*-delayed fission activity, as measured on MG-RAGS. The wheel-stepping time was 2.50 min.

nent. The behavior of the TKE and $\overline{\text{TKE}}$ as a function of mass fraction is shown in the TKE contour[39] plot in Fig. 12. In this case, the $\overline{\text{TKE}}$ at symmetry dips, as expected for light actinides, but again the statistical significance of this point is poor (twenty-six events). This, however, would be the expected behavior if a spherical subshell at $N = 66$ is exerting a strong effect on the fission fragments of ^{232}Pu. Since ^{234}Pu is two neutrons heavier, the symmetric fragments are each further from the subshell. The $\overline{\text{TKE}}$ value of 175 MeV (SKW with Weissenberger constants) for the ^{234}Am εDF is comparable to the predicted $\overline{\text{TKE}}$[40,41] for ground state fission from ^{234}Pu, as shown in Fig. 9. The fission properties of ^{234}Am εDF are summarized in Table III.

37.5.C P_{DF} and σ_ε Results

Am fractions were repeatedly isolated chemically in order to measure the Am *K*-capture x rays. Fission measurements were made on an alternating basis with the chemical separations. The chemically purified Am samples were γ counted repeatedly, and the fission samples were each counted in the proportional counter, and the integrated fissions were recorded. The γ spectra were analyzed using the SAMPO[42] computer code, and half-life analysis was performed with the CLSQ[43] code.

Major contaminants included ^{237}Am and ^{238}Am, probably produced by stripping reactions. A small amount of ^{7}Be, which was produced from the target backings, followed the Am, as did small amounts of ^{28,29}Al and ^{27}Mg. The Al and Mg were most likely produced by scattered beam on the Al target-holder cards. Half-life analysis confirmed the assignment of these peaks.

The initial activities determined for the Am *EC*-decay mode were corrected for detector efficiency, chemical yield, branching ratio, and *K*-fluorescence yield (taken as 97.7%[44]). The resulting initial disintegration rates were used for the calculation of σ_ε and P_{DF}.

The *EC* cross-section was calculated based on the following assumptions. First, the target thickness was estimated the same way as for the apparent fission cross-section, yiel-

Table III. Summary of the Fission Characteristics of the εDF Mode of ^{232}Am and ^{234}Am

	^{232}Am[a]	^{234}Am[a]
Post-neutron $\overline{\text{TKE}}$[b]	173 ± 5 MeV	174 ± 5 MeV
Pre-neutron $\overline{\text{TKE}}$	174 ± 5 MeV	175 ± 5 MeV
Post-neutron $\overline{\text{KE}}$[c] of high-energy fragment	99.4 ± 1.9 MeV	100.1 ± 2.0 MeV
Post-neutron $\overline{\text{KE}}$ of low-energy fragment	73.6 ± 2.0 MeV	73.5 ± 1.5 MeV
Pre-neutron $\overline{\text{KE}}$ of high-energy fragment	100.2 ± 1.9 MeV	101.2 ± 2.0 MeV
Pre-neutron $\overline{\text{KE}}$ of low-energy fragment	74.2 ± 2.0 MeV	74.1 ± 1.5 MeV
Average mass of the light fission fragment	98.7 ± 0.3	99.0 ± 0.1
Average mass of the heavy fission fragment	133.3 ± 0.3	135.0 ± 0.1
Assumed $\overline{\nu}_T$	2.4	2.4

[a] Calculated using the Schmitt, Kiker, and Williams (SKW) [36] method and Weissenberger [37] constants.
[b] Average total kinetic energy.
[c] Average kinetic energy.

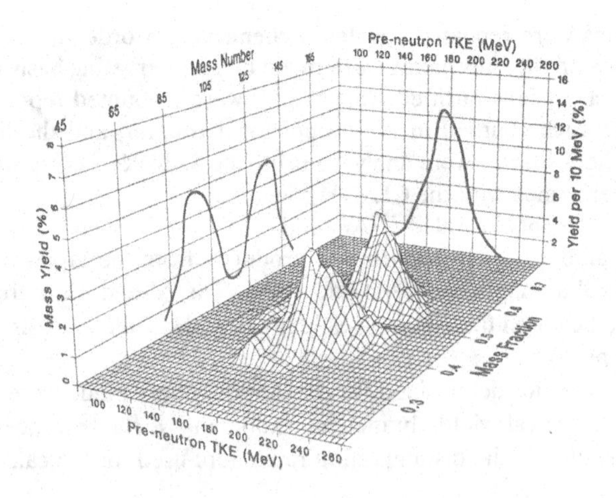

Figure 11. Pre-neutron emission, total kinetic-energy (TKE) distribution of the ^{234}Am εD [illegible]ode and pre-neutron emission mass-yield distribution.

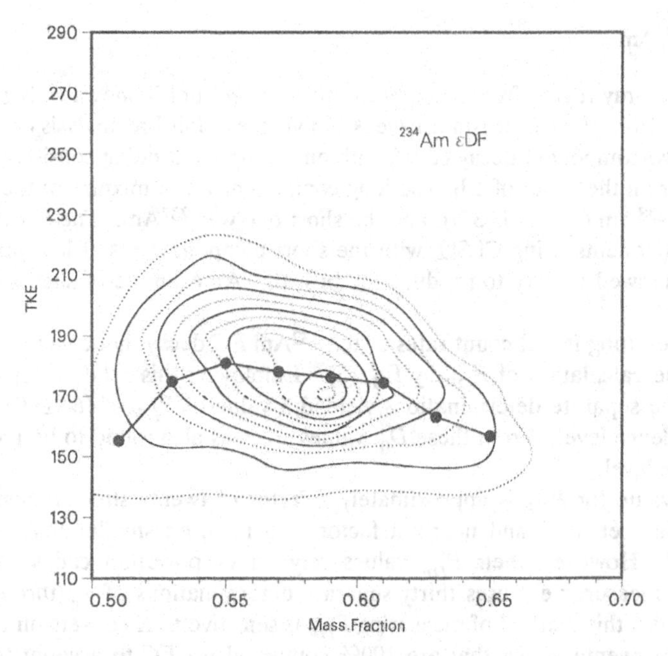

Figure 12. Total kinetic energy and average total kinetic energy of ^{234}Am, as a function of mass fraction.

ding an effective total target thickness of 75 μg/cm^2 per target for ^{234}Am and 100 μg/cm^2 per target for ^{232}Am. Second, the gas-jet yield was assumed to be 100%. Third, because of the lack of discernible γ lines in the spectrum with half-lives consistent with the decay of ^{232}Am or ^{234}Am, it was assumed that the level densities of the Pu daughters were high enough that deexcitation proceeded through a series of high-energy (500–1000 keV) low-multipolarity transitions. Based on this assumption, the *K*–x-ray production from internal conversion was taken as negligible. Of course, the last few transitions could be more highly converted, but without detailed information about the daughter level schemes any estimate on *K* conversion would not be meaningful.

The delayed fission probability was calculated from the *EC* initial activities and the number of fissions observed in the alternating fission samples. By measuring each quantity nearly simultaneously, experimental variables such as the target thickness, the beam flux (since our flux was held constant throughout this measurement, with less than 5% deviation), and the gas-jet yield should all cancel out. This allows us to calculate P_{DF} with a variant of Eq. (1),

$$P_{DF} = \frac{\lambda I_f / [e^{-\lambda t_1} - e^{-\lambda(t_1+t_C)}]}{D_{0,\varepsilon}}, \tag{5}$$

where λ is the decay constant, I_f is the number of fissions observed in a counting time t_C, t_1 is the time from end of bombardment to the start of the fission counting, and $D_{0\varepsilon}$ is the initial activity for *EC*. Employing this relationship, P_{DF} was calculated and an error-weighted average is reported, encompassing all of the separate determinations.

37.5.C(1) ^{232}Am

The K-x-ray region from a representative γ spectrum is shown in Fig. 13. The Pu x rays resulting from the *EC* of Am are weak but visible. Half-life analysis of the Pu *K* x rays revealed a two-component decay curve, with one component being consistent with 1.31 min, and the other on the order of 1 h. The long component was a mixture of the ^{237}Am ($t_{1/2}$ = 73 min) and ^{238}Am ($t_{1/2}$ = 1.63 h), and the short one was ^{232}Am. The *K* x rays were fitted with two components using CLSQ, with the short component set at 1.31 min and the long component allowed to vary to produce the best fit. An example of such a fit is shown in Fig. 14.

The resulting initial count rates of the ^{232}Am *EC* decay mode were converted to D_0 values for the calculation of P_{DF} by Eq. (5). Employing this relationship and averaging over all of the separate determinations yielded a value of P_{DF} of (6.9±1.0)x10^{-4} at the 1σ (68%) confidence level. From these D_0 values, σ_ε was also found to be 1.3±0.2 μb at the 1σ confidence level.

This value for P_{DF} is approximately a factor of twenty smaller than the value reported by Habs et al.,[15] and nearly a factor of a hundred smaller that the estimate of Kuznetsov.[19] However, their P_{DF} values rely on evaportation codes to estimate σ_ε, whereas our measurement uses thirty separate determinations of σ_ε through the Pu *K* x rays. Of course, this method of measuring P_{DF} is sensitive to *K* conversion of γ rays, but it would require twenty γ rays that are 100% converted *per EC* to account for the discrepancy. It seems much more likely that the evaporation codes become unreliable for predicting the magnitude of the cross-section when such a large number of neutrons are evaported [^{232}Am was formed by the ^{237}Np(α,9*n*) reaction] in all of the above experiments.

37.5.C(2) ^{234}Am

The *K*–x-ray region from a representative γ-ray spectrum is shown in Fig. 15. The Pu x rays resulting from the *EC* of Am are clearly visible. The only other peaks in this re-

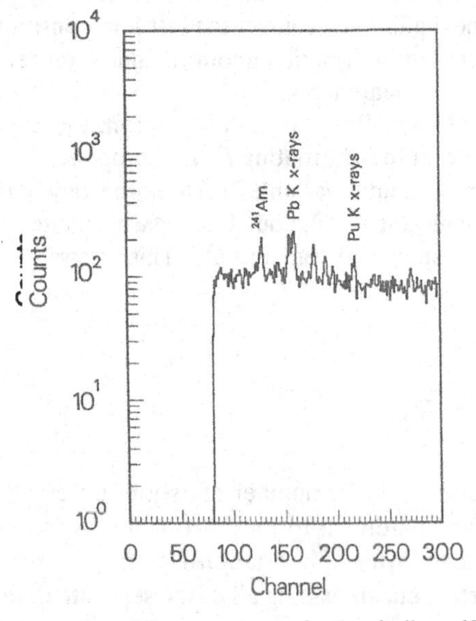

Figure 13. The *K*-x-ray region of the γ spectrum of a chemically purified ^{232}Am sample.

gion are lead-K x rays and the 59.5-keV γ ray from the ^{241}Am yield tracer.

Half-life analysis of the Pu K x rays revealed a two-component decay curve, with one component being short (about 2–3 min.), and the other on the order of 1 h. The long component was a mixture of the ^{237}Am ($t_{1/2}$ = 73 min) and ^{238}Am ($t_{1/2}$ = 1.63 h), and the short one was ^{234}Am. The K x rays were fitted with two components using CLSQ, with the short component being set at 2.32 min and the long component allowed to vary. An example of such a fit is shown in Fig. 16. The resulting initial disintegration rates were determined and used for the calculation of σ_{ε} and P_{DF}.

The EC cross-section, σ_{ε}, was found to be 5.4±1.3 μb at the 1σ confidence level. P_{DF} was calculated and averged over all of the separate determinations. This yielded a value of P_{DF} of (6.6±1.8) x 10^{-5} at the 1σ confidence level. This value is consistent with the value predicted by Kuznetsov,[19] and indicates that the region of unreliability in the evaporation codes are likely to begin after the 7n reaction but before the 9n.

37.5.D X-Ray Fission Results

Samples were collected from the gas-jet system every 4 min for ^{234}Am and at 2 min intervals for ^{232}Am, then these samples were placed in the counting chamber for the correlation studies. Figures 17*A* and 18*A* show the x-ray and the γ-ray spectrum of those events in prompt coincidence with the fission signal. The data in Fig. 17*C* and 18*C* are the logarithms of a maximum-likelihood fit[8] L of an idealized x-ray spectrum (shown in Fig. 18*B*) to the observed data as a function of the $K_{\alpha 1}$ position.

From the likelihood functions, the most probable $K_{\alpha 1}$ energies were found to be 103.8±0.3 keV and 103.6±0.5 keV for ^{232}Am and ^{234}Am, respectively, in excellent agreement with the Pu $K_{\alpha 1}$ energy of 103.76 keV. The total number of K x rays was found to be 42±8 for ^{232}Am and 32±6 for ^{234}Am by allowing the intensity of the ideal spectrum to vary within the maximum-likelihood analysis. Observed and expected x-ray intensities are given in Table IV.

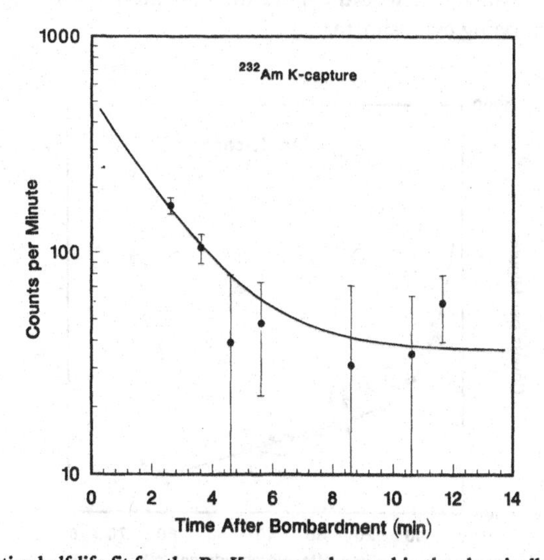

Figure 14. Representative half-life fit for the Pu $K_{\alpha 1}$ x ray observed in the chemically purified ^{232}Am sample.

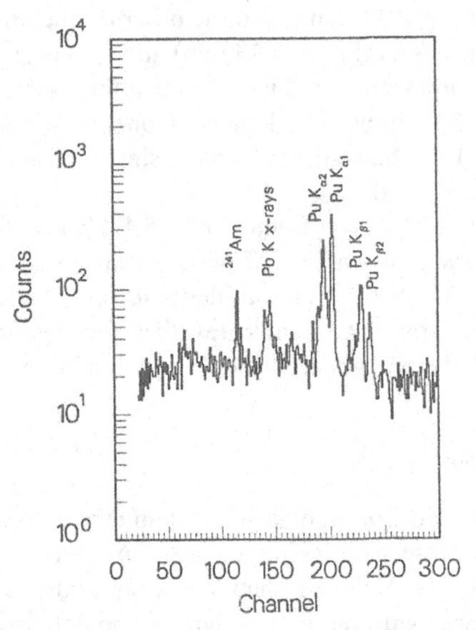

Figure 15. The *K*-x-ray region of the γ spectrum of a chemically purified ^{234}Am sample.

No evidence was observed for fission delay times longer than the best timing resolution of these experiments, about 3–8 ns. The fact that Pu x rays can be seen requires that the lifetime of the fissioning state be longer than the time it takes the orbital electrons to fill a *K* vacancy. The time required for this is on the order of 10^{-17} s.[45] We can therefore set boundaries on the excited states half-lives of 10^{-8} ns < $t_{1/2}$ < 3 ns for both ^{232}Pu and ^{234}Pu. If the nucleus is truly 100% damped in the second well (as is commonly[7,8,15] assumed), then these limits are also limits on the lifetimes of the shape isomers ^{232}Pu*f* and ^{234}Pu*f*. These limits are consistent with the half-life systematics of Pu shape isomers (See Fig. 3 of Ref. 46), from which one would expect the half-life of ^{234}Pu*f* to be in the range of 1 to 100 ps, with ^{232}Am being even shorter.

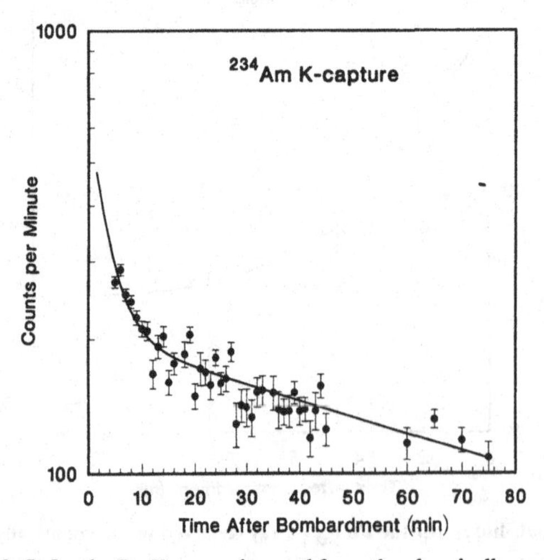

Figure 16. Half-life fit for the Pu *K* x rays observed from the chemically purified ^{234}Am sample.

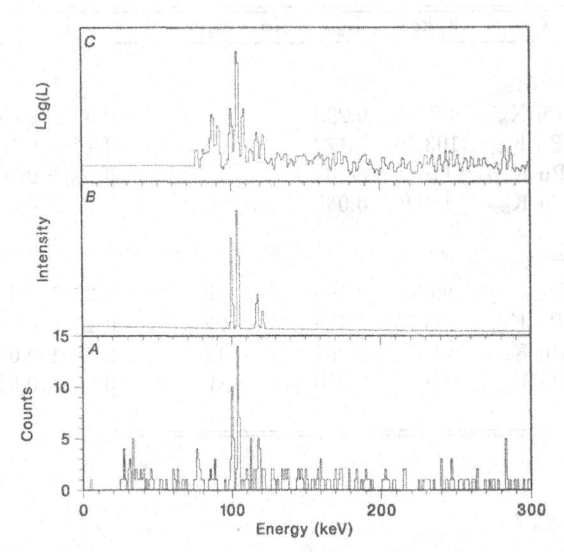

Figure 17. X-ray-fission correlation results for ^{232}Am. *A*) X rays and γ rays in coincidence with delayed fission from ^{232}Am. *B*) An idealized Pu *K* x-ray spectrum, based on the measured detector resolution and prompt γ-ray continuum. *C*) The likelihood function for the position of the ideal spectrum *B*) in the data *A*), as a function of the $K_{\alpha 1}$ position.

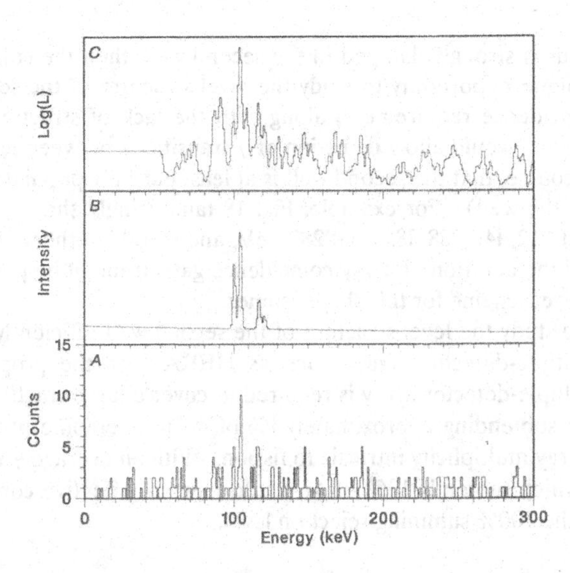

Figure 18. X-ray-fission correlation results for ^{234}Am. *A*) X rays and γ rays in coincidence with delayed fission from ^{234}Am. *B*) An idealized Pu *K* x-ray spectrum, based on the measured detector resolution and prompt γ-ray continuum. *C*) The likelihood function for the position of the ideal spectrum *B*) in the data *A*), as a function of the $K_{\sigma 1}$ position.

Table IV. Observed and Expected X-Ray Intensities from the Correlated X-Ray-Fission Data. Expected X-Ray Intensities are Taken from the "Table of Isotopes". [44]

X-ray	E/keV	I_{theo}	No. Observed[a]	I_{obs}
^{232}Am:				
Pu $K_{\alpha 2}$	99.55	0.299	19	0.33 ± 0.09
Pu $K_{\alpha 1}$	103.76	0.479	23	0.40 ± 0.10
Pu $K_{\beta 1'}$	116.9	0.162	11	0.19 ± 0.06
Pu $K_{\beta 2'}$	120.6	0.060	4	0.07 ± 0.04
^{234}Am:				
Pu $K_{\alpha 2}$	99.55	0.299	10	0.20 ± 0.07
Pu $K_{\alpha 1}$	103.76	0.479	22	0.45 ± 0.12
Pu $K_{\beta 1'}$	116.9	0.162	14	0.29 ± 0.09
Pu $K_{\beta 2'}$	120.6	0.060	3	0.06 ± 0.04

If the nucleus is strongly damped in the second well, then the coincidence γ-fission data provides a unique opportunity to study the level structure of the second well.[47] The highly specific coincidence requirement, along with the lack of structure in the fission—prompt-γ-ray emission, would allow detection of γ transitions between levels in the second well (provided, of course, that the second well is at least partially populated by states above the lowest state in the well). For example, Fig. 18 tantalizingly shows what appear to be true peaks at about 112, 147, 168, 185, and 287 keV, and possibly others. With better statistics in the data and the addition of a γ-γ coincidence gate, it might be possible to construct a fairly complete level scheme for this shape isomer.

However, to study the level structure of the second well efficiently, it will be necessary to use a multiple-detector array such as HERA[48] or the proposed GAMMASPHERE.[49] A multiple-detector array is required to cover a large fraction of 4π with each individual detector subtending approximately 1% of 4π to overcome problems created by the high prompt γ-ray multiplicity intrinsic to fission. With an average γ multiplicity of ten, a single detector which subtends 10% of 4π would have an effective correlation detection rate of 0% due to the 100% summing rejection level.

37.6 CONCLUSIONS

Light Am isotopes were produced using multiple ^{237}Np targets irradiated with α particles. The half-lives of ^{232}Am and ^{234}Am were determined to be 1.31 ± 0.04 min and

2.32±0.08 min, respectively, using a rotating-wheel system. The fission properties of the ϵDF mode in ^{232}Am and ^{234}Am were measured. These are the first delayed-fissile nuclei for which meaurements of the fission properties have been made. These are also the first nuclei for which both the fission and the EC branch leading to the fission have been directly measured.

The highly asymmetric mass-division and symmetric TKE distributions for both ^{232}Am and ^{234}Am show no trace of the Ra anomaly. Therefore, the transition region between "normal" double-humped mass distributions and the triple-humped distribution of the Ra anomaly must begin with lighter elements for this neutron number. Unfortunately, the lighter isotones have considerably smaller Q_ϵ values. This may reduce ϵDF in those nuclei to a level too low to measure their fission properties.

The ϵDF mode provides a mechanism for studying the fission properties of a nucleus far from stability near its ground state. No other technique currently exists which allows the study of near ground-state fission from a specific nucleus this far from β stability.

Finally, the concidence data between the Pu x ray and the fission provides direct proof that the fissions observed in this experiment are the result of K capture in Am followed by fission of excited states in the daughter Pu nucleus. These data also provide the intriguing prospect of studying the level structures of the daughter shape isomers, which are not attainable by other techniques.

ACKNOWLEDGMENTS

This work was supported in part by the Director, Office of Energy Research, Division of Nuclear Physics of the Office of High Energy and Nuclear Physics of the U. S. Department of Energy, under Contract DE-AC03-76SF00098.

The authors wish to thank the staff and crew of the LBL 88-Inch Cyclotron for their assistance.

REFERENCES

1. E. M. Burbidge, G. R. Burbidge, W. A. Fowler, and F. Hoyle, *Rev. Mod. Phys.* **29**, 547 (1957).
2. C. O. Wene and S. A. E. Johansson, *Phys. Scripta* **10A**, 156 (1974).
3. C. O. Wene, *Astron. and Astrophys.* **44**, 233 (1975).
4. H. V. Klapdor, T. Oda, J. Metzinger, W. Hillebrandt, and F. K. Thielman, *Z. Physik* **A299**, 213 (1981).
5. B. S. Meyer, W. M. Howard, G. J. Matthews, K. Takahashi, P. Möller, and G. Leander, *Phys. Rev. C* **39**, 1876 (1989).
6. Yu. A. Lazarev, Yu. Ts. Oganessian, and V. I. Kuznetsov, Joint Institute for Nuclear Research, Dubna, Report No. JINR-E7-80-719 (1980).
7. Yu. P. Gangrskii, M. B. Miller, L. V. Mikhailov, and I. F. Kharisov, *Yad. Fiz.* **31**, 306 (1980) [*Sov. J. Nucl. Phys.* **31**, 162 (1980)].
8. H. L. Hall, K. E. Gregorich, R. A. Henderson, C. M. Gannett, R. B. Chadwick, J. D. Leyba, K. R. Czerwinski, B. Kadkhodayan, S. A. Kreek, D. M. Lee, M. J. Nurmia, D. C. Hoffman, C. E. A. Palmer, and P. A. Baisden, *Phys. Rev. C.* **41**, 618 (1990).
9. H. V. Klapdor, C. O. Wene, I. N. Isosimov, and Yu. W. Naumow, *Z. Phy.* **A292**, 249 (1979).
10. H. L. Hall, K. E. Gregorich, R. A. Henderson, D. M. Lee, D. C. Hoffman, M. E. Bunker, M. M. Fowler, P. Lysaght, J. W. Starner, and J. B. Wilhelmy, *Phys. Rev. C* **39**, 1866 (1989).
11. V. I. Kuznetsov, N. K. Skobelev, and G. N. Flerov, *Yad. Fiz.* **4**, 279 (1966) [*Sov. J. Nucl. Phys.* **4**, 202 (1967)].

12. V. I. Kuznetsov, N. K. Skobelev, and G. N. Flerov, *Yad. Fiz.* **5**, 271 (1967) [*Sov. J. Nucl. Phys.* **5**, 191 (1967)].
13. E. E. Berlovich and Yu. P. Novikov, *Dok. Akad. Nauk SSSR* **185**, 1025 (1969) [*Sov. Physics Doklady* **14**, 349 (1969)].
14. N. K. Skobelev, *Yad. Fiz.* **15**, 444 (1972) [Sov. *J. Nucl. Phys.* **15**, 249 (1972)].
15. D. Habs, H. Klewe-Nebenius, V. Metag, B. Neumann, and H. J. Specht, *Z. Phy.* **A285**, 53 (1978).
16. R. Hingman, W. Kuehn, V. Metag, R. Novotny, A. Ruckelshausen, H. Stroeher, F. Hessberger, S. Hofmann, G. Münzenberg, and W. Reisdorf, GSI, Darmstadt, Report No. GSI **85-1**, 88 (1985).
17. Yu. A. Lazarev, Yu. Ts. Oganessian, I. V. Shirokovsky, S. P. Tretyakova, V. K. Utyonkov, and G. V. Buklanov, *Europhys. Lett.* **4**, 893 (1987).
18. L. P. Somerville, A. Ghiorso, M. J. Nurmia, and G. T. Seaborg, LBL Nuclear Science Division Annual Report, 1976-1977, Report No. LBL-6575, 39 (1977).
19. V. I. Kuznetsov, *Yad. Fiz.* **30**, 321 (1979) [*Sov. J. Nucl. Phys.* **30**, 166 (1979)].
20. R. W. Hoff, in *Weak and Electromagnetic Interactions in Nuclei*, Ed. by H. V. Klapdor, (Springer-Verlag, Heidelburg, 1986), p. 207.
21. R. W. Hoff, *Inst. Phys. Conf. Ser. No. 88/J.Phys. G: Nucl. Phys.* **14** *Suppl.*, S343 (1986).
22. Yu. P. Gangrskii, G. M. Marinescu, M. B. Miller, V. N. Samoyusk, and I. F. Kharisov, *Yad. Fiz.* **27**, 894 (1978) [*Sov. J. Nucl. Phys.* **27**, 475 (1978)].
23. A. Baas-May, J. V. Kratz, and N. Trautmann, *Z. Physik* **A322**, 457 (1985).
24. P. Möller, W. D. Myers, W. J. Swiatecki, and J. Treiner, *At. Data Nucl. Data Tables* **39**, 225 (1988).
25. H. L. Hall, M. J. Nurmia, and D. C. Hoffman, *Nucl. Inst. Meth.* **A276**, 649 (1989).
26. D. C. Hoffman and M. M. Hoffman, *Ann. Rev. Nucl. Sci.* **24**, 151 (1974).
27. H. J. Specht, *Rev. Mod. Phys.* **46**, 773 (1974).
28. H. J. Specht, *Phys. Scripta* **10A**, 21 (1974).
29. R. Vandenbosch and J. R. Huizenga, *Nuclear Fission*, (Academic Press, New York, 1973).
30. E. Konecny, H. J. Specht, and J. Weber, *Phys. Lett.* **45B**, 329 (1973).
31. J. Blons, C. Mazur, D. Paya, M. Ribrag, and H. Weigmann, *Nucl. Phys.* **A414**, 1 (1984).
32. M. G. Itkis, V. N. Okolovich, and A. Ya. Rusanov, *Fiz. Elem. Chastits At. Yadra* **19**, 701 (1988) [*Sov. J. Part. Nucl.* **19**, 301 (1988)].
33. B. S. Bhandari and A. S. Al-Kharam, *Phys. Rev. C* **39**, 917 (1989).
34. D. C. Hoffman and L. P. Somerville, *Charged Particle Emission from Nuclei*, Vol. III, Ed. by D. N. Poenaru and M. Ivascu, (CRC Press, Boca Raton, FL, 1989), p. 1.
35. Y. A. Ellis-Akovali, *Nucl. Data Sheets* **40**, 523 (1983).
36. H. W. Schmitt, W. E. Kiker, and C. W. Williams, *Phys. Rev.* **137**, B837 (1965).
37. E. Weissenberger, P. Geltenbort, A. Oed, F. Gonnenwein, and H. Faust, *Nucl. Inst. Meth.* **A248**, 506 (1986).
38. L. C. Northcliffe and R. F. Schilling, *Nucl. Data Tables* **A7**, 233 (1970).
39. R. Brandt, S. G. Thompson, R. C. Gatti, and L. Phillips, *Phys. Rev.* **131**, 2617 (1963).
40. V. Viola, *Nucl. Data, Sect. B* **1**, 391 (1966).
41. J. P. Unik, J. E. Gindler, L. E. Glendenin, K. F. Flynn, A. Gorski, and R. K. Sjoblom, in *Proceedings of the 3rd IAEA Symposium on the Physics and Chemistry of Fission, 1973*, Vol. 2 (International Atomic Energy Agency, Vienna, 1974), p. 19.
42. J. T. Routti and S. G. Prussin, *Nucl. Inst. Meth.* **72**, 125 (1969).
43. J. B. Cumming, NAS-NS 3107, 25 (1963).
44. C. M. Lederer and V. M. Shirley, *Table of Isotopes*, 7th Ed. (John Wiley & Sons, New York, 1978).
45. J. H. Scofield, At. Data Nucl. *Data Tables* **14**, 121 (1974)
46. D. N. Poenaru, M. S. Ivascu, and D. Mazilu, *Charged Particle Emission from Nuclei, Vol. III*, Ed. by D. N. Poenaru and M. Ivascu, (CRC Press, Boca Raton, FL, 1989), p. 41.
47. H. L. Hall, K. E. Gregorich, R. A. Henderson, C. M. Gannett, R. B. Chadwick, J. D. Leyba, K. R.

Czerwinski, B. Kadkhodayan, S. A. Kreek, D. M. Lee, M. J. Nurmia, and D. C. Hoffman, *Phys. Rev. Lett.* **63**, 2548 (1989).

48. R. M. Diamond, *Nuclei Off the Line of Stability*, Ed. by R. A. Meyer, and D. S. Brenner (ACS Symposium Series, Washington, DC, 1986), p. 341.
49. "GAMMASPHERE - A Proposal for a National Gamma-Ray Facility," Ed. by M.-A. Delaplanque and R. M. Diamond, LBL Publication No. PUB-5202 (1988).

38. High-Spin Spectroscopy For Odd- Z Nuclei With $A \approx 160$

C.-H. Yu
The Niels Bohr Institute
University of Copenhagen
Blegdamsvej 17
Copenhagen Ø, Denmark
and
Department of Physics and Astronomy
The University of Tennessee
Knoxville, TN 37966-1200, USA

J. Gascon* and G. B. Hagemann
The Niels Bohr Institute
University of Copenhagen
Blegdamsvej 17
Copenhagen Ø, Denmark

J. D. Garrett
The Niels Bohr Institute
University of Copenhagen
Blegdamsvej 17
Copenhagen Ø, Denmark
and
Oak Ridge National Laboratory
Oak Ridge, TN 37831, USA

ABSTRACT

Experimental routhians, alignments, band-crossing frequencies, and the $B(M1)/B(E2)$ ratios of the $N = 90$ isotones and several light Lu ($N = 90$–96) isotopes are summarized and discussed in terms of shape changes. This systematic analysis shows a neutron- and proton-number-dependent quadrupole and γ deformations for these light rare-earth nuclei. The stability of the nuclear deformation with respect to β and γ is also found to be particle-number dependent. Such particle-number dependent shapes can be attributed to the different locations of the proton and neutron fermi levels in the Nilsson diagrams. Configuration-dependent shapes are discussed specially concerning the deformation difference between the proton $h_{9/2}1/2^-[541]$ and the high-K $h_{11/2}$ configurations. The observed large neutron band-crossing frequencies in the $h_{9/2}1/2^-[541]$ configuration support the predicted large deformation of this configuration but cannot be reproduced by self-consistent, cranked-shell-model calculations. Lifetime measurement for ^{157}Ho, one of the nuclei that show such a large $\hbar\omega_c$ in the $1/2^-[541]$ band, indicates that deformation difference can only account for 20% of such shift in $\hbar\omega_c$.

*Present address: Laboratoire de Physique Nucléaire, C. P. 612, Succ "A", Université de Montréal, Montréal, Québec H3C 3J7, Canada

38.1 INTRODUCTION

Much spectroscopic information exists for the near-yrast states at high-angular momentum in the rare-earth region. Most of these investigations, however, concentrate on the study of even-even and odd-N nuclei; for example, the series of ytterbium nuclei.[1-5] Such studies provide an understanding of many high-spin phenomena, such as rotational band crossings due to the excitation of pairs of quasi-neutrons and quasi-protons; the quenching of static neutron pair correlations at high angular momentum; and the dependence of the nuclear shape on particle number, configuration, and rotational frequency.

It is important to extend such high-spin studies to odd-Z nuclei in order to establish the spectrum of proton states at large angular momentum. Many phenomena—for example, the variation of the nuclear shape induced by the occupation of various single-proton orbitals with different deformation-driving forces—can be studied and compared with the corresponding neutron effect. Furthermore, because of the large g factors associated with the unpaired proton, it is possible to investigate the details of the nuclear wave-functions by studying the magnetic-dipole transition probabilities between the favoured and unfavoured signature sequences of specific configurations.

A series of experiments have been carried out recently to study the odd-Z light rare-earth nuclei; e.g., ^{161}Lu,[6] ^{165}Lu,[7-9] ^{167}Lu,[10] and ^{157}Ho.[11] Together with the existing data for another odd-A Lu isotope, ^{163}Lu,[12] and the odd-Z, N = 90 isotones, $^{159}Tm_{90}$[13-16] and $^{157}Ho_{90}$,[17-19] the newly-measured data make the odd-A lutetium isotopes and the N = 90 isotones the best studied odd-Z isotopic and isotonic chains at high spin. A systematic analysis on these isotopes and isotones can therefore be made for specific configurations.

The spectroscopy for quasi-proton configurations would be identical for all the Lu isotopes in the absence of mean-field changes, since the proton configurations are the same for all the isotopes. Changes of the single-proton-state spectrum as a function of neutron number are particularly sensitive to changes in the nuclear shape. This sensitivity, combined with the variety of proton orbitals (both down- and up-sloping orbitals on the Nilsson diagrams) in this mass region, leads to a very detailed and interesting spectroscopy for these nuclei. The heaviest Lu isotope, $^{167}Lu_{96}$, is the most stably-deformed nucleus discussed in this work. The rotational effect on single-proton motion is exhibited in this nucleus with the least ambiguity. As a result, this nucleus sets a benchmark for such a study in a stably-deformed system. With the decrease of neutron number, both the magnitude and stability of the nuclear deformation are expected to decrease. The lightest Lu isotope, $^{161}Lu_{90}$, is near the transitional region where the nuclear deformation changes from prolate to spherical shape. Consequently, it is least stable with respect to deformations. In such a "soft" system, configuration- and angular-momentum-dependent nuclear shapes are expected. The rotational modification of single-proton motion in a "soft" system can also be investigated and compared with the more stably-deformed system.

Combined with the data of even-even, N = 90 isotones, ^{162}Hf,[20] ^{160}Yb,[1,4] and ^{158}Er,[21-22] the odd-Z, N = 90 isotones make the N = 90 isotonic chain the best studied isotonic chain at high spin. A systematic study of these isotones allows the investigation of nuclear shapes influenced by the changing mean field due to the change of proton fermi surface.

Figure 1 is a map of the nuclei to be discussed. The contrasting locations of proton and neutron fermi levels on the Nilsson diagrams (see Fig. 2, Refs. 23,24) for these two chains of nuclei (in the upper and lower portions of the shell, respectively) make the comparison of isotopic and isotonic systematics most sensitive to the configuration and particle-number dependent shapes.

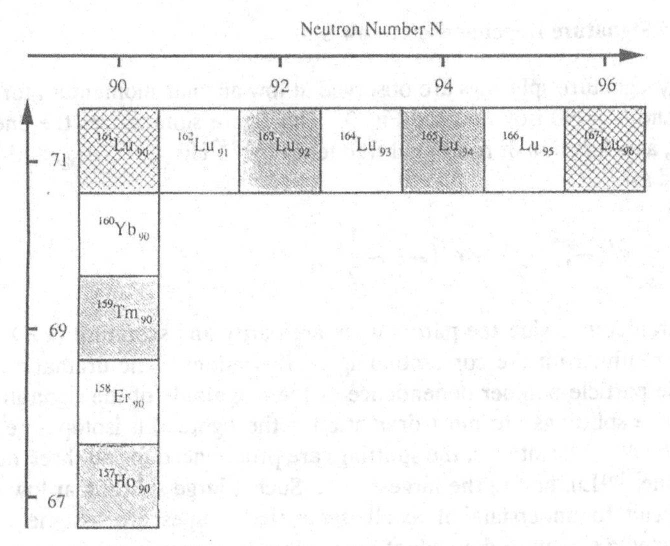

Figure 1. The $N = 90$ isotones and odd-A Lu isotopes discussed in this work.

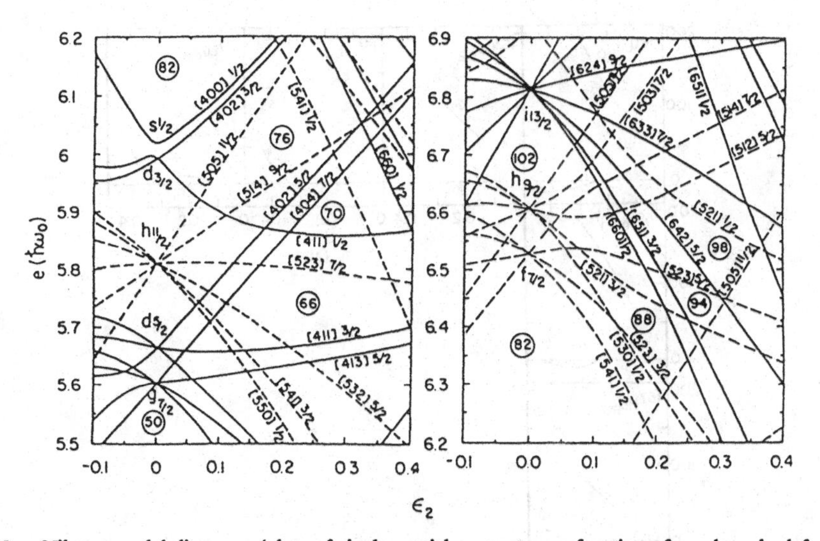

Figure 2. Nilsson-model diagrams (plots of single-particle energy as a function of quadrupole deformation ϵ_2 for protons (left side) and neutrons (right side) calculated using modified-oscillator potentials[23] parametrized as in Ref. 24.

38.2 SYSTEMATIC TREND OF DEFORMATIONS AT LOW ANGULAR MOMENTUM

38.2.A Energy Signature Dependence at Low Spin

Energy signature splittings are observed at low angular momentum for the odd-mass Lu isotopes and $N = 90$ isotones, see Fig. 3. This figure summarizes the energy signature splitting, $\Delta e'$, as a function of rotational frequency for $^{157}Ho_{90}$, $^{159}Tm_{90}$, and $^{161\text{-}167}Lu_{90\text{-}96}$. $\Delta e'$ is defined as

$$\Delta e' = e'(-, +\frac{1}{2}) - e'(-, -\frac{1}{2}) , \tag{1}$$

where the parameters inside the parentheses are parity and signature (π,α), and e' is the experimental routhian for the corresponding configurations. The dramatic feature shown in Fig. 3 is the particle-number dependence of the magnitude of the signature splittings at $\hbar\omega < \hbar\omega_c$. The splittings are most dramatic for the lighter Lu isotopes (e.g., ^{161}Lu and ^{163}Lu). For the $N = 90$ isotones, the splittings are pronounced for all three nuclei, with the heaviest isotone, ^{161}Lu, having the largest $\Delta e'$. Such a large splitting at low spin for these nuclei is difficult to understand if axially-symmetric shapes are assumed. For axially-symmetric nuclei, a signature-dependent decoupling in energy is expected and observed in the $K = 1/2$ rotational sequences. The signature-dependent term in the hamiltonian has different signs for the $\alpha = +1/2$ and $-1/2$ sequences, thus producing the decoupled

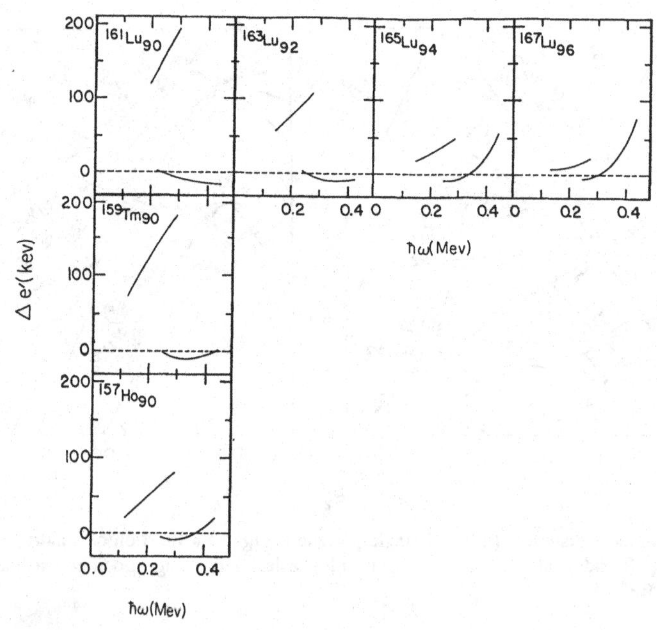

Figure 3. Energy-signature splitting, $\Delta e' \equiv e'(\alpha = +1/2) - e'(\alpha = -1/2)$, as a function of rotational frequency for the decay sequences associated with the proton $h_{11/2}$ yrast configurations in $^{161-167}$Lu, ^{157}Ho, and ^{159}Tm. See Section 38.1 for data sources.

energies for the $K = 1/2$ decay sequences. In a deformed rotating system, K is not a conserved quantum number. The coriolis interaction, which mixes $K = 1/2$ components into the wave-function of bands with $K \neq 1/2$, leads to signature splitting in these sequences. Such a splitting depends on the magnitude of the coriolis matrix elements, which in first order connects states with $\Delta K = 1/2$.[25] Therefore, the splitting for configurations with very large K is inhibited at low spin. The decay sequences shown in Fig. 3 are associated with the orbits in the middle or upper portion of the $h_{11/2}$ "high-j intruder" proton subshell. These orbits (either the $9/2^-[514]$ or the $7/2^-[523]$ Nilsson configuration) have large K values at $\hbar\omega = 0$ ($K = 9/2$ or $7/2$). The rotational frequency at which the large splitting occurs is not high enough to allow sufficient admixture of the low-K components into the wave-function. Thus, large signature splitting is not expected. Indeed, a simple cranked-shell-model calculation, assuming axially-symmetric shapes, predicts no signature splitting in energy for $\hbar\omega < 0.25$ MeV. The large signature splittings for these isotones at low spin can be understood[6,26-28] as the deviation of the nuclear shapes from axial symmetry (i.e., γ deformation). The occupation of a high-j quasi-proton with the favoured signature (the $\alpha = -1/2$ signature for the $h_{11/2}$ shell) strongly polarizes the core shape, especially in the γ degree of freedom. The nuclear shape is "driven" toward $\gamma < 0$ for λ in the lower portion of the shell, not affected for λ in mid-shell, and "driven" toward $\gamma > 0$ for λ in the upper portion of the shell (The Lund convention for the sign of the γ values is assumed; see Ref. 29). In contrast, the polarization effect of occupying a quasi-proton with the unfavoured signature is small. The predicted dependence of the two signatures of the lowest negative-parity quasi-proton configurations on γ deformation is shown for $Z \approx 71$ in Fig. 4. This figure shows that both signature components of the lowest $h_{11/2}$ protons are energetically favoured at negative γ values. The $\alpha = -1/2$ orbit lies lower in energy and has a pronounced minimum at a rather large negative value of γ. Such a non-axially-symmetric nuclear shape and the different shapes for the two signatures give an enhanced energy signature splitting, compared to the axially-symmetric system.

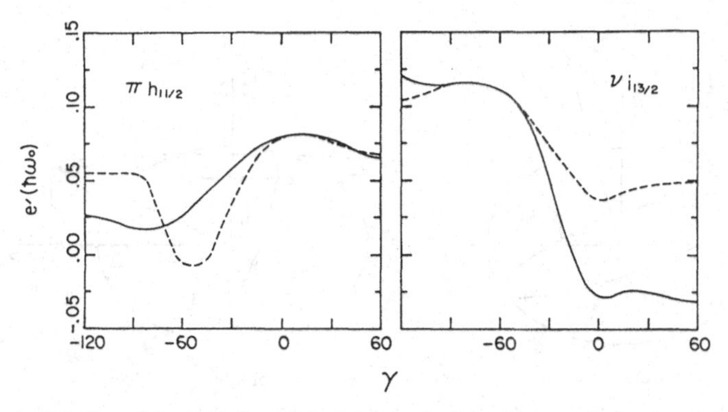

Figure 4. Cranked-shell-model calculation of the γ dependent routhians of the lowest negative-parity quasi-proton (left portion) and positive-parity quasi-neutron (right portion) configurations at $\hbar\omega = 0.03\ \hbar\omega_o$ (0.23 MeV). Solid and dashed curves denote $\alpha = +1/2$ and $\alpha = -1/2$, respectively. The proton and neutron fermi levels [$\lambda_p = 5.88\ \hbar\omega_o$ (for ^{161}Lu, $\hbar\omega_o = 7.240$ and 7.833 MeV for protons and neutrons, respectively), $\lambda_n = 6.46\ \hbar\omega_o$] were fixed at $\gamma = 0$ to correspond to $Z = 71$ and $N = 90$. The deformation parameters, $\varepsilon_2 = 0.2$ and $\varepsilon_4 = 0$, are appropriate for $^{161}Lu_{90}$. The pair-gap parameter, Δ, was taken to be $0.14\ \hbar\omega_o$ (= 1.05 MeV), i.e., approximately equal to 80% and 90% of the odd-even mass differences for protons and neutrons, respectively. The remaining parameters of the oscillator potential were taken from Ref. 24.

Not only can the signature partners of soft nuclei assume different γ deformations, but also such signature-dependent γ deformations will change as a function of the proton and neutron numbers. The nucleus is either driven toward different γ deformations or not influenced by the valence particles, depending on the relative positions of the fermi levels in the shell (see the preceding paragraph). Indeed, the increase of energy signature splitting in the negative-parity decay sequences of the $N = 90$ isotones with increasing Z is attributed to such a change in the γ deformation between the middle and the upper portion of the $h_{11/2}$ shell. For example, a static γ deformation of about $-20°$ is necessary to account for the observed splitting of the negative-parity configurations in $^{161}Lu_{90}$ below the band crossing, and those for $^{159}Tm_{90}$ and $^{157}Ho_{90}$ are $-16°$ and $-10°$, respectively.[30] It is difficult, however, to distinguish the effect of a static γ deformation from a fluctuation with respect to the γ degree of freedom. The magnitude of such fluctuations is inversely proportional to the stiffness of the potential with respect to γ deformation.

The changing of the proton fermi level also affects the β_2 deformation of the nucleus. This can be qualitatively understood from the proton Nilsson diagrams; see Fig. 2. The occupation of an oblate proton orbit* drives the nucleus toward a smaller quadrupole deformation. Such a decrease of quadrupole deformation also gives rise to increased signature splitting, since the proton fermi level is closer to the low-K components of the shell for smaller quadrupole deformation. Hence, the nucleus does not have to rotate so fast to produce the same amount of low-K components in the wave-functions.

The neutron-number dependence of changes in the spectra of quasi-proton states for an odd-Z isotopic chain particularly reflects changes in the nuclear shape, since to first approximation the same spectra are expected for identical quasi-proton configurations, assuming the same shape. Differences in the spectra of quasi-proton states, therefore, are directly attributed to different nuclear shapes. In Fig. 3, the drastic decrease of energy signature splitting at low spin with increasing neutron number for the Lu isotopes is a clear

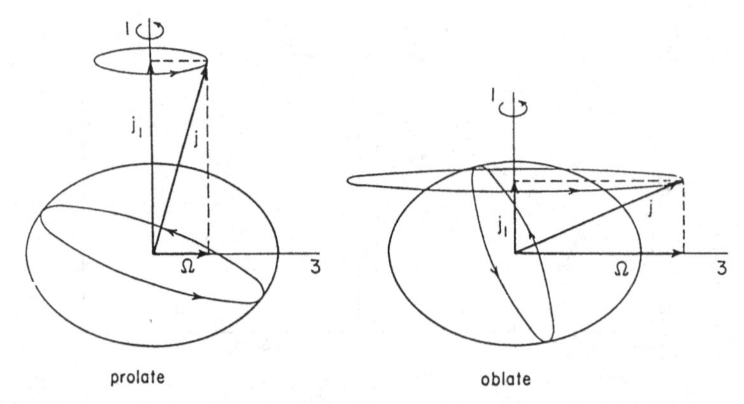

Figure 5. Comparison of the prolate and oblate orbitals in prolately-deformed nuclei. The intrinsic angular momentum, j, for nucleons moving in such orbitals are indicated together with the projections on the nuclear symmetry axis, Ω, and on the axis of rotation, j_1.

*The definition of prolate and oblate orbitals, downward- and upward-sloping, respectively, on the Nilsson diagrams, see Fig. 5, distinguishes the relative projection of the intrinsic angular momentum on the nuclear-symmetry and rotational axes. Oblate and prolate orbitals are illustrated for a prolate nucleus in Fig. 5.

indication of shape changes from the lighter to heavier isotopes. The decrease of splitting with increasing neutron number is the result of both the increased stability of the axially-symmetric shapes and the increased magnitude of the quadrupole deformation associated with the occupation of downward-sloping neutron Nilsson levels between $N = 90$ and 96; see the right side of Fig. 2. In contrast to the case of protons, as discussed in the preceding paragraphs, the occupation of prolate neutron orbitals at the lower portion of the $\nu i_{13/2}$ shell drives the nucleus toward positive γ and larger quadrupole deformations. The almost-vanishing energy signature splitting in ^{167}Lu indicates that this nucleus is most stably deformed and has the largest quadrupole deformation and the least deviation from axial symmetry.

Self-consistent calculations[31] also show consistent deformation systematics for these isotopes. Figure 6 shows the $\beta_2 - \gamma$ dependence of the total routhians for the lowest negative-parity, $\alpha = -1/2$ configurations in the odd-mass $N = 90$ and $Z = 71$ nuclei at $\hbar\omega \sim 0.2$ MeV. Such plots show a particle-number dependence of the stability of the nuclear shapes. The total routhian minimum is best defined on the $\beta_2 - \gamma$ plane for the heaviest Lu isotope, ^{167}Lu. This nucleus has also the largest quadrupole deformation and the smallest deviation from axial symmetry (the energy minimum has the largest β_2, and $\gamma \sim 0$). With the decrease in neutron number, the stability of the nuclear deformation decreases; i.e., the energy minimum becomes less well-defined with respect to β_2 and γ. This "softness" is especially pronounced with respect to γ. For $^{161}Lu_{90}$, the lowest energy-contour line has γ values varying from $<-30°$ to $+10°$. The quadrupole deformation is also significantly reduced from $N = 96$ to 90. For the $N = 90$ isotones, the removal of the upper-shell $h_{11/2}$ protons, by decreasing the proton number from 71 to 67, apparently stabilizes the nuclear shape, especially with respect to γ, and increases the quadrupole deformation β_2. Such systematics are in agreement with the observed systematic trend of the energy signature dependence.

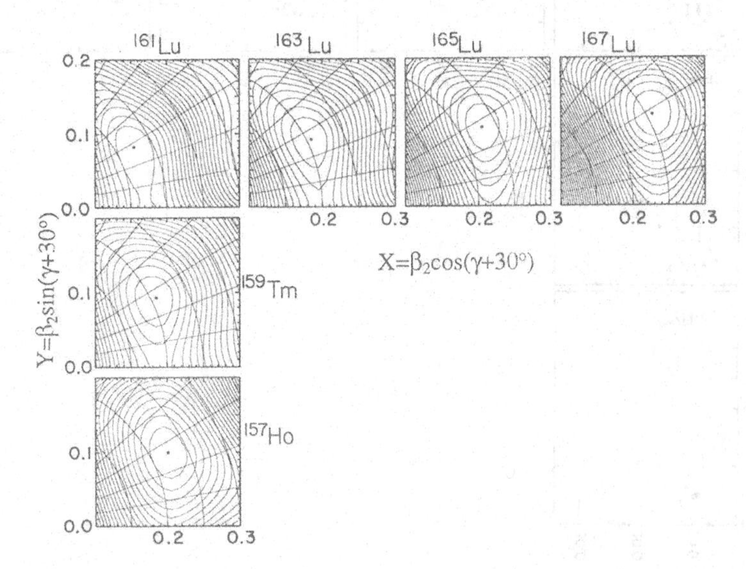

Figure 6. Comparison of calculated[30] potential-energy surfaces for the lowest negative-parity, $\alpha = -1/2$ configurations of ^{161,163,165,167}Lu, ^{157}Ho, and ^{159}Tm. These surfaces were calculated as a function of β_2 and γ at $\hbar\omega \approx 0.2$ MeV, using the Nilsson-Strutinsky procedure. The pairing-gap parameters are fixed at the values calculated with the BCS method at $\hbar\omega = 0$.

38.2.B Signature Dependence of Transition Rates at Low Spin

The large energy signature splitting discussed in the preceding subsection is a clear indication of non-axially-symmetric shapes for the lighter Lu isotopes and $N = 90$ isotones, but it is not definite evidence. Since different models based on varying assumptions and parameters predict different magnitudes of energy splittings, even for the same deformation, it is important to investigate additional experimental quantities that are sensitive to the deformations.

Figure 7 shows the relative ratios, $B(M1; I \to I-1)/B(E2; I \to I-2)$, of the reduced magnetic-dipole and electric-quadrupole transition probabilities of the decay sequences associated with the lowest $\pi h_{11/2}$ configurations for $^{157}Ho_{90}$, $^{159}Tm_{90}$, and $^{161-167}Lu_{90-96}$. In Fig. 7, the $B(M1; I \to I-1)/B(E2; I \to I-2)$ ratios at low spin (below the first band crossing) are characterized by a signature dependence that is smallest for the heaviest Lu isotope, ^{167}Lu. Lifetime measurements for $^{157}Ho_{90}$ (Ref. 17) and $^{159}Tm_{90}$ (Ref. 16) show no signature dependence of $B(E2; I \to I-2)$ values at low spin within the experimental uncertainties of 20% and 15% for $^{159}Tm_{90}$ and $^{157}Ho_{90}$, respectively. The observed signature dependence of the $B(M1/B(E2)$ ratios is therefore attributed to the $B(M1, I \to I-1)$ values.

The signature dependence of $B(M1)$ values is related to the amplitude of $K = 1/2$

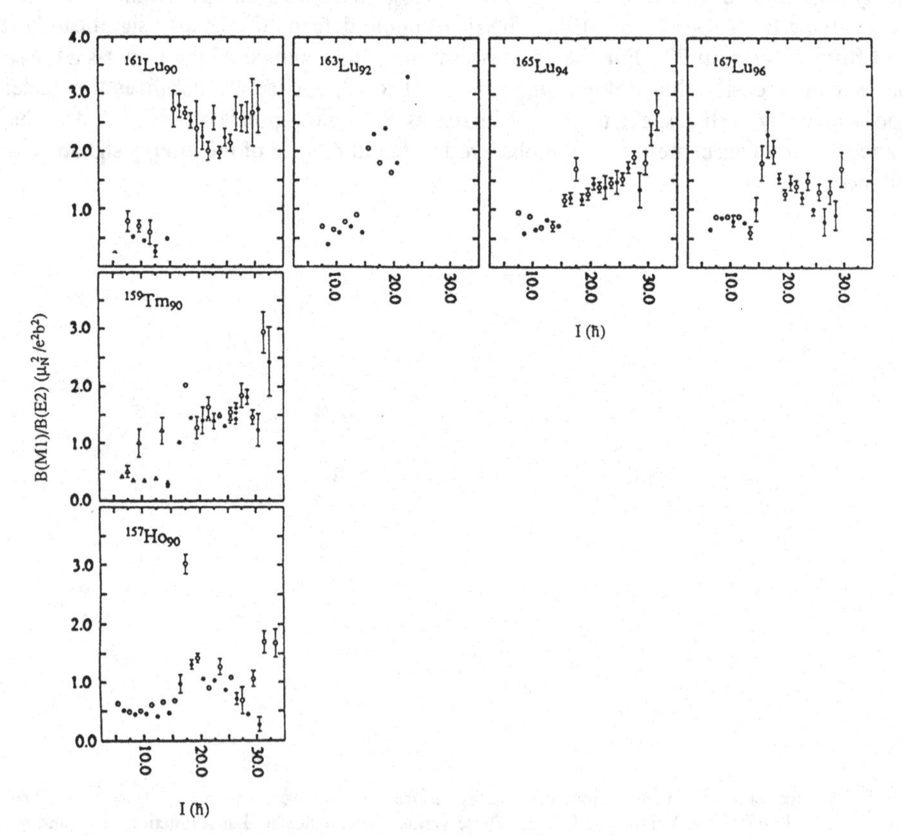

Figure 7. $B(M1, I \to I-1)/B(E2, I \to I-2)$ ratios as a function of spin for $^{157}Ho_{90}$, $^{159}Tm_{90}$, and $^{161-167}Lu_{90}$. See Sect. 38.1 for data sources. For ^{159}Tm, the ratio corresponding to $I = 33.5\hbar$ is 8.25 ± 0.63.

components in the wave-function through a mechanism similar to the decoupling in the excitation energies. Such a signature dependence of $B(M1)$ values does not in itself give definite evidence for the triaxiality of the nuclear shape. However, it is proved[32] that there exists a definite relation between the signature splittings of routhians and the $B(M1)$ values, if the nucleus has an axially-symmetric shape. In a semiclassical approximation[33] based on cranking, if the nucleus is axially symmetric about the z axis (or has a very small γ deformation), the reduced $M1$-transition probability is:

$$B(M1, I \rightarrow I-1) = \frac{3}{8\pi} K^2(\omega) \left\{ (g_p - g_R) \left[\sqrt{1 - \frac{K(\omega)^2}{I^2}} - \frac{i_p}{I} \pm \frac{\Delta e'}{\hbar\omega} \right] - (g_n - g_R)\frac{i_n}{I} \right\}^2 \quad (2)$$

Therefore, below the band crossing (i.e., $i_n = 0$),

$$\frac{B(M1, \alpha = +\frac{1}{2}I \rightarrow \alpha = -\frac{1}{2}I - 1)}{B(M1, \alpha = -\frac{1}{2}I \rightarrow \alpha = +\frac{1}{2}I - 1)} = \frac{\sqrt{1 - \frac{K(\omega)}{I^2}} - \frac{i_p}{I} + \frac{\Delta e'}{\hbar\omega}}{\sqrt{1 - \frac{K(\omega)}{I^2}} - \frac{i_p}{I} - \frac{\Delta e'}{\hbar\omega}} . \quad (3)$$

Particle-rotor calculations[32] also show that a similar relation is valid for axially-symmetric shapes and that it works very well in the deformation-aligned limit. The validity of this relation is, therefore, a test of the axial symmetry of the nuclear shape. From Eq. (3), the relative signature dependence of $B(M1)$ values can be expressed by:

$$\frac{\Delta B(M1)}{B(M1)_{av}} = \frac{4X\frac{\Delta e'}{\hbar\omega}}{X^2 + \left(\frac{\Delta e'}{\hbar\omega}\right)^2}, \quad (4)$$

where

$$X = \sqrt{1 - \frac{K^2(\omega)}{I^2}} - \frac{i_p}{I} . \quad (5)$$

Using experimental values of $\Delta e'/\hbar\omega$, the relative splitting $\Delta B(M1)/B(M1)_{av}$ can be extracted from experimental $B(M1)/B(E2)$ ratios and compared to the empirical values calculated from the right-hand side of Eq. (4). The results for both experimental and empirical values of $\Delta B(M1)/B(M1)$ are shown in Fig. 8 for $^{157}Ho_{90}$, $^{159}Tm_{90}$, and $^{161-167}Lu_{90-96}$ at spin $I \approx 10$. The "expected" $\Delta B(M1)/B(M1)_{av}$ values are overestimated for ^{157}Ho and ^{165}Lu by more than a factor of two and for ^{159}Tm and ^{161,163}Lu by a factor of three to four. For ^{167}Lu, the expected value is close to the experimental value. The large discrepancies between the expected and measured values for ^{161}Lu, ^{163}Lu, and ^{159}Tm suggest sizable triaxial deformations for these nuclei. This conclusion is consistent with the large energy signature splittings at this spin region discussed in the preceding subsection.

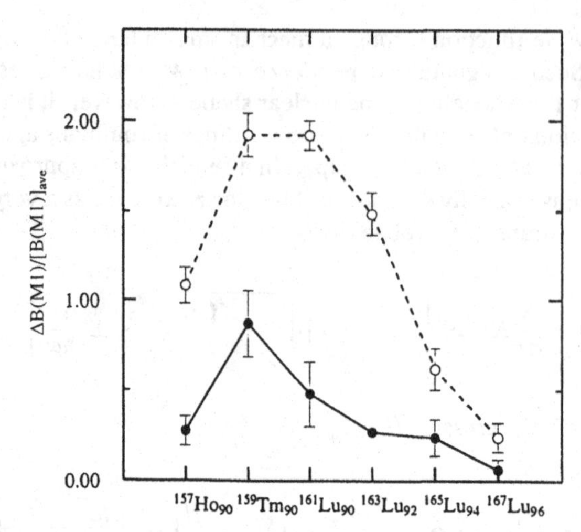

Figure 8. Comparison of "expected" (open ⊙) [see Eq. (4)] and experimentally-extracted (solid ⊙) relative $B(M1)$ signature splittings averaged at spin ≈ 19/2 for the yrast configuration of $^{157}Ho_{90}$, $^{159}Tm_{90}$, and $^{161-167}Lu_{90\text{-}96}$.

38.3 DEFORMATION AND THE *AB* NEUTRON BAND CROSSINGS

Indirect information about the nuclear deformation can be obtained from the band-crossing frequency, $\hbar\omega_c$, which in many cases can be determined accurately from experiment. Theoretically, such a crossing frequency is associated with: 1) the projection, j_x, of the intrinsic angular-momentum component on the rotation axis for the aligning orbitals responsible for the band crossing; and 2) the quasi-particle energies of the aligning orbitals, defined as:

$$E_\nu = \sqrt{\Delta^2 + (\varepsilon_\nu - \lambda)^2} \tag{6}$$

where Δ, ε_ν, and λ are the pair gap parameter, the single-particle energy of state ν taken to be that of the aligning levels, and the fermi level associated with the appropriate particle number. For nuclei with fermi levels close to the aligning orbits [$\varepsilon_\nu - \lambda$ small in Eq. (6)], band-crossing frequencies are dominantly determined by the pair gap, Δ. Band-crossing frequencies are also sensitive to the relative position of the fermi level with respect to the aligning orbitals. This influence is not only derived from the single-particle term, $(\varepsilon_\nu - \lambda)^2$, in Eq. (6), but also associated with the value of j_x. Thus, deformation can influence $\hbar\omega_c$ by changing the position of fermi level relative to the aligning orbitals.

In the past few years, studies on some Ta and Re isotopes[34-35] have shown that the *AB* neutron band crossing is shifted to a larger frequency in the proton $h_{9/2}$ $1/2^-[541]$ Nilsson configuration relative to the other configurations. Such shifts of the *AB* neutron band-crossing frequencies are also observed in some light rare-earth nuclei, e.g., ^{167}Lu,[10] ^{165}Lu,[7] and ^{157}Ho.[19] The magnitudes of these shifts (for this $1/2^-[541]$ Nilsson configuration in ^{157}Ho, ^{165}Lu, ^{167}Lu, ^{169}Ta, $^{171,173}Ta$, ^{171}Re, and ^{177}Re), measured relative to the average crossing frequency in the yrast sequences of the neighbouring even-even isotones are:

$$\delta\hbar\omega_c = \hbar\omega_c(\mathrm{odd}-\mathrm{Z}, \frac{1}{2}^-[541]) - \hbar\omega_c(\mathrm{even}-\mathrm{even},\ \mathrm{yrast})\ , \qquad (7)$$

They are summarized in Fig. 9. Deformation effects qualitatively describe[35] the shift of $\hbar\omega_c$ to larger frequencies in the $1/2^-[541]$ configuration. The strongly down-sloping $1/2^-[541]$ orbit in the Nilsson diagrams (see the left side of Fig. 2) drives the nuclear shapes toward a large quadrupole deformation. As a result, the nuclear deformation is larger when an odd proton occupies this orbit, compared to the occupation of other orbits. Indeed, self-consistent cranking calculations reproduce[35] most of the observed shift for $^{177}Re_{102}$. It should be noted, however, that not only is the observed $\hbar\omega_c$ relatively small in $^{177}Re_{102}$ (e.g., less than half of that in $^{167}Lu_{96}$; see Fig. 9), but also the $i_{13/2}$ quasi-neutron band crossing for this nucleus, for which the fermi level is moved away from the highly-alignable orbits, is more sensitive to the deformation. This sensitivity decreases with decreasing neutron number. For the ninety-six neutrons of $^{167}Lu_{96}$, the neutron fermi level is closer to the highly-alignable $i_{13/2}$ orbits than for $^{177}Re_{102}$. Therefore, the quasi-neutron energy, E_ν, is less sensitive to deformation changes. Self-consistent calculations[39] based on a Woods-Saxon nuclear potential only predict about 50% of the observed shift in crossing frequencies for the $1/2^-[541]$ decay sequence in ^{167}Lu. For the lightest nuclei shown in Fig. 9, $^{157}Ho_{90}$, the predicted shift is less than 20% of the observed value. A recent measurement[11] of the lifetimes for both the $h_{9/2}1/2^-[541]$ and the $h_{11/2}7/2^-[523]$ configurations in ^{157}Ho indicates that the experimentally-determined deformation difference between these two configurations can only account for about 20% of the observed shifts in $\hbar\omega_c$. Such a result agrees with the expected small sensitivity of the *AB* neutron band-crossing frequency to deformations when the neutron fermi level is in the lower portion of the $i_{13/2}$ subshell. The measured large shifts of crossing frequencies in

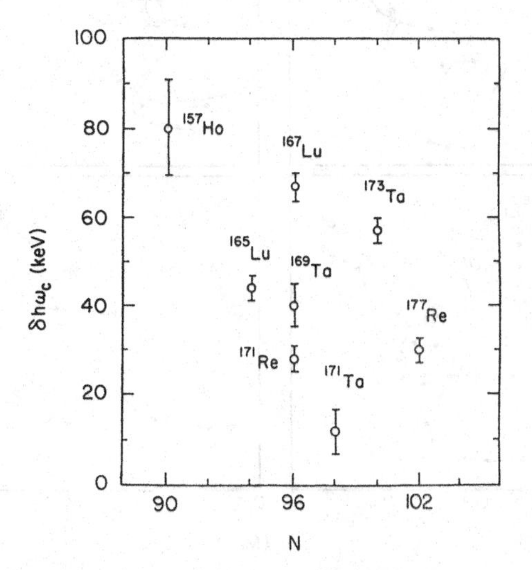

Figure 9. Summary of the shifts, $\delta\hbar\omega_c$, of the *AB* quasi-neutron band crossing frequencies, for the decay sequences based on the $1/2^-[541]$ Nilsson configuration in ^{157}Ho,[19] ^{165}Lu,[7,8] ^{167}Lu,[10] ^{169}Ta,[36] ^{171,173}Ta,[37] ^{171}Re,[38] and ^{177}Re[35] with respect to this crossing in the neighbouring even-even isotones. The precise definition of $\delta\hbar\omega_c$ is given in Eq. (7).

decay sequences associated with the $h_{9/2}1/2^-[541]$ Nilsson configuration vs other configurations for these nuclei is impossible to understand in terms of deformation differences.

38.4 SPECTROSCOPIC PHENOMENA AT HIGHER SPIN

38.4.A Alignment Systematics Above the *AB* Crossing

The systematics of the aligned angular momentum for the N = 90 isotones are summarized in Fig. 10. An interesting feature is observed in the negative-parity decay sequences of $^{161}Lu_{90}$, i. e., the gradual, yet sizable, alignment gain at intermediate

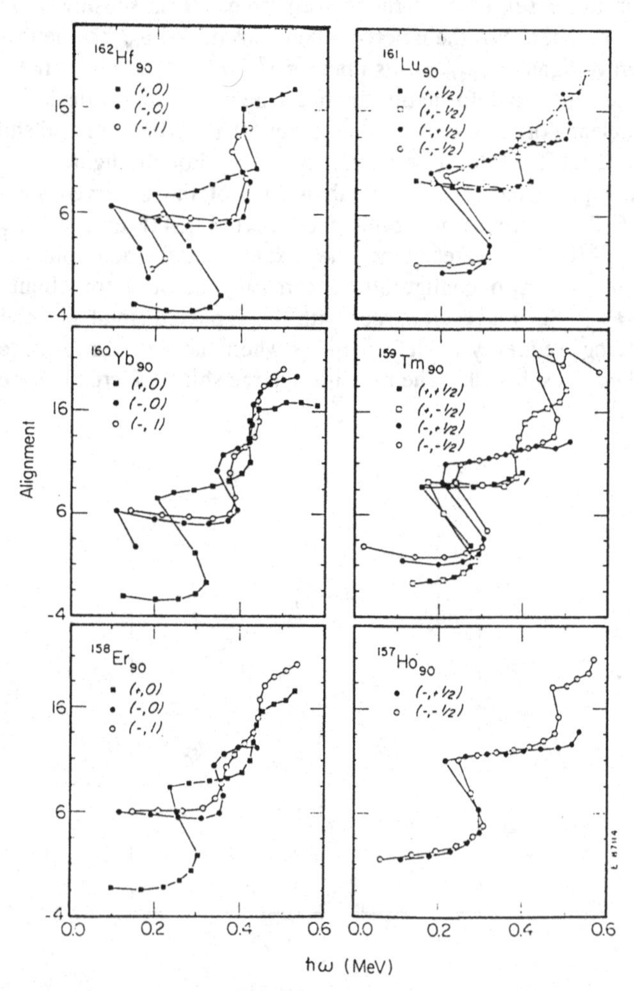

Figure 10. Comparison of the experimental alignments for selected decay sequences of the N = 90 isotones (see Sect. 38.1 for data sources). All the values are referred to a reference configuration parametrized by the Harris[40] formula, $\Im = \Im^{(0)} + \Im^{(1)}\omega^2$ with $\Im^{(0)} = 29$ MeV$^{-1}\hbar^2$ and $\Im^{(1)} = 32$ MeV$^{-3}\hbar^4$.

frequencies ($0.25 < \hbar\omega < 0.40$ MeV) between the neutron and proton band crossings; see Fig. 10. This feature is absent in the positive-parity decay sequences of ^{161}Lu and the other odd-Z, $N = 90$ isotones. However, it is observed, though less pronounced, in the negative-parity decay sequences of $^{159}Tm_{90}$ and is nearly absent in the negative-parity decay sequences of $^{157}Ho_{90}$.

A similar gradual gain in alignment, increasing in magnitude as a function of Z, has been known[4,41,42] at mid frequencies in the (+,0) decay sequences of the even-mass, $N = 90$ isotones; see Fig. 10. Since this feature is absent in the (–,0), (–,1), (–,1/2), and (–,–1/2) sequences (*AF*, *AE*, *ABE*, and *ABF* quasi-neutron excitations, respectively) of the even-A, $N = 90$ and 91 isotones, it was suggested[41,42] to be the result of a band crossing associated with the excitation of the lowest-frequency pair of negative-parity quasi-neutrons (*EF*). This explanation, however, fails to explain the gradual alignments in the negative-parity decay sequences of $^{159}Tm_{90}$ and $^{161}Lu_{90}$, which are absent in the positive-parity sequences of these same nuclei. The positive- and negative-parity decay sequences in these odd-Z isotones should have the same quasi-neutron configuration; thus, any quasi-neutron alignment should either occur, or not occur, in both positive- and negative-parity sequences. (These systematics are also discussed in Refs. 20 and 43.)

The physical basis for the relative alignment gains of the negative-parity decay sequences with respect to the positive-parity sequences in ^{159}Tm and ^{161}Lu remains unexplained. It is noteworthy that this feature is more pronounced for nuclei with a smaller β_2 deformation and also is anti-correlated with the stability of nuclear deformation in both the β_2 and γ degrees of freedom. No such gradual alignment gains are observed for the heavier Lu isotopes. This may be associated with a more stable nuclear shape for these isotopes resulting from the addition of neutrons (see also discussion in Sect. 2).

38.4.B Energy Signature Dependence above the *AB* Band Crossing

The large energy signature splittings at low spin (see Sect. 38.2.*A*), which favours the $\alpha = -1/2$ sequence, in the negative-parity decay sequences of the lighter Lu isotopes and heavier $N = 90$ isotones disappear above the *AB* band crossings (see Fig. 3). The small splitting above the *AB* band crossings for these nuclei favours the $\alpha = +1/2$ sequence. Such a dramatic change of signature splitting is the result of deformation changes due to the occupation of a pair of $i_{13/2}$ quasi-neutrons. The excited low-Ω $i_{13/2}$ quasi-neutron orbitals drive the nuclear shape toward a positive γ deformation (see the right side of Fig. 4), thus cancelling the driving effect of quasi-protons[6] and producing a stable, axially-symmetric nuclear shape for these isotones and isotopes. It should be noted that the relative change of energy-signature splitting, $\Delta e'$, is the largest for ^{161}Lu, indicating a larger change of nuclear shape below and above the *AB* band crossing for this nucleus compared to the other isotones and isotopes. This feature is consistent with the "softness" of the nuclear potential expected in this nucleus.

At the largest rotational frequencies, a sizable signature dependence of energy develops for ^{167}Lu in the negative-parity decay sequences based on the $9/2^-[514]$ Nilsson configuration (see Fig. 3). A similar, though smaller, signature splitting in energy is also observed[8,9] for this configuration in ^{165}Lu (Fig. 3). Such an energy signature splitting at high spin is attributed to the admixture of the low-K $h_{11/2}$ components into the wave-functions.

38.4.C Transition Rates at Higher Spins

Near the $i_{13/2}$ quasi-neutron band crossing, the $B(M1)/B(E2)$ ratio increases for all the nuclei shown in Fig. 7. This feature is expected qualitatively. The $B(M1)$ value is expected to increase because of the increased neutron alignment. However, the magnitude of the increases for the lighter Lu isotopes is considerably larger than those for the other nuclei. Such a large increase cannot be explained entirely by the increased $B(M1)$ values based on the experimental alignment gain if a stable, axially-symmetric shape is assumed. The alignment gains associated with the *AB* neutron band crossing for these isotonic and isotopic chains of nuclei have similar magnitudes. By assuming a constant value of $B(E2)$, the relative increase of $B(M1)$ values at the band crossing can be extracted from the experimental ratios of $B(M1)/B(E2)$. Such experimentally-extracted relative increases of $B(M1)$ values are compared with the "expected" increases according to Eq. (2). Figure 11 shows such a comparison for $^{157}Ho_{90}$, $^{159}Tm_{90}$, and $^{161-167}Lu_{90\text{-}96}$. The experimental relative increases were obtained by assuming a constant value of $B(E2)$ and are defined as:

$$\frac{B(M1,\, I\approx 20)}{B(M1,\, I\approx 10)} = \frac{[B(M1)/B(E2)]_{I\approx 20}}{[B(M1)/B(E2)]_{I\approx 10}} \tag{8}$$

where $[B(M1)/B(E2)]_{I\approx 20}$ and $[B(M1)/B(E2)]_{I\approx 10}$ are the averaged experimental $B(M1)/B(E2)$ ratios at spin $I \approx 20$ and 10, respectively. The expected relative increases of $B(M1)$ values are the ratios of the calculated $B(M1)$ values at $I \approx 20$ and 10, respectively. The proton and neutron gyromagnetic ratios used in the calculation are $g_p = 1.26$ and $g_n = 0.2$ for $h_{11/2}$ protons and $i_{13/2}$ neutrons. The effective g factor for collective rotation, g_R, is

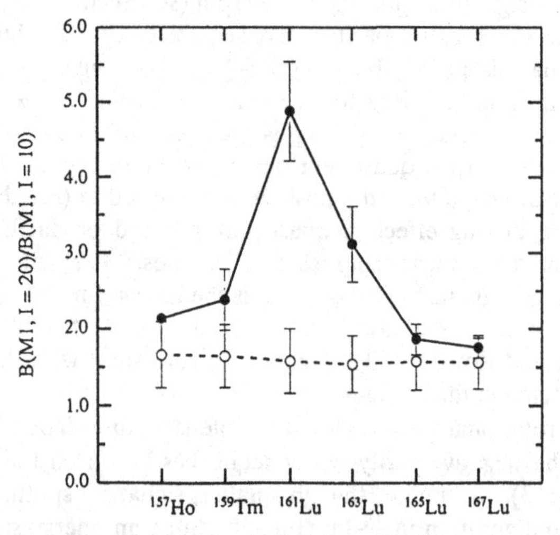

Figure 11. Comparison of experimental (solid points) and predicted (open points) relative increase of the $B(M1)$ values due to the excitation of a pair of $i_{13/2}$ quasi-neutrons for decay sequences associated with the high-K proton $h_{11/2}$ Nilsson state in $^{161,163,165,167}Lu$, ^{157}Ho, and ^{159}Tm. The empirically-predicted relative increases were calculated using Dönau's formula, Eq. (2), and the experimental alignment gains at the band crossing (see text).

assumed to be 0.4 below and 0.3 above the AB neutron band crossing. The alignment gains, Δi, caused by the excitation of a pair of $i_{13/2}$ quasi-neutrons, are taken from the experimental alignments.

The expected increase of $B(M1)$ values for ^{157}Ho, ^{159}Tm, ^{165}Lu, and ^{167}Lu are in good agreement with the experimental values under the assumption of constant $B(E2)$ values. However, the expectations underestimate the increase for ^{161}Lu and ^{163}Lu by a factor of three and two, respectively. Such discrepancies can be the results of either an improper assumption of constant $B(E2)$ values or an abnormal behavior of $B(M1)$ values. Equation (2) is valid only for nuclei with axially-symmetric shape or small triaxiality.[33] Both ^{161}Lu and ^{163}Lu are expected to have considerably large deviations from axial symmetry at low spin. For nuclei with large anisotropy ($|\gamma| \gg 20$), a difference in the principal moments of inertia should be considered.[44] A different coupling scheme for the unfavoured states is also expected when $\gamma \approx -30°$. In this coupling scheme the $B(M1)$ values are nearly quenched.[45] Such a reduction of $B(M1)$ values, due to the non-axial shapes below the AB band crossing, gives rise to the relative increase of $B(M1)$ values at the crossing. A decrease of $B(E2)$ values can also enhance the relative increase of the $B(M1)/B(E2)$ ratios. The observed abnormally-large increase of $B(M1)/B(E2)$ ratios at the AB neutron band crossing most probably are the combination of a reduction of $B(M1)$ values below the band crossing due to the negativ-γ deformation and a decrease of $B(E2)$ values at the crossing.

An increase of $B(M1; I \to I-1)/B(E2; I \to I-2)$ ratios occurs at the largest angular momentum for ^{159}Tm and ^{165}Lu (see Fig. 7). Such an increase may be the result of a decrease of $B(E2)$ values. Detailed discussion for these increases are given in Refs. 9 and 46. For ^{161}Lu, the $B(M1; I \to I-1)/B(E2; I \to I-2)$ ratios above the AB neutron band crossing retain the abnormally-large value up to the rotational frequency where a proton alignment is expected to occur. Such a pattern suggests a possible decrease of $B(E2)$ values, not only at the band crossing, as that discussed in the preceding paragraphs, but also at higher spins, since a decrease in the $B(M1)$ value is expected because of the decrease of the effective K values as a function of $\hbar\omega_c$ (see Ref. 33). At high angular momentum, a signature dependence of $B(M1)/B(E2)$ is observed in ^{167}Lu and ^{157}Ho. Such a signature dependence results from the rotationally-induced coriolis mixing that is also responsible for the signature splitting in energies (see the discussion in Sect. 4.2*A*, 4.2*B*). The signature dependence of $B(M1; I \to I-1)/B(E2; I \to I-2)$ ratios in ^{167}Lu qualitatively agrees with the expectations from the signature dependence in energies.

38.5 ABSOLUTE TRANSITION PROBABILITIES

Direct information about nuclear shapes can be derived from experimentally-measured transition quadrupole moments. Such information is available for most of the N = 90 isotones. Figure 12 summarizes the transition quadrupole moments for the yrast bands of five N = 90 isotones. Two interesting features are observed when odd-A nuclei are compared to their neighbouring even-even isotones: 1) The transition quadrupole moment for ^{157}Ho is smaller than for either of its even-even isotones, ^{156}Dy and ^{158}Er, at low spin. 2) The decrease of Q_t as a function of spin exhibited in the even-even isotones is absent in both ^{157}Ho and ^{159}Tm (except for ^{157}Ho at very low spin).

The first feature indicates that ^{157}Ho has a smaller quadrupole deformation than its neighbouring even-even isotones. The occupation of single-proton states is modified in two ways when an unpaired proton is added to an even-even core: *i*) The single unpaired proton occupies a specific orbit with 100% probability. The quantum numbers for this orbit

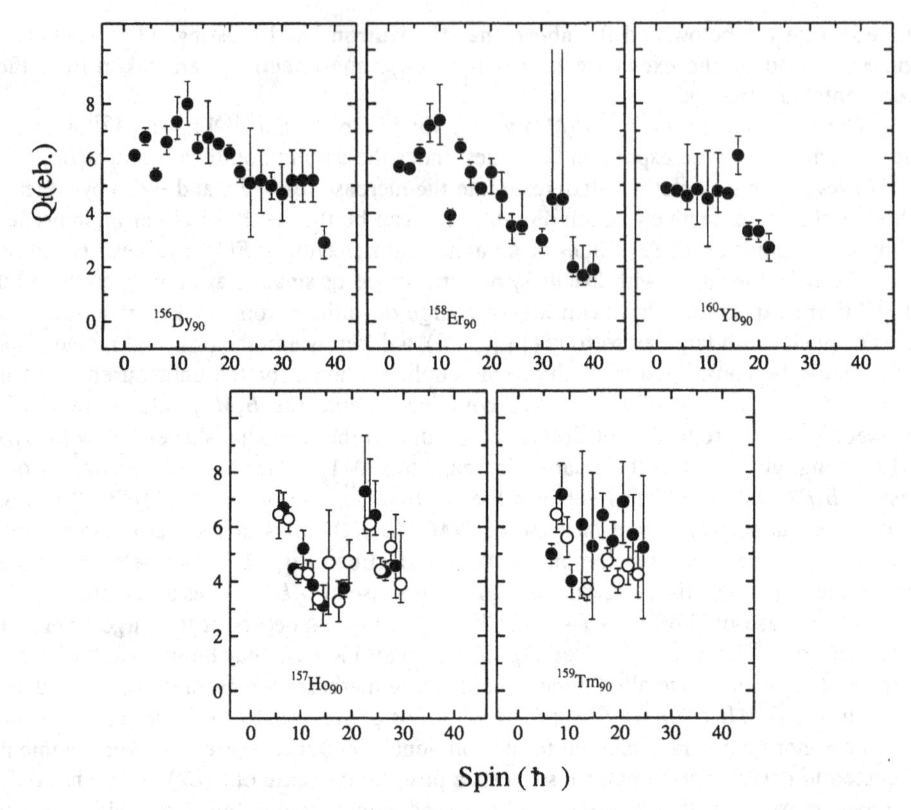

Figure 12. Experimental quadrupole moments for the ΔI = 2 transitions in the yrast decay sequences of ^{156}Dy$_{90}$, ^{157}Ho$_{90}$, ^{158}Er$_{90}$, ^{159}Tm$_{90}$, and ^{160}Yb$_{90}$. Data are taken from Refs. 11, 16, 17, 47, 48, 49, and 50.

define the parity and band head spin of this configuration. *ii*) The addition of a proton "blocks" the pairing contributions associated with the orbit that is occupied. This affect on the proton correlations can be sizeable (a reduction of 20% – 30%) if the configuration is near the fermi surface. Both of these effects influence the nuclear shape. Indeed, a simple cranked-shell model calculation shows that the occupation of the $h_{11/2}$ 7/2$^-$[523] proton orbit at small rotational frequencies "drives" the nucleus toward smaller quadrupole deformations, even though this orbit has a nearly-zero slope in the Nilsson diagrams (see Fig. 2). The occupation of such an orbit in ^{157}Ho may have a stronger influence on the nuclear shape than the occupation of a pair of these orbits in ^{156}Dy and ^{160}Yb. Likewise, the reduction in proton correlation associated with the "blocking" of this orbit reduces the occupation of the strongly prolate-polarizing orbit, i.e., the 1/2$^-$[541] orbit which lies above the proton fermi level (see Fig. 2).

For ^{159}Tm, the Q_t values are nearly halfway in between those of its neighbouring even-even isotones. This is hard to understand in terms of the first cause of the odd-even deformation difference argued for ^{157}Ho (see the preceding paragraph). The second cause argued for ^{157}Ho, however, may not be valid for ^{159}Tm, since the 1/2$^-$[541] orbit is expected to lie further away from the fermi level for the heavier N = 90 isotones. It is therefore possible that no obvious odd-even deformation difference exists for the heavier N = 90 isotones.

The absence of the decrease of Q_t as a function of spin for the odd-A isotones may be partially associated with the odd-even deformation differences. The decrease of transition quadrupole moments [or $B(E2)$ values] as a function of spin were observed for the even-even $N = 90$ isotones[47-50] and for several heavier Yb isotopes.[51,52] This effect has been attributed[53] to a modification of single-particle composition of the intrinsic wavefunction by rotational motion. In the presence of rotationally-induced coriolis and centrifugal forces, the degeneracy of the pairs of strongly-polarizing nucleons (i.e., nucleons moving in a high-j, low-Ω orbits), moving in time-reversed orbits, is removed. At higher frequencies the splitting becomes sufficiently large for the "unfavoured" states (corresponding to the nucleon moving in a direction opposite to the nuclear rotation) to be depopulated. The deoccupation of such strongly shape-driving states causes the decrease of the nuclear quadrupole deformation, thus resulting in a decreased transition quadrupole moment. A smaller quadrupole deformation anticipated for ^{157}Ho (see the preceding paragraph) causes the loss of the high-j, low-Ω orbit to occur at a lower rotational frequency. For a smaller quadrupole deformation, the neutron fermi level is closer to the low-Ω components of the $i_{13/2}$ subshell. As a result, the nucleus does not need to rotate as fast to lose the anti-aligning, high-j orbit. Consequently, no decrease of Q_t is observed for ^{157}Ho in the spin region where the decrease is observed for the even-even isotones (i.e., at or above the AB neutron band crossing; see Fig. 12). For ^{157}Ho, the experimental Q_t decreases by 30% from spin 15/2 to 17/2 (Fig. 12). It should be noted, however, that, at such a low spin, Q_t is sensitive to the effective K values used in the extraction of Q_t from the measured lifetime. The mixing of low-K components, due to a small triaxiality expected in this nucleus, may reduce the amount of the above-mentioned decrease in Q_t. At high spins, the Q_t values show no further decrease in ^{157}Ho. The connection between the lower-frequency (spin) occurrence of the decrease of Q_t and the smaller quadrupole deformation is also observed in a systematic analysis[54] of this phenomenon for several Yb isotopes.

For ^{159}Tm, no decrease of Q_t is observed at low spin. Since ^{159}Tm has a larger quadrupole deformation than ^{160}Yb (the Q_t values for ^{159}Tm are larger than those for ^{160}Yb at low spin; self-consistent calculation[31] also predict a larger quadrupole deformation for ^{159}Tm), the lack of decrease in ^{159}Tm, compared to ^{160}Yb, is difficult to understand in terms of the reasons argued for ^{157}Ho in the preceding paragraph. This indicates a more complicated mechanism behind the decrease of Q_t and the lack or earlier occurrence of this decrease in the odd-Z isotones. More data for the odd-Z nuclei are needed in order to fully understand this odd-even difference of transition quadrupole moments.

38.6 SUMMARY

Several recent high-spin studies on the light Lu isotopes, $^{161-167}$Lu (Refs. 2, 6, 8, 9, 10, and 12) and the odd-Z, $N = 90$ isotones, $^{157}\text{Ho}_{90}$, $^{159}\text{Tm}_{90}$, and $^{161}\text{Lu}_{90}$ (Refs. 6, 11, 15, 16, and 17) are summarized and compared to the relevant even-even isotopes. The systematic analysis concentrates on the study of nuclear shapes and stabilities for the odd-Z, $N = 90$ isotones and the Lu isotopes with respect to the various deformation degrees of freedom.

The results of the systematic analysis of energies and relative transition probabilities indicate that the heaviest Lu isotope, ^{167}Lu, has not only the largest quadrupole deformation but also the best-defined shape with respect to the various deformation degrees of freedom. The nuclear shapes and other properties for the lighter isotopes therefore can be compared to this well-deformed rare-earth nucleus. The results from the comparison show evidence for a decreased quadrupole deformation with decreasing neutron number for the Lu isotopes and with increasing proton number for the N = 90 isotones. Energy signature splittings and their relations to signature-dependent transition rates at low spin indicate that the lightest Lu isotope, ^{161}Lu, has not only the smallest quadrupole deformation, but also the least-defined shape among the nuclei studied. The "softness" of the nuclear potential is associated with both the quadrupole deformation and the γ degree of freedom. The neutron-number dependence of shapes is consistent with the expectation in terms of the polarization effect of the valence particles. The low location of the neutron fermi level in the neutron $i_{13/2}$ subshell for $^{161-169}$Lu suggests a larger and more stable quadrupole deformation and a more stable shape for isotopes occupying a larger number of low-Ω configurations. Such shape systematics are also consistent with the recently-calculated equilibrium shapes for these isotopes. A similar analysis for the N = 90 isotones shows the opposite dependence of nuclear shapes on the proton number. The contrasting locations of the proton fermi levels in the middle or upper portion of the proton shell result in an increased β_2 and a more stable shape for a smaller number of protons.

The large signature splittings in energy at low spin and the inconsistency of these splittings with the signature dependence of $B(M1)$ values expected for an axially-symmetric system also suggest a significant triaxiality for the lighter Lu isotopes and heavier N = 90 isotones. Such a deviation of nuclear shape from axial symmetry is understood as the result of the negative γ-driving force of the unpaired $h_{11/2}$ quasi-proton. However, it is difficult to extract quantitative information about the γ deformation from the experimental data, since the separation of static and dynamic γ deformations is not straightforward.

The polarization effects of different valence quasi-protons on the core are studied by comparing different configurations in the odd-Z, N = 90 isotones and Lu isotopes. The yrast decay sequences of these nuclei are associated with the proton orbits lying in the middle or upper portion of the $h_{11/2}$ subshell. For the heavier Lu isotopes and the lightest N = 90 isotone, ^{157}Ho$_{90}$, level schemes are also established for the decay sequences associated with the 1/2$^-$[541] Nilsson state, the lowest $h_{9/2}$ configuration for prolate deformation. Comparisons of the quasi-neutron band crossing frequencies in the 1/2$^-$[541] band with those in the yrast configurations of the neighbouring even-even nuclei show significant differences. The systematic shifts of the AB quasi-neutron band crossing frequencies to large values in the 1/2$^-$[541] configuration are indicative of a larger quadrupole deformation for this configuration. However, significant discrepancies are observed between the self-consistent, cranked-shell-model calculation and the experimental data, especially for nuclei with the neutron fermi surface low in the $i_{13/2}$ subshell. These sizeable anamalous shifts of the AB neutron band crossing are not understood.

Transition quadrupole moments deduced from lifetime measurements for the N = 90 isotones are also presented and discussed in a systematic way. The smaller average value of the transition quadrupole moments for ^{157}Ho$_{90}$ extracted from the lifetime measurement compared to its neighbouring even-even isotones, ^{156}Dy and ^{158}Er, may be associated with a combination of reduced occupation of prolately-polarizing orbit (i.e., 1/2$^-$[541]) resulting from the decreased proton pair correlations and the occupation of the unpaired, oblate-polarizing 7/2$^-$[523] orbit. The lack of a decrease in Q_t at high spin for ^{157}Ho, compared to the even-even isotones, can be partially attributed to its smaller deformation (see Sect. 38.5). The failure of interpreting the lack of decrease in ^{159}Tm, using the same argument, suggests a more complicated mechanism behind this phenomenon.

ACKNOWLEDGMENTS

The authors wish to thank I. Hamamoto, D. Radford, L. L. Riedinger, and R. Wyss for valuable discussions during this work. This work was supported by the Danish Natural Science Foundation, the U. K. Science and Engineering Research Council, the European Economic Community Stimulation Programme (The ESSA30 Collaboration, contract number ST2J-0205), the NORDBALL Collaboration, the Nordic Council for Accelerator Based Research, the State Education Commission of China, and the Physics Department at the University of Tennessee (supported by the U. S. Department of Energy under contract number DE-SG05-40361, C.-H. Yu), the North Atlantic Treaty Organization (J. Gascon), and Oak Ridge National Laboratory (operated by Martin Marietta Energy Systems, Inc., for the U. S. Department of Energy under contract DE-AC05-840R21400, J. D. Garrett).

REFERENCES

1. J. J. Gaardhoje, Thesis, Univ. of Copenhagen, 1980.
2. S. Jonsson, N. Roy, H. Ryde, W. Walus, J. Kownacki, J. D. Garrett, G. B. Hagemann, B. Herskind, R. Bengtsson, and S. Aberg, *Nucl. Phys.* **A449**, 537 (1986).
3. J. Kownacki, J. D. Garrett, J. J. Gaardhoje, G. B., Hagemann, B, Herskind, S. Jonsson, N. Roy, H. Ryde, and W. Walus, *Nucl. Phys.* **A394**, 269 (1983).
4. L. L. Riedinger, O. Andersen, S. Frauendorf, J. D. Garrett, J. J. Gaardhoje, G. B. Hagemann, B. Herskind, Y. V. Makovetzky, J. C. Waddington, M. Guttormsen, and P. O. Tjøm, *Phys. Rev. Lett.* **44**, 568 (1980).
5. W. Walus, N. Roy, S. Jonsson, J. Carlen, H. Ryde, J. D. Garrett, G. B. Hagemann, B. Herskind, Y. S. Chen, J. Almberger, and G. Leader, *Phys. Scripta* **24**, 324 (1981).
6. C.-H. Yu, M. A. Riley, J. D. Garrett, G. B. Hagemann, J. Simpson, P. D. Forsyth, A. R. Mokhtar, J. D. Morrison, B. M. Nyako, J. F. Sharpey-Shafer, and R. Wyss, *Nucl. Phys.* **A489**, 477 (1988).
7. S. Jonsson, J. Lyttkens, L. Carlen, N. Roy, H. Ryde, W. Walus, J. Kownacki, G. B. Hagemann, B. Herskind, J. D. Garrett, and P. O. Tjøm, *Nucl. Phys.* **A422**, 397 (1984).
8. P. Frandsen, J. D. Garrett, G. B. Hagemann, B. Herskind, M. A. Riley, R. Chapman, J. C. Lisle, J. N. Mo, L. Carlen, J. Lyttkens, H. Ryde, and P. M. Walker, *Phys. Lett.* **B177**, 287 (1986).
9. P. Frandsen, R. Chapman, J. D. Garrett, G. B. Hagemann, B. Herskind, C.-H. Yu, K. Schiffer, D. Clarke, F. Khazaie, J. C. Lisle, J. N. Mo, L. Carlen, P. Ekstrom, and H. Ryde, *Nucl. Phys.* **A489**, 508 (1988).
10. C.-H. Yu, G. B. Hagemann, J. M. Espino, K. Furuno, J. D. Garrett, R. Chapman, D. Clarke, F. Khazaie, J. C. Lisle, J. N. Mo, M. Bergstrom, L. Carlen, P. Ekstrom, J. Lyttkens, and H. Ryde, *Nucl. Phys.* **A511**, 157 (1990).
11. J. Gascon, C.-H. Yu, G. B. Hagemann, M. C. Carpenter, J. M. Espino, Y. Iwata, T. Kamatsubara, J. Nyberg, S. Ogaza, G. Sletten, P. O. Tjøm, D. C. Radford, J. Simpson, M. A. Bentley, P. Fallon, P. D. Forsythand, J. F. Sharpey-Schafer, *Nucl. Phys.* **A513**, 344 (1990).
12. K. Honkanen, H. C. Griffin, D. G. Sarantites, V. Abenante, L. A. Adler, C. Baktash, Y. S. Chen, O. Dietzsch, M. L. Halbert, D. C. Hensley, N. R. Johnson, A. J. Larabee, I. Y. Lee, L. L. Riedinger, J. X. Saladin, T. M. Semkow, and Y. Schutz, in *ACS Symposium Series No. 324, Nuclei off the Line of Stability*, Ed. by R. A. Meyer and D. S. Brenner, (American Chemical Society, 1986), p. 317.
13. A. J. Larabee and J. C. Waddington, *Phys. Rev. C* **24**, 2367 (1981).
14. A. J. Larabee, L. H. Courtney, S. Frauendorf, L. L. Riedinger, J. C. Waddington, M. P. Fewell, N. R. Johnson, I. Y. Lee, and F. K. McGowan, *Phys. Rev. C* 29, 1934 (1984).

15. J. Simpson, B. M. Nyako, A. R. Mokhtar, M. Bentley, H. W. Cranmer-Gordon, P. D. Forsyth, J. D. Morrison, J. F. Sharpey-Schafer, M. A. Riley, J. D. Garrett, C.-H. Yu, A. Johnson, J. Nyberg, and R. Wyss, in *Proceedings of the International Nuclear Physics Conference*, Harrogate, U. K. Aug. 1986, Vol. 1, p. B66, and to be published.
16. J. Gascon, P. Taras, D. C. Radford, D. Ward, H. R. Andrews, and F. Banville, *Nucl. Phys.* **A467**, 539 (1987).
17. G. B. Hagemann, J. D. Garrett, B. Herskind, J. Kownacki, B. M. Nyako, P. L. Nolan, J. F. Sharpey-Schafer, and P. O. Tjøm, *Nucl. Phys.* A424, 365 (1984).
18. J. Simpson, D. V. Elenkov, P. D. Forsyth, D. Howe, B. M. Nyako, M. A. Riley, J. F. Sharpey-Schafer, B. Herskind, A. Holm, and P. O. Tjom, *J. Phys. G*, **12**, L67 (1986).
19. D. Radford, in *Contribution to the Workshop on Nuclear Structure*, The Niels Bohr Institute, May 1988; and D. C. Radford, H. R. Andrews, D. Horn, D. Ward, F. Banville, S. Flibotte, P. Taras, J. Johansen, D. Tuker, and J. D. Waddington, in press.
20. H. Hubel, M. Murzel, E. M. Beck, H. Kluge, A. Kuhnert, K. H. Maier, J. C. Bacelar, M. A. Deleplanque, R. M. Diamond,, and F. S. Stephens, *Z. Phys. A* **329**, 289 (1988).
21. J. Simpson, M. A. Riley, J. R. Cresswell, P. D. Forsyth, D. Howe, B. M. Nyako, J. F. Sharpey-Schafer, J. Bacelar, J. D. Garrett, G. B. Hagemann, B. Herskind, and A. Holm, *Phys, Rev. Lett.* **53**, 648 (1984).
22. P. O. Tjøm, R. M. Diamond, J. C. Bacelar, E. M. Beck, M. A. Deleplanque, J. E. Draper, and F. S. Stephen, *Phys. Rev. Lett.* **55**, 2405 (1985).
23. R. Bengtsson, *J. de Physique* (Coll.) **C10**, 87 (1980); and S. G. Nilsson and C. F. Tsang, *Nucl. Phys.* **A131**, 1 (1969).
24. T. Bengtsson and I. Ragnarsson, *Nucl. Phys.* **A436**, 14 (1985).
25. Aa. Bohr and B. Bottelson, *Nuclear Structure* (Benjamin, Vol. 1 (1969) and Vol. 2 (1975), Reading).
26. R. Bengtsson, H. Frisk, F. R. May, and J. A. Pinston, *Nucl. Phys.* **A415**, 189 (1984).
27. S. Frauendorf and F. R. May, *Phys. Lett.* **105B**, 5 (1981).
28. G. A. Leander, S. Frauendorf, and F. R. May, in *Proceedings of the Conference on High Angular Momentum Properties of Nuclei*, Oak Ridge, p. 281, 1982.
29. G. Andersson, S. E. Larsson, G. Leander, P. Møller, S. G. Nilsson, I. Ragnarsson, S. Aberg, R. Bengtsson, J. Dudek, B. Nerlo-Pomorska, K. Pomorski, and Z. Szymanski, *Nucl. Phys.* **A268**, 205 (1976).
30. I. Hamamoto and H. Sagawa, *Phys. Lett.* **B201**, 415 (1988).
31. R. Wyss, R. Bengtsson, W. Nazarewicz, and J. Nyberg, private communication and to be published.
32. G. B. Hagemann and I. Hamamoto, *Phys. Rev. C* **40**, 2862 (1989).
33. F. Dönau, *Nucl. Phys.* **A471**, 469 (1987).
34. J. C. Bacelar, M. Diebel, O. Andersen, J. D. Garrett, G. B. Hagemann, B. Herskind, J. Kownacki, C.-X. Yang, L. Carlen, J. Lyttkens, H. Ryde, W. Walus, and P. O. Tjøm, *Phys. Lett.* **152B**, 157 (1985).
35. W. Walus, L. Carlen, S. Jonsson, J. Lyttkens, H. Ryde, J. Kownacki, W. Nazarewicz, J. C. Bacelar, J. Dudek, J. D. Garrett, G. B. Hagemann, B. Herskind, and C.-X. Yang, *Phys. Scr.* **34**, 710 (1986).
36. S. G. Li, C.-X. Yang, R. Chapman, F. Khazaie, J. C. Lislie, J. N. Mo, J. D. Garrett, G. B. Hagemann, B. Herskind, and H. Ryde, *Manchester Annual Report*, p. 52 (1987); and G.-J. Yuan, G.-S. Li, S.-X. Wen, S.-G. Li, P.-F. Hua, L. Zhou, P. Wu, L.-K. Zhang, Z.-K. Yu, P.-S. Yu, P.-K. Wen, and C.-X. Yang, *Chinese Journal of Nuclear Physics* **11**, 1 (1989).
37. J. C. Bacelar, R. Chapman, J. R. Leslie, J. C. Lisle, J. N. Mo, E. Paul, A. Simcock, J. C. Willmott, J. D. Garrett, G. B. Hagemann, B. Herskind, A. Holm, and P. M. Walker, *Nucl. Phys.* **A442**, 547 (1985).
38. R. A. Bark, G. D. Dracoulis, A. E. Stuchbery, A. P. Byrne, A. M. Baxter, F. Riess, and P. K. Weng, *J. Phys.* G: **15**, L169 (1989).
39. W. Nazarewicz, J. Dudek, R. Bengtsson, T. Bengtsson, and I. Ragnarsson, *Nucl. Phys.* **A435**, 397 (1985).
40. S. M. Harris, *Phys. Rev.* **138**, B509 (1965).
41. L. L. Riedinger, *Phys. Scr.* **T5**, 36 (1983).

42. S. Frauendorf, L. L. Riedinger, J. D. Garrett, J. J. Gaardhoje, G. B. Hagemann, and B. Herskind, *Nucl. Phys.* **A431**, 511 (1984).
43. C. R. Bingham, L. L. Riedinger, L. H. Courtney, Z. M. Liu, A. J. Larabee, M. Craycraft, D. J. G. Love, P. J. Nolan, A. Kirwan, D. J. Thornley, P. J. Bishop, A. H. Nelson, M. A. Riley, and J. C. Waddington, *J. Phys.* **G14**, L77 (1988).
44. I. Hamamoto in *Nuclear Structure*, Proc. of Niels Bohr Centennial Conf., eds. R. Broglia, G. B. Hagemann, and B. Herskind, p. 129 (North Holland, Amsterdam, 1985).
45. I. Hamamoto, *Phys. Lett.* **B193**, 399 (1987).
46. G. B. Hagemann, R. Chapman, P. Frandsen, A. R. Mokhtar, M. Riley, J. Simpson, and C.-H. Yu, in *Proc. of Int. Conf. on Nuclear Structure Through Static and Dynamic Moments*, p. 313, Melbourne, Australia, Aug. 1987.
47. H. Emling, I. Ahmad, P. J. Daly, B. K. Dichter, M. Drigert, U. Garg, Z. W. Grabowski, R. Holzmann, R. V. F. Janssens, T. L. Khoo, W. C. Ma, M. Piiparinen, M. A. Quader, I. Ragnarsson, and W. H. Trzaska, *Phys. Lett.* **217B**, 33 (1989).
48. M. Oshima, N. R. Johnson, F. R. McGowan, C. Baktash, I. Y. Lee, Y. Schutz, R. V. Ribas, and J. C. Wells, *Phys. Rev. C* 33, 1988 (1986).
49. E. M. Beck, H. Hubel, R. M. Diamond, J. C. Bacelar, M. A. Deleplanque, K. H. Meier, R. J. McDonald, F. S. Stephens, and P. O. Tjøm, *Phys. Lett.* **215B**, 624 (1988).
50. M. P. Fewell, N. R. Johnson, F. K. McGowan, J. S. Hattula, I. Y. Lee, C. Baktash, Y. Schutz, J. C. Wells, L. L. Riedinger, M. W. Guidry, and S. C. Pancholi, *Phys. Rev. Lett.* **37**, 101 (1988).
51. J. C. Bacelar, R. M. Diamond, E. M. Beck, M. A. Deleplanque, J. Draper, B. Herskind, and F. S. Stephens, *Phys. Rev.* **C35**, 1170 (1987).
52. D. Clarke, R. Chapman, F. Khazaie, J. C. Lisle, J. N. Mo, K. Schiffer, J. D. Garrett, G. B. Hagemann, B. Herskind, K. Theine, W. Schmitz, and H. Hubel, *Manchester Annual Report*, 72 (1989).
53. J. D. Garrett, J. Nyberg, C.-H. Yu, J. M. Espino, and M. J. Godfrey, in *Contemporary Topics in Nuclear Structure Physics*, ed. R. F. Casten, *et al.*, p. 699 (World Scientific, Singapore, 1988).
54. J. D. Garrett, contribution to the *Workshop on Nuclear Structure at High Spins*, Bad Honnef, Germany, March 1989, p. 100.

39. Fast Chemical Separations for the Study of Short-Lived Nuclides

Krishnaswamy Rengan
Chemistry Department
Eastern Michigan University
Ypsilanti, MI 48197, USA

ABSTRACT

Study of short-lived nuclides has fascinated nuclear chemists ever since the discovery of radioactivity. Fast-decaying nuclides with half-lives in the range of several seconds to several tens of seconds have been explored in the last ten to twenty years. Automatic batch (autobatch) processes as well as continuous separation procedures are being used for the study of these nuclides. Autobatch processes and continuous production and separation techniques are reviewed in this paper.

39.1 INTRODUCTION

Chemical separations have played an important role in isolating and studying radioactive isotopes of elements ever since the discovery of radioactivity of Becquerel in 1896. Radiochemical Separations provide an unambiguous way of identifying element. The Curies' work leading to the identification of Po and Ra is the earliest radiochemical work reported in the literature.[1,2] Radiochemical separations carried out by Rutherford and Soddy[3,4] led them to conclude that radioactive decay is accompanied by a change of the parent atom.

Interest in the study of short-lived nuclides have led to ingenious fast chemical separation procedures. As early as 1938 Livingood and Seaborg[5] used a precipitation procedure for studying Mn isotopes; the procedure separated Mn from irradiated Fe, Cr, or V metal in a few minutes. The discovery of fission by Hahn and Strassmann in 1939 provided an abundant source of neutron-rich nuclides. As the need and interest for studying shorter and shorter half-lives has grown, new techniques have been developed to the extent that today's separation techniques are challenging chemical reaction rates and the concepts of physical design. Simultaneously, mass separation techniques have improved during the last two decades. Recent developments in ion-source technology have made possible study of many elements; still, for many elements like Se, radiochemical separations provide the only possible means of study.

Several reviews have appeared in the literature. Kusaka and Meinke[6] have reviewed the earlier work and published a volume on "Rapid Radiochemical Separations" as part of the Radiochemistry Monograph series. Herrmann and coworkers publish periodical reviews of radiochemical methods.[7] Autobatch procedures have been reviewed by Meyer.[8] Continuous solvent extraction procedures are reviewed by Skarnemark et al.,[9] while Rengan[10] has reviewed continuous gas phase separations. In this article the general techniques used in fast separation procedures are discussed briefly; autobatch and continuous separation techniques are reviewed.

39.2 TECHNIQUES USED IN FAST-SEPARATION PROCEDURES

Nearly four hundred fast-separation procedures with separation time of 1 min or less (except in the case of lanthanides or actinides and some other elements where procedures require a few minutes) are available in the literature for ninety-two elements. Figure 1 shows the elements for which one or more fast chemical procedures are available in the literature. For each element the figure also shows the separation time for the fastest procedure.

A variety of common analytical techniques like ion exchange, solvent extraction, dis-

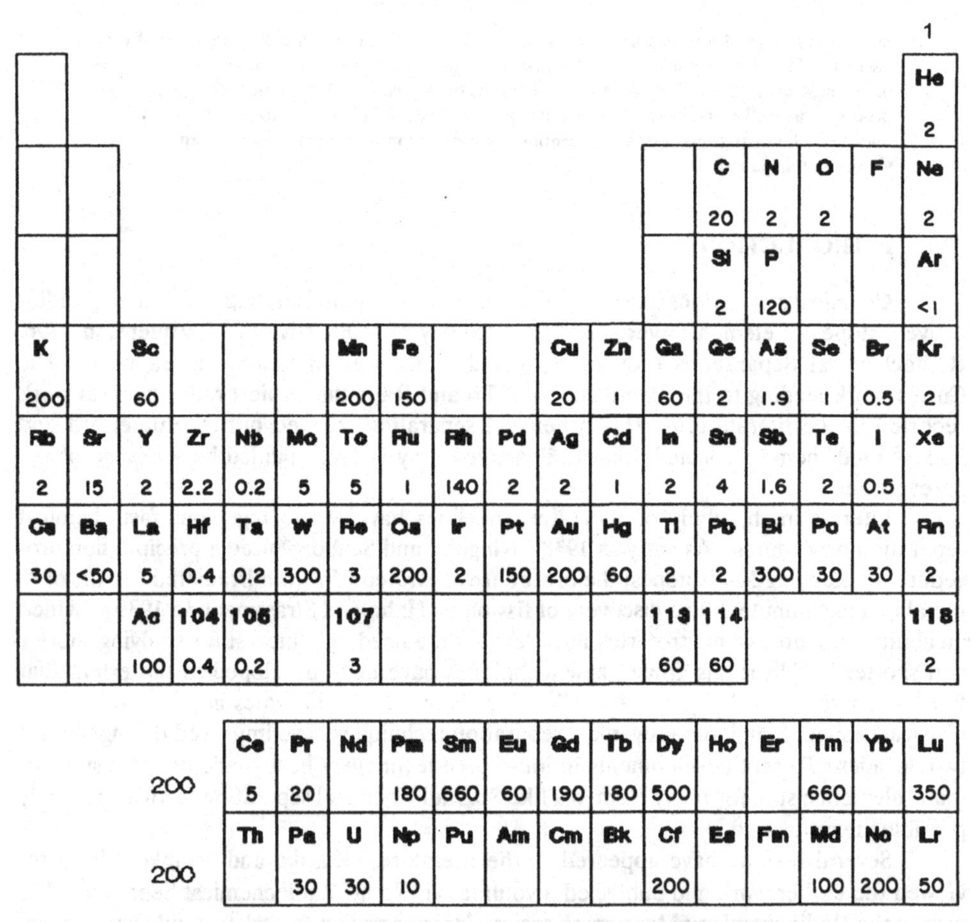

Figure 1. Elements for which fast chemical procedure(s) is (are) available in the literature. The number below the element symbol gives the separation of time (in seconds) for the fastest general procedure. The number on the right of the noble gas group and on the left on the lanthanides and actinides is the time reported for generic procedure for that group of elements.

tillation, precipitation, and volatilization are used in fast chemical separations. Some techniques like exchange, recoil, and hot-atom reactions, which are special for radiochemical separations are also used. Table I lists the techniques used, the number of procedures using the technique, and the elements for which it is used. The complexity of the separation procedure depends on the mode of production of the nuclide to be separated.

Solvent extraction is the most frequently used technique. Table I shows that there are ninety-eight procedures which use this technique. Figure 2 shows the elements for which fast chemical procedures based on solvent extraction are available in the literature. It is evident that the use of this technique is not confined to any particular group or groups of the periodic table. The reason for this is obvious: a variety of extracting agents and the conditions for extraction of different elements are available in the literature. The technique can be easily adapted for the development of a fast procedure to separate a specific element produced in a nuclear reaction.

Volatilization has been used in sixty-nine procedures. Figure 3 shows elements for which volatilization technique is used for the fast separation. This technique is mainly used for elements in groups IV*B*, V*B*, VI*B*, and VII*B* which form volatile hydrides.

Table I. Techniques Used in Fast Chemical Procedures

Technique	No. of Procedures	Elements
Adsorption	10	O,Br,Ne,Nb,Se,Tc,Te
Condensation	3	Ar,Rn,118
Distillation	11	Br,Ge,Os,Rh,Ru
Electrolysis	8	Sb,As
Electrophoresis	16	Ba,Br,Ce,Cs,lanthanides,Se,Sr
Emanation	8	Halogens,He,N,Noble gases
Exchange	12	I,Se,Ag,Sr,Te,Sn
Extraction	98	Ac,Sb,As,At,Ba,Bi,Br,Cd,Ce,Cu, Eu,Ga,Au,In,I,Fe,La,Lr,Pb,Mo, Np,Nb,Pd,Po,Pr,Pa,Rh,Ru,Sc,Ag, Ta,Te,Tl,Sn,V,Zn,Zr,104,113,114
Extraction with other techniques	19	Ac,As,lanthanides,Lr,Md,Np
Gas chromatography	4	Actinides,Br,I,lanthanides
Hot atom reaction	13	Halogens,Se
Ion exchange	21	Ba,Cf,Ir,Fe,Lr,Pb,Md,No,P,Ag, Tl,Y,Zn,104
Plating chemical	9	Sb,Cu,Pd,Ag,Po
Precipitation	38	Ba,Bi,C,Cs,Cu,Fe,La,Mn,Hg,Mo, Pt,Po,K,Rb,Se,Ag,Sr,Tc,Te,Tl,Sn, Y,Zr
Recoil	3	Nb,Rb,Ag
Thermochromatography	40	Cd,Hf,Ir,Hg,Mo,Nb,Os,Rb,Ru, Ta,Tl,W,Zn,Zr,104,105,107
Volatilization	69	Sb,As,At,Br,C,Ge,I,Ir,Kr,Hg,N, Po,Rn,Ru,Se,Tc,Te,Sn,104

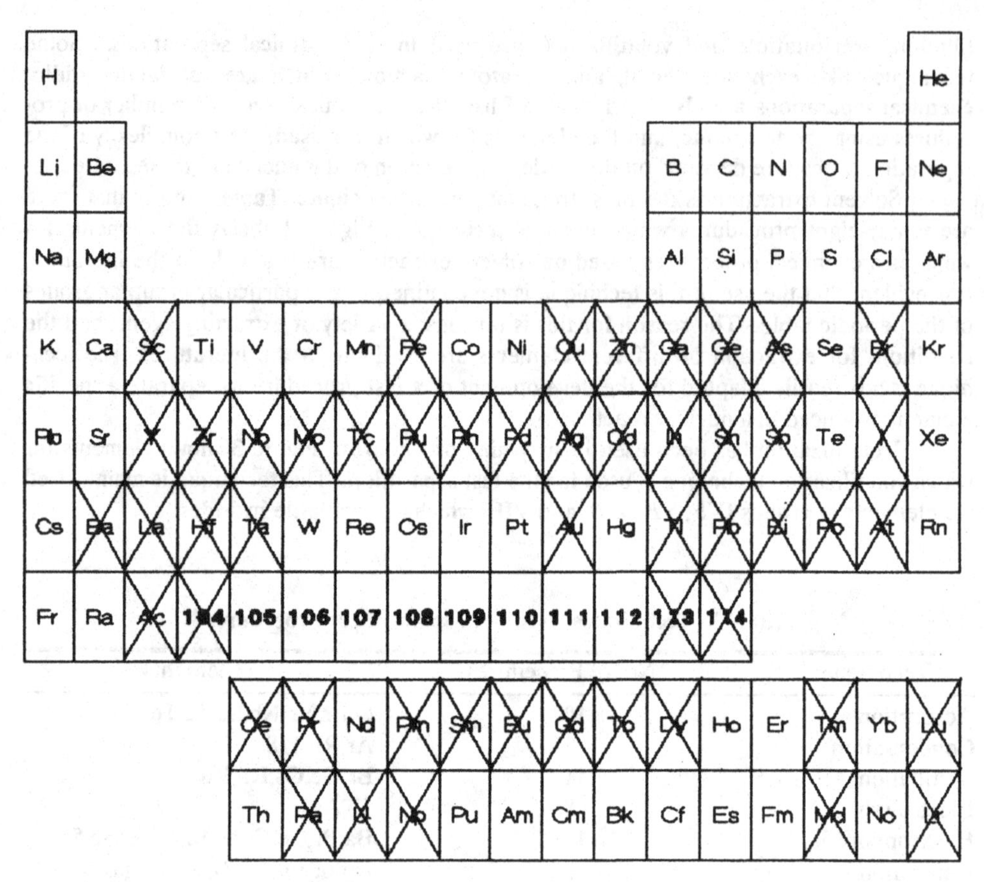

Figure 2. Elements for which fast chemical procedure(s) using solvent extraction techniques is (are) available (marked with an X).

Forty procedures based on thermochromatography have been reported. Figure 4 shows the elements for which thermochromatographic procedure is available. The procedures are available for mostly transition elements which form volatile chlorides, oxychlorides, or bromides; many of the procedures were developed so these could be extended to study transplutonium elements.

Ion-exchange techniques are capable of separating elements with very similar properties, e.g., individual members of lanthanides and actinides from one another. However, the kinetics is slow. Hence, the technique has been used only for a very few elements, as can be seen from Fig. 5. Often it has been utilized for the separation of an element forming an anionic complex from other elements existing as cations under those conditions.

39.3 AUTOMATION OF BATCH-SEPARATION PROCEDURES

As the attention of nuclear chemists shifted to the studies of nuclides with half-lives in the range of several seconds, manual processes became inadequate. Further, in order to

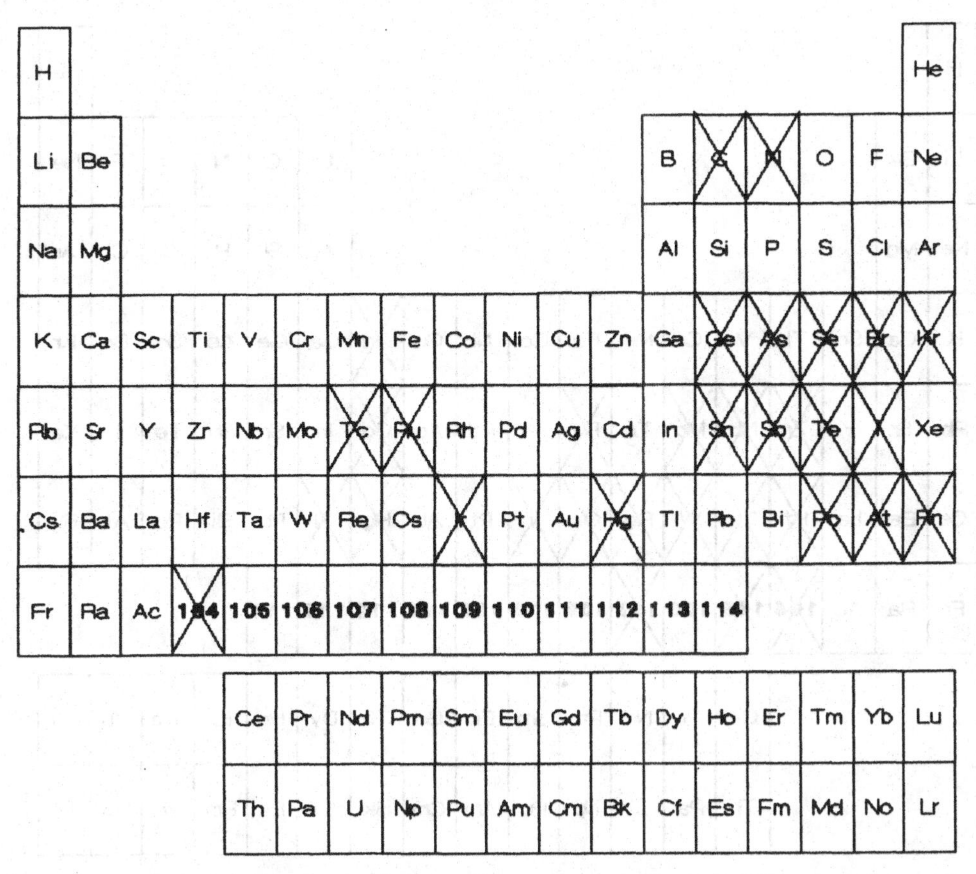

Figure 3. Elements for which fast chemical procedure(s) using volatilization techniques is (are) available (marked with an X).

accumulate statistically significant data for low-intensity γ rays and for coincidence data, several hundreds of separations must be performed. During the last two decades, fully-automated systems have been used. The first direction was to build on the known manual, batch-separation procedures. By appropriate adaptations and modifications of the separation steps, the entire procedure has been carried out by application of pressure or vacuum and opening or closing of valves; the control of pressure/vacuum and the operation of valves were performed by electronic programers in the 70's and by microprocessors in the 80's. There are fifty-one automated batch procedures (autobatch) available in the literature (from '69). Using autobatch techniques, the separations can be repeated more than one thousand times in a typical 12-h day. In the following sections examples of autobatch separation procedures are given.

The autobatch procedure used by Trautmann et al.[11] for the separation of Tc from fission products serves as a good example to illustrate the adaptation of precipitation and solvent extraction steps from batch processes. Figure 6 shows a schematic representation of the procedure. The fission-product solution from an irradiation capsule, smashed by

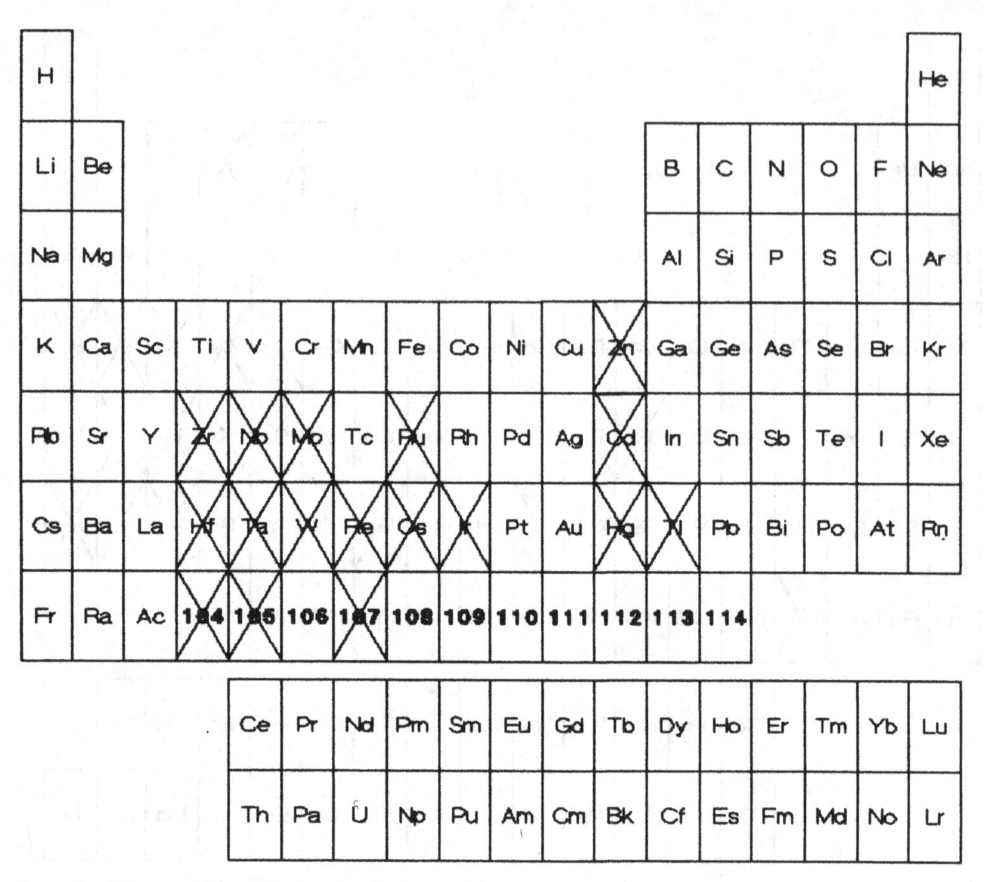

Figure 4. Elements for which fast chemical procedure(s) using thermochromatographic techniques is (are) available (marked with an X).

impact, was passed through two layers of preformed AgCl. Br and I fission products exchange with chloride in AgCl layers and were removed. The exchange took the place of a precipitation step in a batch process. The AgCl layers were then washed with 0.1 *M* HNO_3 containing tartaric acid. The fission-product solution and the washings were collected in a receptacle containing $(NH_4)_2S_2O_8$ and $AgNO_3$; the persulfate oxidized Tc to pertechnetate. Solvent extraction of Tc was achieved by passing the solution through a layer of Chromosorb coated with 0.025-*M* tetraphenyl arsonium chloride (PH_4AsCl) in $CHCl_3$. After washing the layer with 0.1-*M* HNO_3 containing TaA, Tc was stripped by passing 2-*M* HNO_3 containing $(NH_4)_2ReO_4$. Tc was finally coprecipitated with Re as Ph_4AsReO_4. The transfer of liquids and the transfer of the filter containing the final precipitate for counting were achieved by opening and closing of stopcocks and applications of pressure or vacuum at the appropriate time; an electronic programmer controlled the entire operation. Counting of the Tc sample was started 7.5 s after irradiation.

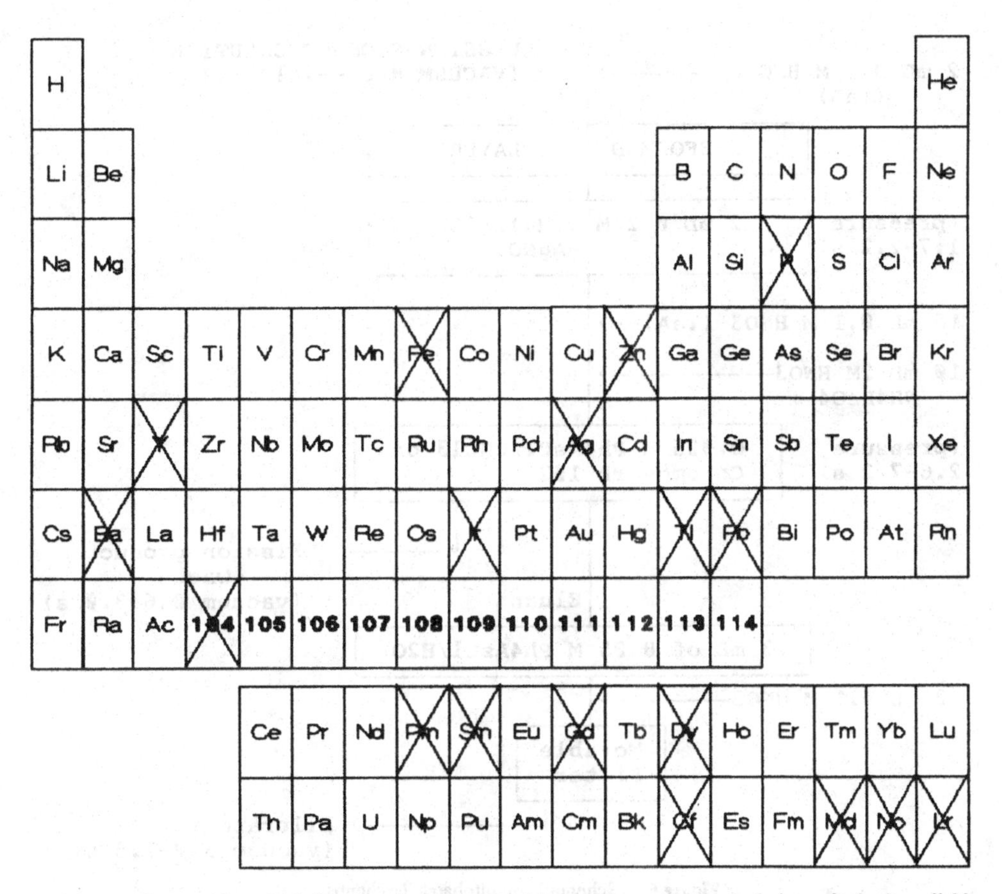

Figure 5. Elements for which fast chemical procedure(s) using ion exchange techniques is (are) available (marked with an X).

Meyer et al.[12,13] developed and used an autobatch procedure for the study of As and Sb nuclides. Figure 7 shows the time sequence of the autobatch As-Sb cycle. The entire process was controlled by a microprocessor. A rabbit containing the U solution was loaded and transferred to an irradiation position through a pneumatic tube. After a preset time the rabbit was returned to a chemistry fume hood through another pneumatic tube. The fission-product solution was transferred to a reaction chamber, purged with N_2 to remove Kr and Xe, and hydrides were generated by the addition of sodium borohydride. Selective decomposition and trapping of hydrides allowed isolation and study of As and Sb nuclides. The separation was achieved in 1.9 s. The counting and the cleaning of the chemistry apparatus started simultaneously. The total cycle time was 30 s. Over 10,000 cycles of As chemistry was performed during the course of 10 d.[14]

Autobatch procedures have made possible study of short-lived nuclides with half-lives as short as a few seconds. Table II gives a list of elements for which autobatch procedures are available in the literature. The technique used for separation as well as the separation time are included in the table.

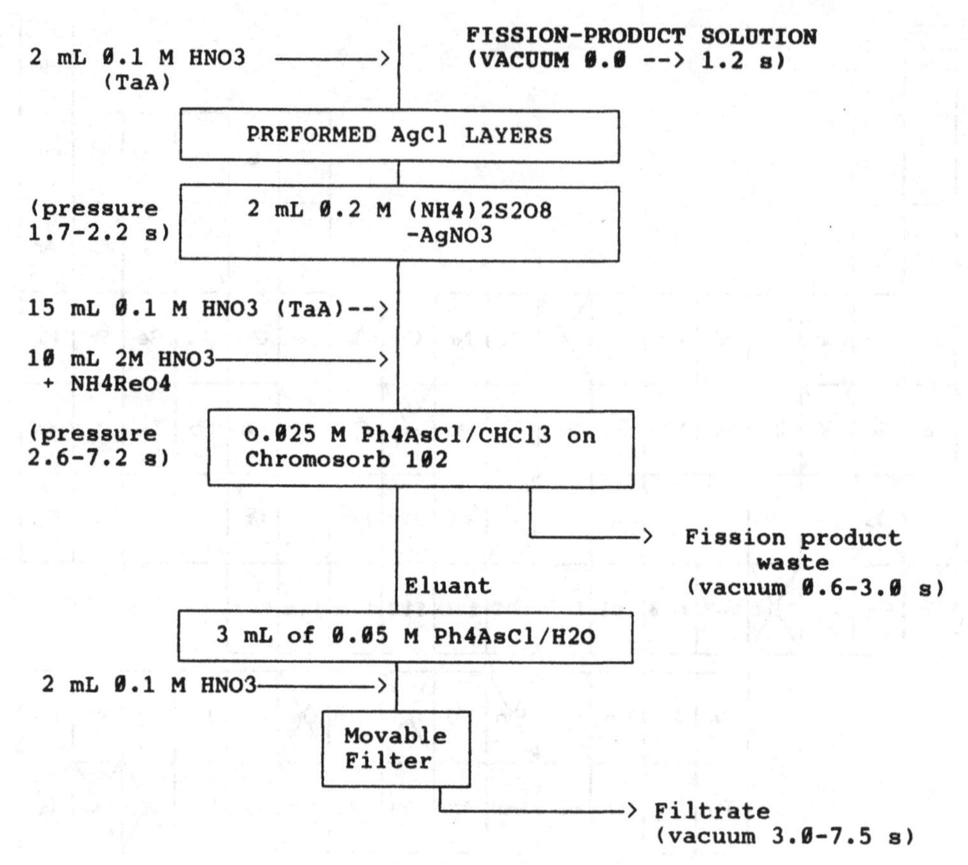

Figure 6. Schematic of autobatch Tc chemistry.

39.4 CONTINUOUS-SEPARATION PROCEDURES

Collection of data using autobatch procedures become tedious, especially if the half-life of the nuclide is 1 s or shorter. An entirely different approach has been to produce and separate reaction products continuously. Such procedures allow data collection in a shorter time. The data collection efficiency of autobatch and continuous processes have been compared by Meyer and coworkers.[12,13] For continuous separation of short-lived nuclides a system should be available for continuous production and transport of reaction products to a chemistry system in a time comparable with the half-life of the nuclide studied. Gas-jet systems have been successfully used in accelerators[15] and in nuclear reactors[16] to carry reaction products from a target chamber to a chemistry site. The typical transport time is 1 s or less. The gas-jet system at the Ford Nuclear Reactor (University of Michigan) delivers fission products in about 0.8 s. He or N_2 gas loaded with aerosol is used as carrier gas. In the study of fission products, reactive gases like HCl and ethylene have also been mixed with N_2 and used as carrier gas.[17]

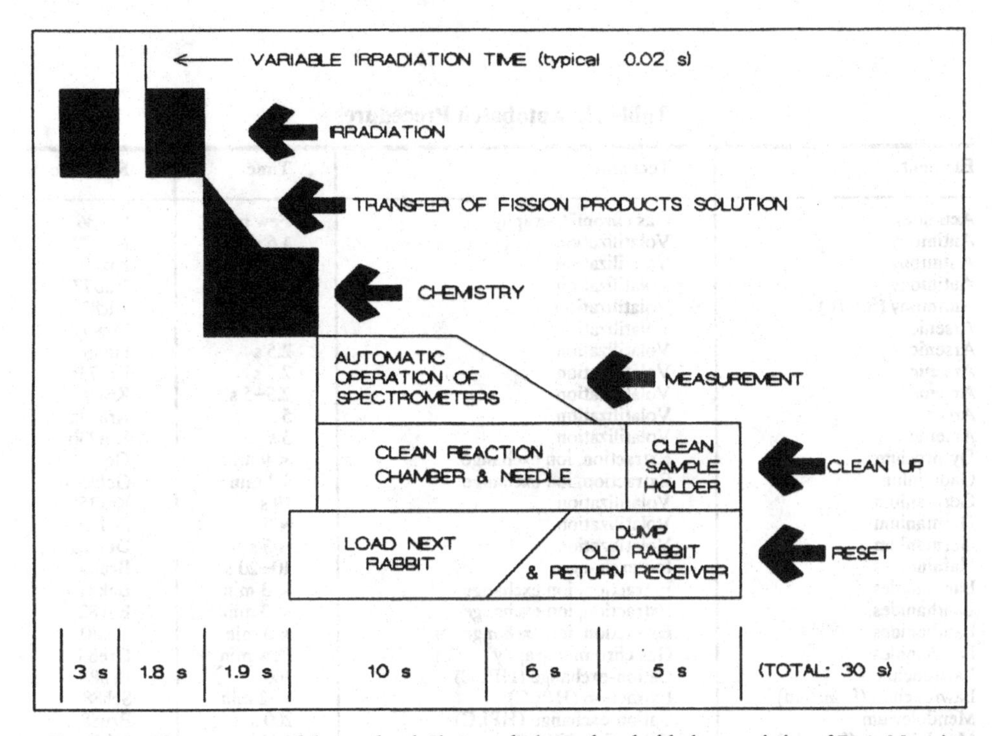

Figure 7. Time sequence of the autobatch Sb-As cycle (reproduced with the permission of R. A. Meyer).

The delivered nuclear-reaction products are subjected to continuous separation in a chemistry system. One of the early attempts to perform continuous chemical separation was reported by Kosanke and coworkers.[18] The recoiling nuclear reaction products carried by the He jet was extracted with HCl (8*M*) and passed through an anion-exchange column. They used the procedure to separate Ga activities from Zn and Cu. In the last two decades several new developments have taken place. Solvent extraction techniques have been adapted with the use of high speed contrifuges to achieve rapid separation of phases. Gas phase chemical reactions have also been utilized for continuous separations. Examples of these types of procedures are given in the following sections.

39.4.A Continuous Solvent-Extraction Procedures

A large volume of literature is available on solvent-extraction systems and their applications for the separation of elements. A number of batch procedures have been developed for the study of short-lived nuclides using solvent extraction. The a slow phase separation was a limiting feature of this technique in its application to achieve separation in a few seconds. The development of high speed centrifuges (H-centrifuges) by Reichert and Rydberg[19] overcame this difficulty. The first version of the H-centrifuge had a hold-up volume of 100 mL. A second version with a hold-up volume of 12 mL and a hold-up time[20,21] of 0.25 s has been used for the development of a number of fast separation procedures. The technique is called SISAK, *S*hort-lived *I*sotopes *S*tudied by the *AK*UFVE

Table II. Autobatch Procedures

Element	Technique	Time	Reference*
Actinides	Gas chromatography	Few min	Gre86
Antimony	Volatilization	1.6 s	Mey79
Antimony	Volatilization	2.7 s	Ktz79
Antimony	Volatilization	2.7 s	Rud77
Antimony (for Te)	Volatilization	7.3 s	Hic83
Arsenic	Volatilization	1.9 s	Mey79
Arsenic	Volatilization	2.5 s	Hen81
Arsenic	Volatilization	2.5 s	Kra73a
Arsenic	Volatilization	2.5–5 s	Kra75
Arsenic	Volatilization	5 s	Kra73a
Arsenic	Volatilization	5 s	Kra73b
Dysprosium	Extraction, ion exchange	< 9 min	Geh82
Gadolinium	Extraction, ion exchange	3.2 min	Gch82a
Germanium	Volatilization	10 s	Kra75
Germanium	Volatilization	≈ 5 s	Del72
Germanium	Volatilization	≈ 5 s	Del72a
Hafnium	Ion exchange	10–20 s	Bru89
Lanthanides	Extraction, ion exchange	< 3 min	Bak81
Lanthanides	Extraction, ion exchange	< 3 min	Bak82
Lanthanides	Extraction, ion exchange	≈ 3 min	Grn81
Lanthanides	Gas chromatography	Few min	Gre86
Lawrencium	Cation-exchange (HPLC)	160 s	Bru88
Lawrencium (from Md)	Extraction (HPLC)	1–2 min	Sch88
Mendelevium	Cation-exchange (HPLC)	200 s	Bru88
Mendelevium (from Lr)	Extraction (HPLC)	1–2 min	Sch88
Molybdenum	Extraction	5 s	Tit77
Neptunium	Extraction, ion exchange	2.7 min	Moo87
Niobium	Adsorption	2.2 s	Ahr76a
Niobium	Adsorption	2.2–2.4 s	Wei81
Niobium	Adsorption	≈ 5 s	Wei87
Palladium	Extraction	135 s	Mei81
Palladium	Extraction	< 1 min	Aro70
Promethium	Extraction, ion exchange	< 3 min	Grn82
Protactinium	Extraction	2.3 min	Baa85
Ruthenium	Extraction	8.3 s	Fra78
Samarium	Extraction, ion exchange	≈ 11 min	Bak80a
Selenium	Volatilization	4.0 s	Fol69
Selenium	Volatilization	5 s	Kra73b
Selenium	Volatilization, extraction	5 s	Kra70
Silver	Exchange	4.1 s	Bru82
Tantalum	Extraction	10–20 s	Bru89
Technetium	Extraction	2.5 s	Tra76
Technetium	Extraction	7.5 s	Kaf73
Technetium	Extraction	7.5 s	Tra72
Technetium	Volatilization	1 s	Mat79
Tellurium	Volatilization	5.0 s	Fol69
Terbium	Extraction, ion exchange	< 3 min	Grn83
Tungsten	Ion exchange	10–20 s	Bru89
Yttrium	Ion exchange	10 s	Kle75
Z = 104	Extraction	2–3 min	Hul80
Zirconium	Extraction	2.2 s	Wei81
Zirconium	Extraction	4.0 s	Tra72

* References are not included in this paper but can be found in Ref. 36.

technique. The following paragraph describes separation of Tc using SISAK.

A schematic of the SISAK Tc chemistry used by Broden and coworkers[22,23] is shown in Fig. 8. N_2 gas containing KCl aerosol carried the fission products from a large chamber. A hot solution of 0.1-*M* HNO_3 – 0.1-*M* $KBrO_3$ dissolved the fission products. The solution was degassed and passed on to the first mixer-centrifugal separator unit (*C*1). Tc in the form of pertechnetate was extracted by 0.05-*M* Alamine-336 in $CHCl_3$, along with Zr, Nb, and Mo. The $CHCl_3$ layer was transferred to the second mixer-centrifugal separator unit (*C*2) and Tc was stripped with 2-*M* HNO_3. The solution contained traces of Zr, Nb and Mo. The γ-ray spectrum of this layer was recorded by a detector (*D*1) as the solution passed on to the third unit (*C*3), where the contaminants were extracted by 0.5-*M* HDEHP (bis-2-ethylhexyl phosphoric acid) in $CHC1_3$. The aqueous phase from *C*3 containing Tc was used for decay studies. The separation was accomplished in 7 s.

Table III lists the elements for which SISAK procedures are available in the literature. The extraction reagent used and the separation time are also included in the table. Recently a mini-H-Centrifuge has been developed by Skarnemark and coworkers.[9,24] This centrifuge has an internal volume of only 0.3 mL and has a hold-up time of about 0.1 s (SISAK 3). It will allow separation to be achieved in even shorter time. Skarnemark and coworkers are planning to study nuclides in the *A* = 110 region and short-lived actinides and transactinides using the SISAK 3 system.

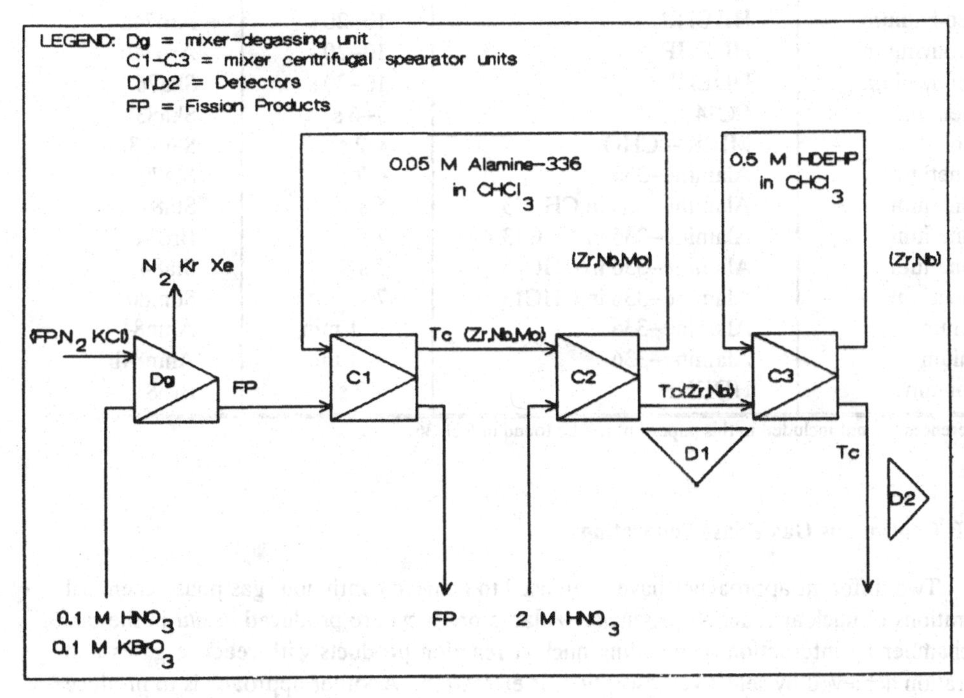

Figure 8. Schematic of the SISAK system used for the separation of Tc from fission products (reproduced with the permission of Professor G. Skarnemark).

Table III. Continuous Solvent-Extraction Procedures SISAK Technique

Element	Reagent	Time	Reference*
Arsenic	CHC13	3 or 4 s	Ska83
Bromine	CHC13	≈ 5 s	Bro81
Cadmium	MIBK – CHO	≈ 2 s	Rob85
Cadmium	MIBK – CHO	≈ 2 s	Sem83
Cerium	HDEHP	10–20 s	Aro74a
Cerium	HDEHP	10–20 s	Aro74b
Cerium	HDEHP	< 5 s	Ska80
Cerium	HDEHP	≈ 5 s	Tra75
Copper	Oxime	20 s	Aro74c
Indium	MIBK – CHO	≈ 2 s	Rob85
Indium	MIBK – CHO	≈ 2 s	Sem83
Indium	MIBK – CHO	≈ 2 s	Sem84
Iodine	CC14	≈ 5 s	Bro81
Lanthanum	HDEHP	10–20 s	Aro74a
Lanthanum	HDEHP	10–20 s	Aro74b
Lanthanum	HDEHP	< 5 s	Ska80
Lanthanum	HDEHP	≈ 4 min	Bjo77
Neptunium	HDEHP	10 s	Tet86
Niobium	HDEHP	≈ 9 s	Bro81
Praseodymium	HDEHP	10–20 s	Aro74a
Praseodymium	HDEHP	10–20 s	Aro74b
Praseodymium	HDEHP	10–20 s	Ska76
Ruthenium	CC14	5–6 s	Ska83
Silver	MIBK – CHO	≈ 2 s	Sem83
Technetium	Alamine–336	≈ 7 s	Sta79
Technetium	Alamine–336 in CHC13	5 s	Sta84
Technetium	Alamine–336 in CHC13	7 s	Bro81
Technetium	Alamine–336 in CHC13	7 s	Sta84
Technetium	Alamine–336 in CHC13	7 s	Sum80
Uranium	Alamine–336	< 1 min	Ahm84
Uranium	Alamine–336	< 1 min	Ahm84b
Zirconium	HDEHP	≈ 7 s	Bro81

* References are not included in this paper but can be found in Ref. 36.

39.4.B Continuous Gas-Phase Separations

Two different approaches have been used to achieve continuous gas-phase chemical separations of nuclear reaction products. Volatile products are produced *in situ* in the target chamber by interaction of recoiling nuclear-reaction products with reactive gases and separation achieved by selective adsorption or extraction. Another approach is to produce volatile products outside the target chamber by allowing nuclear-reaction products to interact chemically with reactive gases at high temperatures; the separation of individual elements is usually achieved by thermochromatography. Both techniques are illustrated in the following sections.

39.4.B(1) Continuous Gas-Phase Separation Using Volatile Products Generated in the Target Chamber

Rengan and Griffin[17] reported a procedure for continuous separation of Br from fission products. A ^{235}U target covered with Al foils allowed light fission products to recoil into the target chamber. A mixture of N_2 and ethylene was used as carrier gas. A schematic of the separation system is shown in Fig. 9. Volatile products were formed by reaction of Se and Br (produced in fission) with ethylene. All other fission products (except Kr and Xe) were carried by ethylene clusters and were retained by quartz wool filter. Volatile Se products were selectively retained by quartz wool coated with silver nitrate. The gas stream passed through an anion-exchange trap which retained Br; Kr and Xe flowed through. Separation of Br was achieved in about 1 s using this procedure.

Continuous gas phase separation procedures were developed for Se[25-27] using N_2-ethylene carrier gas. Using volatile chlorides produced in the target chamber, with HCl as carrier gas, Zendel and coworkers[28] developed separation procedures for As and Ge. Table IV lists the elements for which continuous gas-phase separation procedures based on volatile products generated in the target chamber is available. The table also gives information on the chemical species/reaction used, technique, and time of separation.

The chemical nature of volatile products formed in the target chamber, when ethylene is used as carrier gas, is not known. Se probably forms organoselenides,[27] while Br may form organobromides and HBr. Recent experiments[29] performed in our laboratory indicate that the Se compounds are relatively more nonpolar compared to the Br com-

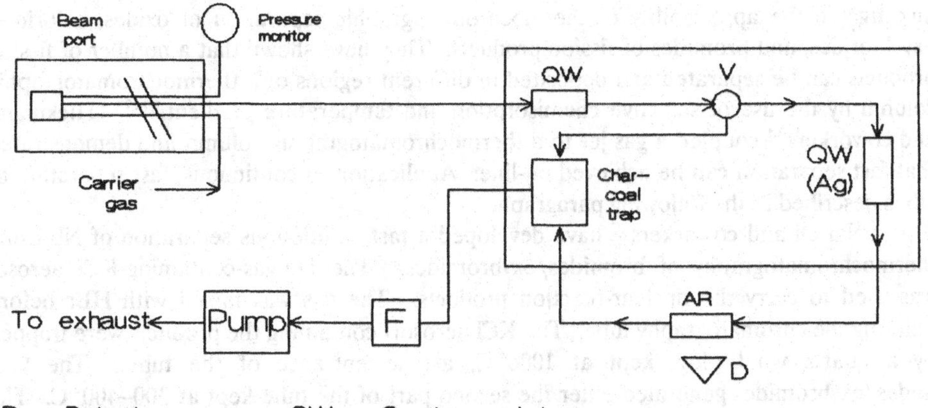

Figure 9. Schematic of continuous gas-phase Br chemistry system.

Table IV. Gas-Phase Chemical Separations where Volatile Species Are Generated in the Target Chamber

Element	Species/Reaction	Technique	Time	Ref.*
Arsenic	Chlorides	Adsorption, Extraction	Few s	Zen81
Bromine	Exchange ?	Volatilization	≈ 1 s	Ray76
Bromine	Reaction with C2H4	Adsorption	≈ 1 s	Ren82b
Bromine	Reaction with CH4	Precipitation	2 s	Lun70a
Germanium	Chlorides	Adsorption, extraction	Few s	Zen81
Iodine	Reaction with CH4	Precipitation	2 s	Lun70a
Radon	Cryosystem	Condensation	Few s	Hil87
Selenium	Reaction with C2H4	Thermal decomposition	1–2 s	Zen78
Selenium	Reaction with C2H4	Thermal decomposition	1–2 s	Zen80
Selenium	Reaction with C2H4/B	Adsorption	≈ 1 s	Ren82a
Silicon	Hydrides	Condensation	Few s	Har75
104	Chlorides	Volatilization	< 1 s	Zva66
118 (112 & 114)	Cryosystem	Condensation	Few s	Hil87

*** References are not included in this paper but can be found in Ref. 36.**

pounds formed in the target chamber. Further experiments are in progress to understand the nature of the chemical reactions taking place in the target chamber between fission fragments and reactive carrier gases. Hopefully, this will help in developing gas phase separation procedures for other elements.

39.4.B(2) Volatile Products Generated Outside the Target Chamber

Zvara and coworkers[30] produced volatile bromides by heating to high temperature nuclear reaction products carried by a stream of He and Br; a thermochromatographic column was used to separate the bromides. Bachmann, Bogl, and coworkers have investigated the applicability of thermochromatographic separation of oxides, chlorides, oxychlorides, and bromides of fission products. They have shown that a number of fission products can be separated and deposited in different regions of a thermochromatographic column by the use of selective chemisorption and temperature gradient.[31-33] Hickmann and coworkers[34] coupled a gas jet to a thermochromatographic column and demonstrated that fast separation can be achieved on-line. Application to continuous, fast separation of Nb is described in the following paragraph.

Nai-Qi and coworkers[35] have developed a fast, continuous separation of Nb using thermochromatography of bromides/oxybromides. The He-gas-containing-KCl aerosol was used to carry the nuclear-reaction products. The gas was mixed with HBr before entering the chromatography tube. The KCl aerosols containing the products were trapped by a quartz wool plug, kept at 1000° C, at the entrance of the tube. The bromides/oxybromides generated enter the second part of the tube kept at 300–400° C. The volatile species leaving the tube were allowed to enter a reclustering chamber; through a capillary N_2 gas containing KCl aerosol carried the volatile products to a glass fiber filter in front of a detector. The average residence time was found to be 18 s, and Nb isotopes with half-lives as low as 7 s was observed.

Table V. Continuous Thermochromatographic Procedures

Element	Chemical Species	Time	Reference*
Cadmium	Mass separated ion	≈ 1 s	Rus81
Hafnium	Chlorides	< 2 s	Zva72
Hafnium	Chlorides	≈ 0.4 s	Zva71
Niobium	Bromides	0.1–0.2 s	Bel75
Niobium	Bromides	0.1–0.2 s	Zva76
Niobium	Bromides/oxybromides	12 s	Nai89
Rhenium	Oxides	< 3 s	Dom83
Rhenium	Oxides	< 3 s	Zva84
Tantalum	Bromides	0.1–0.2 s	Bel75
Tantalum	Bromides	0.1–0.2 s	Zva76
Z = 104	Chlorides	< 2 s	Zva72
Z = 105	Bromides	0.1–0.2 s	Bel75
Z = 105	Bromides	0.1–0.2 s	Zva76
Z = 105	Bromides/oxybromides	12 s	Nai89
Z = 107	Oxides	< 3 s	Dom83
Z = 107	Oxides	< 3 s	Zva84
Zinc	Mass separated ions	≈ 1 s	Rus81

* References are not included in this paper but can be found in Ref. 36.

Table V gives a list of elements for which a fast, continuous procedures based on volatile products generated outside the target chamber is available. The table also gives the separation time and the chemical species used for the separation.

In the last decade, the technology of ion sources has improved tremendously, making it possible to achieve mass separations of many elements. Fast radiochemical separations still provide the only means for studying nuclides with low yields and for certain elements like Se which are not accessible by mass separation. Continuous gas-phase separations and the SISAK technique have provided ways to study nuclides with short half-lives.

ACKNOWLEDGMENTS

The author wishes to thank Eastern Michigan University for the support provided through the Faculty Research Program. The support received through the U. S. Department of Energy Cost Sharing Program (Grant #DEFG0280ER10724) is acknowledged. The material for this article is derived from the data collected for the forthcoming radiochemistry monograph "Ultrafast Radiochemical Separations" to be published by the National Academy of Sciences.[36]

REFERENCES

1. P. Curie and Mme. S. Curie, *Compt. Rend.* **127**, 175 (1898).
2. P. Curie, Mme. S. Curie, and G. Benont, *Compt. Rend.* **128**, 1215 (1898)
3. E. Rutherford and F. Soddy, *Phil. Mag.* **Ser. 5, XIix**, 1 (1900).

4. F. Soddy, *Annual Progress Reports to Chemical Society*, reproduced in *Radioactivity in Atomic Theory*, T. J. Trenn, Vol. 1, (Halsted Press, 1975), p. 263.
5. J. J. Livingood and G. T. Seaborg, *Phys. Rev.* **54**, 391 (1938).
6. Y. Kusaka and W. W. Meinke, *Rapid Radiochemical Separations, National Academy of Sciences Radiochemical Techniques Monograph*, NAS-NS-3104 (1961).
7. G. Herrmann and N. Trautmann, *Ann. Rev. Nucl. Part. Sci.* **32**, 117 (1982).
8. R. A. Meyer, *J. Radioanal. Nucl. Chem.* **142**, xxx (1990).
9. G. Skarnemark, J. Alstad, N. Kaffrell, and N. Trautmann, *J. Radioanal. Nucl. Chem.* **142**, xxx (1990).
10. K. Rengan, *J. Radioanal. Nucl. Chem.* **142**, xxx (1990).
11. N. Trautman, N. Kaffrell, H. W. Behlich, H. Folger, G. Herrmann, D. Hubscher, H. Ahrens, *Radiochim. Acta* **18**, 86 (1972).
12. R. A. Meyer, E. A. Henry, in *Nuclear Spectroscopy of Fission Products*, Ed. by T. von Egidy, (Institute of Physics, Bristol, 1979), p. 59.
13. O. G. Lien III, P. C. Stevenson, E. A. Henry, R. P. Yaffe, R. A. Meyer, *Nucl. Instr. Meth.* **185**, 351 (1981).
14. E. A. Henry, O. G. Lien II, and R. A. Meyer, *Proceedings of the 4th International Conference on Nuclei Far from Stability*, Report CERN 81-09, 339 (1981).
15. R. D. MacFarlane and Wm. C. McHarris, in *Nuclear Spectroscopy and Reactions, Part I*, Ed. by J. Cerny, (Academic Press, New York, 1974), p. 243.
16. K. Rengan and H. C. Griffin, in *Artificial Radioactivity*, K. Eds. Narayana Rao and W. J. Arnikar, (Tata McGraw-Hill, New Delhi), (1985) p. 61.
17. K. Rengan and H. C. Griffin, *J. Radioanal. Nucl. Chem.* **98**, 255 (1986).
18. K. L. Kosanke, Wm. C. McHarris, R. A. Warner, and W. H. Kelly, *Nucl. Inst. Meth.* **115**, 151 (1974).
19. H. Reinhardt, J. Rydberg, *Acta Chem. Scand.* **23**, 2773 (1969).
20. J. Rydberg, H. Persson, P. O. Aronsson, A. Selme, and G. Skarnemark, *Hydrometallurgy* **5**, 273 (1980).
21. G. Skarnemark, P. O. Aronsson, K. Broden, J. Rydberg, T. Bjornstad, N. Kaffrell, E. Stender, and N. Trautmann, *Nucl. Instr. Meth.* **171**, 323 (1980).
22. K. Broden, Ph.D. thesis submitted to Chalmers University of Technology, Goteborg, Sweden (1984).
23. K. Broden, G. Skarnemark, T. Bjornstad, D. Erikson, I. Haldorsen, N. Kaffrell, E. Stender, and N. Trautmann, *J. Inorg. Nucl. Chem.* **43**, 765 (1982).
24. H. Persson, G. Skarnemark, M. Skalberg, J. Alstad, J. O. Liljenzin, G. Bauer, F. Haberberger, N. Kaffrell, J. Rogowski, and N. Trautmann, *Radiochim. Acta* **48**, 177 (1989).
25. M. Zendel, N. Trautmann, and G. Herrman, *J. Inorg. Nucl. Chem.* **42**, 1347 (1980).
26. K. Rengan, J. Lin, T. N. Massey, and R. A. Meyer, *Radiochem. Radioanal. Lett.* **50**, 385 (1982).
27. K. Rengan, J. Lin, M. Zendel, and R. A. Meyer, *Nucl. Instr. Meth.* **197**, 427 (1982).
28. M. Zendel, N. Trautmann, and G. Herrmann, *Radiochim. Acta* **29**, 17 (1981).
29. E. T. Contis, K. Rengan, and H. C. Griffin, private communication (1989).
30. I. Zvara, O. L. Keller, Jr., R. J. Silva, and J. R. Tarrant, *J. Chromatog.* **103**, 77 (1975).
31. K. Bogl and K. Bachmann, *J. Radioanal. Chem.* **34**, 223 (1976).
32. K. Bachmann and J. Rudolph, *J. Radioanal. Chem.* **32**, 243 (1976).
33. K. Bachmann, V. Matschoss, J. Rudolph, A. Steffan, and S. Tsalos, *Nucl. Instr. Meth.* **139**, 343 (1976).
34. V. Hickmann, N. Greulich, N. Trautmann, H. Gaggeler, H. Gaggeler-Koch, B. Eichler, and G. Herrmann, *Nucl. Instr. Meth.* **174**, 507 (1980).
35. Ya Nai-Qi, D. T. Jost, V. Baltensperger, and H. W. Gaggeler, *Radiochim. Acta* **47**, 1 (1989).
36. K. Rengan and R. A. Meyer, "Ultrafast Radiochemical Separations," *National Academy of Sciences Radiochemical Techniques Monograph*, in press (1990).

40. Rotational Population Patterns and Searches for the Nuclear SQUID

L. F. Canto and R. Donangelo
Instituto de Fisica
Universidade Federal do Rio de Janeiro
C. P. 68528 Rio de Janeiro, RJ, Brasil

A. R. Farhan
Department of Physics
Kuwait University
Kuwait

M. W. Guidry
Department of Physics
University of Tennessee
Knoxville, TN 37996
and Oak Ridge National Laboratory
Oak Ridge, TN 37830, USA

J. O. Rasmussen and M. A. Stoyer
Lawrence Berkeley Laboratory
University of California
Berkeley, CA 94720, USA

P. Ring
Physik Department
Technische Universität München
Garching, Germany

ABSTRACT

This paper presents new theoretical results for rotational-population patterns in the nuclear SQUID effect. (The term nuclear SQUID is an analogy to the solid-state *S*uperconducting *QU*antum *I*nterference *D*evices.) The SQUID effect is an interesting new twist to an old quest to understand Coriolis *A*nti-*P*airing (CAP) effects in nuclear rotational bands. Two-neutron transfer-reaction cross-sections among high-spin states have long been touted as more specific CAP probes than other nuclear properties. Heavy projectiles like Sn or Pb generally are recommended to pump the deformed nucleus to as high spin as possible for transfer. The interference and sign reversal of 2*n*-transfer amplitudes at high spin, as predicted in the early SQUID work, impose the difficult requirement of coulomb pumping to near back-bending spins at closest approach. For Pb on rare earths we find a dramatic departure from the sudden approximation, so that the population depression occurs as low as final spin 10$\hbar$.

40.1 INTRODUCTION

The acronym SQUID stands for *S*uperconducting *QU*antum *I*nterference *D*evice, a solid-state device exploiting quantum aspects of superconductivity. Ring and Nikam[1] suggested that a nuclear analogy might exist in the two-neutron-transfer reaction between heavy ions, where one of the nuclei is deformed and excited to a high rotational state by coulomb excitation. In the solid-state case Josephson currents are affected by a superimposed magnetic field. In the nuclear case the pair supercurrent between the nuclei is affected by the rotational field. The idea was developed further by the above authors and others in subsequent papers.[2,3]

It has long been generally believed that the increase of moment of inertia at higher spin in most deformed nuclei is caused mainly by a decrease in the pairing correlation at higher rotational velocities, the so-called *C*oriolis *A*nti-*P*airing (CAP) effect. It was suggested[4] that this CAP effect could be tested by two-neutron transfer reactions between heavy nuclei, the coulomb excitation on the inward path pumping the deformed partner into higher rotational states from which the $2n$ transfer occurs. Pair-transfer is strongly enhanced by the pairing correlation, the enhancement factor being essentially the square of the number of Nilsson orbitals involved in pair configuration mixing. The ground-to-ground enhancement factor is given by the following:

$$F = \sum_{i>0} {}_{A}u_i \; {}_{A+2}v_i \approx \sum - {}_{A}u_i \; {}_{A}v_i = \frac{\Delta}{G} \qquad (1)$$

where v and u are the BCS fullness and emptiness amplitudes, respectively. What was envisioned in this earlier time was that $2n$ transfer would steadily and monotonically weaken with increasing spin.

The new and interesting aspect of the SQUID-effect prediction is that the pair-transfer matrix element decreases through zero and changes sign. To see this effect strongly, it is thought that the coulomb excitation on the inward path must excite to near the backbending spin (diabolic point). The cancellation in $2n$-transfer matrix elements has been related to the Berry phase[5,6] in passing around the diabolic point in spin and particle-number space. Because of the very large coulomb-excitation required, attention has focussed on the heaviest spherical projectiles in the Pb region and targets in the deformed rare-earth region. Semiclassical trajectory theoretical estimates of Landowne et al.[7] make it appear that in the most favorable case the SQUID effect will be a small decrease in the population of the highest rotational states—thus difficult to achieve and to prove. However, as our new method described below shows, the sudden approximation used in Ref. 7 is not applicable for systems with such a high-Z projectile as Pb, and the theoretical rotational-population decrease from the SQUID effect should occur at half the spin given by the sudden approximation.

We have undertaken two new approaches to the transfer theory. First, we have considered the dependence of transfer on the zones of the most-lightly-bound $i_{13/2}$ Nilsson orbitals on the surface of the deformed nucleus. The transfer strength over the surface is taken as a product of factors, the WKB tunneling factor, the absorption factor, and the square of Nilsson wave-functions of the most-lightly-bound orbitals. Second, we have moved a step toward more quantal theory by adapting the coulomb-excitation codes originating with deBoer and Winther,[8] where the collision partners move on a Rutherford hyperbolic orbit during integration of the time-dependent Schrodinger equation for the amplitudes of the various rotational states. In our new approach this integration is paused at closest approach, and the rotational-amplitude vector is matrix-multiplied by a transfer matrix, taking into account transfer-changing angular momentum as well as *S*-wave transfer. The characteristic 2*n*-tunneling length is sufficiently small that this approximation that transfer occurs at the classical turning point should be justified. After matrix multiplication accounting for transfer, tunneling distance, and nuclear-optical-potential absorption and phase shifts, the deBoer-Winther integration continues on the outward path to give the final rotational signature. (We learned recently that Pollarolo, Dasso, and deBoer are also studying the pair-transfer process by similar modifications of the old coulomb-excitation codes. They confine their calculations to energies below the coulomb barrier, where they feel it is safe not to include effects of the nuclear optical potential. They also have not included the effect of the oscillatory Nilsson wave-functions over the nuclear surface.[9])

40.2 E2 COULOMB EXCITATION OF GROUND AND S BAND

We may confidently apply semi-classical trajectory methods to estimate the amount of rotational angular momentum carried in by head-on heavy-ion collisions. The higher the beam energy, the higher the angular momentum up to the limit imposed by the coulomb barrier. Above barrier the collision system predominantly goes into compound nucleus and other complex reaction channels, robbing flux from the simple-transfer and rotational-inelastic channels. For transfer reactions we are concerned with rotational angular momen-

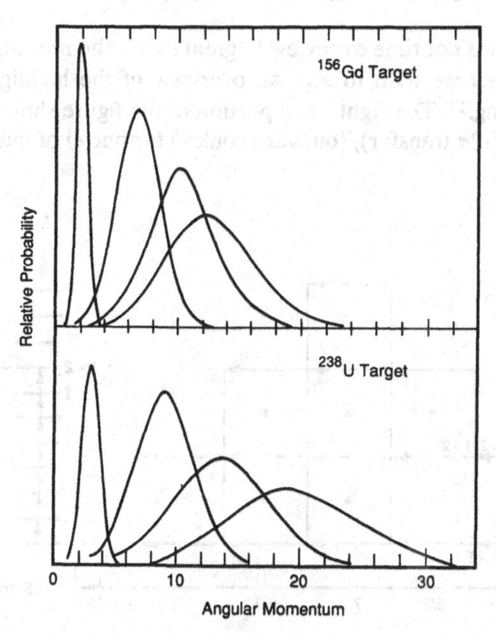

Figure 1. Calculations of the distribution of rotational angular momentum at closest approach for two spheroidal target nuclei and four spherical projectiles, from left to right, ^{16}O, ^{58}Ni, ^{120}Sn, and ^{208}Pb. These calculations are made by classical trajectory methods in head-on collisions of energy causing the nuclei barely to touch. The figure is taken from Ref. 9.

tum pumped in during the inward path, which is half the final coulomb-excited spin in the sudden limit. Guidry et al.[10] made such estimates for two deformed nuclei, ^{156}Gd and ^{238}U, with four different projectiles, ranging from ^{16}O to ^{208}Pb, and these probability distributions are shown in Fig. 1. As backbending can occur in the rare-earth region as low as spin $12\hbar$, we see that with ^{208}Pb as a projectile it should be possible to pump the rotational energy up to this region on the inward path. The nuclear rotor in the semiclassical calculations of Ref. 7 did not deal with coulomb-excitation behavior in the backbending region.

The backbending region in the yrast band may best be thought of as a virtual band crossing, where an upper *S* band with substantial aligned angular momentum crosses below the regular ground band. In the rare-earth region the first backbend is with an *S* band having two $i_{13/2}$ neutrons with angular momentum partially aligned along the rotational-angular-momentum axis and perpendicular to the cylindrical-symmetry axis of the prolate-spheroidal nucleus. In only a few cases has it been possible experimentally to study electric-quadrupole transition rates, i.e., $B(E2)$ values, within and between *S* and ground bands.[11] The $B(E2)$ values evidently are strongest among the yrast levels, with $B(E2)$ values to the next band substantially lower. It can be shown that this behavior is to be expected unless backbending is so sharp that there is a sharp change in the wave-function mixture of yrast levels from one level to the next. Thus, in our work to date we have not explicitly taken into account coulomb excitation into the band above yrast, but it is straightforward with the deBoer-Winther amplitude method to incorporate detailed energies and $B(E2)$ values in the backbending region of interest in our new calculations of diabolic pair-transfer and the SQUID effect.

40.3 DIABOLICAL POINTS AND THE NUCLEAR SQUID

In this paper there is not time to review in great detail the literature of the proposed nuclear SQUID effect, but we wish to give an overview of the highlights. Our Fig. 2 is taken from Fig. 3.2 of Ring.[12] The right-hand portion of the figure shows schematically two paths of (inward coulex)/($2n$ transfer)/(outward coulex) for nuclei of mass numbers *A* and

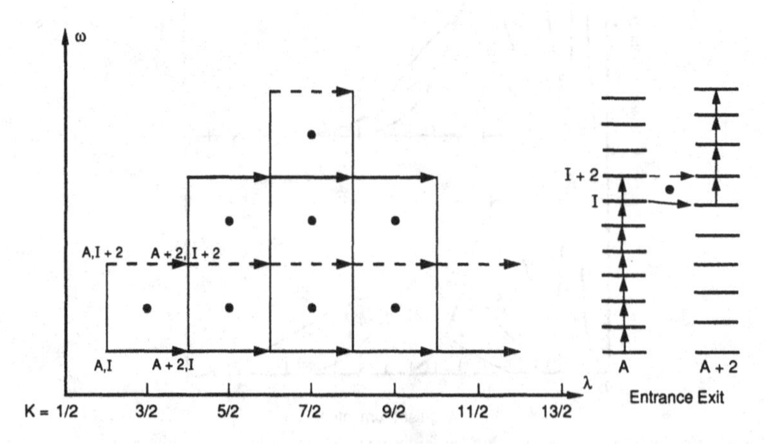

Figure 2. The left-hand portion of the figure is a schematic representation of the diabolic points of the $i_{13/2}$ Nilsson-orbital family in the plane of particle number (chemical potential) and rotational angular momentum. The right-hand rotational-band scheme illustrates the principal interfering paths of coulomb excitation and $2n$ transfer below and above the first diabolic point. This figure is taken from Ref. 2.

$A + 2$, respectively. One path (solid arrow) passes beneath the backbending, or diabolic, point, and the other (dashed arrow) passes above. These paths are supposed to contribute to $2n$ transfer with opposite signs, hence cause a destructive interference in the $2n$-transfer matrix element. The left-hand part of the figure is a diagram showing schematically where the diabolic points for the $i_{13/2}$ shell appear. The abscissa is particle number (or chemical potential λ), and the ordinate is rotational angular velocity ω. Of course, these are really not continuous variables in real nuclei, but they are continuous in some of the general theoretical methods. That is, cranking velocity and angular momentum are continuous variables in deformed-nuclear-potential and cranking models, where spherical symmetry is broken. Likewise, particle number is not conserved in pairing models like the BCS, where gauge symmetry is broken. The lower leftmost dot in our Fig. 2 represents the first band-crossing for spheroidal nuclei with chemical potential close to the $i_{13/2}$, $\Omega = 3/2$ Nilsson orbital. The SQUID interference diagrammed at the right might then be expected for the A nucleus with chemical potential just below the 13/2; 3/2 state and the $A + 2$ nucleus just above. The upper four diabolic points are of purely mathematical interest, as angular momenta in this second 13/2-backbend region could not be realized experimentally in nuclei for transfer studies.

The earlier theoretical work on the SQUID effect focussed on S-wave transfer, in which there is no change in spin intrinsic to the transfer process at closest approach. However, we know from theory of α decay of deformed nuclei that a non-uniform emission or cluster-transfer wave-function over the nuclear surface necessarily implies angular-momentum changes. In particular here, if we can specify the angular function over the surface that transforms incoming waves to outgoing transfer-waves, we can readily convert this to a square matrix transforming incoming deBoer-Winther amplitudes to outgoing transfer-wave amplitudes. The angular transfer function is sandwiched between initial and final rotational functions and integrated over the eulerian angles:

$$T_{ll'} = \int Y^*_{l'0} F(\theta) Y_{l0} \sin\theta \mathrm{d}\phi \tag{2}$$

40.4 SURFACE ANGULAR FORM FACTOR OF PAIR TRANSFER

The concept of the angular form factor $F(x)$ ($x = \cos\theta$) was introduced in Ref. 3, although it was there incorporated into an angular projection integral for the classical limit S matrix in the sudden approximation. In Eq. (4) of that work $F(x)$ was taken to be the product of three real factors, a $2n$-tunneling factor $a_{tunl}(x)$, an absorption factor $a_{abs}(x)$, and the SQUID spectroscopic factor $a_{spec}(I(x))$. The tunneling factor is just the WKB exponential for a two-neutron cluster with experimental separation energy tunneling between surfaces at the classical distance of closest approach at angle χ. Since in Ref. 3 and the present work we do not purport to compute absolute tunneling, only relative rotational-populations, the pre-exponential factor is left at unity. The absorption factor in Ref. 3 is just a semiclassical trajectory time integral over the imaginary part of the optical potential. In the current work we make this factor complex by integrating over the complex optical potential, thus introducing phase shifts coming from the tail of the real part of the nuclear potential. Figures 3 and 4 are examples of tunneling and absorption factors at two bombarding energies for ^{208}Pb on ^{160}Dy. Only the absolute value of the absorption amplitude is plotted. Note that at the higher bombarding energy the absorption becomes very strong as

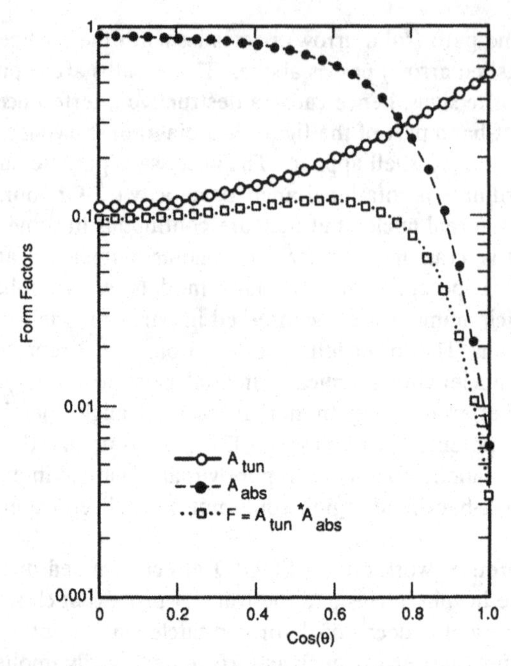

Figure 3. Factors in the transfer-matrix integrand over the spheroidal nuclear surface. The abscissa is cos θ. The solid circles give the absorption factor arising from the tail of the imaginary component of the optical potential. The open circles are the tunneling amplitude factor. The squares represent the product of the two factors. These factors were calculated for ^{208}Pb on ^{160}Dy at a laboratory energy of 1100 MeV.

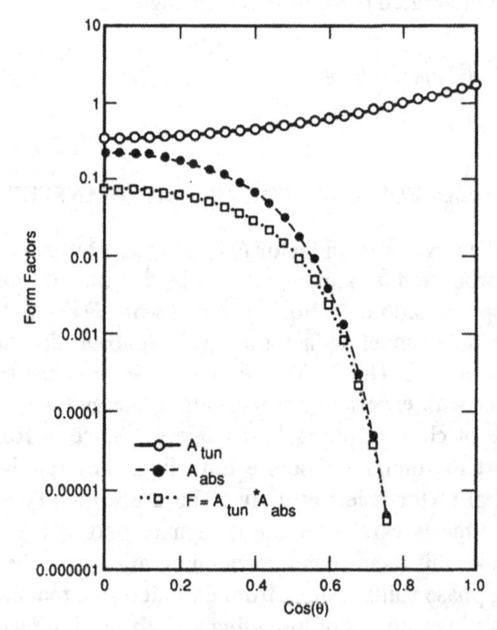

Figure 4. Same as Fig. 3, except calculated for a beam energy of 1200 MeV, sufficiently near the top of the barrier that the absorption amplitude factor is as small as 0.2 (96% absorption) even at the equator, with very high absorption prevailing toward the poles.

one approaches the tip of the prolate nucleus, the right-hand side of the figure, since the abscissa is the cosine of θ. It is implicit in the form of the a_{abs} integral of Eq. (6) in Ref. 3 that the Q value of the reaction is taken to be zero. In fact, non-zero Q values could be incorporated, though we have not yet done so, by including in the time integral the factor $\exp(iQt/\hbar)$; thus,

$$a_{abs} = \exp\left\{-\frac{1}{i\hbar}\int_{-\infty}^{\infty}[V(t)+iW(t)]e^{iQt/\hbar}dt\right\} . \quad (3)$$

Likewise, the tunneling integral of Eq. (5) in Ref. 3 could easily be modified for non-zero Q values by inclusion of such a factor in the time integral for semiclassical tunneling, as done long ago by Breit and Ebel[13]:

$$a_{tunl} = C\exp\left[-\frac{2(2M_n)^{1/2}}{\hbar}(S_{2n})^{1/2}\int_0^{\infty}\frac{dx}{dt}e^{iQt/\hbar}dt\right] , \quad (4)$$

where $x(t) = r(t) - R_p - R_t$, S_{2n} is the two-neutron binding energy, M_n is the neutron mass, C is an arbitrary constant, and Q is the energy release in the transfer reaction.

Next, we address a central matter of this paper, the SQUID factor $a_{spec}(I)$. S-wave matrix elements for pair transfer between cranked-HFB wave-functions were found to cross through zero and go negative at about the spin where backbending or upbending occurs in the energy levels. The S-wave $2n$-transfer matrix elements are as follows:

$$P = \langle A+2|S^{+}|A\rangle , \quad (5)$$

where the pair creation operator is

$$S^{+} = (a^{+}a^{+})_{I=0} , \quad (6)$$

with a^{+} the neutron creation operators.

We show in Fig. 5 the results from Ref. 3 for the deformed nucleus ^{160}Dy. The solid line gives the pair-transfer matrix elements, and the dashed line is the pairing-gap parameter, which only gradually decreases. As mentioned in the introduction, the oscillatory behavior of the transfer matrix elements shows up with realistic cranked-HFB calculations comprising more than one oscillator shell, but it is the high-j intruder orbital $i_{13/2}$ that is mainly responsible, and cranked-HFB solutions with just this orbital show qualitatively the same behavior.

It is possible to understand microscopically the origin of the effect in terms of just the three $i_{13/2}$ Nilsson states nearest the fermi energy in initial and final nuclei. At zero spin the pairing-force makes the ground states a coherent mixture of the various pair arrangements in the three orbitals, and the well-known superfluid pair-transfer enhancement occurs. At increasing cranking velocity a second-order coriolis term through the $K = 1^{+}$ intermediate states begins to oppose the pairing-force mixture. The S band in a weak-pairing limit at low spin is just the band derived from ground by promotion of one pair in the $i_{13/2}$ family from the Nilsson level just below the fermi energy to the Nilsson level just above. (For strong pairing the S-band structure is a little more complicated, derivable in the num-

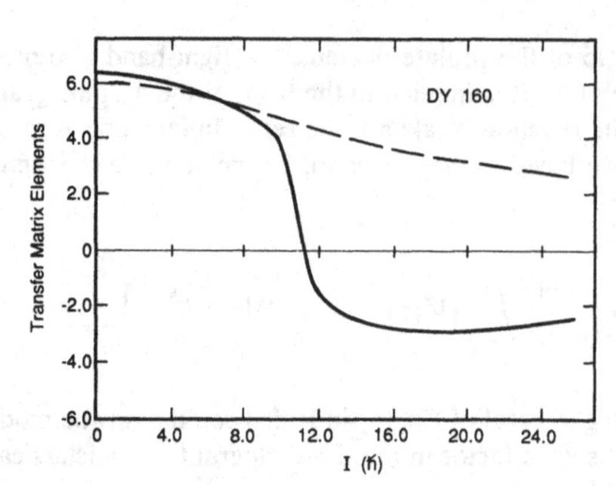

Figure 5. Calculated neutron pair-transfer ($L = 0$) matrix elements (solid line) as a function of spin. The traditional pair-transfer sum of uv products (dashed line). This calculation and figure is from Ref. 3.

ber-conserving space by matrix diagonalization. It is expressible in HFB by a linear combination of quasiparticle operators mainly operating on the Nilsson orbitals nearest the fermi energy.) With increasing spin the $i_{13/2}$ pairing coherence in ground is steadily reduced by the coriolis interaction until the principal pairing-admixed term goes through zero and changes sign, usually before the virtual band crossing. (In sharp backbending cases the pairing mixing cancellation comes at about the same angular velocity as the band crossing.) The overall pair-transfer matrix element is still positive when the admixed term first goes through zero. As the angular velocity of band crossing is approached, the negative admixed terms become comparable to the formerly dominant term, and the pair-transfer for S waves goes through zero, as shown in Fig. 5. Although the S-wave transfer goes through zero near band-crossing, that does not mean that the transfer amplitude is everywhere zero over the nuclear surface. The simple model discussed above suggests indeed a strongly-oscillatory transfer amplitude over the nuclear surface near the diabolic point, since there is a subtraction of the pair-transfer from successive Nilsson $i_{13/2}$ states. At the deadline time of this chapter we have not yet completed checking the surface angular form factors for transfer using cranked-HFB $i_{13/2}$ wave functions, but it is clear that there are strong oscillations near the diabolic region. The results of theoretical rotational-population patterns we present in the next section are preliminary in the sense that we approximate the angular-transfer form factor by the product of the a_{spec} of Fig. 5 and Ref. 3 times the square of a spherical harmonic $|Y_{61}|^2$. The oscillatory form factor produces considerable angular-momentum change intrinsic to the transfer at closest approach.

40.5 NUMERICAL RESULTS—SQUID SIGNATURE IN ROTATIONAL POPULATION

When the form factors above are used with our modified deBoer-Winther transfer code, we find a remarkable effect of the SQUID sign reversal at unexpectedly low spin for ^{210}Pb on ^{160}Dy at E_{lab} = 1200 MeV. Fig. 6 shows the rotational population pattern with and without the sign reversing a_{spec}. With the SQUID-effect factor there is a considerable suppression in the population around spin 10ħ. We may gain an understanding of the reasons by examining the semiclassical trajectory quantum-number functions, even though

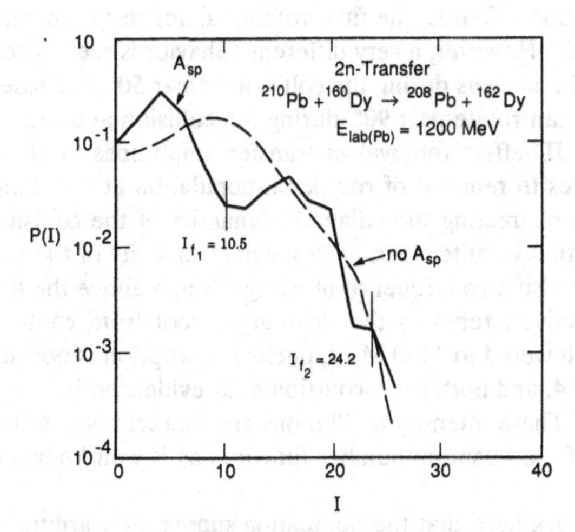

Figure 6. Calculated yrast rotational-transfer-population patterns for ^{210}Pb on ^{160}Dy at 1200 MeV. The dashed line is a traditional calculation with the dashed spectroscopic factor of Fig. 5, and the solid line is the SQUID-effect calculation with the SQUID spectroscopic factor of Fig. 5. Note the SQUID suppression of population around spin $10\hbar$.

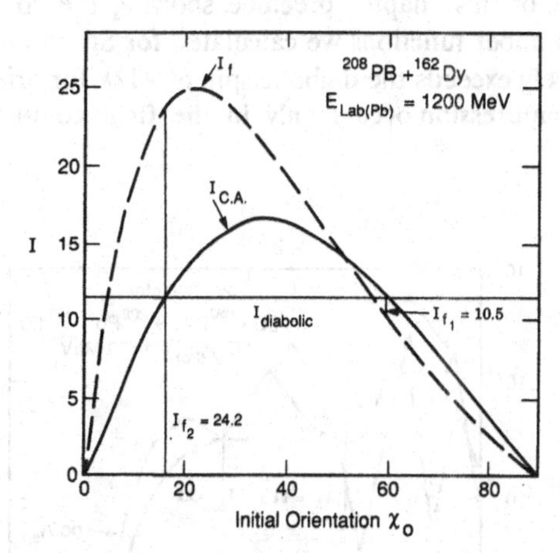

Figure 7. Classical quantum-number functions at closest approach (solid line) and at the end of the collision (dashed line) for Pb on Dy at 1200 MeV. The spin at the diabolic (band-crossing) region is shown by the horizontal line labeled $I_{diabolic}$. The two roots for the diabolic spin at closest approach are indicated by vertical lines. Note that the forward root has a final spin nearly twice that at closest approach (as in the sudden approximation), whereas the back root has a final spin slightly lower than at closest approach, a dramatic consequence of the finite moment of inertia and rotation of the nucleus during the time of the coulomb torque.

they are not now used in the population calculation. Fig. 7 shows for head-on collisions the spin at closest approach (solid line) and at infinite time (dashed line) as a function of initial orientation of the deformed nucleus. For small initial angles we see nearly the behavior of the sudden approximation. That is, the final rotational angular momentum is nearly twice that at closest approach. However, a very different behavior is seen beyond 20° because of the finite rotation of the nucleus during the collision. Near 50° I_f crosses under $I_{c.a.}$, since the deformed nucleus can rotate past 90° during the collision and thus reverse the sign of the torque. The SQUID-effect removal of transfer amplitudes at closest approach near spins 10 to 12 translates to removal of rotational-population at the same or slightly lower spin. The importance of treating the adiabatic dynamics of the collision instead of using the sudden approximation is quite clear. The special simplicity of the rotational pattern at 1200 MeV is evidently also a consequence of being enough above the barrier that the a_{abs} absorption factor effectively removes the small-angle root from contributing. When the bombarding energy is lowered to 1100 MeV, nuclear absorption is not so strong, as seen by comparing Figs. 3 and 4, and both roots contribute, as evidenced by the oscillating population pattern of Fig. 8. These intensity oscillations are characteristic of the interference between the two roots of the quantum-number function, as is well known in simple multiple coulomb excitation.

We should remark here that the population suppression around spin 10$\hbar$ seems to require an oscillatory angular form factor near band crossing, though the final results do not depend much on whether m = 0, 1, or 2 is taken for the spherical harmonic. If the spherical-harmonic function is omitted, the population suppression is barely evident at all. The large size of the spherical collision partner is expected to smear a sharply-oscillating angular-transfer strength and hence somewhat reduce the L transfer intrinsic to the transfer process, but we are still studying these modifications.

Space limitations of this chapter preclude showing the corresponding population patterns and quantum-number functions we calculated for Sn on Dy systems. Suffice it to say that the rotation barely exceeds the diabolic spin of ~12$\hbar$ for orientations in the 25-45° range. The population suppression occurs only in the final coulomb-excitation rainbow

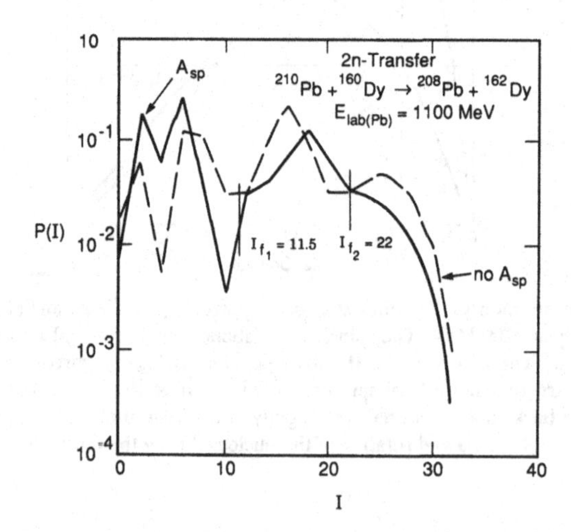

Figure 8. Same as Fig. 6, except at the lower beam energy of 1100 MeV.

maximum around spin 16ħ.

In another chapter in this book by one of us (M. W. Guidry) a puzzling new inhibition of high-spin 2*n* transfer was presented.[14] This inhibition for Ni on Dy begins to cut in around spin 4 and reduces the population pattern to quite small by spin 10. We are trying to understand the origin of this inhibiting factor, which does not appear in the cranked HFB. If this FDSM inhibiting factor must be included, then the experimental proof of the nuclear SQUID effect will be harder, though perhaps still possible by careful comparison of transfer population patterns with a range of projectiles, such as, Ni, Sn, and Pb.

We hope soon to submit a short paper presenting the cranked-HFB results with these new methods. Our preliminary calculations of HFB population patterns appear similar to those presented here.

ACKNOWLEDGMENTS

The main support for this research came from the Director, Office of Energy Research, Division of Nuclear Physics of the Office of High Energy and Nuclear Physics of the U. S. Department of Energy under contract DE-AC03-76SF00098. We acknowledge also the essential travel support of U. S. – Brazil cooperative-research grant INT-8302853 of the U. S. National Science Foundation and the Brazilian Conselho Nacional de Pesquisas e Desenvolvimento Cientifico. Travel support of the Kuwait Institute for Scientific Research is also acknowledged.

REFERENCES

1. P. Ring and R. S. Nikam, in *Proc. Intern. Conf. on Microscopic Theory of Nuclear Structure* (Sorrento, Italy, May 1986).
2. R. S. Nikam, P. Ring, and L. F. Canto, *Z. Phys.* **324**, 241 (1986); *Phys. Lett.* **B185**, 269 (1987); P. Ring, *Proc. IX Workshop in Nuclear Physics, Buenos Aires* (June, 1986), Ed. by Macchiavelli, Sofia, and Ventura, (World Scientific, Singapore, 1987), p. 143; P. Ring, *Proc. Workshop on Microscopic Models in Nuclear Structure Physics, Oak Ridge Oct. 3-6, 1988*, (World Scientific, Singapore), p. 298.
3. L. F. Canto, R. Donangelo, R. S. Nikam, and P. Ring, *Phys. Lett.* **B192**, 4 (1987).
4. M. W. Guidry, T. L. Nichols, R. E. Neese, J. O. Rasmussen, L. F. Oliveira, and R. J. Donangelo, *Nucl. Phys.* **A361**, 274, (1981).
5. P. Ring, in *Proc. Workshop on Microscopic Models in Nuclear Structure Physics, Oak Ridge Oct. 3-6, 1988*, Ed. by M. W. Guidry, J. H. Hamilton, D. H. Feng, N. R. Johnson, and J. B. McGrory, (World Scientific, Singapore, 1989).
6. M. V. Berry, *Proc. Roy. Soc.* **A392**, 45 (1984).
7. C. Price, H. Esbenson, and S. Landowne, *Phys. Lett.* **B197**, 15 (1987).
8. A. Winther and J. deBoer, in K. Alder and A. Winther, *Coulomb-Excitation* (Acad. Press, NY 1966), p. 303.
9. J. deBoer, C. H. Dasso, and G. Pollarolo, preprint, 1989.
10. M. W. Guidry, R. W. Kincaid, and R. Donangelo, *Phys. Lett.* **B150**, 265 (1985).
11. H. Emling, E. Grosse, R. Kulessa, D. Schwalm, and H. J. Wallersheim, *Nucl. Phys.* **A419**, 187 (1984).
12. P. Ring, in *Proceedings of the IV Jorge Andre Swieca Summer School in Nuclear Physics*, Caxambu, Brazil, February 1989.
13. G. Breit and M. E. Ebel, *Phys. Rev.* **103**, 679 (1956); *Phys. Rev.* **104**, 1030 (1956).
14. M. W. Guidry, This Book, Chapter 29.

41. Time-Resolved and Time-Integral Studies of Nuclear Relaxation: An Extension of the On-Line Nuclear Orientation Technique to Shorter Half-Lives

N. J. Stone
Clarendon Laboratory
Parks Road
Oxford, UK

ABSTRACT

A new method of on-line measurement of nuclear orientation and spin-lattice relaxation is presented which allows separation of decay and relaxation effects in cases where relaxation time and nuclear lifetime are comparable. The decay and orientation of activity implanted into a cold (≈ 10 mK) ferromagnetic lattice in a short pulse is observed as a function of time, the implantation/relaxation sequence being repeated many times to improve statistics. Both the relaxation rate and the fully-relaxed orientation of the nuclei can be extracted, yielding data which assist the determination of the hyperfine interaction strength and spin of the oriented nuclei. Examples of application of the method to Cs and Au isotopes are presented.

41.1 INTRODUCTION

The study of anisotropic emissions from radioactive Nuclei Oriented at Low Temperatures (LTNO) is well established as a method of measuring the hyperfine interaction of the oriented parent and transition multipolarities, plus level spins and parities involved in the decay process. Full formalism for analysis yielding these parameters exists for cases where the oriented parent is in thermal equilibrium with its solid host lattice at a known temperature T_L.[1] Recent years have seen the advent of On-Line Nuclear Orientation (OLNO) facilites. Shorter-lived radioactive isotopes are produced by reaction in a target which forms part of an isotope-separator ion source. The desired nuclear species are selected and implanted at ≈60 keV into a suitable host lattice at ≈10 mK. The OLNO method has greatly extended the range of isotopes accessible to the nuclear orientation technique.[2]

For each element studied by OLNO, as we move further from stability, the accessible isotopes are limited by the nuclear spin-lattice relaxation time with which the nuclei, initially unpolarised upon implantation, achieve thermal equilibrium with their cold host lattice—which has to date always been metallic. The dominant relaxation mechanism is the Korringa interaction between the nuclei and the conduction electrons close to the fermi surface. Other interactions such as magnon effects are usually negligible.

This chapter considers the ways in which the effects of incomplete nuclear spin-lattice relaxation can be included in the analysis of OLNO experiments. A new method of pulsed-source measurement—Time-Resolved On-Line Nuclear Orientation (TR-OLNO)

Exotic Nuclear Spectroscopy, Edited by W. C. McHarris
Plenum Press, New York, 1990

—which allows separate measurement of relaxation and equilibrium orientation parameters, is introduced and discussed.

The ability of TR-OLNO to extract both types of information constitutes a valuable extension of the range of the OLNO method to include nuclei with lifetimes comparable to their spin-lattice relaxation time. Furthermore, the differing dependence of orientation and relaxation parameters upon the hyperfine interaction strength allows more definitive determination of the oriented nuclear spin and hyperfine splitting in pulsed-source measurements than can be obtained from normal continuous-orientation studies.

An outline of the formalism used to calculate relaxation is given in the limit of very low concentration of implant nuclei (single-impurity limit). More complete treatments are those of Klein[3] and Shaw and Stone.[4]

Time-averaged attenuation factors relevant to the analysis of continuous-implantation experiments on isotopes exhibiting incomplete relaxation are briefly discussed, followed by a more extended description of the new pulsed-source method. Application of this method in situations where relaxation of the implant itself is studied and to more complex cases involving a cascade sequence of isotopes, several of which may exhibit incomplete relaxation, is considered in some detail. Examples are given from the few TR-OLNO experiments published to date and from computer simulation.

41.2 TRANSITION PROBABILITIES AND RELAXATION IN THE SINGLE-IMPURITY LIMIT

The angular distribution of γ radiation from an ensemble of nuclei oriented by an axially symmetric static hyperfine interaction is given by[1]

$$W(\theta) = 1 + \sum_{\lambda \mathrm{even}} B_\lambda[p_m(t)] A_\lambda U_\lambda Q_\lambda P_\lambda(\cos\theta) \,, \tag{1}$$

where all terms have their usual meaning. The $p_m(t)$ refer to the nuclear sub-level ($M_I = m$ state) populations at time t, which, in the absence of external perturbations and for constant lattice temperature T_L will always tend towards the boltzmann distribution characteristic of that temperature. The anisotropy ε measured in OLNO experiments is defined as

$$\varepsilon = [W(0)/W(90)] - 1 \,, \tag{2}$$

a quantity which is independent of the instantaneous source strength.

For on-line implantation, the implanted nuclei, given existing isotope-separator yields, come to rest on average tens to hundreds of lattice spacings apart. For example, at ISOLDE-3, although yields close to stability reach $10^{10}\ \mathrm{s}^{-1}$, those of on-line orientation isotopes are typically below $10^8\ \mathrm{s}^{-1}$ so that a 1-at % local concentration takes many hours to produce. At such dilution the nuclei relax independently of each other by coupling with the conduction electrons, and mutual interactions within the nuclear spin system can be neglected. As a result the p_m are not, in general, to be described by a boltzmann distribution during relaxation, i.e., the implanted nuclear ensemble cannot be described in terms of a temperature.

The time evolution of the p_m can be shown to follow a simple gain/loss (master) equation of the form[3]

$$\frac{dp_m}{dt} = \sum_n W_{n,m}p_n - W_{m,n}p_m \,, \tag{3}$$

where $W_{m,n}$ is the time-independent transition probability from sublevel $|m\rangle$ to sublevel $|n\rangle$. For relaxation of nuclei in states $|m\rangle$ via magnetic-dipole interaction with conduction electrons $|ks\rangle$ conserving energy and angular momentum, the transition $|mks\rangle \rightarrow |nk's'\rangle$ involves the nuclear raising/lowering operators,

$$\langle n|I_{\pm}|m\rangle = \sqrt{[I(I+1) - m(m\pm 1)]}\delta_{n,m\pm 1} \,. \tag{4}$$

Summation over electron initial and final states gives

$$W_{m,n} = \sum_{k's',ks} W_{mks,nk's'} f(E_{ks})[1 - f(E_{k's'})] \,, \tag{5}$$

where $f(E_{ks})$ is the fermi function, giving the probability that the initial electron state is occupied, and $[1 - f(E_{k's'})]$ is the probability that the final electron state is vacant. The result of evaluating this summation is

$$W_{m+1,m} = \frac{h\nu}{2kC_K}\frac{[I(I+1) - m(m+1)]}{1 - e^{-h\nu/kT_L}} \,; \tag{6}$$

and

$$W_{m,m+1} = \frac{h\nu}{2kC_K}\frac{[I(I+1) - m(m+1)]}{e^{-h\nu/kT_L} - 1} \,, \tag{7}$$

where $h\nu = |E_{m+1} - E_m|$ (assumed single valued; i.e., a pure magnetic nuclear zeeman splitting) and C_K, the Korringa constant, is a host-dependent constant resulting from the integral over electron states.

It is clear from the master equation and the equations for $W_{m,n}$ that given knowledge of $h\nu$, C_K, T_L, I (parent spin), and of the initial conditions $p_m(0)$, it is straightforward to evaluate $p_m(t)$ and hence model the approach of the nuclei to thermal equilibrium with the host lattice. Mathematical techniques for this operation, developed by Klein, have been followed by Shaw and Stone in dealing with relaxation in the on-line implantation situation and in the present work. Decay of the orienting nuclei is incorporated by adding the term, $-p_m/\tau$, where τ is the mean lifetime ($= t_{1/2}/\ln 2$), to the master Eq. 3.

In conventional NMR, the signal approaches equilibrium with a single exponential time constant T_1 given by

$$T_1 = \frac{2C_K}{T_{int}}\tanh\frac{T_{int}}{2T_L} \,. \tag{8}$$

In the limit $T_L >> T_{int}$, this yields the simple Korringa relation $T_1 T_L = C_K$. The fact that this single exponential relaxation does not apply in nuclear orientation measurements has been emphasized in many places.[3,4] To give an order of magnitude for the time taken to approach equilibrium in nuclear orientation, Klein introduced an effective time constant τ_{SLR} defined by

$$B_2(\tau_{SLR}) - B_2(\infty) = [B_2(0) - B_2(\infty)]/e\,. \tag{9}$$

τ_{SLR} has been shown empirically[5] to be close to

$$\tau_{htl} = \frac{4C_K}{3T_L} \qquad T_L \gg T_{int}\,, \tag{10}$$

and

$$\tau_{ltl} = \frac{3.3C_K}{(I+1/2)T_{int}} \qquad T_L \ll T_{int}\,. \tag{11}$$

Solid-state aspects of the behaviour of C_K for impurities in Fe have been recently discussed by Akai.[6]

For different isotopes of the same element in a given host, the C_K values are related by

$$\gamma_N^2 C_K = \text{const}, \tag{12}$$

Where γ_N is the nuclear magneto-gyric ratio. Thus, when C_K has been measured for one isotope of an element in a particular host lattice, relaxation for any other isotope of that element in the same host can be calculated as a function of the nuclear g factor; C_K = const/g^2. Experimental determination of C_K is therefore a sensitive measure of the nuclear g factor. (These remarks neglect any effect of the hyperfine anomaly, which makes the hyperfine field B_{hf} weakly isotope dependent; however, the anomaly is usually $\leqslant 1\%$.)

In this sense, we see that, for those nuclei for which the limitation of OLNO $\tau_{SLR} \approx t_{1/2}$ becomes critical, there is a valuable new experimental possibility to study the relaxation process directly and from such measurements to determine the nuclear hyperfine interaction. This chapter discusses various ways in which relaxation studies can be made with emphasis on their potential for extending the range of application of the OLNO technique to measuring nuclear moments further from stability.

41.2.A Time-Integral Experiments

i) In a continuous implantation OLNO experiment the implant lifetime is comparable with τ_{SLR}.

ii) In an experiment on a long-lived-parent isotope, decay feeds an isomeric daughter state for which the lifetime is comparable with τ_{SLR}.

In these two situations, relaxation is not directly observed but is deduced from the observation of time-averaged, decay-weighted, attenuation of anisotropy, as compared with full-thermal-equilibrium effects. The attenuation coefficients give information concerning the relaxation process. Such experiments have been fully described theoretically by Shaw and Stone[4] and are included here briefly in Section 3.

41.2.B Time-Resolved Experiments

These are a new type of experiment which are described here generally for the first time. Examples considered specifically in the paper are:

iii) In the simplest pulsed-implantation experiment, following a short implantation the implanted nuclei are observed to orient and decay, as a function of time, the process being repeated cyclically to achieve statistically-adequate data.

iv) Pulsed implantation of a spin $I = 0$ nucleus is followed by a daughter for which the lifetime is comparable with τ_{SLR}. Orientation and decay of the daughter are observed.

v) Pulsed implantation feeds a sequence of even-*A* isotopes for which the lifetime is comparable with τ_{SLR}. The complications of isomerism are briefly considered.

vi) Pulsed implantation feeds a sequence of odd-*A* isotopes for which the lifetime is comparable with τ_{SLR}.

These form the new class of *Time-Resolved* OLNO experiments.

Situations *iii)* and *iv)* have been experimentally demonstrated at the DOLIS-COLD facility of the NSF, SERC Daresbury Laboratory, UK,[7] and by the NICOLE Collaboration at ISOLDE-3 CERN, Geneva,[8] respectively. *iv)* is a relatively trivial extension of *iii)*, but *v)* and *vi)* can be complex and interesting in the variety of cases which arise. These ideas and the published experiments to date are described in Sections 4 and 5 of this paper. Section 7 summarises the ways in which TR-OLNO experiments enjoy advantage over continuous implantation studies in their ability to yield more specific results through separate determination of relaxation and static-equilibrium hyperfine-interaction parameters.

41.3 RELAXATION IN CONTINUOUS-SOURCE-PRODUCTION EXPERIMENTS—TIME-INTEGRAL ATTENUATION FACTORS

Situation i)

When the mean life of the nuclei under study is comparable with the relaxation time, the average anisotropy is given by

$$\overline{W(\theta)} = \frac{1}{\tau}\int W(\theta,t)e^{-t/\tau}dt\,, \tag{13}$$

Here, $W(\theta, t)$ is obtained from the solution of the $p_m(t)$, using the master equation and the usual factors in Eq. 1. For situation *i*), continuous implantation of randomly-oriented nuclei, the master equation becomes

$$\frac{dp_m}{dt} = \sum_n(W_{n,m}p_n - W_{m,n}p_m) - \frac{p_m}{\tau} + \frac{I_0(t)}{2I+1}\,, \tag{14}$$

where I is the spin of the parent and $I_0(t)$ is the implantation rate at time t. For constant plantation rate (or rate varying slowly over time τ_{SLR}, we can define the secular equilibrium condition, somewhere between the initial unoriented system and full thermal equilibrium, for which $dp_m/dt = 0$, and the sublevel populations $p_m(\infty)$, can be calculated. This situation can be described by introducing attentuation factors,

$$\rho_\lambda(I\,, T_{int}/T_L\,, \tau T_{int}/C_K) = \overline{B_\lambda}/B_\lambda(\mathrm{eq})\;, \tag{15}$$

and these factors have been tabulated for a wide range of experimental conditions and nuclear parameters by Shaw and Stone.[4] An example of the B_2 attentuation factors for $I = 2,5$ as a function of interaction strength is given in Fig. 1.

Experimentally, this situation arose in the study of OLNO of $^{118}Cs^m$ ($t_{1/2}$ = 17 s) in iron by the DOLIS-COLD group. Figure 2 shows the fit to the temperature-dependence data, using hyperfine field and C_K values obtained from measurements on other Cs isotopes (see in part below). The dotted line is the unattenuated orientation assuming instantaneous relaxation, with the same interaction parameters.

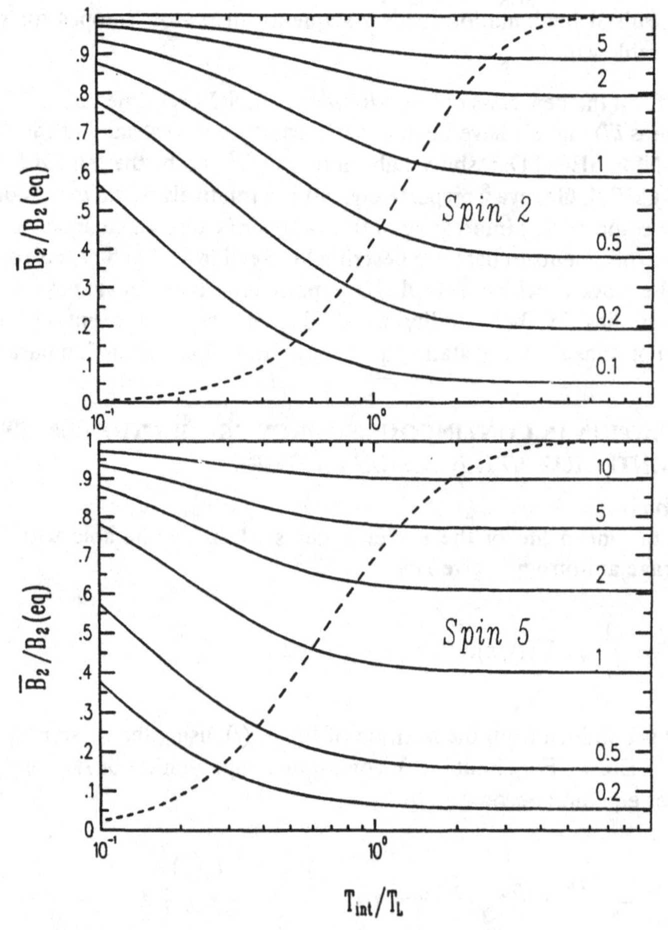

Figure 1. $\rho_2 = \overline{B}_2/B_2$ (eq) as a function of T_{int}/T_L for $I = 2,5$. The curves correspond to different values of the parameter $\tau\, T_{int}/C_K$, which is approximately the ratio of mean lifetime to relaxation time. The dashed lines represent the temperature dependence of $B2$(eq), normalised to its full saturation value.[5]

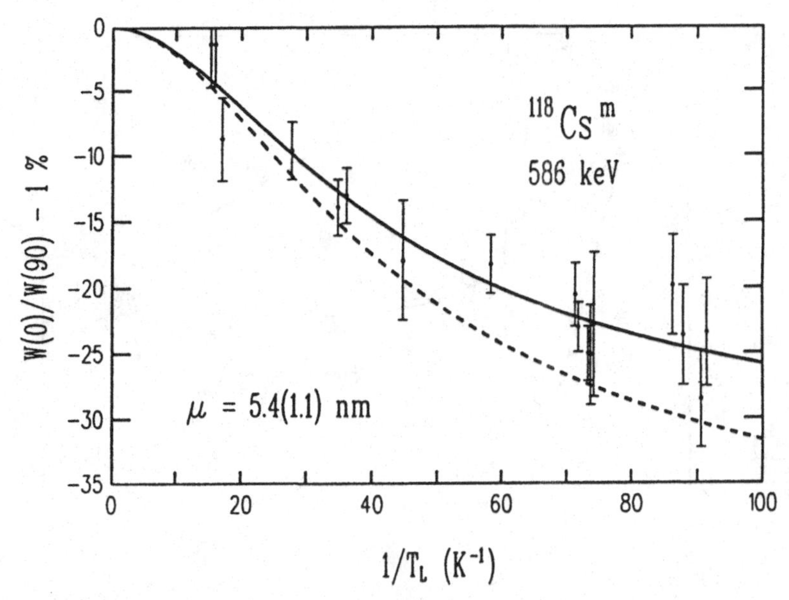

Figure 2. Anisotropy data and fits for the 586-keV transition in the decay of $^{118}Cs^m$.[7]

Situation ii)

Although not an OLNO experiment, precisely the same situation arises if a long-lived but unoriented (perhaps $I = 0$) parent feeds an isomer for which $\tau \approx \tau_{SLR}$. For the related case that the long-lived parent is itself oriented, the last term in the master equation, Eq. 14, is replaced by

$$p_m = (-1)^{I+m}(2I+1)^{-1/2}\sum_{\lambda}^{2I}(2\lambda+1)^{1/2}\begin{pmatrix} I & I & \lambda \\ -m & m & 0 \end{pmatrix} U_\lambda B_\lambda(I_{\text{parent}}) , \qquad (16)$$

normalised to

$$\sum_m p_m = \frac{1}{\tau} . \qquad (17)$$

As an example of this situation, Fig. 3 shows the anisotropy measured by Fahad et al.[9] of radiation from the short-lived isomer $^{89}Y^m$ ($t_{1/2} = 16$ s), populated *in situ* in iron following the decay of recoil-implanted ^{89}Zr ($t_{1/2} = 78$ h), fitted using calculations by Shaw. If the $^{89}Y^m$ were to show only the "inherited" orientation from the Zr parent, the data would follow the "unrelaxed" curve, whilst instantaneous relaxation would give the "relaxed" curve. It is clear from these examples that incomplete relaxation can be included into fitting procedures either to extract the nuclear-hyperfine-interaction strength (as was done in the case of $^{118}Cs^m$Fe) or to determine the Korringa constant (which was the case for $^{89}Y^m$Fe). Nevertheless, the introduction of this additional complication adds to the uncertainty of the result in the former case, and it would be desirable to have a method in which the relaxation and equilibrium orientation effects could be separately determined.

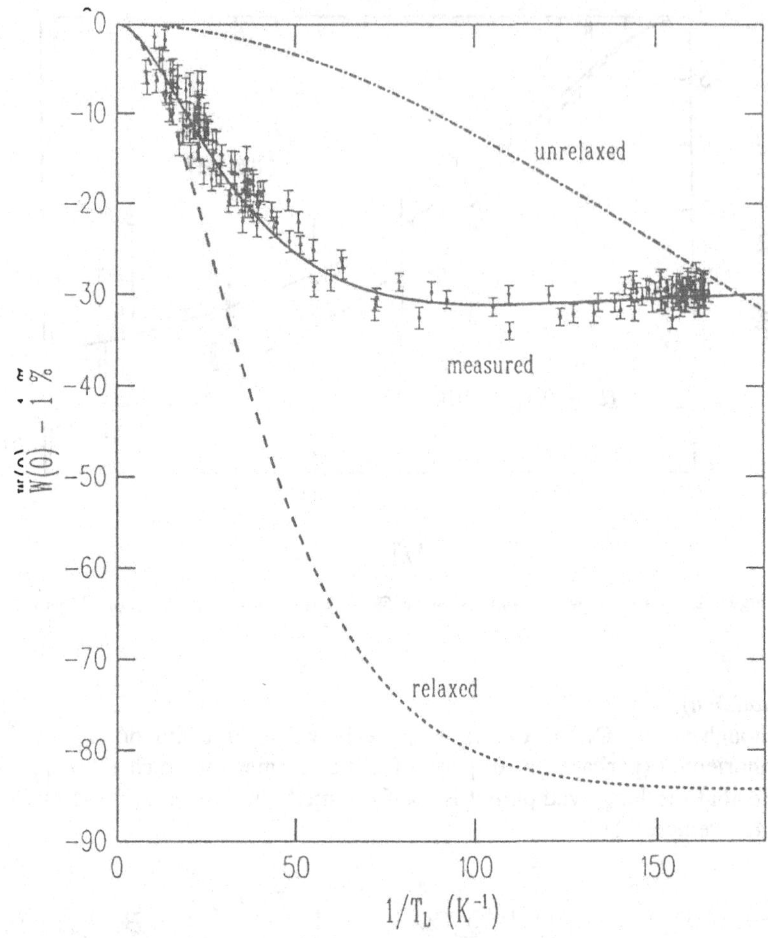

Figure 3. Measured (fitted) and calculated anisotropy for the 909-keV isomeric decay of $^{89}Y^m$ plotted vs inverse temperature.[9]

41.4 PULSED-SOURCE TIME-RESOLVED OLNO EXPERIMENTS—STUDY OF A SINGLE PARTIALLY-RELAXED ISOTOPE

Situation iii)

TR-OLNO allows experimental separation of relaxation effects from equilibrium orientation in a way quite inaccessible to conventional long-lived NO studies. Simply stated, a small sample activity, implanted unoriented in a short pulse, is observed by taking spectra over a succession of short time intervals as the nuclei simultaneously become polarised and decay. The sequence is repeated many times, spectra taken at the same time following each implantation being summed, to build up data of adequate statistical quality.

To describe such a time-resolved experiment, the master equation is modified to allow for source production over time t_{on} and decay and relaxation over time t_{off}. The choice of these times is governed by the desirable conditions $t_{on} \ll \tau_{SLR}$ and $t_{off} \gg \tau$.

When these can both be well satisfied, relaxation is observed with each cycle dominated by freshly-implanted nuclei and little carryover of more relaxed nuclei from previous cycles. In a computer simulation Shaw[5] investigated the behaviour of source strength and of magnetic sub-level population and orientation parameter ratio $B_2/B_2(eq)$ in such a multi-cycle experiment. Figure 4 shows the growth of the source $n(t)$ from the start of the first cycle $n(0) = 0$, with particular attention to the values n^i_{on}, n^i_{off} at the start and end of the ith implantation cycle. The approach to limiting values n_{on}^{∞}, n_{off}^{∞} is clearly seen, for the case that the implant rate I_0 is set equal to $1/\tau$, in the expressions:

$$n_{\text{on}}^{\infty} = n_{\text{off}}^{\infty} e^{-t_{\text{off}}/\tau} \tag{18}$$

$$n_{\text{off}}^{\infty} = \frac{1 - e^{-t_{\text{on}}/\tau}}{1 - e^{(-t_{\text{on}}+t_{\text{off}})/\tau}} \tag{19}$$

Figure 5 shows, for the case $C_K = T_{int} = T_L$, $t_{1/2}/t_{on} = 10$, and $t_{off}/t_{on} = 5$ (dotted lines) and 20 (dashed lines), how the source strengths vary relative to that for continuous implantation at the start and end of each implantation (lower plot) and how the B_2 orientation-parameter ratio at the end of the implantation (source just made) compares with the value just before implantation (source at its oldest). It is clear that cyclic equilibrium for both variables is achieved after a relatively small number of cycles for both parameter sets. The choice of timings will vary with individual experiments; a longer cycle time giving cyclic equilibrium in fewer cycles, but at the expense of a lower overall count rate.

These calculations illustrate the design of the first TR-OLNO experiment, carried out at the DOLIS-COLD facility at the SERC Daresbury Laboratory, to determine the

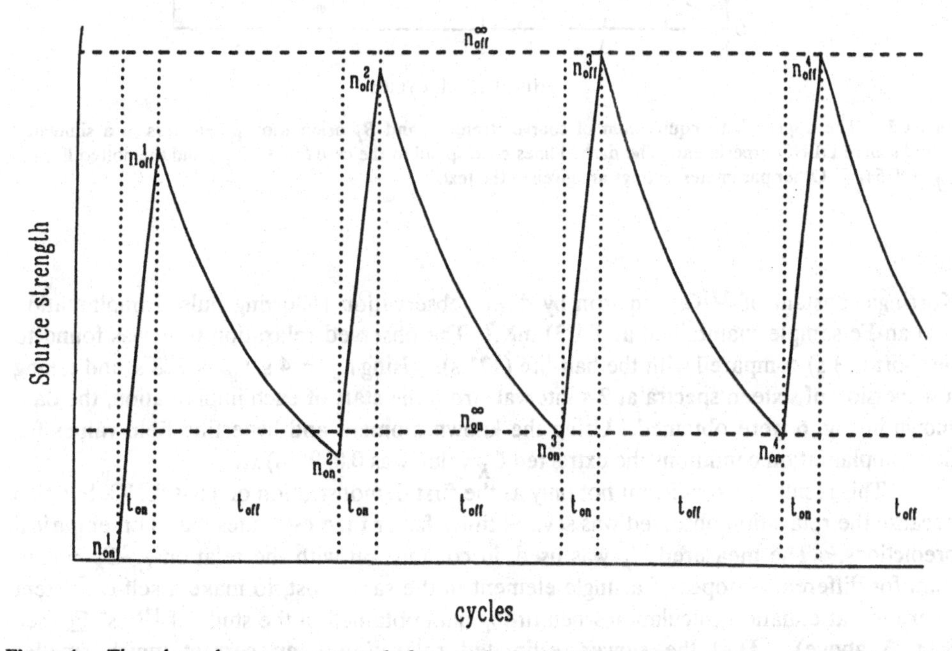

Figure 4. Time-dependent source strength for a pulsed-relaxation experiment, where implantation of I_0 atoms/s for a time t_{on} is performed at intervals (t_{on} + t_{off}). For other symbols, see text.[5]

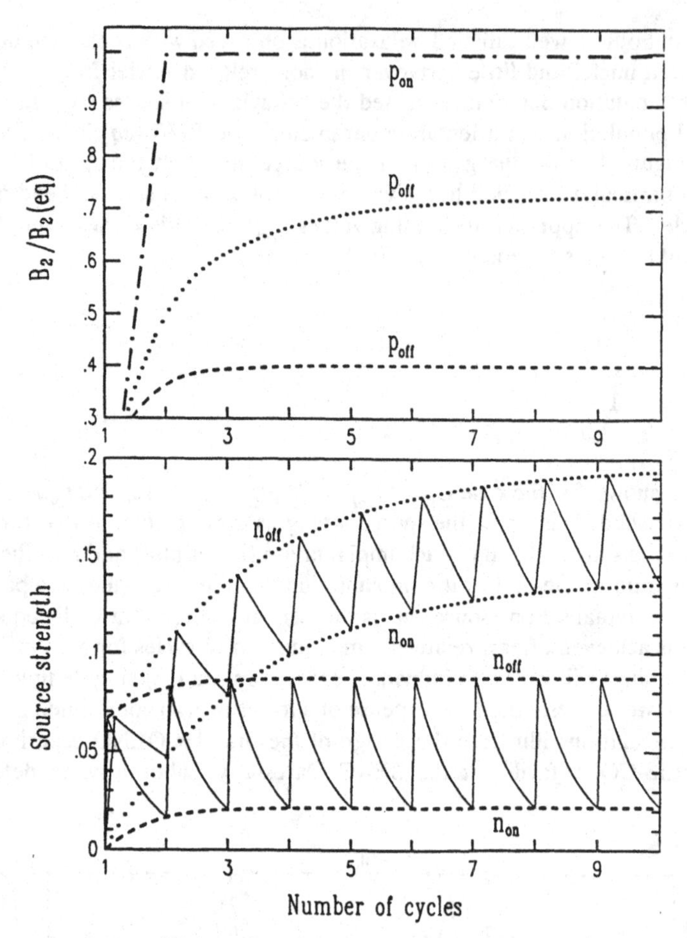

Figure 5. The approach to equilibrium of source strengths and B_2 orientation parameters in a simulated pulsed source OLNO experiment. The dashed lines correspond to the case $t_{off} = 2t_{1/2}$ and the dotted lines to $t_{off} = 0.5\,t_{1/2}$. Other parameter settings are given in the text.[5]

Korringa constant of $^{121}Cs^m$ in iron by direct observation following pulsed implantation into an Fe sample maintained at 7.4(3) mK.[7] The observed relaxation time was found to be short (≈3 s) compared with the half-life (121 s). Using t_{on} = 4 s, t_{off} = 228 s, and taking a succession of sixteen spectra at 2-s intervals from the start of each implantation, the data shown in Fig. 6 were obtained. Using the known moment and hyperfine field values for these implantation conditions the extracted C_K value was 0.059(16) *s*K.

This result was significant not only as the first demonstration of TR-OLNO, but also because the relaxation observed was several times faster than estimates based on empirical predictions.[2] The measured C_K was used, in conjunction with the relation $\gamma_n{}^2C_K$ = constant for different isotopes of a single element in the same host, to make a self-consistent relaxation-attenuation calculation when fitting data obtained in the study of $^{118}Cs^m\underline{Fe}$ (see Sect. 3 above). Had the slower, estimated relaxation been correct, much smaller anisotropies would have been observed.

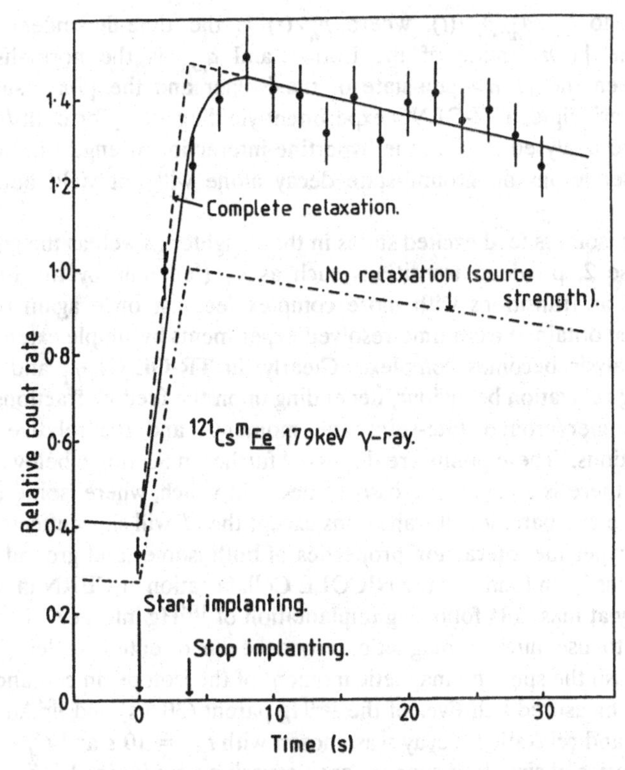

Figure 6. Observed relaxation in the first TR-OLNO experiment at the DOLIS-COLD facility. The anisotropy of the 179-keV transition in the decay of $^{121}Cs^m$ implanted in Fe is plotted as a function of time after the activity pulse.[7]

41.5 TIME-INTEGRAL AND TIME-RESOLVED OLNO EXPERIMENTS IN EVEN-*A* DECAY SEQUENCES

Situations iv) and v)

Where no isomerism is involved, the analysis of even-*A* decay sequences is straightforward even if several of the odd-odd $I \neq 0$ members of the cascade are incompletely oriented during their lifetime. The simplicity follows from the fact that each intervening even-even isotope, with $I = 0$, "resets" the polarisation to zero so that there is no "history" passing down the cascade. In a TI-OLNO experiment the ρ_2, ρ_4 attenuation-factor treatment of Section 3 and Ref. 4 is fully applicable for each odd-odd isotope in the sequence. In TR-OLNO the only difference for members of the sequence other than the first is that the source-strength evolution is smeared out by intervening decays. This means that, even if $\tau_{SLR} < \tau$, the fully-relaxed anisotropy will not be observed unless the cycle length exceeds the summed lifetimes of the preceding decays.

Three situations involving isomerism, starting from an $I = 0$ parent, are shown schematically in Fig. 7. In the first the isomer decays solely by an internal transition. Provided this transition can be observed, the orientation "inherited" by the ground state can be allowed for in the simple two-step decay.[10] The calculation involves changing the feeding term in Eq. 14 for the ground state from $I_0(t)/(2I+1)$, corresponding to unoriented random feeding, to $\sum_n a_{nm} p_{n'}(t)$, where $p_{n'}(t)$ is the time-dependent value of the population of the $|I'n\rangle$ state of the isomer and a_{nm} is the normalised transition probability between the $|I'n\rangle$ sub-state of the isomer and the $|Im\rangle$ sub-state of the ground state. In principle, a TI-OLNO experiment yielding ρ_2,ρ_4 for both *IT* and ground-state decay can be analysed to yield the hyperfine-interaction strengths in both levels, but time-integral observations on ground-state decay alone will not yield both interactions uniquely.

Where the isomers feed excited states in the daughter as well as the ground state via an *IT*, as in Case 2, provided transitions such as γ_1 (fed only by the isomer) can be observed as well as transitions with more complex feeding, once again both hyperfine interactions can be obtained from time-resolved experiments by simple extension of Eq. 14. Time-integral analysis becomes complex. Clearly, in TR-OLNO γ_1 and γ_2 may show markedy differing relaxation behaviour, depending upon the feeding fractions of the decay, the respective isomer/ground-state-spins and moments, and the relative sign of their hyperfine interactions. These points are discussed further in Section 6 below.

In Case 3 there is a complex crossover decay in which, where isomeric decay by *IT* and β emission are comparable, all transitions except the *IT* will show effects influenced by the nuclear and hyperfine interaction properties of both isomer and ground state. An example of Case 3 has been found by the NICOLE Collaboration at CERN in the TR-OLNO study of the decay at mass 184 following implantation of ^{184}Hg into Fe.[8] The design of the experiment was to use nuclear magnetic resonance on oriented nuclei (NMR/ON) of ^{184}Au<u>Fe</u> to establish the spin and magnetic moment of the isotope, in conjunction with TR-OLNO. With the measured half-lives of the ^{184}Hg parent (30.6 s) and ^{184}Au (53 s), a cycle for implantation and relaxation/decay was chosen with t_{on} = 10 s and t_{off} = 310 s. Data were taken as a series of sixty-four spectra, each covering a 5-s interval. Figure 8 shows the observed TR-OLNO anisotropy of the 272-keV ($4^+ \rightarrow 2^+$) transition in the decay of ^{184}Au at 13 mK, and it is clear that relaxation is complete after ~ 20 x 5 = 100 s. However, data taken with the same time cycle on warm nuclei (~1 K) gave the first direct evidence for the existence of isomerism in ^{184}Au and led to the proposal of a decay scheme in which this

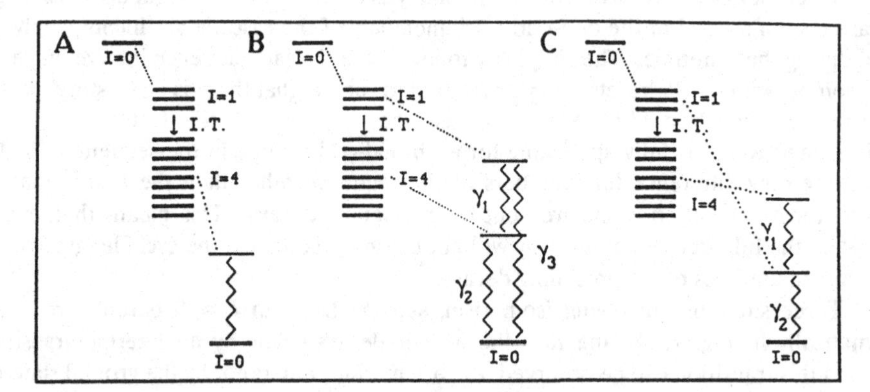

Figure 7. Schematic decay schemes involving isomerism with decay of increasing complexity. For details, see text.

isotope has two isomers, the higher in energy having low spin (I = 2,3) and half-life 53 s, and the lower, the true ground state, higher spin (5,6) and half-life ≈ 12 s, the spin difference ΔI = 3 being extracted from the isomeric half-life. In subsequent experiments the existence of a prominent $M3$ transition at 67 keV, seen only in conversion electrons, confirmed the presence of these two isomers, as indicated by the pulsed source measurements. This elaboration of the mass-184 Hg-Au-Pt decay chain complicates analysis of the relaxation/orientation data on the prominent 362- ($6^+ \rightarrow 4^+$), 272-, and 162-keV ($2^+ \rightarrow 0^+$) transitions, as all are fed in varying degrees via both isomer and ground state of ^{184}Au. This analysis is still in progress. It is clear, however, that the time-resolved experiment prevented continued erroneous interpretation of the properties of this decay.

41.6 TR-OLNO EXPERIMENTS ON ODD-*A* DECAY SEQUENCES

Situation vi)

For odd-A decay and magnetic hyperfine interaction, there is no equivalent of the I = 0 polarisation "reset to zero." Polarisation occurs for all non-zero nuclear spin values, including I = 1/2 for magnetic interactions (however, it is clear that in the case of pure electric-quadrupole interaction there is no alignment for I + 1/2, and "reset to zero" does occur for these nuclei). Although the primary nuclei implanted into ferromagnetic host metals are unoriented, each decay product will "inherit" some degree of polarisation from its parent, followed by relaxation towards its own equilibrium orientation. As is shown below, the relaxation behaviour can be characteristic of the precursors in the decay chain, as

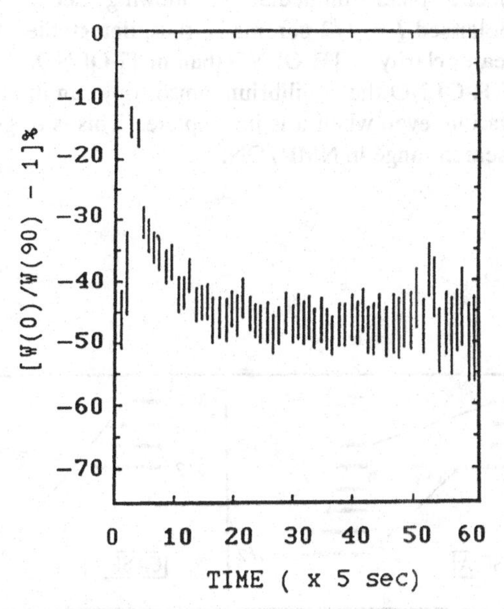

Figure 8. Observed relaxation in the TR-OLNO experiment on ^{184}Au<u>Fe</u> performed at the NICOLE facility, ISOLDE-3, CERN. The anisotropy of the 272-keV $4^+ \rightarrow 2^+$ transition is shown as a function of time after the activity pulse.[8]

well as of the daughter at different stages of the implant-relaxation/decay cycle, so that analysis of the whole chain may be interlinked.

For a sequence of partially-relaxed decays, TI-OLNO analysis presents major difficulty and may give little information on the hyperfine interactions of individual isotopes in the chain. Naturally, the NMR/ON technique is available to obtain $|g_I B_{hf}|$ when resonance is observed, but care is needed to etablish which isotope is being resonated if there is strong "inherited" polarisation! Integral OLNO is of uncertain value in predicting the appropriate NMR/ON frequency search range under circumstances involving partial relaxation.

By contrast, TR-OLNO has much to offer in complex odd-*A* decay chains. Direct relaxation observation in each isotope of the sequence allows the full history to be analysed by straightforward theoretical extension of Eq. 14. Values of the Korringa constant C_K can be extracted for each nucleus which, using the relation $C_K = \text{const}(I/\mu)^2$, with known constant for many implant/host combinations, yield the *g* factor of the relaxing isotope with useful precision.

A significant additional feature, not available in any conventional orientation experiment, is that the relaxation is sensitive to the *relative* sign of the magnetic hyperfine interaction in successive members of the decay chain. This is illustrated in Fig. 9. In Case *A* the sequence of nuclear substates $|I\,m\rangle$ is the same in both parent and daughter, and the sense of polarisation is maintained during decay involving a small change in magnetic quantum number *m* (typically $\Delta m = \pm 1,0$). In Case *B*, for which in the daughter either the hyperfine field (usually known) or the nuclear magnetic moment is of the opposite sign to the parent, decay "inverts" the polarisation. Examples of time evolution of the daughter p_m and B_2 orientation parameter are shown in Fig. 10, clearly demonstrating the sensitivity of TR-OLNO to such a sign change. The computer simulation was done by Harding.[10]

A third point is that "inherited" polarisation is clearly visible in TR-OLNO, since it will dominate spectra taken immediately following decay. Gamma-ray anisotropy originating from polarised $I = 1/2$ parents is, thus, detectable in $I > 1/2$ daughter-decay transitions with greater clarity in TR-OLNO than in TI-OLNO.

Finally, in TR-OLNO the equilibrium anisotropy can in principle be extracted from the observed relaxation, even when it is incomplete. This is of great assistance in predicting the frequency search range in NMR/ON.

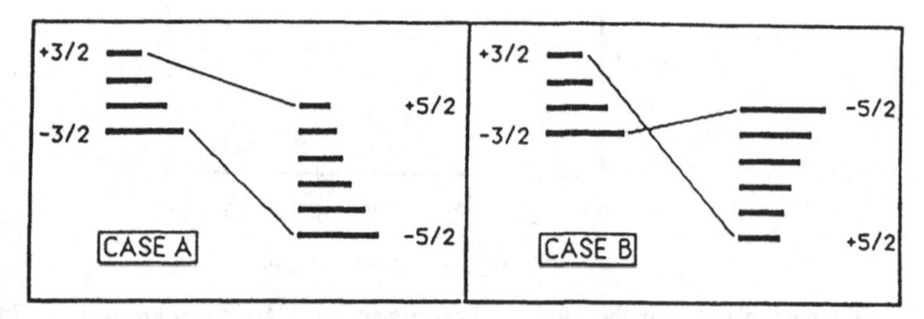

Figure 9. Illustrating the differing initial conditions for relaxation which give sensitivity of TR-OLNO experiments to the relative sign of hyperfine interaction of successive members of an odd-*A* decay chain.

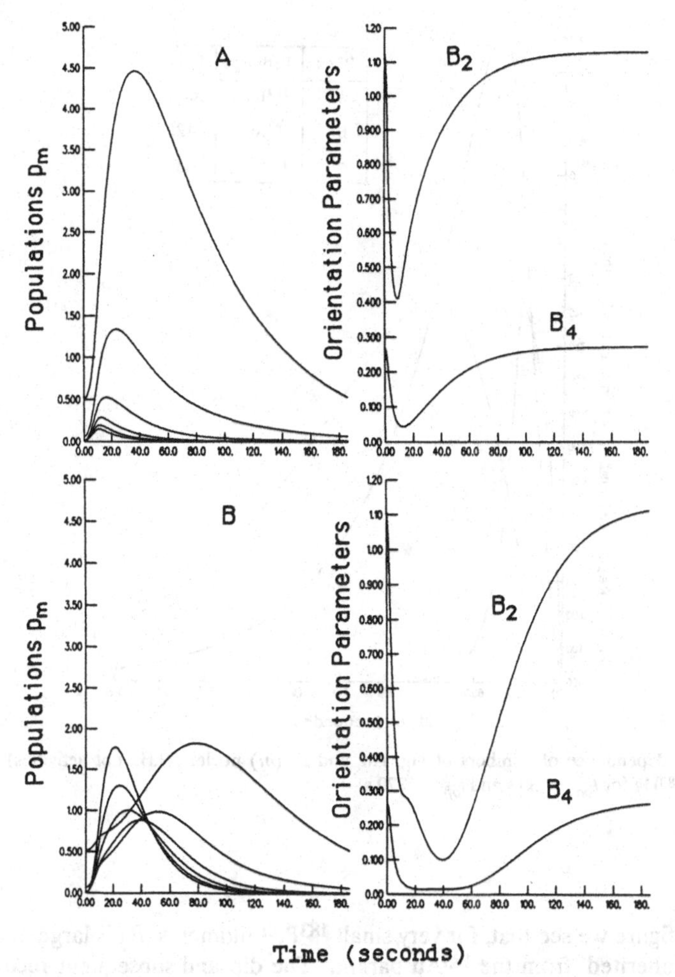

Figure 10. A simulated example of the time evolution of populations p_m and the B_2 and B_4 orientation parameters for the case of a spin-5/2 daughter of a spin-3/2 parent. *A*) with the same sign *hfi* in both parent and daughter; *B*) with opposite sign.

As an example of such a decay chain, we show in Fig. 11 the calculated variation with time of the number of ^{183}Hg, ^{183}Au, and ^{183}Pt^m nuclei present in a cyclic TR-OLNO experiment at mass 183, starting with implanted ^{183}Hg. Since the decay chain is long, the γ ray spectra are complex and the distinctive time dependence of intensity from different isotopes is of value in assigning transitions to their correct decay. Figure 12 gives the calculated B_2 orientation parameters as a function of time over the cycle, after cyclic equilibrium has been achieved, for a transition in the decay of the $7/2^-$ isomer of ^{183}Pt, the third element in the chain. The parameters used in the calculation are given in Table I. The cycle has t_{on} = 30 s and t_{off} = 270 s, and the calculation is for a lattice temperature T_L of 25 mK. The magnetic dipole moment of the Pt isomer is a variable in the calculation.

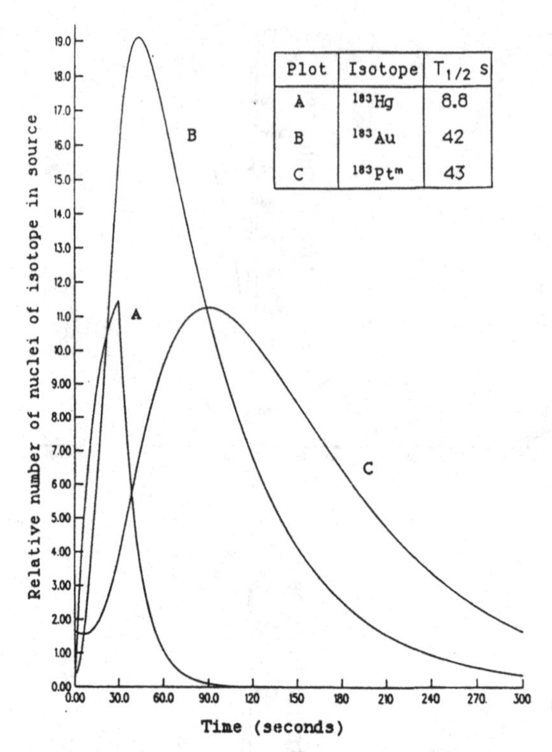

Figure 11. Time dependence of numbers of Hg, Au, and Pt (*m*) nuclei (N.B., not activities) in a cycle with implantation of ^{183}Hg for t_{on} = 30 s and t_{off} = 270 s.

In the figure we see that, for very small ^{183}Pt^m moments, B_2 is large at all times, being entirely "inherited" from the ^{183}Au parent. The dip and subsequent recovery in B_2 at early times reflects the relaxation of the *parent* ^{183}Au and is not related to the Pt isomer. With increasing Pt moment, the ^{183}Pt relaxation time shortens, which initially leads to a *reduction* in B_2 at later times in the cycle, as the ^{183}Pt equilibrium polarisation is lower than the "inherited" value. With further moment increase, the relaxation time becomes shorter than the cycle time, and a full equilibrium B_2 characteristic of the Pt $7/2^-$ isomer is observed in the latter part of the cycle. For the higher moments the initial time dependence becomes a compound of Au-parent and Pt-isomer relaxation. It is clear that over a wide range of moments the TR-OLNO signal shows characteristic features which can be fitted to data of good statistical quality to yield the magnetic moment to useful accuracy. Measurements as a function of lattice temperature are of value, the temperature dependence being fully calculable.

41.7 THE VALUE OF TR-OLNO MEASUREMENTS

In this chapter, we have surveyed the ways in which relaxation times can be extracted from orientation studies of nuclei for which the lifetime is comparable to the spin-lattice relaxation time. We argue that the relaxation time constant, in metals the Kor-

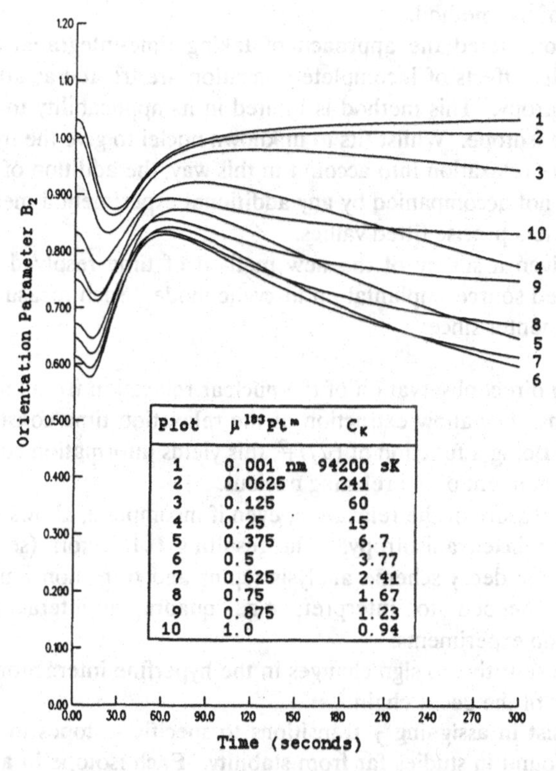

Figure 12. Simulated TR-OLNO B_2 versus time for $^{183}Pt^m$ produced following decay of ^{183}Hg implanted into Fe at $T_L = 25$ mK. The calculation simplifies decay as a sequence of direct ground state-ground state β transitions with angular momentum equal to the difference of the spins involved. The assumed magnetic moments of $^{183}Pt^m$ are given in the figure. Other parameters used are given in Table I. For discussion, see text.

Table I. Parameters Used in the TR-OLNO Simulation of $^{183}Pt^m$ Produced Following Decay of ^{183}Hg Implanted into Iron at T_L = 25 mK

Isotope	$t_{1/2}$ (s)	I	μ (μ_N)	μB_{hf} ($\mu_N T$)	C_K (sK)
^{183}Hg	8.8	1/2	+0.524	−59.7	0.135
^{183}Au	42	5/2	+2.2	−250.8	0.121
^{183}Pt^m	43	7/2	variable subject to $\nu^2 C_K = 8 \times 10^4$ (MHz^{2}sK), where ν is the calculated NMR frequency in Fe		

ringa constant C_K, is a valuable measure of the nuclear-hyperfine-interaction strength. This means that measurements which reveal the effects of incomplete relaxation are of direct value. As experiments extend further from the region of stability, falling lifetime is a limiting parameter for the OLNO method, and the development of a new experimental technique, applicable particularly to shorter-lived isotopes, is most valuable and directly extends the range of the method.

We have considered the approach of taking time-integrated measurements, TI-OLNO, in which the effects of incomplete relaxation are treated as attenuation factors in the observed anisotropy. This method is limited in its applicability to simple cases effectively involving one isotope. Whilst fits to unknown nuclei to give the hyperfine interaction can take incomplete relaxation into account in this way, the addition of the relaxation constant into the fit is not accompanied by any additional experimental measurable and hence leads in general to less precise fitted values.

We have given a survey of the new method of time resolved measurement, TR-OLNO, using pulsed source implantation in cyclic mode. Such measurements have much to offer the experimenter since:

a) By giving direct observation of the nuclear relaxation from known starting conditions, they allow extraction of the relaxation time constant for each isotope. Being a function of $(\mu/I)^2$, this yields information concerning the spin and moment of the relaxing nucleus.
b) A good measure of the relaxation, even if incomplete, allows extraction of the fully relaxed anisotropy. The resulting *UA* factors (see Eq. 1) are valuable for decay-scheme analysis (spins and transition multipolarities) and are needed to interpret; e.g., quadrupole-interaction nuclear-orientation experiments.
c) They are sensitive to sign changes in the hyperfine interaction in differing members of the decay chain.
d) They assist in assigning γ transitions to specific isotopes in the complex spectra found in studies far from stability. Each isotope in a decay chain has a specific time evolution.
e) They can reveal isomerism in nuclei where lifetimes are more than a few seconds.

The benefits of TR-OLNO are not without their cost, the most significant of which is that the requirement of pulsed implantation means a reduction in average counting rate. However, the time-integral data available with continuous implantation in complex decay chains is quite inaccessible to interpretation. The only apparent value of TI-OLNO arises where it may afford improved statistical data quality during NMR/ON searches.

ACKNOWLEDGMENTS

The author acknowledges the contributions to the ideas of TR-OLNO by other members of the Oxford nuclear orientation group—in particular T. L. Shaw, D. E. Brown, P. A. Harding, and V. R. Green—and also the work of the DOLIS-COLD and NICOLE OLNO groups in realising the first TR-OLNO experiments.

REFERENCES

1. K. S. Krane, in *Low Temperature Nuclear Orientation*, Ed. by Stone and Postma (North Holland, Amsterdam 1986), Chapter 2.
2. N. J. Stone, *Hyp. Int.* **22**, 3 (1985).
3. E. Klein, op.cit., Ref. 1, Chap. 12
4. T. L. Shaw and N. J. Stone, *Atom. Nucl. Data Tables* **42**, 339 (1989); *Hyp. Int.* **43**, 299 (1988).
5. T. L. Shaw, Ph.D. Thesis, Oxford, 1987.
6. H. Akai, *Hyp. Int.* **43**, 255 (1988).
7. T. L Shaw, V. R. Green, C. J. Ashworth, J. Rikovska, J. Stone, P. M. Walker, and I. S. Grant, *Phys. Rev. C* **36**, 41 (1987).
8. R. Eder et al., to appear in *Proc. 8th Int. Conf. on Hyperfine Interactions, Prague*, (1989); *Hyp. Int.*, to be published.
9. M. Fahad, I. Berkes, R. Hassani, M. Massaq, T. L. Shaw, and N. J. Stone, *Hyp. Int.* **43**, 307 (1988).
10. P. A. Harding, to be published.

42. Exotic Nuclear Spectroscopy—Remembrance of Past Futures

Wm. C. McHarris
National Superconducting Cyclotron Laboratory
and Departments of Chemistry and Physics/Astronomy
Michigan State University
East Lansing, MI 48824

ABSTRACT

Many of the developments presented in *Exotic Nuclear Spectroscopy* are realizations of dreams and goals of long ago. Radioactive beams, 4π high-multiplicity γ-ray spheres, on-line recoil mass spectrometers, nuclei ever further from stability, peculiar nuclear shapes and transitions, insight into details of nuclear structure unthought of several decades ago—all these are now being realized in nuclear spectroscopy. And theory is keeping pace with experiment—more elaborate shell-model calculations, partial details of the *p-n* residual interaction, successful group-theoretical models, feedback from quarks and elementary-particle theory into models of nuclear structure. It is an exciting time to be working in nuclear spectroscopy, for we are producing far-out but concrete results. In this summary chapter, I shall try to tie together some of the new findings and ideas presented in the book, relate some of these to "past futures," and even make a few cautious predictions about the state of nuclear spectroscopy in future futures.

42.1 EXOTIC NUCLEAR SPECTROSCOPY

What exactly is exotic nuclear spectroscopy? This book and the symposium that generated it concerned themselves with this question, so it is unlikely that the reader has reached this chapter without having a pretty good notion as to the answer. However, it still behooves us to investigate the question once again in summary, if only to tie nuclear spectroscopy of the present into predictions made in the past and to make some attempt to envision directions it might take in the future.

In a sense, essentially *all* nuclear spectroscopy—certainly that being performed in major laboratories around the world today—is exotic. Gone are the days when simple desk-top instrumentation was used to measure half-lives and spectra, then work out the straightforward decay schemes of nuclei close to stability or available conveniently, say, as fission products. Gone, too, are the days when nuclear experimentalists and theorists were distinct, seldom-communicating personages.

Those were simpler but not necessarily better days. A spectroscopist could measure the shape of a β spectrum, make a Kurie plot (or, if old enough, perhaps even a Feather plot), apply a few tabulated corrections, determine the end-point energy, and look it all up in a table of selection rules and log*ft* values. He/she could then publish a quick paper,

Exotic Nuclear Spectroscopy, Edited by W. C. McHarris
Plenum Press, New York, 1990

declaring, "Aha, a first-forbidden unique transition, which means the J^π value of the parent level is $3/2^\pm$, $5/2^\pm$, or $7/2^\pm$!" There could also be art as well as science involved—if a γ-ray spectrum had been taken with a NaI(Tl) detector, with or without manually-set coincidence gates, then the ensuing number of transitions depended at least in part on the imagination of the person who resolved the spectrum. As to insight into structure, one rarely bothered with that—at most a comparison with the calculations of Kisslinger and Sorensen.[1] At that, a referee was likely to criticize the authors for "dabbling in theory and adding unnecessary theoretical complications to an essentially experimental paper."[2]

Times have changed. During the past two decades, nuclear spectroscopy has undergone a quiet revolution and has again become a forefront science.

This has come about partly through fortuitous developments in experimental techniques: 1) Order-of-magnitude better detectors; in particular, the development of Ge γ-ray detectors. 2) Large-scale integrated circuits and their application to nuclear electronics, leading to coincidence logic and timing capabilities unheard of ten years ago. 3) The low(er)-cost availability of large-scale computing power to handle and digest the proliferation of data produced by high-resolution, multi-coincident detection systems. 4) Last but far from least, the construction of heavy-ion and higher-energy accelerators needed to produce nuclei far from stability.

Theoretical tools have improved significantly at the same time: 1) Huge shell-model calculations make possible meaningful predictions and interpretations of experimental data. 2) Group-theoretical interpretations complement the more geometrical models of nuclei. 3) Feedback from quark theory and quantum chromodynamics is just beginning to usher in some understanding of nuclear structure perhaps on a deeper level.

But the most significant changes have come about in nuclear spectroscopists themselves. There are fewer of us, we tend to work cooperatively in larger groups, and more and more we attack a problem from all sides, experimentally and theoretically. Chemists and physicists alike, we band together to perform long-range, massive experiments using specialized equipment, such as a multi-detector array, at one of the major laboratories. Familiarity with intricate electronics and extensive computer analysis of the data are considered only part of standard operating procedure. Such a large array of data is expected to produce more than mere numerology—even a so-called experimental paper nowadays is considered superficial and incomplete unless it makes a concerted effort at a theoretical understanding of what the data mean in terms of nuclear structure. This is usually accomplished by a collaboration of experimentalists and/or theorists—sometimes the distinction is difficult to discern, for many of us experimentalists now dabble rather deeply in theory (and even do a pretty good job of it!).

So, what is exotic nuclear spectroscopy? Originally, "exotic" nuclei referred to nuclei far from stability, those that could be produced by unusual nuclear reactions. They might even exhibit some rather unusual properties.

I remember when our group got all excited about the β decay of the then-exotic nucleus, $^{139}Nd^m$.[3,4] Produced by a straightforward (p, $3n$) reaction, its only exotic behavior was β^+ decay to a multiplet of high-lying (1.6–2.2 MeV) states rather than to states near the ground state. This could be explained quite readily, if one looked beyond the tables of empirical solution rules: Although $^{139}Nd^m \rightarrow {}^{139}Pr$ goes from an odd-n to an odd-p nucleus, this is β^+ decay, and one of the paired p's must decay to a similar n state. An examination of the available shell-model states showed the most likely final states to be three-quasiparticle states. This was dabbling in theory in the most superficial way; yet, it explained a minor puzzle that had existed for a few years.

Nowadays, of course, the term "exotic" is usually reserved for nuclei much further from stability, those near the limits of particle binding or super-heavy elements or for nuclear states having extreme properties, such as superdeformed bands, octupole bands, or decoupled bands.

The theory is correspondingly more elaborate: Large-scale shell-model calculations (Brown, Chapter 20), the Fermion-Dynamical-Symmetry Model (Guidry, Chapter 29), the Interacting-Boson Model (Chou, McHarris, and Scholten, Chapter 34), and elaborate extensions of the geometrical models (Kreiner, Chapter 26; Yu, Gascon, Hagemann, and Garrett, Chapter 38). Although, I was pleased to find that certain more-empirical approaches (Sood and Sheline, Chapter 31; Brenner, Casten, Wesselborg, Warner, and Zhang, Chapter 33) still produce worthwhile results. Nuclear spectroscopy retains an element of intrigue in its exotic renaissance.

42.2 PAST FUTURES

What were a few of those past predictions that have come to pass?

First of all, the early 1970's was the heyday of interest and speculation about superheavy elements.[5,6] All sorts of predictions were made about their nuclear and chemical properties, and proposals abounded for constructing accelerators to produce them or for instituting ingenious searches for them in nature. Since "doubly-magic" element 114 should be eka-Pb, a number of searches were made for it by dismantling old stained-glass windows—the idea was that some of this glass had been in contact with Pb for centuries; hence, it could record fission tracks coming from trace amounts of element-114 mixed in with the Pb. Also, there was a period when it was thought that superheavy elements might have been produced in the beam stops of high-energy *p* accelerators through secondary heavy-ion collisions induced by the *p*'s. For a few months, there was a flurry of activity in dissolving and performing radiochemical separations on beam stops.

What has come of all this? Well, superheavy elements, as such—an "island of stability separated from the main peninsula of elements by a sea of instability"—have not been discovered. However, by using a careful, systematic approach, scientists at GSI Darmstadt have mapped out a portion of the mass surface for trans-actinide nuclei and have produced new elements up through element 109, with the distinct possibility of proceeding to even higher elements (Münzenberg and Armbruster, Chapter 11). Not quite so romantic as shooting directly for element 114, but far better science.

Just as important, the drive toward superheavy elements helped promote the construction of a new generation of higher-energy heavy-ion accelerators, from the Super HI-LAC at Berkeley through Unilac at GSI to Holifield (Oak Ridge), Atlas (Argonne), NSCL (Michigan State), and GANIL (Caen), to name only a few. Add these accelerators to increased availability of high-energy physics accelerators (for spallation reactions), and we now have facilities for producing all sorts of exotic, far-out nuclei.

Second, what about the possibility of exotic beams? At the 1970 Leysin Conference,[5] Flerov gave a short talk on the possibility of developing an ^{8}He accelerator at Dubna. The concept was straightforward: Produce ^{8}He in a spallation reaction, and then accelerate it as a secondary beam in a linear accelerator. His talk met with skepticism. The consensus was that radioactive beams in usable intensities were not likely to become available in the foreseeable near-future.

This, too, has changed. Radioactive beams are now a hot item (D'Auria, Chapter 6; Haustein, Chapter 1), and construction of radioactive-beam facilities is underway at many laboratories.[7] This should have an imminent effect on the production of exotic nuclei, especially opening up new regions of very neutron-rich nuclei.

Third, large detector arrays. Actually, these were almost beyond the dreams of nuclear spectroscopists in the 1970's, having become available only within the past decade. Experiments performed using existing, rather large arrays have provided much of the "hard" science in this book: Ahmad et al., Chapter 12; Cline, Chapter 14; Chowdhury et al., Chapter 17; Deleplanque, Chapter 21; Johnson, Chapter 22; Janzen et al., Chapter 23; Paul et al., Chapter 24; Olivier et al., Chapter 25; Sarantites et al., Chapter 30. None of these experiments would have been possible without relatively large detector arrays.

And the really big arrays are yet to come—GAMMASPHERE (Lee, Chapter 15) and EUROGAM. These arrays, with their hundreds of detectors, will increase the sensitivities for observing exotic nuclear states by orders of magnitude.

Fourth, on-line isotope separators. For many years, ISOLDE (CERN, Geneva) was the preeminent "on-line" isotope separator. Good as it was, it had a slow interface between production and mass separation; also, it was more or less limited to volatile species. Recent and proposed systems use recoil mass separators to overcome this slow interface and to extend the range of applicability (Cole, Chapter 2; Mohar et al., Chapter 7; Hill et al., Chapter 8; Reeder et al., Chapter 35; Moltz et al., Chapter 36). It is now possible to separate—and perform *clean* experiments—on virtually any exotic species that can be produced.

Fifth, the willingness to study unusual nuclei. There are many examples of this, from superdeformed nuclei (Deleplanque, Chapter 21; Janzen et al., Chapter 23) to investigations of the moments of "strangely-shaped" light Hg isotopes (Rikovska, Chapter 10). One of the clearest examples is the new surge of interest in odd-odd nuclei. As recently as ten years ago, odd-odd nuclei were considered "too difficult" and "not worth the effort for the amount of information gleaned." That, too, has changed—Paul et al., Chapter 24; Olivier et al., Chapter 25; Kreiner, Chapter 26; Hoff et al., Chapter 27; Meyer, Morrison, and Walters, Chapter 28. Of particular interest are the "doubly-decoupled" bands (Chapters 25 and 26), which hold out the hope of providing some sort of insight into the *p-n* residual interaction in heavier nuclei with minimal interference from the nuclear core. Spectacular work has also been done on ϵ -delayed fission (Hall and Hoffman, Chapter 37). And we should not forget the really far-out effects—such as searches for the nuclear SQUID (Canto et al., Chapter 40). These are all active fields of nuclear research that would have been considered highly impracticable just a short age ago.

42.3 FUTURE FUTURES

What about the future? Making predictions about what will come to pass is always a dangerous business, but I am going to be brave (foolhardy?) and make a few predictions about "hot" topics in nuclear spectroscopy around the year 2000. Some are relatively straightforward to envision, others not—and some downright contradictory, but here goes:

First, large detector arrays at heavy-ion and high-energy facilities. The use of these should become more and more routine, and electronic/computer developments should do away with much of the drudge work now associated with analyzing the massive amounts of data they can and will produce. I imagine that accelerators used for nuclear spectroscopy will develop in two seemingly contradictory directions—higher energies (especially for spallation and the production of radioactive beams) and yet higher resolution/lower energies for direct heavy-ion spectroscopy. A few years ago, great things seemed about to happen with cluster-transfer reactions, and, with improved spectrometers, this may actually come about. And extremely heavy beams, such as Pb and U, should be widely available. The bulk of the action will still be in "classical" in-beam γ-ray and particle-transfer spectroscopy, but there should be added interest in exotic processes, such as the nuclear Josephson effort or SQUID (Chapter 40). In short, nuclear spectroscopy should develop into a more subtle, less brute-force science.

Second, a major breakthrough in detectors, especially for γ-ray detectors. As good as they are, we are reaching diminishing returns with Ge γ-ray detectors. Most of the progress yet to come lies in the development of large detector arrays. After all, there is a practical limit to just how big one can construct a high-resolution Ge detector and how much one can suppress the compton background. Side by side with Ge detectors, we should see new detectors constructed out of more esoteric materials with intrinsically better efficiency, resolution, and peak-to-compton ratios. One such possibility is InSb,[8] a difficult and treacherous material with which to work but one that may become amenable to fast-developing improvements in materials-science techniques.

The ultimate improvement in γ-ray spectroscopy would be to heterodyne γ rays, much as is now done with visible light for laser Raman spectroscopy.[9] Unfortunately, a minimum requirement for this is a coherent source of γ rays, and γ-ray lasers remain at the threshold of science fiction. (No one is likely to want to perform spectroscopy based on a single "super-radiant" burst in the Nevada desert!) However, some progress has been made in using heavy-ion channeling to produce coherent nuclear effects in crystals; so, it is conceivable that coherent radiation in the γ-ray range could become available as a reference beam with which to beat nuclear γ rays and extract the difference signal. Perhaps ten years is overly optimistic, but eventually we can expect to see γ-ray spectroscopy with resolutions in the eV range!

Third, feedback from high-energy nuclear science and from elementary-particle physics. "Bevelac, CERN, and, someday, RHIC experiments are East, and nuclear spectroscopy and structure are West, and ne'er the twain shall meet!" Such has been the accepted dogma until comparatively recently. However, we find that, even in this book, a number of authors are starting to try to bridge the gap. Nuclear structure studies can certainly profit from high-energy feedback, and I am sure there will be more and more of it—in fields starting with giant resonances and continuing through π interactions and Δ spectroscopy.

Why stop there? Feedback from elementary-particle theory and quantum chromodynamics is just around the corner. Indeed, neutral weak currents are already the subject of intense, successful investigations in atoms, so can nuclei be far behind? One possible example,[12] still at the stage of speculation, concerns the possible contribution of neutral weak currents to the so-called anomalous $E1$ transitions in actinide nuclei. All the requisites seem to be there—highly-hindered γ transitions plus relativistically enhanced β processes. Neutral β decay would masquerade as internal conversion, so perhaps the excess electrons are a result not of anomalous γ transitions but of neutral β transitions?! Whether or not this turns out to be the particular case, many more illuminating examples should turn up during the next decade.

Fourth, improved and unusual nuclear models. Powerful group-theoretical models should become more the vogue. IBA and FDSM models are already paving the path, and I, for one, expect great things from the FDSM model (Chapter 29). Another direction group-theoretical models can proceed is through higher-order "geometrical" groups, such as $R(4)$, which has already been developed to an elegant degree of precision for the spectra of diatomic molecules.[13] This is not to say that we shall or should give up on the true geometrical models—Chapter 38 gives the lie to this. But, during the next decade, the exact correspondences between the group-theoretical and geometrical parameters should be worked out, giving us the advantages of both approaches. The whole will be greater than the sum of its parts.

In addition to these models, which use more or less standard mathematics, I expect an upsurge of interest in chaos-based models, for, after all, nuclei are rather nonlinear systems. At this juncture, nuclear scientists are just beginning to understand chaotic systems in general, but there have been some semi-successful attempts to apply them to nuclei.[14] Another possibility along this line is the use of fuzzy sets, which have already been applied to radionuclide transfer predictions.[15]

Fifth, applications. As nuclear scientists, we have been lax in selling ourselves and our work to the general public. And yet we complain about their lack of interest (and funding!). I expect this to change somewhat with the dynamic new generation of nuclear chemists, in particular. If I may be forgiven one more personal example: About a year ago, our group serendipitously stumbled into the field of tribology.[16] We developed a technique for implanting ^{7}Be and ^{22}Na from fragmentation reactions into a surface (piston, engine wall, rocket ceramic, etc.) so that the wear of this surface could be monitored by γ-ray spectroscopy. Interest from industry, the State of Michigan, and even the federal funding agencies was immediate, gratifying, and surprising (to me, at least). But, I think it behooves all of us to demonstrate from time to time that ours can be a practical as well as an elegant, aesthetically-pleasing science. Our "pure-science" productivity—and most certainly our egos—cannot help but profit from it.

Such are a few of my predictions about the future of our science. What can be said with assurance is that nuclear spectroscopy—exotic or not—is alive and kicking. And we are probably on the threshold of an important renaissance. The exact form it will take, of course, must remain speculative, for ours is not really an exact science. (That is why it is so much fun.) I would like to close with a "song" from my fantasy book, *Into the Atom*. It attempts to explain the Uncertainty Principle in a grown-up childish way, but it may also have relevance as to predictions about the future.

Uncertainty

You toss a tack,
It bounces back,
Whatever does that prove?
It shows you where,
Behind the chair,
That book, it did not move.

You throw a rock
To find the clock,
Whose time is not too good;
Here's what you've proved:
That clock has moved
Two feet from where it stood.

This ten-pound ball
Rolls down the hall
To find out where's the broom;
It can but tell,
The broom must dwell
Somewhere inside the room.

Chorus: Uncertainty, uncertainty—
It gives us thinkers pause—
To realize that this Effect
May have no clear-cut Cause.

REFERENCES

1. L. S. Kisslinger and R. A. Sorensen, *Mat.-Fys. Medd. Dan. Vid. Selsk.* **32**, No. 9 (1960).
2. A direct quote from a referee about a paper of mine published in *Phys. Rev. C* in the early 1970's.
3. Wm. C. McHarris, D. B. Beery, and W. H. Kelly, *Phys. Rev. Lett.* **22**, 1191 (1969).
4. Wm. C. McHarris, R. E. Eppley, R. A. Warner, R. R. Todd, and W. H. Kelly, *Proceedings of the International Conference on the Properties of Nuclei Far from the Region of Beta-Stability, Leysin, Switzerland, 31 August-4 September 1970*, CERN Report CERN 70-30, 1, 435.
5. Various articles in *Proceedings* of Ref. 4.
6. *Proceedings of the Robert A. Welch Foundation Conferences on Chemical Research - XIII. The Transuranic Elements*, Ed. by W. O. Milligan (Robert A. Welch Foundation, Houston, 1970).
7. *Radioactive Nuclear Beams - The First International Conference, Berkeley, Calif., 16-18 October 1989*, Ed. by W. O. Myers, J. M. Nitschke, and E. B .Norman (World Scientific, Singapore, 1990).
8. Wm. C. McHarris, *Nucl. Instr. Meth.* **A242**, 373 (1986).
9. A. T. Forrester, *J. Opt. Soc. Am.* **51**, 253 (1961).
10. V. V. Okorokov and S. V. Proshin, Inst. of Theor. and Exp. Physics, Moscow, Report ITEP-13 (1980), and letter of intent for LBL Bevelac proposal.
11. E.g., E. O. Cummins and P. H. Bucksbaum, *Weak Interactions of Leptons and Quarks* (Cambridge Univ. Press, 1983), Chaps. 4 and 9.
12. Wm. C. McHarris, *Proceedings of Second International Conference on Nucleus-Nucleus Collisions, Visby, Sweden, 10-14 June 1985*, p. 121.
13. B. R. Judd, *Angular Momentum Theory for Diatomic Molecules* (Academic Press, New York, 1975).
14. M. Baranger, *Proceedings of the 1989 International Nuclear Physics Conference, Sao Paulo* (World Scientific, Singapore, 1990), p. 247.
15. W. Shaw and P. Grindrod, Commission of the European Communities, Nuclear Science and Technology Report No. EUR 12499 (1989).
16. M. L. Mallory, R. M. Ronningen, Wm. C. McHarris, B. Sherrill, Y. X. Dardenne, and H. J. Schock, *Nucl. Instr. Meth.* **B40/41**, 579 (1989).

Author Index

Abenante, V. 457
Ahmad, I. 205, 293
Andrews, H. R. 271
Armbruster, P. 181
Aryaeinejad, R. 379
Baktash, C. 457
Baum, E. M. 259
Beard, K. B. 293
Beausang, C. W. 369
Beene, J. R. 457
Belgya, T. 259
Benenson, W. 101
Benet, Ph. 205, 293
Blomqvist, J. 23
Brenner, D. S. 511
Brown, B. A. 295
Canto, L. F. 625
Carpenter, M. P. 205
Carter, H. K. 495
Casten, R. F. 511
Cerny, J. 549
Chasteler, R. M. 57
Chishti, A. A. 23
Chou, W.-T. 379, 523
Chowdhury, P. 271
Cline, D. 229
Cole, J. D. 11
Cormier, T. M. 11
Crowell, B. 271
D'Auria, J. M. 83
Daly, P. J. 23, 293
Deleplanque, M.-A. 321
Donangelo, R. J. 625
Drigert, M. W. 205, 293
Durell, J. L. 205
Emling, H. 205
Ennis, P. J. 271
Farhan, A. R. 625
Fazekas, B. 259
Firestone, R. B. 57
Fitzgerald, J. B. 205
Ford, C. E. 495
Fossan, D. B. 369
Garcia-Bermudez, G. 457
Garg, U. 293
Garrett, J. D. 587
Gascon, J. 587
Gatenby, R. A. 259
Gelletly, W. 23
Gilat, J. 57
Gill, R. L. 111
Göres, J. 101
Grabowski, Z. 293
Griffin, H. C. 457
Guidry, M. 437, 625
Hagemann, G. B. 587
Halbert, M. L. 457
Hall, H. L. 561
Hamilton, J. H. 11, 283
Haustein, P. E. 1
Hensley, D. C. 457
Hensley, W. K. 535
Hill, J. C. 111
Hoff, R. W. 413
Hoffman, D. C. 561
Holzmann, R. 205, 293
Horn, D. 271
Hotchkis, M. A. C. 23
Jain, A. K. 413
Janssens, R. V. F. 205, 293
Janzen, V. P. 355
Johansson, J. K. 271, 355
Johnson, E. L. 259
Johnson, N. R. 339, 457
Khoo, T. L. 205, 293
Kreiner, A. J. 393
Kuehner, J. A. 355
Kvasil, J. 413
Lang, T. F. 549
Lee, I.-Y. 245, 457
Liang, Y. 369
Lister, C. J. 271
Ma, R. 369
Majka, Z. 457
Mantica, P. F. 495
McGowan, F. K. 457
McHarris, Wm. C. 379, 523, 657
McNeill, J. H. 23
Meyer, R. A. 427
Mohar, M. F. 101
Molnar, G. 259
Moltz, D. M. 549
Moore, E. F. 205, 293
Morrison, I. 427
Morrissey, D. J. 101
Morse, L. R. 205
Mowbary, A. S. 205
Münzenberg, G. 181

Nicolis, N. G. 457
Nitschke, J. M. 57, 127
Olivier, W. A. 379
Paul, E. S. 369
Phillips, W. R. 205
Piel, W. F. 369
Piiparinen, M. 23
Pilotte, S. 271
Radford, D. C. 271, 355
Ramayya, A. V. 283
Rasmussen, J. O. 625
Reeder, P. L. 535
Reiff, J. E. 549
Rengan, K. 609
Rikovska, J. 139, 495
Riley, M. A. 457
Ring, P. 625
Rios, A. 379
Robertson, J. D. 217, 549
Ronningen, R. M. 101
Rupnik, D. 495
Sarantites, D. G. 457
Scholten, O. 523
Schuhmann, R. B. 111
Schwarzenberg, J. 39
Semkow, T. M. 457
Sheline, R. K. 413, 473
Sherrill, B. 101
Shi, S. 369
Shihab-Eldin, A. A. 57
Sood, P. C. 413, 473
Stevenson, J. 101
Stone, N. J. 495, 637
Stoyer, M. A. 625
Subotic, K. 101
Toth, K. S. 127
Vanhoy, J. R. 259
Varley, B. J. 23
Vieira, D. J. 535
Vierinen, K. S. 57, 127
Virtanen, A. 457
Waddington, J. C. 271, 355
Walters, W. B. 217, 427, 495
Warburton, E. K. 111, 295
Ward, D. 271, 355
Warner, D. D. 511
Warner, R. A. 535
Wesselborg, C. 511
Wildenthal, B. H. 295
Wilmarth, P. A. 57, 127
Winfield, J. S. 101
Winger, J. A. 111
Winter, Ch. 271
Wohn, F. K. 111
Wolfs, F. L. H. 293
Wood, J. L. 39
Woods, P. J. 23
Wouters, J. M. 535
Xu, N. 369
Yates, S. W. 259
Ye, D.-Z. 205, 293
Yu, C.-H. 587
Yurkon, J. 101
Zganjar, E. F. 39, 495
Zhang, J.-Y. 511
Zimmerman, B. E. 495

Nuclide Index

NOTE: Important references are given in boldface. Passing references are given in lightface. For each chapter, only the first listing is given.

H-2 114

He-2 554
He-3 114, 314, 551
He-4 **297**
He-5 **297**
He-6 **297**
He-7 **297**
He-8 **297**, 659
He-9 **297**
He-10 **297**

Li-6 316
Li-7 84, 316
Li-8 84, 316
Li-9 97, 316, **535**
Li-10 316, 537
Li-11 97, 313, **536**

Be-7 316, 575
Be-8 84, 316
Be-9 84, 316
Be-10 316
Be-11 84, 310
Be-12 310, **535**
Be-13 537
Be-14 315, **543**
Be-15 537

B-8 551
B-11 88
B-13 **535**
B-14 **535**
B-15 535, 536
B-16 537
B-17 **536**
B-18 537

C-9 314, 550
C-12 17, 272
C-13 222, 281, 499
C-14 118, 540
C-16 **537**
C-17 **535**
C-18 **543**
C-19 **543**
C-21 537
C-23 537

N-13 84
N-17 517, **535**
N-18 **535**
N-19 **543**
N-21 **535**
N-24 537

O-14 88
O-15 88
O-16 46, 118, 281, 287, **298**, 381, 627
O-17 **298**, 517
O-18 118, **298**, 517, 540
O-19 **298**, 540
O-20 **298**
O-21 **298**
O-22 **298**
O-23 **298**
O-24 **298**
O-25 537
O-26 **298**
O-27 306
O-28 305
O-29 306
O-30 306
O-31 306

F-17 88
F-18 88
F-19 371
F-22 313
F-23 313
F-24 313
F-25 **535**
F-27 306
F-28 301
F-29 301
F-30 306
F-31 306
F-32 306

Ne-18 88, 517
Ne-19 88
Ne-20 118, 309, 381, 459
Ne-22 326, 381
Ne-23 100
Ne-24 100
Ne-28 306
Ne-29 306
Ne-30 301
Ne-31 306
Ne-32 301
Ne-33 306
Ne-34 301
Na-19 88
Na-20 88, 551
Na-21 88
Na-24 95
Na-25 97
Na-26 97
Na-27 97
Na-29 306
Na-30 306
Na-31 306
Na-32 306
Na-33 306
Na-34 306

Mg-22 88, 551
Mg-24 226, 309
Mg-27 575
Mg-30 306
Mg-31 306
Mg-32 112, 296
Mg-33 306
Mg-34 306
Mg-35 306

Al-22 **551**
Al-26 88

Al-27 **303**
Al-28 97, 575
Al-29 97, 575
Al-31 306
Al-32 306
Al-33 306
Al-34 306
Al-35 306
Al-36 306

Si-25 556
Si-27 88
Si-28 226, 281, 309, 558
Si-29 275
Si-30 276, 358
Si-32 306
Si-33 306
Si-34 305
Si-35 306
Si-36 7, 305
Si-37 306

P-26 552
P-33 306
P-34 306
P-35 306
P-36 306
P-37 306
P-38 306

S-32 281
S-32 226, 309, 340, 558
S-34 362, 457
S-36 281
S-40 7

Cl-34 100
Cl-35 273, 289
Cl-38 100

Ar-31 552
Ar-32 40, 314
Ar-33 314
Ar-35 100
Ar-36 101, 103, 309
Ar-40 182, 381
Ar-41 100
Ar-43 100

K-36 97
K-38 97
K-40 303
K-42 97
K-43 97
K-44 97
K-45 97
K-46 97
K-47 97

Ca-35 552
Ca-40 5, 89, 101, 236, 281, 289, 303
Ca-48 7, 182, 313
Ca-50 199
Ca-52 199
Ca-60 301
Ca-70 301

Sc-48 313

Ti-39 554
Ti-44 8
Ti-50 182, 184, 459
Ti-52 199

Cr-48 8
Cr-52 340, 457
Cr-54 182
Cr-56 199

Fe-54 25
Fe-58 182
Fe-60 199

Co-53 549
Co-56 264

Ni-56 8
Ni-58 5, 127, 233, 287, 627
Ni-60 275
Ni-64 459
Ni-66 122
Ni-68 112
Ni-78 **111**

Cu-64 108
Cu-67 122
Cu-68 122
Cu-69 122
Cu-71 118
Cu-72 118
Cu-73 118
Cu-74 **111**
Cu-75 118
Cu-76 **111**
Cu-77 118
Cu-78 118

Zn-64 127
Zn-68 122
Zn-69 122
Zn-70 122
Zn-72 122
Zn-74 **111**
Zn-76 **111**
Zn-80 112

Ga-63 89
Ga-64 102

Ge-61 550
Ge-64 102
Ge-65 551
Ge-70 273
Ge-72 240
Ge-76 118
Ge-82 112
Ge-83 **111**

As-65 102, 552
As-83 **111**

Se-65 550
Se-69 286
Se-70 **284**
Se-71 **286**
Se-72 **283**
Se-76 240
Se-78 240
Se-80 240, 283
Se-82 118

Br-69 552
Br-73 12, 289
Br-85 114
Br-92 430

Kr-73 **289**, 551
Kr-74 283
Kr-75 283
Kr-86 106, 114, 184, 432
Kr-87 432

Rb-75 98
Rb-76 40, 98
Rb-77 98
Rb-78 98
Rb-79 98
Rb-80 98
Rb-81 98
Rb-82 98
Rb-83 98
Rb-84 98
Rb-86 98
Rb-87 114
Rb-88 98
Rb-89 98
Rb-90 98
Rb-91 98
Rb-92 99
Rb-93 99
Rb-94 99
Rb-95 99
Rb-96 99
Rb-97 99
Rb-98 99

Sr-77 98, 551
Sr-82 **457**
Sr-83 98
Sr-85 98
Sr-87 98
Sr-88 116, 117, 432
Sr-89 432
Sr-91 98
Sr-96 430, 433
Sr-97 436
Sr-98 431

Y-89 116, **643**

Zr-80 8, 275
Zr-82 275
Zr-84 **277, 284**
Zr-86 **271**
Zr-88 275
Zr-89 643
Zr-90 116, 184, 275, 432
Zr-91 **432**
Zr-92 184, 436
Zr-94 184, 436
Zr-96 8, 184, **259**
Zr-98 **435**
Zr-99 436
Zr-100 430
Zr-102 436

Nb-93 127

Mo-92 127, 184, 272, 432
Mo-93 432
Mo-94 127, 184
Mo-95 127
Mo-96 127, 184
Mo-98 184
Mo-100 8, 184, 431
Mo-101 436
Mo-102 431
Mo-104 431
Mo-106 431
Mo-108 431, 506

Ru-94 116, 432
Ru-95 432
Ru-96 127
Ru-103 436
Ru-104 8, 184
Ru-106 8
Ru-110 505

Pd-96 432
Pd-97 432
Pd-100 272
Pd-102 25
Pd-105 436
Pd-110 184, 234
Pd-112 503
Pd-185 **44**

Ag-97 272
Ag-101 **271**
Ag-103 272

Cd-102 272
Cd-106 499
Cd-107 436
Cd-111 499
Cd-114 **240, 504**
Cd-180 499

In-99 116
In-122 99
In-124 99
In-126 99
In-128 99

Sn-100 116
Sn-109 433
Sn-110 459
Sn-111 433
Sn-113 433
Sn-114 **458**
Sn-115 433
Sn-116 **503**, 512
Sn-117 371, 433
Sn-119 433
Sn-120 627
Sn-121 433
Sn-122 340
Sn-123 433
Sn-124 184, 340, 358
Sn-125 433
Sn-127 433
Sn-129 433
Sn-131 112, **432**
Sn-132 112, 219

Sb-111 433
Sb-113 433
Sb-115 433
Sb-117 433
Sb-119 433
Sb-121 433
Sb-123 433
Sb-125 433
Sb-127 433
Sb-129 433
Sb-131 433
Sb-133 112, 433

Te-118 496
Te-120 496
Te-122 496
Te-130 429
Te-132 427
Te-136 427

I-109 549
I-119 507

Xe-114 **499**
Xe-116 **499**
Xe-118 **499**
Xe-120 **495**
Xe-122 505
Xe-123 **498**
Xe-124 **497**
Xe-125 **497**
Xe-126 **497**
Xe-127 **497**
Xe-128 **497**
Xe-130 **497**
Xe-131 **497**
Xe-132 429, **497**
Xe-133 **497**
Xe-134 429
Xe-136 232, **498**
Xe-137 221
Xe-138 209, 428
Xe-139 221
Xe-140 209, 428
Xe-141 209, 221
Xe-142 **212**, 428
Xe-143 221
Xe-144 428

Cs-113 549
Cs-118 **642**
Cs-119 221
Cs-120 **495**
Cs-121 221, 507, **646**
Cs-124 99
Cs-126 99
Cs-128 99
Cs-130 99
Cs-133 500
Cs-137 221
Cs-139 99, 221
Cs-140 99
Cs-141 99, 221
Cs-142 99
Cs-143 99, 220
Cs-144 99
Cs-145 99, 220

Ba-119 68
Ba-122 505
Ba-133 500
Ba-134 429

Ba-136 429
Ba-139 221
Ba-140 428
Ba-141 221
Ba-142 213, 428
Ba-143 **217**
Ba-144 **212**, 220, 428
Ba-145 219
Ba-146 **209**, 220, 428

La-120 69
La-122 69
La-128 373
La-130 376
La-131 376
La-132 **371**
La-133 78
La-139 381, 381
La-147 **217**

Ce-123 68
Ce-124 506
Ce-125 68
Ce-132 332
Ce-136 429
Ce-138 429
Ce-142 428
Ce-144 215, 220, 428
Ce-146 **213**, 220, 428
Ce-147 220
Ce-148 213, 220, 428
Ce-150 213

Pr-124 69
Pr-126 69
Pr-128 69
Pr-132 **371**
Pr-134 **371**
Pr-139 658

Nd-127 68
Nd-129 68
Nd-131 68
Nd-135 **330**
Nd-136 **334**
Nd-138 429, 459
Nd-139 658
Nd-140 429
Nd-143 221
Nd-144 340, 428
Nd-145 221
Nd-146 215, 428
Nd-147 221
Nd-148 215, **229**, 428
Nd-149 221
Nd-150 215, 428, 459
Nd-151 **485**
Nd-154 475

Pm-130 69
Pm-132 69
Pm-134 69
Pm-135 69
Pm-136 69, **371**
Pm-142 63
Pm-153 488
Pm-154 475

Sm-131 68
Sm-133 68
Sm-134 69
Sm-135 68
Sm-136 69
Sm-140 221, 429
Sm-141 69, 131, 221
Sm-142 63, 429
Sm-144 68, 70, **259**
Sm-146 428
Sm-148 428
Sm-149 221
Sm-150 428
Sm-151 221
Sm-152 428
Sm-153 488
Sm-155 488

Eu-134 69
Eu-135 69
Eu-136 69
Eu-138 **371**
Eu-140 66
Eu-141 69
Eu-142 63
Eu-144 70
Eu-145 70
Eu-147 358
Eu-152 264
Eu-154 415
Eu-155 488
Eu-156 419

Gd-110 499
Gd-137 69
Gd-139 69
Gd-140 66
Gd-141 69, **71**
Gd-142 63, 429
Gd-143 131
Gd-144 358, 429
Gd-145 66
Gd-146 24, 70, 259, 357, **474**
Gd-147 259, 489
Gd-148 259, 357, 428, 489
Gd-149 **357**, 489
Gd-150 357, 428, 489
Gd-151 489
Gd-152 428, 475
Gd-154 428
Gd-156 17, 234, 627
Gd-161 488

Tb-140 66
Tb-141 69, **71**
Tb-142 63, **65**
Tb-144 67
Tb-145 69, 133
Tb-146 70, 129
Tb-147 70, 489
Tb-148 34, 489
Tb-149 136, 489
Tb-150 357, 489
Tb-151 357, 489
Tb-152 475
Tb-155 488
Tb-156 **413**
Tb-157 488
Tb-158 419
Tb-159 381, 488
Tb-160 415
Tb-161 488
Tb-163 484
Tb-164 486

Dy-141 69, 72
Dy-142 63
Dy-143 69
Dy-144 69, 75
Dy-145 **127**
Dy-146 75, 129
Dy-147 130
Dy-148 **24**, 70, 80, 489
Dy-149 **31**, 69, 489
Dy-150 489
Dy-151 357, 489
Dy-152 246, **321**, 355, 489
Dy-153 **357**, 489
Dy-154 475
Dy-155 488
Dy-156 602
Dy-157 **488**
Dy-158 483
Dy-159 488
Dy-160 **629**
Dy-162 **633**
Dy-163 484
Dy-164 482
Dy-165 484
Dy-168 69
Dy-169 69

Ho-144 69
Ho-145 67, 129
Ho-146 69, 129
Ho-147 129
Ho-148 70, 129
Ho-149 **24**, 62, 129, 135, 489

Ho-150 34, 489
Ho-151 489
Ho-152 478
Ho-153 136, 489, 358
Ho-154 475
Ho-157 488, **588**
Ho-158 478
Ho-159 488
Ho-161 488
Ho-163 488
Ho-164 **481**
Ho-165 **232**, 381
Ho-165 483
Ho-166 415
Ho-167 488
Ho-169 69
Ho-171 69

Er-145 69, 129
Er-147 **127**
Er-148 69, 129
Er-149 62, 129
Er-150 **24**, 80, 129, 489
Er-151 **31**, 69, 489
Er-152 69, 351, 489
Er-153 136, 489
Er-154 489
Er-157 488
Er-158 588
Er-159 488
Er-161 488
Er-163 484
Er-164 482
Er-165 484
Er-166 234
Er-167 488
Er-168 **234**
Er-170 58
Er-174 69

Tm-147 **127**, 549
Tm-149 69, 129
Tm-150 66, 129
Tm-151 **24**, 69, 129, 489
Tm-152 34, 129, 489
Tm-153 69, 129, 489
Tm-154 489
Tm-155 129, 489
Tm-156 489
Tm-159 **588**
Tm-163 484
Tm-164 **481**
Tm-165 484, 488
Tm-167 488
Tm-168 **413**
Tm-170 415
Tm-172 419
Tm-174 69
Tm-175 484

Yb-151 66, 129, 130
Yb-152 **24**, 80, 129, 489
Yb-153 **24**, 66, 129
Yb-154 67, 129, 489
Yb-155 69, 129
Yb-156 489
Yb-157 136
Yb-160 588
Yb-163 484
Yb-164 17, 459
Yb-165 484, 488
Yb-167 488
Yb-170 **458**
Yb-172 242
Yb-175 484
Yb-176 58, 326
Yb-177 485

Lu-151 549
Lu-152 66, 129
Lu-153 **24**, 69, **127**
Lu-154 **24**, 67, 129
Lu-155 69, **127**
Lu-157 69, **127**
Lu-161 **588**
Lu-162 588
Lu-163 **588**
Lu-164 588
Lu-165 484, **588**
Lu-166 69, **482**, 588
Lu-167 **588**
Lu-168 69
Lu-169 69
Lu-170 419
Lu-171 69
Lu-172 69, 419
Lu-173 484
Lu-174 415
Lu-176 **413**
Lu-177 485
Lu-180 483
Lu-183 485

Hf-154 **24**, 80
Hf-156 25, 490
Hf-162 588
Hf-170 **247**, **402**
Hf-171 **402**
Hf-172 **402**
Hf-173 **402**
Hf-176 **486**
Hf-178 482
Hf-180 483
Hf-183 485

Ta-156 490
Ta-169 596
Ta-171 **399**, 596
Ta-172 **399**
Ta-173 **399**, 596
Ta-174 **399**
Ta-175 **399**
Ta-176 419, **487**
Ta-177 485
Ta-178 482
Ta-181 46
Ta-182 **413**
Ta-184 419

W-170 **339**
W-172 **340**
W-174 381, **402**
W-175 **408**, 530
W-176 381, 402
W-177 402, 485, 530
W-178 402
W-179 485, 530
W-182 234
W-184 **232**
W-190 475

Re-171 596
Re-173 **398**
Re-174 **379**, **398**
Re-175 **379**, **398**, 528
Re-176 **379**, **398**, **527**
Re-177 381, **398**, 528, 596
Re-178 **398**, **527**
Re-179 **398**, 485, 528
Re-180 **527**
Re-181 **526**
Re-182 **413**, **523**
Re-183 **523**
Re-185 **526**
Re-186 415
Re-188 419
Re-190 419, 475

Os-172 **340**
Os-176 526
Os-177 530
Os-178 526
Os-179 530
Os-180 **402**, **525**
Os-181 **402**, **527**
Os-182 402, **525**
Os-183 402, **524**
Os-184 **394**, **523**
Os-185 527
Os-186 **236**, 526
Os-188 **236**
Os-190 **239**, 475
Os-191 236
Os-192 **236**

Ir-181 **397**

Ir-182 **397**
Ir-183 **397**
Ir-184 **397**
Ir-185 **397**
Ir-186 **394**
Ir-187 **397**
Ir-190 475
Ir-191 475

Pt-129 **139**
Pt-176 140, 158
Pt-177 **139**
Pt-178 140, 158
Pt-180 140
Pt-181 **139**
Pt-182 140
Pt-183 **52, 139, 651**
Pt-184 **53**, 140, **649**
Pt-185 **139**
Pt-186 140
Pt-187 49, **139**
Pt-188 140
Pt-189 51, **139**
Pt-190 140
Pt-191 **139**
Pt-192 140
Pt-193 **139**
Pt-194 140, **236**
Pt-195 **139**
Pt-196 140

Au-183 **52, 651**
Au-184 **53, 648**
Au-185 **44**, 176
Au-187 **40**
Au-189 **46**

Hg-181 148
Hg-183 **651**
Hg-184 **648**
Hg-185 **45**
Hg-186 51
Hg-187 **40**
Hg-189 **46**
Hg-190 **47**
Hg-191 293, 324, 358
Hg-192 **326**, 358

Tl-180 564
Tl-187 **40**
Tl-189 46
Tl-190 **48**

Pb-195 49
Pb-206 253
Pb-207 500
Pb-208 182, 223, 232, 251, 259, **627**
Pb-210 **632**

Bi-207 500
Bi-209 184

Ra-224 225
Ra-225 **208**
Ra-226 264

Ac-223 209
Ac-225 **208**
Ac-227 209

Th-222 **206**
Th-224 **206**
Th-226 206
Th-228 206
Th-231 248
Th-232 535

Pa-229 **208**
Pa-234 **419**
Pa-236 **419, 564**
Pa-238 **564**

U-234 206, 253
U-235 112, 253, 621
U-238 627

Np-228 564
Np-229 564
Np-230 564
Np-231 564
Np-232 564
Np-233 564
Np-234 564
Np-235 209
Np-237 209, 567
Np-238 **415**
Np-239 209

Pu-229 564
Pu-230 564
Pu-231 564
Pu-232 564
Pu-233 564
Pu-234 564
Pu-235 564

Am-230 564
Am-231 564
Am-232 **563**
Am-233 564
Am-234 **564**
Am-235 564
Am-236 564
Am-237 575
Am-238 575, 579
Am-240 419
Am-241 569
Am-242 415
Am-243 209
Am-244 415

Cm-248 **209**

Bk-240 564
Bk-250 415

Cf-180 187
Cf-246 191
Cf-249 187
Cf-252 **209**, 569
Cf-254 **209**

Es-242 563
Es-244 564
Es-246 564
Es-248 564
Es-256 **563**

Fm-250 191
Fm-254 454
Fm-256 454, 563
Fm-258 197

Md-250 564

No-254 192

Lr-254 192

104-254 **187**
104-255 **187**
104-256 **187**
104-257 **187**
104-258 **187**
104-259 **187**
104-260 **187**, 219
104-261 **187**
104-262 **187**

105-256 **187**
105-257 **187**
105-258 **187**
105-260 **187**
105-261 **187**
105-262 **187**

106-259 **187**
106-260 **187**
106-261 **187**
106-263 **187**

107-261 **187**
107-262 **187**

108-264 **187**
108-265 **187**

109-266 **187**

Subject Index

Abnormal-parity orbitals 451
Absolute transition probabilities 601
Alaga selection rules 473
ALICE (computer code) 58, 103
Alignment (nuclear) 389, 402, 587
Alignment mechanisms 339
Allowed unhindered (au) decays 473
Alpha decay 127, 192, 208
Alpha emission 455
Alpha particles 457
Alternating parity 208
Angular correlation (γ rays) 327
Angular distributions 261
Angular frequency 278
Anisotropy 638
Anisotropy coefficients 461
Anisotropy ratios 469
Anomalous $E1$ transitions 661
Antiparallel coupling 398
Argonne National Laboratory 14, 210, 293
Argonne National Laboratory - Notre Dame 357
Astrophysics (nuclear) 87
Asymetric fission 219
Asymmetric shapes 205, 217
Asymptotic quantum numbers 474
ATLAS 293
Attenuation factors 641
Autobatch techniques (chemical) 609
Axially-symmetric oscillator 323
$B(E2)$ values 344, 374, 437, 511, 523, 628
$B(M1)$ values 347, 374, 523, 594
Backbending 340, 387, 407, 598, 628
Band crossing 339, 387, 407
Band mixing 236
Band-crossing frequencies 587
Barrier (fission) 562
Bass barrier 183
BCS calculations 145
Beta decay 495
Beta decay (neutral) 661
Beta transitions 473
Beta-decay rates 312
Beta-delayed fission 561
Beta-delayed neutron emission 535
Beta-delayed α decay 551
Beta-delayed particle emission 313
Beta-delayed proton emission 549
Beta-delayed two-proton emission 550
$Beta^+/EC$ ratios 62
Binding energies 297, 511
Bosons 437
Brookhaven National Laboratory 112
CASCADE (computer code) 26, 103
Casimir operators 441
Central zero-range force 416
Chalk River Nuclear Laboratories 271, 355
Chemical separations 568, 609
Cluster deformation 222
Cluster structures 219
Coherent γ rays 661
Cold-fusion reactions 183
Cold-temperature techniques 637
Collective oblate states 376
Collectivity (nuclear) 339
Compton-Suppression Spectrometer System (ORNL) 339
Continuous gas-phase separations 621
Continuous solvent extraction 617
Continuum analysis (γ rays) 280
Conversion coefficients 27, 501
Conversion-electron spectroscopy 495
Coriolis antipairing 455, 625
Coriolis calculations 395
Coriolis coupling 413
Coriolis generator 395

Coriolis matrix elements 208
Coulomb break-up 315
Coulomb excitation 229, 627
Cranked-HFB calculations 340, 635
Cranked shell-model calculations 340, 370, 389, 393, 439, 457, 587
Cross-shell excitations 303
CSI(Tl) detectors 461
Damping (of rotational motion) 248
Daresbury 14, 23
Decay-scheme normalization 60
Decoupling parameter 395
Deformation (nuclear) 427, 473, 495
Deformation (γ) 139
Deformation (octupole) 207
Deformation (onset of) 70
Deformation (spin-stabilized) 271
Deformation changes 283
Deformation effects (on particle emission) 459
Deformation parameters 140
Deformations (quadrupole) 207
Delayed fission 561
Delayed neutron emission 535
Delayed-particle emission 296
Detector telescopes 107
Diabolical points 628
Diagonal figure of merit 358
Differences (in binding energies) 302
Diffusion (for producing radioactive beams) 84
Dipole moments (electric) 214
Dipole moments (magnetic) 141, 221
Directional correlations 53
Dispersion (in mass spectrometers) 17
Divergences (in mass predictions) 4
Doppler shifts 262
Doppler-shift attenuation method 262, 336, 358
Doppler-shift recoil-distance technique 339
Doubly-decoupled bands 376, 379, 393, 527
Doubly-magic region 111
Doubly-odd nuclei 369, 379, 393, 413, 523
Dwarf Ball 461
Dynamic moment of inertia 284, 327, 364
Dynamical Pauli effect 437
Dynamical symmetries 437
*E*0 transitions 45, 495
*E*1 matrix elements 241
*E*1 transitions 44, 89, 208, 220, 259
*E*2 coulomb excitation 627
*E*2 isomers 24
*E*2 matrix elements 229
*E*2 transition rates 25
*E*2 transitions 80, 265, 276, 344, 358
*E*3 isomers 33
*E*3 matrix elements 229
*E*3 transitions 79, 265
EC-delayed fission 561
ECR 93
Effective charge 25
Eight-π spectrometer 271, 355
Electric dipole moment 214
Electromagnetic moments 141, 374
Electron-cyclotron resonance 93
Electronics systems 256
Energy surface 435
Enhanced transitions (*E*0) 49
Excitation functions 182, 261
Extra push (in fusion reactions) 184
Fast β transitions 473
Feather plot 657
Fermion-Dynamical-Symmetery model 437
Fermions 437
Fission (delayed) 561
Fission (spontaneous) 192
Fission barriers 195, 562
Fission yields 210, 217
Forbidden transitions (*M*1) 29
Four-quasiparticle states 483
Frequency (of transitions) 329
Frequency shifts 597
G matrix 300
Gallagher-Moszkowski splittings 413
Gamma band 237
Gamma deformation 139, 284, 587
Gamma-gamma coincidence matrix 360
Gamma-gamma matrix 465
Gamma-ray spectrometer 25, 41, 59, 113, 131, 209, 230, 245, 260, 271, 287, 293, 324
Gamma-ray transition probabilities (reduced) 75

GAMMASPHERE 16, 230, 245, 365
Gamow-Teller strength 295
Gesellschaft fur Schwionenforschung 13, 182
Giant resonances 251
Ginnochio model 442
Glauber model 315
GOSIA (computer code) 232
Ground-state masses 447
Group generators 444
Group theoretical structures 444
Group theory 437
Group-theoretical decompositions 445
GSI 13, 182
Half-life 27, 572
Half-lives 191, 540
Hamiltonian 524
Hartree-Fock calculations 295, 324
Helium-jet recoil transport method 554, 569
HERA 324, 357
Heritage isomers 455
HHIRF 11, 39, 339, 459, 499
High-*j* orbitals 369
High-spin states 371, 587
Holifield Heavy-Ion Research Facility 11, 39, 245, 284, 339, 459, 499
Hyperdeformation 246, 366, 457
i-active group structures 442
IBA calculations 524
IBFA calculations 524
IBFFA calculations 523
Implantation 637
InSb detectors 661
Inelastic neutron scattering 259
Inertia paramaters 402
Interacting-boson calculations 391, 523
Interacting-boson model 222, 236, 440
Interaction energies (*p-n*) 515
Internal-conversion coefficients 63
Internal-conversion electron spectrometers 43
Intruder states 495
Ion exchange 615
Ion sources 93
Ion yields 538
ISAC facility 89
Isobaric analog states 314
Isobaric-multiplet mass equation 314
Isomeric transitions 24
Isospin-forbidden decay 295
Isotope separators 111, 499
j-j coupling 313
Josephson effect (nuclear) 661
k-active group structures 442
Kentucky (University of) 260
Korringa constant 639
Kurie plot 657
LAMPF 536
Large radii (in nuclei) 315
Lawrence Berkeley Laboratory 57, 127, 324, 357, 554, 567
Legnaro 14
Lifetime measurements 262
Light output (of detectors) 464
LIM target system 567
Lineshapes (γ ray) 330
Liquid-drop calculations 457
Liquid-drop model 217
log*ft* values 72, 484
*M*1 forbidden transition 29
*M*1 transitions 208
*M*2 transitions 78
*M*4 transitions 75
Magnetic dipole moments 221
Magnetic spectrograph 103
Magnetic spectrometer 14
Mass formulas 196, 455
Mass measurements 1
Mass models 2
Mass predictions 3
Mass separators 57
Mass spectrometer (recoil) 11, 23, 39
Mass surfaces 196
Mass-yield distributions 561
Matrix elements ($E\lambda$, $M\lambda$) 232
Mean-field calculations 209, 295
Mean-field theory 458
Michigan State University 13, 101, 379
Mixing ratios 374, 587
Mixing ratios ($M1/E2$) 53
Model spaces (shell model) 296
Moment of inertia 139, 277, 284, 327, 358
Moment of inertia (rigid body) 285
Moments (nuclear) 139
Moments of inertia 598

n-p interaction 33, 413, 427, 444, 511
Neutral β decay 661
Neutral weak currents 661
Neutron counters 539
Neutron emission (delayed) 535
Newby shifts 413
Nilsson diagrams 72, 587
NMR 639
N_nN_p scheme 512
Nuclear alignment 339, 389, 402, 587
Nuclear astrophysics 87
Nuclear collectivity 339
Nuclear deformation 427
Nuclear moments 139
Nuclear orientation 637
Nuclear orientation (on-line) 53
Nuclear relaxation 637
Nuclear resonance fluorescence 267
Nuclear shapes 139, 205, 229, 283, 321, 369
Nuclear temperature 562
OASIS 57, 127
Oblate states 376
Octupole deformations 207, 220
Octupole minima 209
Octupole shapes 205, 229
Octupole vibrations 205
Octupole-octupole correlations 215
Octupole-octupole excitations 259
Octupole-quadrupole excitations 259
Odd-even staggering (in band spacings) 422
Odd-odd nuclei 369, 379, 393, 413, 523
Orientation (nuclear) 637
ORNL 459, 499
OSIRIS 357
p-n interaction 33, 413, 427, 444, 511
Pair-transfer matrix elements 632
Pairing force 324
Parallel coupling 398
Parity doublets 208
Particle-hole configurations 51
Particle-plus-rotor calculations 393
Particle-rotor model 439
Particle-triaxial-rotor calculations 142
Pauli effect (dynamical) 437
Phase-space effects 468
Potential-energy surface 249, 322, 339, 593
Pre-neutron emission energy 576
Production cross-section 26
Prolate-oblate differences 592
Proton decay (β-delayed) 64, 127
Proton drip line 101, 127, 549
Proton emission 457
Proton-evaporation spectra 461
Pseudo-orbit 442
Pseudo-spin 442
Q_{EC} determination 62
Q_α 193
Quadrupole deformation 207, 323, 398, 587
Quadrupole moments 141, 330, 344, 374, 523, 601
Quadrupole moments (electric) 221
Quadrupole shapes 229
Quantum numbers (asymptotic) 474
Radioactive beams 1, 83, 199
Radioactive targets 7
Radioactive-beams facilities (table) 86
RAMA 555
Random-phase approximation 316
Reactor 112
Recoil mass spectrometer 11, 23, 39
Recoil techniques (for producing radioactive beams) 84
Recoil-distance measurements 340
Reduced transition probabilities (γ-ray) 75
Reflection assymetry 205, 217
Relaxation (nuclear) 637
RHIC 661
Rigidity (magnetic) 108
Rochester (University of) 14, 230
Rotational frequencies 339
Rotational population patterns 625
Routhians 287, 350, 370, 403, 587
RP process 88, 102
Seniority (quantum number) 24
Shape changes 293, 457
Shape coexistence 47, 240, 283, 339, 495
Shape evolution 283, 339
Shape transitions 236, 283
Shapes (nuclear) 139, 205, 229, 283, 321

Shell-model calculations 115, 143, 295
Shell-model orbitals 73
Shell-model states 131, 205
SHIP 13, 182
Side-feeding (in rotational bands) 332
Signature (in rotational bands) 276
Signature dependence 594
Signature splittings 594
Single-neutron energies 432
Single-particle energies 300, 432
Single-proton energies 433
Skyrme interactions 295
Solvent extraction 612
Spin Spectrometer (ORNL) 343
Spin-flip transitions 473
Spin-orbit partners 435
Spin-stabilized deformation 271
Spontaneous-fission decay 192
SQUID 625, 661
Staggering (in band spacings) 422
Statistical-model calculations 458
Strength measurements (β and γ) 81
Strutinsky calculations 457
Sub-barrier decay 457
Sub-barrier fusion 225
Subshell at $Z = 64$ 427
SUNY Stony Brook 369, 379
Super deformation 246
Super HILAC 57
Superconducting Quantum Interference Device (SQUID) 625
Superdeformation 293, 321, 355, 455, 457
Superheavy elements 181, 217
TASCC 355
Temperature (high) 293
Temperature-induced rotation 457
TESSA 14
TESSA 3 357
Thermochromatography 614
Three-particle states 658
Three-phonon states 259
Three-quasiparticle states 483
Time-integral experiments (orientation) 640
Time-of-flight spectrometer 536
Time-of-flight techniques 105
Time-resolved experiments (orientation) 641
Time-Resolved On-Line Nuclear Orientation (TR-OLNO) 637
TISOL 83
TOFI Spectrometer 536
Total-kinetic-energy distributions 561
TR-OLNO 637
Transfer reactions 252
Transfer strength 626
Transition probabilities 149
Transitional nuclei 271
Triaxial deformation 287
Triaxial rotor 142
Triaxial shapes 369
Tribology 662
TRISTAN 111
TRIUMF 83
Tunneling factor 629
Two-body matrix elements 116, 296
Two-dimensional correlation (array) 280
Two-neutron emission 544
Two-neutron transfer 625
Two-phonon states 260
Two-proton radioactivity 552
Two-quasiparticle-plus-rotor model 393
Uncertainty 662
UNILAC accelerator 188
Unique-parity orbitals 24
UNISOR 39, 499
Velocity filter 13, 188
Vibrations 445
Vibrations (octupole) 205
Volatilization 613
Weak coupling model 304
Wear studies 662
Weisskopf estimates 75
Weisskopf units 208, 265
Wien filter 13
WKB tunneling factor 629
Woods-Saxon potential 197, 323
X rays 569
X-ray intensities 61
Yale University 271
Yrast sequence 324
Yukawa potential 197
Zero-energy-dispersion spectrometer 14

Printed in the United States
by Baker & Taylor Publisher Services

Printed in the United States
by Baker & Taylor Publisher Services